Intelligent Systems, Control and Automation: Science and Engineering

Volume 69

For further volumes:
http://www.springer.com/series/6259

José M. Maestre · Rudy R. Negenborn
Editors

Distributed Model Predictive Control Made Easy

Springer

Editors

José M. Maestre
Department of Systems and Automation Engineering
University of Seville
Seville
Spain

Rudy R. Negenborn
Department of Marine and Transport Technology
Delft University of Technology
Delft
The Netherlands

ISSN 2213-8986 ISSN 2213-8994 (electronic)
ISBN 978-94-007-7005-8 ISBN 978-94-007-7006-5 (eBook)
DOI 10.1007/978-94-007-7006-5
Springer Dordrecht Heidelberg New York London

Library of Congress Control Number: 2013944513

Printed on acid-free paper

Springer is part of Springer Science+Business Media (www.springer.com)

We are driven by self-interest, it is necessary to survive. But we need wise self-interest that is generous and co-operative, taking others' interests into account.

Dalai Lama

A small boat with only one pole cannot sail across a sea; Paddling many quants together can move a huge carrack.
A small tree standing alone cannot resist a storm; Acres of forest can survive extremely cold winters.
One chopstick by itself can be broken easily; A dozen of chopsticks bundled together are very strong.
One hand by itself cannot make any sound; Applause of many hands together can be very loud.

Freely translated lines from Chinese song "Paddling together moves a huge carrack"

Disheng Fu

Preface

In the beginning of 2012, we were discussing the value that automatic control could have for a type of system that had not been considered by the control community before, viz. large-scale intermodal transport networks, typically being considered the interconnection of road, rail, and water infrastructures. We soon got to the question about what kind of control architecture we could propose.

Information and communication technology is rapidly being introduced for this type of system, facilitating data collection and remote actuation. At the same time, domain specific scientists are working on making models that could explain the behavior of such a system using measured data. Such models can in the future be used for making predictions of dynamic behavior. Then, after defining optimal behavior from the perspective of the different actors, we could assess the performance of different predicted behaviors. It is therefore reasonable to explore the development of model-based controllers further, in particular distributed model-based controllers, as we are facing a system of a very large scale, that moreover involves many different actors. Now, the question is: What distributed model-based control structure would be the most appropriate? This is not an easy question to answer.

Over the years, a large group of research groups has focused on advancing distributed model-based control techniques. We found that a coherent overview that introduces the reader to many techniques available was missing. Hence, we brought the idea to create such an overview to the control community. This book in front of you is the result. For us, the book will directly contribute to answering the question what kind of distributed model-based control technique could be appropriate for control of large-scale intermodal transport networks, or parts thereof. Apart from this, we hope that this book will also help other researchers' to find those schemes that could be useful and provide them with insight in what has been done, and what has not been done.

This book features 35 contributions, each describing noncentralized model-based predictive control techniques. The authors of the contributions have structured their contributions along a common framework as much as possible. The intended audience of this book consists of researchers, (technically oriented) control practitioners, M.Sc. and Ph.D. candidates. The readers aimed at have backgrounds in optimization and control.

We thank all authors for their high-quality contributions and the reviewers for their constructive remarks and suggestions; without them this book would not exist. We also thank Ms. Jacobs, Ms. Feenstra, and Ms. Pot for their guidance in the publication process at Springer. Moreover, we acknowledge the financial support for this project via the VENI project "Intelligent Multi-agent Control for Flexible Coordination of Transport Hubs" (project 11210) of the Dutch Technology Foundation STW, a subdivision of The Netherlands Organisation for Scientific Research (NWO), and the projects "Model Predictive Techniques for Efficient Management of Renewable Emicro-Networks" (project DPI2010-21589-C05-01) and "Networked Model Predictive Control" (project DPI2008-05818) of the Spanish Ministry of Economy and Competitiveness. Finally, financial support from the EU Network of Excellence Highly Complex and Networked Control Systems (HYCON2) under grant agreement no. 257462 is also gratefully acknowledged.

We hope that you as reader reading this book will appreciate the aggregated efforts of the 88 authors that contributed. Let this book inspire you in your professional activities and stimulate you to explore to use of distributed model-based control techniques.

Suggestions are welcome, so please do not hesitate to contact us via: http://distributedmpc.net/.

Sevilla & Delft
February 2013

José M. Maestre
Rudy R. Negenborn

Contents

1 **On 35 Approaches for Distributed MPC Made Easy** 1
R. R. Negenborn and J. M. Maestre

Part I From Small-Scale to Large-Scale: The Group of Autonomous Systems Perspective

2 **Bargaining Game Based Distributed MPC**. 41
F. Valencia, J. D. López, J. A. Patiño and J. J. Espinosa

3 **Cooperative Tube-Based Distributed MPC for Linear Uncertain Systems Coupled Via Constraints** 57
P. A. Trodden and A. G. Richards

4 **Price-Driven Coordination for Distributed NMPC Using a Feedback Control Law**. 73
R. Martí, D. Sarabia and C. de Prada

5 **Distributed MPC for Consensus and Synchronization** 89
M. A. Müller and F. Allgöwer

6 **Distributed MPC Under Coupled Constraints Based on Dantzig-Wolfe Decomposition**. 101
R. Bourdais, J. Buisson, D. Dumur, H. Guéguen and P.-D. Moroşan

7 **Distributed MPC Via Dual Decomposition and Alternative Direction Method of Multipliers** 115
F. Farokhi, I. Shames and K. H. Johansson

8 **D-SIORHC, Distributed MPC with Stability Constraints Based on a Game Approach** 133
J. M. Lemos and J. M. Igreja

9 **A Distributed-in-Time NMPC-Based Coordination Mechanism for Resource Sharing Problems** 147
M. Y. Lamoudi, M. Alamir and P. Béguery

10 **Rate Analysis of Inexact Dual Fast Gradient Method for Distributed MPC** 163
I. Necoara

11 **Distributed MPC Via Dual Decomposition** 179
B. Biegel, J. Stoustrup and P. Andersen

12 **Distributed Optimization for MPC of Linear Dynamic Networks** 193
E. Camponogara

13 **Adaptive Quasi-Decentralized MPC of Networked Process Systems** 209
Y. Hu and N. H. El-Farra

14 **Distributed Lyapunov-Based MPC** 225
R. Hermans, M. Lazar and A. Jokić

15 **A Distributed Reference Management Scheme in Presence of Non-Convex Constraints: An MPC Based Approach** 243
F. Tedesco, D. M. Raimondo and A. Casavola

16 **The Distributed Command Governor Approach in a Nutshell** . . . 259
A. Casavola, E. Garone and F. Tedesco

17 **Mixed-Integer Programming Techniques in Distributed MPC Problems** 275
I. Prodan, F. Stoican, S. Olaru, C. Stoica and S.-I. Niculescu

18 **Distributed MPC of Interconnected Nonlinear Systems by Dynamic Dual Decomposition** 293
A. Grancharova and T. A. Johansen

19 **Generalized Accelerated Gradient Methods for Distributed MPC Based on Dual Decomposition** 309
P. Giselsson and A. Rantzer

20 Distributed Multiple Shooting for Large Scale Nonlinear Systems 327
A. Kozma, C. Savorgnan and M. Diehl

21 Nash-Based Distributed MPC for Multi-Rate Systems 341
S. Roshany-Yamchi, R. R. Negenborn and A. A. Cornelio

Part II From Large-Scale to Small-Scale: The Decomposed Monolithic System Perspective

22 Cooperative Dynamic MPC for Networked Control Systems 357
I. Jurado, D. E. Quevedo, K. H. Johansson and A. Ahlén

23 Parallel Implementation of Hybrid MPC 375
D. Axehill and A. Hansson

24 A Hierarchical MPC Approach with Guaranteed Feasibility for Dynamically Coupled Linear Systems 393
M. D. Doan, T. Keviczky and B. De Schutter

25 Distributed MPC Based on a Team Game 407
J. M. Maestre, F. J. Muros, F. Fele, D. Muñoz de la Peña and E. F. Camacho

26 Distributed MPC: A Noncooperative Approach Based on Robustness Concepts 421
G. Betti, M. Farina and R. Scattolini

27 Decompositions of Augmented Lagrange Formulations for Serial and Parallel Distributed MPC 437
R. R. Negenborn

28 A Hierarchical Distributed MPC Approach: A Practical Implementation 451
A. Zafra-Cabeza and J. M. Maestre

29 Distributed MPC Based on Agent Negotiation 465
J. M. Maestre, D. Muñoz de la Peña and E. F. Camacho

30 Lyapunov-based Distributed MPC Schemes: Sequential and Iterative Approaches 479
J. Liu, D. Muñoz de la Peña and P. D. Christofides

31 **Multi-layer Decentralized MPC of Large-Scale Networked Systems** 495
C. Ocampo-Martinez, V. Puig, J. M. Grosso and S. Montes-de-Oca

32 **Distributed MPC Using Reinforcement Learning Based Negotiation: Application to Large Scale Systems** 517
B. Morcego, V. Javalera, V. Puig and R. Vito

33 **Hierarchical MPC for Multiple Commodity Transportation Networks** 535
J. L. Nabais, R. R. Negenborn, R. B. Carmona-Benítez, L. F. Mendonça and M. A. Botto

34 **On the Use of Suboptimal Solvers for Efficient Cooperative Distributed Linear MPC** 553
G. Pannocchia, S. J. Wright and J. B. Rawlings

35 **Cooperative Distributed MPC Integrating a Steady State Target Optimizer** 569
A. Ferramosca, D. Limon and A. H. González

36 **Cooperative MPC with Guaranteed Exponential Stability** 585
A. Ferramosca

Contributors

A. Ahlén Department of Engineering Sciences, Signals and Systems, Uppsala University, Uppsala, Sweden, e-mail: Anders.Ahlen@signal.uu.se

M. Alamir Gipsa-Lab, CNRS-University of Grenoble, Saint Martin d'Hères, France, e-mail: Mazen.alamir@gipsa-lab.grenoble-inp.fr

F. Allgöwer Institute for Systems Theory and Automatic Control, University of Stuttgart, Stuttgart, Germany, e-mail: frank.allgower@ist.uni-stuttgart.de

P. Andersen Department of Electronic Systems, Automation and Control, Aalborg University, Aalborg, Denmark, e-mail: pa@es.aau.dk

D. Axehill Division of Automatic Control, Linköping University, Linköping, Sweden, e-mail: axehill@isy.liu.se

G. Betti Dipartimento di Elettronica e Informazione, Politecnico di Milano, Milan, Italy, e-mail: betti@elet.polimi.it

B. Biegel Department of Electronic Systems, Automation and Control, Aalborg University, Aalborg, Denmark, e-mail: bbi@es.aau.dk

M. A. Botto IDMEC-Instituto Superior Técnico, Technical University of Lisbon, Lisboa, Portugal, e-mail: ayalabotto@ist.utl.pt

R. Bourdais IETR (UMR-CNRS 6164), SUPELEC, Cesson-Sévigné, France, e-mail: roman.bourdais@supelec.fr

J. Buisson IETR (UMR-CNRS 6164), SUPELEC, Cesson-Sévigné, France, e-mail: jean.buisson@supelec.fr

E. F. Camacho Departamento de Sistemas y Automática, Universidad de Sevilla, Sevilla, Spain, e-mail: eduardo@cartuja.us.es

E. Camponogara Department of Automation and Systems Engineering, Federal University of Santa Catarina, Florianópolis, Brazil, e-mail: camponog@das.ufsc.br

R. B. Carmona-Benítez School of Business and Economics, Universidad Anáhuac México Norte, Huixquilucan, México, e-mail: rafael.carmona@anahuac.mx

A. Casavola DEIS-Università degli Studi della Calabria, Rende, Italy, e-mail: casavola@deis.unical.it

A. A. Cornelio Department of Electrical Engineering, Mercedes-Benz R&D, Bangalore, India, e-mail: avi_anthony.cornelio@daimler.com

P. D. Christofides Department of Chemical and Biomolecular Engineering, Department of Electrical Engineering, University of California, Los Angeles, CA, USA, e-mail: pdc@seas.ucla.edu

C. de Prada Department of Systems Engineering and Automatic Control, School of Industrial Engineering, University of Valladolid, Valladolid, Spain, e-mail: prada@autom.uva.es

B. De Schutter Delft Center for Systems and Control, Delft University of Technology, Delft, The Netherlands, e-mail: b.deschutter@tudelft.nl

M. Diehl Department of Electrical Engineering (ESAT/SCD), Optimization in Engineering Center, Leuven, Belgium, e-mail: Moritz.Diehl@esat.kuleuven.be

M. D. Doan Delft Center for Systems and Control, Delft University of Technology, Delft, The Netherlands, e-mail: m.d.doan@tudelft.nl

D. Dumur E3S, SUPELEC, Gif sur Yvette, France, e-mail: didier.dumur@supelec.fr

P. Béguery Schneider-Electric Industries, Strategy and Innovation, Grenoble, France, e-mail: patrick.beguery@fr.schneider-electric.com

N. H. El-Farra University of California, Davis, California, e-mail: nhelfarra@ucdavis.edu

J. J. Espinosa Departamento de Energía Eléctrica y Automática, Facultad de Minas, Universidad Nacional de Colombia, Medellín, Colombia, e-mail: jjespino@unal.edu.co

M. Farina Dipartimento di Elettronica e Informazione, Politecnico di Milano, Milan, Italy, e-mail: farina@elet.polimi.it

F. Farokhi ACCESS Linnaeus Centre, School of Electrical Engineering, KTH Royal Institute of Technology, Stockholm, Sweden, e-mail: farokhi@ee.kth.se

F. Fele Departamento de Sistemas y Automática, Universidad de Sevilla, Sevilla, Spain, e-mail: filiberto@cartuja.us.es

A. Ferramosca Institute of Technological Development for the Chemical Industry (INTEC), CONICET-Universidad Nacional del Litoral, Santa Fe, Argentina, e-mail: ferramosca@santafe-conicet.gov.ar

E. Garone Universite Libre de Bruxelles, Bruxelles, Belgium, e-mail: egarone@ulb.ac.be

P. Giselsson Department of Automatic Control, Lund University, Lund, Sweden, e-mail: pontusg@control.lth.se

A. H. González Institute of Technological Development for the Chemical Industry (INTEC), CONICET-Universidad Nacional del Litoral (UNL), Santa Fe, Argentina, e-mail: alejgon@santafe-conicet.gov.ar

A. Grancharova Institute of System Engineering and Robotics, Bulgarian Academy of Sciences, Sofia, Bulgaria; Department of Industrial Automation, University of Chemical Technology and Metallurgy, Sofia, Bulgaria, e-mail: alexandra.grancharova@abv.bg

J. M. Grosso Institut de Robòtica i Informàtica Industrial, Technical University of Catalonia, Barcelona, Spain, e-mail: jgrosso@iri.upc.edu

H. Guéguen IETR (UMR-CNRS 6164), SUPELEC, Cesson-Sévigné, France, e-mail: herve.gueguen@supelec.fr

A. Hansson Division of Automatic Control, Linköping University, Linköping, Sweden, e-mail: hansson@isy.liu.se

R. Hermans Department of Electrical Engineering, Eindhoven University of Technology, Eindhoven, The Netherlands, e-mail: r.m.hermans@tue.nl

Y. Hu University of California, Davis, California, e-mail: yehhu@ucdavis.edu

J. M. Igreja INESC-ID/ISEL/IPL, Lisboa, Portugal, e-mail: jigreja@deea.isel.ipl.pt

V. Javalera Institut de Robòtica i Informàtica Industrial (CSIC-UPC), Technical University of Catalonia, Terrassa, Spain, e-mail: vjavaler@iri.upc.edu

T. A. Johansen Department of Engineering Cybernetics, Norwegian University of Science and Technology, Trondheim, Norway, e-mail: Tor.Arne.Johansen@itk.ntnu.no

K. H. Johansson School of Electrical Engineering, ACCESS Linnaeus Centre, KTH Royal Institute of Technology, Stockholm, Sweden, e-mail: kallej@ee.kth.se

A. Jokić Faculty of Mechanical Engineering and Naval Architecture, University of Zagreb, Zagreb, Croatia, e-mail: andrej.jokic@fsb.hr

I. Jurado Departamento de Ingeniería de Sistemas y Automática, Escuela Superior de Ingenieros, Universidad de Sevilla, Seville, Spain, e-mail: ijurado@cartuja.us.es

T. Keviczky Delft Center for Systems and Control, Delft University of Technology, Delft, The Netherlands, e-mail: t.keviczky@tudelft.nl

A. Kozma Department of Electrical Engineering (ESAT/SCD), Optimization in Engineering Center, Leuven, Belgium, e-mail: Attila.Kozma@esat.kuleuven.be

M. Y. Lamoudi Schneider-Electric Industries, Strategy and Innovation, Grenoble, France, e-mail: mohamed-yacine.lamoudi@fr.schneider-electric.com

M. Lazar Department of Electrical Engineering, Eindhoven University of Technology, Eindhoven, The Netherlands, e-mail: m.lazar@tue.nl

J. M. Lemos INESC-ID/IST/UTL, Lisboa, Portugal, e-mail: jlml@inesc-id.pt

D. Limon Departamento de Ingeniería de Sistemas y Automática, Escuela Superior de Ingenieros, Universidad de Sevilla, Sevilla, Spain, e-mail: limon@cartuja.us.es

J. Liu Department of Chemical and Materials Engineering, University of Alberta, Edmonton, Canada, e-mail: jinfeng@ualberta.ca

J. D. López Departamento de Energía Eléctrica y Automática, Facultad de Minas, Universidad Nacional de Colombia, Medellín, Colombia, e-mail: jodlopezhi@unal.edu.co

J. M. Maestre Departamento de Sistemas y Automática, Universidad de Sevilla, Sevilla, Spain, e-mail: pepemaestre@cartuja.us.es

R. Martí Department of Systems Engineering and Automatic Control, School of Industrial Engineering, University of Valladolid, Valladolid, Spain, e-mail: ruben@autom.uva.es

L. F. Mendonça Department of Marine Engineering, Escola Superior Naútica Infante D. Henrique, PaÇo d'Arcos, Portugal, e-mail: luismendonca@enautica.pt

S. Montes-de-Oca Automatic Control Department, Technical University of Catalonia, Barcelona, Spain, e-mail: saul.montes.de.oca@upc.edu

B. Morcego Advanced Control Systems Group, Terrassa, Spain, e-mail: bernardo.morcego@upc.edu

P.-D. Moroşan Acsystème, Rennes, France, e-mail: petru-daniel.morosan@acsysteme.com

M. A. Müller Institute for Systems Theory and Automatic Control, University of Stuttgart, Stuttgart, Germany, e-mail: matthias.mueller@ist.uni-stuttgart.de

D. Muñoz de la Peña Departamento de Sistemas y Automática, Universidad de Sevilla, Sevilla, Spain, e-mail: davidmps@cartuja.us.es

F. J. Muros Departamento de Sistemas y Automática, Universidad de Sevilla, Sevilla, Spain, e-mail: franmuros@gmail.com

J. L. Nabais IDMEC-Department of Informatics and Systems Engineering, Setúbal School of Technology, Polytechnical Institute of Setúbal, Setúbal, Portugal, e-mail: joao.nabais@estsetubal.ips.pt

I. Necoara Automation and Systems Engineering Department, University Politehnica Bucharest, Bucharest, Romania, e-mail: ion.necoara@acse.pub.ro

R. R. Negenborn Department of Marine and Transport Technology, Delft University of Technology, Delft, The Netherlands, e-mail: r.r.negenborn@tudelft.nl

S.-I. Niculescu Signals and Systems Laboratory, SUPELEC Systems Sciences (E3S), Gif Sur Yvette, France, e-mail: silviu.niculescu@lss.supelec.fr

C. Ocampo-Martinez Institut de Robòtica i Informàtica Industrial (CSIC-UPC), Technical University of Catalonia, Barcelona, Spain, e-mail: cocampo@iri.upc.edu

S. Olaru Automatic Control Department, SUPELEC Systems Sciences (E3S), Gif Sur Yvette, France, e-mail: sorin.olaru@supelec.fr

J. A. Patiño Departamento de Energía Eléctrica y Automática, Facultad de Minas, Universidad Nacional de Colombia, Medellín, Colombia, e-mail: japatin0@unal.edu.co

G. Pannocchia Deparartment of Civil and Industrial Engineering, University of Pisa, Pisa, Italy, e-mail: g.pannocchia@diccism.unipi.it

I. Prodan Automatic Control Department, SUPELEC Systems Sciences (E3S), Gif Sur Yvette, France; Signals and Systems Laboratory, SUPELEC Systems Sciences (E3S), Gif Sur Yvette, France, e-mail: ionela.prodan@supelec.fr; ionela.prodan@lss.supelec.fr

V. Puig Institut de Robòtica i Informàtica Industrial (CSIC-UPC), Technical University of Catalonia, Barcelona, Spain, e-mail: vpuig@iri.upc.edu

D. E. Quevedo School of Electrical Engineering and Computer Science, The University of Newcastle, Newcastle, Australia, e-mail: dquevedo@ieee.org

A. Rantzer Department of Automatic Control, Lund University, Lund, Sweden, e-mail: rantzer@control.lth.se

D. M. Raimondo Dipartimento di Informatica e Sistemistica, Università degli Studi di Pavia, Pavia, Italy, e-mail: davide.raimondo@unipv.it

J. B. Rawlings Chemical and Biological Engineering Department, University of Wisconsin, Madison, Wisconsin, USA, e-mail: rawlings@engr.wisc.edu

A. G. Richards Department of Aerospace Engineering, University of Bristol, Bristol, UK, e-mail: arthur.richards@bris.ac.uk

S. Roshany-Yamchi NIMBUS Center for Embedded Research, Cork Institute of Technology, Cork, Ireland, e-mail: samira.roshany@cit.ie

D. Sarabia Department of Systems Engineering and Automatic Control, School of Industrial Engineering, University of Valladolid, Valladolid, Spain, e-mail: dsarabia@autom.uva.es

C. Savorgnan Department of Electrical Engineering (ESAT/SCD), Optimization in Engineering Center, Leuven, Belgium, e-mail: Carlo.Savorgnan@esat.kuleuven.be

R. Scattolini Dipartimento di Elettronica e Informazione, Politecnico di Milano, Milano, Italy, e-mail: riccardo.scattolini@elet.polimi.it

I. Shames Department of Electrical and Electronic Engineering, University of Melbourne, Melbourne, Australia, e-mail: iman.shames@unimelb.edu.au

C. Stoica Automatic Control Department, SUPELEC Systems Sciences (E3S), Gif Sur Yvette, France, e-mail: cristina.stoica@supelec.fr

F. Stoican Department of Engineering Cybernetics, Norwegian University of Science and Technology, Trondheim, Norway, e-mail: florin.stoican@itk.ntnu.no

J. Stoustrup Department of Electronic Systems, Automation and Control, Aalborg University, Aalborg, Denmark, e-mail: jakob@es.aau.dk

F. Tedesco Università degli Studi della Calabria, DEIS, Rende, Italy, e-mail: tedesco@deis.unical.it

P. A. Trodden Department of Automatic Control and Systems Engineering, University of Sheffield, Sheffield, UK, e-mail: p.trodden@sheffield.ac.uk

F. Valencia Departamento de Energía Eléctrica y Automática, Facultad de Minas, Universidad Nacional de Colombia, Medellín, Colombia, e-mail: fvalenc@unal.edu.co

R. Vito Institut de Robòtica i Informàtica Industrial (CSIC-UPC), Technical University of Catalonia, Terrassa, Spain, e-mail: rvito@iri.upc.edu

S. J. Wright Computer Sciences Department, University of Wisconsin, Madison, WI, USA, e-mail: mail: swright@cs.wisc.edu

A. Zafra-Cabeza Department of Automatic Control and Systems Engineering, University of Seville, Seville, Spain, e-mail: asunzafra@us.es

Chapter 1
On 35 Approaches for Distributed MPC Made Easy

R. R. Negenborn and J. M. Maestre

Abstract In this chapter the motivation for developing a comprehensive overview of distributed MPC techniques such as presented in this book is discussed. Understanding the wide range of techniques available becomes easier when a common structure and notation is adopted. Therefore, a list of questions is proposed that can be used to obtain a structured way in which such techniques can be described, and a preferred notation is suggested. This chapter concludes with an extensive categorization of the techniques described in this book, and compact representations of the properties of each individual technique. As such, this chapter serves as a starting point for further developing understanding of the various particularities of the different techniques.

1.1 Introduction

1.1.1 From Centralized to Distributed Control

The evolution of computer science and information technology has made possible the application of control techniques to systems that were beyond the possibilities of control theory just a decade ago. The size of the problems faced today by control engineers has grown enormously as the limitations imposed by the communication and computational capabilities decrease. In this sense, there are strong incentives to be ambitious: society heavily depends on infrastructure systems, such as road-traffic networks, water networks, electricity networks, intermodal transport networks, etc.

R. R. Negenborn (✉)
Department of Marine and Transport Technology, Delft University of Technology, Delft, The Netherlands
e-mail: r.r.negenborn@tudelft.nl

J. M. Maestre
Departamento de Sistemas y Automática, Universidad de Sevilla, Sevilla, Spain
e-mail: pepemaestre@cartuja.us.es

J. M. Maestre and R. R. Negenborn (eds.), *Distributed Model Predictive Control Made Easy*, Intelligent Systems, Control and Automation: Science and Engineering 69,
DOI: 10.1007/978-94-007-7006-5_1, © Springer Science+Business Media Dordrecht 2014

These are examples of large-scale, networked systems, that everybody makes use of on a daily basis, and any performance improvement would have a great direct impact on the society. However, traditional centralized control approaches cannot be used with these kind of systems due to, e.g., centralized computational issues or issues with centralized modeling, data collection and actuation.

It is at this point where distributed controllers come into play. The idea behind distributed control approaches is simple: the centralized problem is divided in several different parts whose control is assigned to a certain number of local controllers or *agents*. Therefore, each agent does not have a global vision of the problem. Depending on the degree of interaction that exists between the local subsystems, the agents may need to communicate so that they can coordinate themselves.

Distributed approaches have important advantages that justify their use. The first advantage is that in general these schemes are easier to implement. Their computational requirements are lower because a difficult problem is substituted by several smaller problems. In addition, these schemes are scalable. Their inherent modularity simplifies the system maintenance and the possible expansions of the control system. Moreover, the modularity provides robustness in comparison with a centralized controller. A possible failure does not have to affect the overall system. For this reason, distributed systems have a greater tolerance to failures. Nevertheless, these systems have also several drawbacks that have to be taken into account, being the main one the loss of performance in comparison with a centralized controller. This loss depends on the degree of interaction between the local subsystems and the coordination mechanisms between the agents.

During the last years many distributed control approaches using the control strategy model predictive control (MPC) have been proposed. MPC is a popular control strategy for the design of high performance model-based process control systems because of its ability to handle multi-variable interactions, constraints on control (manipulated) inputs and system states, and optimization requirements in a systematic manner. These features are essential in this context because they allow the control engineer to handle explicitly the interactions between the different subsystems.

1.1.2 Need for an Overview

Although many approaches for distributed MPC have been proposed, a coherent and easily accessible overview of the components of these approaches is lacking. Having such an overview would facilitate making the approaches easier known to a wider community, and help students, researchers and practitioners in choosing the approach most suitable for their particular situation.

This book is the result of the efforts of creating such an overview. For researchers and students, the book can provide a state-of-the-art overview, while at the same time making clear directions of research that deserve more attention. The goal that was kept in mind while developing the book was to make available to a wide audience in a systematic, practical, and accessible way the available approaches for distributed (and hierarchical) MPC. To make this possible, in each chapter of the book one particular approach is described, including:

1. the rationale and background of the approach;
2. the assumptions made on the system dynamics, control objectives and constraints;
3. a step-by-step description of the computations/equations;
4. the availability of theoretical results;
5. the availability of real (or simulated) applications.

The consistent description of all approaches enables readers to compare the approaches and assess the usefulness of these approaches for their respective applications.

This remainder of this chapter is structured as follows. In Sect. 1.2 we present the list of questions that served as basis for structuring the descriptions of the various schemes and the notation suggested to be used is presented. In Sect. 1.3, specific properties of distributed MPC schemes are described and a naming convention for distributed MPC schemes is proposed. Section 1.4 provides concluding remarks. Finally, Sect. 1.4 gives a categorization of schemes according to values of the properties and summaries of the specific values of a property per scheme.

1.2 Distributed MPC Schemes Commonalities

In order to obtain coherence in the description of distributed MPC schemes, we propose to structure the description of a scheme using a pre-specified set of questions and a common notation. These questions and notation suggestions were also used in the preparation of the schemes described in this book.

1.2.1 Description Questions

The most important and distinguishing elements of each distributed MPC scheme are captured by describing a scheme along the following line:

1. Short introduction
   - What is the rationale behind the approach?
   - Where did it originate from?
   - What makes this approach interesting?
2. Boundary conditions on considered system, control objectives, and constraints
   - What kind of system partition and type of dynamical model is assumed?
   - What kind of control problem is being solved for this kind of system?
   - What kind of communication architecture is assumed available?

3. Step-by-step description of how the approach works
   - What initialization is required?
   - What equation/optimization is used when?
   - When does what communication take place with which agent?
   - When do agents agree on/decide to take an action?
4. Availability of theoretical results
   - What theoretical properties have been investigated?
   - How does the approach relate to other existing approaches?
   - Where have results of this been published?
5. Availability of actual applications of the approach
   - What are the systems in which your approach has been tested?
   - Where have results of this been published?

1.2.2 Consistent Notation

Considering a common notation increases the coherence among the chapters and facilitates comparing the various techniques discussed. The following notational guidelines are suggested:

- Number types:
  - $\mathbb{R}$ for real numbers.
  - $\mathbb{Z}$ for integers.
- Vectors:
  - Boldface and small characters, e.g., $\mathbf{x}$.
  - n_x for dimension of a vector $\mathbf{x}$.
  - $\mathbf{x}(k+1:k+N_\mathrm{p})$ represents $[\mathbf{x}^\mathrm{T}(k+1),\ldots,\mathbf{x}^\mathrm{T}(k+N_\mathrm{p})]^\mathrm{T}$.
- Matrices: Boldface and capital characters, e.g., $\mathbf{A}$, $\mathbf{B}$.
- Sets: Calligraphic characters, e.g., $\mathcal{N}$.
- Model description:
  - $\mathbf{x}$ for states
  - $\mathbf{u}$ for inputs/actions
  - $\mathbf{y}$ for outputs/measurements
  - $\mathbf{d}$ for disturbances.
- Functions:
  - f, g, h for functions returning scalars; boldface versions for functions returning a column vector.
  - $\oplus$ for Minkowski sum, i.e., $\mathcal{A}\oplus\mathcal{B}\triangleq\{a+b : a\in\mathcal{A}, b\in\mathcal{B}\}$.
  - $\sim$ for the Pontryagin difference, i.e., $\mathcal{A}\sim\mathcal{B}\triangleq\{a : a+b\in\mathcal{A}, \forall b\in\mathcal{B}\}$.

- Subsystems:
  - $\mathcal{N}$ for the set of subsystems.
  - $|\mathcal{N}|$ for the cardinality of $\mathcal{N}$, i.e., the number of subsystems.
  - Subscript i for a variable of subsystem i, e.g., $\mathbf{x}_i$.
  - Subscript j for a variable of another subsystem (e.g., neighboring), e.g., $\mathbf{x}_j$.
- Network: Graph $(\mathcal{N}, \mathcal{E})$, where $\mathcal{N}$ is the set of subsystems and $\mathcal{E} \subset \mathcal{N} \times \mathcal{N}$ is the set of edges/links.
- Time:
  - Discrete time index k.
  - Continuous time t.
  - Time index as parameter of variable: in brackets behind a variable, e.g., $\mathbf{x}(k)$.
- Model predictive controller:
  - Performance index: J.
  - Penalty matrices: $\mathbf{Q}$ for the state, $\mathbf{R}$ for the actuation.
  - Target/references values for $\mathbf{x}$: $\mathbf{x}_{\text{ref}}$.
  - Target/references values for subsystem i: $\mathbf{x}_{\text{ref},i}$.
  - Prediction horizon length: N_{p}.
  - Control horizon length: N_{c}.
  - Time step within prediction horizon: l.
  - The value of a variable x at iteration p will be denoted using a superscript: x^p.
- Math:
  - Transpose: Superscript T, e.g., $\mathbf{x}^{\text{T}}$.
  - Min/max: Subscripts min and max of a variable, e.g., $x_{\min}$ and $x_{\max}$, represent the minimum and maximum value of that variable, respectively.
  - $\|\mathbf{z}\|_{\mathbf{M}}$ is the weighted Euclidean norm, i.e., $\sqrt{\mathbf{z}^{\text{T}}\mathbf{M}\mathbf{z}}$

1.3 Properties for Categorization

There exist many different features that can be used to described a distributed control scheme. We consider the following main sets of properties as basis for comparison: process properties, control architecture properties, and theoretical properties.

1.3.1 Process Properties

This set of properties is related with the specifications of the physical system that is going to be controlled using a distributed scheme. This is an important point since it determines what schemes can be used to control the system. The most important process features are:

- **System type**, or the way in which the scheme is derived: either starting from a group of autonomous systems and then introducing communication to obtain coordination, or from a (hypothetical) monolithic system decomposed into subsystems that are coordinated taking into account limitations in communication/processing power. This book itself has been partitioned in two parts according to this feature.
- **Process type**, or the kind of dynamics that better capture the behavior of the system: linear, non-linear, or hybrid.
- **Type of model**, or the way in which the system model is described mathematically: transfer function or state-space.
- **Randomness**, i.e., whether the process shows a deterministic or non-deterministic behavior.
- **Type of control**, i.e., the control goal: regulation, tracking, or economic.
- **Coupling source**, or whatever makes the overall optimization problem become non-separable: inputs, outputs, state, objective, or constraints. Notice that some schemes may deal with more than one coupling source and consequently they appear in more than one column.

We present a classification of the schemes contained in the book according to this feature in Table 1.1. Table 1.2 shows a classification of the schemes in the book according to this feature. The classification of schemes according to the type of model is presented in Table 1.3. Table 1.4 presents a classification of the schemes according to the randomness feature. Table 1.5 shows a classification of the schemes in the book according to the type of control. In Table 1.6 we present the schemes that deal with each of the five possible coupling sources.

1.3.2 Control Architecture Properties

This set of properties describe the essence of the schemes presented in the book. These features are important from a practical point of view, as there may be real situations in which a scheme cannot be applied due to, for example, communicational or informational constraints. The most important features are:

- **Architecture**, or how the coordination between local controllers is structured: decentralized, distributed, and hierarchical. In general, the controllers can be classified depending on how many of them participate in the solution of the control problem and the relative importance between them. We say that a control system is centralized if there is a single controller that solves the plant-wide problem. The control is decentralized when there are local controllers in charge of the local subsystems of the plant that require no communication among them. When the local controllers communicate in order to find a cooperative solution for the overall control problem the control system is distributed. Finally, if there are different control layers coordinated to take care of the process the control system is hierarchical. In this case, upper layers manage the global objectives of the process and provide references for the lower layers, which control directly the plant.

- **Controller knowledge**, which measures the degree of global information that a a controller has: strictly local or partially global. Naturally, the amount of information that each agent has about the overall system has a great impact in the performance and ease of implementation of the scheme.
- **Computation type**, or how the joint control actions are calculated, in an iterative or non-iterative fashion.
- **Controller's attitude**, i.e., the degree in which an agent takes into account other agents' objectives: non-cooperative, or cooperative. In general, attitude is related with the will of collaboration between subsystems. We say that a controller has a noncooperative attitude if it behaves selfishly, i.e., it only seeks the maximization of its own objective function. On the other hand, the controllers's attitude is cooperative when the it minimizes not only its cost but the cost of its neighbors. Hence, it may make sacrifices in terms of its own welfare to help the overall system attain a better global situation. Notice that there are some schemes that are really in between these two categories as their local controllers have both cooperative and noncooperative features. In this case, we have approximated the controller to the category in which it fits better.
- **Communication**, or if there is a sequence in which the agents transmit and receive information: serial and parallel. In particular, under the serial communication paradigm only one controller can communicate at the same time in contrast with parallel communication, where several controllers are allowed to communicate at the same time.
- **Timing**, or whether there is or not a strict schedule in the communication process that determines when controllers can communicate: synchronous or asynchronous.
- **Optimization variables**, i.e., the nature of the variables in the optimization problem: real or integer. Notice that some schemes are in both columns since they use both these type of variables, i.e., they solve a mixed-integer optimization problem.

The classification of schemes according to this feature is presented in Table 1.7. The classification of schemes according to the controller knowledge is shown in Table 1.8. Table 1.9 shows the classification of schemes according to the computation type. Table 1.10 presents a classification of the schemes contained in the book according to the controller's attitude. In Table 1.11 we present the classification of the schemes according to the communication. A classification of the chapters of the book according to timing can be seen in Table 1.12. Table 1.13 presents the schemes that use each type of optimization variable.

1.3.3 Theoretical Properties

This set of properties has to do with the availability of mathematical results that provide a certain guarantee regarding the scheme's performance. The following set of properties have been considered:

- **Optimality**, i.e., if the scheme provides the same result that the corresponding centralized optimization problem.
- **Suboptimality bounds**, i.e., if the scheme provides a measurement of the distance with respect to the optimum of the corresponding centralized optimization problem.
- **Stability**, i.e., if the the scheme guarantees a non-divergent evolution of the state and the output of the system.
- **Robustness**, if the scheme is able to reject external unknown disturbances.

Table 1.14 shows a classification of the schemes according to optimality. A classification of the schemes of the book according to the suboptimality bounds property is presented in Table 1.15. Table 1.16 shows a classification of the schemes according to the stability property. We present a classification of the schemes according to the robustness property in Table 1.17.

1.3.4 Naming Convention

Using the values of the properties of the distributed MPC schemes, we propose a naming convention, to able to refer to different schemes in a standardized way. The short name that we propose is derived from three of the aforementioned features, the key features that are used to structure the distributed MPC schemes in this book. In particular, each short name is composed of:

- Two letters indicating if the scheme is tailored for monolithical systems (Mo) or multiple independent systems (Ma).
- Two or three letters indicating if the local controllers use a sensible amount of global information (Glo) or mainly local information (Lo).
- Two or three letters indicating whether the scheme is iterative (It) or not (Nit).
- A slash, followed by the first letters of the first author of the chapter.

This short naming convention provides the essence of each scheme at a glance. Naturally, there may be other short naming options. We propose this way for the following reasons.

- We consider the main property the *System type* feature. Undoubtedly, this is the most important feature since it has to do with the perspective from which the distributed control scheme has been designed, bottom-up or top-down. Likewise, there is also a sensitive difference in the target applications of these two families of schemes. We observe two different perspectives: *The Group of Autonomous Systems Perspective* and *The Decomposed Monolithic System Perspective*.
- The second property has to do with the degree of global information that local controllers have. On the one have we have local controllers that only have strict local information and on the other we have controllers with a significant amount of global information. While the first approach could be favorable for systems with

a dynamic structure (e.g. plug-and-play networks), the second can simplify the coordination due to the additional knowledge about the overall system.
- Finally, we have chosen as third distinguishing property in the short naming the *Computation type*, i.e., if the scheme is iterative or not.

This book is structured using the same criteria on which the short naming convention is based. The chapters in the book are ordered from most distributed to most centralized, as shown in Table 1.1. Indeed, the first chapter discussed scheme MaLoNit, i.e., multiple subsystems controlled by agents that only have local knowledge and that do not iterate in order to find a solution. This contrasts the last chapter of the book, which is MoGloIte, i.e, a scheme considering local controllers that have global system information and exchange information iteratively in order to find a control action for the decomposed monolithical system they are controlling.

1.4 Concluding Remarks

The techniques presented in the book, introduced in accessible and standardized way, constitute a true handbook of distributed model predictive control. In this sense, we believe that this book will become a valuable aid for those readers that are beginning their research careers (e.g.: master and PhD students). Likewise, researchers in the field will find in this book a valuable survey of the state-of-the-art methods carefully explained by their original contributors. But the contribution of the book goes far beyond theory. The current compilation of hierarchical, decentralized and specially distributed model predictive control schemes provides a potential solution to almost any imaginable application. Vehicle formation, irrigation canals, chemical processes, energy networks, among many others, are used as benchmarks in the book. For this reason, practitioners will also find the tools and *recipes* they need to face their practical challenges.

As editors of this compilation of schemes, we do not expect this first edition of the book to be the last one. Despite that the field of distributed MPC has registered a strong research activity during the last decade, there are still new schemes being proposed every now and then. On top of that, it is also easy to find enhancements of previous schemes that guarantee new theoretical properties. Nevertheless, we can say the basis of this young research field is already settled.

What will the future bring? We envision several potential research lines. We expect a refinement of the most popular schemes aiming to a commercial implementation. So far, most distributed control schemes have only been tested via simulations or with lab benchmarks. A real implementation will demand to discard some of the simplifications made during the research, e.g: common distributed computing fallacies as assuming that the network is reliable. We also expect a proliferation of dynamic distributed control schemes, able to adapt the degree of coordination to the circumstances. For example, one may think of a traffic network. When the system is close to congestion, a high degree of coordination and communication is necessary between the agents. On the contrary, if the roads are almost empty (e.g., at night), such coordination is not needed and agents can work in a decentralized fashion. Furthermore, the development of the field will also be linked to the evolution of paradigms that may become strong research topics such as the domain of systems of systems.

Acknowledgments This research is supported by the VENI project "Intelligent multi-agent control for flexible coordination of transport hubs" (project 11210) of the Dutch Technology Foundation STW, a subdivision of The Netherlands Organisation for Scientific Research (NWO), and the projects "Model predictive techniques for efficient management of renewable energy micro-networks" (project DPI2010-21589-C05-01) and "Networked Model Predictive Control" (project DPI2008-05818), from the Spanish Ministry of Economy and Competitiveness.

Appendix: Comparing the Approaches

Below follows in Tables 1.1–1.17 per property a categorization per value of the described schemes. After this follows per scheme a summary of the particular values of a property of that scheme. The short names of the distributed MPC schemes falling in particular categories are mentioned, as well as references to the related chapters.

Table 1.1 Scheme classification with respect to the system type

System type			
Multiple autonomous systems		Decomposed monolithical system	
MaLoNit-Val (Chap. 2; [34])	MaLoNit-Tro (Chap. 3; [33])		
MaLoNit-Mar (Chap. 4; [22])	MaLoNit-Mul (Chap. 5; [24])	MoLoIt-Jur (Chap. 22; [1])	MoLoIt-Axe(Chap. 23; [2])
MaLoIt-Bou (Chap. 6; [5])	MaLoIt-Far (Chap. 7; [9])	MoLoIt-Doa (Chap. 24; [8])	MoLoNit-Mae (Chap. 25; [21])
MaLoIt-Lem (Chap. 8; [18])	MaLoIt-Lam (Chap. 9; [17])	MoLoNit-Bet (Chap. 26; [3])	MoLoIt-Neg (Chap. 27; [27])
MaLoIt-Nec (Chap. 10; [26])	MaLoIt-Bie (Chap. 11; [4])	MoLoIt-Zaf (Chap. 28; [35])	MoLoIt-MaeMun (Chap. 29; [20])
MaLoIt-Cam (Chap. 12; [6])	MaGloNit-Hu (Chap. 13; [15])	MoLoIt-Liu (Chap. 30; [19])	MoGloNit-Oca (Chap. 31; [28])
MaGloNit-Her (Chap. 14; [14])	MaGloNit-Ted (Chap. 15; [32])	MoGloNit-Mor (Chap. 32; [23])	MoGloNit-Nab (Chap. 33; [25])
MaGloNit-Cas (Chap. 16; [7])	MaGloIt-Pro (Chap. 17; [30])	MoGloIt-Pan (Chap. 34; [29])	MoGloIt-Fer (Chap. 35; [11])
MaGloIt-Gra (Chap. 18; [13])	MaGloIt-Gis (Chap. 19; [12])	MoGloIt-Fer2 (Chap. 36; [10])	
MaGloIt-Koz (Chap. 20; [16])	MaGloIt-Ros (Chap. 21; [31])		

Table 1.2 Scheme classification with respect to the process dynamics

Process type		
Linear	Nonlinear	Hybrid
MaLoNit-Val (Chap. 2; [34])		
MaLoNit-Tro (Chap. 3; [33])		
MaLoIt-Bou (Chap. 6; [5])		
MaLoIt-Lem (Chap. 8; [18])		
MaLoIt-Nec (Chap. 10; [26])		
MaLoIt-Bie (Chap. 11; [4])		
MaLoIt-Cam (Chap. 12; [6])		
MaGloNit-Her (Chap. 14; [14])		

(continued)

Table 1.2 (continued)

Process type		
Linear	Nonlinear	Hybrid
MaGloNit-Ted (Chap. 15; [32])		
MaGloNit-Cas (Chap. 16; [7])	MaLoNit-Mar (Chap. 4; [22])	
MaGloIt-Pro (Chap. 17; [30])	MaLoNit-Mul (Chap. 5; [24])	
MaGloIt-Gis (Chap. 19; [12])	MaLoIt-Far (Chap. 7; [9])	
MaGloIt-Ros (Chap. 21; [31])	MaLoIt-Lam (Chap. 9; [17])	MoLoIt-Axe (Chap. 23; [2])
MoLoIt-Jur (Chap. 22; [1])	MaGloNit-Hu (Chap. 13; [15])	
MoLoIt-Doa (Chap. 24; [8])	MaGloIt-Gra (Chap. 18; [13])	
MoLoNit-Mae (Chap. 25; [21])	MaGloIt-Koz (Chap. 20; [16])	
MoLoNit-Bet (Chap. 26; [3])	MoLoIt-Liu (Chap. 30; [19])	
MoLoIt-Neg (Chap. 27; [27])		
MoLoIt-Zaf (Chap. 28; [35])		
MoLoIt-Mae2 (Chap. 29; [20])		
MoGloNit-Oca (Chap. 31; [28])		
MoGloNit-Mor (Chap. 32; [23])		
MoGloNit-Nab (Chap. 33; [25])		
MoGloIt-Pan (Chap. 34; [29])		
MoGloIt-Fer (Chap. 35; [11])		
MoGloIt-Fer2 (Chap. 36; [10])		

Table 1.3 Scheme classification with respect to the type of model used

Type of model		
Transfer function	State space	
	MaLoNit-Val (Chap. 2; [34])	MaLoNit-Tro (Chap. 3; [33])
	MaLoNit-Mar (Chap. 4; [22])	MaLoNit-Mul (Chap. 5; [24])
	MaLoIt-Bou (Chap. 6; [5])	MaLoIt-Far (Chap. 7; [9])
	MaLoIt-Lem (Chap. 8; [18])	MaLoIt-Lam (Chap. 9; [17])
	MaLoIt-Nec (Chap. 10; [26])	MaLoIt-Bie (Chap. 11; [4])
	MaLoIt-Cam (Chap. 12; [6])	MaGloNit-Hu (Chap. 13; [15])
	MaGloNit-Her (Chap. 14; [14])	MaGloNit-Ted (Chap. 15; [32])
	MaGloNit-Cas (Chap. 16; [7])	MaGloIt-Pro (Chap. 17; [30])
MoGloIt-Pan (Chap. 34; [29])	MaGloIt-Gra (Chap. 18; [13])	MaGloIt-Gis (Chap. 19; [12])
	MaGloIt-Koz (Chap. 20; [16])	MaGloIt-Ros (Chap. 21; [31])
	MoLoIt-Jur (Chap. 22; [1])	MoLoIt-Axe (Chap. 23; [2])
	MoLoIt-Doa (Chap. 24; [8])	MoLoNit-Mae (Chap. 25; [21])
	MoLoNit-Bet (Chap. 26; [3])	MoLoIt-Neg (Chap. 27; [27])
	MoLoIt-Zaf (Chap. 28; [35])	MoLoIt-Mae2 (Chap. 29; [20])
	MoLoIt-Liu (Chap. 30; [19])	MoGloNit-Oca (Chap. 31; [28])
	MoGloNit-Mor (Chap. 32; [23])	MoGloNit-Nab (Chap. 33; [25])
	MoGloIt-Fer (Chap. 35; [11])	MoGloIt-Fer2 (Chap. 36; [10])

Table 1.4 Scheme classification with respect to the randomness of the process

Randomness			
Deterministic		Non-deterministic	
MaLoNit-Val (Chap. 2; [34])	MaLoNit-Mar (Chap. 4; [22])		
MaLoNit-Mul (Chap. 5; [24])	MaLoIt-Bou (Chap. 6; [5])		
MaLoIt-Lem (Chap. 8; [18])	MaLoIt-Lam (Chap. 9; [17])		
MaLoIt-Nec (Chap. 10; [26])	MaLoIt-Bie (Chap. 11; [4])		
MaLoIt-Cam (Chap. 12; [6])	MaGloNit-Hu (Chap. 13; [15])		
MaGloNit-Her (Chap. 14; [14])	MaGloNit-Ted (Chap. 15; [32])		
MaGloNit-Cas (Chap. 16; [7])	MaGloIt-Gra (Chap. 18; [13])	MaLoNit-Tro (Chap. 3; [33])	MaLoIt-Far (Chap. 7; [9])
MaGloIt-Gis (Chap. 19; [12])	MaGloIt-Koz (Chap. 20; [16])	MaGloIt-Pro (Chap. 17; [30])	MoLoIt-Jur (Chap. 22; [1])
MaGloIt-Ros (Chap. 21; [31])	MoLoIt-Axe (Chap. 23; [2])	MoLoIt-Zaf (Chap. 28; [35])	
MoLoIt-Doa (Chap. 24; [8])	MoLoNit-Mae (Chap. 25; [21])		
MoLoNit-Bet (Chap. 26; [3])	MoLoIt-Neg (Chap. 27; [27])		
MoLoIt-Mae2 (Chap. 29; [20])	MoLoIt-Liu (Chap. 30; [19])		
MoGloNit-Oca (Chap. 31; [28])	MoGloNit-Mor (Chap. 32; [23])		
MoGloNit-Nab (Chap. 33; [25])	MoGloIt-Pan (Chap. 34; [29])		
MoGloIt-Fer (Chap. 35; [11])	MoGloIt-Fer2 (Chap. 36; [10])		

Table 1.5 Scheme classification with respect to the objective of type of control that the operation of the process demands

Type of Control		
Regulation	Tracking	Economic
MaLoNit-Tro(Chap. 3; [33])		
MaLoNit-Mul (Chap. 5; [24])		
MaLoIt-Far (Chap. 7; [9])		
MaLoIt-Nec (Chap. 10; [26])	MaLoNit-Val (Chap. 2; [34])	
MaLoIt-Cam (Chap. 12; [6])	MaLoNit-Mar (Chap. 4; [22])	
MaGloNit-Hu (Chap. 13; [15])	MaLoIt-Bou (Chap. 6; [5])	MaLoNit-Val (Chap. 2; [34])
MaGloNit-Her (Chap. 14; [14])	MaLoIt-Lem (Chap. 8; [18])	MaLoIt-Bou (Chap. 6; [5])
MaGloIt-Pro (Chap. 17; [30])	MaGloNit-Ted (Chap. 15; [32])	MaLoIt-Lam (Chap. 9; [17])
MaGloIt-Gra (Chap. 18; [13])	MaGloNit-Cas (Chap. 16; [7])	MaLoIt-Bie (Chap. 11; [4])
MaGloIt-Gis (Chap. 19; [12])	MaGloIt-Ros (Chap. 21; [31])	MaGloIt-Koz (Chap. 20; [16])
MoLoIt-Jur (Chap. 22; [1])	MoLoIt-Axe (Chap. 23; [2])	MoLoIt-Neg (Chap. 27; [27])
MoLoIt-Doa (Chap. 24; [8])	MoLoIt-Neg (Chap. 27; [27])	MoGloNit-Oca (Chap. 31; [28])
MoLoNit-Mae (Chap. 25; [21])	MoGloNit-Nab (Chap. 33; [25])	
MoLoNit-Bet (Chap. 26; [3])	MoGloIt-Pan (Chap. 34; [29])	
MoLoIt-Neg (Chap. 27; [27])	MoGloIt-Fer (Chap. 35; [11])	
MoLoIt-Zaf (Chap. 28; [35])		
MoLoIt-Mae2 (Chap. 29; [20])		
MoLoIt-Liu (Chap. 30; [19])		
MoGloNit-Mor (Chap. 32; [23])		
MoGloIt-Fer2 (Chap. 36; [10])		

Table 1.6 Scheme classification with respect to the possible sources of coupling

Coupling source Constraints		Objective	
MaLoNit-Val (Chap. 2; [34])	MaLoNit-Tro (Chap. 3; [33])		
MaLoNit-Mar (Chap. 4; [22])	MaLoNit-Mul (Chap. 5; [24])		
MaLoIt-Bou (Chap. 6; [5])	MaLoIt-Lam (Chap. 9; [17])	MaLoNit-Val (Chap. 2; [34])	MaLoNit-Mul (Chap. 5; [24])
MaLoIt-Nec (Chap. 10; [26])	MaLoIt-Cam (Chap. 12; [6])	MaLoIt-Far (Chap. 7; [9])	MaLoIt-Nec (Chap. 10; [26])
MaGloNit-Hu (Chap. 13; [15])	MaGloNit-Ted (Chap. 15; [32])	MaGloIt-Pro (Chap. 17; [30])	MaGloIt-Gra (Chap. 18; [13])
MaGloNit-Cas (Chap. 16; [7])	MaGloIt-Pro (Chap. 17; [30])	MoLoIt-Jur (Chap. 22; [1])	MoLoIt-Doa (Chap. 24; [8])
MaGloIt-Gis (Chap. 19; [12])	MoLoIt-Doa (Chap. 24; [8])	MoLoIt-Liu (Chap. 30; [19])	MoGloNit-Oca (Chap. 31; [28])
MoLoNit-Bet (Chap. 26; [3])	MoGloNit-Oca (Chap. 31; [28])	MoGloIt-Pan (Chap. 34; [29])	
MoGloNit-Nab (Chap. 33; [25])	MoGloIt-Pan (Chap. 34; [29])		
MoGloIt-Fer2 (Chap. 36; [10])			

Input	**Output**	**State**
MaLoNit-Val (Chap. 2; [34])		
MaLoNit-Mar (Chap. 4; [22])		
MaLoIt-Lem (Chap. 8; [18])		MaLoNit-Val (Chap. 2; [34])
MaLoIt-Lam (Chap. 9; [17])		MaLoIt-Far (Chap. 7; [9])
MaLoIt-Bie (Chap. 11; [4])		MaLoIt-Bie (Chap. 11; [4])
MaLoIt-Cam (Chap. 12; [6])		MaGloNit-Hu (Chap. 13; [15])
MaGloNit-Cas (Chap. 16; [7])		MaGloNit-Her (Chap. 14; [14])
MaGloIt-Gis (Chap. 19; [12])	MaLoNit-Val (Chap. 2; [34])	MaGloNit-Cas (Chap. 16; [7])
MaGloIt-Ros (Chap. 21; [31])	MaLoIt-Cam (Chap. 12; [6])	MaGloIt-Gra (Chap. 18; [13])
MoLoIt-Doa (Chap. 24; [8])	MaGloIt-Koz (Chap. 20; [16])	MaGloIt-Gis (Chap. 19; [12])
MoLoNit-Mae (Chap. 25; [21])	MoLoNit-Bet (Chap. 26; [3])	MaGloIt-Koz (Chap. 20; [16])
MoLoNit-Bet (Chap. 26; [3])	MoLoIt-Neg (Chap. 27; [27])	MaGloIt-Ros (Chap. 21; [31])
MoLoIt-Neg (Chap. 27; [27])		MoLoIt-Jur (Chap. 22; [1])
MoLoIt-Zaf (Chap. 28; [35])		MoLoIt-Doa (Chap. 24; [8])
MoLoIt-MaeMun (Chap. 29; [20])		MoLoNit-Bet (Chap. 26; [3])
MoGloNit-Oca (Chap. 31; [28])		MoLoIt-Neg (Chap. 27; [27])
MoGloNit-Mor (Chap. 32; [23])		MoLoIt-Zaf (Chap. 28; [35])
MoGloIt-Pan (Chap. 34; [29])		MoLoIt-Liu (Chap. 30; [19])
MoGloIt-Fer (Chap. 35; [11])		MoGloNit-Nab (Chap. 33; [25])
MoGloIt-Fer2 (Chap. 36; [10])		

Table 1.7 Scheme classification with respect to the type of control architecture that the scheme implements

Architecture		
Decentralized	Distributed	Hierarchical
	MaLoNit-Val (Chap. 2; [34])	
	MaLoNit-Tro(Chap. 3; [33])	
	MaLoNit-Mul (Chap. 5; [24])	
	MaLoIt-Bou (Chap. 6; [5])	
	MaLoIt-Far (Chap. 7; [9])	
	MaLoIt-Lam (Chap. 9; [17])	
	MaLoIt-Nec (Chap. 10; [26])	
	MaLoIt-Bie (Chap. 11; [4])	
	MaLoIt-Cam (Chap. 12; [6])	
	MaGloNit-Hu (Chap. 13; [15])	
	MaGloNit-Ted (Chap. 15; [32])	MaLoNit-Mar (Chap. 4; [22])
	MaGloNit-Cas (Chap. 16; [7])	MaGloIt-Pro (Chap. 17; [30])
	MaGloIt-Gra (Chap. 18; [13])	MaGloIt-Koz (Chap. 20; [16])
MaLoIt-Lem (Chap. 8; [18])	MaGloIt-Gis (Chap. 19; [12])	MoLoIt-Doa (Chap. 24; [8])
MoGloNit-Oca (Chap. 31; [28])	MaGloIt-Koz (Chap. 20; [16])	MoLoIt-Zaf (Chap. 28; [35])
	MaGloIt-Ros (Chap. 21; [31])	MoGloNit-Oca (Chap. 31; [28])
	MoLoIt-Jur (Chap. 22; [1])	MoGloNit-Nab (Chap. 33; [25])
	MoLoIt-Axe (Chap. 23; [2])	
	MoLoIt-Doa (Chap. 24; [8])	
	MoLoNit-Mae (Chap. 25; [21])	
	MoLoNit-Bet (Chap. 26; [3])	
	MoLoIt-Neg (Chap. 27; [27])	
	MoLoIt-Mae2 (Chap. 29; [20])	
	MoLoIt-Liu (Chap. 30; [19])	
	MoGloNit-Mor (Chap. 32; [23])	
	MoGloIt-Pan (Chap. 34; [29])	
	MoGloIt-Fer (Chap. 35; [11])	
	MoGloIt-Fer2 (Chap. 36; [10])	

Table 1.8 Scheme classification with respect to the controller knowledge

Controller knowledge			
Strictly local		Partially global	
MaLoNit-Val (Chap. 2; [34])	MaLoNit-Tro (Chap. 3; [33])		
MaLoNit-Mar (Chap. 4; [22])	MaLoNit-Mul (Chap. 5; [24])	MaGloNit-Hu (Chap. 13; [15])	MaGloNit-Her (Chap. 14; [14])
MaLoIt-Bou (Chap. 6; [5])	MaLoIt-Far (Chap. 7; [9])	MaGloNit-Ted (Chap. 15; [32])	MaGloNit-Cas (Chap. 16; [7])
MaLoIt-Lem (Chap. 8; [18])	MaLoIt-Lam (Chap. 9; [17])	MaGloIt-Pro (Chap. 17; [30])	MaGloIt-Gra (Chap. 18; [13])
MaLoIt-Nec (Chap. 10; [26])	MaLoIt-Bie (Chap. 11; [4])	MaGloIt-Gis (Chap. 19; [12])	MaGloIt-Koz (Chap. 20; [16])
MaLoIt-Cam (Chap. 12; [6])	MoLoIt-Jur (Chap. 22; [1])	MaGloIt-Ros (Chap. 21; [31])	MoLoIt-Axe (Chap. 23; [2])
MoLoIt-Axe (Chap. 23; [2])	MoLoIt-Doa (Chap. 24; [8])	MoGloNit-Oca (Chap. 31; [28])	MoGloNit-Mor (Chap. 32; [23])
MoLoNit-Mae (Chap. 25; [21])	MoLoNit-Bet (Chap. 26; [3])	MoGloNit-Nab (Chap. 33; [25])	MoGloIt-Pan (Chap. 34; [29])
MoLoIt-Neg (Chap. 27; [27])	MoLoIt-Zaf (Chap. 28; [35])	MoGloIt-Fer (Chap. 35; [11])	MoGloIt-Fer2 (Chap. 36; [10])
MoLoIt-Mae2 (Chap. 29; [20])	MoLoIt-Liu (Chap. 30; [19])		

Table 1.9 Scheme classification with respect to the type of computation used by the scheme

Computation type			
Iterative		Non-iterative	
MaLoIt-Bou (Chap. 6; [5])	MaLoIt-Far (Chap. 7; [9])		
MaLoIt-Lem (Chap. 8; [18])	MaLoIt-Lam (Chap. 9; [17])		
MaLoIt-Nec (Chap. 10; [26])	MaLoIt-Bie (Chap. 11; [4])	MaLoNit-Val (Chap. 2; [34])	MaLoNit-Tro (Chap. 3; [33])
MaLoIt-Cam (Chap. 12; [6])	MaGloIt-Pro (Chap. 17; [30])	MaLoNit-Mar (Chap. 4; [22])	MaLoNit-Mul (Chap. 5; [24])
MaGloIt-Gra (Chap. 18; [13])	MaGloIt-Gis (Chap. 19; [12])	MaGloNit-Hu (Chap. 13; [15])	MaGloNit-Her (Chap. 14; [14])
MaGloIt-Koz (Chap. 20; [16])	MaGloIt-Ros (Chap. 21; [31])	MaGloNit-Ted (Chap. 15; [32])	MaGloNit-Cas (Chap. 16; [7])
MoLoIt-Jur (Chap. 22; [1])	MoLoIt-Axe (Chap. 23; [2])	MoLoNit-Mae (Chap. 25; [21])	MoLoNit-Bet (Chap. 26; [3])
MoLoIt-Doa (Chap. 24; [8])	MoLoIt-Neg (Chap. 27; [27])	MoGloNit-Oca (Chap. 31; [28])	MoGloNit-Mor (Chap. 32; [23])
MoLoIt-Zaf (Chap. 28; [35])	MoLoIt-Mae2 (Chap. 29; [20])	MoGloNit-Nab (Chap. 33; [25])	
MoLoIt-Liu (Chap. 30; [19])	MoGloIt-Pan (Chap. 34; [29])		
MoGloIt-Fer (Chap. 35; [11])	MoGloIt-Fer2 (Chap. 36; [10])		

Table 1.10 Scheme classification with respect to the attitude shown by the controllers

Controller attitude			
Cooperative		Non-cooperative	
MaLoNit-Val (Chap. 2; [34])	MaLoNit-Tro (Chap. 3; [33])		
MaLoNit-Mul (Chap. 5; [24])	MaLoIt-Bou (Chap. 6; [5])		
MaLoIt-Far (Chap. 7; [9])	MaLoIt-Lem (Chap. 8; [18])		
MaLoIt-Lam (Chap. 9; [17])	MaLoIt-Nec (Chap. 10; [26])		
MaLoIt-Bie (Chap. 11; [4])	MaLoIt-Cam (Chap. 12; [6])	MaLoNit-Mar (Chap. 4; [22])	MaGloNit-Hu (Chap. 13; [15])
MaGloIt-Pro (Chap. 17; [30])	MaGloIt-Gra (Chap. 18; [13])	MaGloNit-Her (Chap. 14; [14])	MaGloNit-Ted (Chap. 15; [32])
MaGloIt-Gis (Chap. 19; [12])	MaGloIt-Koz (Chap. 20; [16])	MaGloNit-Cas (Chap. 16; [7])	MaGloIt-Ros (Chap. 21; [31])
MoLoIt-Jur (Chap. 22; [1])	MoLoIt-Axe (Chap. 23; [2])	MoLoNit-Bet (Chap. 26; [3])	MoLoIt-MaeMun (Chap. 29; [20])
MoLoIt-Doa (Chap. 24; [8])	MoLoNit-Mae (Chap. 25; [21])	MoGloNit-Oca (Chap. 31; [28])	
MoLoIt-Neg (Chap. 27; [27])	MoLoIt-Liu (Chap. 30; [19])		
MoGloNit-Mor (Chap. 32; [23])	MoGloNit-Nab (Chap. 33; [25])		
MoGloIt-Pan (Chap. 34; [29])	MoGloIt-Fer (Chap. 35; [11])		
MoGloIt-Fer2 (Chap. 36; [10])			

Table 1.11 Scheme classification with respect to the way in which the communication takes place

Communication			
Serial		Parallel	
		MaLoNit-Val (Chap. 2; [34])	MaLoNit-Tro (Chap. 3; [33])
		MaLoNit-Mar (Chap. 4; [22])	MaLoIt-Bou (Chap. 6; [5])
		MaLoIt-Far (Chap. 7; [9])	MaLoIt-Lem (Chap. 8; [18])
		MaLoIt-Lam (Chap. 9; [17])	MaLoIt-Nec (Chap. 10; [26])
		MaLoIt-Bie (Chap. 11; [4])	MaLoIt-Cam (Chap. 12; [6])
		MaGloNit-Hu (Chap. 13; [15])	MaGloNit-Her (Chap. 14; [14])
		MaGloNit-Cas (Chap. 16; [7])	MaGloIt-Pro (Chap. 17; [30])
MaLoNit-Mul (Chap. 5; [24])	MaLoIt-Cam (Chap. 12; [6])	MaGloIt-Gra (Chap. 18; [13])	MaGloIt-Gis (Chap. 19; [12])
MaGloNit-Ted (Chap. 15; [32])	MoLoIt-Neg (Chap. 27; [27])	MaGloIt-Koz (Chap. 20; [16])	MaGloIt-Ros (Chap. 21; [31])
MoGloNit-Oca (Chap. 31; [28])	MoGloNit-Nab (Chap. 33; [25])	MoLoIt-Jur (Chap. 22; [1])	MoLoIt-Axe (Chap. 23; [2])
		MoLoIt-Doa (Chap. 24; [8])	MoLoNit-Mae (Chap. 25; [21])
		MoLoNit-Bet (Chap. 26; [3])	MoLoIt-Neg (Chap. 27; [27])
		MoLoIt-Zaf (Chap. 28; [35])	MoLoIt-MaeMun (Chap. 29; [20])
		MoLoIt-Liu (Chap. 30; [19])	MoGloNit-Oca (Chap. 31; [28])
		MoGloNit-Mor (Chap. 32; [23])	MoGloIt-Pan (Chap. 34; [29])
		MoGloIt-Fer (Chap. 35; [11])	MoGloIt-Fer2 (Chap. 36; [10])

Table 1.12 Scheme classification with respect to the timing employed in the communication

Timing			
Synchronous		Asynchronous	
MaLoNit-Val (Chap. 2; [34])	MaLoNit-Tro (Chap. 3; [33])		
MaLoNit-Mar (Chap. 4; [22])	MaLoNit-Mul (Chap. 5; [24])		
MaLoIt-Bou (Chap. 6; [5])	MaLoIt-Far (Chap. 7; [9])		
MaLoIt-Lem (Chap. 8; [18])	MaLoIt-Lam (Chap. 9; [17])		
MaLoIt-Nec (Chap. 10; [26])	MaLoIt-Bie (Chap. 11; [4])		
MaLoIt-Cam (Chap. 12; [6])	MaGloNit-Her (Chap. 14; [14])		
MaGloNit-Ted (Chap. 15; [32])	MaGloNit-Cas (Chap. 16; [7])		
MaGloIt-Pro (Chap. 17; [30])	MaGloIt-Gra (Chap. 18; [13])		
MaGloIt-Gis (Chap. 19; [12])	MaGloIt-Koz (Chap. 20; [16])	MaGloNit-Hu (Chap. 13; [15])	MaGloIt-Ros (Chap. 21; [31])
MaGloIt-Ros (Chap. 21; [31])	MoLoIt-Jur (Chap. 22; [1])	MoLoIt-Axe (Chap. 23; [2])	MoLoIt-Zaf (Chap. 28; [35])
MoLoIt-Axe (Chap. 23; [2])	MoLoIt-Doa (Chap. 24; [8])		
MoLoNit-Mae (Chap. 25; [21])	MoLoNit-Bet (Chap. 26; [3])		
MoLoIt-Neg (Chap. 27; [27])	MoLoIt-MaeMun (Chap. 29; [20])		
MoLoIt-Liu (Chap. 30; [19])	MoGloNit-Oca (Chap. 31; [28])		
MoGloNit-Mor (Chap. 32; [23])	MoGloNit-Nab (Chap. 33; [25])		
MoGloIt-Pan (Chap. 34; [29])	MoGloIt-Fer (Chap. 35; [11])		
MoGloIt-Fer2 (Chap. 36; [10])			

Table 1.13 Scheme classification with respect to the nature of the optimization variables used in the optimization problem

Optimization variables			
Real		Integer	
MaLoNit-Val (Chap. 2; [34])	MaLoNit-Tro (Chap. 3; [33])		
MaLoNit-Mar (Chap. 4; [22])	MaLoNit-Mul (Chap. 5; [24])		
MaLoIt-Bou (Chap. 6; [5])	MaLoIt-Far (Chap. 7; [9])		
MaLoIt-Lem (Chap. 8; [18])	MaLoIt-Lam (Chap. 9; [17])		
MaLoIt-Nec (Chap. 10; [26])	MaLoIt-Bie (Chap. 11; [4])		
MaLoIt-Cam (Chap. 12; [6])	MaGloNit-Hu (Chap. 13; [15])		
MaGloNit-Her (Chap. 14; [14])	MaGloNit-Ted (Chap. 15; [32])		
MaGloNit-Cas (Chap. 16; [7])	MaGloIt-Pro (Chap. 17; [30])		
MaGloIt-Gra (Chap. 18; [13])	MaGloIt-Gis (Chap. 19; [12])	MaGloNit-Ted (Chap. 15; [32])	MaGloIt-Pro (Chap. 17; [30])
MaGloIt-Koz (Chap. 20; [16])	MaGloIt-Ros (Chap. 21; [31])	MoLoIt-Axe (Chap. 23; [2])	MoLoIt-Zaf (Chap. 28; [35])
MoLoIt-Jur (Chap. 22; [1])	MoLoIt-Axe (Chap. 23; [2])		
MoLoIt-Doa (Chap. 24; [8])	MoLoNit-Mae (Chap. 25; [21])		
MoLoNit-Bet (Chap. 26; [3])	MoLoIt-Neg (Chap. 27; [27])		
MoLoIt-Zaf (Chap. 28; [35])	MoLoIt-MaeMun (Chap. 29; [20])		
MoLoIt-Liu (Chap. 30; [19])	MoGloNit-Oca (Chap. 31; [28])		
MoGloNit-Mor (Chap. 32; [23])	MoGloNit-Nab (Chap. 33; [25])		
MoGloIt-Pan (Chap. 34; [29])	MoGloIt-Fer (Chap. 35; [11])		
MoGloIt-Fer2 (Chap. 36; [10])			

Table 1.14 Scheme classification with respect to the availability of optimality guarantees

Optimality			
Yes		No	
		MaLoNit-Tro (Chap. 3; [33])	MaLoNit-Mar (Chap. 4; [22])
MaLoNit-Val (Chap. 2; [34])	MaLoIt-Bou (Chap. 6; [5])	MaLoNit-Mul (Chap. 5; [24])	MaLoIt-Far (Chap. 7; [9])
MaLoIt-Bie (Chap. 11; [4])	MaLoIt-Cam (Chap. 12; [6])	MaLoIt-Lem (Chap. 8; [18])	MaLoIt-Lam (Chap. 9; [17])
MaGloNit-Her (Chap. 14; [14])	MaGloNit-Cas (Chap. 16; [7])	MaLoIt-Nec (Chap. 10; [26])	MaGloNit-Hu (Chap. 13; [15])
MaGloIt-Gis (Chap. 19; [12])	MaGloIt-Koz (Chap. 20; [16])	MaGloNit-Ted (Chap. 15; [32])	MaGloIt-Pro (Chap. 17; [30])
MaGloIt-Ros (Chap. 21; [31])	MoLoIt-Axe (Chap. 23; [2])	MaGloIt-Gra (Chap. 18; [13])	MoLoIt-Jur (Chap. 22; [1])
MoLoIt-Neg (Chap. 27; [27])	MoLoIt-Liu (Chap. 30; [19])	MoLoIt-Doa (Chap. 24; [8])	MoLoNit-Mae (Chap. 25; [21])
MoGloIt-Pan (Chap. 34; [29])	MoGloIt-Fer (Chap. 35; [11])	MoLoNit-Bet (Chap. 26; [3])	MoLoIt-Zaf (Chap. 28; [35])
MoGloIt-Fer2 (Chap. 36; [10])		MoLoIt-Mae2 (Chap. 29; [20])	MoGloNit-Oca (Chap. 31; [28])
		MoGloNit-Mor (Chap. 32; [23])	MoGloNit-Nab (Chap. 33; [25])

Table 1.15 Scheme classification with respect to the existence of suboptimality bounds

Suboptimality bounds			
Yes		No	
		MaLoNit-Tro (Chap. 3; [33])	MaLoNit-Mar (Chap. 4; [22])
MaLoNit-Val (Chap. 2; [34])	MaLoIt-Far (Chap. 7; [9])	MaLoNit-Mul (Chap. 5; [24])	MaLoIt-Bou (Chap. 6; [5])
MaLoIt-Nec (Chap. 10; [26])	MaLoIt-Bie (Chap. 11; [4])	MaLoIt-Lem (Chap. 8; [18])	MaLoIt-Lam (Chap. 9; [17])
MaLoIt-Cam (Chap. 12; [6])	MaGloNit-Ted (Chap. 15; [32])	MaGloNit-Hu (Chap. 13; [15])	MaGloNit-Her (Chap. 14; [14])
MaGloIt-Gra (Chap. 18; [13])	MaGloIt-Gis (Chap. 19; [12])	MaGloNit-Cas (Chap. 16; [7])	MaGloIt-Pro (Chap. 17; [30])
MaGloIt-Ros (Chap. 21; [31])	MoLoIt-Axe (Chap. 23; [2])	MaGloIt-Koz (Chap. 20; [16])	MoLoIt-Jur (Chap. 22; [1])
MoLoIt-Doa (Chap. 24; [8])	MoLoIt-Liu (Chap. 30; [19])	MoLoNit-Mae (Chap. 25; [21])	MoLoNit-Bet (Chap. 26; [3])
MoGloNit-Oca (Chap. 31; [28])	MoGloIt-Fer (Chap. 35; [11])	MoLoIt-Neg (Chap. 27; [27])	MoLoIt-Zaf (Chap. 28; [35])
MoGloIt-Fer2 (Chap. 36; [10])		MoLoIt-Mae2 (Chap. 29; [20])	MoGloNit-Mor (Chap. 32; [23])
		MoGloNit-Nab (Chap. 33; [25])	MoGloIt-Pan (Chap. 34; [29])

Table 1.16 Scheme classification with respect to the availability of stability guarantees

Stability			
Yes		No	
MaLoNit-Val (Chap. 2; [34])	MaLoNit-Tro (Chap. 3; [33])		
MaLoNit-Mul (Chap. 5; [24])	MaLoIt-Nec (Chap. 10; [26])		
MaGloNit-Hu (Chap. 13; [15])	MaGloNit-Her (Chap. 14; [14])	MaLoNit-Mar (Chap. 4; [22])	MaLoIt-Bou (Chap. 6; [5])
MaGloNit-Ted (Chap. 15; [32])	MaGloNit-Cas (Chap. 16; [7])	MaLoIt-Far (Chap. 7; [9])	MaLoIt-Lem (Chap. 8; [18])
MaGloIt-Pro (Chap. 17; [30])	MaGloIt-Gra (Chap. 18; [13])	MaLoIt-Lam (Chap. 9; [17])	MaLoIt-Bie (Chap. 11; [4])
MaGloIt-Gis (Chap. 19; [12])	MoLoIt-Axe (Chap. 23; [2])	MaLoIt-Cam (Chap. 12; [6])	MaGloIt-Koz (Chap. 20; [16])
MoLoIt-Doa (Chap. 24; [8])	MoLoNit-Mae (Chap. 25; [21])	MaGloIt-Ros (Chap. 21; [31])	MoLoIt-Jur (Chap. 22; [1])
MoLoNit-Bet (Chap. 26; [3])	MoLoIt-Zaf (Chap. 28; [35])	MoLoIt-Neg (Chap. 27; [27])	MoGloNit-Oca (Chap. 31; [28])
MoLoIt-Mae2 (Chap. 29; [20])	MoLoIt-Liu (Chap. 30; [19])	MoGloNit-Mor (Chap. 32; [23])	MoGloNit-Nab (Chap. 33; [25])
MoGloIt-Pan (Chap. 34; [29])	MoGloIt-Fer (Chap. 35; [11])		
MoGloIt-Fer2 (Chap. 36; [10])			

Table 1.17 Scheme classification with respect to the availability of robustness guarantees

Robustness			
Yes		No	
MaLoNit-Tro (Chap. 3; [33])	MaLoNit-Mul (Chap. 5; [24])	MaLoNit-Val (Chap. 2; [34])	MaLoNit-Mar (Chap. 4; [22])
MaLoIt-Nec (Chap. 10; [26])	MaGloNit-Hu (Chap. 13; [15])	MaLoIt-Bou (Chap. 6; [5])	MaLoIt-Far (Chap. 7; [9])
MaGloNit-Ted (Chap. 15; [32])	MaGloNit-Cas (Chap. 16; [7])	MaLoIt-Lem (Chap. 8; [18])	MaLoIt-Lam (Chap. 9; [17])
MaGloIt-Pro (Chap. 17; [30])	MaGloIt-Gra (Chap. 18; [13])	MaLoIt-Bie (Chap. 11; [4])	MaLoIt-Cam (Chap. 12; [6])
MaGloIt-Gis (Chap. 19; [12])	MoLoIt-Axe (Chap. 23; [2])	MaGloNit-Her (Chap. 14; [14])	MaGloIt-Koz (Chap. 20; [16])
MoLoNit-Mae (Chap. 25; [21])	MoLoNit-Bet (Chap. 26; [3])	MaGloIt-Ros (Chap. 21; [31])	MoLoIt-Jur (Chap. 22; [1])
MoLoIt-Zaf (Chap. 28; [35])	MoLoIt-Mae2 (Chap. 29; [20])	MoLoIt-Doa (Chap. 24; [8])	MoLoIt-Neg (Chap. 27; [27])
MoLoIt-Liu (Chap. 30; [19])	MoGloNit-Oca (Chap. 31; [28])	MoGloNit-Mor (Chap. 32; [23])	MoGloNit-Nab (Chap. 33; [25])
MoGloIt-Pan (Chap. 34; [29])		MoGloIt-Fer (Chap. 35; [11])	MoGloIt-Fer2 (Chap. 36; [10])

Chapter 2	**Bargaining game based distributed MPC MaLoNit-Val [34]**		
	Process		
System Type	Multiple Autonomous Systems		Decomposed Monolithical System
Process Type	Linear	Nonlinear	Hybrid
Type of Model	Transfer function		State Space
Randomness	Deterministic		Non-deterministic
Type of Control	Regulation	Tracking	Economic
Coupling Source	Constraints		Objective
	Input	Output	State
	Control Architecture		
Architecture	Decentralized	Distributed	Hierarchical
Controller Knowledge	Strictly Local		Partially Global
Computation Type	Iterative		Non-iterative
Controller Attitude	Cooperative		Noncooperative
Communication	Serial		Parallel
Timing	Synchronous		Asynchronous
Optimization Variables	Real		Integer
	Theoretical Properties		
Optimality	Yes		No
Suboptimality Bounds	Yes		No
Stability	Yes		No
Robustness	Yes		No

Chapter 3	**Cooperative tube-based distributed MPC for linear uncertain systems coupled via constraints MaLoNit-Tro [33]**		
	Process		
System Type	Multiple Autonomous Systems		Decomposed Monolithical System
Process Type	Linear	Nonlinear	Hybrid
Type of Model	Transfer function		State Space
Randomness	Deterministic		Non-deterministic
Type of Control	Regulation	Tracking	Economic
Coupling Source	Constraints		Objective
	Input	Output	State
	Control Architecture		
Architecture	Decentralized	Distributed	Hierarchical
Controller Knowledge	Strictly Local		Partially Global
Computation Type	Iterative		Non-iterative
Controller Attitude	Cooperative		Noncooperative
Communication	Serial		Parallel
Timing	Synchronous		Asynchronous
Optimization Variables	Real		Integer
	Theoretical Properties		
Optimality	Yes		No
Suboptimality Bounds	Yes		No
Stability	Yes		No
Robustness	Yes		No

Chapter 4	Price-driven coordination for distributed NMPC using a feedback control law MaLoNit-Mar [22]		
	Process		
System Type	Multiple Autonomous Systems		Decomposed Monolithical System
Process Type	Linear	Nonlinear	Hybrid
Type of Model	Transfer function		State Space
Randomness	Deterministic		Non-deterministic
Type of Control	Regulation	Tracking	Economic
Coupling Source	Constraints		Objective
	Input	Output	State
	Control Architecture		
Architecture	Decentralized	Distributed	Hierarchical
Controller Knowledge	Strictly Local		Partially Global
Computation Type	Iterative		Non-iterative
Controller Attitude	Cooperative		Noncooperative
Communication	Serial		Parallel
Timing	Synchronous		Asynchronous
Optimization Variables	Real		Integer
	Theoretical Properties		
Optimality	Yes		No
Suboptimality Bounds	Yes		No
Stability	Yes		No
Robustness	Yes		No

Chapter 5	Distributed MPC for consensus and synchronization MaLoNit-Mul [24]		
	Process		
System Type	Multiple Autonomous Systems		Decomposed Monolithical System
Process Type	Linear	Nonlinear	Hybrid
Type of Model	Transfer function		State Space
Randomness	Deterministic		Non-deterministic
Type of Control	Regulation	Tracking	Economic
Coupling Source	Constraints		Objective
	Input	Output	State
	Control Architecture		
Architecture	Decentralized	Distributed	Hierarchical
Controller Knowledge	Strictly Local		Partially Global
Computation Type	Iterative		Non-iterative
Controller Attitude	Cooperative		Noncooperative
Communication	Serial		Parallel
Timing	Synchronous		Asynchronous
Optimization Variables	Real		Integer
	Theoretical Properties		
Optimality	Yes		No
Suboptimality Bounds	Yes		No
Stability	Yes		No
Robustness	Yes		No

Chapter 6	Distributed MPC under coupled constraints based on Dantzig-Wolfe decomposition MaLoIt-Bou [5]		
	Process		
System Type	Multiple Autonomous Systems		Decomposed Monolithical System
Process Type	Linear	Nonlinear	Hybrid
Type of Model	Transfer function		State Space
Randomness	Deterministic		Non-deterministic
Type of Control	Regulation	Tracking	Economic
Coupling Source	Constraints		Objective
	Input	Output	State
	Control Architecture		
Architecture	Decentralized	Distributed	Hierarchical
Controller Knowledge	Strictly Local		Partially Global
Computation Type	Iterative		Non-iterative
Controller Attitude	Cooperative		Noncooperative
Communication	Serial		Parallel
Timing	Synchronous		Asynchronous
Optimization Variables	Real		Integer
	Theoretical Properties		
Optimality	Yes		No
Suboptimality Bounds	Yes		No
Stability	Yes		No
Robustness	Yes		No

Chapter 7	Distributed MPC via dual decomposition and alternative direction method of multipliers MaLoIt-Far [9]		
	Process		
System Type	Multiple Autonomous Systems		Decomposed Monolithical System
Process Type	Linear	Nonlinear	Hybrid
Type of Model	Transfer function		State Space
Randomness	Deterministic		Non-deterministic
Type of Control	Regulation	Tracking	Economic
Coupling Source	Constraints		Objective
	Input	Output	State
	Control Architecture		
Architecture	Decentralized	Distributed	Hierarchical
Controller Knowledge	Strictly Local		Partially Global
Computation Type	Iterative		Non-iterative
Controller Attitude	Cooperative		Noncooperative
Communication	Serial		Parallel
Timing	Synchronous		Asynchronous
Optimization Variables	Real		Integer
	Theoretical Properties		
Optimality	Yes		No
Suboptimality Bounds	Yes		No
Stability	Yes		No
Robustness	Yes		No

Chapter 8	D-SIORHC, distributed MPC with stability constraints based on a gameapproach MaLoIt-Lem [18]		
	Process		
System Type	Multiple Autonomous Systems		Decomposed Monolithical System
Process Type	Linear	Nonlinear	Hybrid
Type of Model	Transfer function		State Space
Randomness	Deterministic		Non-deterministic
Type of Control	Regulation	Tracking	Economic
Coupling Source	Constraints		Objective
	Input	Output	State
	Control Architecture		
Architecture	Decentralized	Distributed	Hierarchical
Controller Knowledge	Strictly Local		Partially Global
Computation Type	Iterative		Non-iterative
Controller Attitude	Cooperative		Noncooperative
Communication	Serial		Parallel
Timing	Synchronous		Asynchronous
Optimization Variables	Real		Integer
	Theoretical Properties		
Optimality	Yes		No
Suboptimality Bounds	Yes		No
Stability	Yes		No
Robustness	Yes		No

Chapter 9	A distributed-in-time NMPC-based coordination mechanism for resource sharing problems MaLoIt-Lam [17]		
	Process		
System Type	Multiple Autonomous Systems		Decomposed Monolithical System
Process Type	Linear	Nonlinear	Hybrid
Type of Model	Transfer function		State Space
Randomness	Deterministic		Non-deterministic
Type of Control	Regulation	Tracking	Economic
Coupling Source	Constraints		Objective
	Input	Output	State
	Control Architecture		
Architecture	Decentralized	Distributed	Hierarchical
Controller Knowledge	Strictly Local		Partially Global
Computation Type	Iterative		Non-iterative
Controller Attitude	Cooperative		Noncooperative
Communication	Serial		Parallel
Timing	Synchronous		Asynchronous
Optimization Variables	Real		Integer
	Theoretical Properties		
Optimality	Yes		No
Suboptimality Bounds	Yes		No
Stability	Yes		No
Robustness	Yes		No

Chapter 10	**Rate analysis of inexact dual fast gradient method for distributed MPC MaLoIt-Nec [26]**		
	Process		
System Type	Multiple Autonomous Systems		Decomposed Monolithical System
Process Type	Linear	Nonlinear	Hybrid
Type of Model	Transfer function		State Space
Randomness	Deterministic		Non-deterministic
Type of Control	Regulation	Tracking	Economic
Coupling Source	Constraints		Objective
	Input	Output	State
	Control Architecture		
Architecture	Decentralized	Distributed	Hierarchical
Controller Knowledge	Strictly Local		Partially Global
Computation Type	Iterative		Non-iterative
Controller Attitude	Cooperative		Noncooperative
Communication	Serial		Parallel
Timing	Synchronous		Asynchronous
Optimization Variables	Real		Integer
	Theoretical Properties		
Optimality	Yes		No
Suboptimality Bounds	Yes		No
Stability	Yes		No
Robustness	Yes		No

Chapter 11	**Distributed MPC via dual decomposition MaLoIt-Bie [4]**		
	Process		
System Type	Multiple Autonomous Systems		Decomposed Monolithical System
Process Type	Linear	Nonlinear	Hybrid
Type of Model	Transfer function		State Space
Randomness	Deterministic		Non-deterministic
Type of Control	Regulation	Tracking	Economic
Coupling Source	Constraints		Objective
	Input	Output	State
	Control Architecture		
Architecture	Decentralized	Distributed	Hierarchical
Controller Knowledge	Strictly Local		Partially Global
Computation Type	Iterative		Non-iterative
Controller Attitude	Cooperative		Noncooperative
Communication	Serial		Parallel
Timing	Synchronous		Asynchronous
Optimization Variables	Real		Integer
	Theoretical Properties		
Optimality	Yes		No
Suboptimality Bounds	Yes		No
Stability	Yes		No
Robustness	Yes		No

Chapter 12	Distributed optimization for MPC of linear dynamic networks MaLoIt-Cam [6]		
	Process		
System Type	Multiple Autonomous Systems		Decomposed Monolithical System
Process Type	Linear	Nonlinear	Hybrid
Type of Model	Transfer function		State Space
Randomness	Deterministic		Non-deterministic
Type of Control	Regulation	Tracking	Economic
Coupling Source	Constraints		Objective
	Input	Output	State
	Control Architecture		
Architecture	Decentralized	Distributed	Hierarchical
Controller Knowledge	Strictly Local		Partially Global
Computation Type	Iterative		Non-iterative
Controller Attitude	Cooperative		Noncooperative
Communication	Serial		Parallel
Timing	Synchronous		Asynchronous
Optimization Variables	Real		Integer
	Theoretical Properties		
Optimality	Yes		No
Suboptimality Bounds	Yes		No
Stability	Yes		No
Robustness	Yes		No

Chapter 13	Adaptive quasi-decentralized MPC of networked process systems MaGloNit-Hu [15]		
	Process		
System Type	Multiple Autonomous Systems		Decomposed Monolithical System
Process Type	Linear	Nonlinear	Hybrid
Type of Model	Transfer function		State Space
Randomness	Deterministic		Non-deterministic
Type of Control	Regulation	Tracking	Economic
Coupling Source	Constraints		Objective
	Input	Output	State
	Control Architecture		
Architecture	Decentralized	Distributed	Hierarchical
Controller Knowledge	Strictly Local		Partially Global
Computation Type	Iterative		Non-iterative
Controller Attitude	Cooperative		Noncooperative
Communication	Serial		Parallel
Timing	Synchronous		Asynchronous
Optimization Variables	Real		Integer
	Theoretical Properties		
Optimality	Yes		No
Suboptimality Bounds	Yes		No
Stability	Yes		No
Robustness	Yes		No

Chapter 14	**Distributed Lyapunov-based MPC MaGloNit-Her [14]**		
	Process		
System Type	Multiple Autonomous Systems		Decomposed Monolithical System
Process Type	Linear	Nonlinear	Hybrid
Type of Model	Transfer function		State Space
Randomness	Deterministic		Non-deterministic
Type of Control	Regulation	Tracking	Economic
Coupling Source	Constraints		Objective
	Input	Output	State
	Control Architecture		
Architecture	Decentralized	Distributed	Hierarchical
Controller Knowledge	Strictly Local		Partially Global
Computation Type	Iterative		Non-iterative
Controller Attitude	Cooperative		Noncooperative
Communication	Serial		Parallel
Timing	Synchronous		Asynchronous
Optimization Variables	Real		Integer
	Theoretical Properties		
Optimality	Yes		No
Suboptimality Bounds	Yes		No
Stability	Yes		No
Robustness	Yes		No

Chapter 15	**A distributed reference management scheme in presence of non-convex constraints: an MPC based approach MaGloNit-Ted [32]**		
	Process		
System Type	Multiple Autonomous Systems		Decomposed Monolithical System
Process Type	Linear	Nonlinear	Hybrid
Type of Model	Transfer function		State Space
Randomness	Deterministic		Non-deterministic
Type of Control	Regulation	Tracking	Economic
Coupling Source	Constraints		Objective
	Input	Output	State
	Control Architecture		
Architecture	Decentralized	Distributed	Hierarchical
Controller Knowledge	Strictly Local		Partially Global
Computation Type	Iterative		Non-iterative
Controller Attitude	Cooperative		Noncooperative
Communication	Serial		Parallel
Timing	Synchronous		Asynchronous
Optimization Variables	Real		Integer
	Theoretical Properties		
Optimality	Yes		No
Suboptimality Bounds	Yes		No
Stability	Yes		No
Robustness	Yes		No

Chapter 16	The distributed command governor approach in a nutshell MaGloNit-Cas [7]		
	Process		
System Type	Multiple Autonomous Systems		Decomposed Monolithical System
Process Type	Linear	Nonlinear	Hybrid
Type of Model	Transfer function		State Space
Randomness	Deterministic		Non-deterministic
Type of Control	Regulation	Tracking	Economic
Coupling Source	Constraints		Objective
	Input	Output	State
	Control Architecture		
Architecture	Decentralized	Distributed	Hierarchical
Controller Knowledge	Strictly Local		Partially Global
Computation Type	Iterative		Non-iterative
Controller Attitude	Cooperative		Noncooperative
Communication	Serial		Parallel
Timing	Synchronous		Asynchronous
Optimization Variables	Real		Integer
	Theoretical Properties		
Optimality	Yes		No
Suboptimality Bounds	Yes		No
Stability	Yes		No
Robustness	Yes		No

Chapter 17	Mixed-integer programming techniques in distributed MPC problems MaGloIt-Pro [30]		
	Process		
System Type	Multiple Autonomous Systems		Decomposed Monolithical System
Process Type	Linear	Nonlinear	Hybrid
Type of Model	Transfer function		State Space
Randomness	Deterministic		Non-deterministic
Type of Control	Regulation	Tracking	Economic
Coupling Source	Constraints		Objective
	Input	Output	State
	Control Architecture		
Architecture	Decentralized	Distributed	Hierarchical
Controller Knowledge	Strictly Local		Partially Global
Computation Type	Iterative		Non-iterative
Controller Attitude	Cooperative		Noncooperative
Communication	Serial		Parallel
Timing	Synchronous		Asynchronous
Optimization Variables	Real		Integer
	Theoretical Properties		
Optimality	Yes		No
Suboptimality Bounds	Yes		No
Stability	Yes		No
Robustness	Yes		No

Chapter 18	Distributed MPC of interconnected nonlinear systems by dynamic dualdecomposition MaGloIt-Gra [13]		
	Process		
System Type	Multiple Autonomous Systems		Decomposed Monolithical System
Process Type	Linear	Nonlinear	Hybrid
Type of Model	Transfer function		State Space
Randomness	Deterministic		Non-deterministic
Type of Control	Regulation	Tracking	Economic
Coupling Source	Constraints		Objective
	Input	Output	State
	Control Architecture		
Architecture	Decentralized	Distributed	Hierarchical
Controller Knowledge	Strictly Local		Partially Global
Computation Type	Iterative		Non-iterative
Controller Attitude	Cooperative		Noncooperative
Communication	Serial		Parallel
Timing	Synchronous		Asynchronous
Optimization Variables	Real		Integer
	Theoretical Properties		
Optimality	Yes		No
Suboptimality Bounds	Yes		No
Stability	Yes		No
Robustness	Yes		No

Chapter 19	Generalized accelerated gradient methods for distributed MPC based on dual decomposition MaGloIt-Gis [12]		
	Process		
System Type	Multiple Autonomous Systems		Decomposed Monolithical System
Process Type	Linear	Nonlinear	Hybrid
Type of Model	Transfer function		State Space
Randomness	Deterministic		Non-deterministic
Type of Control	Regulation	Tracking	Economic
Coupling Source	Constraints		Objective
	Input	Output	State
	Control Architecture		
Architecture	Decentralized	Distributed	Hierarchical
Controller Knowledge	Strictly Local		Partially Global
Computation Type	Iterative		Non-iterative
Controller Attitude	Cooperative		Noncooperative
Communication	Serial		Parallel
Timing	Synchronous		Asynchronous
Optimization Variables	Real		Integer
	Theoretical Properties		
Optimality	Yes		No
Suboptimality Bounds	Yes		No
Stability	Yes		No
Robustness	Yes		No

Chapter 20	Distributed multiple shooting for large scale nonlinear systems MaGloIt-Koz [16]		
	Process		
System Type	Multiple Autonomous Systems		Decomposed Monolithical System
Process Type	Linear	Nonlinear	Hybrid
Type of Model	Transfer function		State Space
Randomness	Deterministic		Non-deterministic
Type of Control	Regulation	Tracking	Economic
Coupling Source	Constraints		Objective
	Input	Output	State
	Control Architecture		
Architecture	Decentralized	Distributed	Hierarchical
Controller Knowledge	Strictly Local		Partially Global
Computation Type	Iterative		Non-iterative
Controller Attitude	Cooperative		Noncooperative
Communication	Serial		Parallel
Timing	Synchronous		Asynchronous
Optimization Variables	Real		Integer
	Theoretical Properties		
Optimality	Yes		No
Suboptimality Bounds	Yes		No
Stability	Yes		No
Robustness	Yes		No

Chapter 21	Nash-based distributed MPC for multi-rate systems MaGloIt-Ros [31]		
	Process		
System Type	Multiple Autonomous Systems		Decomposed Monolithical System
Process Type	Linear	Nonlinear	Hybrid
Type of Model	Transfer function		State Space
Randomness	Deterministic		Non-deterministic
Type of Control	Regulation	Tracking	Economic
Coupling Source	Constraints		Objective
	Input	Output	State
	Control Architecture		
Architecture	Decentralized	Distributed	Hierarchical
Controller Knowledge	Strictly Local		Partially Global
Computation Type	Iterative		Non-iterative
Controller Attitude	Cooperative		Noncooperative
Communication	Serial		Parallel
Timing	Synchronous		Asynchronous
Optimization Variables	Real		Integer
	Theoretical Properties		
Optimality	Yes		No
Suboptimality Bounds	Yes		No
Stability	Yes		No
Robustness	Yes		No

Chapter 22	Cooperative dynamic MPC for networked control systems MoLoIt-Jur [1]		
	Process		
System Type	Multiple Autonomous Systems		Decomposed Monolithical System
Process Type	Linear	Nonlinear	Hybrid
Type of Model	Transfer function		State Space
Randomness	Deterministic		Non-deterministic
Type of Control	Regulation	Tracking	Economic
Coupling Source	Constraints		Objective
	Input	Output	State
	Control Architecture		
Architecture	Decentralized	Distributed	Hierarchical
Controller Knowledge	Strictly Local		Partially Global
Computation Type	Iterative		Non-iterative
Controller Attitude	Cooperative		Noncooperative
Communication	Serial		Parallel
Timing	Synchronous		Asynchronous
Optimization Variables	Real		Integer
	Theoretical Properties		
Optimality	Yes		No
Suboptimality Bounds	Yes		No
Stability	Yes		No
Robustness	Yes		No

Chapter 23	Parallel implementation of hybrid MPC MoLoIt-Axe [2]		
	Process		
System Type	Multiple Autonomous Systems		Decomposed Monolithical System
Process Type	Linear	Nonlinear	Hybrid
Type of Model	Transfer function		State Space
Randomness	Deterministic		Non-deterministic
Type of Control	Regulation	Tracking	Economic
Coupling Source	Constraints		Objective
	Input	Output	State
	Control Architecture		
Architecture	Decentralized	Distributed	Hierarchical
Controller Knowledge	Strictly Local		Partially Global
Computation Type	Iterative		Non-iterative
Controller Attitude	Cooperative		Noncooperative
Communication	Serial		Parallel
Timing	Synchronous		Asynchronous
Optimization Variables	Real		Integer
	Theoretical Properties		
Optimality	Yes		No
Suboptimality Bounds	Yes		No
Stability	Yes		No
Robustness	Yes		No

Chapter 24	A hierarchical MPC approach with guaranteed feasibility for dynamically coupled linear systems MoLoIt-Doa [8]		
	Process		
System Type	Multiple Autonomous Systems		Decomposed Monolithical System
Process Type	Linear	Nonlinear	Hybrid
Type of Model	Transfer function		State Space
Randomness	Deterministic		Non-deterministic
Type of Control	Regulation	Tracking	Economic
Coupling Source	Constraints		Objective
	Input	Output	State
	Control Architecture		
Architecture	Decentralized	Distributed	Hierarchical
Controller Knowledge	Strictly Local		Partially Global
Computation Type	Iterative		Non-iterative
Controller Attitude	Cooperative		Noncooperative
Communication	Serial		Parallel
Timing	Synchronous		Asynchronous
Optimization Variables	Real		Integer
	Theoretical Properties		
Optimality	Yes		No
Suboptimality Bounds	Yes		No
Stability	Yes		No
Robustness	Yes		No

Chapter 25	Distributed MPC based on a team game MoLoNit-Mae [21]		
	Process		
System Type	Multiple Autonomous Systems		Decomposed Monolithical System
Process Type	Linear	Nonlinear	Hybrid
Type of Model	Transfer function		State Space
Randomness	Deterministic		Non-deterministic
Type of Control	Regulation	Tracking	Economic
Coupling Source	Constraints		Objective
	Input	Output	State
	Control Architecture		
Architecture	Decentralized	Distributed	Hierarchical
Controller Knowledge	Strictly Local		Partially Global
Computation Type	Iterative		Non-iterative
Controller Attitude	Cooperative		Noncooperative
Communication	Serial		Parallel
Timing	Synchronous		Asynchronous
Optimization Variables	Real		Integer
	Theoretical Properties		
Optimality	Yes		No
Suboptimality Bounds	Yes		No
Stability	Yes		No
Robustness	Yes		No

Chapter 26	Distributed MPC: A noncooperative approach based on robustness concepts MoLoNit-Bet [3]		
	Process		
System Type	Multiple Autonomous Systems		Decomposed Monolithical System
Process Type	Linear	Nonlinear	Hybrid
Type of Model	Transfer function		State Space
Randomness	Deterministic		Non-deterministic
Type of Control	Regulation	Tracking	Economic
Coupling Source	Constraints		Objective
	Input	Output	State
	Control Architecture		
Architecture	Decentralized	Distributed	Hierarchical
Controller Knowledge	Strictly Local		Partially Global
Computation Type	Iterative		Non-iterative
Controller Attitude	Cooperative		Noncooperative
Communication	Serial		Parallel
Timing	Synchronous		Asynchronous
Optimization Variables	Real		Integer
	Theoretical Properties		
Optimality	Yes		No
Suboptimality Bounds	Yes		No
Stability	Yes		No
Robustness	Yes		No

Chapter 27	Decompositions of augmented Lagrange formulations for serial and parallel distributed MPC MoLoIt-Neg [27]		
	Process		
System Type	Multiple Autonomous Systems		Decomposed Monolithical System
Process Type	Linear	Nonlinear	Hybrid
Type of Model	Transfer function		State Space
Randomness	Deterministic		Non-deterministic
Type of Control	Regulation	Tracking	Economic
Coupling Source	Constraints		Objective
	Input	Output	State
	Control Architecture		
Architecture	Decentralized	Distributed	Hierarchical
Controller Knowledge	Strictly Local		Partially Global
Computation Type	Iterative		Non-iterative
Controller Attitude	Cooperative		Noncooperative
Communication	Serial		Parallel
Timing	Synchronous		Asynchronous
Optimization Variables	Real		Integer
	Theoretical Properties		
Optimality	Yes		No
Suboptimality Bounds	Yes		No
Stability	Yes		No
Robustness	Yes		No

Chapter 28	**A hierarchical distributed MPC approach: A practical implementation MoLoIt-Zaf [35]**		
	Process		
System Type	Multiple Autonomous Systems		Decomposed Monolithical System
Process Type	Linear	Nonlinear	Hybrid
Type of Model	Transfer function		State Space
Randomness	Deterministic		Non-deterministic
Type of Control	Regulation	Tracking	Economic
Coupling Source	Constraints		Objective
	Input	Output	State
	Control Architecture		
Architecture	Decentralized	Distributed	Hierarchical
Controller Knowledge	Strictly Local		Partially Global
Computation Type	Iterative		Non-iterative
Controller Attitude	Cooperative		Noncooperative
Communication	Serial		Parallel
Timing	Synchronous		Asynchronous
Optimization Variables	Real		Integer
	Theoretical Properties		
Optimality	Yes		No
Suboptimality Bounds	Yes		No
Stability	Yes		No
Robustness	Yes		No

Chapter 29	**Distributed MPC Based on Agent Negotiation MoLoIt-Mae2 [20]**		
	Process		
System Type	Multiple Autonomous Systems		Decomposed Monolithical System
Process Type	Linear	Nonlinear	Hybrid
Type of Model	Transfer function		State Space
Randomness	Deterministic		Non-deterministic
Type of Control	Regulation	Tracking	Economic
Coupling Source	Constraints		Objective
	Input	Output	State
	Control Architecture		
Architecture	Decentralized	Distributed	Hierarchical
Controller Knowledge	Strictly Local		Partially Global
Computation Type	Iterative		Non-iterative
Controller Attitude	Cooperative		Noncooperative
Communication	Serial		Parallel
Timing	Synchronous		Asynchronous
Optimization Variables	Real		Integer
	Theoretical Properties		
Optimality	Yes		No
Suboptimality Bounds	Yes		No
Stability	Yes		No
Robustness	Yes		No

Chapter 30	Lyapunov-based distributed MPC schemes: Sequential and iterative approaches MoLoIt-Liu [19]		
	Process		
System Type	Multiple Autonomous Systems		Decomposed Monolithical System
Process Type	Linear	Nonlinear	Hybrid
Type of Model	Transfer function		State Space
Randomness	Deterministic		Non-deterministic
Type of Control	Regulation	Tracking	Economic
Coupling Source	Constraints		Objective
	Input	Output	State
	Control Architecture		
Architecture	Decentralized	Distributed	Hierarchical
Controller Knowledge	Strictly Local		Partially Global
Computation Type	Iterative		Non-iterative
Controller Attitude	Cooperative		Noncooperative
Communication	Serial		Parallel
Timing	Synchronous		Asynchronous
Optimization Variables	Real		Integer
	Theoretical Properties		
Optimality	Yes		No
Suboptimality Bounds	Yes		No
Stability	Yes		No
Robustness	Yes		No

Chapter 31	Multi-layer decentralized MPC of large-scale networked systems MoGloNit-Oca [28]		
	Process		
System Type	Multiple Autonomous Systems		Decomposed Monolithical System
Process Type	Linear	Nonlinear	Hybrid
Type of Model	Transfer function		State Space
Randomness	Deterministic		Non-deterministic
Type of Control	Regulation	Tracking	Economic
Coupling Source	Constraints		Objective
	Input	Output	State
	Control Architecture		
Architecture	Decentralized	Distributed	Hierarchical
Controller Knowledge	Strictly Local		Partially Global
Computation Type	Iterative		Non-iterative
Controller Attitude	Cooperative		Noncooperative
Communication	Serial		Parallel
Timing	Synchronous		Asynchronous
Optimization Variables	Real		Integer
	Theoretical Properties		
Optimality	Yes		No
Suboptimality Bounds	Yes		No
Stability	Yes		No
Robustness	Yes		No

Chapter 32	**Distributed MPC using reinforcement learning based negotiation: Application to large scale systems MoGloNit-Mor [23]**		
		Process	
System Type	Multiple Autonomous Systems		Decomposed Monolithical System
Process Type	Linear	Nonlinear	Hybrid
Type of Model	Transfer function		State Space
Randomness	Deterministic		Non-deterministic
Type of Control	Regulation	Tracking	Economic
Coupling Source	Constraints		Objective
	Input	Output	State
		Control Architecture	
Architecture	Decentralized	Distributed	Hierarchical
Controller Knowledge	Strictly Local		Partially Global
Computation Type	Iterative		Non-iterative
Controller Attitude	Cooperative		Noncooperative
Communication	Serial		Parallel
Timing	Synchronous		Asynchronous
Optimization Variables	Real		Integer
		Theoretical Properties	
Optimality	Yes		No
Suboptimality Bounds	Yes		No
Stability	Yes		No
Robustness	Yes		No

Chapter 33	**Hierarchical MPC for multiple commodity transportation networks MoGloNit-Nab [25]**		
		Process	
System Type	Multiple Autonomous Systems		Decomposed Monolithical System
Process Type	Linear	Nonlinear	Hybrid
Type of Model	Transfer function		State Space
Randomness	Deterministic		Non-deterministic
Type of Control	Regulation		Economic
Coupling Source	Constraints		Objective
	Input	Output	State
		Control Architecture	
Architecture	Decentralized	Distributed	Hierarchical
Controller Knowledge	Strictly Local		Partially Global
Computation Type	Iterative		Non-iterative
Controller Attitude	Cooperative		Noncooperative
Communication	Serial		Parallel
Timing	Synchronous		Asynchronous
Optimization Variables	Real		Integer
		Theoretical Properties	
Optimality	Yes		No
Suboptimality Bounds	Yes		No
Stability	Yes		No
Robustness	Yes		No

Chapter 34	On the use of suboptimal solvers for efficient cooperative distributed linear MPC MoGloIt-Pan [29]		
	Process		
System Type	Multiple Autonomous Systems		Decomposed Monolithical System
Process Type	Linear	Nonlinear	Hybrid
Type of Model	Transfer function		State Space
Randomness	Deterministic		Non-deterministic
Type of Control	Regulation	Tracking	Economic
Coupling Source	Constraints		Objective
	Input	Output	State
	Control Architecture		
Architecture	Decentralized	Distributed	Hierarchical
Controller Knowledge	Strictly Local		Partially Global
Computation Type	Iterative		Non-iterative
Controller Attitude	Cooperative		Noncooperative
Communication	Serial		Parallel
Timing	Synchronous		Asynchronous
Optimization Variables	Real		Integer
	Theoretical Properties		
Optimality	Yes		No
Suboptimality Bounds	Yes		No
Stability	Yes		No
Robustness	Yes		No

Chapter 35	Cooperative distributed MPC integrating a steady state target optimizer MoGloIt-Fer [11]		
	Process		
System Type	Multiple Autonomous Systems		Decomposed Monolithical System
Process Type	Linear	Nonlinear	Hybrid
Type of Model	Transfer function		State Space
Randomness	Deterministic		Non-deterministic
Type of Control	Regulation	Tracking	Economic
Coupling Source	Constraints		Objective
	Input	Output	State
	Control Architecture		
Architecture	Decentralized	Distributed	Hierarchical
Controller Knowledge	Strictly Local		Partially Global
Computation Type	Iterative		Non-iterative
Controller Attitude	Cooperative		Noncooperative
Communication	Serial		Parallel
Timing	Synchronous		Asynchronous
Optimization Variables	Real		Integer
	Theoretical Properties		
Optimality	Yes		No
Suboptimality Bounds	Yes		No
Stability	Yes		No
Robustness	Yes		No

Chapter 36	Cooperative MPC with guaranteed exponential stability MoGloIt-Fer2 [10]		
	Process		
System Type	Multiple Autonomous Systems		Decomposed Monolithical System
Process Type	Linear	Nonlinear	Hybrid
Type of Model	Transfer function		State Space
Randomness	Deterministic		Non-deterministic
Type of Control	Regulation	Tracking	Economic
Coupling Source	Constraints		Objective
	Input	Output	State
	Control Architecture		
Architecture	Decentralized	Distributed	Hierarchical
Controller Knowledge	Strictly Local		Partially Global
Computation Type	Iterative		Non-iterative
Controller Attitude	Cooperative		Noncooperative
Communication	Serial		Parallel
Timing	Synchronous		Asynchronous
Optimization Variables	Real		Integer
	Theoretical Properties		
Optimality	Yes		No
Suboptimality Bounds	Yes		No
Stability	Yes		No
Robustness	Yes		

References

1. I. Jurado, D.E. Quevedo, K.H. Johansson, A. Ahlén, Cooperative dynamic MPC for networked control systems, in *Distributed Model Predictive Control Made Easy*, ed. by J.M. Maestre, R.R. Negenborn (Springer, Dordrecht, 2013), pp. 357–373
2. D. Axehill, A. Hansson, Parallel implementation of hybrid MPC, in *Distributed Model Predictive Control Made Easy*, ed. by J.M. Maestre, R.R. Negenborn (Springer, Dordrecht, 2013), pp. 375–392
3. G. Betti, M. Farina, R. Scattolini, Distributed MPC: A noncooperative approach based on robustness concepts, in *Distributed Model Predictive Control Made Easy*, ed. by J.M. Maestre, R.R. Negenborn (Springer, Dordrecht, 2013), pp. 421–435
4. B. Biegel, J. Stoustrup, P. Andersen, Distributed MPC via dual decomposition, in *Distributed Model Predictive Control Made Easy*, ed. by J.M. Maestre, R.R. Negenborn (Springer, Dordrecht, 2013), pp. 179–192
5. R. Bourdais, J. Buisson, D. Dumur, H. Guéguen, P.-D. Moroşan, Distributed MPC under coupled constraints based on Dantzig-Wolfe decomposition, in *Distributed Model Predictive Control Made Easy*, ed. by J.M. Maestre, R.R. Negenborn (Springer, Dordrecht, 2013), pp. 101–114
6. E. Camponogara, Distributed optimization for MPC of linear dynamic networks, in *Distributed Model Predictive Control Made Easy*, ed. by J.M. Maestre, R.R. Negenborn (Springer, Dordrecht, 2013), pp. 193–208
7. A. Casavola, E. Garone, F. Tedesco, The distributed command governor approach in a nutshell, in *Distributed Model Predictive Control Made Easy*, ed. by J.M. Maestre, R.R. Negenborn (Springer, Dordrecht, 2013), pp. 259–274

8. M.D. Doan, T. Keviczky, B. De Schutter, A hierarchical MPC approach with guaranteed feasibility for dynamically coupled linear systems, in *Distributed Model Predictive Control Made Easy*, ed. by J.M. Maestre, R.R. Negenborn (Springer, Dordrecht, 2013), pp. 393–406
9. F. Farokhi, I. Shames, K.H. Johansson, Distributed MPC via dual decomposition and alternative direction method of multipliers, in *Distributed Model Predictive Control Made Easy*, ed. by J.M. Maestre, R.R. Negenborn (Springer, Dordrecht, 2013), pp. 115–131
10. A. Ferramosca, Cooperative mpc with guaranteed exponential stability, in *Distributed Model Predictive Control Made Easy*, ed. by J.M. Maestre, R.R. Negenborn (Springer, Dordrecht, 2013), pp. 585–600
11. A. Ferramosca, D. Limon, A.H. González, Cooperative distributed MPC integrating a steady sate target optimizer, in *Distributed Model Predictive Control Made Easy*, ed. by J.M. Maestre, R.R. Negenborn (Springer, Dordrecht, 2013), pp. 569–584
12. P. Giselsson, A. Rantzer, Generalized accelerated gradient methods for distributed MPC based on dual decomposition, in *Distributed Model Predictive Control Made Easy*, ed. by J.M. Maestre, R.R. Negenborn (Springer, Dordrecht, 2013), pp. 309–325
13. A. Grancharova, T.A. Johansen, Distributed MPC of interconnected nonlinear systems by dynamic dual decomposition, in *Distributed Model Predictive Control Made Easy*, ed. by J.M. Maestre, R.R. Negenborn (Springer, Dordrecht, 2013), pp. 293–308
14. R. Hermans, M. Lazar, A. Jokić, Distributed Lyapunov-based MPC, in *Distributed Model Predictive Control Made Easy*, ed. by J.M. Maestre, R.R. Negenborn (Springer, Dordrecht, 2013), pp. 225–241
15. Y. Hu, N.H. El-Farra, Adaptive quasi-decentralized MPC of networked process systems, in *Distributed Model Predictive Control Made Easy*, ed. by J.M. Maestre, R.R. Negenborn (Springer, Dordrecht, 2013), pp. 209–223
16. A. Kozma, C. Savorgnan, M. Diehl, Distributed multiple shooting for large scale nonlinear systems, in *Distributed Model Predictive Control Made Easy*, ed. by J.M. Maestre, R.R. Negenborn (Springer, Dordrecht, 2013), pp. 327–340
17. M.Y. Lamoudi, M. Alamir, P. Béguery, A distributed-in-time NMPC-based coordination mechanism for resource sharing problems, in *Distributed Model Predictive Control Made Easy*, ed. by J.M. Maestre, R.R. Negenborn (Springer, Dordrecht, 2013), pp. 147–162
18. J.M. Lemos, J.M. Igreja, D-SIORHC, Distributed MPC with stability constraints based on a game approach, in *Distributed Model Predictive Control Made Easy*, ed. by J.M. Maestre, R.R. Negenborn (Springer, Dordrecht, 2013), pp. 133–146
19. J. Liu, D. Muñoz de la Peña, P.D. Christofides, Lyapunov-based distributed MPC schemes: Sequential and iterative approaches, in *Distributed Model Predictive Control Made Easy*, ed. by J.M. Maestre, R.R. Negenborn (Springer, Dordrecht, 2013), pp. 479–494
20. J.M. Maestre, D. Muñoz de la Peña, E.F. Camacho, Distributed MPC based on agent negotiation, in *Distributed Model Predictive Control Made Easy*, ed. by J.M. Maestre, R.R. Negenborn (Springer, Dordrecht, 2013), pp. 465–477
21. J.M. Maestre, F.J. Muros, F. Fele, D. Muñoz de la Peña, E.F. Camacho, Distributed MPC based on a team game, in *Distributed Model Predictive Control Made Easy*, ed. by J.M. Maestre, R.R. Negenborn (Springer, Dordrecht, 2013), pp. 407–419
22. R. Martí, D. Sarabia, C. de Prada, Price-driven coordination for distributed NMPC using a feedback control law, in *Distributed Model Predictive Control Made Easy*, ed. by J.M. Maestre, R.R. Negenborn (Springer, Dordrecht, 2013), pp. 73–88
23. B. Morcego, V. Javalera, V. Puig, R. Vito, Distributed MPC using reinforcement learning based negotiation: Application to large scale systems, in *Distributed Model Predictive Control Made Easy*, ed. by J.M. Maestre, R.R. Negenborn (Springer, Dordrecht, 2013), pp. 517–533
24. M.A. Müller, F. Allgöwer, Distributed MPC for consensus and synchronization, in *Distributed Model Predictive Control Made Easy*, ed. by J.M. Maestre, R.R. Negenborn (Springer, Dordrecht, 2013), pp. 89–100
25. J.L. Nabais, R.R. Negenborn, R.B. Carmona-Benítez, L.F. Mendonşa, M.A. Botto, Hierarchical MPC for multiple commodity transportation networks, in *Distributed Model Predictive Control Made Easy*, ed. by J.M. Maestre, R.R. Negenborn (Springer, Dordrecht, 2013), pp. 535–552

26. I. Necoara, Rate analysis of inexact dual fast gradient method for distributed MPC, in *Distributed Model Predictive Control Made Easy*, ed. by J.M. Maestre, R.R. Negenborn (Springer, Dordrecht, 2013), pp. 163–178
27. R.R. Negenborn, Decompositions of augmented Lagrange formulations for serial and parallel distributed MPC, in *Distributed Model Predictive Control Made Easy*, ed. by J.M. Maestre, R.R. Negenborn (Springer, Dordrecht, 2013), pp. 437–450
28. C. Ocampo-Martinez, V. Puig, J.M. Grosso, S. Montes-de-Oca, Multi-layer decentralized MPC of large-scale networked systems, in *Distributed Model Predictive Control Made Easy*, ed. by J.M. Maestre, R.R. Negenborn (Springer, Dordrecht, 2013), pp. 495–515
29. G. Pannocchia, S.J. Wright, J.B. Rawlings, On the use of suboptimal solvers for efficient cooperative distributed linear MPC, in *Distributed Model Predictive Control Made Easy*, ed. by J.M. Maestre, R.R. Negenborn (Springer, Dordrecht, 2013), pp. 553–568
30. I. Prodan, F. Stoican, S. Olaru, C. Stoica, S.-I. Niculescu, Mixed-integer programming techniques in distributed MPC problems, in *Distributed Model Predictive Control Made Easy*, ed. by J.M. Maestre, R.R. Negenborn (Springer, Dordrecht, 2013), pp. 275–291
31. S. Roshany-Yamchi, R.R. Negenborn, A.A. Cornelio, Nash-based distributed MPC for multirate systems, in *Distributed Model Predictive Control Made Easy*, ed. by J.M. Maestre, R.R. Negenborn (Springer, Dordrecht, 2013), pp. 341–353
32. F. Tedesco, D.M. Raimondo, A. Casavola, A distributed reference management scheme in presence of non-convex constraints: An MPC based approach, in *Distributed Model Predictive Control Made Easy*, ed. by J.M. Maestre, R.R. Negenborn (Springer, Dordrecht, 2013), pp. 243–257
33. P.A. Trodden, A.G. Richards, Cooperative tube-based distributed MPC for linear uncertain systems coupled via constraints, in *Distributed Model Predictive Control Made Easy*, ed. by J.M. Maestre, R.R. Negenborn (Springer, Dordrecht, 2013), pp. 57–72
34. F. Valencia, J.D. López, J.A., Pati no, J.J. Espinosa, Bargaining game based distributed MPC, in *Distributed Model Predictive Control Made Easy*, ed. by J.M. Maestre, R.R. Negenborn (Springer, Dordrecht, 2013), pp. 41–56
35. A. Zafra-Cabeza, J.M. Maestre, A hierarchical distributed MPC approach: A practical implementation, in *Distributed Model Predictive Control Made Easy*, ed. by J.M. Maestre, R.R. Negenborn (Springer, Dordrecht, 2013), pp. 451–464

Part I
From Small-Scale to Large-Scale: The Group of Autonomous Systems Perspective

Chapter 2
Bargaining Game Based Distributed MPC

F. Valencia, J. D. López, J. A. Patiño and J. J. Espinosa

Abstract Despite of the efforts dedicated to design methods for distributed model predictive control (DMPC), the cooperation among subsystems still remains as an open research problem. In order to overcome this issue, game theory arises as an alternative to formulate and characterize the DMPC problem. Game theory is a branch of applied mathematics used to capture the behavior of the players (agents or subsystems) involved in strategic situations where the outcome of a player is function not only of his choices but also depends on the choices of others. In this chapter a bargaining game based DMPC scheme is proposed; roughly speaking, a bargaining game is a situation where several players jointly decide which strategy is best with respect to their mutual benefit. This allows to deal with the cooperation issues of the DMPC problem. Additionally, the bargaining game framework allows to formulate solutions where the subsystems do not have to solve more than one optimization at each time step. This also reduces the computational burden of the local optimization problems.

F. Valencia (✉) · J. D. López · J. A. Patiño · J. J. Espinosa
Departamento de Energía Eléctrica y Automática, Universidad Nacional de Colombia, Facultad de Minas, Medellín, Colombia
e-mail: fvalenc@unal.edu.co

J. D. López
e-mail: jodlopezhi@unal.edu.co

J. A. Patiño
e-mail: japatino@unal.edu.co

J. J. Espinosa
e-mail: jjespino@unal.edu.co

J. M. Maestre and R. R. Negenborn (eds.), *Distributed Model Predictive Control Made Easy*, Intelligent Systems, Control and Automation: Science and Engineering 69, DOI: 10.1007/978-94-007-7006-5_2,

2.1 Introduction

The main goal of the DMPC is to achieve some degree of coordination among subsystems that are solving local MPC problems with locally relevant variables, costs, and constraints, without solving the centralized MPC problem [4, 6, 10]. However, despite of the efforts dedicated to the formulation of the DMPC schemes, the paradigm to formulate DMPC approaches often requires an iterative procedure to compute the control actions to be applied to the controlled system, while stability and controllability properties of such process must be guaranteed. Most approaches also force the subsystems to cooperate without taking into account whether the cooperative behavior gives some operational advantage to the subsystems, and might steer the subsystems to operating points where they do not perceive any benefit. In the DMPC case, the cooperation is not always an advantage because, as it will be shown in the next sections, there exists a compromise between the local and the global behavior. Thus, if the local performance is not good enough, the whole system behavior is not good either. Considering all these issues, game theory arises as an alternative to formulate and characterize the DMPC problem.

Game theory is a branch of applied mathematics attempting to capture behaviors in strategic situations or games, where the outcome of a player (subsystem in the DMPC case) is function not only of his choices but also of the choices of others [8]. Game theory defines a game as formed by the set of all rules used to describe the situation as a mathematical description or model. Every particular instance of the game is called a play, at each play each player performs a move; formally, a move is a choice performed by a player between various alternatives (under conditions precisely prescribed by the mandatory rules of the game that cannot be infringed), following an strategy decided by the player [9].

If the DMPC problem can be viewed as calculated circumstances where the success of each subsystem is based upon the choices of the other subsystems, then it can be viewed as a game. Besides, if this situation can be described, analyzed and solved as a game where the subsystems are able to share information with each other and have a common goal, the DMPC problem can be described as a **bargaining game**. A bargaining situation involves a group of subsystems having the opportunity to collaborate for mutual benefit in more than one way. If an agreement is not achieved, each subsystem selects a course of action according to the available information [9]. In the DMPC case, such course of action is determined by a threshold of maximum lost of performance. This guarantees that the performance of each subsystem is better or at least equal to such threshold. In general terms, all bargaining situations share the following five elements:

1. A group of players involved in the bargaining.
2. A mutual benefit which is the objective of the bargaining, often defined as a profit or cost function.
3. A decision space composed by all the available choices of the subsystems.
4. A disagreement point defined by the minimum expected satisfaction for the bargaining.

5. An utopia point defined by the set of choices where all players involved in the bargaining achieve at the same time their maximum benefit.

In game theory a solution means a determination of the amount of satisfaction each player should expect to get from the situation. In specific games like bargaining ones, the concept of solution is associated with the determination of how much it should be worth to each of these players the right to have a chance for bargaining.

The selection of the bargaining approach is made because its main insight is focusing in others, i.e., "to assess your added value, you have to ask not only what other players can bring to you but what you can bring to other players" [3]. This approach adds flexibility to the DMPC schemes, and taking as a mathematical framework the theory presented by [9] about bargaining games, a bargaining-game-based DMPC scheme can be proposed with the following features:

- An iterative procedure is not required to compute the control actions to be applied to the controlled system.
- The subsystems are not forced to cooperate, i.e., each subsystem is able to decide whether to cooperate or not with other subsystems depending on the benefit received from the cooperative behavior.
- The paradigm to determine the interaction among subsystems is: *Focusing on others*.
- Each subsystem does not have to compute more than one optimization problem at each time step.
- The solution of the game is Pareto optimal.

2.2 Mathematical Formulation of the DMPC as a Game

Mathematically, a game G can be defined as a tuple $G = (\mathcal{N}, \{\Omega_i\}_{i \in \mathcal{N}}, \{\phi_i\}_{i \in \mathcal{N}})$ where:

- $\mathcal{N} = \{1, \ldots, N\}$ is the set of players.
- Ω_i is the decision space of player i determined by the game, i.e., is the set of feasible decisions of player i.
- $\phi_i : \Omega_1 \times \Omega_N \mapsto \mathbb{R}$ is the profit function of the i-th player.

Often, ϕ_i quantifies the preferences of player i (and determines its strategy), and gives to each player some degree of rationality [1]. In this case, it is assumed that the players are able to communicate and "bargain" with each other in order to achieve a common goal. Mathematically, a bargaining game for $\mathcal{N}$ is a pair $(\mathcal{S}, d)$ where [11]:

1. $\mathcal{S}$ denotes the decision space of the game defined as a nonempty closed subset of $\mathbb{R}^N$, i.e., the feasible set of profit functions.
2. $d \in \text{int}(\mathcal{S})$, d being the disagreement point.
3. $\zeta_i(\mathcal{S})$ the maximum profit available in $\mathcal{S}$ for the i-th player, i.e., it is the utopia point of the i-th subsystem: $\zeta_i(\mathcal{S}) := \max\{\phi_i : (\phi_i)_{i \in \mathcal{N}} \in \mathcal{S}\}$ exists for every $i \in \mathcal{N}$.

In general, the outcome of a game $(\mathcal{S}, d)$ is a tuple $\varphi(\mathcal{S}, d) = (\phi_1, \ldots, \phi_N)$ of profits received by the players. If any player does not cooperate, its corresponding position in $\varphi(\mathcal{S}, d)$ is replaced by its disagreement point. Hence, if all subsystems decide not to cooperate: $\varphi(\mathcal{S}, d) = (d_1, \ldots, d_N)$.

Game theory can be applied to the DMPC problem with the following considerations [14]:

- The rules are provided by the physical and operational constraints of the whole system (the DMPC game).
- Each time step k corresponds to an instance at which the optimal control action should be computed (a play of the DMPC game).
- Each time step k is an opportunity for each subsystem to choose a local control action between various alternatives (the choice in a DMPC game). If an iterative procedure is carried out to compute the local control actions, the iterations are the choices in such DMPC game.
- The moves are determined by the procedure to solve the game, often called negotiation model.

Based on these considerations, the DMPC game can be formulated in its strategic form as $G_{\text{DMPC}} = (\mathcal{N}, \{\Omega_i\}_{i\in\mathcal{N}}, \{\phi_i(\widetilde{\mathbf{u}}(k))\}_{i\in\mathcal{N}})$ where $\mathcal{N} = \{1, \ldots, N\}$ is the set of subsystems, Ω_i and $\phi_i(\widetilde{\mathbf{u}}(k))$ are the set of feasible control actions and the cost function of the i-th subsystem respectively, and $\widetilde{\mathbf{u}}(k) = [\mathbf{u}^T(k), \ldots, \mathbf{u}^T(k + N_c), \ldots, \mathbf{u}^T(k + N_p)]^T$. Note that the unique requirement to formulate the DMPC problem as G_{DMPC} is the dependence of each local cost function in the decisions of the other subsystems.

Moreover, as it was stated before, the game G_{DMPC} is cooperative since it comes from a distributed control situation. Additionally, in G_{DMPC} all subsystems have a common goal. Then, this game satisfies the conditions of a bargaining game $(\mathcal{S}, d)$. However, the game G_{DMPC} has infinite plays and its decision environment has a dynamic evolution. These facts obligate to extend the original bargaining game theory to discrete-time dynamic bargaining games.

2.3 Description of the Approach

In this section the DMPC problem is formulated as a discrete-time dynamic bargaining game (see [14] for details about the definition of discrete-time dynamic bargaining games). With this purpose we recall that (same assumptions were made in [2, 15, 16]):

- $\phi_i(\widetilde{\mathbf{u}}(k); \mathbf{x}(k))$ denotes the cost function for subsystem i, $i = 1, \ldots, N$, where the notation $(\widetilde{\mathbf{u}}(k); \mathbf{x}(k))$ indicates that the function ϕ_i depends on $\widetilde{\mathbf{u}}(k)$, and $\mathbf{x}(k)$ is a time variant parameter.
- It is assumed that the evolution of $\mathbf{x}(k)$ is given by the linear state update equation

$$\mathbf{x}(k+1) = \mathbf{A}\mathbf{x}(k) + \mathbf{B}\mathbf{u}(k)$$

where $\mathbf{A}$ and $\mathbf{B}$ are obtained by linearization and discretization of the model describing the behavior of the whole system.

- $\phi_i(\widetilde{\mathbf{u}}(k))$ is a quadratic positive convex function of $\widetilde{\mathbf{u}}(k)$ for $i = 1, \ldots, N$.
- It is assumed that all subsystems are able to "bargain" in order to achieve a common goal: to maintain both the local and the global system performances by driving the system states to their reference values.

Let $\Upsilon(k) := \{\phi_i(\widetilde{\mathbf{u}}(k)) \colon \widetilde{\mathbf{u}}(k) \in \Omega,\ \forall i \in \mathcal{N}\}$ be a feasible set of cost functions. Since Ω is time-invariant for $i = 1, \ldots, N$, the feasible set $\Upsilon(k)$ is also time-invariant, i.e., $\Upsilon(1) = \Upsilon(2) = \ldots = \Upsilon$. Moreover, since Ω is closed and convex, and by the continuity and convexity of $\phi_i(\widetilde{\mathbf{u}}(k))$ with respect to $\widetilde{\mathbf{u}}(k)$, the set Υ is closed and convex. Note that Υ defines a set of possible values of the cost function of every subsystem given the set Ω. Then, Υ is **the decision space** in the DMPC problem treated in this chapter.

Once defined the decision space in the DMPC problem, it is required to define the disagreement point. The disagreement point is the benefit perceived by the player when an agreement is not possible. Such benefit is associated with an alternative plan carried out by the player in this situation, which is determined by the locally available information [9]. Based on the definition of bargaining game, in [9] the author establishes that the disagreement point should give to the players a strong incentive to increase their demands as much as possible without losing collaboration. Therefore, following these statements let us define **the disagreement point** for the DMPC problem as [14]

$$\eta_i(k+1) = \begin{cases} \eta_i(k) - \alpha(\eta_i(k) - \phi_i(\widetilde{\mathbf{u}}(k))) & \text{if } \eta_i(k) \geq \phi_i(\widetilde{\mathbf{u}}(k)) \\ \phi_i(\widetilde{\mathbf{u}}(k)) & \text{if } \eta_i(k) < \phi_i(\widetilde{\mathbf{u}}(k)) \end{cases} \tag{2.1}$$

$\forall i \in \mathcal{N}$, with $0 < \alpha < 1$.

With this definition of the disagreement point, if the i-th subsystem decides to cooperate it can improve its expected performance by reducing the disagreement point in a factor $\alpha(\eta_i(k) - \phi_i(\widetilde{\mathbf{u}}(k)))$. But if the i-th subsystem decides not to cooperate, its expected performance is increased by a factor $(\phi_i(\widetilde{\mathbf{u}}(k)) - \eta_i(k))$, resulting in a disagreement point equal to $\phi_i(\widetilde{\mathbf{u}}(k))$, allowing the i-th subsystem to cooperate few time steps ahead. For both cooperating and non-cooperating subsystems the disagreement point tends to the optimal expected value of the cost function $\phi(\widetilde{\mathbf{u}}(k))$, given the behavior of the states $\mathbf{x}(k)$ [14].

The definition of the DMPC problem as a discrete-time dynamic bargaining is completed by defining the utopia point. Let $\zeta_i(\Upsilon)$ denote **the utopia point** of the i-th subsystem (in a DMPC problem) defined as $\zeta_i(\Upsilon) := \min\{\phi_i(\widetilde{\mathbf{u}}(k)) \colon \phi_i(\widetilde{\mathbf{u}}(k)) \in \Upsilon\}$. From such a definition $\zeta_i(\Upsilon)$ exist for every $i \in \mathcal{N}$. Then the DMPC problem can be analyzed as a discrete-time dynamic bargaining game $\{(\Upsilon, \eta(k))\}_{k=0}^{\infty}$, where

1. $\Upsilon := \{\phi_i(\widetilde{\mathbf{u}}(k)) \colon \widetilde{\mathbf{u}}(k) \in \Omega,\ \forall i \in \mathcal{N}\}$ is **the decision space**.
2. $\eta(k) := \{\eta_1(k), \ldots, \eta_N(k)\}$ is **the disagreement point**, whose evolution is determined by (2.1).
3. $\zeta_i(\Upsilon) := \min\{\phi_i(\widetilde{\mathbf{u}}(k)) \colon \phi_i(\widetilde{\mathbf{u}}(k)) \in \Upsilon\}$ is **the utopia point**.
4. The dynamic evolution of the decision environment is given by the dynamic model of the controlled system.

Note that only the disagreement point depends on the time step k in $\{(\Upsilon, \eta(k))\}_{k=0}^{\infty}$.

2.3.1 Symmetric and Non-symmetric Games

It is possible to define (into a mathematical model) the desire of each player of maximizing his own gain in bargaining. However, depending of the complexity of the analyzed situation the resultant model can be intractable. Then, some ideal assumptions must be considered for the players [9]: They are highly rational, can accurately compare their desires for various aspects, are equal in bargaining skill, and have full knowledge of the tastes and preferences of the others.

All the situations satisfying these requirements are known as symmetric bargaining games. If all players have the same characteristics the expected satisfaction after a play should be the same. Note that symmetry conditions imply that all players have the same disagreement point in order to achieve the same utility at the end of the play. However, these symmetry conditions can be heavily restrictive in real applications because often players think different, they do not have the same bargaining skills, or simply the players do not share the same interests. As a consequence of these differences among players the bargaining game becomes non-symmetric [11].

The symmetry conditions of discrete-time dynamic bargaining games establish that the game $\{(\Upsilon, \eta(k))\}_{k=0}^{\infty}$ is symmetric if $\eta_1(k) = \ldots = \eta_N(k)$ for $k = 0, 1, 2, \ldots, \infty$. From the symmetry conditions it is possible to conclude that (naturally) the DMPC problem arising from the distributed control of homogeneous systems is an example of symmetric discrete-time dynamic bargaining games. Here the expression homogeneous systems refers to those systems composed by several subsystems with the same characteristic. Although there exist several real systems satisfying the symmetry conditions of discrete-time dynamic bargaining games, these conditions are heavily restrictive, mainly because real large-scale systems are composed by several different subsystems with different time evolution equations. As a consequence, a DMPC game $\{(\Upsilon, \eta(k))\}_{k=0}^{\infty}$ is non-symmetric in general.

The non-symmetric bargaining solution of a DMPC game $\{(\Upsilon, \eta(k))\}_{k=0}^{\infty}$ at time step k can be computed in a centralized way as a solution of the maximization problem [2, 14, 15]:

Algorithm 2.1 Negotiation model for DMPC games

1: At time step k, each subsystem sends the values of $\mathbf{x}_i(k)$, $\eta_i(k)$ to the remaining subsystems.
2: With the information received, each subsystem solves local optimization problem (2.3).
3: Let $\widetilde{\mathbf{u}}_i^*(k)$ denote optimal control actions for subsystem i. If (2.3) is feasible, the subsystem i selects the first control action of $\widetilde{\mathbf{u}}_i^*(k)$. Otherwise, the subsystem i selects the first control action of $\widetilde{\mathbf{u}}_i(k)$, where $\widetilde{\mathbf{u}}_i(k)$ is the initial condition of the subsystem i at time step k.
4: Each subsystem updates its disagreement point. If (2.3) is feasible the update of the disagreement point of the subsystem i is given by $\eta_i(k+1) = \eta_i(k) - \alpha(\eta_i(k) - \phi_i(\widetilde{\mathbf{u}}(k)))$. Otherwise, the update of the disagreement point of the subsystem i is given by $\eta_i(k+1) = \eta_i(k) + (\phi_i(\widetilde{\mathbf{u}}(k)) - \eta_i(k))$ or $\eta_i(k+1) = \phi_i(\widetilde{\mathbf{u}}(k))$.
5: Each subsystem communicates its updated control action and its updated disagreement point to the others.
6: Go to step 1.

$$\begin{aligned}
&\max_{\widetilde{\mathbf{u}}(k)} \sum_{i=1}^{N} w_i \log(\eta_i(k) - \phi_i(\widetilde{\mathbf{u}}(k))) \\
&\text{Subject to:} \\
&\qquad \eta_i(k) > \phi_i(\widetilde{\mathbf{u}}(k)) \\
&\qquad \widetilde{\mathbf{u}}(k) \in \Omega
\end{aligned} \tag{2.2}$$

Note that in the optimization problem of (2.2), $w = \{w_1, \ldots, w_N\}$ requires the selection of the weights for each subsystem. However, there are not guidelines for choosing their values. Often, in control theory field they can be arbitrarily selected as $w_i = \frac{1}{N}$, $i = 1, \ldots, N$ (such a selection is made in [5, 16]). However, performing controllability and/or sensitivity analysis or prior knowledge of the system could help to derive guidelines for the selection of the weights. Also note that for $w_i = w_j = \frac{1}{N}$, $\forall i, j \in \mathcal{N}$, a symmetric game solution is obtained, i.e., the proposed negotiation model can be used for both symmetric and non-symmetric DMPC games.

Let $\phi_i(\widetilde{\mathbf{u}}(k)) = \sigma_r(\widetilde{\mathbf{u}}_i(k), \widetilde{\mathbf{u}}_{-i}(k))$ with $\widetilde{\mathbf{u}}_{-i}(k)$ denoting the inputs of all subsystems except the i-th one. Then, the maximization problem of (2.2) is still centralized, but it can be solved in a distributed way by locally solving the system-wide control problem [2, 14, 15]

$$\begin{aligned}
&\max_{\widetilde{\mathbf{u}}_i(k)} \sum_{r=1}^{N} w_r \log(\eta_r(k) - \sigma_r(\widetilde{\mathbf{u}}_i(k), \widetilde{\mathbf{u}}_{-i}(k))) \\
&\text{Subject to:} \\
&\qquad \eta_r(k) > \sigma_r(\widetilde{\mathbf{u}}_i(k), \widetilde{\mathbf{u}}_{-i}(k)),\ r = 1, \ldots, N \\
&\qquad \widetilde{\mathbf{u}}_i(k) \in \Omega_i
\end{aligned} \tag{2.3}$$

This formulation, for both symmetric and non-symmetric games, allows each subsystem (see [14] for details):

- To take into account the effect of its decisions in the behavior of the remaining subsystems.
- To include the effect of the local decisions in the profit of the remaining subsystems and in their decision about whether to cooperate or not.
- Combined with the negotiation model, to decide whether or not to cooperate depending on the utility perceived by the cooperative behavior.

These aspects reflect the paradigm underlying the proposed control scheme: *on others*.

2.3.2 Negotiation Model

A negotiation model is a sequence of steps for computing the outcome of a game. The negotiation model proposed to solve DMPC games $\{(\Upsilon, \eta(k))\}_{k=0}^{\infty}$ in a distributed way is based on the model proposed in [9] for two-player games. Such model adapted for solving the DMPC game has the steps as given in Algorithm 1.

The initial condition for solving (2.3) at time step $k+1$ is given by the shifted control input: $\bar{\mathbf{u}}(k+1:k+N_{\mathrm{p}})$, where the bar denotes the selected control input.

As in the case of the negotiation model proposed in [9], the negotiation model to solve the DMPC game in a distributed way represents a two moves game where the decisions are taken in steps 3 and 4. In this negotiation model each subsystem is:

- Fully informed on the structure of the game.
- Fully informed on the utility function of the remaining subsystems.
- Assumed intelligent (intelligence is given by the optimization procedure).
- Assumed rational (rationality is given by the decision procedure).

Additionally, it is assumed that the communication architecture allows each subsystem to communicate with the remaining subsystems in order to transmit their disagreement point and their local measurements of the states and inputs.

2.4 Theoretical Results

Most of the DMPC schemes based on Game theory require to perform two or more optimizations. These optimizations are focused on computing the local control actions of each subsystem and suggesting the other subsystems which control actions to use, and/or on creating a matrix of costs used for each subsystem to select which control action apply (see [7, 13] for examples of these kind of approaches). However, in the proposed negotiation model only one optimization problem should be solved. This allows to reduce the computational burden of the DMPC scheme associated with the communications among subsystems.

In addition, only local functions that depend from decisions of the other subsystems are required. This makes more flexible the bargaining approach to the DMPC problem than almost all the DMPC schemes presented in the literature. This statement is validated in [12], where a nonlinear DMPC is formulated for traffic control based on the approach described in this chapter. Moreover, in [15] the proposed control scheme was also used to formulate a DMPC scheme for the chain of reactors followed by a flash separator proposed in [16]. Besides, a comparison of several DMPC schemes (including the control scheme described in this chapter) using a quadruple tank process as a testbed is presented in [2].

The stability of the proposed DMPC method depends on the decision of each subsystem about cooperation. In [14] the case when some subsystems initially do not cooperate, but few steps ahead they start cooperating has been considered in order to demonstrate the stability of the closed-loop system.

2.5 Application Results

In this section the results of the application of the bargaining game based DMPC to an HPV are presented. Here, two scenarios were considered: a power tracking scenario and a price based operation scenario. The system description and model can be found in [14]. The HPV considered in this Chapter is presented in Fig. 2.1 and the full mathematical deployment of both examples is shown in [14].

As in almost all large-scale systems, the most common control scheme employed in practice for controlling the HPV comprises a PI controller with disturbance feed-

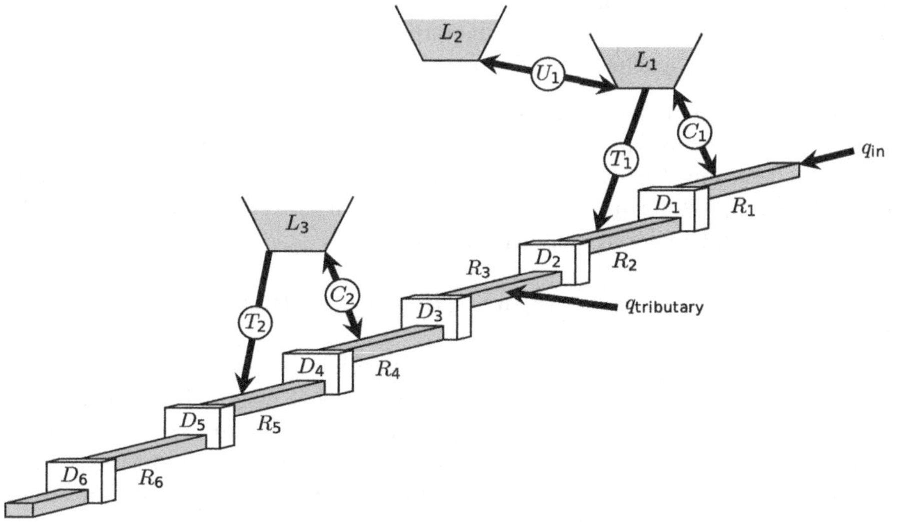

Fig. 2.1 Hydro-power valley used as case of study [14]

forward installed on each individual power plant. However, the use of local PI controllers does not guarantee an efficient use of the stored water, and in presence of disturbances the performance of the entire system could be compromised. For tackling these issues, multivariable control structures have been proposed for controlling HPV systems. Often, these are centralized optimal control schemes. But, since an HPV is a large scale system, a centralized MPC is inflexible and unsuitable. Therefore distributed and/or hierarchical MPC controllers are required (several references regarding the implementation of centralized optimal control schemes in HPV systems can be found in [14]).

2.5.1 The Power Tracking Scenario

In this scenario the power output of the system should follow a given reference while keeping the water levels in the lakes and at the dams as constant as possible. Such power reference is determined by the expected daily demand in the zone fed by the HPV. So, the global cost function considered for the DMPC is composed by two terms [14]:

- The first term penalizes the 1-norm of the power tracking error.
- The second term penalizes the 2-norm of the deviations of the levels in the lakes and dams from their steady state values.

The power reference to be followed by the entire system is known 24 h in advance and the inputs of the system can be changed every 30 min.

From [14], it is possible to divide the HPV of Fig. 2.1 into eight subsystems:

- Subsystem 1: lakes L_1 and L_2, turbine T_1, and turbine-pump C_1
- Subsystem 2: lake L_3, turbine T_2, and turbine-pump C_2
- Subsystems 3–8: reaches R_1 to R_6 respectively

Based on this system decomposition, the local cost function for each of the $\mathcal{N} = \{1, \dots, 8\}$ subsystems can be obtained. For each subsystem, there exists a decision space Ω_i determined by the state and input constraints, and a performance index $\sigma_i(\widetilde{\mathbf{u}}_i(k), \widetilde{\mathbf{u}}_{-i}(k))$ indicating the preferences of each subsystem. Hence, the DMPC for the HPV can be viewed as a strategic game

$$G_{\text{HPV}} = \{\mathcal{N}, \{\sigma_i(\widetilde{\mathbf{u}}_i(k), \widetilde{\mathbf{u}}_{-i}(k))\}_{i \in \mathcal{N}}, \{\Omega_i\}_{i \in \mathcal{N}}\}.$$

Since all subsystems have the same goal: to minimize the power tracking error while keeping the levels in the lakes and dams as close as possible to their steady state values; the game G_{HPV} can be analyzed and solved as a discrete-time dynamic bargaining game $\{(\Upsilon, \eta(k))\}_{k=0}^{\infty}$. In addition, due to the diversity of the physical phenomena involved in an HPV, it is expected that the game G_{HPV} belongs to the class of non-symmetric bargaining games.

Since the power produced by the HPV at time step k is equal to the sum of the powers generated by all subsystems, and assuming that each one communicates the value of its states and inputs to the remaining subsystems, it is possible to locally compute the power produced by the others. Hence, a reference value for each subsystem is not required. This also reduces the computational and communication burden of the proposed DMPC scheme.

2.5.1.1 Simulation Results

A closed-loop simulation of the HPV was performed along 24 h (simulation time). In this simulation the sampling time was $T_s = 1{,}800$ s (30 min), $N_\mathrm{p} = 48$ (corresponding to a day), $N_\mathrm{c} = 32$, $w_{1,2} = \frac{0.4}{2}$, $w_{3-8} = \frac{0.6}{6}$ (the weights of subsystems 1 to 8), $\eta_i(0) = 1 \times 10^5$ (the initial disagreement point of subsystems 1–8), $\gamma = 50$, and $\mathbf{Q} = \mathbf{I}$ ($\mathbf{I}$ being the identity matrix). The values of the parameters as well as the lower and upper limits of the inputs and the states were taken as proposed in [14].

Figure 2.2 shows the comparison between the power produced by the HPV and the power reference when the proposed DMPC scheme computes the inputs of each subsystem. This Fig. 2.2 shows how the power produced by the HPV followed the power reference, satisfying one of the objectives proposed for the control scheme. However, there was an oscillation at the beginning of the experiment due to the transient generated by the change of power from 175 MW (equilibrium power) to the initial required power 150 MW. Recall that in the proposed control scheme there is not a power reference for each subsystem; hence, the initial change requires a negotiation among subsystems in order to decide the amount of power delivered by each power plant to supply the demanded power. The evolution of the disagreement points is presented in Fig. 2.3. In this Figure, the disagreement started at the same point but as they were evolving each subsystem had its own value, indicating the non-symmetry of the game G_HPV. Figure 2.3 also shows a zoom between 4×10^4 s and 7.5×10^4 s, note that all disagreement points decreased with low frequency oscillations; such oscillations were associated to the decision process of each subsystem.

2.5.2 *The Economic Scenario*

The economic scenario concerns the operation of the HPV based on the electricity price. Here, the objective is to compute the optimal control actions such that the profit of the HPV is maximized. Following the economic scenario proposed by [14], the profit maximization relates two important elements:

- The amount of money perceived by the production of electricity in the HPV.
- The opportunity cost determined by the water remaining in the system at the end of the prediction horizon.

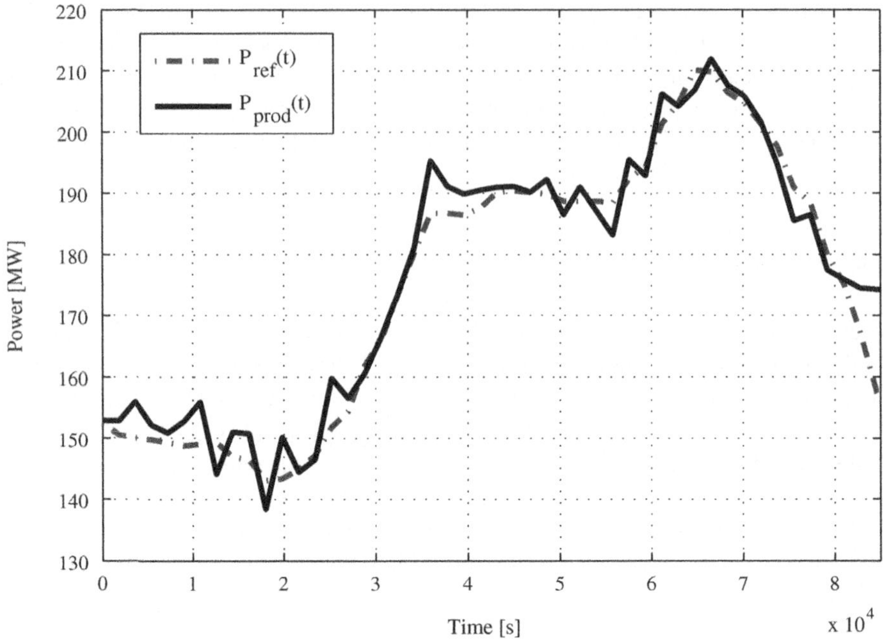

Fig. 2.2 Comparison between the power produced by the HPV and the power reference, when the proposed game-theory-based DMPC is used for computing the inputs of the subsystems

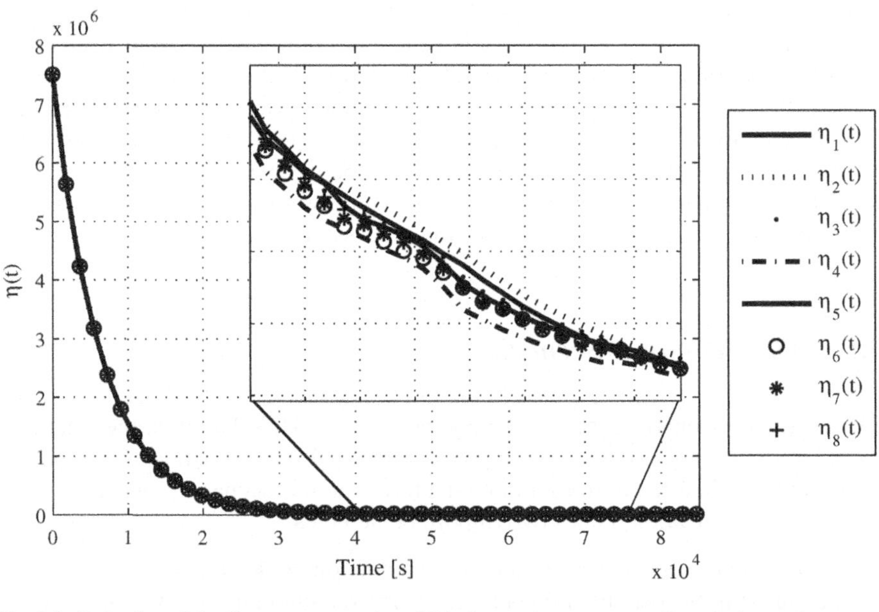

Fig. 2.3 Behavior of the disagreement points. This figure shows an overall evolution and presents a detailed view that allows to evidence the non-symmetry of the game

In order to solve this problem it is assumed that the electricity prices are known 24 h in advance and change every hour, they have the same behavior during the next day, and the prices of the remaining water in the system are constant.

As in the power tracking scenario, there exists a set of subsystems $\mathcal{N} = \{1, \ldots, 8\}$ in the price based operation of the HPV, each one with a decision space Ω_i determined by the input and state constraints, and with a profit function $\sigma_i(\widetilde{\mathbf{u}}_i(k), \widetilde{\mathbf{u}}_{-i}(k))$ indicating their preferences. Hence the DMPC for the price based operation of the HPV can be viewed as an strategic game

$$G_{\text{ec}} = \{\mathcal{N}, \{\sigma_i(\widetilde{\mathbf{u}}_i(k), \widetilde{\mathbf{u}}_{-i}(k))\}_{i\in\mathcal{N}}, \{\Omega_i\}_{i\in\mathcal{N}}\}.$$

Furthermore, since the value of $\sigma_i(\widetilde{\mathbf{u}}_i(k), \widetilde{\mathbf{u}}_{-i}(k))$ depends of the decisions of the other subsystems, all subsystems are able to communicate with each other in order to decide the local control action to be applied. All of them have the same goal: to maximize the profit perceived by the electricity production in the HPV. The game G_{ec} can be analyzed and solved as a discrete-time dynamic bargaining game $\{(\Upsilon, \eta(k))\}_{k=0}^{\infty}$ (due to the diversity of the physical phenomena involved in the HPV, it is expected that the game G_{ec} belongs to the class of non-symmetric bargaining games).

There must be noted that in the power tracking scenario the objective is to minimize a cost function, while the objective in the economic scenario is to maximize a profit function. As a consequence the value of the disagreement point $\eta(k)$ in the economic scenario indicates the minimum profit expected by each subsystem for the production of electricity in the HPV.

2.5.2.1 Simulation Results

In order to test the performance of the proposed control scheme in the economic scenario, a closed-loop simulation of the HPV was carried out along 24 h (simulation time). The sampling time, prediction horizon, control horizon, and weights of the subsystems were the same of the power tracking scenario. However, in the economic scenario the disagreement point of all subsystems was started in zero. The values of c_l, $c_{f,i}$, and the lower and upper limits of the inputs and the states were taken from [14].

Figure 2.4 shows the behavior of the prices and the power produced by the HPV along the simulation. In this Fig. 2.4, despite of the changes of the prices of the electricity, the power produced by the HPV remained almost constant along the simulation. Such behavior was determined by the prices of changing the levels of the reaches. If the use of the stored water to produce energy provided a higher profit than keeping the water at the same levels while producing the same amount of power, then the water levels were changed; otherwise, the levels at the lakes and reaches were kept at a constant value. Recall that the profit of each subsystem is given by the difference between its disagreement point $\eta_i(k)$ and its local profit function $\sigma_i(\cdot)$.

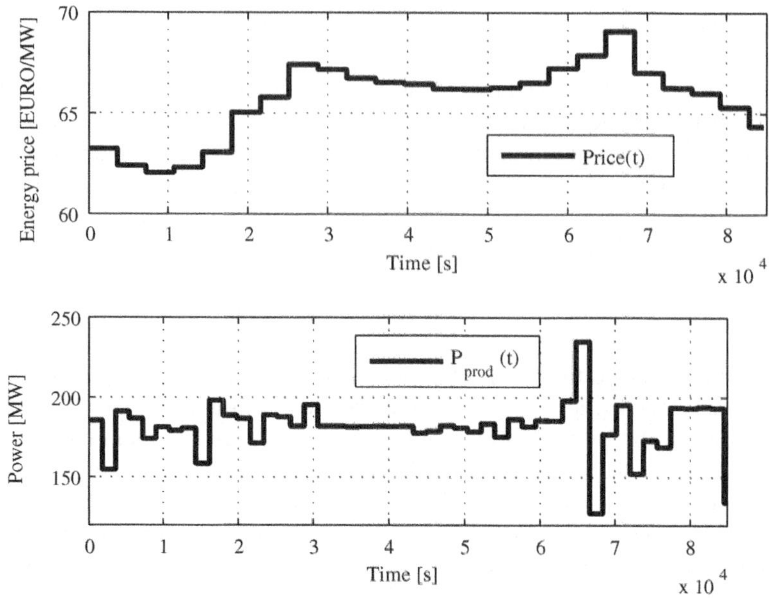

Fig. 2.4 Economic operation of the HPV. The *top panel* shows the behavior of the electricity prices along the simulation. The *bottom panel* shows the power produced by the HPV. Despite of the changes in the electricity prices the power produced by the HPV remains constant along the simulation

From [14], the water of the lake 1 provides more profit than the water in lakes 2 and 3, being the cheapest water the water stored in lake 2. Furthermore, the water stored at the dam of the reach 3 provides more profit than the water stored at the dams in the remaining reaches. Since the price of the water is higher (in all the cases) than the price of the electricity, the results shown in Fig. 2.4 were the expected: a constant production of electricity in order to maximize the water stored in the lakes and reaches.

Figure 2.5 presents the evolution of the disagreement points. The cooperation among subsystems was evident, all disagreement points started at the same value but their increase was cooperative (with different rates, which means that the game G_{ec} is non-symmetric) until 6×10^4 s where a non-cooperative behavior appeared. At that moment, some subsystems did not perceived any benefit by the cooperation, i.e., there were no feasible control actions such that the minimum expected profit was achieved. As a consequence, a decrease of the disagreement point of these subsystems was done in order to promote the cooperation of these subsystems several time steps ahead (note that the cooperation started again at the next time step).

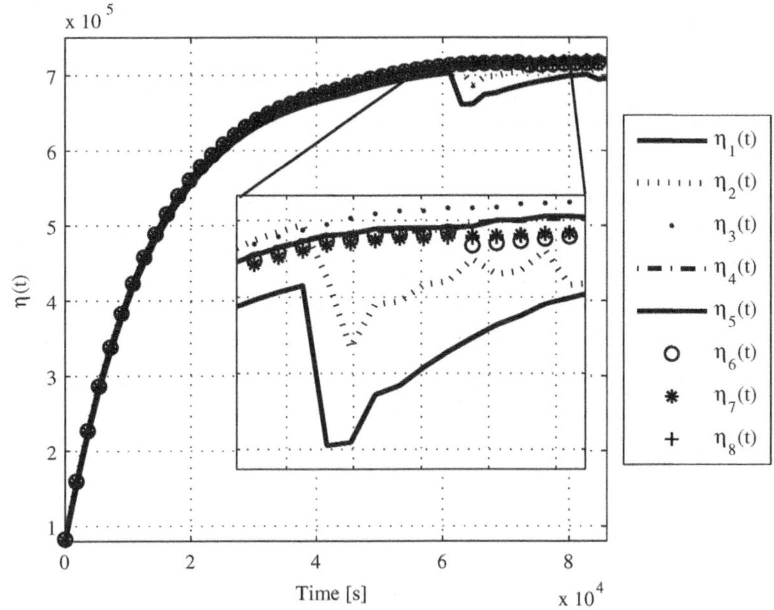

Fig. 2.5 Evolution of the disagreement points. Here, the cooperative behavior of the subsystems remained during 6×10^4 s where the non-cooperative behavior emerged. This is reflected in the decrease of the value of the disagreement point of some subsystems

2.6 Conclusions

In this chapter, the formulation of the distributed model predictive control as a bargaining game was presented. With this purpose, the concept of discrete-time dynamic bargaining game was considered. Moreover, a negotiation model to solve the distributed model predictive control game was proposed. Properties like closed-loop stability of the system when the control actions are computed by the proposed control scheme were also discussed. The application of the DMPC scheme presented in this chapter to an HPV was also included. In this application two scenarios were considered: a power tracking scenario and a price based operation scenario. In both cases the original control problem was formulated and solved as a bargaining game.

Acknowledgments This research has been supported by the European 7th framework STREP project "Hierarchical and distributed model predictive control (HD-MPC)", contract number INFSO-ICT-223854.

References

1. O. Akira. A noncooperative approach to general N-person cooperative games. Discussion Papers 2005–01, Graduate School of Economics, Hitotsubashi University, 2005
2. I. Alvarado, D. Limon, D. Munoz de la Pena, J.M. Maestre, R.R. Negenborn, F. Valencia, H. Scheu, M.A. Ridao, B. De Schutter, J. Espinosa, and W. Marquardt. A comparative analysis of distributed [mpc] techniques applied to the [hd-mpc] four-tank benchmark. J. Process Control **21**, 800–815 2010
3. A.M. Brandenburger, B.J. Nalebuff, *The Right Game: Use the Game Theory to Shape Strategy* (Harvard Bussines Review, July-August, 1995)
4. E. Camponogara, D. Jia, B.H. Krogh, S. Talukdar, Distributed model predictive control. IEEE Control Syst. Mag. **22**(1), 44–52 (2002)
5. D. Doan, T. Keviczky, I. Necoara, and M. Diehl. A Jacobi algorithm for distributed model predictive control of dynamically coupled systems. Technical Report arXiv:0809.3647, Delf University of Technology, Sep 2008. Comments: Shorter version submitted to 2009 American Control Conference
6. D. Jia and B.H. Krogh. Distributed Model Predictive Control. *Proceedings of the 2001 American Control Conference*, pages 2767–2772, 2001
7. J. M. Maestre, D. Muñoz de la Peña, and E. F. Camacho. Distributed MPC Based on a Cooperative Game. *48th IEEE Conference on Decision and Control and 28th Chinese Control Conference*, 2009
8. R.B. Myerson. *Game Theory: Analysis of Conflict*. Harvard University Press, 1991
9. J. Nash, Two-person cooperative games. Econometrica **21**(1), 128–140 (1953)
10. I. Necoara, D. Doan, and J.A.K. Suykens. Application of the proximal center decomposition method to distributed model predictive control. *Proceedings of the 2008 IEEE Conference on Decision and, Control*, pp. 2900–2905, 2008
11. H.J. Peters, *Axiomatic Bargaining Game Theory* (Kluwer Academic Publishers, Dordrecht, 1992)
12. C. Roviro, F. Valencia, J.D. López, J.J. Espinosa, A. Nú nez, and B. De Schutter. Non linear model predictive control based on game theory for traffic control on highways. In *IFAC Conference on Nonlinear Model Predictive, Control*, 2012
13. P.A. Trodden, D. Nicholson, and A.G. Richards. Distributed model predictive control as a game with coupled contraints. *Proceedings of the European Control Conference*, 2009
14. F. Valencia. Game Theory Based Distributed Model Predictive Control: An Approach to Large-Scale Systems Control. Ph.D thesis, Facultad de Minas, 2012
15. F. Valencia, J. Espinosa, B. De Schutter, and K. Stavnková. Feasible-cooperation distributed model predictive control scheme based on game theory. In 18th IFAC world congress, 2010
16. A.N. Venkat, J.B. Rawlings, and S.J. Wright. Stability and optimality of distributed, linear model predictive control. Part I: State feedback. Texas-Wisconsin Modeling and Control Consortium Technical, Report, 3, 2006

Chapter 3
Cooperative Tube-based Distributed MPC for Linear Uncertain Systems Coupled Via Constraints

P. A. Trodden and A. G. Richards

Abstract This chapter presents a robust form of distributed model predictive control for multiple, dynamically decoupled subsystems subject to bounded, persistent disturbances. Control agents make decisions locally and exchange plans; satisfaction of coupling constraints is ensured by permitting only non-coupled subsystems to update simultaneously. Robustness to disturbances is achieved by use of the tube MPC concept, in which a local control agent designs a tube, rather than a trajectory, for its subsystem to follow. Cooperation between agents is promoted by a local agent, in its optimization, designing hypothetical tubes for other subsystems, and trading local performance for global. Uniquely, robust feasibility and stability are maintained without the need for negotiation or bargaining between agents.

3.1 Introduction

This chapter presents a distributed form of MPC for systems defined by the following characteristics: the overall system is composed of, or may be decomposed to, a number of dynamically decoupled subsystems. Each has linear, time-invariant dynamics, and is subject to local constraints and persistent, bounded disturbances. The subsystems are coupled via constraints, and should coordinate decision-making to satisfy these constraints robustly and also to minimize some system-wide cost.

In the described approach, the distributed control agents exchange plans to achieve constraint satisfaction. Key features are that (i) coupled subsystems may not update

P. A. Trodden (✉)
Department of Automatic Control and Systems Engineering, University of Sheffield, Sheffield, UK
e-mail: p.trodden@sheffield.ac.uk

A. G. Richards
Department of Aerospace Engineering, University of Bristol, Bristol, UK
e-mail: arthur.richards@bris.ac.uk

J. M. Maestre and R. R. Negenborn (eds.), *Distributed Model Predictive Control Made Easy*, Intelligent Systems, Control and Automation: Science and Engineering 69,
DOI: 10.1007/978-94-007-7006-5_3,

their plans simultaneously; (ii) robust stability is guaranteed for any choice of update sequence; (iii) an agent communicates only when strictly necessary, and (iv) cooperation between agents is promoted by a local agent considering the objectives of, and designing hypothetical plans for, other subsystems. The resulting algorithm offers flexibility in communication and computation, and requires no inter-agent negotiation, iteration or bargaining.

The approach, which first appeared in its non-cooperative form [10], uses the concept of tube MPC [6], a form of robust MPC that guarantees feasibility and stability despite the action of an unknown but bounded disturbance. The approach shares similarities with the 'sequential' DMPC method of Richards and How [7], in that robust feasibility and stability of the overall system is guaranteed by local agents updating plans one at a time, without negotiation. However, tube DMPC permits a flexible order of updating, in contrast to a fixed, pre-determined sequence. Thus, this approach combines guaranteed robust feasibility and convergence, in the presence of a persistent disturbance, with flexible communication.

In the cooperative form of the algorithm [11], a local agent designs not only its own tube, but also hypothetical tubes for other subsystems in the problem. The idea is that an agent may now consider the objectives and intentions of others in order to arrive at a more cooperative solution. Here, cooperation is taken to mean the improvement of system-wide performance through the avoidance of greedy behaviour by individual agents. Coupled constraint satisfaction is, however, maintained without the need for inter-agent negotiation or bargaining. In comparison, approaches to cooperation based on inter-agent iteration or bargaining [4, 8, 15], require multiple and repeated information exchanges at each time step in order to achieve constraint satisfaction and stability. Thus, the approach combines robust satisfaction of coupled constraints with cooperation. Overall, cooperation offers performance close to that of centralized MPC but with less computation and communication.

The chapter begins with a formal statement of the problem. In Sect. 3.3, the distributed MPC approach is described, including local optimization problems, algorithms and communication requirements. Theoretical results are summarized in Sect. 3.4. Finally, Sect. 3.5 discusses applications of the approach.

3.2 Problem Statement

The system under consideration consists of a set $\mathcal{N}$ of dynamically decoupled subsystems. Subsystem dynamics are linear and time invariant (LTI):

$$\mathbf{x}_i(k+1) = \mathbf{A}_i\mathbf{x}_i(k) + \mathbf{B}_i\mathbf{u}_i(k) + \mathbf{d}_i(k), \quad \forall i \in \mathcal{N}, k \in \mathbb{N}.$$

For a subsystem i, $\mathbf{x}_i \in \mathbb{R}^{n_{x_i}}$, $\mathbf{u}_i \in \mathbb{R}^{n_{u_i}}$ and $\mathbf{d}_i \in \mathbb{R}^{n_{x_i}}$ are, respectively, the state, control input and disturbance. The latter is unknown *a priori*, but is assumed to lie in a known independent, bounded, compact set that contains the origin in its interior:

$$\mathbf{d}_i(k) \in \mathcal{D}_i \subset \mathbb{R}^{n_{x_i}}, \quad \forall i \in \mathcal{N}, k \in \mathbb{N}.$$

Each subsystem $i \in \mathcal{N}$ is subject to local constraints on an output $\mathbf{y}_i \in \mathbb{R}^{n_{y_i}}$:

$$\mathbf{y}_i(k) = \mathbf{C}_i\mathbf{x}_i(k) + \mathbf{D}_i\mathbf{u}_i(k) \in \mathcal{Y}_i,$$

where $\mathcal{Y}_i \subset \mathbb{R}^{n_{y_i}}$ is closed and contains the origin in its interior. The subsystems are coupled via constraints, constructed as follows. Define a coupling output $\mathbf{z}_{ci} \in \mathbb{R}^{n_{z_c}}$ for a constraint $c \in \mathcal{C}$ and a subsystem $i \in \mathcal{N}$. The sum of coupling outputs for a constraint c must lie in a closed set $\mathcal{Z}_c$ that contains the origin:

$$\mathbf{z}_{ci}(k) = \mathbf{E}_{ci}\mathbf{x}_i(k) + \mathbf{F}_{ci}\mathbf{u}_i(k), \text{ and} \sum_{i \in \mathcal{N}} \mathbf{z}_{ci}(k) \in \mathcal{Z}_c.$$

The following definitions identify structure in these coupling constraints, and are used later in the requirements for communication. Let $\mathcal{N}_c$ be the set of subsystems involved in constraint c, and $\mathcal{C}_i$ the set of constraints involving subsystem i:

$$\mathcal{N}_c \triangleq \{i \in \mathcal{N} : [\mathbf{E}_{ci}\ \mathbf{F}_{ci}] \neq \mathbf{0}\}, \tag{3.1}$$

$$\mathcal{C}_i \triangleq \{c \in \mathcal{C} : [\mathbf{E}_{ci}\ \mathbf{F}_{ci}] \neq \mathbf{0}\}. \tag{3.2}$$

Then the set of all other subsystems coupled to a subsystem i is

$$\mathcal{Q}_i = \left(\bigcup_{c \in \mathcal{C}_i} \mathcal{N}_c \right) \setminus \{i\}. \tag{3.3}$$

The control objective is, without loss of generality, to steer each subsystem state $\mathbf{x}_i$ to the origin, while satisfying constraints. To this end, assume that each $(\mathbf{A}_i, \mathbf{B}_i)$ is controllable, and that the state $\mathbf{x}_i$ is available to the control agent for subsystem i at each sampling instant. Define $\mathbf{K}_i \in \mathbb{R}^{n_{u_i} \times n_{x_i}}$ as a stabilizing controller for each $i \in \mathcal{N}$, let $\mathcal{R}_i$ be a disturbance-invariant set [3] for the resulting controlled system. That is, $(\mathbf{A}_i + \mathbf{B}_i\mathbf{K}_i)\mathbf{x}_i + \mathbf{d}_i \in \mathcal{R}_i$ for all $\mathbf{x}_i \in \mathcal{R}_i$ and $\mathbf{d}_i \in \mathcal{D}_i$; equivalently, $(\mathbf{A}_i + \mathbf{B}_i\mathbf{K}_i)\mathcal{R}_i \oplus \mathcal{D}_i \subseteq \mathcal{R}_i$. It is assumed that the disturbance sets $\mathcal{D}_i$ are sufficiently small such that $(\mathbf{C}_i + \mathbf{D}_i\mathbf{K}_i)\mathcal{R}_i \subset \text{interior}(\mathcal{Y}_i)$, $\forall i \in \mathcal{N}$, and $\bigoplus_{i \in \mathcal{N}}(\mathbf{E}_{ci} + \mathbf{F}_{ci}\mathbf{K}_{ci})\mathcal{R}_i \subset \text{interior}(\mathcal{Z}_c)$, $\forall c \in \mathcal{C}$. This latter assumption is not unusual and represents a mild condition for many practical constraints and disturbances.

3.3 Distributed MPC Using Tubes

Tube MPC was introduced by Mayne et al. in [6]. Instead of optimizing a sequence of states, i.e. points in the state space, it optimizes for a *tube*, or a sequence of sets, within the state space. Drawing on invariance concepts, one can prove that once in the tube, the system state can stay in the tube despite the disturbance. A consequence of this is that the initial centre of the tube is variable. One can envisage this as the problem of catching a speck of dust with a vacuum cleaner: the centre of the hose can be placed anywhere, provided the speck is within the hose opening. The internal dynamics of the hose will then get the dust to the bag, without having to move the hose further. It is this invariance that makes the tube approach so attractive for DMPC: if each subsystem has a tube that is invariant under local control, the tube does not need to be updated to accommodate the disturbance. Hence, communication is necessary only when an agent chooses to change the tube. Coupling can be captured by ensuring that the tubes are consistent across the system.

This section begins with a review of centralized tube MPC, also serving to introduce the relevant notation. Then two different forms of the distributed approach are described: one that leads to a non-cooperative form of DMPC, and a second that uses a conceptual extension to promote cooperation between agents.

3.3.1 Review of Centralized Tube MPC

The tube MPC approach [6] uses the nominal system dynamics to design a sequence of disturbance-invariant state sets. The decision variable includes not only the control sequence for each subsystem i, $\bar{\mathbf{u}}_i(k : k+N_p-1) \triangleq \left[\bar{\mathbf{u}}_i^{\mathrm{T}}(k), \bar{\mathbf{u}}_i^{\mathrm{T}}(k+1), \ldots, \bar{\mathbf{u}}_i^{\mathrm{T}}(k+N-1)\right]^{\mathrm{T}}$, but also the initial state predictions, $\bar{\mathbf{x}}_i(k)$ for all i, which correspond to the tube centres. As the optimization involves only nominal terms, complexity is comparable to standard MPC. Robustness to disturbances is guaranteed by use of a feedback law to keep the state of each subsystem around its tube centre.

For the system state $\{\mathbf{x}_i(k)\}_{i\in\mathcal{N}}$, the centralized optimal control problem is

$$\min_{\{\bar{\mathbf{x}}_i(k),\bar{\mathbf{u}}_i(k:k+N_p-1)\}_{i\in\mathcal{N}}} \sum_{i\in\mathcal{N}} J_i\left(\bar{\mathbf{x}}_i(k), \bar{\mathbf{u}}_i(k : k+N_p-1)\right) \tag{3.4}$$

subject to, $\forall l \in \{1, \ldots, N_p-1\}, i \in \mathcal{N}$,

$$\bar{\mathbf{x}}_i(k+l+1) = \mathbf{A}_i\bar{\mathbf{x}}_i(k+l) + \mathbf{B}_i\bar{\mathbf{u}}_i(k+l), \tag{3.5a}$$

$$\bar{\mathbf{y}}_i(k+l) = \mathbf{C}_i\bar{\mathbf{x}}_i(k+l) + \mathbf{D}_i\bar{\mathbf{u}}_i(k+l), \tag{3.5b}$$

$$\bar{\mathbf{z}}_{ci}(k+l) = \mathbf{E}_{ci}\bar{\mathbf{x}}_i(k+l) + \mathbf{F}_{ci}\bar{\mathbf{u}}_i(k+l), \forall c \in \mathcal{C}, \tag{3.5c}$$

$$\mathbf{x}_i(k) - \bar{\mathbf{x}}_i(k) \in \mathcal{R}_i, \tag{3.5d}$$

$$\bar{\mathbf{x}}_i(k+N_p) \in \mathcal{X}_i^{\mathrm{f}}, \tag{3.5e}$$

$$\bar{\mathbf{y}}_i(k+l) \in \tilde{\mathcal{Y}}_i, \tag{3.5f}$$

$$\sum_{i\in\mathcal{N}} \bar{\mathbf{z}}_{ci}(k+l) \in \tilde{\mathcal{Z}}_c, \forall c \in \mathcal{C}. \tag{3.5g}$$

Here, the local cost function for each subsystem is

$$J_i\big(\bar{\mathbf{x}}_i(k), \bar{\mathbf{u}}_i(k:k+N_p-1)\big) = F_i\big(\bar{\mathbf{x}}_i(k+N)\big) + \sum_{l=0}^{N_p-1} L_i\big(\bar{\mathbf{x}}_i(k+l), \bar{\mathbf{u}}_i(k+l)\big), \quad (3.6)$$

where the stage cost $L_i : \mathbb{R}^{n_{x_i}} \times \mathbb{R}^{n_{u_i}} \mapsto \mathbb{R}_{0+}$, and $F_i : \mathbb{R}^{n_{xp}} \mapsto \mathbb{R}_{0+}$ is a terminal cost.

The constraint sets in the problem, $\tilde{\mathcal{Y}}_i$, $\tilde{\mathcal{Z}}_c$ are tightened versions of the original sets, in order to provide a margin for uncertainty:

$$\tilde{\mathcal{Y}}_i = \mathcal{Y}_i \sim (\mathbf{C}_i + \mathbf{D}_i\mathbf{K}_i)\mathcal{R}_i, \forall i \in \mathcal{N}, \quad (3.7a)$$

$$\tilde{\mathcal{Z}}_c = \mathcal{Z}_i \sim \bigoplus_{i\in\mathcal{N}} (\mathbf{E}_{ci} + \mathbf{F}_{ci}\mathbf{K}_i)\mathcal{R}_i, \forall c \in \mathcal{C}. \quad (3.7b)$$

The sets $\mathcal{R}_i$ are the cross-sections of the tubes, and satisfy the assumptions in the previous section; the tube itself is given by $\{\bar{\mathbf{x}}_i(k) \oplus \mathcal{R}_i, \ldots, \bar{\mathbf{x}}_i(k+N_p) \oplus \mathcal{R}_i\}$ for a subsystem i. The tightening is necessary to accommodate the tube approach: instead of dealing with the exact output values, the constraints act upon the centres of the tubes, and the actual outputs could be anywhere within the cross-section.

The sets $\mathcal{X}_i^{\mathrm{f}}$ for all $i \in \mathcal{N}$ are terminal constraint sets, and each is assumed to be an admissible control invariant set [3]. That is, there is assumed to exist a control law $\mathbf{u}_i = \kappa_i^{\mathrm{f}}(\mathbf{x}_i)$ such that, for all $\mathbf{x}_i \in \mathcal{X}_i^{\mathrm{f}}$,

$$\mathbf{A}_i\mathbf{x}_i + \mathbf{B}_i\kappa_i^{\mathrm{f}}(\mathbf{x}_i) \in \mathcal{X}_i^{\mathrm{f}}, \quad (3.8a)$$

$$\mathbf{C}_i\mathbf{x}_i + \mathbf{D}_i\kappa_i^{\mathrm{f}}(\mathbf{x}_i) \in \tilde{\mathcal{Y}}_i, \quad (3.8b)$$

$$\sum_{i\in\mathcal{N}} \mathbf{E}_{ci}\mathbf{x}_i + \mathbf{F}_{ci}\kappa_i^{\mathrm{f}}(\mathbf{x}_i) \in \tilde{\mathcal{Z}}_c, \quad \forall c \in \mathcal{C}. \quad (3.8c)$$

It is often assumed that $\kappa_i^{\mathrm{f}}(\mathbf{x}_i) = \mathbf{K}_i\mathbf{x}_i$, but this is not necessary, and an alternative choice of κ_i^{f} may simplify the determination of the sets $\mathcal{X}_i^{\mathrm{f}}$. Here, we leave the problem in its most flexible form, and the reader is referred to the examples cited in Sect. 3.5 for further options.

Assuming that a feasible solution, $\{\bar{\mathbf{x}}_i^*(k), \bar{\mathbf{u}}_i^*(k:k+N_p-1)\}_{i\in\mathcal{N}}$, is available to this problem at time k, the following control is applied to each subsystem

$$\mathbf{u}_i(k) = \bar{\mathbf{u}}_i^*(k) + \mathbf{K}_i\big(\mathbf{x}_i(k) - \bar{\mathbf{x}}_i^*(k)\big). \quad (3.9)$$

The resulting controlled system, under the assumptions described, is recursively feasible, despite the actions of the persistent disturbances [6]. This is because, given a feasible solution $\{\bar{\mathbf{x}}_i^*(k), \bar{\mathbf{u}}_i^*(k:k+N_p-1)\}_{i\in\mathcal{N}}$ for time k, the candidate or "tail" solution $\{\bar{\mathbf{x}}_i^*(k+1), \bar{\mathbf{u}}_i^*(k+1:k+N_p)\}_{i\in\mathcal{N}}$, where

$$\bar{\mathbf{x}}_i^*(k+1) = \mathbf{A}_i \bar{\mathbf{x}}_i^*(k) + \mathbf{B}_i \bar{\mathbf{u}}_i^*(k), \tag{3.10a}$$

$$\bar{\mathbf{u}}_i^*(k+1:k+N_p) = \left[\bar{\mathbf{u}}_i^{*\mathrm{T}}(k+1), \ldots, \bar{\mathbf{u}}_i^{*\mathrm{T}}(k+N_p-1), \kappa_i^{\mathrm{f}}(\bar{\mathbf{x}}_i^*(k+N_p))^{\mathrm{T}}\right]^{\mathrm{T}}, \tag{3.10b}$$

is a feasible solution to the centralized problem at $k+1$. Furthermore, with standard assumptions [5] on the stage and terminal costs, asymptotic or exponential convergence of the states of the system to the sets $\mathcal{R}_i$ is guaranteed [6].

In the sequel, let $\mathbf{U}_i(k) = \{\bar{\mathbf{x}}_i(k), \bar{\mathbf{u}}_i(k:k+N_p-1)\}$. $\mathbf{U}_i^*(k)$ is a feasible solution for step k, and $\tilde{\mathbf{U}}_i(k+1)$ denotes the candidate solution for $k+1$, as formed by (3.10).

3.3.2 Distributed Tube MPC

In the distributed MPC approach described in the remainder of this section, the centralized problem is distributed among the subsystem control agents as local optimization problems. In order to maintain coupled constraint satisfaction, only a subset of agents solve their optimizations for a new plan. Meanwhile, the other agents 'freeze' their plans by adopting the tail solution (3.10) unchanged.

In the non-cooperative form, a control agent for subsystem i minimizes only its local share of the system-wide cost (3.4), which is J_i, as defined by (3.6). The local optimization problem, $P_i^{\mathrm{dmpc}}(\mathbf{x}_i(k), \mathbf{Z}_i^*(k))$, for i is

$$\min_{\mathbf{U}_i(k)} J_i(\mathbf{U}_i(k)) \tag{3.11}$$

subject to local constraints (3.5a)–(3.5f) for i, and the coupling constraint

$$\bar{\mathbf{z}}_{ci}(k+l) + \sum_{j \in \mathcal{N}_c \setminus \{i\}} \bar{\mathbf{z}}_{cj}^*(k+l) \in \tilde{\mathcal{Z}}_c, \forall l \in \{1, \ldots, N_p-1\}, c \in \mathcal{C}_i. \tag{3.12}$$

where * denotes a fixed, previously published output of a coupled subsystem. $\mathbf{Z}_i^*(k)$ denotes the aggregate information required by i to evaluate the coupling constraints.

The optimization is employed in Algorithm 3.1, to be executed by all agents in parallel.

Though the algorithm is executed by all agents in parallel, only agents in a set $\mathcal{N}_k \subseteq \mathcal{N}$ are permitted to update by optimization at a step k. All other agents $j \notin \mathcal{N}_k$ renew their current plans, as per (3.10), each by shifting in time the tail of its previous, feasible plan and augmenting with a step of terminal control.

Having obtained a plan $\mathbf{U}^*(k)$ at step k, by either optimization or renewal, each agent unilaterally applies the first control of the planned sequence. No negotiation or iterative refinement of solutions takes place during a time step.

The order in which the subsystems' plans are optimized is determined by the *update sequence*, $\{\mathcal{N}_1, \ldots, \mathcal{N}_k, \mathcal{N}_{k+1}, \ldots\}$. This is to be chosen by the designer,

Algorithm 3.1 Distributed MPC for a subsystem i

1: Wait for an initial feasible plan $\mathbf{U}_i^*(0) = \{\bar{\mathbf{x}}_i^*(0), \bar{\mathbf{u}}_i^*(0 : N_p - 1)\}$ and information, including $\mathbf{Z}_i^*(0)$. Set $k = 0$.
2: Apply control $\mathbf{u}_i(k) = \bar{\mathbf{u}}_i^*(k) + \mathbf{K}_i(\mathbf{x}_i(k) - \bar{\mathbf{x}}_i^*(k))$. Wait one time step, increment k.
3: Measure state $\mathbf{x}_i(k)$.
4: Update plan. If $i \in \mathcal{N}_k$,

1. Propagate intentions of other agents, $\bar{\mathbf{z}}_{cj}^*(\cdot)$, to the current planning horizon.
2. Obtain new plan $\mathbf{U}_i^{\text{opt}}(k)$ by solving the local problem $P_i^{\text{dmpc}}(k)$.
3. Set $\mathbf{U}_i^*(k) = \mathbf{U}_i^{\text{opt}}(k)$.
4. If necessary, transmit new information to other agents.

Else, renew existing plan: form $\tilde{\mathbf{U}}_i(k)$ according to (3.10) and set $\mathbf{U}_i^*(k) = \tilde{\mathbf{U}}_i(k)$.
5: Go to step 2.

and may be a static (*i.e.*, pre-determined) or dynamic sequence. For a particular time step k, the only criterion for the selection of the updating set of agents, $\mathcal{N}_k$, is that no two optimizing subsystems may share coupled constraints. That is, $(i, j) \in \mathcal{N}_k$ only if $j \notin \mathcal{Q}_i$. In the limiting case of coupling between all pairs of subsystems, this reduces the maximum size of $\mathcal{N}_k$ to a single agent. Note, in addition, that the empty set is always a valid choice for $\mathcal{N}_k$, such that no optimization is solved at a time k.

3.3.2.1 Initialization of DMPC

It is assumed that all of the local subsystem information required to formulate the local problem is made available off-line to agent i.

Requirement 3.1 *[Local subsystem information] The matrices* $\mathbf{A}_i$, $\mathbf{B}_i$, $\mathbf{C}_i$, $\mathbf{D}_i$, $\mathbf{E}_{ci}$, $\mathbf{F}_{ci}$, $\forall c \in \mathcal{C}_i$, *stabilizing controllers* $\mathbf{K}_i$ *and* κ_i^{f}, *sets* $\mathcal{R}_i$, $\mathcal{X}_i^{\text{f}}$, $\tilde{\mathcal{Y}}_i$, $\tilde{\mathcal{Z}}_c$, $\forall c \in \mathcal{C}_i$, *and cost function* J_i *shall be available to the control agent for subsystem* i.

The computation of $\mathbf{K}_i$, κ_i^{f}, $\mathcal{R}_i$ and $\mathcal{X}_i^{\text{f}}$—as well as the tightening of sets according to (3.7)—need be done only once, off-line, at initialization. It is assumed that this is done centrally, with the results communicated to agents.

Also, since each agent needs to be able to deduce its neighbors' current intentions based on their last published plans, information is required on the dynamics and constraints of those neighbours:

Requirement 3.2 *[Coupled subsystem information] The matrices* $\mathbf{A}_j$, $\mathbf{B}_j$, $\mathbf{E}_{cj}$, $\mathbf{F}_{cj}$, $\forall c \in \mathcal{C}_i$ *and terminal controller* κ_i^{f} *shall be available to the control agent for subsystem* i *for each coupled subsystem* $j \in \mathcal{Q}_i$.

Finally, it is required that a feasible initial plan is available to each control agent.

Requirement 3.3 *[Initial plan] A feasible local plan* $\mathbf{U}_i^*(0)$ *shall be available to the control agent for subsystem* i *at time* $k = 0$.

This is a common assumption of many DMPC approaches (e.g. [1, 7]). Note that this does not imply that the centralized problem must be solved to optimality; often a simple feasible solution is available, such as all subsystems remaining stationary. No further centralized processing is required after this initialization step.

3.3.2.2 Inter-agent Communication

An updating agent $i \in \mathcal{N}_k$ must have received sufficient information from each coupled agent $j \in \mathcal{Q}_i$ so as to enforce constraints (3.12). Specifically, this requires the construction of signals $\bar{\mathbf{z}}^*_{cj}(k+l)\ \forall l \in \{1, \ldots, N_p - 1\}, \forall j \in \mathcal{N}_c, c \in \mathcal{C}_i$.

It is not necessary to obtain the whole *plan* $\mathbf{U}^*_j(k)$ from some coupled $j \in \mathcal{Q}_i$. Instead, define a *message* vector for subsystem i regarding constraint c at time k as

$$\mathbf{m}_{ci}(k) \triangleq \left[\bar{\mathbf{z}}^{*\mathrm{T}}_{ci}(k) \ldots \bar{\mathbf{z}}^{*\mathrm{T}}_{ci}(k+N_p-1)\ \bar{\mathbf{x}}^{*\mathrm{T}}_{i}(k+N_p)\right]^{\mathrm{T}}, \tag{3.13}$$

which includes the terminal state, and where * denotes a feasible solution.

Put simply, each updating agent must have the latest information available about every other coupled agent. More formally:

Requirement 3.4 *[Information exchange] Consider any two coupled agents i and $j \in \mathcal{Q}_i$ and any two time steps k_i and $k_j > k_i$ such that $i \in \mathcal{N}_{k_i}$ and $j \in \mathcal{N}_{k_j}$, i.e. agents i and j updated at times k_i and k_j respectively. Then message $\mathbf{m}_{ci}(k)$ for every coupling constraint $c \in \mathcal{C}_i \bigcap \mathcal{C}_j$ must have been sent from i to j at least once during time steps $k \in [k_i, k_j]$.*

A sufficient means of achieving this is for an agent $i \in \mathcal{N}_k$, following update, to transmit $\mathbf{m}_{ci}(k)$ regarding constraints $c \in \mathcal{C}_i \bigcap \mathcal{C}_j$ to each agent $j \in \mathcal{Q}_j$.

Note that information in $\mathbf{m}_{cj}(k)$ may be "out of date" if sent at an earlier time, in the sense that it may not include outputs for all times in the current planning horizon. However, since it is required that coupled agents must adopt their tail plans unless publishing otherwise, plans for others can be brought up to date by the propagation step 4.1 in Algorithm 3.1. The knowledge of others' dynamics from Requirement 3.2 and the inclusion of the terminal state $\bar{\mathbf{x}}^*_i(k+N_p)$ in the message $\mathbf{m}_{cj}(k)$ ensures that this can be done.

Minimal communication strategies vary significantly with different update sequences. The reader is referred to [14] for a full formal coverage of both communication and propagation mathematics.

3.3.3 Cooperative DMPC

A shortcoming of the method described in the previous section, and of many DMPC methods, is that 'greedy' local decision making can lead to poor system-wide performance [15]. Even if the dynamics and objectives are decoupled, the closed-loop performance of subsystems coupled via the constraints is coupled. Consequently, the solutions applied by agents can be severely sub-optimal—even with inter-agent iteration—and hence cooperation is required to obtain good performance.

A logical approach to promoting cooperation is for distributed control agents to consider, in addition to their own objectives, the objectives or intentions of other agents in the system. To help illustrate this meaning of cooperation, consider an analogy of driving in a long congested stream of traffic. From our car, we observe another car waiting to turn on to our road. Considering only our local objective, we wish to get to our destination as quickly as possible, and we are not constrained to give way to the waiting car, so we continue. However, we could consider the objective of the waiting car as well as our own. A small sacrifice of our objective—the time to let the waiting car pull out—saves a long wait for the other car, and hence an improvement of the global objective. We don't need to instruct the other car to pull out though: merely by slowing to create the opportunity, it is natural for them to take it. This is the key to cooperation: considering the objectives of others can improve global performance without increasing communication requirements. Cooperative control could also be thought of as "considerate control".

This approach to cooperation has been shown to work well for subsystems coupled only via dynamics or objectives. However, it is generally incompatible with maintaining coupled constraint satisfaction. For example, in the method of [2] a local control agent designs, in addition to its own plan, *hypothetical* plans for directly-coupled subsystems. That is, the local problem for subsystem i is

$$\min_{\mathbf{U}_i(k), \hat{\mathbf{U}}^i_j(k), j \in \mathcal{Q}_i} J_i\big(\mathbf{U}_i(k)\big) + \sum_{j \in \mathcal{Q}_i} J_j\big(\hat{\mathbf{U}}^i_j(k)\big)$$

subject to (3.5a)–(3.5f) on $\mathbf{U}_i(k)$, similar local constraints on each $\hat{\mathbf{U}}^i_j(k)$, $j \in \mathcal{Q}_i$, and the coupling constraints

$$\bar{\mathbf{z}}_{ci}(k+l) + \sum_{j \in \mathcal{N}_c \setminus \{i\}} \hat{\mathbf{z}}^i_{cj}(k+l), \quad \forall c \in \mathcal{C}_i, l \in \{1, \ldots, N_p - 1\},$$

where $\{\cdot\}^i_j$ denotes a variable for an agent j that has been computed by agent i.

A crucial detail is that various representations of a plan for a subsystem i might exist at any instance. Firstly, a subsystem $j \in \mathcal{N}$ has the plan $\mathbf{U}^*_j(k)$ for time k, which it is currently following. In addition, an $i \in \mathcal{Q}_j$ has, as part of its own decision-making process, designed a hypothetical plan $\hat{\mathbf{U}}^i_j(k)$ for j that is not necessarily equal to $\mathbf{U}^*_j(k)$. In doing so, it has ensured satisfaction of the coupling constraints by its optimized, local plan $\mathbf{U}^{\text{opt}}_i(k)$ when taken together with the plans $\hat{\mathbf{U}}^i_j(k)$, $j \in \mathcal{Q}_i$. However, satisfaction of the constraints by $\mathbf{U}^{\text{opt}}_i(k)$ together with the actual plans $\mathbf{U}^*_j(k)$, $j \in \mathcal{Q}_i$—and therefore feasibility of the overall, closed-loop system—is not assured, even if only a single agent updates at a time step.

In the approach described in this section, a local agent i designs hypothetical plans for others, yet now this set of other subsystems is an arbitrary *cooperating set* $\mathcal{S}_i(k)$. The local problem is to minimize a weighted sum of local costs by designing a local plan $\mathbf{U}_i(k)$ and a hypothetical plan $\hat{\mathbf{U}}^i_j(k)$ for each $j \in \mathcal{S}_i(k)$. The problem is solved subject to local constraints on $\mathbf{U}_i(k)$ and each $\hat{\mathbf{U}}^i_j(k)$, and coupling constraints on $\mathbf{U}_i(k)$ together with (i) fixed $\mathbf{U}^*_j(k)$ for all coupled $j \in \mathcal{Q}_i$ and (ii) the hypothetical plans $\hat{\mathbf{U}}^i_j(k)$ for all $j \in \mathcal{S}_i$ and the fixed plans $\mathbf{U}^*_m(k)$ for all coupled $m \in \mathcal{Q}_i \setminus \mathcal{S}_i$. The additional decision variables $\{\hat{\mathbf{U}}^i_j(k)\}$ are internal to agent i's decision making and will not be communicated to other agents. Following the optimization, i communicates information about only its own plan, $\mathbf{U}^{\text{opt}}_i(k)$, as before. Moreover, there is no obligation for a cooperating subsystem $j \in \mathcal{S}_i$ to itself optimize at the next step or indeed ever adopt the plan $\hat{\mathbf{U}}^i_j(k)$. The main point is that the an agent i, in determining its own plan, considers what others may be able to achieve.

The presence of *two* sets of coupling constraints in the optimization is crucial in the development here. Effectively, two different representations of a plan for a cooperating subsystem $j \in \mathcal{S}_i$ appear in the local optimization for i: firstly, a previously published plan, $\mathbf{U}^*_j(k)$, originating from the last time step at which j optimized, and the plan that subsystem is currently following; secondly, a hypothetical plan, $\hat{\mathbf{U}}^i_j(k)$, designed locally by agent i. This leads to a key feature of the method; that of promoting inter-agent cooperation yet maintaining robust feasibility of all local decisions.

The local optimization problem, $P^{\text{cdmpc}}_i(\mathbf{x}_i(k), \check{\mathbf{Z}}^*_i(k))$, where $\check{\mathbf{Z}}^*_i(k)$ denotes the extended information required, for subsystem i is formally defined as

$$\min_{\mathbf{U}_i(k), \hat{\mathbf{U}}^i_j(k), j \in \mathcal{S}_i} \quad J_i\big(\mathbf{U}_i(k)\big) + \sum_{j \in \mathcal{S}_i(k)} \alpha^i_j J_j\big(\hat{\mathbf{U}}^i_j(k)\big) \tag{3.14}$$

subject to local constraints (3.5a)–(3.5f) on $\mathbf{U}_i(k)$, (3.5a)–(3.5c), (3.5e), (3.5f) on $\hat{\mathbf{U}}^i_j(k)$, $\forall j \in \mathcal{S}_i(k)$, the initial constraints

$$\hat{\mathbf{x}}^i_j(k) = \bar{\mathbf{x}}^*_j(k), \tag{3.15a}$$

$$\hat{\mathbf{u}}^i_j(k) = \bar{\mathbf{u}}^*_j(k), \tag{3.15b}$$

for all $j \in \mathcal{S}_i$, and, for prediction steps $l \in \{1, \ldots, N_p - 1\}$, the coupling constraints

$$\bar{\mathbf{z}}_{ci}(k+l) + \sum_{j \in \mathcal{N}_c \setminus \{i\}} \bar{\mathbf{z}}^*_{cj}(k+l) \in \tilde{\mathcal{Z}}_c, \forall c \in \mathcal{C}_i, \tag{3.15c}$$

$$\bar{\mathbf{z}}_{ci}(k+l) + \sum_{j \in \mathcal{S}_i(k)} \hat{\mathbf{z}}^i_{cj}(k+l) + \sum_{m \in \mathcal{N}_c \setminus \{i, \mathcal{S}_i(k)\}} \bar{\mathbf{z}}^*_{cm}(k+l) \in \tilde{\mathcal{Z}}_c, \forall c \in \mathcal{C}_{\mathcal{S}_i(k)} \triangleq \bigcup_{j \in \mathcal{S}_i(k)} \mathcal{C}_j, \tag{3.15d}$$

Algorithm 3.2 Cooperative distributed MPC for a subsystem i

1: Wait for an initial feasible plan $\mathbf{U}_i^*(0) = \{\bar{\mathbf{x}}_i^*(0), \bar{\mathbf{u}}_i^*(0 : N_p - 1)\}$ and information, including $\check{\mathbf{Z}}_i^*(0)$. Set $k = 0$.
2: Apply control $\mathbf{u}_i(k) = \bar{\mathbf{u}}_i^*(k) + \mathbf{K}_i(\mathbf{x}_i(k) - \bar{\mathbf{x}}_i^*(k))$. Wait one time step, increment k.
3: Measure state $\mathbf{x}_i(k)$.
4: Update plan. If $i \in \mathcal{N}_k$,

1. Choose cooperating set $\mathcal{S}_i(k)$ and weights α_j^i for each $j \in \mathcal{S}_i(k)$.
2. Obtain new plan $\mathbf{U}_i^{\text{opt}}(k)$ by solving the local problem $P_i^{\text{cdmpc}}(\mathbf{x}_i(k), \check{\mathbf{Z}}_i^*(k))$.
3. Set $\mathbf{U}_i^*(k) = \mathbf{U}_i^{\text{opt}}(k)$.
4. Transmit new information to other agents.

Else, renew existing plan: form $\tilde{\mathbf{U}}_i(k)$ according to (3.10) and set $\mathbf{U}_i^*(k) = \tilde{\mathbf{U}}_i(k)$.
5: Go to step 2.

The initial constraints (3.15a) and (3.15b), which replace (3.15d), provide the starting point of the hypothetical trajectory $\hat{\mathbf{U}}_j^i(k)$ for each $j \in \mathcal{S}_i(k)$. These constraints act on the assumption that any $j \in \mathcal{S}_i(k)$ can not optimize its own plan until, at the earliest, the next step $k+1$. Hence, these predicted trajectories may begin to diverge from the previously published trajectories, $\mathbf{U}_j^*(k)$, only at the $k+1$ prediction time. The cooperative problem is solved in the Algorithm 3.2.

The cooperating set $\mathcal{S}_i(k)$ and the scalar weightings α_j^i are essentially tuning parameters for the level of cooperation. The parameter $\alpha_j^i \geq 0$ is the weighting applied to the local subsystem cost J_j for $j \in \mathcal{S}_i(k)$; smaller values ($\alpha_j^i < 1$) place more emphasis on i's own objective and self interest, while larger values ($\alpha_j^i > 1$) have the opposite effect. The size of the cooperating set maps to what portion of the system-wide cost is considered in the local optimization. If $\mathcal{S}_i(k)$ is empty, the problem reverts simply to the non-cooperative problem P_i^{dmpc}. Conversely, as $\mathcal{S}_i(k) \to \mathcal{N} \setminus \{i\}$, the local optimization attempts to solve a problem more closely resembling the system-wide, centralized problem, but with modified constraints.

As before, having obtained a plan $\mathbf{U}^*(k)$ at step k, by either optimization or renewal, each agent unilaterally applies the first control of the planned sequence. No negotiation or iterative refinement of solutions takes place during a time step, and hypothetical plans are never exchanged or compared.

3.3.3.1 Initialization of Cooperative DMPC

As before, it is assumed that each agent has the information required to formulate its local optimization problem (Requirements 3.1 and 3.2). It is also assumed that initial plans are available for each subsystem (Requirement 3.3).

3.3.3.2 Inter-agent Communication

The additional coupling constraints (3.15d) and initial intention constraints (3.15a) and (3.15b) both demand information beyond that required for non-cooperative DMPC.

Considering first the coupling constraints, constraint (3.15c) matches the non-cooperative counterpart (3.12), and hence Requirement 3.4 still applies. For the new constraint (3.15d), an updating agent further requires coupling information from any agents coupled to those in its cooperating set $\mathcal{S}_i(k)$. Formally:

Requirement 3.5 *[Additional coupling information exchange] Consider any three agents i, j and h such that $i \in \mathcal{Q}_h$ and $h \in \mathcal{S}_j$ and any two time steps k_i and $k_j > k_i$ such that $i \in \mathcal{N}_{k_i}$ and $j \in \mathcal{N}_{k_j}$, i.e. agents i and j updated at times k_i and k_j respectively. Then message $\mathbf{m}_{ci}(k)$ for every coupling constraint $c \in \mathcal{C}_i \bigcap \mathcal{C}_h$ must have been sent from i to j at least once during time steps $k \in [k_i, k_j]$.*

For the initial intent constraints (3.15a) and (3.15b), full state and control outputs are required but might not be included in the coupling messages. Thus the requirement is for the latest complete plan information from any agents in the cooperating set:

Requirement 3.6 *[Cooperating plan exchange] Consider an updating agent j and another agent $i \in \mathcal{S}_j$, (i.e. agent j wants to cooperate with i) and any two time steps k_i and $k_j > k_i$ such that $i \in \mathcal{N}_{k_i}$ and $j \in \mathcal{N}_{k_j}$, i.e. agent i updated at time k_i and agent j updated at k_j. Then the plan $\mathbf{U}_i^*(k)$ must have been sent from i to j at least once during time steps $k \in [k_i, k_j]$.*

A sufficient, yet conservative, means of meeting these requirements is for the communication step in Algorithm 3.2 to specify transmission of the full plan to all other subsystems following update. While this may seem significant, it should be noted that to meet the requirement it is sufficient for one agent to transmit its plan to others only after that plan has changed, i.e., as a result of optimization. Moreover, it is not necessary for an agent to update at every time step, and robust coupled constraint satisfaction and stability are guaranteed for any choices of update sequence and cooperating sets. Thus, data exchanges need not occur at every time step, and the cooperating set and update sequence may be tailored to exploit this flexibility, as has been shown in [14] for the non-cooperative form.

3.4 Theoretical Results

3.4.1 Robust Constraint Satisfaction and Feasibility

With no assumptions extra to those already stated, the system controlled according to Algorithm 3.1 or 3.2 attains the properties of guaranteed robust constraint satisfaction and robust feasibility. This result, established in [14], relies on the observation that, given a feasible solution at time k_0, the candidate solution for each subsystem at the next time step $k_0 + 1$—as defined by (3.10)—is feasible for all possible realizations of the disturbances, $\mathbf{d}_i \in \mathcal{D}_i$, and any choice of update sequence.

In [11], this result is extended to the cooperative form of the algorithm by noting that, at k_0, an updating local agent i has available both a feasible local plan $\tilde{\mathbf{U}}_i(k_0)$ (*i.e.*, the candidate plan) and feasible hypothetical plans $\hat{\mathbf{U}}^i_j(k_0) = \tilde{\mathbf{U}}_j(k_0)$ for $j \in \mathcal{S}_i(k_0)$, and for any choice of cooperating set $\mathcal{S}_i(k_0)$. Viewed differently, the cooperative form of DMPC is equivalent to the non-cooperative form with only a modified cost, which happens to include the evaluation of options for other agents. Since constraint satisfaction and feasibility depend only on the constraints of each optimization and not on its cost, it is logical that cooperative DMPC inherits the feasibility properties of its non-cooperative counterpart.

3.4.2 Robust Convergence and Stability

Under further assumptions, asymptotic convergence of the states of the system to a neighbourhood of the origin is guaranteed.

(A1) The stage cost $L_i(\mathbf{x}_i, \mathbf{u}_i) \geq c\|(\mathbf{x}_i, \mathbf{u}_i)\|$, for $c > 0$, and $L_i(\mathbf{0}, \mathbf{0}) = 0$, $\forall i \in \mathcal{N}$.
(A2) The terminal cost is a local Lyapunov function: for all $\mathbf{x}_i \in \mathcal{X}_i^{\mathrm{f}}$ and $i \in \mathcal{N}$, $F_i(\mathbf{A}_i\mathbf{x}_i + \mathbf{B}_i\kappa_i^{\mathrm{f}}(\mathbf{x}_i)) - F_i(\mathbf{x}_i) \leq -L_i(\mathbf{x}_i, \kappa_i^{\mathrm{f}}(\mathbf{x}_i))$.
(A3) The local cost of an adopted plan $\mathbf{U}_i^*(k)$ for any $i \in \mathcal{N}_k$ updating at k satisfies $J_i(\mathbf{U}_i^*(k)) \leq J_i(\tilde{\mathbf{U}}_i(k)) + \sum_{j \in \mathcal{N}} \epsilon_j L_j(\bar{\mathbf{x}}_j^*(k-1), \bar{\mathbf{u}}_j^*(k-1))$ for some chosen $0 \leq \epsilon_j < 1$, $\forall j \in \mathcal{N}$, where $\tilde{\mathbf{U}}_i(k)$ is the candidate plan (3.10).

Together with the assumptions on $\mathbf{K}_i$, κ_i^{f}, $\mathcal{R}_i$, and $\mathcal{X}_i^{\mathrm{f}}$, Assumptions (A1) and (A2) represent a specific case of the standard assumptions (A1)–(A4) in [5].

The non-cooperative approach requires only Assumptions (A1) and (A2) [14]. It may be established that each $\mathbf{x}_i(k) \to \mathcal{R}_i$ and $\mathbf{u}_i \to \mathbf{K}_i\mathbf{u}_i$ as $k \to \infty$. This holds for all realizations of the disturbances $\mathbf{d}_i \in \mathcal{D}_i$, and for any choice of update sequence.

Robust convergence and stability of the cooperative form is established with the help of Assumption (A3). This limits the amount by which the local cost J_i of an agent's solution is permitted to *increase* over that of the candidate plan in order to benefit other agents. Intuitively, an unbounded increase may lead to instability if repeated by many agents over time. However, if (A3) holds, monotonic descent of the *global* cost, $\sum_{i \in \mathcal{N}} J_i(\mathbf{U}_i^*(k))$ is assured, and each $\mathbf{x}_i(k) \to \mathcal{R}_i$ and $\mathbf{u}_i \to \mathbf{K}_i\mathbf{u}_i$ as $k \to \infty$. This holds regardless of disturbances, choice of update sequence, and choices of cooperating sets. Although difficult to prove, (A3) often holds anyway. Since it can be shown that it is always possible to satisfy (A3), it may be enforced by a direct constraint, though resulting in a more complex and constrained problem.

3.4.3 The Benefit of Cooperation

The problem of how the cooperating sets $\mathcal{S}_i$ are chosen to obtain the maximal benefit to system-wide performance is studied in [13]. It is proven that, depending on the coupling structure, it is not necessary to cooperate with all others in the problem, yet not sufficient to cooperative only with directly-coupled subsystems. An adaptive form of cooperation between agents is proposed, in which agents cooperate with others connected by *paths* in a graph of active coupling constraints

In [9], the DMPC approach is studied in a game-theoretical framework. Under assumptions milder than those required for asymptotic convergence of the states to a neighbourhood of the origin, (i.e., (A1) and (A2)), the states of the controlled system converge to some limit set. The system is in such a limit set if and only if the control agents are continually playing Nash solutions. Relating the Nash solutions to the cooperation set choices, it may be proven that increasing inter-agent cooperation does not enlarge the set of Nash solutions. Thence, it follows that increasing the size of cooperating sets does not enlarge the set of state limit sets for the system.

3.5 Applications of the Approach

Figure 3.1 shows results from an example, taken from [10], concerning the control of a group of point masses, coupled by the requirement to stay close together. The trade between communication and computation is shown for both centralized and tube DMPC. Centralized cannot reduce communication without lowering its update rate. Tube DMPC can manipulate the update sequence to change communication, and hence can out-perform centralized at low communication levels.

Figure 3.2 illustrates the benefits of cooperation when applied to a system of five point masses. The objective is for the masses to reach the origin, yet the coupled constraint on positions, $\sum_{i \in \mathcal{N}} x_{i,1} \geq 1$, means that this cannot be achieved by all masses simultaneously. Cooperative DMPC clearly delivers a fairer outcome.

In [13], the coupling structure is exploited, with the cooperating set for the point-mass system being chosen on-line, according to the active coupling constraints.

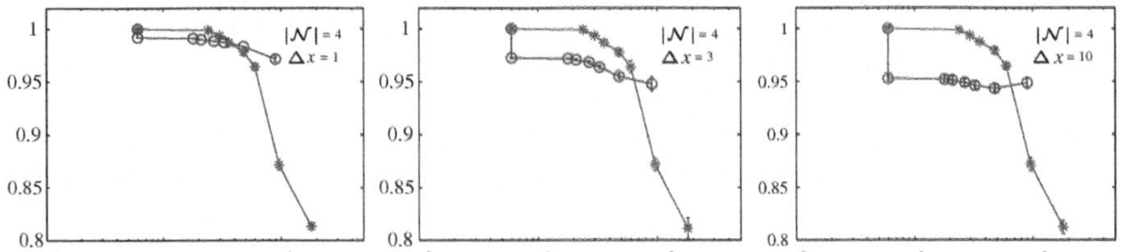

Fig. 3.1 Cost versus number of data exchanges for (i) DMPC (○) and (ii) CMPC (∗). Each mass is required to remain within Δx of the others, with (l to r) $\Delta x = 1$, 3, and 10. Reproduced from [10]

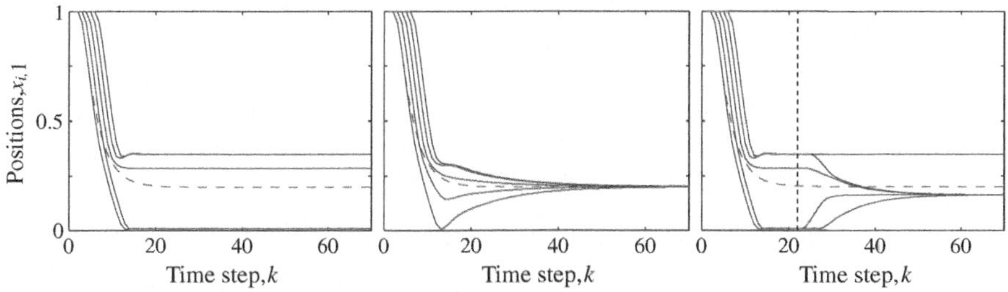

Fig. 3.2 Convergence of a five-mass system when controlled by (*left* to *right*) non-cooperative DMPC; cooperative DMPC with one other agent in the cooperating set $\mathcal{S}_i$; firstly non-cooperative DMPC, and then cooperative DMPC. Reproduced from [13]

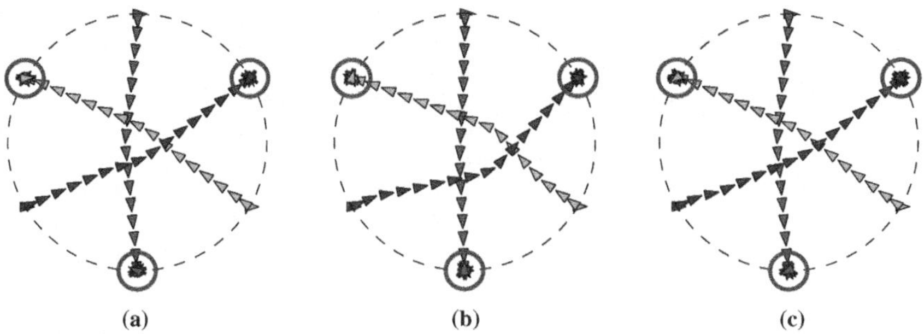

Fig. 3.3 Three vehicles traversing a circle when controlled by **a** centralized MPC, **b** non-cooperative DMPC and **c** cooperative DMPC with the next-to-plan vehicle in the cooperating set for vehicle i. Reproduced from [11]

Simulations show that it is beneficial for an agent to cooperate with non-directly coupled agents, but not necessary to include all other masses in the cooperating set.

The distributed MPC algorithms described in this chapter are applicable to those systems comprising dynamically decoupled, LTI subsystems that share coupling constraints. A natural application is guidance and control of multiple vehicles, for which collision avoidance can be enforced via coupled constraints. In [11], a problem similar to air traffic control is simulated, and it is demonstrated that the cooperative approach leads to a more equitable arrangement of flight paths (Fig. 3.3). In [12], multiple vehicles are given the shared objective of achieving complete coverage or search of an area in minimum time.

References

1. W.B. Dunbar, Distributed receding horizon control of dynamically coupled nonlinear systems. *IEEE Trans. Autom. Control* **52(7)**, 1249–1263 (2007)
2. T. Keviczky, F. Borrelli, G.J. Balas, Decentralized receding horizon control for large scale dynamically decoupled systems. *Automatica* **42**(12), 2105–2115 (2006)

3. I. Kolmanovsky, E.G. Gilbert, Theory and computation of disturbance invariant sets for discrete-time linear systems. *Mathematical Problems in Engineering* **4**, 317–367 (1998)
4. Y. Kuwata, J.P. How, Cooperative distributed robust trajectory optimization using receding horizon MILP. *IEEE Transactions on Control Systems Technology* **19**(2), 423–431 (2011)
5. D.Q. Mayne, J.B. Rawlings, C.V. Rao, P.O.M. Scokaert, Constrained model predictive control: Stability and optimality. *Automatica* **36**, 789–814 (2000)
6. D.Q. Mayne, M.M. Seron, S.V. Raković, Robust model predictive control of constrained linear systems with bounded disturbances. *Automatica* **41**(2), 219–224 (2005)
7. A.G. Richards, J.P. How, Robust distributed model predictive control. *International Journal of Control* **80**(9), 1517–1531 (2007)
8. B.T. Stewart, A.N. Venkat, J.B. Rawlings, S.J. Wright, G. Pannocchia, Cooperative distributed model predictive control. *Systems & Control Letters* **59**, 460–469 (2010)
9. P. A. Trodden, D. Nicholson, A. G. Richards, Distributed model predictive control as a game with coupled constraints, In *Proceedings of the European Control Conference*, pp. 2996–3001, 2009
10. P. A. Trodden, A. G. Richards, Robust distributed model predictive control using tubes. In *Proceedings of the American Control Conference*, pp. 2034–2039, 2006
11. P. A. Trodden, A. G. Richards. Robust distributed model predictive control with cooperation. In *Proceedings of the European Control Conference*, pp. 2172–2178, 2007
12. P. A. Trodden and A. G. Richards, Multi-vehicle cooperative search using distributed model predictive control, In *AIAA Guidance, Navigation, and Control Conference*, 2008
13. P.A. Trodden, A.G. Richards, Adaptive cooperation in robust distributed model predictive control, In *Proceedings of the IEEE Multi-conference on Systems and Control*, 2009
14. P.A. Trodden, A.G. Richards, Distributed model predictive control of linear systems with persistent disturbances. *International Journal of Control* **83**(8), 1653–1663 (2010)
15. A.N. Venkat, I.A. Hiskens, J.B. Rawlings, S.J. Wright, Distributed MPC strategies with application to power system automatic generation control. *IEEE Transaction on Control System Technology* **16**(6), 1192–1206 (2008)

Chapter 4
Price-driven Coordination for Distributed NMPC Using a Feedback Control Law

R. Martí, D. Sarabia and C. de Prada

Abstract This chapter presents a distributed coordinated control algorithm based on a hierarchical scheme for systems consisting of nonlinear subsystems coupled by input constraints: the bottom layer is composed of several non-linear model predictive controllers (NMPC) working in parallel, and in a top layer, a price-driven coordination technique is used to coordinate these controllers. The price coordination problem is formulated as a feedback control law to fulfill the global constraints that affect all NMPC controllers. To illustrate this approach, the price-driven coordination method is used to control a four-tank process in a distributed manner and is compared with centralized and fully decentralized approaches.

4.1 Introduction

The main goal of the distributed MPC is to overcome computational and communication limitations of centralized architectures. There are many approaches to this problem, following the lines of hierarchical, distributed and price coordination: acting on the setpoints, interchanging information or modifying the cost functions as different ways of coordinating the actions of the distributed controllers, an excellent review of these techniques can be seen in [9]. In particular, in price coordination, several MPC controllers, each one acting on a subset of the process, reach a global optimum using a market-like mechanism for coordination: each controller modifies

R. Martí (✉) · D. Sarabia · C. de Prada
Department of Systems Engineering and Automatic Control,
School of Industrial Engineering, University of Valladolid, Valladolid, Spain
e-mail: ruben@autom.uva.es

D. Sarabia
e-mail: dsarabia@autom.uva.es

C. de Prada
e-mail: prada@autom.uva.es

J. M. Maestre and R. R. Negenborn (eds.), *Distributed Model Predictive Control Made Easy*, Intelligent Systems, Control and Automation: Science and Engineering 69,
DOI: 10.1007/978-94-007-7006-5_4,

the cost function that defines its control or optimization aims according to a set of prices assigned from an upper layer, as in [5] and [4]. The advantages of this approach are the ease of application, because it is only necessary to implement a simple coordinator and to add just one extra term on the objective function. Therefore, the existing local industrial model predictive controllers can still be used without any problem. Several approaches based on price coordination methods have been proposed for large-scale and networked systems, some of them being based on price coordination methods [3, 7] and [10].

The purpose of this chapter is to show the principal ideas of the price-driven coordination approach in the coordination of coupled dynamic subsystems and its application in a simulation benchmark. Sections 4.2 and 4.3 show how to solve dynamic shared resource allocation where several consumer units are fed by a common resource, implying inequality global constraints. This approach has been successfully implemented in [8] and it will be described in the first part of Sect. 4.4. In the second part of Sect. 4.4, the method is particularized to solve subsystems which are coupled by interprocess streams which imply equality global constraints.

4.2 Boundary Conditions

In this section, the generic formulation of the NMPC problem for large-scale systems, where there are some shared resources, is explained. Consider a system composed by a set of interconnected subsystems $\mathcal{N} = \{S_1, S_2, \ldots, S_{|\mathcal{N}|}\}$ where $|\mathcal{N}|$ corresponds to the total number of subsystems. The associated optimization problem (4.1) implies minimizing an objective function which can include both control and economic aims subject to nonlinear models $\mathbf{h_i}(.)$ and $\mathbf{g_i}(.)$ and $i \in \mathcal{N}$ corresponding to the dynamics and local constraints of each subsystem. On the other hand, the optimization problem (4.1) has some global constraints related to common shared resources (see constraint type 1 for a specific resource on Fig. 4.1).

$$\begin{aligned} \min_{\{\mathbf{u},\mathbf{v}\}} J &= \sum_{i=1}^{|\mathcal{N}|} J_i\left(\mathbf{v_i}, \mathbf{u_i}, \mathbf{x_i}\right) \qquad (4.1)\\ \text{s.t.} & \\ &\mathbf{h_i}\left(\dot{\mathbf{x}}_\mathbf{i}, \mathbf{x_i}, \mathbf{v_i}, \mathbf{u_i}\right) = 0 \;\; \forall i \in \mathcal{N} \\ &\mathbf{g_i}\left(\dot{\mathbf{x}}_\mathbf{i}, \mathbf{x_i}, \mathbf{v_i}, \mathbf{u_i}\right) \le 0 \;\; \forall i \in \mathcal{N} \\ \sum_{i=1}^{|\mathcal{N}|} & r_{ji}\left(u_{ji}\right) \le r_{T_j} \;\; \forall j \in \mathcal{K} \end{aligned}$$

In this expression, $\mathcal{K} = \{1, 2, 3, \ldots\}$ represents the set of shared resources and $|\mathcal{K}|$ is the number of shared resources, r_{T_j} is the availability of the shared resource j, and $r_{ji}(u_{ji})$ represents the consumption of the shared resource j in each subsystem i,

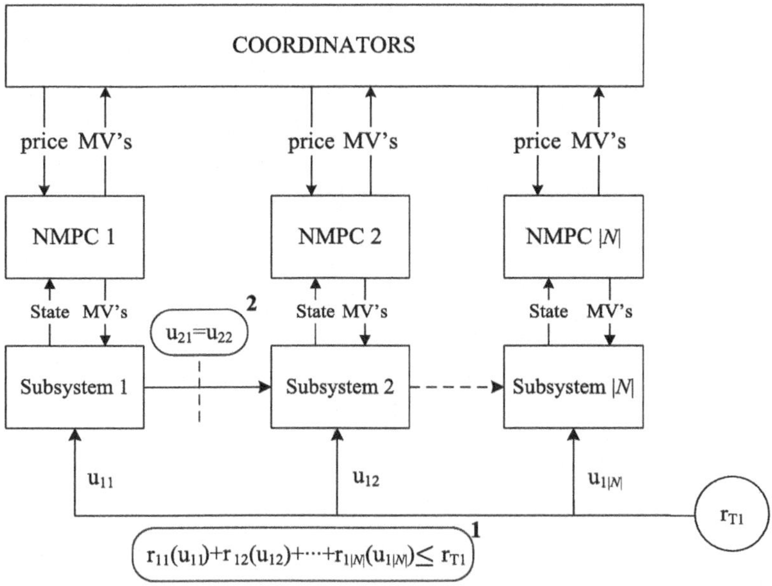

Fig. 4.1 Hierarchical scheme for using price-driven coordination and subsystem interconnections. Type 1: Shared resource constraint and type 2: interprocess stream constraints

being a function of the corresponding manipulated variable u_{ji}. So, u_{ji} is the use of resource j in the plant i, and these variables can be written in a compact form $\mathbf{u_i} = [u_{1i}, u_{2i}, \ldots, u_{|\mathcal{K}|i}]$. In addition, each subsystem can have its own independent manipulated variables $\mathbf{v_i} = [v_{1i}, v_{2i}, \ldots, v_{n_{v_i}i}]$ and states $\mathbf{x_i} = [x_{1i}, x_{2i}, \ldots, x_{n_{x_i}i}]$, where n_{v_i} and n_{x_i} are the corresponding dimension of the vectors in each subsystem i.

Now we consider (4.1) under the relaxed assumption that the objective functions $J_i(\mathbf{v_i}, \mathbf{u_i})$ are not strictly increasing, then the associated Lagrangian is:

$$\mathcal{L} = \sum_{i=1}^{|\mathcal{N}|} J_i\,(\mathbf{v_i}, \mathbf{u_i}, \mathbf{x_i}) + \sum_{i=1}^{|\mathcal{N}|} \lambda_{\mathbf{i}}{}^T \mathbf{h_i} + \sum_{i=1}^{|\mathcal{N}|} \mu_{\mathbf{i}}^T\, \mathbf{g_i} + \sum_{j=1}^{|\mathcal{K}|} p_j \left(\sum_{i=1}^{|\mathcal{N}|} r_{ji}(u_{ji}) - r_{T_j} \right) \quad (4.2)$$

For simplicity, and without loss of generality, in (4.1), we are going to consider only the manipulated variables related to shared resources, then the Lagrangian of the optimization problem is:

$$\mathcal{L} = \sum_{i=1}^{|\mathcal{N}|} J_i\,(\mathbf{u_i}, \mathbf{x_i}) + \sum_{i=1}^{|\mathcal{N}|} \lambda_{\mathbf{i}}{}^T h_i + \sum_{i=1}^{|\mathcal{N}|} \mu_{\mathbf{i}}{}^T \mathbf{g_i} + \sum_{j=1}^{|\mathcal{K}|} p_j \left(\sum_{i=1}^{|\mathcal{N}|} r_{ji}(u_{ji}) - r_{T_j} \right) \quad (4.3)$$

The conditions for optimality for (4.3) (which are necessary for the existence of an optimum) are:

$$\nabla_{\mathbf{u_i}} J_i(\mathbf{u_i}, \mathbf{x_i}) + \nabla_{\mathbf{u_i}} \mathbf{h_i^T} \lambda_i^* + \nabla_{\mathbf{u_i}} \mathbf{g_i}^{\mathbf{T}} \mu_{\mathbf{i}}^* + \sum_{\mathbf{j=1}}^{|\mathcal{K}|} \frac{\mathbf{dr_{ji}(u_{ji})}}{\mathbf{du_{ji}}} \mathbf{p_j^*} = 0 \ \forall \mathbf{i} \in \mathcal{N} \tag{4.4}$$

$$\mathbf{h_i}(\dot{\mathbf{x}}_\mathbf{i}, \mathbf{x_i}, \mathbf{u_i}) = 0 \ \forall i \in \mathcal{N}$$

$$\mathbf{g_i}(\dot{\mathbf{x}}_\mathbf{i}, \mathbf{x_i}, \mathbf{u_i}) \leq 0 \ \forall i \in \mathcal{N}$$

$$\mathbf{g_i}(\dot{\mathbf{x}}_\mathbf{i}, \mathbf{x_i}, \mathbf{u_i})^T \mu_{\mathbf{i}}^* = 0 \ \forall i \in \mathcal{N}$$

$$\mu_{\mathbf{i}}^* \geq 0 \ \forall i \in \mathcal{N}$$

$$\sum_{i=1}^{|\mathcal{N}|} r_{ji}(u_{ji}) - r_{T_j} \leq 0 \ \forall j \in \mathcal{K}$$

$$p_j^* \left(\sum_{i=1}^{|\mathcal{N}|} r_{ji}(u_{ji}) - r_{T_j} \right) \leq 0 \ \forall j \in \mathcal{K}$$

$$p_j^* \geq 0 \ \forall j \in \mathcal{K}$$

Notice that, KKT conditions (4.4) are equivalent to the ones that result from solving $|\mathcal{N}|$ independent optimization problems (4.5) for given Lagrangian multipliers p_j, plus a coordinator level to fulfill the global conditions (4.6).

$$\min_{\{\mathbf{u_i}\}} J = J_i(\mathbf{u_i}, \mathbf{x_i}) + \sum_{j=1}^{|\mathcal{K}|} \left(p_j \, r_{ji}(u_{ji}) \right) \tag{4.5}$$

s.t.

$$\mathbf{h_i} \ (\dot{\mathbf{x}}_\mathbf{i}, \mathbf{x_i}, \mathbf{u_i}) = 0$$

$$\mathbf{g_i} \ (\dot{\mathbf{x}}_\mathbf{i}, \mathbf{x_i}, \mathbf{u_i}) \leq 0$$

$$\sum_{i=1}^{|\mathcal{N}|} r_{ji}(u_{ji}) - r_{T_j} \leq 0 \ \forall j \in \mathcal{K} \tag{4.6}$$

$$p_j^* \left(\sum_{i=1}^{|\mathcal{N}|} r_{ji}(u_{ji}) - r_{T_j} \right) = 0 \ \forall j \in \mathcal{K}$$

$$p_j^* \geq 0 \forall j \in \mathcal{K}$$

Therefore, the principle of the price coordination method is based on assigning a price p_j to the resource j consumed in each individual subsystem i, in such a way that each subsystem tries to optimize its objective function by accepting an amount of the resources at a certain price such that the global constraint is satisfied. When the price meets (4.6), then the individual optimal solutions of (4.5) are equal to the centralized optimal solution of problem (4.1) and the sum of the local cost function is equal to the global cost function. An extended demonstration can be seen in [5].

The mechanisms for updating the price p until it satisfies Eq. (4.6) can be considered as the coordinator (see Fig. 4.1) in price-driven approaches [3]. Different policies to fix p have given rise to different price-coordination methods, for instance, in [3], the authors use a price-adjustment algorithm based on Newton's method, in which sensitivity analysis and active set change identification techniques are employed. In addition, [10] uses the same technique, but in this case, Broyden's algorithm [6], which does not require the calculation of the Jacobian matrix, is implemented. All these techniques use the Eq. (4.7) for updating the price, which is evaluated iteratively each sampling time. On the other hand, in [2], the authors try to overcome the problematic of the interaction balance. For this reason, they show three algorithms based on the theoretical aspects of the interaction balance theorem and their application to the optimum allocation problem.

$$p_j(k+1) = p_j(k) - \left(\frac{\sum_{i=1}^{|\mathcal{N}|} dr_{ji}(u_{ji})}{d\, p_j} \right)^{-1} \left(\sum_{i=1}^{|\mathcal{N}|} r_{ji}(u_{ji}) - r_{T_j} \right) \tag{4.7}$$

Finally, the distributed coordinated control algorithm to solve (4.1) will be explained in the next sections and is based on the following main assumptions:

1. The objective function is separable and strictly convex.
2. The nonlinear internal model is separable in several subsystems $i \in \mathcal{N}$.
3. The global constraints are formulated in terms of manipulated variables $\mathbf{u_i}$ and they are written as inequality constraints (see the lower part of Fig. 4.1). This means that the subsystems are only coupled by the inputs.
4. The local NMPC controllers solve all these dynamic optimization problems using an SQP algorithm based on a sequential approach, where each manipulated variable $\mathbf{u_i}$ and $\mathbf{v_i}$ is discretized using a constant parameterization. On the other hand, the NMPC controllers are well-tunned to achieve their objectives, obtaining a stable solution.
5. The data exchange is done once per sampling time between each NMPC controller and the coordinator or coordinators following the topology shown in Fig. 4.1. The NMPC controllers communicate the computed optimal manipulated variable to an upper layer and they receive the Lagrangian multipliers or prices.

Of particular interest are those kinds of subsystem interconnections common in chemical processes, electric power distribution networks, water supply networks, etc., that represent global equality constraints (see Fig. 4.1). This kind of constraints are related to interprocess streams. For instance, one way to decompose these global equality constraints is to apply a standard Lagrangian decomposition, but as it is argued in [5], this technique has convergence problems because the pricing interprocess streams are not well-defined. To overcome this problem, it is possible to rewrite the equality constraints into shared resource constraints or inequality constraints, and then subsystems coupled by equality constraints can be addressed first using shared resource decomposition. How to deal with these problems and what kind of mechanism is used for updating the price in the coordinator layer will be explained in the next section.

Algorithm 4.1 Price adjustment

1: **Initialization:** An arbitrary price p_j for each shared resource is established and sends that information to every subsystem.
2: **Optimization performance:** Each sampling time, based on the price provided by the coordinator, each subsystem i solves its own optimization problem (4.5) using an NMPC$_i$ controller and calculates the consumed resources r_{ji}. This information is communicated back to the coordinator.
3: **Implementation of control action:** The first predicted values are applied in the process.
4: **Price update:** The coordinators gather the information from each subsystem; it calculates $\sum_{i=1}^{|\mathcal{N}|}(r_{ji})$. Then, the coordinators update the prices p_j using a non-linear controller implementing (4.6). In this case a PID controller (4.8) is used and the new prices are sent to each subsystem. In the next sampling time the algorithm starts in step 2.

$$p_{aux,j}(s) = K_c\left(1 + \frac{1}{T_i s} + \frac{T_d s}{\beta f_d s + 1}\right) E(s)\ with\ E(s) = \sum_{i=1}^{|\mathcal{N}|}\left(r_{ji}(u_{ji}) - r_{T_j}\right) \quad (4.8)$$

$$\textbf{If} p_{aux,j} \geq 0 \rightarrow p_j = p_{aux,j}$$
$$\textbf{If} p_{aux,j} < 0 \rightarrow p_j = 0$$

4.3 Description of the Approach

The price-Driven Coodination NMPC scheme (PCNMPC) uses an upper layer where the coordinator is seen as a control problem (see Fig. 4.2). A non-linear SISO controller is defined for each common resource and gathers the value of the total demanded resource as the sum of all r_{ji} and compares it with the maximum r_{T_j}. If it is below this total, the price is assigned to zero, otherwise, a control law is enforced to maintain the total resource consumed as less than or equal to r_{T_j} using the price p_j of each shared resource as the manipulated variable.

The implementation of the PCNMPC scheme is carried out according to Algorithm 4.1. The coordinator (4.8) is executed continuously along the time of the process, but the values of u_{ji} and p_j are exchanged at each sampling time.

On the other hand, to deal with interprocess streams and use the decomposition (4.5) it is necessary to rewrite the equality constraints in a form of a resource constraint

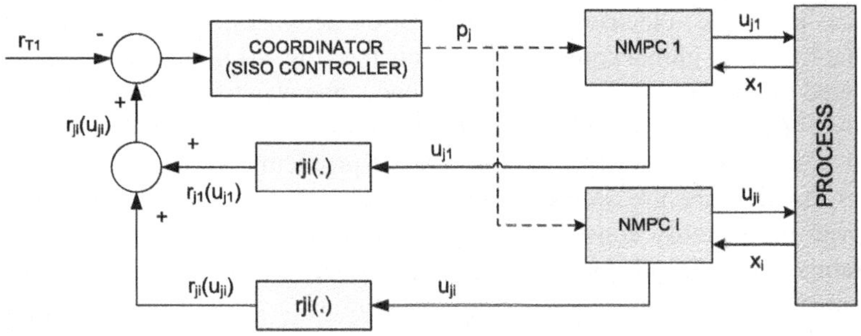

Fig. 4.2 Structure of price-driven resource allocation for solving shared resource constraints

(4.1). For instance in Fig. 4.1, an interprocess streams $u_{21} = u_{22}$ is equivalent to:

$$u_{21} = u_{22} \Leftrightarrow u_{21} - u_{22} = 0 \Leftrightarrow |u_{21} - u_{22}| \leq 0 \Leftrightarrow (u_{21} - u_{22})^2 \leq 0 \tag{4.9}$$
$$(u_{21} - u_{22})^2 \leq 0 \Leftrightarrow \underbrace{(u_{21} - u_{22})^2}_{r_{21}()} + \underbrace{(u_{22} - u_{22})^2}_{r_{22}()} \leq \underbrace{0}_{r_{T_2}}$$

In the last inequality of (4.9), the lefthand side can be interpreted as an aggregate demand for common resource, and the righthand side can be interpreted as an available resource r_{T_2}. Notice, the first term of the left hand side in this inequality depends on a variable u_{21} of subsystem 1, but also on a variable u_{22} of subsystem 2. However, the second term depends only on a variable of the subsystem 2. This particular way of rewriting the interprocess stream has been chosen because it is easy to transfer to the standard MPC controller's cost function in order to solve the problem (4.5).

4.4 Application Results

In this section, two applications of price-driven coordination are presented. The first one is a dynamic resource allocation problem corresponding to the problem denominated as 1 in Fig. 4.1. Meanwhile, the second one corresponds to the problem denominated as 2, where subsystems are coupled by interprocess streams. In addition, the last one is compared to centralized and decentralized schemes.

4.4.1 Oxygen Supply Network. Preliminary Results

In [8], the approach showed in Algorithm 4.2 was directly applied in a simulation of an oxygen distribution network (Fig. 4.3) which includes two oxygen generators (P_1 and P_2): each of them produces oxygen, which are the shared resources r_{T_1} and r_{T_2}, at different purities, capacities and production costs (r_{T_1} is cheaper than r_{T_2}). There are also two different collectors in charge of driving the oxygen from each generator to each consumer unit (C_1, C_2 and C_3). Every consumer unit is a chemical reactor where a reaction that consumes oxygen takes place, where the manipulated variables are the feed flows from both collectors and the controlled variable is the concentration of dissolved oxygen. In addition, each one has a NMPC controller to maintain a certain level of dissolved oxygen minimizing, at the same time, operational costs. This implies that each NMPC tries to use only the cheapest resource. On the other hand, three kinds of laws for updating prices have been implemented: market policy, PI and PID controllers.

The solid line in the upper picture of Fig. 4.4 shows the availability of the resource, which is the total production of resource r_{T_1}. In the same picture, each dashed line corresponds to the needed oxygen demand ($\sum_{i=1}^{3} r_{1i}(u_{1i})$) by all NMPC controllers

Algorithm 4.2 Price-driven coordination approach

1: **Initialization:** An initial price p_j for each interprocess stream is established and sends that information to every subsystem.

2: **Optimization performance:** Every sampling time, $NMPC_1$ and $NMPC_2$ gather the measured variables and solve independently the following optimization problems, based on the prices p_1 and p_2 provided by the coordinators:

$$\begin{aligned}
\min_{\left\{\begin{smallmatrix} q_{A1}(0),\dots,q_{A1}(N_c-1) \\ q_{B1}(0),\dots,q_{B1}(N_c-1)\end{smallmatrix}\right\}} J_1 &= \int_{t=0}^{N_p l} (h_1(t) - h_{1\text{ref}}(t))^2\, dt \\
&+ \sum_{k=0}^{N_c-1} \left(p_1 (\Delta q_{A1}(k))^2\right) + \sum_{k=0}^{N_c-1} \left(p_2 (\Delta q_{B1}(k))^2\right)
\end{aligned} \tag{4.10}$$

$$\begin{aligned}
&\text{s.t.} \\
&\qquad \text{Equations (4.10, 4.12) model subsystem 1} \\
h_{j\,min} &\leq h_j(t) \leq h_{j\,max} \;\; \forall j \in \{1, 3\} \;\; t \in [0, N_p] \\
q_{min} &\leq q_{i,1}(k) \leq q_{i,max} \;\; \forall i \in \{A, B\} \;\; k = 0, \dots, N_c - 1 \\
q_{i1}(k) &= q_{i1}(N_c - 1) \;\; \forall i \in \{A, B\} \;\; k = N_c, \dots, N_p - 1 \\
\Delta q_{i1}(k) &= q_{i1}(k) - q_{i1}(k-1) \;\; \forall i \in \{A, B\} \;\; k = 0, \dots, N_c - 1 \\
q_{A1}(-1) &= q^*_{A2}, \;\; q_{B1}(-1) = q^*_{B1}
\end{aligned}$$

$$\begin{aligned}
\min_{\left\{\begin{smallmatrix} q_{A2}(0),\dots,q_{A2}(N_c-1) \\ q_{B2}(0),\dots,q_{B2}(N_c-1)\end{smallmatrix}\right\}} J_2 &= \int_{t=0}^{N_p l} (h_2(t) - h_{2\text{ref}}(t))^2\, dt \\
&+ \sum_{k=0}^{N_c-1} \left(p_1 (\Delta q_{A2}(k))^2\right) + \sum_{k=0}^{N_c-1} \left(p_2 (\Delta q_{B2}(k))^2\right)
\end{aligned} \tag{4.11}$$

$$\begin{aligned}
&\text{s.t.} \\
&\qquad \text{Equations (4.11, 4.13) model subsystem 2} \\
h_{j\,min} &\leq h_j(t) \leq h_{j\,max} \;\; \forall j \in \{1, 3\} \;\; t \in [0, N_p] \\
q_{min} &\leq q_{i,2}(k) \leq q_{i,max} \;\; \forall i \in \{A, B\} \;\; k = 0, \dots, N_c - 1 \\
q_{i2}(k) &= q_{i2}(N_c - 1) \;\; \forall i \in \{A, B\} \;\; k = N_c, \dots, N_p - 1 \\
\Delta q_{i2}(k) &= q_{i2}(k) - q_{i2}(k-1) \;\; \forall i \in \{A, B\} \;\; k = 0, \dots, N_c - 1 \\
q_{A2}(-1) &= q^*_{A2}, \;\; q_{B2}(-1) = q^*_{B1}
\end{aligned}$$

3: **Solution implementation:** The first predicted value $q_{A2}(0)$ and $q_{B1}(0)$ are applied in the process.

4: **Price update:** The excess or lack of the resources is used for updating price (p_1 and p_2), using a PID controller law (4.8). In the next sampling time the algorithm starts in step 2.

for the same resource using different price-adjustment policies. The optimal solution, from the point of view of the centralized approach, is to use as much cheap resource as possible. This solution is addressed in a distributed price-coordination scheme when the optimal price is achieved (see lower picture of Fig. 4.4). Therefore, an affective coordination of all NMPC controllers has been obtained. In [8] it can be seen how this

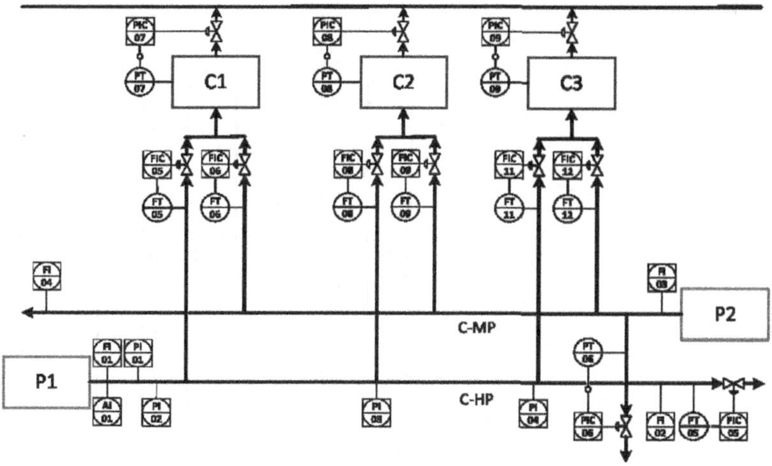

Fig. 4.3 The oxygen benchmark process diagram

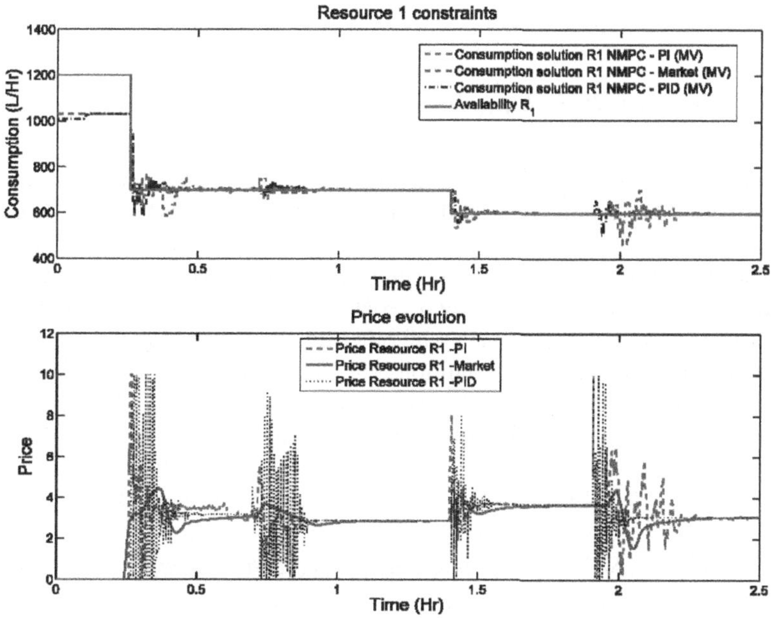

Fig. 4.4 Dynamic resource allocation results and price evolution

optimal solution cannot be achieved with a fully decentralized architecture. Notice, during the transition of the prices, the resource constraints are not fulfilled, but when the steady state is achieved, then the optimal allocation of the resource is equal to the allocation achieved using the centralized scheme.

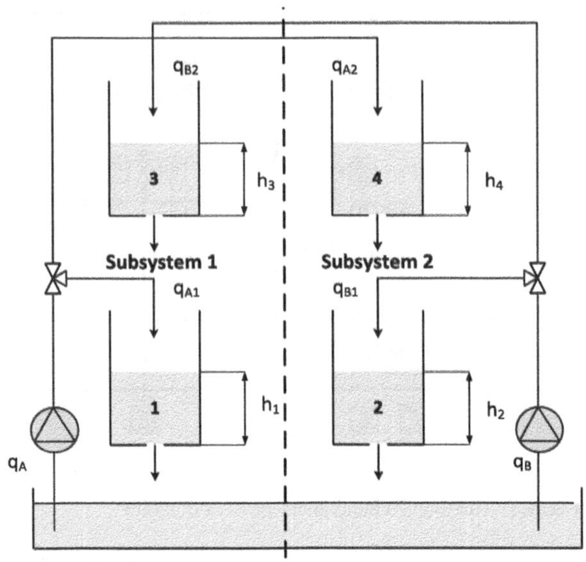

Fig. 4.5 The four-tank process diagram

4.4.2 Four-Tank Benchmark

The plant chosen (Fig. 4.5) is based on the one used by Alvarado et al. in [1]. It consists of four tanks interconnected in such a way that the flow q_A fills tanks 1 and 4, whereas flow q_B fills tanks 2 and 3. On the other hand, tank 3 empties its content into tank 1, and tank 4 does the same into tank 2. So, two manipulated variables q_A and q_B are available to control two variables h_1 and h_2. In order to compare the price-driven coordination approach with other schemes such as the distributed MPC ones presented in [1], the same conditions, the same tracking experiment (references change in the levels of tank 1 and 2) and the same performance index J_{per} has been used. The model is given by the following differential equations:

$$Sdh_1/dt = -a_1\sqrt{2gh_1} + a_3\sqrt{2gh_3} + \gamma_a q_A \qquad (4.12)$$

$$Sdh_2/dt = -a_2\sqrt{2gh_2} + a_4\sqrt{2gh_4} + \gamma_a q_B \qquad (4.13)$$

$$Sdh_3/dt = -a_3\sqrt{2gh_3} + (1-\gamma_b)q_B \qquad (4.14)$$

$$Sdh_4/dt = -a_4\sqrt{2gh_4} + (1-\gamma_a)q_A \qquad (4.15)$$

where h_i and a_i are the water level and the constant discharge of tank $i \in \{1, 2, 3, 4\}$, S is the cross section of the tanks, q_j and γ_j denote the flow and the ratio of the three-way valve of pump $j \in \{A, B\}$, and g is the gravitational acceleration.

4.4.3 Price Coordination Non-Linear Model Predictive Control

The plant is split into two subsystems: subsystem 1 includes tank 1 and tank 3, while subsystem 2 comprises tanks 2 and 4, as indicated in Fig. 4.5. One NMPC controller is assigned to each subsystem, but both of them are able to manipulate flows q_A and q_B, denoting as q_{A1}, q_{B1} or q_{A2}, q_{B2} the ones seen by each controller. This means that two global constraints related to interprocessing streams have to be satisfied; $q_{A1} = q_{A2}$ and $q_{B1} = q_{B2}$. So, it is necessary to rewrite these equality constraints in the form of a resource constraint (4.1) as follows:

$$q_{A1} = q_{A2} \Leftrightarrow q_{A1} - q_{A2} = 0 \Leftrightarrow |q_{A1} - q_{A2}| \leq 0 \Leftrightarrow (q_{A1} - q_{A2})^2 \leq 0 \quad (4.16)$$
$$(q_{A1} - q_{A2})^2 \leq 0 \Leftrightarrow \underbrace{(q_{A1} - q_{A2})^2}_{r_{11}()} + \underbrace{(q_{A2} - q_{A2})^2}_{r_{12}()} \leq \underbrace{0}_{r_{T_1}}$$

$$q_{B1} = q_{B2} \Leftrightarrow q_{B1} - q_{B2} = 0 \Leftrightarrow |q_{B1} - q_{B2}| \leq 0 \Leftrightarrow (q_{B1} - q_{B2})^2 \leq 0 \quad (4.17)$$
$$(q_{B1} - q_{B2})^2 \leq 0 \Leftrightarrow \underbrace{(q_{B1} - q_{B1})^2}_{r_{21}()} + \underbrace{(q_{B1} - q_{B2})^2}_{r_{22}()} \leq \underbrace{0}_{r_{T_2}}$$

The price coordinated optimization problem is then composed of two layers: In the lower one, two separate NMPC; NMPC_1 (4.17) and NMPC_2 (4.18), perform the control level of the corresponding subsystem. The first one, NMPC_1, manipulates the variables q_{A1} and q_{B1}, and the second one, NMPC_2, manipulates q_{A2} and q_{B2}. In the upper coordinating layer, two controllers assign prices p_1 and p_2 to the NMPCs, as in Fig. 4.6. Then, the term $p_1 r_{11}$ is added to NMPC_1's cost function, while the term $p_1 r_{12}$ is added to NMPC_2's cost function. Similar terms are added for the second resource.

Two SISO PID controllers are used for updating the prices, selecting one of the equivalent statements in equations (4.16) and (4.17). The controlled variable is the absolute difference between the flows demanded, which are the solutions obtained by each NMPC controller in the previous sampling time $|q^*_{A1} - q^*_{A2}|$ or $|q^*_{B1} - q^*_{B2}|$. The setpoints correspond to the total shared resource available (which is equal to zero because it is an interprocess stream), and the manipulated variables are the prices that are communicated to every NMPC controller in order to modify their cost functions. This means that, when the optimal prices are found, then the global equality constraints $q_{A1} = q_{A2}$ and $q_{B1} = q_{B2}$ are satisfied.

Notice that the first NMPC computes q_{A1} and q_{B1}, and the second one computes q_{A2} and q_{B2}, but only a single q_A and q_B can be applied physically to the process. The coordination layer changes the prices p_1 and p_2 to equate them, but perfect equality cannot be guaranteed at every sampling time. So, an implementation policy must be designed. For this purpose, the relative gain array (RGA) of the process can be computed as in [1]. The RGA, calculated from the linearized process modelat

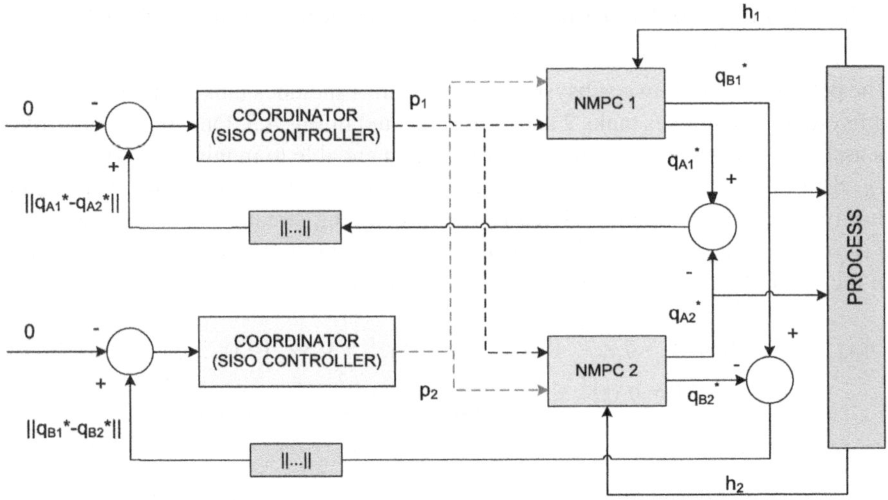

Fig. 4.6 Coordination feedback control scheme using a SISO controller for each interprocess stream

steady state, is the following one:

$$RGA = \begin{pmatrix} q_A & q_B & \\ -0.4 & 1.38 & h_1 \\ 1.38 & -0.4 & h_2 \end{pmatrix} \tag{4.18}$$

This matrix indicates that, for subsystem 1 (tanks 1 and 3) level h_1 should be controlled with q_B, and for subsystem 2 (tanks 2 and 4), h_2 should be controlled with q_B. Then, the implemented flow q_A is the solution obtained by NMPC$_2$ (q^*_{A2}) and the implemented flow q_B is the solution obtained by NMPC$_1$ (q^*_{B1}).

For both NMPC controllers, the cost function in (4.17) and (4.18) includes penalty terms in manipulated variables in order to penalize excessive control changes, i.e. $\Delta q_{A1}(k) = q_{A1}(k) - q_{A1}(k-1)$, the first term of this sequence being ($k = 0$) $\Delta q_{A1}(0) = q_{A1}(0) - q_{A1}(-1)$. However, $q_{A1}(k-1)$ is the implemented solution in the process for the flow q_A, that is, the solution obtained by NMPC$_2$ (q^*_{A2}). In this way, the first term of the sequence is $\Delta q_{A1}(0) = q_{A1}(0) - q^*_{A2}$, corresponding just with the term r_{11} in (4.16), q_{A1} is a decision variable of the controller NMPC$_1$, but q^*_{A2} is data read from the process. A similar penalty term is included in the cost function of the NMPC$_2$: $\Delta q_{A2}(k) = q_{A2}(k) - q_{A2}(k-1)$ and for $k = 0$, $\Delta q_{A2}(0) = q_{A2}(0) - q_{A2}(-1) = q_{A2}(0) - q^*_{A2}$, where q_{A2} is a decision variable of the controller NMPC$_2$, and q^*_{A2} is the solution obtained before for this same controller, being equivalent to the term r_{12} of Eq. (4.16). In this way, and during the process time, the individual different flow solutions for the same stream are penalized.

The results which have been obtained using the PCNMPC are shown in Fig. 4.7 and they illustrate the fact that the scheme can fulfill all the global constraints and achieve

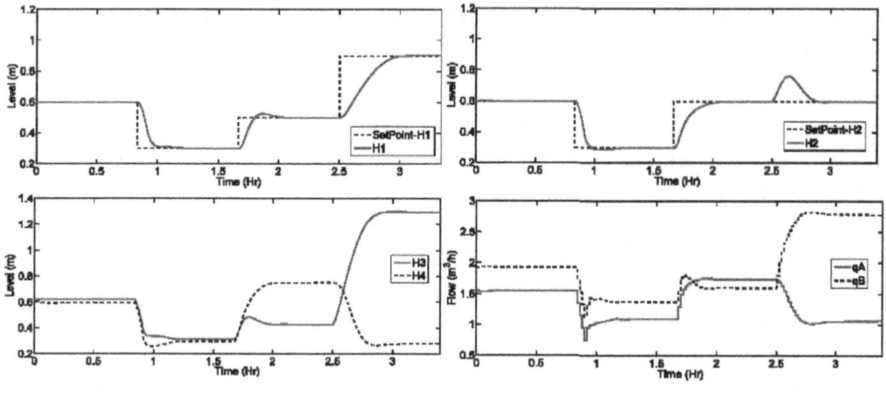

Fig. 4.7 Manipulated and controlled variables using a price-driven coordination method

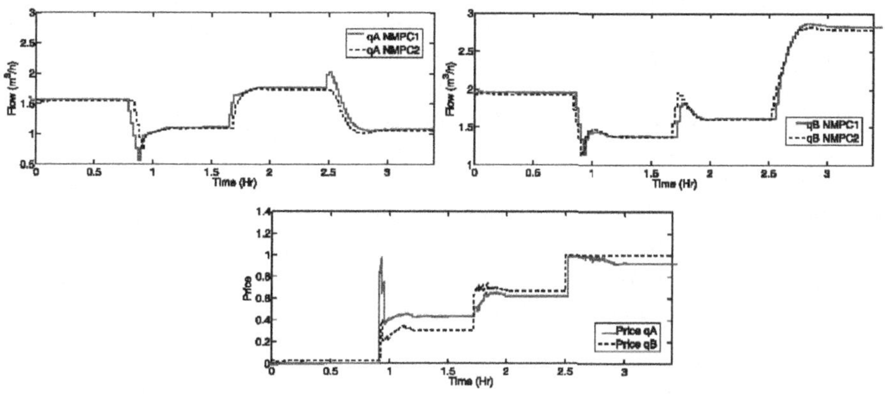

Fig. 4.8 Evolution of the interprocess stream flow and prices

a good tracking of the different setpoints, the performance index being $J_{per} = 32.4$. The lower picture in Fig. 4.8 shows the evolution of the prices, i.e. when the price of q_A achieves an optimal steady value, because Eq. 4.6 is met, (4.6), then both flows, q_{A1} calculated by $NMPC_1$ and q_{A2} calculated by $NMPC_2$, are equal,which means that an effective coordination of both controllers has been achieved (see also the upper pictures in Fig. 4.8). Local solutions of each NMPC controller are the optimal solution of the optimization of the entire process, but of course, during the transient equality constraints are not satisfied.

4.4.4 Comparative Test with Centralized and Decentralized Schemes

The centralized NMPC architecture implies solving a global optimization problem by taking into account the complete nonlinear model of the process, q_A and q_B being the manipulated variables. The results are shown in Fig. 4.9 and the performance index was $J_{per} = 23.83$.

The decentralized non-linear model predictive architecture involves using two independent NMPC controllers. Subsystem 1 is controlled by NMPC controller 1, where q_B is the manipulated variable and q_A is treated as a measured disturbance. On the other hand, subsystem 2 is controlled by NMPC controller 2, where q_A is the manipulated variable and q_B is treated as a measured disturbance. The pairing between manipulated and controlled variables has been done using RGA analysis (4.16). In this architecture, there is no kind of communication between controllers and the results are shown in Fig. 4.10. The performance index $J_{per} = 34.63$ was the worst of all approaches, the value of the offset between setpoint and controlled variable in subsystem 2 being significantly.

The results show that this technique achieves the same steady state solution which has been obtained using the centralized approach. They also match the optimal analytic steady state (see Table 4.1), so a correct decomposition of the global dynamical problem has been achieved. In addition, all control objectives are satisfied, but the response is slower than the centralized architecture compare the controlled variables in Fig. 4.7 with the controlled ones in Fig. 4.9. This effect can seen in the performance indexes which are different (see Table 4.1).

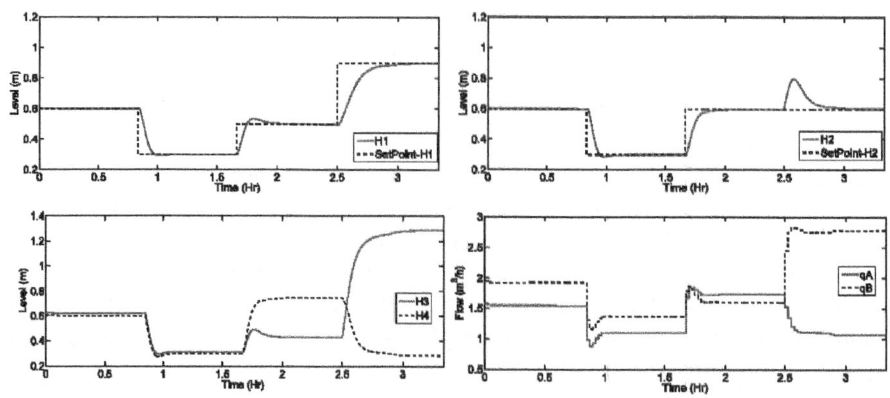

Fig. 4.9 Manipulated and controlled variables using a centralized NMPC scheme

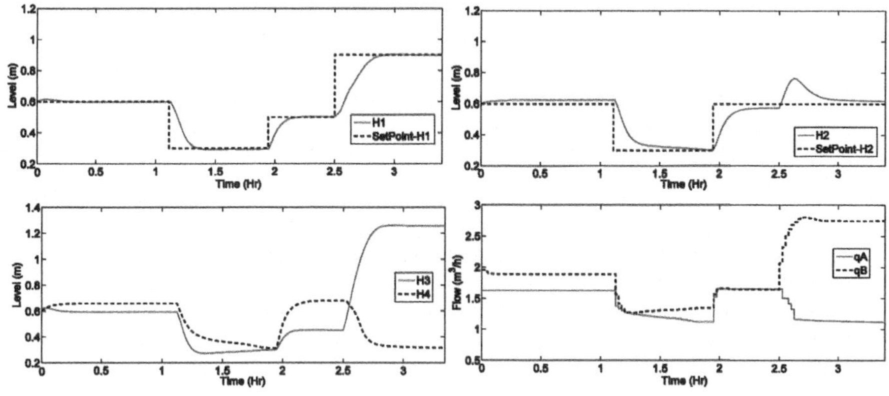

Fig. 4.10 Manipulated and controlled variables using a decentralized NMPC scheme

Table 4.1 Steady state achieved by the different approaches

Time(h)	Steady state		Centralized scheme		Decentralized scheme		Price-driven coordination	
	q_A^s	q_B^s	q_A	q_B	q_A	q_B	q_A	q_B
[0 − 0.84]	1.55	1.93	1.55	1.93	1.63	1.89	1.55	1.93
[0.84 − 1.67]	1.09	1.36	1.09	1.36	1.12	1.34	1.09	1.36
[1.67 − 2.5]	1.74	1.60	1.74	1.60	1.65	1.65	1.74	1.60
[2.5 − 3.34]	1.06	2.78	1.06	2.78	1.12	2.75	1.06	2.78
J_{per}				23.83		34.63		32.40

4.5 Conclusions

In this chapter, a method has been developed to coordinate several NMPC controllers working in parallel in a price-driven scheme. Each NMPC controls a subsystem being all subsystems coupled by the manipulated variables and the coordinator is formulated as a control problem, that uses the prices as manipulated variables and try to fulfill (4.6) as target. This method presents several advantages. It is easy to implement in existing MPC industrial controllers and less data interchange is needed to coordinate the NMPC controllers. Moreover, if a PID controller is used inside the non-linear controller, to find the optimal price, it is possible to determine, in a well known way, the behavior of the fulfillment of the global constraints in terms of the tuning parameters of that PID.

Acknowledgments The research leading to these results has received funding from the European Union Seventh Framework Programme [FP7/2007-2013]. The financial support of EMECW Lot 17 program and the project "MICINN DPI 2009-12805", is also greatly appreciated.

References

1. I. Alvarado, D. Limón, D. Muñoz de la Peña, J.M. Maestre, M.A. Ridao, H. Scheu, W. Marquardt, R.R. Negenborn, B. De Schutter, F. Valencia, J. Espinosa, A comparative analysis of distributed MPC techniques applied to the HD-MPC four-tank benchmark. Journal of Process Control **21**(5), 800–815 (June 2011)
2. M. Cepak and V. Havlena. New price coordination methods solving optimum allocation problem. Power India Conference, pp. 536–542, 2006
3. R. Cheng and F. Forbes, Price-driven Coordination for Solving Plantwide MPC Problems. In 16th IFAC World Congress, 17:429–438, 2007
4. R.A. Jose and L.H. Ungar, Auction-driven coordination for plantwide optimization. 3rd International Conference on Foundations of Computer-Aided Process Operations, pp 152–157, 1998
5. R.A. Jose, L.H. Ungar, Pricing interprocess streams using slack auctions. AIChE Journal **46**(3), 575–587 (2000)
6. G.D. *Luengerber*, Introduction to linear and nonlinear programming (Addison-Wesley Publishing Company, Reading, 1973)
7. N.I. Marcos, J.F. Forbes, M. Guay, Coordination of distributed model predictive controllers for constrained dynamic processes, in ADCHEM 2009 (International Symposium on Advanced Control of Chemical Processes), 2009
8. R. Martí, D. Sarabia, and C. de Prada, Shared Resources Management by Price Coordination. ESCAPE22 (European Symposium on Computer Aided Process Engineering), pp 902–906, 2012
9. Riccardo Scattolini, Architectures for distributed and hierarchical Model Predictive Control - A review. J. Process Control **19**, 723–731 (2009)
10. Holger Voos. Resource Allocation in Continuous Production using Market-Based Multi-Agent Systems, in 5th IEEE International Conference on Industrial Informatics, pages 1085–1090, 2007

Chapter 5
Distributed MPC for Consensus and Synchronization

M. A. Müller and F. Allgöwer

Abstract In this chapter, we describe a distributed MPC algorithm for cooperative control of a network of systems which are coupled by constraints and pursue a common, cooperative control objective. The proposed DMPC algorithm cannot only be used for classical control objectives such as set point stabilization, but also for more general cooperative control tasks such as consensus and synchronization problems. Possible application fields include teams of mobile robots, formation flight of aircrafts, as well as satellite control.

5.1 Introduction

Most of the distributed MPC algorithms described in this book consider the problem of solving a centralized problem for an overall large-scale system by distributing it into several coupled subproblems. These subproblems are then assigned to local controllers or agents, which together compute a (approximate) solution to the original centralized problem in a distributed or decentralized fashion. In this chapter, we look at distributed MPC from a different point of view. Namely, we consider the setup of a team of physically (i.e., dynamically) decoupled systems which pursue a common, cooperative control task and which additionally may have to satisfy certain coupling constraints. This setup includes, for example, formation flight of a team of aircrafts subject to collision avoidance constraints, or remote sensing of a team of satellites with connectivity maintenance constraints. In such a setting, more general cooperative control objectives are of great importance, compared to classical ones

M. A. Müller (✉) · F. Allgöwer
University of Stuttgart, Institute for Systems Theory and Automatic Control,
Stuttgart, Germany
e-mail: matthias.mueller@ist.uni-stuttgart.de

F. Allgöwer
e-mail: frank.allgower@ist.uni-stuttgart.de

J. M. Maestre and R. R. Negenborn (eds.), *Distributed Model Predictive Control Made Easy*, Intelligent Systems, Control and Automation: Science and Engineering 69,
DOI: 10.1007/978-94-007-7006-5_5,

such as the stabilization of an a priori known set point. In particular, consensus and synchronization problems are of interest (see, e.g., [3, 10, 11] and the references therein), where the systems have to agree on a common trajectory online, in contrast to following an a priori specified reference trajectory.

In this chapter, which is based on the results presented in [8, 9], we describe a distributed MPC algorithm which can be used for a variety of cooperative control tasks for dynamically independent systems including the above mentioned. The proposed algorithm is non-iterative and the systems optimize their performance criteria in a specific order, similar to [13], where such an idea was used for the robust stabilization of a set point for systems only coupled via constraints. The advantage of such an approach is that less communication between the systems is needed in comparison to iterative schemes and that the control action is computed in a truly distributed way, in the sense that only neighboring systems have to communicate with each other and no centralized optimization problem is solved iteratively.

The remainder of this chapter is structured as follows. In Sect. 5.2, we more formally state the considered problem setup, the control objective and the communication structure between the systems. Section 5.3 describes in detail the proposed distributed MPC algorithm, before we discuss its various properties in Sect. 5.4. In contrast to the approach presented here, the vast majority of available distributed MPC schemes in the literature only consider the stabilization of an a priori known set point or reference trajectory, with only very few exceptions considering other cooperative control tasks [2, 4, 6]. In Sect. 5.4, we will also discuss and compare these results to the distributed MPC algorithm proposed in Sect. 5.3. Section 5.5 describes possible applications of the proposed algorithm, in particular consensus and synchronization problems. Finally, we give a short conclusion in Sect. 5.6.

5.2 Problem Setup

5.2.1 System and Network Structure

We consider a set $\mathcal{N}$ of discrete-time dynamically decoupled systems s_i of the form

$$s_i: \quad \mathbf{x}_i(k+1) = \mathbf{f}_i(\mathbf{x}_i(k), \mathbf{u}_i(k)), \qquad k \in \mathbb{Z}_+, \tag{5.1}$$

where $\mathbf{x}_i \in \mathcal{X}_i \subseteq \mathbb{R}^{n_{x_i}}$ denotes the state and $\mathbf{u}_i \in \mathcal{U}_i \subseteq \mathbb{R}^{m_{u_i}}$ the control input to system i, and $i \in \mathcal{I} := \{1, 2, \ldots, |\mathcal{N}|\}$. The input constraint sets $\mathcal{U}_i$ are assumed to be compact and to contain the origin in their interior. Denote by $\mathbf{x}$ and $\mathbf{u}$ the state and input of the overall system, i.e., $\mathbf{x} := [\mathbf{x}_1^T, \ldots, \mathbf{x}_{|\mathcal{N}|}^T]^T$ and $\mathbf{u} := [\mathbf{u}_1^T, \ldots, \mathbf{u}_{|\mathcal{N}|}^T]^T$, and let $\mathcal{X} := \mathcal{X}_1 \times \cdots \times \mathcal{X}_{|\mathcal{N}|}$ and $\mathcal{U} := \mathcal{U}_1 \times \cdots \times \mathcal{U}_{|\mathcal{N}|}$. Furthermore, denote by $\mathbf{f}$ the overall system dynamics such that

$$\mathbf{x}(k+1) = \mathbf{f}(\mathbf{x}(k), \mathbf{u}(k)). \tag{5.2}$$

As already mentioned in the Introduction, while in our setup the systems (5.1) are dynamically decoupled, i.e., only its own state $\mathbf{x}_i$ and input $\mathbf{u}_i$ appear in the dynamics (5.1) of each system i, they are coupled with each other through common constraints and a common control objective (to be made precise later). Hence for each system i, we can define system j to be a neighbor of system i if the two systems are subject to coupled constraints or a coupled objective function. More formally, we identify each of the systems s_i with a vertex of a graph $\mathcal{G} = (\mathcal{S}, \mathcal{E})$, where $\mathcal{S} = \{s_1, ..., s_{|\mathcal{N}|}\}$ is the set of vertices, and the set of edges $\mathcal{E} \subseteq \{(s_i, s_j) \in \mathcal{S} \times \mathcal{S} | i \neq j\}$ describes the interconnection topology of the systems, induced by the couplings in the constraints and the common objective. Let $\mathcal{N}_i := \{j | (s_i, s_j) \in \mathcal{E}\}$ be the set of indices of the neighbors of system i, and $d_i := |\mathcal{N}_i|$ its cardinality. According to the above definition of a neighbor, system i is a neighbor of system j if and only if also system j is a neighbor of system i, as coupling constraints and a coupled objective function affect both systems. This means that we model the graph $\mathcal{G}$ to be undirected, i.e., for any $i, j \in \mathcal{I}$, $(s_i, s_j) \in \mathcal{E}$ if and only if also $(s_j, s_i) \in \mathcal{E}$. In order to be able to achieve the common control objective and to satisfy the coupling constraints, each system must be able to communicate with each of its neighbors. In particular, we assume that the systems can exchange predicted trajectories with their neighbors, which we will make preciser later.

5.2.2 Cost Functions and Optimal Control Problem

As stated above, our control objective is the achievement of some common, cooperative task. We will translate this into the stabilization of some closed set $\mathcal{X}^0 \subseteq \mathcal{X}$. This rather general problem formulation includes as two special cases the situations where $\mathcal{X}^0$ is (i) a prespecified set point; and (ii) where $\mathcal{X}^0$ is the consensus subspace $\mathcal{X}^0 = \{\mathbf{x} \in \mathcal{X} : \mathbf{x}_1 = \mathbf{x}_2 = \cdots = \mathbf{x}_{|\mathcal{N}|}\}$. In order to achieve the stabilization of $\mathcal{X}^0$ via distributed MPC, we associate a stage cost function L_i and a terminal cost function F_i to each of the systems. For each system s_i, these cost functions can also depend on neighboring system states according to the considered cooperative control objective, and we assume that they exhibit the following separable structure:

$$
\begin{aligned}
L_i(\mathbf{x}_i, \{\mathbf{x}_j\}_{j\in\mathcal{N}_i}, \mathbf{u}_i) &= L_{ii}(\mathbf{x}_i, \mathbf{u}_i) + \sum_{j\in\mathcal{N}_i} L_{ij}(\mathbf{x}_i, \mathbf{x}_j), \\
F_i(\mathbf{x}_i, \{\mathbf{x}_j\}_{j\in\mathcal{N}_i}) &= F_{ii}(\mathbf{x}_i) + \sum_{j\in\mathcal{N}_i} F_{ij}(\mathbf{x}_i, \mathbf{x}_j).
\end{aligned} \tag{5.3}
$$

This means that the stage and terminal cost functions of each system are separated into a part L_{ii} (respectively, F_{ii}) consisting of its own state and input, and parts L_{ij} (respectively, F_{ij}) consisting of its own and *one* of its neighbors' states. Furthermore, as is standard in tracking MPC, we assume that both the stage cost L_i and the terminal cost F_i are such that their respective sum is positive definite with respect to the target set $\mathcal{X}^0$, i.e., there exist $\alpha_1, \alpha_2 \in \mathcal{K}_\infty$ such that $\sum_{i=1}^N L_i(\mathbf{x}, \mathbf{u}) \geq \alpha_1(|\mathbf{x}|_{\mathcal{X}^0})$ for all $\mathbf{u} \in \mathcal{U}$ and $\sum_{i=1}^N F_i(\mathbf{x}) \geq \alpha_2(|\mathbf{x}|_{\mathcal{X}^0})$, respectively, where $|\mathbf{x}|_{\mathcal{X}^0}$ denotes the point to set distance from x to $\mathcal{X}^0$.

Remark 5.1 In this chapter, we restrict ourselves to cost functions exhibiting the above structure as it results in less communication effort in the distributed MPC algorithm later on. Note that such an assumption is in general not very restrictive and is fulfilled in various applications such as, e.g., multi-vehicle formation stabilization, synchronization of oscillators and others (see, e.g., [1, 5, 8]). Furthermore, in [8], we presented a second DMPC algorithm where this assumption is not needed, but more communication between the systems is required. However, we do not elaborate this second algorithm in more detail here in order to streamline the presentation. □

Now consider the following finite horizon open-loop optimal control problem for system i with prediction horizon N_p:

Problem $\mathcal{P}_i$: At time instant k,

$$\begin{aligned}
&\underset{\mathbf{u}_i(k:k+N_p-1)}{\text{minimize}} \; J_i(\mathbf{x}_i(k), \tilde{\mathbf{x}}_{-i}(k:k+N_p), \mathbf{u}_i(k:k+N_p-1)) \\
&= J_{ii}\big(\mathbf{x}_i(k), \tilde{\mathbf{x}}_{-i}(k:k+N_p), \mathbf{u}_i(k:k+N_p-1)\big) \\
&\quad + \sum_{j\in\mathcal{N}_i} J_{ji}\big(\mathbf{x}_i(k), \tilde{\mathbf{x}}_{-i}(k:k+N_p), \mathbf{u}_i(k:k+N_p-1)\big) \\
&:= \sum_{l=0}^{N_p-1} L_i\big(\mathbf{x}_i(k+l|k), \tilde{\mathbf{x}}_{-i}(k+l|k), \mathbf{u}_i(k+l|k)\big) + F_i\big(\mathbf{x}_i(k+N_p|k), \tilde{\mathbf{x}}_{-i}(k+N_p|k)\big) \\
&\quad + \sum_{j\in\mathcal{N}_i} \left\{ \sum_{l=0}^{N_p-1} L_{ji}(\tilde{\mathbf{x}}_j(k+l|k), \mathbf{x}_i(k+l|k)) + F_{ji}(\tilde{\mathbf{x}}_j(k+N_p|k), \mathbf{x}_i(k+N_p|k)) \right\}
\end{aligned} \tag{5.4a}$$

subject to

$$\mathbf{x}_i(k+l+1|k) = \mathbf{f}_i\big(\mathbf{x}_i(k+l|k), \mathbf{u}_i(k+l|k)\big) \tag{5.4b}$$

$$\mathbf{x}_i(k|k) = \mathbf{x}_i(k) \tag{5.4c}$$

$$\mathbf{u}_i(k+l|k) \in \mathcal{U}_i \tag{5.4d}$$

$$\mathbf{x}_i(k+l+1|k) \in \mathcal{X}_i \tag{5.4e}$$

$$\mathbf{x}_i(k+N_p|k) \in \mathcal{X}_i^f(k) \subseteq \mathcal{X}_i \tag{5.4f}$$

$$c_q(\mathbf{x}_i(k+l+1|k), \tilde{\mathbf{x}}_{-i}(k+l+1|k)) \in \mathcal{C}_q \tag{5.4g}$$

for all $l \in \{0, 1, \ldots, N_p-1\}$.

First, note that in Problem $\mathcal{P}_i$, the cost function J_i which is minimized is comprised of the sum of a cost function J_{ii} consisting of system i's own state and terminal cost functions L_i and F_i, respectively, and cost functions J_{ji} which consist of those terms in the cost functions of system i's neighbors which involve its own state $\mathbf{x}_i$, i.e., L_{ji}

and F_{ji}. This special structure of the cost function J_i together with the distributed MPC algorithm proposed in Sect. 5.3 will ensure a decrease of a specific Lyapunov function and hence the achievement of the cooperative control goal. Furthermore, in Problem $\mathcal{P}_i$, $\mathbf{u}_i(k : k + N_p - 1)$ consists of the sequence of input values $\mathbf{u}_i(k : k + N_p - 1) := [\mathbf{u}_i(k|k)^T, \ldots, \mathbf{u}_i(k + T - 1|k)^T]^T$ predicted at time k, and $\mathbf{x}_i(k : k + N_p) := [\mathbf{x}_i(k|k)^T, \ldots, \mathbf{x}_i(k + T|k)^T]^T$ is the corresponding predicted state sequence with initial condition equal to the actual system state at time k according to (5.4c). Furthermore, $\tilde{\mathbf{x}}_{-i}(k : k + N_p)$ consists of the predicted state sequences of system i's neighbors, i.e.,

$$\tilde{\mathbf{x}}_{-i}(k : k + N_p) = [\tilde{\mathbf{x}}_{-i}(k|k)^T, \ldots, \tilde{\mathbf{x}}_{-i}(k + N_p|k)^T]^T$$

with

$$\tilde{\mathbf{x}}_{-i}(l|k) = \left[\tilde{\mathbf{x}}_{i_1}(l|k)^T, \ldots, \tilde{\mathbf{x}}_{i_{d_i}}(l|k)^T\right]^T$$

for all $k \leq l \leq k + N_p$, where $\{i_1, \ldots, i_{d_i}\}$ is an ordered sequence of the elements of the set $\mathcal{N}_i$, i.e. $i_1 < \cdots < i_{d_i}$. As system i optimizes only over its local input trajectory $\mathbf{u}_i(k : k + N_p - 1)$, the predicted trajectories of its neighbors, $\tilde{\mathbf{x}}_{-i}$, are held as constant parameters. We will specify later in the distributed MPC algorithm which predicted sequences $\tilde{\mathbf{x}}_{-i}(k : k + N_p)$ system i assumes for its neighbors. The functions c_q and the sets $\mathcal{C}_q \subseteq \mathbb{R}^{\ell_q}$ with $q \in \{1, ..., Q\}$ define the Q coupling constraints the systems have to fulfill. Clearly, for Problem $\mathcal{P}_i$, only those of the Q coupling constraints are relevant where $\mathbf{x}_i$ appears explicitly. Finally, the terminal region $\mathcal{X}_i^f(k)$ in (5.4f) is time-varying, and it will be defined in each step of the distributed MPC algorithm in Sect. 5.3. Before stating the proposed distributed MPC algorithm, we shortly discuss the necessary assumptions concerning the terminal cost and terminal region approach which we use here.

5.2.3 Terminal Regions and Controllers

When using a terminal cost and terminal region approach in MPC, the crucial assumption in order to establish stability is that the terminal region can be made invariant under a local auxiliary control law, and that the terminal cost function can be used as a control Lyapunov function inside the terminal region [12]. We will make a similar assumption here.

Assumption 5.1 There exists a terminal region $\mathcal{X}^f \subseteq \mathcal{X}$, and for each system $i \in \mathcal{I}$ an auxiliary local control law $\mathbf{u}_i = \mathbf{k}_i^{loc}(\mathbf{x}_i, \{\mathbf{x}_j\}_{j\in\mathcal{N}_i})$, such that the terminal region $\mathcal{X}^f$ is invariant with respect to the overall nominal closed-loop system $\mathbf{x}(k+1) = \mathbf{f}(\mathbf{x}(k), \mathbf{k}^{loc}(\mathbf{x}))$ with $\mathbf{k}^{loc} := [(\mathbf{k}_1^{loc})^T, \ldots, (\mathbf{k}_{|\mathcal{N}|}^{loc})^T]^T$, and the following holds for all $\mathbf{x} \in \mathcal{X}^f$ and for all $i \in \mathcal{I}$:

$$\mathbf{k}_i^{loc}(\mathbf{x}_i, \{\mathbf{x}_j\}_{j\in\mathcal{N}_i}) \in \mathcal{U}_i \tag{5.5a}$$

$$c_q(\mathbf{x}_i, \{\mathbf{x}_j\}_{j\in\mathcal{N}_i}) \in \mathcal{C}_q \qquad \forall q \in \{1, \ldots, Q\} \tag{5.5b}$$

$$\sum_{i=1}^{N} F_i^+(\mathbf{x}_i, \{\mathbf{x}_j\}_{j\in\mathcal{N}_i}) - F_i(\mathbf{x}_i, \{\mathbf{x}_j\}_{j\in\mathcal{N}_i}) + L_i(\mathbf{x}_i, \{\mathbf{x}_j\}_{j\in\mathcal{N}_i}, \mathbf{k}_i^{loc}(\mathbf{x}_i, \{\mathbf{x}_j\}_{j\in\mathcal{N}_i})) \leq 0 \tag{5.5c}$$

In Assumption 4.1, (5.5a) ensures the satisfaction of the input constraints when applying the auxiliary local control laws, (5.5b) the satisfaction of the coupling constraints inside the terminal region, and (5.5c) the decayrate of the sum of the terminal cost functions. The invariance condition for $\mathcal{X}^f$ is implied by (5.5c), if $\mathcal{X}^f$ is chosen as a sublevel set of the sum of the terminal cost functions. Note that the local controllers are only allowed to depend on neighboring system states, whereas we assume the existence of a "centralized" terminal region $\mathcal{X}^f$ for the overall system. We use such a centralized terminal region as in typical cooperative control tasks such as consensus problems, only the decaying of the sum of the terminal cost functions (5.5c) can be ensured, but not the decaying of the single terminal cost functions. Hence only invariance of a centralized terminal region can be ensured, but not invariance of some distributed terminal regions. However, due to the separable structure of the cost function assumed in (5.3), it turns out that the required centralized terminal constraint

$$\mathbf{x}(k+T|k) \in \mathcal{X}^f \tag{5.6}$$

can be distributed into time-varying terminal regions as used in Problem $\mathcal{P}_i$. The exact definition of these time-varying terminal regions will be given in the distributed MPC algorithm proposed in Sect. 5.3.

5.3 A Distributed MPC Algorithm for Cooperative Control

We are now ready to state the proposed distributed MPC algorithm for cooperative control tasks as Algorithm 5.1.

5.4 Theoretical Properties of the Proposed Distributed MPC Algorithm

In this section, we discuss various properties of the proposed distributed MPC algorithm and the resulting closed-loop system. We start with some remarks on the communication requirements and scalability of the algorithm, before discussing recursive feasibility of the optimization problems, asymptotic stability of the

Algorithm 5.1 Distributed MPC for cooperative control

Initialization: Let $k = 0$ and $\mathbf{x}(0) := \mathbf{x}_0$ be the initial condition for system (5.2). For all systems i, find a feasible solution $\hat{\mathbf{u}}_i(0 : N_p - 1)$ with corresponding state sequence $\hat{\mathbf{x}}_i(0 : N_p)$, such that the constraints (5.4b)–(5.4e), (5.4g) and (5.6) are satisfied. Each system transmits $\hat{\mathbf{x}}_i(0 : N_p)$ to its neighbors. Each system initializes $\tilde{\mathbf{x}}_{-i}(0 : N_p)$ with $\tilde{\mathbf{x}}_{-i}(l|0) := [\hat{\mathbf{x}}_{i_1}(l|0)^T \ldots \hat{\mathbf{x}}_{i_{d_i}}(l|0)^T]^T$ for all $0 \le l \le N_p$. Go to step 3.

1: At time instant k, each system i computes the candidate input sequence

$$\begin{aligned}\hat{\mathbf{u}}_i(k : k + N_p - 1) := \Big\{&\mathbf{u}_i^*(k|k-1), \ldots, \mathbf{u}_i^*(k + N_p - 1|k-1),\\ &\mathbf{k}_i^{loc}\big(\mathbf{x}_i^*(k + N_p - 1|k-1), \tilde{\mathbf{x}}_{-i}(k + N_p - 1|k-1)\big)\Big\}\end{aligned} \tag{5.7}$$

by taking the remaining part of the old optimal input sequence $\mathbf{u}_i^*$ and adding the auxiliary local control law $\mathbf{k}_i^{loc}$, as well as the corresponding candidate state sequence $\hat{\mathbf{x}}_i(k : k + N_p)$. Each system sends $\hat{\mathbf{x}}_i(k + N_p|k)$ to all of its neighbors. (Note that it is sufficient to only send the last state $\hat{\mathbf{x}}_i(k + N_p|k)$ of the sequence $\hat{\mathbf{x}}_i(k : k + N_p)$ to the neighbors. Namely, for $k \le l \le k + N_p - 1$, we have $\hat{\mathbf{x}}_i(l|k) = \mathbf{x}_i^*(l|k-1)$, and these states (i.e., the previously optimal sequence) were already sent to the neighbors in Step 3c at time $k - 1$.)

2: After receiving $\hat{\mathbf{x}}_j(k + N_p|k)$ from all of its neighbors $j \in \mathcal{N}_i$, each system i initializes $\tilde{\mathbf{x}}_{-i}(k : k + N_p)$ with $\tilde{\mathbf{x}}_{-i}(l|k) := [\hat{\mathbf{x}}_{i_1}(l|k)^T \ldots \hat{\mathbf{x}}_{i\,d_i}(l|k)^T]^T$ for all $k \le l \le k + N_p$.

3: For $i = 1$ to N, system i

1. defines the terminal region

$$\begin{aligned}\mathcal{X}_i^f(k) := \Big\{\mathbf{y} \in \mathcal{X}_i : {}& F_i(\mathbf{y}, \tilde{\mathbf{x}}_{-i}(k + N_p|k)) + \sum_{j \in \mathcal{N}_i} F_{ji}(\tilde{\mathbf{x}}_j(k + N_p|k), \mathbf{y})\\ &\le F_i(\hat{\mathbf{x}}_i(k + T|k), \tilde{\mathbf{x}}_{-i}(k + T|k))\\ &+ \sum_{j \in \mathcal{N}_i} F_{ji}(\tilde{\mathbf{x}}_j(k + N_p|k), \hat{\mathbf{x}}_i(k + N_p|k))\Big\},\end{aligned} \tag{5.8}$$

2. solves Problem $\mathcal{P}_i$ with $\mathcal{X}_i^f(k)$ given by (5.8), and denotes the solution by $\mathbf{u}_i^*(k : k + N_p - 1)$ and the corresponding state sequence by $\mathbf{x}_i^*(k : k + N_p)$,
3. sends $\mathbf{x}_i^*(k : k + N_p)$ to its neighbors $j \in \mathcal{N}_i$, who update $\tilde{\mathbf{x}}_{-j}(k : k + N_p)$ by replacing $\hat{\mathbf{x}}_i(l|k)$ with $\mathbf{x}_i^*(l|k)$ for all $k \le l \le k + N_p$.

4: Each system applies $\mathbf{u}_i(k) = \mathbf{k}_i^{MPC} := \mathbf{u}_i^*(k|k)$.

5: Increment k and go to Step 1).

closed-loop system with respect to the target set $\mathcal{X}^0$ as well as robustness issues. After that, we shortly compare our algorithm to other existing distributed MPC approaches.

First, note that Algorithm 5.1 is non-iterative, i.e., each system solves its optimization problem $\mathcal{P}_i$ only once during each time step (in Step 3b of the algorithm). The advantage of such an approach is that less communication and computational power is needed than if the systems had to exchange predicted trajectories and solve their respective optimization problems multiple times. Namely, the communication requirements in Algorithm 5.1 are such that each system has to send information to each of its neighbors two times at each time step, once a whole predicted state sequence and once only one element of a predicted state sequence: In Step 1, the last element $\hat{\mathbf{x}}_i(k+N_p|k)$ of the feasible sequence $\hat{\mathbf{x}}_i(k:k+N_p)$, and in Step 3c the whole newly calculated optimal sequence $\mathbf{x}_i^*(k:k+N_p)$ have to be sent to the neighbors. On the other hand, while with the proposed non-iterative algorithm we can ensure recursive feasibility and the achievement of the cooperative control objective (see below), we cannot expect to achieve a performance which is as good as if the centralized problem for the overall system was solved iteratively.

For clarity of presentation, in Step 3 of Algorithm 5.1 we required the systems to solve their optimization problem in a specified, sequential order. This special sequential structure helps to ensure recursive feasibility as well as asymptotic stability of the closed-loop to the target set $\mathcal{X}^0$, as discussed in more detail later. However, if the systems have to execute Step 3 in a sequential order during one sampling interval, this implies that the algorithm is not very well scalable with the number of systems $|\mathcal{N}|$. Nevertheless, it turns out that the algorithm enjoys the same properties if non-neighboring systems execute Step 3 in parallel. This means that the scalability of the algorithm depends on the underlying graph of the network. For example, if the underlying graph is a tree (i.e., does not contain cycles), then the systems can be divided into two disjoint independent sets (i.e., sets which contain only non-neighboring nodes), independent of the number of systems $|\mathcal{N}|$. This means that only two optimization problems have to be solved sequentially during each sampling time k independent of the number of systems $|\mathcal{N}|$, i.e., the algorithm is scalable. For general graphs, d optimization problems have to be solved sequentially during each sampling interval, where d is the minimum number of independent sets of the graph such that their union contains all nodes. However, determining independent sets for a given graph, and hence also the number d, can in general be a difficult problem.

Next, we discuss the issues of recursive feasibility as well as stability and robustness when applying Algorithm 5.1. In [8], it was shown that if a feasible solution can be found in Step 0, then the optimization problems $\mathcal{P}_i$ in Step 3b remain recursively feasible, and the calculated optimal sequences $\mathbf{x}_i^*$ together also satisfy the centralized terminal constraint (5.6). The proof of this relies on the fact that neighboring systems do not solve their respective optimization problems in parallel, so that the assumed predicted trajectories for the neighbors, $\tilde{\mathbf{x}}_{-i}$, stay constant while system i solves its optimization problem, and hence fulfillment of the coupling constraints and the (time-varying) terminal constraint can be ensured. Furthermore, the closed-loop system is asymptotically stable with respect to the target set $\mathcal{X}^0$. An interesting feature of the respective proof in [8] is the fact that not the sum over the optimal value functions J_i (5.4a), but only the sum over the parts J_{ii} of the optimal value functions J_i is used as a Lyapunov function. Finally, in [9] we examined robustness properties of Algorithm 5.1. Namely, for the case of linear system dynamics subject to additive unknown but bounded disturbances, we modified Algorithm 5.1 based on the tube MPC approach initially proposed in [7]. Again, if a feasible solution can be found in Step 0, recursive feasibility of the optimization problems $\mathcal{P}_i$ in Step 3b can be ensured, as well as asymptotic stability of the set $\mathcal{X}^0 \oplus \mathcal{Z}$, where $\mathcal{Z}$ is a robust positively invariant set for the overall system.

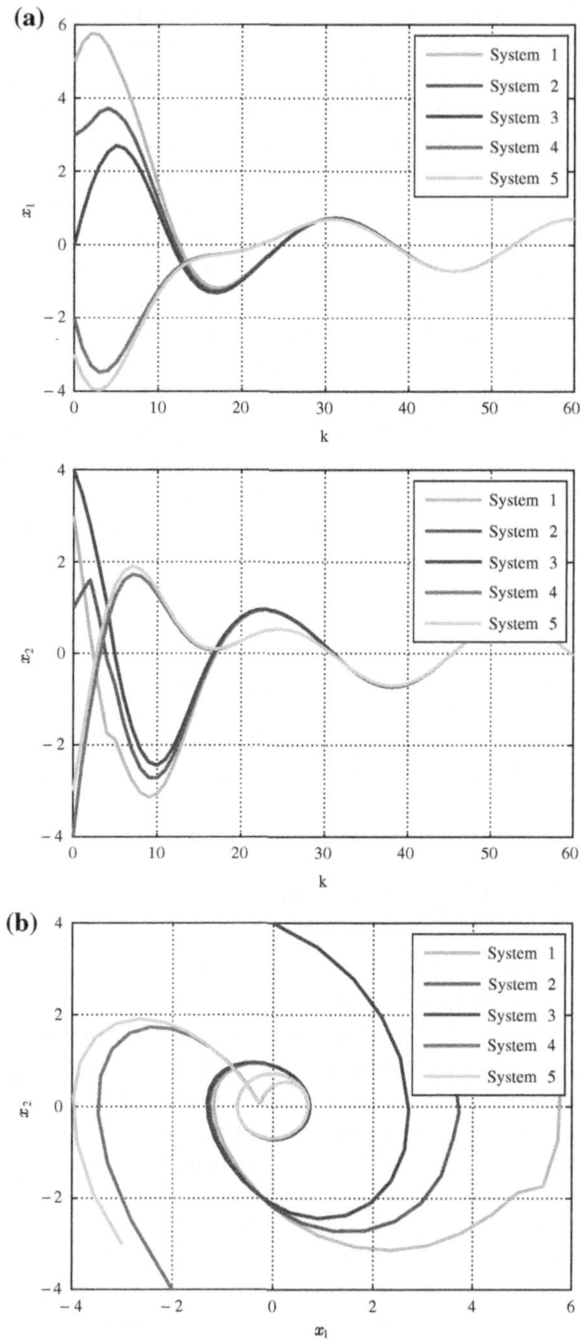

Fig. 5.1 Synchronization of linear oscillators via distributed MPC [8]

Compared to other existing distributed Fig. 5.1 MPC algorithms in the literature, our approach is closest related to and based on the one presented in [13], where the idea of sequential optimization was first used in a distributed MPC setup. However, in contrast to the distributed MPC algorithm presented here, in [13] only linear (uncertain) systems coupled via constraints (but not through the objective function) were considered, and also only the control objective of set point stabilization was treated there. Other distributed MPC approaches for dynamically decoupled systems considering the control objective of set point stabilization such as [1, 5, 14] establish stability via consistency constraints, meaning that at each time step, the newly calculated optimal trajectories of each system must not deviate too much from the ones calculated at the previous time step [1] or from the ones the neighboring systems assumed [5], or via an additional constraint requiring that the MPC control law asymptotically converges to some fixed local control law [14]. As already stated in the introduction, there are only very few distributed MPC algorithms available in the literature which also treat more general cooperative control tasks than set point stabilization. In [2], a distributed MPC approach was developed for consensus of single and double integrator dynamics. Furthermore, in [4, 6] a distributed MPC algorithm was proposed for linear systems, where an optimal consensus point is calculated first at each time step by iteratively solving a centralized optimization problem, which is then used as a set point in the MPC formulation. Furthermore, convergence to a common consensus point corresponding to an equilibrium of the system was established. Compared to this algorithm, in our approach we do not need to determine an optimal consensus point first at each time step in order to design the cost functional accordingly; also, the trajectory along which the systems synchronize does not necessarily have to be an equilibrium point (compare Sect. 5.5).

5.5 Applications of the Proposed Distributed MPC Algorithm

As already highlighted before, a special feature of our algorithm is that is not only suited for set point stabilization, but also for more general cooperative control tasks such as consensus and synchronization problems. In [8], we showed for all these cases how the terminal region and terminal controllers can be computed such that Assumption 5.1 is satisfied. Specifically, in the case of state synchronization of identical linear systems, the terminal region and terminal controllers can be calculated by solving a specific LMI which depends on certain algebraic properties of the underlying graph of the network, in particular on its incidence matrix. Furthermore, we extended this approach to nonlinear systems based on linearization techniques. Furthermore, this technique of calculating the terminal region and terminal controllers can also be applied to the disturbed case [9]. The following simulation results, taken from [8], show an example of five identical linear oscillators. As can be seen, the system states synchronize to a common trajectory, which is not given a priori. As mentioned before, cooperative control tasks such as consensus and synchronization problems arise in many different application settings, such as teams of mobile robots, formation flight of aircrafts, or satellite control, and hence the presented distributed MPC algorithm enjoys as a wide range of possible applications.

5.6 Conclusions

In this chapter, we proposed a distributed MPC algorithm for a network of systems subject to coupling constraints and a common, cooperative control objective. The algorithm is non-iterative and requires neighbor-to-neighbor communication only. One special feature of this algorithm is that it can be used for both set point stabilization as well as more general cooperative control tasks such as consensus and synchronization problems.

Acknowledgments This work was supported by the German Research Foundation (DFG) within the Priority Programme 1305 "Control Theory of Digitally Networked Dynamical Systems" and within the Cluster of Excellence in Simulation Technology (EXC 310/1) at the University of Stuttgart.

References

1. W.B. Dunbar, R.M. Murray, Distributed receding horizon control for multi-vehicle formation stabilization. Automatica **42**, 549–558 (2006)
2. G. Ferrari-Trecate, L. Galbusera, M.P.E. Marciandi, R. Scattolini, Model predictive control schemes for consensus in multi-agent systems with single- and double-integrator dynamics. IEEE Trans. Autom. Control **54**(11), 2560–2572 (2009)
3. A. Jadbabaie, J. Lin, A.S. Morse, Coordination of groups of mobile autonomous agents using nearest neighbor rules. IEEE Trans. Autom. Control **48**(6), 988–1001 (2003)
4. B.J. Johansson, A. Speranzon, M. Johansson, K.H. Johansson, Distributed model predictive consensus, in *Proceedings of the 17th International Symposium on Mathematical Theory of Networks and Systems*, pp. 2438–2444 (2006)
5. T. Keviczky, F. Borrelli, G.J. Balas, Decentralized receding horizon control for large scale dynamically decoupled systems. Automatica **42**, 2105–2115 (2006)
6. T. Keviczky, K.H. Johansson, A study on distributed model predictive consensus, in *Proceedings of the 17th IFAC World Congress*, pp. 1516–1521 (2008)
7. D.Q. Mayne, M.M. Seron, S.V. Rakovic, Robust model predictive control of constrained linear systems with bounded disturbances. Automatica **41**, 219–224 (2005)
8. M.A. Müller, M. Reble, F. Allgöwer, Cooperative control of dynamically decoupled systems via distributed model predictive control. Int. J. Robust Nonlinear Control **22**(12), 1376–1397 (2012)
9. M.A. Müller, B. Schürmann, F. Allgöwer, Robust cooperative control of dynamically decoupled systems via distributed MPC, in *Proceedings of the IFAC Conference on Nonlinear Model Predictive, Control*, pp. 412–417 (2012)
10. R. Olfati-Saber, J.A. Fax, R.M. Murray, Consensus and cooperation in networked multi-agent systems. Proc. IEEE **95**(1), 215–233 (2007)
11. R. Olfati-Saber, R.M. Murray, Consensus problems in networks of agents with switching topology and time-delays. IEEE Trans. Autom. Control **49**(9), 1520–1533 (2004)
12. J.B. Rawlings, D.Q. Mayne, *Model Predictive Control: Theory and Design* (Nob Hill Publishing, Madison, 2009)

13. A. Richards, J.P. How, Robust distributed model predictive control. Int. J. Control **80**(9), 1517–1531 (2007)
14. C. Wang, C.-J. Ong, Distributed model predictive control of dynamically decoupled systems with coupled cost. Automatica **46**, 2053–2058 (2010)

Chapter 6
Distributed MPC Under Coupled Constraints Based on Dantzig-Wolfe Decomposition

R. Bourdais, J. Buisson, D. Dumur, H. Guéguen and P-D. Moroşan

Abstract In this chapter, we propose a distributed model predictive control scheme based on the Dantzig-Wolfe decomposition to control a collection of linear dynamical systems coupled by linear global constraints. The resulting structure is composed of one optimization agent for each system, and another one that has to ensure that the global constraints are fulfilled. The global solution of the problem is found in a finite number of iterations.

6.1 Short Introduction

This chapter deals with large-scale systems, composed with independent subsystems that share common resources. A typical example of that kind of system is the problem of energy management system for a residential house: heating systems, refrigerant systems, hot-water tank, washing-machine are independent equipments but the global available power is limited by a subscripted contract. A global objective is to minimize the total consumption of the building, while fulfilling the local objectives such as a

R. Bourdais · J. Buisson · H. Guéguen
IETR (UMR-CNRS 6164), SUPELEC, Cesson-Sévigné, France
e-mail: roman.bourdais@supelec.fr

J. Buisson
e-mail: jean.buisson@supelec.fr

H. Guéguen
e-mail: herve.gueguen@supelec.fr

D. Dumur
E3S, SUPELEC, Gif sur Yvette, France
e-mail: didier.dumur@supelec.fr

P-D. Moroşan
Acsystème, Rennes, France
e-mail: petru-daniel.morosan@acsysteme.com

J. M. Maestre and R. R. Negenborn (eds.), *Distributed Model Predictive Control Made Easy*, Intelligent Systems, Control and Automation: Science and Engineering
DOI: 10.1007/978-94-007-7006-5_6, © Springer Science+Business Media Dordrecht 2014

good ambient air temperature, ... Other examples of such a system could be a local residential area in which each house has its own behavior, or systems connected by smart grid.

The treatment of the optimization problem could be too complex with a global approach. Moreover, each time another subsystem is added, the whole controller synthesis has to be done again. For these reasons, a distributed structure is interesting. It leads to define smaller and less complex optimization problems as well as to ensure a modularity of the structure.

In the following, the whole optimization problem is linear: the local models and their constraints as well as the global objective and the global constraints. The special structure of all the constraints and the linear aspect of the problem allow to use the Dantzig-Wolfe decomposition. In a few words, the Dantzig-Wolfe decomposition is an algorithm which is well-known to simplify the complexity of a linear optimization problem when the matrix of constraints is block-angular. The initial problem is decomposed into several smaller linear problems, and an coordination mechanism is used such that the solution of the initial problem is found. In this chapter, as it will be explained, the distributed control structure will be based on this decomposition, which will be used not only to reduce the complexity of the global problem, but also to ensure the modularity.

6.2 Boundary Conditions

6.2.1 System Description

We suppose in this chapter that the studied system is composed of a collection of q independent discrete-time dynamical systems (S_i) that are interconnected by different input constraints. Their dynamical equation can be described as follows:

$$\begin{cases} \mathbf{x}_i(k+1) = \mathbf{A}_i\mathbf{x}_i(k) + \mathbf{B}_i\mathbf{u}_i(k), \\ \mathbf{y}_i(k) \quad\;\; = \mathbf{C}_i\mathbf{x}_i(k). \end{cases} \tag{6.1}$$

In this equation, the local state $\mathbf{x}_i$ is in $\mathbb{R}^{n_{x_i}}$, the local control input $\mathbf{u}_i$ is in $\mathcal{U}_i \subset \mathbb{R}^{n_{u_i}}$ and the local output $\mathbf{y}_i$ is in $\mathbb{R}^{n_{y_i}}$. We suppose that the local input constraint set is defined by n_{c_i} linear constraints:

$$\mathcal{U}_i := \left\{\mathbf{u}_i \in \mathbb{R}^{n_{u_i}} \quad \backslash \quad \forall n \in \{1, \ldots, n_{c_i}\}, \quad \mathbf{e}_{i_n}^{\mathrm{T}}\mathbf{u}_i \le d_{i_n}\right\} \tag{6.2}$$

This set can be also described by the aggregated notation:

$$\mathcal{U}_i := \left\{\mathbf{u}_i \in \mathbb{R}^{n_{u_i}} \quad \backslash \quad \mathbf{E}_i\mathbf{u}_i \le \mathbf{d}_i\right\}$$

Though all these systems are dynamically independent, we suppose that they are connected by n_G global linear input constraints. The global constraint set is defined as follows:

$$\mathcal{U}_G := \left\{ (\mathbf{u}_i)_{i=1,q} \in \mathbb{R}^{n_{u_1}+\cdots+n_{u_q}} \ \setminus \ \forall n \in \{1,\ldots,n_G\}, \ \sum_{i=1}^{q} \mathbf{h}_{i_n}^{\mathrm{T}} \mathbf{u}_i \le g_n \right\}. \tag{6.3}$$

For the different inputs, we suppose that they fulfill the conditions given in Assumption 6.1, which are not restrictive as soon as we consider engineering systems. Besides, we suppose that there exists at least a collection of local inputs that satisfies the global constraints, this point is given in Assumption 6.1:

Assumption 6.1 Boundedness and existence of a global solution. *We suppose that:*

- *all the sets $\mathcal{U}_i$ are nonempty bounded convex sets.*
- *there exists $\mathbf{v} = (\mathbf{v}_i)_{i=1,q}$ such that $\mathbf{v} \in \mathcal{U}_G$ and for all $i \in \{1,\ldots,q\}$, $\quad \mathbf{v}_i \in \mathcal{U}_i$.*

The objective in this paper is to implement a model predictive control for each systems, while taking in consideration the global constraints. This problem is formalized in the next section.

6.2.2 Control Objective

In predictive control, most of the results are provided for a quadratic cost which is interesting for different reasons such as the explicit solution or stability analysis. Here, the problem is tackled with the l_1-norm which is more representative of real costs (price, power consumption, ...).

For a given prediction horizon N_p, we denote by $\mathbf{u}_i(k : k+N_\mathrm{p}-1)$ the vector of all the control inputs over the prediction horizon. $\mathbf{y}_i(k+1 : k+N_\mathrm{p})$ is the predicted output, constructed from the state $\mathbf{x}_i(k)$ and according to the input vector $\mathbf{u}_i(k : k+N_\mathrm{p}-1)$. $\mathbf{y}_i(k+1 : k+N_\mathrm{p})$ can be defined recursively by:

$$\begin{aligned} \mathbf{y}_i(k+l) &= \mathbf{C}_i \mathbf{x}_i(k+l) \\ &= \mathbf{C}_i \left(\mathbf{A}_i^{(l)} \mathbf{x}_i(k) + \sum_{j=1}^{l} \mathbf{A}_i^{(l-j)} \mathbf{B}_i \mathbf{u}_i(k+j-1) \right) \end{aligned} \tag{6.4}$$

Then, this equation can be aggregated as follows:

$$\mathbf{y}_i(k+1 : k+N_\mathrm{p}) = \mathbf{M}_i \mathbf{x}_i(k) + \mathbf{N}_i \mathbf{u}_i(k : k+N_\mathrm{p}-1) \tag{6.5}$$

The expressions of the matrices $\mathbf{M}_i$ and $\mathbf{N}_i$ are directly derived from equation (6.4). Let us denote with $\mathbf{y}_{\mathrm{ref,i}}(k+1 : k+N_\mathrm{p})$ the vector of the future reference.

The objective function for each system (S_i) can be stated as follows:

$$J_i(\mathbf{x}_i(k), \mathbf{u}_i(k : k + N_\text{p} - 1)) = \left\|\mathbf{y}_{\text{ref,i}}(k + 1 : k + N_\text{p}) - \mathbf{y}_i(k + 1 : k + N_\text{p})\right\|_1 + \alpha_i \left\|\mathbf{u}_i(k : k + N_\text{p} - 1)\right\|_1 . \quad (6.6)$$

α_i is a positive weight factor. From these local objective functions, the global problem can be derived, integrating the global constraints. This leads to:

Problem 6.1 *Centralized Problem*. Given all the current states and the future references $\left(\mathbf{x}_i(k), \mathbf{y}_{\text{ref,i}}(k + 1 : k + N_\text{p})\right)_{i=1,q}$

$$\min_{(\mathbf{u}_i(k:k+N_\text{p}-1))_{i=1,q}} \sum_{i=1}^{q} J_i(\mathbf{x}_i(k), \mathbf{u}_i(k : k + N_\text{p} - 1)), \quad (6.7)$$

subject to:

- the local dynamical equations (6.1),
- the local constraints, $\forall i \in \{1, \ldots, q\}$:

$$\forall l \in \{1, \ldots, N_\text{p}\}, \quad \mathbf{u}_i(k + l - 1) \in \mathcal{U}_i,$$

- and the global constraints,

$$\forall l \in \{1, \ldots, N_\text{p}\}, \quad (\mathbf{u}_i(k + l - 1))_{i=1,q} \in \mathcal{U}_G.$$

6.2.3 Control Structure

Because of the very particular structure of the global system (all the local systems are independent), the distribution of the optimization problem can be very interesting. The resulting control structure is based on $q + 1$ optimization agents. One will be dedicated to each system (S_i), and one which will be in charge of the fulfillment of the global constraints, called the coordinator agent. Concerning the communication protocols, the coordinator agent will exchange data with all the other agents, but the other agents will not communicate with each other. The control structure is presented in Fig. 6.1. All the interactions will be formalized in the next section.

In this chapter, we suppose that all the exchanges are perfect between the agents, this leads to the following assumption:

Assumption 6.2 *Communication and data exchange*. For the communication protocol, we suppose that:

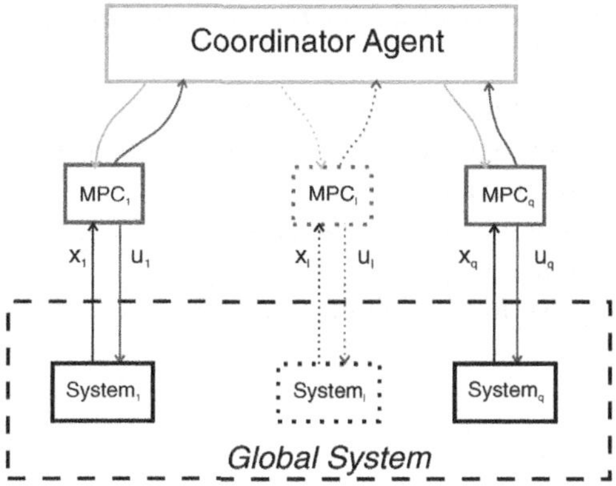

Fig. 6.1 Distributed structure

- there is no delay induced by the communications.
- there is no loss of information during the exchanges.

Now that the global problem and the control structure are presented, the different steps of the resolution will be specified in the following section.

6.3 Description of the Approach

The principle of the decomposition is to use an iterative procedure at each time step (k). Then, for each iterative step, three actions are sequentially achieved:

- A first one is done by the coordinator agent which has to solve its own optimization problem. Its solution is then sent to each other agent.
- A second action in which all the local agents solve, in parallel, their optimization problem. Their solution is sent to the coordinator agent.
- A last step, in which a decision is taken by the coordinator agent. Two cases can occur: either the convergence is reached and the procedure is stopped, or the convergence is not reached, and then a new iteration has to be done.

In the first two steps, different optimizations have to be solved. The distribution of the initial problem relies on linear decomposition technique, and all the agents solve their problem using a linear solver. As it can be noticed, the local cost function (6.6) is the sum of l_1-norm functions. It has to be rewritten into a linear form. This point is developed in the next paragraph, before the formalization of all the optimization problems and the interactions between the agents.

6.3.1 Exact Linearization of the Local Problem

As the function (6.6) is a sum of l_1-norm functions, it is also a sum of absolute values. To linearize the problem, the idea is to introduce two auxiliary variables for each absolute value.

For a better understanding, let us consider the case where $n_{u_i} = n_{y_i} = 1$ and the prediction horizon $N_\text{p} = 1$.

The local minimization problem is the following one:

$$\min_{u_i(k)} |y_{ref,i}(k+1) - y_i(k+1)| + \alpha_i |u_i(k)|, \tag{6.8}$$

subject to

$$u_i(k) \in \mathcal{U}_i. \tag{6.9}$$

Then by introducing 4 positive auxiliary variables $\underline{\mu_i}(k), \overline{\mu_i}(k), \underline{\eta_i}(k), \overline{\eta_i}(k)$, the previous problem is equivalent to:

$$\min_{u_i(k), \underline{\mu_i}(k), \overline{\mu_i}(k), \underline{\eta_i}(k), \overline{\eta_i}(k)} \left(\underline{\eta_i}(k) + \overline{\eta_i}(k)\right) + \alpha_i \left(\underline{\mu_i}(k) + \overline{\mu_i}(k)\right), \tag{6.10}$$

subject to

$$\mathbf{E}_i u_i(k) \le \mathbf{d}_i \tag{6.11}$$

$$y_i(k+1) + \underline{\eta_i}(k) - \overline{\eta_i}(k) = y_{ref,i}(k+1) \tag{6.12}$$

$$u_i(k) + \underline{\mu_i}(k) - \overline{\mu_i}(k) = 0 \tag{6.13}$$

$$\underline{\mu_i}(k), \overline{\mu_i}(k), \underline{\eta_i}(k), \overline{\eta_i}(k) \ge 0 \tag{6.14}$$

Sketches of proof can be found in [3], in which another formulation is proposed to linearize a l_1-norm predictive control.

6.3.2 Local Agent Optimization Problem

This transformation is exactly the same in the general case. Just to have an estimation of the number of variables involved for each subsystem optimization, the initial problem has $n_{u_i} \cdot N_\text{p}$ variables and $n_{c_i} \cdot N_\text{p}$ constraints with a nonlinear cost function whereas the rewritten problem has $(3n_{u_i} + 2n_{y_i}) \cdot N_\text{p}$ variables and $(n_{c_i} + n_{u_i} + n_{y_i}) \cdot N_\text{p}$ constraints but with a linear cost function.

Then, let us denote with $\mathbf{w}_i(k : k + N_\text{p} - 1)$ the vector of all the variables of the rewritten local optimization problem, with

$$\mathbf{w}_i(k:k+N_\mathrm{p}-1) = \begin{pmatrix} \mathbf{u}_i(k:k+N_\mathrm{p}-1) \\ \underline{\mu_i}(k:k+N_\mathrm{p}-1) \\ \overline{\mu_i}(k:k+N_\mathrm{p}-1) \\ \underline{\eta_i}(k:k+N_\mathrm{p}-1) \\ \overline{\eta_i}(k:k+N_\mathrm{p}-1) \end{pmatrix}$$

As we will see in the following, from an iteration to the next one during the distribution process, all the constraints of the local problem remain the same, but the linear cost function can change. At an iteration p, let us denote with ${\mathbf{k}_i^{(p)}}^\mathrm{T}\mathbf{w}_i(k:k+N_\mathrm{p}-1)$ this cost function.

Then, the local problem can be stated as follows:

Problem 6.2 *Rewritten Local Optimization Problem.* For a time step k, and at an iteration p, the local optimization problem is:

$$\min_{\mathbf{w}_i(k:k+N_\mathrm{p}-1)} {\mathbf{k}_i^{(p)}}^\mathrm{T}\mathbf{w}_i(k:k+N_\mathrm{p}-1), \tag{6.15}$$

subject to

$$\overline{\mathbf{E}}_i\mathbf{w}_i(k:k+l-1) \leq \overline{\mathbf{d}}_i(k). \tag{6.16}$$

It is important to notice that in this local problem, the matrix $\overline{\mathbf{E}}_i$ is independent of the time contrary to the vector $\overline{\mathbf{d}}_i(k)$, which has to be calculated at each time step and depends on the future target $\mathbf{y}_{\mathrm{ref,i}}(k+1:k+N_\mathrm{p})$ and the current state $\mathbf{x}_i(k)$. Their expressions are the following:

$$\overline{\mathbf{E}}_i = \left[\begin{array}{c|cccc} diag(\mathbf{E}_i) & \mathbf{0} & \mathbf{0} & \mathbf{0} & \mathbf{0} \\ \hline \mathbf{N}_i & \mathbf{0} & \mathbf{0} & \mathbf{I} & -\mathbf{I} \\ -\mathbf{N}_i & \mathbf{0} & \mathbf{0} & -\mathbf{I} & \mathbf{I} \\ \hline \mathbf{I} & \mathbf{I} & -\mathbf{I} & \mathbf{0} & \mathbf{0} \\ -\mathbf{I} & -\mathbf{I} & \mathbf{I} & \mathbf{0} & \mathbf{0} \\ \hline \mathbf{0} & -\mathbf{I} & \mathbf{0} & \mathbf{0} & \mathbf{0} \\ \hline \mathbf{0} & \mathbf{0} & -\mathbf{I} & \mathbf{0} & \mathbf{0} \\ \hline \mathbf{0} & \mathbf{0} & \mathbf{0} & -\mathbf{I} & \mathbf{0} \\ \hline \mathbf{0} & \mathbf{0} & \mathbf{0} & \mathbf{0} & -\mathbf{I} \end{array}\right], \quad \overline{\mathbf{d}}_i(k) = \left[\begin{array}{c} \mathbf{d}_i \\ \vdots \\ \mathbf{d}_i \\ \hline \mathbf{y}_{\mathrm{ref,i}}(k+1:k+N_\mathrm{p}) - \mathbf{M}_i\mathbf{x}_i(k) \\ -\mathbf{y}_{\mathrm{ref,i}}(k+1:k+N_\mathrm{p}) + \mathbf{M}_i\mathbf{x}_i(k) \\ \hline \mathbf{0} \\ \mathbf{0} \\ \hline \mathbf{0} \\ \hline \mathbf{0} \\ \hline \mathbf{0} \\ \hline \mathbf{0} \end{array}\right] \tag{6.17}$$

During the iteration process, the cost function ${\mathbf{k}_i^{(p)}}^\mathrm{T}\mathbf{w}_i(k:k+N_\mathrm{p}-1)$ will change from an iteration to another one. This is the way that the global constraints are going to be taken into account. When there is no global constraint, the cost function of the rewritten local optimization problem would be ${\mathbf{k}_{i,0}}^\mathrm{T}\mathbf{w}_i(k:k+N_\mathrm{p}-1)$, with:

$$\mathbf{k}_{i,0}^\mathrm{T} = \left[\mathbf{0}^\mathrm{T}\middle|\alpha_i\mathbf{1}^\mathrm{T}\middle|\alpha_i\mathbf{1}^\mathrm{T}\middle|\mathbf{1}^\mathrm{T}\middle|\mathbf{1}^\mathrm{T}\right] \tag{6.18}$$

As it will be explained in the following, this vector will be used during the exchanges from the local agents to the coordinator agent.

6.3.3 The Coordinator Agent Problem

The main objective of the coordinator agent is to guarantee that the global constraints are fulfilled. This is ensured by different optimizations that will be done iteratively. From an iteration to the next one, a new optimization variable is added. The way this new optimization variable is integrated, on the cost function and on the constraints, only depends on the solutions of the local optimizations that have been done at the previous iteration. But before starting, the initialization requires N_0 different admissible points for the local agents. More precision on the initialization will be given in a next paragraph.

Then, at iteration p, the coordinator agent will have $\upsilon^{(p)}$ optimization variables, with $\upsilon^{(p)} = N_0 + p - 1$. For $s = \{1, \ldots, \upsilon^{(p)}\}$, let us denote with $\mathbf{w}_i^{(s)}(k : k + N_\mathrm{p} - 1)$ the s-th admissible point of the local agent.[1]

Then, we define the following variables, $\forall s \in \{1, \ldots, \upsilon^{(p)}\}$,

$$z^{(s)}(k) = \sum_{i=1}^{q} \mathbf{k}_{i,0}^{\mathrm{T}} \mathbf{w}_i^{(s)}(k : k + N_\mathrm{p} - 1) \tag{6.19}$$

$$r_{n,l}^{(s)}(k) = \sum_{i=1}^{q} \mathbf{h}_{i_n}^{\mathrm{T}} \mathbf{u}_i^{(s)}(k + l - 1), \quad \forall n \in \{1, \ldots, n_G\}, \forall l \in \{1, \ldots, N_\mathrm{p}\} \tag{6.20}$$

From this, the coordinator agent has to solve its own problem, which can be stated as follows:

Problem 6.3 *Coordinator Optimization Problem.* At an iteration p, given the variables $z^{(s)}(k)$ and $r_{n,l}^{(s)}(k)$, with $s \in \{1, \ldots, \upsilon^{(p)}\}$, the optimization problem is:

$$\min_{(v_s)_{s=1,\upsilon^{(p)}}} \sum_{s=1}^{\upsilon^{(p)}} z^{(s)}(k) v_s \tag{6.21}$$

with respect to

[1] For $s \in \{1, \ldots, N_0\}$, this point is given by the initialization, and for $s \in \{N_0 + 1, \ldots, N_0 + p - 1\}$ the point is the result of a local optimization problem.

$$\forall n \in \{1,\ldots,n_G\}, \forall l \in \{1,\ldots,N_\mathrm{p}\}, \sum_{s=1}^{\upsilon^{(p)}} r_{n,l}^{(s)}(k) v_s \le g_n \quad : \lambda_{n,l} \tag{6.22}$$

$$\sum_{s=1}^{\upsilon^{(p)}} v_s = 1 \quad : \sigma \tag{6.23}$$

$$\forall s \in \{1,\ldots,\upsilon^{(p)}\}, \quad v_s \ge 0 \tag{6.24}$$

The coordinator agent has to find not only the solution v_s but also the dual variable values $\lambda_{n,l}$, associated to the linear constraints.[2] (6.22) and the dual variable value σ associated to the linear constraint (6.23).

6.3.4 Iterative Mechanism

Each iteration consists in the following three steps:

1. The coordination agent optimization. At iteration p, it has to solve its problem 6.3. Let us denote with $(v_s^{(p)})_{s=1,\upsilon^{(p)}}$ the solution of this problem, and $\lambda_{n,l}^{(p)}$ and $\sigma^{(p)}$ the associated optimal dual variable values.
   The values of the dual variables $\lambda_{n,l}^{(p)}$ are then sent to each local agent, and will be used to define a new cost function $\mathbf{k}_i^{(p)^\mathrm{T}} \mathbf{w}_i(k : k + N_\mathrm{p} - 1)$, which is calculated using:

$$\mathbf{k}_i^{(p)} = \mathbf{k}_{i,0} + \begin{bmatrix} -\left(\sum\limits_{n=1}^{n_G} \lambda_{n,1}^{(p)} \mathbf{h}_{i_n}\right) \\ \vdots \\ -\left(\sum\limits_{n=1}^{n_G} \lambda_{n,N_\mathrm{p}}^{(p)} \mathbf{h}_{i_n}\right) \\ \hline \mathbf{0} \\ \hline \mathbf{0} \\ \hline \mathbf{0} \\ \hline \mathbf{0} \end{bmatrix} \tag{6.25}$$

   The important point is that the second term of this sum can be calculated by coordinator agent and given numerically to the local agent. This term is a penalization cost vector whose objective is to take into account the global constraints in the local optimization.

---

[2] $\forall n \in \{1,\ldots,n_G\}, \forall l \in \{1,\ldots,N_\mathrm{p}\}$.

2. In parallel, each optimization agent has to solve its own problem 6.2 with the new cost function $\mathbf{k}_i^{(p)^{\mathrm{T}}}\mathbf{w}_i(k : k + N_{\mathrm{p}} - 1)$ to find its solution $\mathbf{w}_i^{(\upsilon^{(p)}+1)}(k : k + N_{\mathrm{p}} - 1)$, which has to be sent to the coordinator agent.
3. The convergence checking is done after these two steps by comparing two values:
   - the value of the dual variable $\sigma^{(p)}$
   - the value of the sum of all the local optimization values, calculated as follows:

$$\psi^{(p)} = \sum_{i=1}^{q} \mathbf{k}_i^{(p)^{\mathrm{T}}} \mathbf{w}_i^{(\upsilon^{(p)}+1)}(k : k + N_{\mathrm{p}} - 1) \tag{6.26}$$

   The value $\psi^{(p)}$ obtained by Eq. (6.26) can be seen as the cost function resulting from a relaxed optimization problem with less constraint than the initial one. Two cases can occur:

3a. If $\psi^{(p)} < \sigma^{(p)}$, a new iteration has to be done.The coordinator agent will use the local solutions $\mathbf{w}_i^{(\upsilon^{(p)}+1)}(k : k + N_{\mathrm{p}} - 1)$ to calculate new variable values $z^{(\upsilon^{(p)}+1)}(k)$ and $r_{n,l}^{(\upsilon^{(p)}+1)}(k)$, according to Eqs. (6.19) and (6.20).

3b. In the other case, (i.e. $\psi^{(p)} \geq \sigma^{(p)}$), the optimal solution of the initial problem has been achieved. More precisely, the local optimal solution can be computed as follows:

$$\mathbf{w}_i^{\star}(k : k + N_{\mathrm{p}} - 1) = \sum_{s=1}^{\upsilon^{(p)}} \nu_s^{(p)} \mathbf{w}_i^{(s)}(k : k + N_{\mathrm{p}} - 1), \tag{6.27}$$

   which is a convex combination of all the previous solutions done for the different cost function.

   As usual in model predictive control, only the first input of the control sequence is brought to the system, which is $\mathbf{u}_i^{\star}(k)$.

6.3.5 Initialization

This point is maybe the main difficulty and drawback of the method. To initialize the algorithm, it is necessary to provide enough different solutions to the master problem such that it can be feasible. From a theoretical point of view, only two different solutions are enough, but only if they are good ones ...they cannot be chosen randomly, but such that the resulting master problem is feasible. From a practical point of view, they can be provided using engineering considerations, but also could be the results of different local optimization done with arbitrary cost functions. But anyway, there exist different techniques such as the one proposed in [5] to develop an always feasible coordinator agent problem.

6.3.6 Overview of the Procedure

From all the previous considerations, the iterative procedure of the distributed MPC based on the Dantzig-Wolfe decomposition can be recapitulated as in Algorithm 6.1. This algorithm will be called at each time step k.

6.3.7 A Remark on Modularity

Among the mechanisms used during the distribution procedure, it is important to notice that each local agent only interacts with the coordinator agent but not with other local agents. Then, adding a new subsystem implies a new local agent which is integrated in the structure, but it only communicates with the coordinator agent without any modification on the other agents. Moreover the complexity of the coordinator optimization problem is not increasing, because it only changes the way the coefficients $z^{(s)}(k)$ and $r_{n,l}^{(v^{(s)})}(k)$ are calculated. All these considerations ensure the modularity of the approach.

6.4 Theoretical Results Availability

The two major advantages of this method are that:

- The global solution will be found
- It requires a finite number of iterations.

The presented methodology is very close to the Dantzig-Wolfe decomposition procedure, which has been firstly proposed in [6]. The conditions of the convergence is linked to the fact that at each step, all the solutions of the optimization problems are finite, and there is no unfeasible problem.

For all the local agent, as from an optimization to the next one, only the cost function changes, it is always possible to solve the problem and the solution is finite. This point is ensured by the Assumption 6.1. For the coordinator agent problem, as soon as the initialization is well tuned, the problem cannot be unfeasible anymore. Because of Assumption 6.2, in which all the communications are supposed to be perfect, the iterative process is then ensured to converge to the optimal solution.

All the mathematical concepts that are used to prove the convergence of the decomposition can be found in the different following references [1, 2, 4], in which many different analyzes have been provided.

Algorithm 6.1 dMPC Algorithm

1: **Initialization:**

Step 1a: Local agent, in parallel:

- Acquisition of $\mathbf{x}_i(k)$, $\mathbf{y}_{\text{ref,i}}(k)$
- Calculation of $\overline{\mathbf{d}}_i(k)$
- Determination of N_0 admissible points $\mathbf{w}_i^{(s)}(k : k + N_\text{p} - 1)$
- Transmission of the data to the coordinator agent.

Step 1b: Coordination agent:

- $p = 1, \upsilon^{(p)} = N_0$.
- Calculation of $z^{(s)}(k)$ and $r_{n,l}^{(\upsilon^{(s)})}(k)$ for $s \in \{1, \ldots, \upsilon^{(p)}\}$

2: **Iteration process: Coordinator Agent Optimization**

- Solving the coordinator agent problem, to get $(v_s^{(p)})_{s=1,\upsilon^{(p)}}$ the solution of this problem, and $\lambda_{n,l}^{(p)}$ and $\sigma^{(p)}$ the associated optimal dual variable values.
- Calculation of the cost penalization vector for each agent, using the $\lambda_{n,l}^{(p)}$ values, and transmission of this vector.

3: **Iteration process: Local Agent Optimizations, in parallel**

- Solving the local optimization problem, to get the solution $\mathbf{w}_i^{(\upsilon^{(p)}+1)}(k : k + N_\text{p} - 1)$.
- Transmission of the solution to the coordinator agent.

4: **Iteration process: Coordinator agent, Convergence checking**

- Calculation of $\psi^{(p)}$.
- Test $\psi^{(p)} < \sigma^{(p)}$

Case 1: YES. Then:

- Calculation of $z^{(\upsilon^{(p)}+1)}(k)$ and $r_{n,l}^{(\upsilon^{(p)}+1)}(k)$
- $p \leftarrow p + 1, \upsilon^{(p)} \leftarrow \upsilon^{(p)} + 1$
- GO BACK to Step 2.

Case 2: NO. Then:

- Send $(v_s^{(p)})_{s=1,\upsilon^{(p)}}$ to all the agent.
- GO TO Step 5.

5: **Local Agent, Execution, in parallel**

- Calculation of $\mathbf{w}_i^\star(k : k + N_\text{p} - 1)$
- Applying the first control input $\mathbf{u}_i^\star(k)$
- END OF Procedure

6.5 Application Results Availability

In this chapter, we propose a distributed structure to control different systems when they are connected by global input constraints. This is the easiest case to use the Dantzig-Wolfe decomposition. This can be extended to a more general class of systems, when there exist global constraints on the outputs or on the states. In this case, the main difficulty is linked to the feasibility of the problem, which is not ensured anymore if the constraints are hard one. Different methods can be used to soften these constraints and then to use the decomposition.

In this chapter, all the systems are dynamically independent. In a more general case, if the systems are connected by the inputs, states or outputs, the methodology can be adapted to take into account these interactions. This can be done using nested iterations and a communication-based algorithm, such the one proposed in [9]. The major drawback of these nested iterations is the fact that the algorithm converges but not to the optimal solution of the problem. Anyway, this method has been used with success on building temperature regulation, where it has been shown that the suboptimality is not really significant.

In a case where all the systems are connected by common input variables, the problem can also be distributed using the dual of the Dantzig-Wolfe decomposition, the Benders' decomposition. This has been used to regulation temperature in building in the multi-source case [10].

If the first applications of the method dealt with temperature regulation in buildings [8], it also has been extended to energy management system in residential house to develop an adjustable structure to optimize the energy bill [7].

6.6 Conclusions

In this chapter, a distributed structure has been proposed to control a collection of linear dynamical systems coupled by linear global constraints. The particular structure of the constraints set allows the use of the Dantzig-Wolfe decomposition to define the control architecture. The two major advantages of this method are first the fact that the solution found during the iteration process is the global solution. The second one is link to the modularity induced by the control structure which allow adding or removing subsystems in the structure without any modification for the other local controllers. The methodology is proposed for a collection of independent subsystems, a current direction for future works is how to generalize this method to interconnected subsystems.

References

1. M.S. Bazaraa, J.J. Jarvis, H.D. Sherali. Linear Programming and Network Flows, (Wiley, New York, 1990)
2. S.P. Bradley, A.C. Hax, T.L. Magnanti, *Applied Mathematical Programming* (Addison-Wesley Publishing Company, Reading, 1997)
3. E.F. Camacho, C. Bordons, *Model Predictive Control in the Process Industry*, 2nd edn. (Springer-Verlag, London, England, 2004)
4. V. Chvátal, *Linear Programming* (Freeman and Company, New York, 1983)
5. A.J. Conejo, E. Castillo, R. Mínguez, R. García-Bertrand. *Decomposition Techniques in Mathematical Programming*. (Springer, London, 2006)
6. G.B. Dantzig. *Linear Programming and Extensions*. (Princeton University Press, NJ, 1963)
7. A. Lefort, R. Bourdais, G. Ansannay-Alex, H. Guéguen. Planification de la consommation énergétique d'un bâtiment par une méthode d'optimisation linéaire distribuée, *In Proceedings of CIFA 2012*, (2012)
8. P-D. Moroşan, R. Bourdais, D. Dumur, J. Buisson. Distributed model predictive control for building temperature regulation. In *American Control Conference*, pp. 3174–3179, Jun. 2010
9. P.-D. Moroşan, R. Bourdais, D. Dumur, J. Buisson. In Distributed MPC for multi-zone temperature regulation with coupled constraints, *IFAC World Congress*, Aug, 2011
10. P.-D. Moroşan, R. Bourdais, D. Dumur, J. Buisson, A distributed MPC strategy based on Benders' decomposition applied to multi-source multi-zone temperature regulation. J. Process Control **21**(5), 729–737 (2011)

Chapter 7
Distributed MPC Via Dual Decomposition and Alternative Direction Method of Multipliers

F. Farokhi, I. Shames and K. H. Johansson

Abstract A conventional way to handle model predictive control (MPC) problems distributedly is to solve them via dual decomposition and gradient ascent. However, at each time-step, it might not be feasible to wait for the dual algorithm to converge. As a result, the algorithm might be needed to be terminated prematurely. One is then interested to see if the solution at the point of termination is close to the optimal solution and when one should terminate the algorithm if a certain distance to optimality is to be guaranteed. In this chapter, we look at this problem for distributed systems under general dynamical and performance couplings, then, we make a statement on validity of similar results where the problem is solved using alternative direction method of multipliers.

7.1 Introduction

Model predictive control (MPC) can be used to control dynamical systems with input and output constraints while ensuring the optimality of the performance of the system with respect to cost functions [5, 15, 19]. Typically, the way that the

---

This work was supported in part by the Swedish Research Council, the Swedish Foundation for Strategic Research, and the Knut and Alice Wallenberg Foundation.

F. Farokhi (✉) · K. H. Johansson
School of Electrical Engineering, ACCESS Linnaeus Centre, KTH Royal Institute of Technology, Stockholm, Sweden
e-mail: farokhi@ee.kth.se

K. H. Johansson
e-mail: kallej@ee.kth.se

I. Shames
Department of Electrical and Electronic Engineering, University of Melbourne, Melbourne, Australia
e-mail: iman.shames@unimelb.edu.au

J. M. Maestre and R. R. Negenborn (eds.), *Distributed Model Predictive Control Made Easy*, Intelligent Systems, Control and Automation: Science and Engineering 69, DOI: 10.1007/978-94-007-7006-5_7,

control input is calculated at each time-step is via applying the first control in a sequence obtained from solving an optimal control problem over a finite or infinite horizon. The optimal problem is reformulated at each time step based on the available measurements at that time step. Traditionally, a full model of the system is required to solve the MPC problem and all the control inputs are calculated centrally. However, in large-scale interconnected systems, such as power systems [12, 18], water distribution systems [18], transport systems [17], manufacturing systems [8], biological systems [10], and irrigation systems [11], the assumption on knowing the whole model and calculating all the inputs centrally is often not realistic. Recently, much attention has been paid to solve MPC problems in a distributed way [6, 9, 16, 22, 23, 25]. The problem of distributed model predictive control using dual decomposition was considered in [25]. However, in solving any optimization problem when using dual decomposition methods the convergence behaviors of dual iterations does not necessarily coincides to that of the primal formulation. Hence, the authors in [9] presented a distributed MPC algorithm using dual decomposition accompanied with a stopping criterion to guarantee a pre-specified level of performance. The authors only addressed linear coupled dynamics with separable cost functions.

In this chapter, specifically, we formulate the problem of achieving a control objective cooperatively by a network of dynamically coupled systems under constraints using MPC. We consider discrete-time nonlinear control systems. We are interested in casting the problem in a distributed way and we consider the case where the cost function associated with each system is not necessarily decoupled from the rest. Additionally, we are not limiting our formulation to the case where the coupling in the cost function is the same as the coupling in the dynamics [9, 25]. We note that a natural method to solve such problems is to use dual-decomposition at each time-step and solve the problem iteratively. However, a problem that in implementing the dual solution iterations is that generally one cannot make any statement on how close the solution is to the optimum if the dual algorithm is terminated prematurely. That is, there is no termination guideline to ensure that the variables obtained from the dual algorithm are within an acceptable bound for the primal problem. In this chapter, we propose such termination guidelines that indicate how many iterations are needed to ensure a certain suboptimality guarantee, i.e., distance to optimality. We extend the results of [9] and present a more general frameworks, i.e., nonlinear interconnected dynamics and cost functions. A way to achieve better numerical properties for solving distributed MPC is to apply *alternating direction method of multipliers* (ADMM) [3]. ADMM is a powerful algorithm for solving structured convex optimization problems. Combining the strong convergence properties of the method of multipliers and the decomposability property of dual ascent, the method is particularly applicable to large-scale decision problems. In particular, recently, optimal control synthesis and MPC via ADMM has gained some attention [14, 24]. However, to the best of our knowledge, no attention has been paid to distributed MPC using ADMM. Hence, we show how to address distributed MPC via ADMM.

The origins of ADMM can be traced back to the alternating direction implicit (ADI) techniques for solving elliptic and parabolic partial difference equations. In the 1970s, see [3] and references therein, ADMM was first introduced for solving optimization problems and enjoyed much attention in the following years. However, the main advantage of applying ADMM in solving optimization problems, its ability to deal with very large problem through its superior stability properties and its decomposability, remained largely untapped due to the lack of ubiquity of very large scale problems. Nevertheless, the technique has again raised to prominence in the last few years as there are many applications, e.g. in financial or biological data analysis, that are too large to be handled by generic optimization solvers.

The outline of this chapter is as follows. In Sect. 7.2 we formally define the problem of interest in this chapter and particularly present the plant model and the performance criteria we consider. In Sect. 7.3 the suboptimality guarantee for the dually decomposed MPC is presented. (For the sake of brevity the proofs of the theorems are omitted and can be found in [7].) Additionally, we make some comments on finding similar guarantee when the problem is solved via ADMM. We illustrate the applicability of our results on a formation of nonholonomic agents which employ distributed MPC to acquire a desired formation. Section 7.4. Finally, some concluding remarks are presented in Sect. 7.5.

7.1.1 Notation

The sets of real and integer numbers are denoted by $\mathbb{R}$ and $\mathbb{Z}$, respectively. For any $n_1, n_2 \in \mathbb{Z} \cup \{\pm\infty\}$, we define $\mathbb{Z}_{\geq n_1}^{\leq n_2} = \{n \in \mathbb{Z} \mid n_1 \leq n \leq n_2\}$. When $n_2 = +\infty$, we use $\mathbb{Z}_{\geq n_1}$. For any $x \in \mathbb{R}$, we also define $\mathbb{R}_{\geq x} = \{y \in \mathbb{R} \mid y \geq x\}$. Other sets are denoted by calligraphic letters, such as $\mathcal{X}$ and $\mathcal{E}$. Each (directed) graph is a pair of sets as $\mathcal{G} = (\mathcal{V}, \mathcal{E})$, where $\mathcal{V}$ is the vertex set and $\mathcal{E}$ is the edge set. Each edge in the edge set $\mathcal{E}$ is an ordered pair of vertices, e.g., $(v_1, v_2) \in \mathcal{E}$.

7.2 Problem Formulation

7.2.1 Plant Model

Let a directed graph $\mathcal{G}^P = (\{1, \ldots, N\}, \mathcal{E}^P)$ be given. Consider a discrete-time nonlinear control system composed of N subsystems, where, for each $1 \leq i \leq N$, subsystem i can be described in state-space form as

$$\mathbf{x}_i[k+1] = \mathbf{f}_i(\mathbf{x}_i[k], \mathbf{v}_i[k]; \mathbf{u}_i[k]), \tag{7.1}$$

with state vector $\mathbf{x}_i[k] \in \mathcal{X}_i \subseteq \mathbb{R}^{n_i}$ and control input $\mathbf{u}_i[k] \in \mathcal{U}_i \subseteq \mathbb{R}^{m_i}$ for given integers $n_i, m_i \geq 1$. In addition, let $\mathbf{v}_i[k] = (\mathbf{x}_j[k])_{(j,i)\in\mathcal{E}^P} \in \mathbb{R}^{\sum_{(j,i)\in\mathcal{E}^P} n_j}$ denote the tuple of the state vector of all the subsystems that can influence subsystem i through its dynamics. For each $1 \leq i \leq N$, mapping $\mathbf{f}_i : \mathcal{X}_i \times \prod_{(j,i)\in\mathcal{E}^P} \mathcal{X}_j \times \mathcal{U}_i \to \mathcal{X}_i$ determine the trajectory of subsystem i given the initial condition $x_i[0] \in \mathcal{X}_i$ and the inputs.

7.2.2 Performance Criterion

Let a directed graph $\mathcal{G}^C = (\{1, \ldots, N\}, \mathcal{E}^C)$ be given. For each time-instance $k \in \mathbb{Z}_{\geq 0}$, we introduce the running cost function

$$J_k\big((\mathbf{x}_i[k])_{i=1}^N; (\mathbf{u}_i[k:+\infty])_{i=1}^N\big) = \sum_{t=k}^{\infty}\sum_{i=1}^{N} \boldsymbol{\ell}_i(\mathbf{x}_i[t], \mathbf{w}_i[t]; \mathbf{u}_i[t]),$$

where $\mathbf{w}_i[k] = (\mathbf{x}_j[k])_{(j,i)\in\mathcal{E}^C} \in \mathbb{R}^{\sum_{(j,i)\in\mathcal{E}^C} n_j}$ denotes the tuple of the state vector of all the subsystems that can influence subsystem i through its cost. Note that for the described dynamical system, given the control sequence $(\mathbf{u}_i[k:+\infty])_{i=1}^N$ and boundary condition $(\mathbf{x}_i[k])_{i=1}^N$, the trajectory of the system $(\mathbf{x}_i[k:+\infty])_{i=1}^N$ is uniquely determined by the described system dynamics in (7.1). Hence, we do not explicitly show the dependency of the cost function $J_k((\mathbf{x}_i[k])_{i=1}^N; (\mathbf{u}_i[k:+\infty])_{i=1}^N)$ to the trajectory $(\mathbf{x}_i[k+1:+\infty])_{i=1}^N$. We make the following standing assumption concerning the cost function which is crucial for proving stability of the origin for the closed-loop system with a MPC controller in feedback interconnection.

Assumption 7.1 *For each* $1 \leq i \leq N$, $\boldsymbol{\ell}_i : \mathcal{X}_i \times \prod_{(j,i)\in\mathcal{E}^C} \mathcal{X}_j \times \mathcal{U}_i \to \mathbb{R}_{\geq 0}$ *is a mapping such that (a)* $\boldsymbol{\ell}_i(\mathbf{x}_i, \mathbf{w}_i; \mathbf{u}_i)$ *is continuous in* $\mathbf{x}_i$ *for all* $\mathbf{x}_i \in \mathcal{X}_i$ *and (b)* $\boldsymbol{\ell}_i(\mathbf{x}_i, \mathbf{w}_i; \mathbf{u}_i) = 0$ *if and only if* $\mathbf{x}_i = 0$.

7.2.3 MPC

In each time instance $k \in \mathbb{Z}_{\geq 0}$, the objective of the designer is to solve an infinite-horizon optimal control problem given by

$$\begin{aligned}
(\hat{\mathbf{u}}_i^*[k:+\infty])_{i=1}^N &= \operatorname{argmin}_{(\hat{\mathbf{u}}_i[k:+\infty])_{i=1}^N} J_k\big((\mathbf{x}_i[k])_{i=1}^N; (\hat{\mathbf{u}}_i[k:+\infty])_{i=1}^N\big),\\
\text{subject to}\quad & \hat{\mathbf{x}}_i[t+1] = \mathbf{f}_i(\hat{\mathbf{x}}_i[t], \hat{\mathbf{v}}_i[t]; \hat{\mathbf{u}}_i[t]),\ 1 \leq i \leq N, \forall\, t \in \mathbb{Z}_{\geq k},\\
& \hat{\mathbf{x}}_i[k] = \mathbf{x}_i[k],\ \ 1 \leq i \leq N,\\
& \hat{\mathbf{x}}_i[t] \in \mathcal{X}_i, \hat{\mathbf{u}}_i[t] \in \mathcal{U}_i,\ \ 1 \leq i \leq N, \forall\, t \in \mathbb{Z}_{\geq k},
\end{aligned} \tag{7.2}$$

where $(\hat{\mathbf{x}}_i[k:+\infty])_{i=1}^N$ is the state estimate initialized with the state measurement $\hat{\mathbf{x}}_i[k] = \mathbf{x}_i[k]$, for all $1 \leq i \leq N$. Note that we use $\hat{\mathbf{x}}_i$ and $\hat{\mathbf{u}}_i$ to emphasize the fact that these variables are forecast variables and are predicted using the systems model. We relax the infinite-horizon optimal control problem in (7.2) into a finite-horizon optimal control problem given by

$$(\hat{\mathbf{u}}_i^*[k:k+T])_{i=1}^N = \arg\min_{(\hat{\mathbf{u}}_i[k:k+T])_{i=1}^N} J_k^{(T)}\big((\mathbf{x}_i[k])_{i=1}^N; (\hat{\mathbf{u}}_i[k:k+T])_{i=1}^N\big),$$
$$\text{subject to} \quad \hat{\mathbf{x}}_i[t+1] = \mathbf{f}_i(\hat{\mathbf{x}}_i[t], \hat{\mathbf{v}}_i[t]; \hat{\mathbf{u}}_i[t]),\ 1 \le i \le N, \forall\, t \in \mathbb{Z}_{\ge k}^{\le k+T},$$
$$\hat{\mathbf{x}}_i[k] = \mathbf{x}_i[k],\ 1 \le i \le N,$$
$$\hat{\mathbf{x}}_i[t] \in \mathcal{X}_i, \hat{\mathbf{u}}_i[t] \in \mathcal{U}_i,\ 1 \le i \le N, \forall\, t \in \mathbb{Z}_{\ge k}^{\le k+T}, \tag{7.3}$$

where $J_k^{(T)}((\mathbf{x}_i[k])_{i=1}^N; (\hat{\mathbf{u}}_i[k:k+T])_{i=1}^N) = \sum_{t=k}^{k+T}\sum_{i=1}^N \boldsymbol{\ell}_i(\hat{\mathbf{x}}_i[t], \hat{\mathbf{w}}_i[t]; \hat{\mathbf{u}}_i[t])$, and $T \in \mathbb{Z}_{\ge 0}$ denotes the horizon of estimation and control. After solving this optimization problem, subcontroller i implements $\mathbf{u}_i[k] = \hat{\mathbf{u}}_i^*[k]$, for each $1 \le i \le N$. Doing so, the overall cost of the system equals

$$J_0\big((\mathbf{x}_i[0])_{i=1}^N; (\mathbf{u}_i[0:+\infty])_{i=1}^N\big) = \sum_{t=0}^{\infty}\sum_{i=1}^{N} \boldsymbol{\ell}_i(\mathbf{x}_i[t], \mathbf{w}_i[t]; \mathbf{u}_i[t]),$$

where the control sequence $(\mathbf{u}_i[0:+\infty])_{i=1}^N$, as described earlier, is extracted step-by-step from the optimization problem in (7.3). For the MPC problem to be well-posed, we make the following standing assumption:

Assumption 7.2 *The optimization problem*

$$\arg\min_{(\hat{\mathbf{u}}_i[k:+\infty])_{i=1}^N} \sum_{t=k}^{k+T}\sum_{i=1}^{N} \boldsymbol{\ell}_i(\hat{\mathbf{x}}_i[t], \hat{\mathbf{w}}_i[t]; \hat{\mathbf{u}}_i[t]),$$
$$\text{subject to} \quad \hat{\mathbf{x}}_i[t+1] = \mathbf{f}_i(\hat{\mathbf{x}}_i[t], \hat{\mathbf{v}}_i[t]; \hat{\mathbf{u}}_i[t]),\ 1 \le i \le N, \forall\, t \in \mathbb{Z}_{\ge k}^{\le k+T},$$
$$\hat{\mathbf{x}}_i[k] = \mathbf{x}_i[k],\ 1 \le i \le N,$$
$$\hat{\mathbf{x}}_i[t] \in \mathcal{X}_i, \hat{\mathbf{u}}_i[t] \in \mathcal{U}_i,\ 1 \le i \le N, \forall\, t \in \mathbb{Z}_{\ge k}^{\le k+T},$$

admits a unique global minimizer for all time horizon $T \in \mathbb{Z}_{\ge 0} \cup \{\infty\}$.

Assumption 7.2 is evidently satisfied if, for each $1 \le i \le N$, (a) mapping $\boldsymbol{\ell}_i : \mathcal{X}_i \times \prod_{(j,i)\in\mathcal{E}^C} \mathcal{X}_j \times \mathcal{U}_i \to \mathbb{R}_{\ge 0}$ is quadratic, and (b) mapping $\mathbf{f}_i : \mathcal{X}_i \times \prod_{(j,i)\in\mathcal{E}^P} \mathcal{X}_j \times \mathcal{U}_i \to \mathcal{X}_i$ is linear [1]. We can also consider strictly convex mappings $\boldsymbol{\ell}_i : \mathcal{X}_i \times \prod_{(j,i)\in\mathcal{E}^C} \mathcal{X}_j \times \mathcal{U}_i \to \mathbb{R}_{\ge 0}$ when working with finite-horizon cases [4].

7.3 Main Results

Formulating a constrained optimization problem as a dual problem, in some cases, enables us to solve it in a decentralized manner across a network of agents. Typically, each iteration for solving the dual problem involves broadcasting and receiving variables for each agent. The variables that need to be communicated between the agents are the variables appearing in the cost function and the variables (Lagrange multipliers) used to enforce the constraints. In the rest of this section, we first cast the MPC problem in a dual decomposition framework and then introduce our result on guaranteeing the performance of the iterations to solve the decomposed problem distributedly.

7.3.1 Dual Decomposition

Let us, for each $1 \leq i \leq N$, introduce slack variables $\bar{\mathbf{v}}_i[k] \in \mathbb{R}^{\sum_{(j,i)\in\mathcal{E}^P} n_j}$ and $\bar{\mathbf{w}}_i[k] \in \mathbb{R}^{\sum_{(j,i)\in\mathcal{E}^C} n_j}$. Doing so, we can rewrite the finite-horizon problem in (7.3) as

$$
\begin{aligned}
(\hat{\mathbf{u}}_i^*[k:k+T])_{i=1}^N = & \underset{(\hat{\mathbf{u}}_i[k:k+T])_{i=1}^N}{\arg\min} \sum_{t=k}^{k+T}\sum_{i=1}^{N} \boldsymbol{\ell}_i(\hat{\mathbf{x}}_i[t], \bar{\mathbf{w}}_i[t]; \hat{\mathbf{u}}_i[t]),\\
\text{subject to } & \hat{\mathbf{x}}_i[t+1] = \mathbf{f}_i(\hat{\mathbf{x}}_i[t], \bar{\mathbf{v}}_i[t]; \hat{\mathbf{u}}_i[t]),\ 1 \leq i \leq N, \forall\, t \in \mathbb{Z}_{\geq k}^{\leq k+T},\\
& \hat{\mathbf{x}}_i[k] = \mathbf{x}_i[k],\ \ 1 \leq i \leq N,\\
& \hat{\mathbf{x}}_i[t] \in \mathcal{X}_i, \hat{\mathbf{u}}_i[t] \in \mathcal{U}_i,\ \ 1 \leq i \leq N, \forall\, t \in \mathbb{Z}_{\geq k}^{\leq k+T},\\
& \bar{\mathbf{w}}_i[t] = \hat{\mathbf{w}}_i[t], \bar{\mathbf{v}}_i[t] = \hat{\mathbf{v}}_i[t],\ 1 \leq i \leq N, \forall\, t \in \mathbb{Z}_{\geq k}^{\leq k+T},
\end{aligned}
$$

We can incorporate the set of constraints $\bar{\mathbf{v}}_i[t] = \hat{\mathbf{v}}_i[t]$ and $\bar{\mathbf{w}}_i[t] = \hat{\mathbf{w}}_i[t]$ into the cost function as

$$
\begin{aligned}
\max_{(\boldsymbol{\lambda}_i,\boldsymbol{\mu}_i)_{i=1}^N}\ \min_{(\hat{\mathbf{u}}_i,\bar{\mathbf{v}}_i,\bar{\mathbf{w}}_i)_{i=1}^N} \sum_{t=k}^{k+T}\sum_{i=1}^{N} \Big[& \boldsymbol{\ell}_i(\hat{\mathbf{x}}_i[t], \bar{\mathbf{w}}_i[t]; \hat{\mathbf{u}}_i[t])\\
& + \boldsymbol{\lambda}_i[t]^\top(\bar{\mathbf{v}}_i[t] - \hat{\mathbf{v}}_i[t]) + \boldsymbol{\mu}_i[t]^\top(\bar{\mathbf{w}}_i[t] - \hat{\mathbf{w}}_i[t])\Big],
\end{aligned} \tag{7.4}
$$

where, for each $1 \leq i \leq N$, variables $(\boldsymbol{\lambda}_i[k:k+T], \boldsymbol{\mu}_i[k:k+T])_{i=1}^N$ denote the Lagrange multipliers $\boldsymbol{\lambda}_i[t] = (\boldsymbol{\lambda}_{i,j}[t])_{(j,i)\in\mathcal{E}^P} \in \mathbb{R}^{\sum_{(j,i)\in\mathcal{E}^P} n_j}$ and $\boldsymbol{\mu}_i[t] = (\boldsymbol{\mu}_{i,j}[t])_{(j,i)\in\mathcal{E}^C} \in \mathbb{R}^{\sum_{(j,i)\in\mathcal{E}^C} n_j}$, for all $k \leq t \leq k+T$. Note that in (7.4) we dropped the time index of the variables in the subscripts of the minimization and maximization operators to simplify the presentation.

We can rearrange the cost function in (7.4) as

Algorithm 7.1 Distributed algorithm for solving MPC problem (7.3)

Input: $\mathbf{x}_i[k]$, $1 \le i \le N$, and as parameters Iteration numbers $\{S_k\}_{k=0}^{\infty}$ and gradient ascent step sizes $\{h_i^{(s)}, g_i^{(s)}\}_{i,s=0}^{\infty}$
Output: $\mathbf{u}_i[k]$, $1 \le i \le N$

for $k = 1, 2, \ldots$ **do**
  Initialize Lagrange multipliers $(\boldsymbol{\lambda}_i^{(0)}[k : k+T], \boldsymbol{\mu}_i^{(0)}[k : k+T])_{i=1}^{N}$.
  for $s = 1, 2, \ldots, S_k$ **do**
    for $i = 1, 2, \ldots, N$ **do**
      Solve the optimization problem

$$
\begin{aligned}
&(\hat{\mathbf{u}}_i^{(s)}[k:k+T], \bar{\mathbf{v}}_i^{(s)}[k:k+T], \bar{\mathbf{w}}_i^{(s)}[k:k+T]) \\
&\quad = \arg\min_{\hat{\mathbf{u}}_i, \bar{\mathbf{v}}_i, \bar{\mathbf{w}}_i} \quad L_i(\hat{\mathbf{u}}_i[k:k+T], \bar{\mathbf{v}}_i[k:k+T], \bar{\mathbf{w}}_i[k:k+T]) \\
&\qquad \text{subject to} \quad \hat{\mathbf{x}}_i[t+1] = \mathbf{f}_i(\hat{\mathbf{x}}_i[t], \bar{\mathbf{v}}_i[t]; \hat{\mathbf{u}}_i[t]), \forall\, t \in \mathbb{Z}_{\ge k}^{\le k+T}, \\
&\qquad\qquad \hat{\mathbf{x}}_i[k] = \mathbf{x}_i[k], \\
&\qquad\qquad \hat{\mathbf{x}}_i[t] \in \mathcal{X}_i, \hat{\mathbf{u}}_i[t] \in \mathcal{U}_i, \ \forall\, t \in \mathbb{Z}_{\ge k}^{\le k+T}, \\
&\qquad\qquad \bar{\mathbf{v}}_i[t] \in \prod_{(j,i)\in\mathcal{E}^P} \mathcal{X}_j, \bar{\mathbf{w}}_i[t] \in \prod_{(j,i)\in\mathcal{E}^C} \mathcal{X}_j, \ \forall\, t \in \mathbb{Z}_{\ge k}^{\le k+T},
\end{aligned}
$$

      where

$$
\begin{aligned}
L_i(\hat{\mathbf{u}}_i[k:k+T], \bar{\mathbf{v}}_i[k:k+T], \bar{\mathbf{w}}_i[k:k+T]) = \sum_{t=k}^{k+T} \Big[\ell_i(\hat{\mathbf{x}}_i[t], \bar{\mathbf{w}}_i[t]; \hat{\mathbf{u}}_i[t]) + \boldsymbol{\lambda}_i^{(s)}[t]^\top \bar{\mathbf{v}}_i[t] \\
+ \boldsymbol{\mu}_i^{(s)}[t]^\top \bar{\mathbf{w}}_i[t] - \sum_{(i,j)\in\mathcal{E}^P} \boldsymbol{\lambda}_{j,i}^{(s)}[t]^\top \hat{\mathbf{x}}_i[t] - \sum_{(i,j)\in\mathcal{E}^C} \boldsymbol{\mu}_{j,i}^{(s)}[t]^\top \hat{\mathbf{x}}_i[t] \Big].
\end{aligned}
$$

    end for
    $\boldsymbol{\lambda}_i^{(s+1)}[t] = \boldsymbol{\lambda}_i^{(s)}[t] + h_i^{(s)}(\bar{\mathbf{v}}_i^{(s)}[t] - \hat{\mathbf{v}}_i^{(s)}[t])$, $1 \le i \le N, \forall\, t \in \mathbb{Z}_{\ge k}^{\le k+T}$.
    $\boldsymbol{\mu}_i^{(s+1)}[t] = \boldsymbol{\mu}_i^{(s)}[t] + g_i^{(s)}(\bar{\mathbf{w}}_i^{(s)}[t] - \hat{\mathbf{w}}_i^{(s)}[t])$, $1 \le i \le N, \forall\, t \in \mathbb{Z}_{\ge k}^{\le k+T}$.
  end for
  $\mathbf{u}_i[k] = \hat{\mathbf{u}}_i^{(S_k)}[k]$, $1 \le i \le N$.
end for

$$
\begin{aligned}
\max_{(\boldsymbol{\lambda}_i, \boldsymbol{\mu}_i)_{i=1}^N} \sum_{i=1}^{N} \min_{\hat{\mathbf{u}}_i, \bar{\mathbf{v}}_i, \bar{\mathbf{w}}_i} \sum_{t=k}^{k+T} \Big[& \ell_i(\hat{\mathbf{x}}_i[t], \bar{\mathbf{w}}_i[t]; \hat{\mathbf{u}}_i[t]) + \boldsymbol{\lambda}_i[t]^\top \bar{\mathbf{v}}_i[t] + \boldsymbol{\mu}_i[t]^\top \bar{\mathbf{w}}_i[t] \\
& - \sum_{(i,j)\in\mathcal{E}^P} \boldsymbol{\lambda}_{j,i}[t]^\top \hat{\mathbf{x}}_i[t] - \sum_{(i,j)\in\mathcal{E}^C} \boldsymbol{\mu}_{j,i}[t]^\top \hat{\mathbf{x}}_i[t] \Big].
\end{aligned}
\tag{7.5}
$$

Using (7.5), we can separate subsystem cost functions, which allows us to develop a distributed scheme for solving the finite-horizon MPC problem in (7.3). This distributed scheme is presented in Procedure 7.1. We note that the communication graph, $\mathcal{G}$, considered here is the union of the plant graph $\mathcal{G}^P$ and the cost graph $\mathcal{G}^C$, viz. $\mathcal{G} = (\{1, \ldots, N\}, \mathcal{E}^C \cup \mathcal{E}^P)$.

7.3.2 From Infinite to Finite Horizon

Let us introduce the notations $V_k((\mathbf{x}_i[k])_{i=1}^N) = \min_{(\hat{\mathbf{u}}_i[k:+\infty])_{i=1}^N} J_k\big((\mathbf{x}_i[k])_{i=1}^N; (\hat{\mathbf{u}}_i[k : +\infty])_{i=1}^N\big)$, and $V_k^{(T)}((\mathbf{x}_i[k])_{i=1}^N) = \min_{(\hat{\mathbf{u}}_i[k:k+T])_{i=1}^N} J_k^{(T)}\big((\mathbf{x}_i[k])_{i=1}^N; (\hat{\mathbf{u}}_i[k : k + T])_{i=1}^N\big)$, subject to the constraints introduced the infinite-horizon optimal control problem in (7.2) and the finite-horizon optimal control problem in (7.3), respectively.

Theorem 7.1 *Assume that there exist an a priori given constant* $\alpha \in [0, 1]$ *and controllers* $\boldsymbol{\phi}_i : \prod_{i=1}^N \mathcal{X}_i \to \mathcal{U}_i$, *such that, for all* $(\mathbf{x}_i[k])_{i=1}^N \in \prod_{i=1}^N \mathcal{X}_i$, *we have*

$$V_k^{(T)}((\mathbf{x}_i[k])_{i=1}^N) \geq V_{k+1}^{(T)}((\mathbf{x}_i[k+1])_{i=1}^N) + \alpha \sum_{i=1}^N \boldsymbol{\ell}_i(\mathbf{x}_i[k], \mathbf{w}_i[k]; \boldsymbol{\phi}_i((\mathbf{x}_i[k])_{i=1}^N)), \tag{7.6}$$

where $\mathbf{x}_i[k+1] = \mathbf{f}_i(\mathbf{x}_i[k], \mathbf{v}_i[k]; \boldsymbol{\phi}_i((\mathbf{x}_i[k])_{i=1}^N))$, *for* $1 \leq i \leq N$. *Then,* $\alpha V_k((\mathbf{x}_i[k])_{i=1}^N) \leq V_k^{(T)}((\mathbf{x}_i[k])_{i=1}^N)$ *for all* $(\mathbf{x}_i[k])_{i=1}^N \in \prod_{i=1}^N \mathcal{X}_i$.

This theorem illustrates that by solving the finite-horizon optimal control problem in (7.3), we get a sub-optimal solution, which is in a vicinity of the solution of the infinite-horizon optimal control problem in (7.2) if α is chosen close to one. Hence, in the chapter, we assume that the horizon T is chosen such that it satisfies (7.6). In that way, we do not lose much by abandoning the infinite-horizon optimal control problem for the finite-horizon one.

7.3.3 Convergence

Generically, in solving any optimization problem, if one resorts to use dual decomposition methods the convergence behaviors of dual iterations does not necessarily coincides to that of the primal formulation. In other words, if one terminates the dual iterations after S_k steps and obtains the decision variables, one cannot make a statement on how close is the primal cost function evaluated at the obtained variable to its optimal value, i.e., the optimality gap cannot be determined. However, for model predictive control one can find a bound on such a distance. We aim to propose a way to calculate the optimality gap for general distributed MPC problems based on the results proposed by [9]. Let us introduce the notation

$$V_k^{(T),(s)}((\mathbf{x}_i[k])_{i=1}^N) = \sum_{t=k}^{k+T} \sum_{i=1}^N \Big[\boldsymbol{\ell}_i(\hat{\mathbf{x}}_i[t], \bar{\mathbf{w}}_i^{(s)}[t]; \hat{\mathbf{u}}_i^{(s)}[t]) + \boldsymbol{\lambda}_i^{(s)}[t]^\top (\bar{\mathbf{v}}_i^{(s)}[t] - \hat{\mathbf{v}}_i[t]) + \boldsymbol{\mu}_i^{(s)}[t]^\top (\bar{\mathbf{w}}_i^{(s)}[t] - \hat{\mathbf{w}}_i[t]) \Big],$$

where $(\hat{\mathbf{u}}_i^{(s)}[k : k+T], \bar{\mathbf{v}}_i^{(s)}[k : k+T], \bar{\mathbf{w}}_i^{(s)}[k : k+T])_{i=1}^N$ is extracted from Procedure 7.1.

Theorem 7.2 *Let $\{\tilde{V}_k\}_{k=0}^{\infty}$ be a given family of mappings, such that, for each $k \in \mathbb{Z}_{\geq 0}$, $\tilde{V}_k : \prod_{i=1}^N \mathcal{X}_i \to \mathbb{R}_{\geq 0}$ satisfies*

$$\tilde{V}_k((\mathbf{x}_i[k])_{i=1}^N) \geq V_k^{(T)}((\mathbf{x}_i[k])_{i=1}^N), \tag{7.7}$$

for all $(\mathbf{x}_i[k])_{i=1}^N \in \prod_{i=1}^N \mathcal{X}_i$. In addition, let iteration number S_k in Procedure 7.1, in each time-step $k \in \mathbb{Z}_{\geq 0}$, be given such that

$$V_k^{(T),(S_k)}((\mathbf{x}_i[k])_{i=1}^N) - \tilde{V}_{k+1}((\mathbf{x}_i[k+1])_{i=1}^N) \geq e[k] + \alpha \sum_{i=1}^N \boldsymbol{\ell}_i(\mathbf{x}_i[k], \mathbf{w}_i[k]; \hat{\mathbf{u}}_i^{(S_k)}[k]) \tag{7.8}$$

for a given constant $\alpha \in [0, 1]$, where $\mathbf{x}_i[k+1] = \mathbf{f}_i(\mathbf{x}_i[k], \mathbf{v}_i[k]; \hat{\mathbf{u}}_i^{(S_k)}[k])$, for each $1 \leq i \leq N$. The sequence $\{e[k]\}_{k=0}^{\infty}$ is described by the difference equation

$$e[k] = e[k-1] + \alpha \sum_{i=1}^N \boldsymbol{\ell}_i(\mathbf{x}_i[k-1], \mathbf{w}_i[k-1]; \mathbf{u}_i[k-1]) + \tilde{V}_k((\mathbf{x}_i[k])_{i=1}^N) - \tilde{V}_{k-1}((\mathbf{x}_i[k-1])_{i=1}^N),$$

for all $k \in \mathbb{Z}_{\geq 2}$ and

$$e[1] = \alpha \sum_{i=1}^N \boldsymbol{\ell}_i(\mathbf{x}_i[0], \mathbf{w}_i[0]; \mathbf{u}_i[0]) + \tilde{V}_1((\mathbf{x}_i[1])_{i=1}^N) - V_0^{(T),(S_0)}((\mathbf{x}_i[0])_{i=1}^N),$$

and $e[0] = 0$. Then, $\alpha J_0((\mathbf{x}_i[0])_{i=1}^N; (\mathbf{u}_i[0 : +\infty])_{i=1}^N) \leq V_0((\mathbf{x}_i[0])_{i=1}^N)$ for any initial condition $(\mathbf{x}_i[0])_{i=1}^N \in \prod_{i=1}^N \mathcal{X}_i$. In addition, if $V_0((\mathbf{x}_i[0])_{i=1}^N) < \infty$ for any initial condition $(\mathbf{x}_i[0])_{i=1}^N \in \prod_{i=1}^N \mathcal{X}_i$, then $\lim_{k\to\infty} \mathbf{x}_i[k] = 0$ for $1 \leq i \leq N$.

Theorem 7.2 shows that, provided that $\{S_k\}_{k=0}^{\infty}$ guarantees (7.8), the cost of the sub-optimal controller extracted from Procedure 7.1 is in a close vicinity of the global optimal controller, i.e., the cost of the sub-optimal controller is never worse than $1/\alpha$ times the cost of the global optimal controller. In addition, the closed-loop system is stable.

Now, we only need to present a mapping $\tilde{V}_k(\cdot)$ that satisfies condition (7.7). We use the method presented in [9] for generating a reasonable upper bound. Let us introduce the one-step forward shift operator $\mathbf{q}_T : (\prod_{i=1}^N \mathcal{U}_i)^{T+1} \to (\prod_{i=1}^N \mathcal{U}_i)^{T+1}$, so that for any control sequence $(\mathbf{u}_i[0 : T])_{i=1}^N \in (\prod_{i=1}^N \mathcal{U}_i)^{T+1}$ we have $\mathbf{q}_T((\mathbf{u}_i[0 : T])_{i=1}^N) = (\mathbf{u}'_i[0 : T])_{i=1}^N$, where for all $1 \leq i \leq N$, $\mathbf{u}'_i[t] = \mathbf{u}_i[t+1]$ for $0 \leq t \leq T-1$ and $\mathbf{u}'_i[T] = 0$. Now, for any time-step $k \in \mathbb{Z}_{\geq 1}$, we can define

$$\tilde{V}_k((\mathbf{x}_i[k])_{i=1}^N) = J_k\big((\mathbf{x}_i[k])_{i=1}^N; \mathbf{q}_T(\hat{\mathbf{u}}_i^{(S_{k-1})}[k-1:T+k-1])\big),$$

where the control sequence $\hat{\mathbf{u}}_i^{(S_{k-1})}[k-1:T+k-1]$ denotes the control actions of step $k-1$ extracted from Procedure 7.1. For this described function, we get

$$\tilde{V}_k((\mathbf{x}_i[k])_{i=1}^N) = J_k^{(T)}\big((\mathbf{x}_i[k])_{i=1}^N; \mathbf{q}_T(\hat{\mathbf{u}}_i^{(S_{k-1})}[k-1:T+k-1])\big) \geq V_k^{(T)}((\mathbf{x}_i[k])_{i=1}^N),$$

because the control sequence $\mathbf{q}_T(\{\hat{\mathbf{u}}_i^{(S_{k-1})}[t]\}_{t=k-1}^{T+k-1})$ might not be optimal for time-step k. Hence, we have proposed a suitable mapping for Theorem 7.2.

7.3.4 ADMM Formulation

A way to achieve better numerical properties for solving distributed MPC is to apply ADMM, which retains the decomposability of the dual formulation while ensuring better convergence properties in terms of speed and stability [3]. Recently, solving MPC via ADMM has gained some attention [24]. In what comes next, we cast the problem introduced in this chapter in an ADMM framework and give a sub-optimality guarantee for this scenario.

We rewrite the MPC problem proposed in Sect. 7.2.3:

$$\begin{aligned}(\hat{\mathbf{u}}_i^*[k:k+T])_{i=1}^N &= \arg\min_{(\hat{\mathbf{u}}_i[k:k+T])_{i=1}^N} \sum_{t=k}^{k+T}\sum_{i=1}^{N} \boldsymbol{\ell}_i(\mathbf{y}_i[t]),\\ &\text{subject to } \mathbf{y}_i[t] \in \mathcal{C}_i[t],\ 1 \leq i \leq N,\ \forall\, t \in \mathbb{Z}_{\geq k}^{\leq k+T},\end{aligned} \tag{7.9}$$

where, for each $k \leq t \leq k+T$, $\mathbf{y}_i[t] = [\hat{\mathbf{x}}_i[t]^\top, \bar{\mathbf{w}}_i[t]^\top, \hat{\mathbf{u}}_i[t]^\top, \bar{\mathbf{v}}_i[t]]^\top$, and

$$\begin{aligned}\mathcal{C}_i[t] = \big\{&[\hat{\mathbf{x}}_i[t]^\top, \bar{\mathbf{w}}_i[t]^\top, \hat{\mathbf{u}}_i[t]^\top, \bar{\mathbf{v}}_i[t]]^\top \,\big|\, \hat{\mathbf{x}}_i[s+1] = \mathbf{f}_i(\hat{\mathbf{x}}_i[s], \bar{\mathbf{v}}_i[s]; \hat{\mathbf{u}}_i[s]),\\ &\hat{\mathbf{x}}_i[s] \in \mathcal{X}_i, \hat{\mathbf{u}}_i[s] \in \mathcal{U}_i,\ \bar{\mathbf{w}}_i[s] = \hat{\mathbf{w}}_i[s],\ \bar{\mathbf{v}}_i[s] = \hat{\mathbf{v}}_i[s],\ \forall\, s \in \mathbb{Z}_{\geq k}^{\leq t},\ \hat{\mathbf{x}}_i[k] = \mathbf{x}_i[k]\big\}.\end{aligned} \tag{7.10}$$

Provided that $\mathcal{C}_i[t]$ is convex for all $t \in \mathbb{Z}_{\geq k}^{\leq k+T}$, then (7.9) is equivalent to

$$\begin{aligned}(\hat{\mathbf{u}}_i^*[k:k+T])_{i=1}^N &= \arg\min_{(\hat{\mathbf{u}}_i[k:k+T])_{i=1}^N} \sum_{t=k}^{k+T}\sum_{i=1}^{N} \big[\boldsymbol{\ell}_i(\mathbf{y}_i[t]) + \psi_i(\boldsymbol{\zeta}_i[t])\big],\\ &\text{subject to}\quad \mathbf{y}_i[t] = \boldsymbol{\zeta}_i[t],\ 1 \leq i \leq N,\ \forall\, t \in \mathbb{Z}_{\geq k}^{\leq k+T},\end{aligned}$$

where $\psi_{\mathcal{C}_i[t]}(\cdot)$ is an indicator for $\mathcal{C}_i[t]$, viz. $\psi_{\mathcal{C}_i[t]}(z) = 0$ if $z \in \mathcal{C}_i[t]$ and $\psi_{\mathcal{C}_i[t]}(z) = +\infty$ otherwise. The augmented Lagrangian for this problem is

---

Algorithm 7.2 Distributed algorithm for solving MPC problem (7.3) via ADMM

Input: $\mathbf{x}_i[k]$, $1 \leq i \leq N$
Output: $\mathbf{u}_i[k]$, $1 \leq i \leq N$
1: **for** $k = 1, 2, \ldots$ **do**
2: - Initialize scaled dual variables $(\gamma_i^{(0)}[t])_{i=1}^N$.
3: **for** $s = 1, 2, \ldots, S_k$ **do**
4: **for** $i = 1, 2, \ldots, N$ **do**
5: - Solve the optimization problem
6: $\mathbf{y}_i^{(s)}[k:k+T] = \arg\min_{\mathbf{y}_i[k:k+T]} \sum_{t=k}^{k+T} \left[\ell_i(\mathbf{y}_i[t]) + \frac{\rho}{2}\|\mathbf{y}_i[t] - \zeta_i^{(s)}[t] - \gamma_i^{(s)}[t]\|^2\right]$.
7: **end for**
8: - $\zeta_i^{(s+1)}[t] = \Pi_{\mathcal{C}_i[t]}(\mathbf{y}_i[t] + \gamma_i^{(s)}[t])$, $1 \leq i \leq N, \forall\, t \in \mathbb{Z}_{\geq k}^{\leq k+T}$. $\{\Pi_{\mathcal{C}_i[t]}(\cdot)$ is a projection onto $\mathcal{C}_i[t]\}$
9: - $\gamma_i^{(s+1)}[t] := \gamma_i^{(s)}[t] + (\mathbf{y}_i[t] - \zeta_i^{(s+1)}[t])$, $1 \leq i \leq N, \forall\, t \in \mathbb{Z}_{\geq k}^{\leq k+T}$.
10: **end for**
11: - $\mathbf{u}_i[k] = \hat{\mathbf{u}}_i^{(S_k)}[k]$, $1 \leq i \leq N$.
12: **end for**

$$L((\mathbf{y}_i[k:k+T])_{i=1}^N, (\zeta_i[k:k+T])_{i=1}^N, (\gamma_i[k:k+T])_{i=1}^N)$$
$$= \sum_{t=k}^{k+T}\sum_{i=1}^{N}\left[\ell_i(\mathbf{y}_i[t]) + \psi_i(\zeta_i[t]) + \frac{\rho}{2}\|\mathbf{y}_i[t] - \zeta_i[t] - \gamma_i[t]\|^2\right], \tag{7.11}$$

where $\gamma_i[t]$, $1 \leq i \leq N$, are the scaled dual variables. We outline a distributed procedure that solves the problem in Procedure 7.2. Now, we reintroduce the following function for solving the MPC problem via ADMM:

$$V^{(T),(s)}((\mathbf{x}_i[k])_{i=1}^N) = \sum_{t=k}^{k+T}\sum_{i=1}^{N}\left[\ell_i(\mathbf{y}_i^{(s)}[t]) + \frac{\rho}{2}\left\|\mathbf{y}_i^{(s)}[t] - \zeta_i^{(s)}[t] - \gamma_i^{(s)}[t]\right\|^2\right].$$

For the case that $\|\mathbf{y}_i^{(s)}[t] - \zeta_i^{(s)}[t] - \gamma_i^{(s)}[t]\|^2$ is less than a given threshold $\varepsilon \ll 1$, one might be able to follow the same line of reasoning as in Theorem 7.2 (see [7] for a more detailed discussion).

7.4 Simulations

In this section, we portray the applicability of the algorithm to a formation acquisition problem. We assume that the nonholonmic vehicle i for each $1 \leq i \leq N$, can be described in state-space representation as

$$\mathbf{x}_i[k+1] = \mathbf{x}_i[k] + \begin{bmatrix} \mathbf{v}_i[k]\cos(\theta_i[k]) \\ \mathbf{v}_i[k]\sin(\theta_i[k]) \end{bmatrix},$$

where $\mathbf{x}_i[k] = \begin{bmatrix} x_{i,1}[k] & x_{i,2}[k] \end{bmatrix}^\top \in \mathbb{R}^2$ is the position of the vehicle, $\mathbf{v}_i[k] \in \mathbb{R}$ is its velocity, and $\theta_i[k] \in \mathbb{R}$ is its steering-wheel angle. Because of the vehicles' mechanical constraints, i.e., bounded speed and steering angle, we assume that the control inputs should always remain bounded as $0 \leq \mathbf{v}_i[k] \leq 0.5$ and $|\theta_i[k]| \leq \pi/6$ for all $k \in \mathbb{Z}_{\geq 0}$. We define each vehicle control input as $\mathbf{u}_i[k] = \begin{bmatrix} \mathbf{v}_i[k] & \theta_i[k] \end{bmatrix}^\top$. Let us start with two vehicles. At each time-step $k \in \mathbb{Z}_{\geq 0}$, these vehicles are interested in minimizing the cost function

$$J_k^{(T)}((\mathbf{x}_i[k])_{i=1}^2; (\mathbf{u}_i[k:k+T])_{i=1}^2) = \sum_{t=k}^{k+T} \left[2\|\mathbf{x}_1[t] - \mathbf{x}_2[t] - \mathbf{d}_{12}\|_2^2 + 10(v_1^2[t] + v_2^2[t]) \right],$$

where $\mathbf{d}_{12} = \begin{bmatrix} 2 & 1 \end{bmatrix}^\top$. Let us fix the starting points of the vehicles as $\mathbf{x}_1[0] = \begin{bmatrix} +4.0 & -1.0 \end{bmatrix}^\top$ and $\mathbf{x}_2[0] = \begin{bmatrix} +1.0 & -5.0 \end{bmatrix}^\top$. We also fix the planning horizon $T = 5$. We use Procedure 7.1 to calculate the sequence of sub-optimal control signals when the termination law (i.e., iteration number S_k) is given by Theorem 7.2. Figure 7.1 illustrates the trajectory and the control signals of both vehicles with finite-horizon planning when the sub-optimality parameter is fixed at $\alpha = 0.5$. To be precise, Fig. 7.1a, b portray different coordinates of the vehicle position while Fig. 7.1c, d illustrate the velocities and steering-wheel angels of the vehicles, respectively. The red color denotes the first vehicle and the blue color denotes the second one. The portrayed simulation is done for 200 time-steps. It is interesting to note

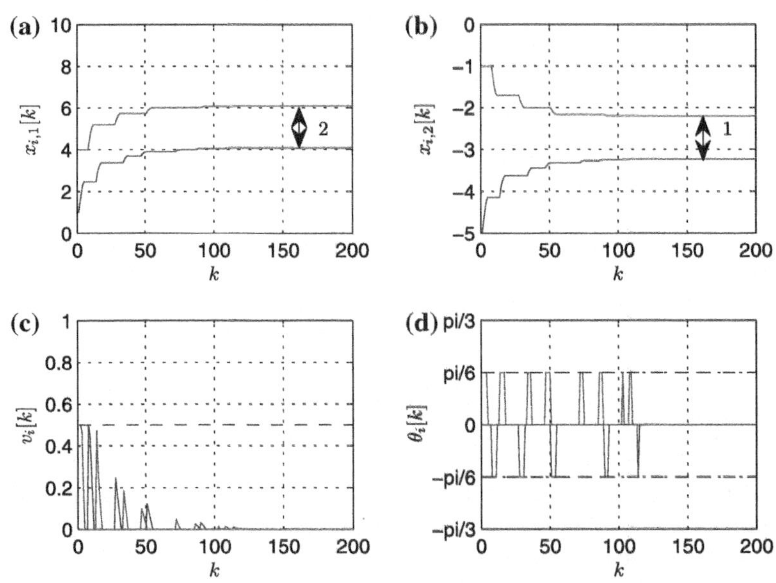

Fig. 7.1 Trajectory and control signal of two vehicles when using Procedure 7.1 and termination law described in Theorem 7.2.

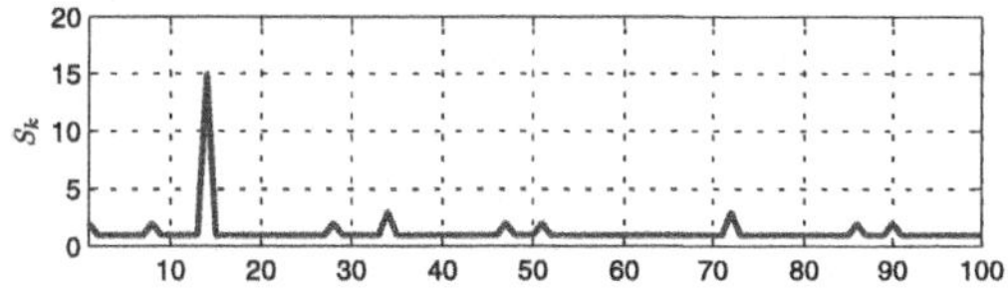

Fig. 7.2 Iteration numbers Sk versus time-step k for the simulation results in Fig. 7.1

that over the first 100 time-steps, in average 1.25 iterations per time-step were used in Procedure 1 to extract the sub-optimal control signal (see Fig. 7.2). Figure 7.3 illustrates the trajectory and the control signals of both vehicles with finite-horizon planning using a centralized optimization algorithm as a reference. We also check the influence of α. To do so, we introduce some notations. For each time-step k, we define $(\mathbf{u}_i^{\text{Primal}}[k])_{i=1}^2 = (\hat{\mathbf{u}}_i[k])_{i=1}^2$ where

$$(\hat{\mathbf{u}}_i[k:k+T])_{i=1}^2 = \underset{(\hat{\mathbf{u}}_i[k:k+T]\in\mathcal{U}_i)_{i=1}^2}{\arg\min} \quad J_k^{(T)}((\mathbf{x}_i[k])_{i=1}^2; (\hat{\mathbf{u}}_i[k:k+T])_{i=1}^2).$$

Similarly, we define $(\mathbf{u}_i^{\text{Dual}}[k])_{i=1}^2 = (\hat{\mathbf{u}}_i^{(S_k)}[k])_{i=1}^2$ where, for each vehicle, $\hat{\mathbf{u}}_i^{(S_k)}[k]$ is calculated using Procedure 7.1 when the dual decomposition iteration numbers $\{S_k\}_{k=0}^{\infty}$ is extracted from Theorem 7.2. Now, we define the ratio

$$\rho = J_0^{(H)}\left((\mathbf{x}_i[0])_{i=1}^2; (\mathbf{u}_i^{\text{Dual}}[0:H])_{i=1}^2\right) / J_0^{(H)}\left((\mathbf{x}_i[0])_{i=1}^2; (\mathbf{u}_1^{\text{Primal}}[0:H])_{i=1}^2\right),$$

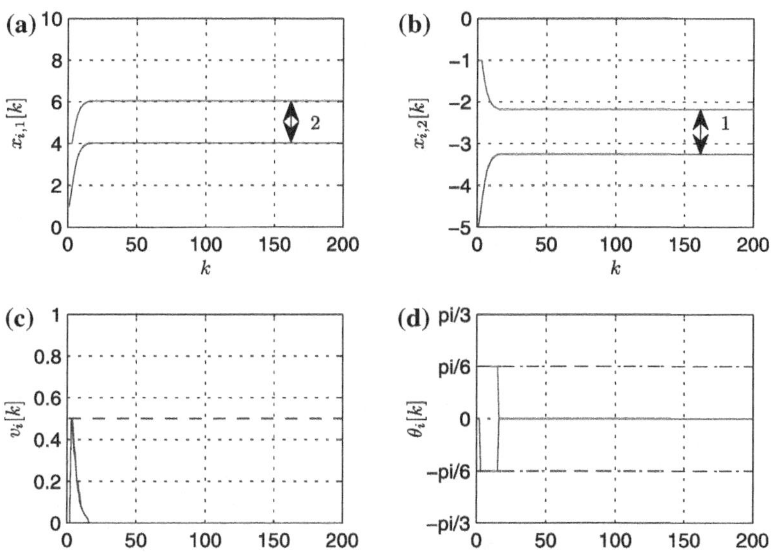

Fig. 7.3 Trajectory and control signal of two vehicles when using a centralized optimization algorithm

Table 7.1 Sub-optimality ratio as function of α

α	0.1	0.3	0.5	0.7
ρ	9.7875	3.2725	1.9684	1.4198

where H is the simulation horizon. Table 7.1 shows ρ as a function of α for $H = 1000$. Based on this table, we can numerically verify the claim of Theorem 7.2 that using Procedure 7.1 when the dual decomposition iteration numbers is extracted from (7.8) provides a suboptimality ratio ρ that is inversely proportional to α.

These simulations can be readily extended to more vehicles. Figure 7.4 illustrates the trajectory and the control signals of three vehicles when trying to minimize the cost function

$$J_k^{(T)}((\mathbf{x}_i[k])_{i=1}^3; (\mathbf{u}_i[k:k+T])_{i=1}^3) = \sum_{t=k}^{k+T} \Big[2\|\mathbf{x}_1[t] - \mathbf{x}_2[t] - \mathbf{d}_{12}\|_2^2 + 2\|\mathbf{x}_1[t] - \mathbf{x}_3[t] - \mathbf{d}_{13}\|_2^2 + 2\|\mathbf{x}_2[t] - \mathbf{x}_3[t] - \mathbf{d}_{23}\|_2^2 + 10(v_1^2[t] + v_2^2[t] + v_2^2[t]) \Big],$$

with $\mathbf{d}_{12} = \begin{bmatrix} +1 & -5 \end{bmatrix}^\top$, $\mathbf{d}_{13} = \begin{bmatrix} -3 & +2 \end{bmatrix}^\top$, and $\mathbf{d}_{23} = \begin{bmatrix} -4 & +7 \end{bmatrix}^\top$. Let us fix the starting points of the vehicles as $\mathbf{x}_1[0] = \begin{bmatrix} +4.0 & -1.0 \end{bmatrix}^\top$, $\mathbf{x}_2[0] = \begin{bmatrix} +1.0 & -3.0 \end{bmatrix}^\top$, and $\mathbf{x}_3[0] = \begin{bmatrix} -2.0 & +3.0 \end{bmatrix}^\top$. We consider planning horizon $T = 3$. As before, we use Procedure 7.1 to calculate the sequence of sub-optimal control signals when

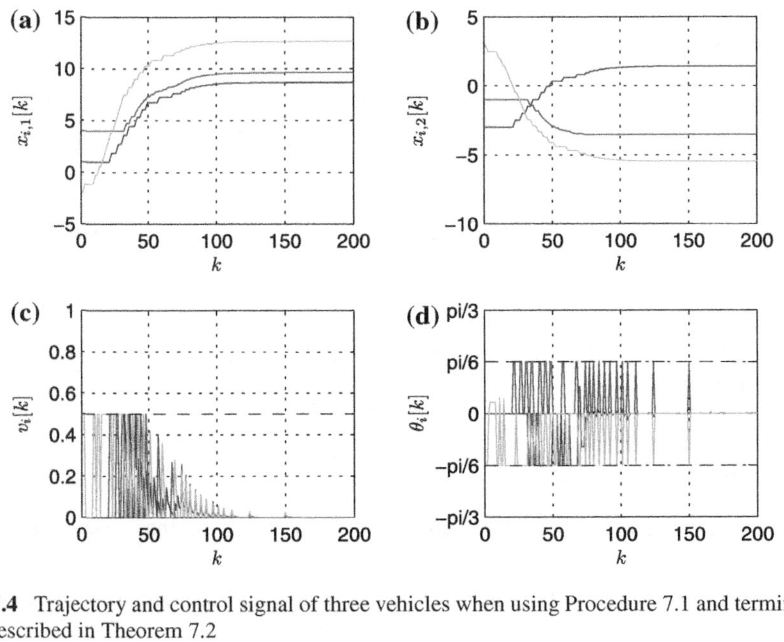

Fig. 7.4 Trajectory and control signal of three vehicles when using Procedure 7.1 and termination law described in Theorem 7.2

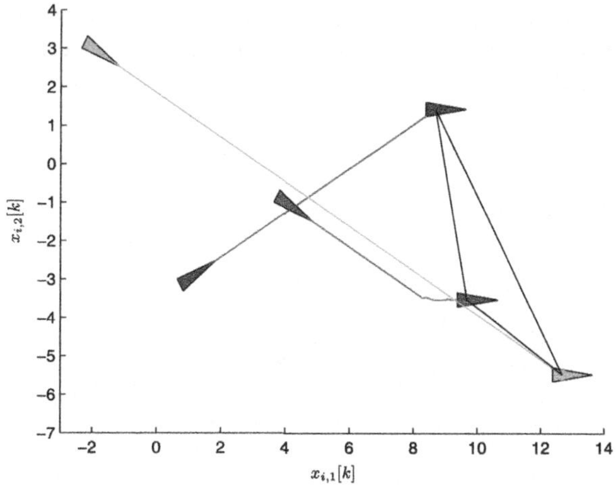

Fig. 7.5 Trajectory of the vehicles in the 2-D plane

the termination law (i.e., iteration number S_k) is given by Theorem 7.2 and the sub-optimality parameter is $\alpha = 0.2$. The red, blue, and green colors denotes the first, second, and third vehicle, respectively. Figure 7.5 portrays the trajectory of the vehicles in the 2-D plane. The final formation is illustrated by the triangle in black color. The codes to generate the results of this section can be found at [21].

We conclude this section by briefly noting that the system consisting of N agents under the aforementioned cost function converges to the desired formation if and only if $\mathcal{G}^C$ is connected. This is a direct consequence of the structural rigidity of the desired formation, for more information see [20]. Other cost functions, such as

$$J_k^{(T)}((\mathbf{x}_i[k])_{i=1}^N; (\mathbf{u}_i[k:k+T])_{i=1}^N) = \sum_{t=k}^{k+T}\left[\sum_{(i,j)\in\mathcal{E}^C}(\|\mathbf{x}_i[t]-\mathbf{x}_j[t]\|_2^2 - \|\mathbf{d}_{ij}\|_2^2)^2 + \sum_{i=1}^{N} v_i^2[t]\right],$$

can be considered as well. In this case, the system converges to the desired formation if and only if the formation is globally rigid with $N \geq 4$, see [2, 13] and references therein.

7.5 Conclusions

In this chapter, we considered the dual decomposition formulation of a distributed MPC problem for systems with arbitrary dynamical couplings. More specifically, we studied the problem of calculating performance bounds on the solution obtained from iteratively solving the dual problem in a distributed way when the iterations are terminated after S_k steps at time-step k. Later, we commented on how the problem can be cast in an ADMM setting. We demonstrated the validity of the proposed performance bound through simulations on formation acquisition by a group of nonholonomic agents. As a future research direction, one might consider providing better performance bounds for the case where ADMM is implemented to solve the MPC problem.

References

1. B.D.O. Anderson, J.B. Moore, *Linear Optimal Control* (Prentice-Hall, Inc., Englewood Cliffs, 1971)
2. B.D.O. Anderson, I. Shames, G. Mao, B. Fidan, Formal theory of noisy sensor network localization. SIAM J. Discret. Math. **24**(2), 684–698 (2010)
3. S. Boyd, N. Parikh, E. Chu, B. Peleato, J. Eckstein, Distributed optimization and statistical learning via the alternating direction method of multipliers. Found. Trends. Mach. Learn. **3**(1), 1–122 2011
4. S. P. Boyd, L. Vandenberghe, *Convex Optimization*. (Cambridge University Press, Cambridge, 2004)
5. E. Camponogara, D. Jia, B.H. Krogh, S. Talukdar, Distributed model predictive control. IEEE Control Syst. Mag. **22**(1), 44–52 (2002)
6. W.B. Dunbar, Distributed receding horizon control of dynamically coupled nonlinear systems. IEEE Trans. Autom. Control **52**(7), 1249–1263 (2007)
7. F. Farokhi, I. Shames, K. H. Johansson, Distributed MPC via dual decomposition and alternative direction method of multipliers (2012). arXiv:1207.3178 [math.OC], http://arxiv.org/abs/1207.3178
8. C.E. Garcia, D.M. Prett, M. Morari, Model predictive control: theory and practice-a survey. Automatica **25**(3), 335–348 (1989)
9. P. Giselsson, A. Rantzer, Distributed model predictive control with suboptimality and stability guarantees, in *Proceedings of the 49th IEEE Conference on Decision and Control*, (2010), pp. 7272–7277
10. R. Hovorka, V. Canonico, L.J. Chassin, U. Haueter, M. Massi-Benedetti, M.O. Federici, T.R. Pieber, H.C. Schaller, L. Schaupp, T. Vering, M.E. Wilinska, Nonlinear model predictive control of glucose concentration in subjects with type 1 diabetes. Physiol. Meas. **25**, 905 (2004)
11. M. Kearney, M. Cantoni, P.M. Dower, Model predictive control for systems with scheduled load and its application to automated irrigation channels, in *Proceedings of the IEEE International Conference on Networking, Sensing and Control*, (2011), pp. 186–191
12. S. Kouro, P. Cortes, R. Vargas, U. Ammann, J. Rodriguez, Model predictive control: A simple and powerful method to control power converters. EEE Trans. Ind. Electron. **56**(6), 1826–1838 (2009)
13. L. Krick, M.E. Broucke, B.A. Francis, Stabilisation of infinitesimally rigid formations of multi-robot networks. Int. J. Control **82**(3), 423–439 (2009)
14. F. Lin, M. Fardad, M.R. Jovanović, Design of optimal sparse feedback gains via the alternating direction method of multipliers, in *Proceedings of the American Control Conference* (2012)
15. J. Mattingley, Y. Wang, S. Boyd, Receding horizon control. IEEE Control Syst. **31**(3), 52–65 (2011)
16. N. Motee, B. Sayyar-Rodsari, Optimal partitioning in distributed model predictive control, in *Proceedings of the American Control Conference*, (2003), pp. 5300–5305
17. R.R. Negenborn, B. De Schutter, J. Hellendoorn, Multi-agent model predictive control for transportation networks: Serial versus parallel schemes. Eng. Appl. Artif. Intell. **21**(3), 353–366 (2008)
18. R.R. Negenborn, Z. Lukszo, H. Hellendoorn (eds.), *Intelligent Infrastructures* (Springer, Berlin, 2010)
19. J.B. Rawlings, Tutorial overview of model predictive control. IEEE Control Syst. **20**(3), 38–52 (2000)

20. B. Servatius, W. Whiteley, Constraining plane configurations in CAD: combinatorics of directions and lengths. SIAM J. Discret. Math. **12**, 136–153 (1999)
21. I. Shames http://eemensch.tumblr.com/post/25506192175/mpc-grnt-code
22. A. Venkat, J. Rawlings, S. Wright, Distributed model predictive control of large-scale systems, in *Assessment and Future Directions of Nonlinear Model Predictive, Control*, (2007), p. 591–605
23. A.N. Venkat, J.B. Rawlings, S.J. Wright, Stability and optimality of distributed model predictive control, in *Proceedings of the IEEE Conference on Decision and Control and European Control Conference*, (2005), pp. 6680—6685
24. B. Wahlberg, S. Boyd, M. Annergren, Y. Wang, An ADMM algorithm for a class of total variation regularized estimation problems, in *Proceedings of the 16th IFAC Symposium on System Identification* (2012)
25. Y. Wakasa, M. Arakawa, K. Tanaka, T. Akashi, Decentralized model predictive control via dual decomposition, in *Proceedings of the 47th IEEE Conference on Decision and Control*, (2008), pp. 381–386

Chapter 8
D-SIORHC, Distributed MPC with Stability Constraints Based on a Game Approach

J. M. Lemos and J. M. Igreja

Abstract This chapter describes D-SIORHC, a distributed MPC algorithm with stability constraints based on a game approach. This controller is designed for chained linear systems, in which each local subsystem interacts only with their neighbors. At the beginning of each sampling interval, each local controller agent computes the value of the corresponding manipulated variable in an iterative process where, in each iteration, it optimizes a quadratic cost by assuming that the neighbor controllers will use for their manipulated variables the value which they have computed in the previous iteration. Therefore, in a game theory framework, if this coordination procedure converges, a Nash equilibrium is reached. The use of linear plant models and the absence of inequality operational constraints allows to compute the manipulated variables in an explicit way, in each iteration of the coordination procedure, thereby reducing the computational load. This approach differs from other distributed MPC algorithms based on linear models in the inclusion of stability constraints in the local controllers that leads to a different control law. The controller usage is illustrated through its application to a water delivery canal.

8.1 Introduction

8.1.1 The Rationale Behind the Approach Proposed

The approach to distributed MPC described in this chapter addresses chained linear systems. The class of plants considered is such that they can be decomposed in local subsystems, each one interacting only with their neighbors. To each of these so called

J. M. Lemos (✉)
INESC-ID/IST/UTL, Lisboa, Portugal
e-mail: jlml@inesc-id.pt

J. M. Igreja
INESC-ID/ISEL/IPL,Lisboa, Portugal
e-mail: jigreja@deea.isel.ipl.pt

J. M. Maestre and R. R. Negenborn (eds.), *Distributed Model Predictive Control Made Easy*, Intelligent Systems, Control and Automation: Science and Engineering 69, DOI: 10.1007/978-94-007-7006-5_8, © Springer Science+Business Media Dordrecht 2014

local subsystems, a local control agent is connected that computes the value of its manipulated variable based on the information available locally, given by the output variables of the local system and the manipulated variables of its neighbors.

Following a game theory approach to distributed MPC [6], in order to act in a coordinated way, local controllers play a game among themselves. This game consists in minimizing the cost function that results from the minimization by each controller of a local objective function. For that sake, at the beginning of each sampling interval, each local controller agent computes the value of the corresponding manipulated variable in an iterative process where, in each iteration, it optimizes a quadratic cost that assumes that the neighbor controllers will use for their own manipulated variables the value that they have computed in the previous iteration.

Therefore, in a game theory framework, if the above coordination procedure converges, a Nash equilibrium is reached. As is known [1], the Nash equilibrium is a suboptimal approximation of the optimum of the overall objective function given by the sum of the local objective functions. One needs thus to check *a posteriori* whether the control solution obtained is acceptable and suits the objectives of the application being considered.

As will be explained in Sect. 8.3, in order to overcome the limitations inherent to a game that converges to a Nash equilibrium, it is remarked that the local costs depend not only of the tracking error of each node, but on the tracking errors of its neighbors as well. The resulting "altruistic" action results in a contribution to the coordination between the different local control agents.

8.1.2 Interest of the Approach

Despite the above mentioned limitations, it is possible to find examples of practical interest in which the approach proposed provides a solution that yields the desired performance with a low computational load. A major advantage of the method proposed for obtaining a distributed version of MPC is its simplicity with respect to other approaches, for instance those based on dual optimization [2, 10, 11]. Although these other methods are in general more powerful, yielding tighter approximations to the overall optimum, they require more complicated algorithms and the tuning of more parameters. The low computational load of the algorithm described in this chapter stems from the fact that, in each round of the negotiation procedure, a closed-form optimization formula is used and no numeric iterative method is required.

8.1.3 Origin of the Method

In [12] the same coordination procedure is employed to obtain a distributed MPC for serially chained systems. With respect to [12], the algorithm proposed in this chapter (originally published in [3]) has the non-trivial difference of designing the local controllers by embedding a stability constraint that consists in imposing the value of the output at the final part of the prediction horizon for a number of coincidence points. This constraint ensures the stability of the local control loops if taken in isolation (i.e. with the manipulated variables of the other subsystems frozen) for a finite value of the horizon. This is the SIORHC algorithm [8, 9] and, hence, the structure proposed in this chapter is called D-SIORHC or "Distributed SIORHC".

Although the above stability constraints do not, just by themselves, ensure stability of the overall system, they are an important ingredient in a stability proof. Furthermore, the number of coincidence points provides an extra knob that can be used to influence the resulting performance.

8.2 Boundary Conditions

8.2.1 System Partition and Plant Dynamic Model

Consider the plant described by the linear state-space model

$$\mathbf{x}(k+1) = \mathbf{A}\mathbf{x}(k) + \mathbf{B}\Delta\mathbf{u}(k) \tag{8.1}$$

and the output equation

$$\mathbf{y}(k) = \mathbf{C}\mathbf{x}(k), \tag{8.2}$$

where $\mathbf{x} \in \mathbb{R}^n$ is the overall plant state, $\Delta\mathbf{u} \in \mathbb{R}^m$ is the vector of increments of the manipulated inputs, $\mathbf{y} \in \mathbb{R}^p$ is the vector of all the measured outputs, k is an integer number that denotes discrete time and $\mathbf{A} \in \mathbb{R}^{n\times n}$, $\mathbf{B} \in \mathbb{R}^{n\times m}$, and $\mathbf{C} \in \mathbb{R}^{p\times n}$ are matrices.

The first step in designing a distributed controller consists in decomposing the plant (8.1) in a number of subsystems and defining the interactions among them. In this chapter we consider plants made of serially connected subsystems, as shown in Fig. 8.1. Let $\mathcal{N}$ denote the set of subsystems in which the plant is decomposed and assume that there are $N_s = |\mathcal{N}|$ subsystems and that these are numbered sequentially by $1, 2, \ldots, N_s$. Subsystem number i interacts only with subsystems $i-1$ and $i+1$. Therefore, the matrices $\mathbf{A}$ and $\mathbf{B}$ in (8.1) have the tri-diagonal block structure

$$\mathbf{A} = \begin{bmatrix} \mathbf{A}_{11} & \mathbf{A}_{12} & \mathbf{0} & \mathbf{0} & \ldots & \mathbf{0} \\ \mathbf{A}_{21} & \mathbf{A}_{22} & \mathbf{A}_{23} & \mathbf{0} & \ldots & \mathbf{0} \\ \mathbf{0} & \mathbf{A}_{32} & \mathbf{A}_{33} & \mathbf{A}_{34} & \ldots & \mathbf{0} \\ \vdots & \ddots & \ddots & \ddots & \ddots & \vdots \\ \mathbf{0} & \ldots & \ldots & \mathbf{0} & \mathbf{A}_{N_sN_s-1} & \mathbf{A}_{N_sN_s} \end{bmatrix}, \tag{8.3}$$

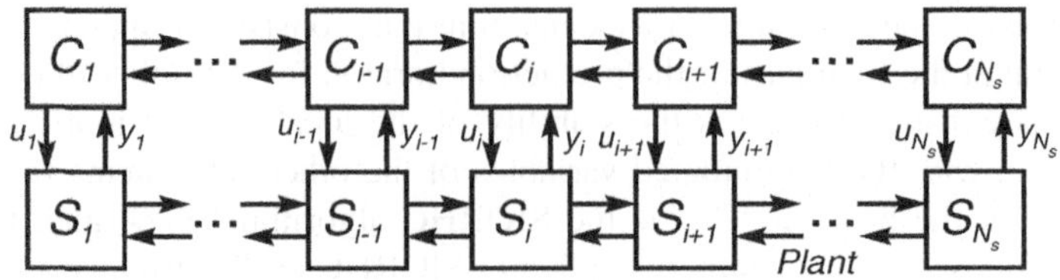

Fig. 8.1 Plant decomposition and structure of the distributed MPC controller

$$\mathbf{B} = \begin{bmatrix} \mathbf{B}_{11} & \mathbf{B}_{12} & \mathbf{0} & \mathbf{0} & \dots & \mathbf{0} \\ \mathbf{B}_{21} & \mathbf{B}_{22} & \mathbf{B}_{23} & \mathbf{0} & \dots & \mathbf{0} \\ \mathbf{0} & \mathbf{B}_{32} & \mathbf{B}_{33} & \mathbf{B}_{34} & \dots & \mathbf{0} \\ \vdots & \ddots & \ddots & \ddots & \ddots & \vdots \\ \mathbf{0} & \dots & \dots & \mathbf{0} & \mathbf{B}_{N_s N_s - 1} & \mathbf{B}_{N_s N_s} \end{bmatrix}, \tag{8.4}$$

and

$$\mathbf{C} = \begin{bmatrix} \mathbf{C}_{11} & \mathbf{0} & \dots & \mathbf{0} \\ \mathbf{0} & \mathbf{C}_{22} & \ddots & \vdots \\ \vdots & \ddots & \ddots & \mathbf{0} \\ \mathbf{0} & \dots & \mathbf{0} & \mathbf{C}_{N_s N_s} \end{bmatrix}. \tag{8.5}$$

Each subsystem $\mathcal{S}_i$, $i = 1, \dots, N_s$, is described by the linear state-space models

$$\mathbf{x}_i(k+1) = \mathbf{A}_{ii}\mathbf{x}_i(k) + \mathbf{B}_{ii}\Delta u_i(k) + \Gamma \mathbf{d}_i, \tag{8.6}$$

such that the concatenation of the $\mathbf{x}_i$ for all $i = 1, \dots, N_s$ yields the state $\mathbf{x}$ of the overall plant, Γ_i is a matrix and the accessible disturbance $\mathbf{d}_i$ is

$$\mathbf{d}_1 = \mathbf{A}_{1,2}\mathbf{x}_2(k) + \mathbf{B}_{12}\Delta u_2(k), \tag{8.7}$$

$$\mathbf{d}_i = \mathbf{A}_{i,i-1}\mathbf{x}_{i-1}(k) + \mathbf{A}_{i,i+1}\mathbf{x}_{i+1}(k) + \mathbf{B}_{i,i-1}\Delta u_{i-1}(k) + \mathbf{B}_{i,i+1}\Delta u_{i+1}(k), \tag{8.8}$$

for $i = 2, \dots, N_s - 1$ and

$$\mathbf{d}_{N_s} = \mathbf{A}_{N_s,N_s-1}\mathbf{x}_{N_s}(k) + \mathbf{B}_{N_s,N_s-1}\Delta u_{N_s}(k). \tag{8.9}$$

The output equation for the local subsystem $\mathcal{S}_i$ is

$$y_i(k) = \mathbf{C}_{ii}\mathbf{x}_i(k). \tag{8.10}$$

We consider the case in which both the Δu_i and the y_i are scalar, although the method can be extended to multivariable local subsystems.

8.2.2 The Control Problem

Let $\mathbf{Y}_{i,N}$, $\mathbf{Y}^*_{i,N}$ and $\Delta\mathbf{U}_i$ denote vectors of samples of the output, of the reference and of the manipulated variable along the prediction horizon of length N_p for subsystem $\mathcal{S}_i$, written as

$$\mathbf{Y}_{i,N_p} \triangleq \left[\, y_i(k+1) \,\ldots\, y_i(k+N_p) \,\right]^T, \tag{8.11}$$

$$\mathbf{Y}^*_{i,N_p} \triangleq \left[\, y^*_i(k+1) \,\ldots\, y^*_i(k+N_p) \,\right]^T, \tag{8.12}$$

$$\Delta\mathbf{U}_i \triangleq \left[\, \Delta u_i(k) \,\ldots\, \Delta u_i(k+N_p-1) \,\right]^T \tag{8.13}$$

where the $y^*_i(\cdot)$ represent the reference to be tracked by $y_i(\cdot)$.

To each one of the subsystems $\mathcal{S}_i$, $i = 1, \ldots, N_s$, a local control agent $\mathcal{C}_i$ is associated that computes the manipulated variable u_i by minimizing in a receding horizon sense the quadratic cost

$$J_1 \triangleq \|\mathbf{Y}^*_{1,N} - \mathbf{Y}_{1,N}\|^2_{\mathbf{Q}_1} + \beta_1 \|\mathbf{Y}^*_{2,N} - \mathbf{Y}_{2,N}\|^2_{\mathbf{Q}_2} + \|\Delta\mathbf{U}\|^2_{\mathbf{R}_1} \tag{8.14}$$

for $i = 1$,

$$\begin{aligned} J_i \triangleq\ & \alpha_i \|\mathbf{Y}^*_{i-1,N_p} - \mathbf{Y}_{i-1,N_p}\|^2_{\mathbf{Q}_{i-1}} + \|\mathbf{Y}^*_{i,N_p} - \mathbf{Y}_{i,N_p}\|^2_{\mathbf{Q}_i} \\ & + \beta_i \|\mathbf{Y}^*_{i+1,N_p} - \mathbf{Y}_{i+1,N_p}\|^2_{\mathbf{Q}_{i+1}} + \|\Delta\mathbf{U}\|^2_{\mathbf{R}_i} \end{aligned} \tag{8.15}$$

for $2 \le i \le N_s - 1$, and

$$J_{N_s} \triangleq \alpha_{N_s} \|\mathbf{Y}^*_{N_s-1,N_p} - \mathbf{Y}_{N_s-1,N_p}\|^2_{\mathbf{Q}_{N_s-1}} + \|\mathbf{Y}^*_{N_s,N_p} - \mathbf{Y}_{N_s,N_p}\|^2_{\mathbf{Q}_{N_s}} + \|\Delta\mathbf{U}\|^2_{\mathbf{R}_{N_s}} \tag{8.16}$$

for $i = N_s$.

In (8.14–8.16), α_i and β_i denote weights that can be positive or zero. If these weights are selected to be zero, then the corresponding cost will depend only on the tracking error of that control agent. Otherwise, the cost will also depend on the tracking errors of its neighbors.

The minimization of each of the costs in (8.14–8.16) is done while satisfying the constraints

$$\mathbf{Y}_{i,P} = \mathbf{Y}^*_{i,P} \tag{8.17}$$

with

$$\mathbf{Y}_{i,P} \triangleq \left[\, y_i(k+N_p+1) \,\ldots\, y_i(k+N_p+P) \,\right]^T, \tag{8.18}$$

$$\mathbf{Y}^*_{i,P} \triangleq \left[\, y^*_i(k+N_p+1) \,\ldots\, y^*_i(k+N_p+P) \,\right]^T, \tag{8.19}$$

where P is the number of coincidence points.

As mentioned in the introduction, the equality constraints (8.17) concern stability. If the value of P is at least the order of the local plant, the local closed-loop system, assuming that there are no interactions with the neighbors, is asymptotically stable [8]. If the horizon N_p is large enough so that stability is achieved with lower values of P, the number of coincidence points can be used as a tuning knob that influences the performance obtained. In this case, no method other then trial and error is available to select convenient values for P.

8.2.3 Communication Architecture

In order to minimize the local costs in a coordinated way, each local controller interchanges with its neighbors information about the decision it takes on its own manipulated variable. The structure of the interconnection of the plant with the local control agents is thus as shown in Fig. 8.1, with each control agent able to communicate only with their immediate neighbors.

8.3 Description of the Approach

The approach followed combines the synthesis of the local control agents that use predictive models to minimize local costs in a receding horizon sense, while imposing stability equality constraints, with an algorithm that coordinates the different coordinates. Both these things are detailed hereafter.

8.3.1 Coordination Algorithm

At the beginning of each sampling interval, a coordination recursive procedure is run a sufficient number of times so as to ensure convergence of the decisions taken by the local control agents $\mathcal{C}_i$ about the value of their manipulated variables. In each step of this recursion, for each value of i, the cost J_i is optimized with respect to u_i, assuming knowledge of the values of Δu_{i-1} and Δu_{i+1} (whenever these variables are defined) computed in the previous iteration. Following [12], this negotiation takes place in a recursive way, shown in the time diagram of Fig. 8.2, and according to the Algorithm 8.1.

Figure 8.2 shows the time progression of the control moves computed by the coordination procedure. Each local control agent does not have access to full plant knowledge but, instead, exchanges information only with its neighbors. As mentioned in step 3 of the coordination algorithm, each control agent knows the vector of virtual control actions computed in the previous iteration by its neighbors that is made of

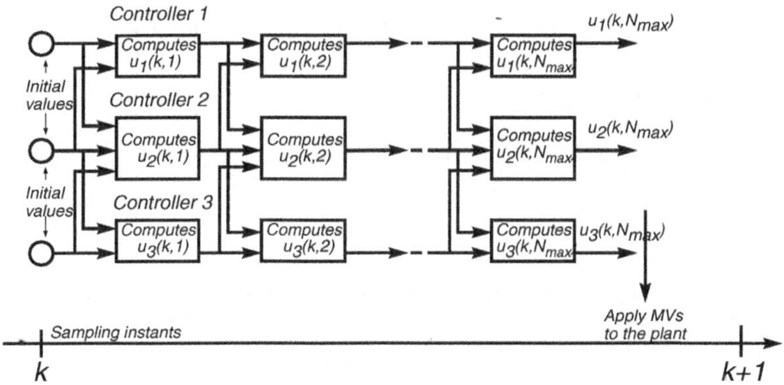

Fig. 8.2 Schematic view of the time progression of the control moves computed by the coordination procedure for 3 local control agents, assuming that the maximum number of iterations N_{max} is performed

---

Algorithm 8.1 Coordination algorithm

1: Let q be the negotiation step index and $\Delta\mathbf{U}_i(k, q)$ be the vector of increments of the manipulated variable of local controller i from sampling time k until $k + N_p - 1$ and after performing q steps of this algorithm.
2: Set the counter $q = 1$.
3: Assume that each local control agent C_i, $i = 1, 2, 3$, at time k and after performing q steps, knows $\Delta\mathbf{U}_{i-1}(k, q-1)$ and $\Delta\mathbf{U}_{i+1}(k, q-1)$, *i. e.*, each local control agent knows the previous iteration of the neighbor control agents. Update the control increment of each local controller by
$$\Delta\mathbf{U}_i(k, q) = \mathscr{F}(\Delta\mathbf{U}_{i-1}(k, q-1), \Delta\mathbf{U}_{i+1}(k, q-1)), \tag{8.20}$$
where $\mathscr{F}$ denotes the optimization procedure used, to be detailed in Sect.8.3..
4: If convergence is reached or if $q = N_{max}$, set $\Delta\mathbf{U}_i(k) = \Delta\mathbf{U}_i(k, q)$ and go to step 5. Otherwise, increment the counter $l \rightarrow l + 1$ and go to step 2.
5: According to a receding horizon strategy, apply to the system S_i the first entry of $\Delta\mathbf{U}_i(k) = \Delta\mathbf{U}_i(k, q)$.

---

control moves that the neighbors would apply along the control horizon in case there were no no further coordination updates (according to a receding horizon strategy only the first element of this vector would be applied).

8.3.2 Local Controllers

The minimization of (8.14–8.16) with the constraints (8.17) implements $\mathscr{F}$ in the coordination procedure described in Sect. 8.3.1. Performing this minimization requires output predictors for the local subsystems and uses the method of Lagrange multipliers [3].

For the system given by (8.1) the predicted outputs at $k + j$ given observations up to time k are given by

$$\hat{y}(k + j) = \sum_{i=0}^{j-1} C\, \mathbf{A}^{j-i-1} B \Delta \mathbf{u}(k + i) + \hat{y}_0(k + j)$$
$$\hat{y}_0(k + j) = \mathbf{C}\, \mathbf{A}^j \hat{x}(k), \tag{8.21}$$

where $\hat{y}_0$ is the output prediction of y without control moves (the system "free response") and $\hat{x}$ denotes either the plant state or its estimate obtained with a suitable observer. For $j = 1 \ldots N_p, N_p + 1, \ldots, N_p + P$ (8.1) the predictors can be written in a compact way as

$$\hat{Y}_{N_p} = \mathbf{G}_{N_p} \Delta \mathbf{U} + \hat{Y}_{0N_p}$$
$$\hat{Y}_{P_p} = \mathbf{G}_P \Delta \mathbf{U} + \hat{Y}_{0P} \tag{8.22}$$

where $\mathbf{G}_{N_p}$ and $\mathbf{G}_P$ are matrices computed from the parameters of (8.1),

$$\hat{Y}_{N_p} = \left[y^T(k + 1 \cdots y^T(k + N_p) \right]^T$$
$$\hat{Y}_P = \left[y^T(k + N_p + 1) \cdots y^T(k + N_p + P) \right]^T. \tag{8.23}$$

and

$$\hat{Y}_{0,N_p} = \left[\hat{y}_0^T(k + 1) \cdots \hat{y}_0^T(k + N_p) \right]^T$$
$$\hat{Y}_{0,P} = \left[\hat{y}_0^T(k + N_p + 1) \cdots \hat{y}_0^T(k + N_p + P) \right]^T. \tag{8.24}$$

For the sake of simplicity consider the case $N_s = 3$. Equation (8.22) is then approximated by considering only interactions between neighboring serially connected systems, yielding

$$\begin{aligned}
\hat{Y}_{1N_p} &= \hat{Y}_{10N_p} + G_{11N_p} \Delta U_1 + G_{12N_p} \Delta U_2, \\
\hat{Y}_{1P} &= \hat{Y}_{10P} + G_{11P} \Delta U_1 + G_{12P} \Delta U_2, \\
\hat{Y}_{2N_p} &= \hat{Y}_{20N_p} + G_{21N_p} \Delta U_1 + G_{22N_p} \Delta U_2, + G_{23N_p} \Delta U_3 \\
\hat{Y}_{2P} &= \hat{Y}_{20P} + G_{21P} \Delta U_1 + G_{22P} \Delta U_2, + G_{23P} \Delta U_3 \\
\hat{Y}_{3N_p} &= \hat{Y}_{30N_p} + G_{32N_p} \Delta U_2 + G_{33N_p} \Delta U_3, \\
\hat{Y}_{3P} &= \hat{Y}_{30P} + G_{32P} \Delta U_2 + G_{33P} \Delta U_3.
\end{aligned}$$

where the $G_{i,j}$ are matrices that are computed from the matrices that parameterize (8.1).

The local control agent that optimizes J_1 is given by the solution with respect to $\Delta\mathbf{U}_1$ of

$$M_1 \Delta U_1 = \left(I - G_{11P}^T W_1 G_{11P} M_1^{-1}\right)\left(G_{11N_p}^T Q_1 E_1 + G_{21N_p}^T Q_2 E_2\right) + W_1 F_1 \tag{8.25}$$

with

$$W_1 = \left(G_{11P} M_1^{-1} G_{11P}^T\right)^{-1}, \tag{8.26}$$

$$M_1 = G_{11N_p}^T Q_1 G_{11N_p} + G_{21N_p}^T Q_2 G_{11N_p} + R_1, \tag{8.27}$$

and, for $i = 1, \ldots, N_s = 3$

$$E_i = Y_{i,N_p}^* - Y_{i,N_p}, \tag{8.28}$$

$$F_i = Y_{i,P}^* - Y_{i,P}. \tag{8.29}$$

Using analogous procedures, another two equations are obtained for the controllers associated with $\mathcal{C}_2$ and $\mathcal{C}_3$ by minimizing the local functionals

$$\min_{\Delta U_2} \|Y_{1N_p}^* - \hat{Y}_{1N_p}\|_{Q_1}^2 + \|Y_{2N_p}^* - \hat{Y}_{2N_p}\|_{Q_2}^2 + \|Y_{3N_p}^* - \hat{Y}_{3N_p}\|_{Q_3}^2 + \|\Delta U_2\|_{R_2}^2 \tag{8.30}$$

$$\text{subject to} \quad \hat{Y}_{2P} = Y_{2P}^*$$

and

$$\min_{\Delta U_3} \|Y_{2N_p}^* - \hat{Y}_{2N_p}\|_{Q_2}^2 + \|Y_{3N_p}^* - \hat{Y}_{3N_p}\|_{Q_3}^2 + \|\Delta U_3\|_{R_3}^2 \tag{8.31}$$

$$\text{subject to} \quad \hat{Y}_{3P} = Y_{3P}^*$$

yielding

$$\begin{aligned} M_2 \Delta U_2 = &\left(I - G_{22P}^T W_2 G_{22P} M_2^{-1}\right)\Big(G_{12N_p}^T Q_1 E_1 \\ &+ G_{22N_p}^T Q_2 E_2 + G_{32N_p}^T Q_3 E_3\Big) + G_{22P}^T W_2 F_2 \end{aligned} \tag{8.32}$$

and

$$\begin{aligned} M_3 \Delta U_3 = &\left(I - G_{33P}^T W_3 G_{33P} M_3^{-1}\right)\Big(G_{23N_p}^T Q_2 E_2 \\ &+ G_{33N_p}^T Q_3 E_3\Big) + G_{33P}^T W_3 F_3 \end{aligned} \tag{8.33}$$

The distributed SIORHC solution for the serially connected sub-systems can be obtained, using the procedure in Sect. 8.3.1, from the matrix algebraic equations system

$$\Phi \Delta U = \Psi, \tag{8.34}$$

where the Φ matrix building blocks are:

$$\begin{aligned}
\Phi_{11} &= M_1 \\
\Phi_{12} &= S_1 \left(G_{11N_p}^T Q_1 G_{12N_p} + G_{21N_p}^T Q_2 G_{22N_p} \right) + G_{11P}^T W_1 G_{12P} \\
\Phi_{13} &= S_1 \left(G_{21N_p}^T Q_2 G_{23N_p} \right) \\
\Phi_{21} &= S_2 \left(G_{12N_p}^T Q_1 G_{11N_p} + G_{22N_p}^T Q_2 G_{21N_p} \right) + G_{22P}^T W_2 G_{21P} \\
\Phi_{22} &= M_2 \\
\Phi_{23} &= S_2 \left(G_{22N_p}^T Q_2 G_{23N_p} + G_{32N_p}^T Q_3 G_{33N_p} \right) + G_{22P}^T W_2 G_{23P} \\
\Phi_{31} &= S_3 \left(G_{23N_p}^T Q_2 G_{21N_p} \right) \\
\Phi_{32} &= S_3 \left(G_{23N_p}^T Q_2 G_{22N_p} + G_{33N_p}^T Q_3 G_{32N_p} \right) + G_{33P}^T W_3 G_{32P} \\
\Phi_{33} &= M_3
\end{aligned}$$

with $S_i = I - G_{iiP}^T W_i G_{iiP} M_i^{-1}$, $i = 1, 2, 3$, the entries of Ψ being

$$\begin{aligned}
\Psi_1 &= S_1 \left(G_{11N_p}^T Q_1 E_1 + G_{21N_p}^T Q_2 E_2 \right) + G_{11P}^T W_1 F_1 \\
\Psi_2 &= S_2 \left(G_{12N_p}^T Q_1 E_1 + G_{22N_p}^T Q_2 E_2 + G_{32N_p}^T Q_3 E_3 \right) + G_{22P}^T W_2 F_2 \\
\Psi_3 &= S_3 \left(G_{23N_p}^T Q_2 E_2 + G_{33N_p}^T Q_3 E_3 \right) + G_{33P}^T W_3 F_3
\end{aligned} \tag{8.35}$$

and

$$\Delta U = \begin{bmatrix} \Delta U_1 & \Delta U_2 & \Delta U_3 \end{bmatrix}^T. \tag{8.36}$$

To apply the iterative procedure described in Sect. 8.3.1, write (8.34) as

$$\Phi_d \Delta U(k, l+1) + \Phi_{nd} \Delta U(k, l) = \Psi, \tag{8.37}$$

where

$$\Phi_d = \begin{bmatrix} \Phi_{11} & 0 & 0 \\ 0 & \Phi_{22} & 0 \\ 0 & 0 & \Phi_{11} \end{bmatrix} \tag{8.38}$$

and

$$\Phi_{nd} = \begin{bmatrix} 0 & \Phi_{12} & \Phi_{13} \\ \Phi_{21} & 0 & \Phi_{23} \\ \Phi_{31} & \Phi_{32} & 0 \end{bmatrix}. \tag{8.39}$$

8.4 Available Theoretical Results

8.4.1 Stability of the Local Controllers

The main theoretical results concern the stability of the local controllers if taken in isolation, that have been published in [9] and [7], page 114. This result states that the SIORHC will stabilize the closed loop when applied to local system $\mathcal{S}_i$ provided that $N_p \geq n_i$, where n_i is the dimension of the local state $\mathbf{x}_i$.

8.4.2 Coordination Convergence

Another result concerns the convergence of the coordination procedure of Algorithm 8.1. From (8.37) it is concluded that the coordination algorithm will converge provided that the spectral radius

$$\lambda_{max} := \max\ eig\left(\Phi_d^{-1}\Phi_{nd}\right) \tag{8.40}$$

verifies

$$|\lambda_{max}| < 1. \tag{8.41}$$

The verification of condition 8.41 is plant dependent. In [3] this condition is checked, for different values of the manipulated variable weights given by the entries of $\mathbf{R}_i$, for the water delivery canal application described in the next section.

8.5 Application to a Water Delivery Canal

The D-SIORHC algorithm has been applied to the control of local upstream water level in a water delivery canal. The results presented hereafter were originally published in [3].

The work reported hereafter was performed at the experimental canal of *Núcleo de Hidráulica e Controlo de Canais* (Universidade de Évora, Portugal), described in [4]. The canal, of which Fig. 8.3 shows a schematic view, has four pools with a length of $35m$ each, separated by three undershoot gates, with the last pool ended by an overshoot gate. In this work only the first three gates are used, gate number 4 being kept at a constant position. The water enters the canal by gravity with a flow that can be imposed by a controller. There are water off-takes downstream from each branch connected to orifices in the channel walls, that are used to generate disturbances corresponding to water usage by farmers.

Water level sensors are installed downstream of each pool, each one close to the gate that delimits the pool downstream. The water level sensors allow to measure

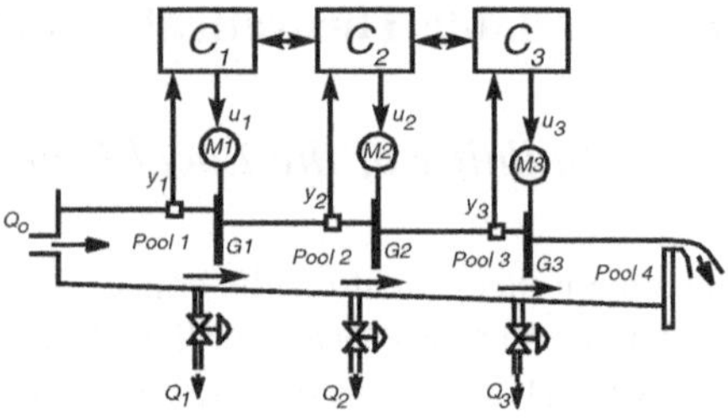

Fig. 8.3 Schematic view of the canal used to test D-SIORHC

values between 0 mm and 900 mm, a value that corresponds to the canal bank. For pool number i, $i = 1, \ldots, 3$, the downstream level is denoted y_i and the opening of gate i is denoted u_i. Pool number i ends with gate number i.

In the example shown, local controllers are connected according to the so called close upstream control structure. For that sake, the canal is decomposed in subsystems, each made of one pool and the gate that limits this pool downstream. The manipulated variable is the gate position, the output is the pool downstream water level and the main disturbances are the offtake flows. To each instance $\mathcal{S}_i$ of these subsystems, associate a local controller $\mathcal{C}_i$ that computes the gate position u_i so that the water level y_i tracks the reference r_i. As explained in Sect. 8.3, each control agent negotiates the value of its gate position with its neighbors. This approach results in the plant/controller interconnection shown in Fig. 8.3.

In the experiment reported, the water flow entering the canal is 0.025 $m^3 s^{-1}$. The number of iterations of the coordination cycle is $N_c = 20$ a value for which it was concluded from simulations that convergence is attained. The weighting matrices are $Q_1 = Q_2 = Q_3 = I$ and $R_1 = R_2 = 100$, $R_3 = 1000$. The sampling interval is 5 s. The prediction horizon is $N = 25$ and the number of coincidence points for the stability constraint is $P = 1$.

Figures 8.4 and 8.5 show the results obtained for a sequence of reference steps in an experiments that lasts about three and a half hour. In the first part of the experiment the reference of pool level 1 is raised and then, after a while, reduced. With the reference kept constant, a disturbance is created by opening the offtake of pool 1 at the time instant corresponding to the vertical line labeled "A". This offtake is then closed again, at the time instant corresponding to the vertical line labeled "B".

Although these maneuvers are performed in direct relation to controller $\mathcal{C}_1$, they also affect the other local control loops as well. In the absence of a coordination mechanism the blind interaction between the different controllers may lead to instability. For instance, if isolated PI controllers are used in this same example, the system is likely to become unstable [5].

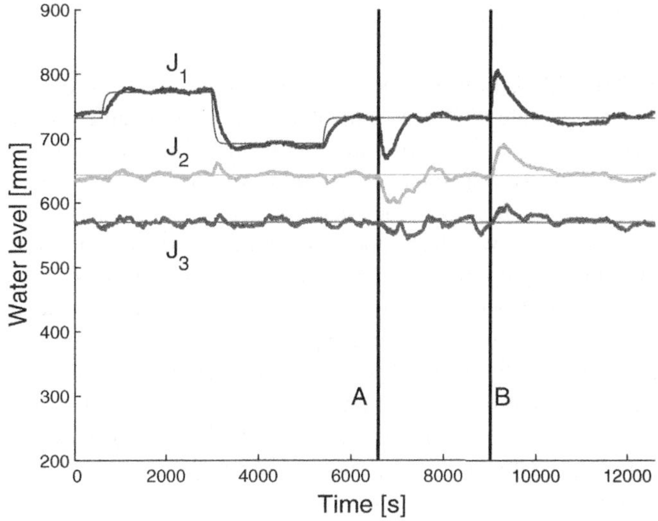

Fig. 8.4 Experimental results with a water delivery canal: downstream water levels for pools 1 up to 3

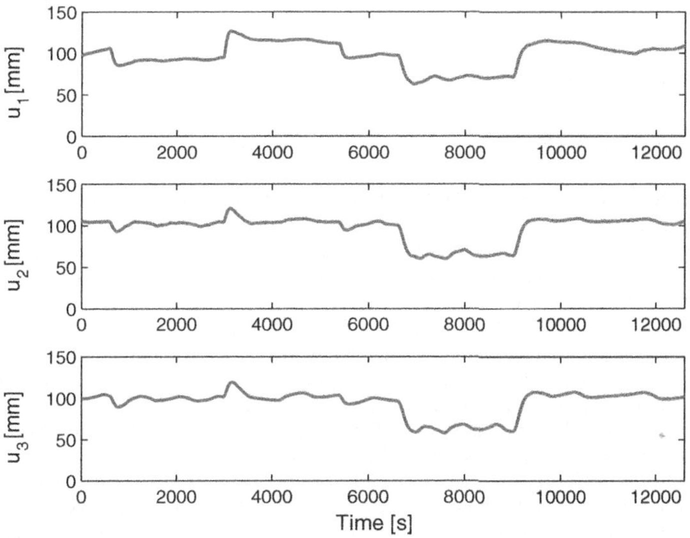

Fig. 8.5 Experimental results with a water delivery canal. Gate positions

8.6 Conclusions

A distributed MPC algorithm for linear systems is described in this chapter. Local controllers consist of state feedback control laws that are obtained by the minimization, in a receding horizon sense, of a moving horizon quadratic cost, subject to the constraint that the output vanishes for a number of so called coincidence points at the end of the prediction horizon. The quadratic cost penalizes not only the tracking error of the subsystem to which the local controller is connected, but also the ones of their neighbors. The coordination between local control agents is obtained by alternating cost optimization that amounts to a negotiation among neighbor control agents. If the negotiation procedure converges, a Nash equilibrium is approximated. The main advantage of this method is its computational simplicity, that depend on the fact that the coordination relies on a closed form formula for the controllers. The incorporation of operational constraints must be done by reference manipulation with a command governor. The control algorithm has been demonstrated in the regulation of pool water level in a water delivery canal.

Acknowledgments This work was supported by FCT—Fundação para a Ciência e a Tecnologia, Portugal, under project *AQUANET: Decentralized and reconfigurable control for water delivery multipurpose canal systems, contract PTDC/EEA-CRO/102102/2008*, and INESC-ID multi-annual funding through PEst-OE/EEI/LA0021/2011.

References

1. T. Başar, G.J. Olsder, *Dynamic Noncooperative Game Theory*, 2nd edn. (SIAM, Philadelphia, 1999)
2. D. Georges, Decentralized adaptive control for a water distribution system, in *Proceedings of the 3rd IEEE Conference on Control Applications*, pp. 1411–1416 (1994)
3. J.M. Igreja, J.M. Lemos, F.M. Cadete, L.M. Rato, M. Rijo, Control of a water delivery canal with cooperative distributed mpc, in *Proceedings of the 2012 Americain Control Conference*, pp. 1961–1966, Montreal, Canada (2012)
4. J.M. Lemos, F. Machado, N. Nogueira, L. Rato, M. Rijo, Adaptive and non-adaptive model predictive control of an irrigation channel. Netw. Heterog. Media **4**(2), 303–324 (2009)
5. J.M. Lemos, L.F. Pinto, Distributed linear-quadratic control of serially chained systems: application to a water delivery canal. IEEE Control Syst. **32**(6) (2012)
6. J.M. Maestre, D. Munoz de la Peña, E.F. Camacho, T. Alamo, Distributed model predictive control based on agent negotiation. J. Proc. Control **21**, 685–697 (2011)
7. E. Mosca, *Optimal, Predictive, and Adaptive Control* (Prentice Hall, Englewood Cliffs, 1995)
8. E. Mosca, J.M. Lemos, J. Zhang, Stabilizing i/o receding horizon control, in *Proceedings of the 29th Conference on Decision and Control*, pp. 2518–2523, Honolulu, Hawai (1990)
9. E. Mosca, J. Zhang, Stable redesign of predictive control. Automatica **28**(6), 1229–1233 (1992)
10. I. Necoara, V. Nedelcu, I. Dumitrache, Parallel and distributed optimization methods for estimation and control in networks. J. Proc. Control **21**, 756–766 (2011)
11. A.N. Venkat, I.A. Hiskens, J.B. Rawlings, S.J. Wright, Distributed mpc strategies with application to powet system generation control. IEEE Trans. Control Syst. Technol. **16**, 1192–1206 (2008)
12. Y. Zhang, S.Y. Li, Networked model predictive control based on neighbourhood optimization for serially connected large-scale processes. J. Process Control **17**(1), 37–50 (2007)

Chapter 9
A Distributed-in-Time NMPC-Based Coordination Mechanism for Resource Sharing Problems

M. Y. Lamoudi, M. Alamir and P. Béguery

Abstract In this chapter, a hierarchical model predictive control framework is presented for a network of subsystems that are submitted to general resource sharing constraints. The method is based on a primal decomposition of the centralized open-loop optimization problem over several subsystems. A coordinator is responsible of adjusting the parameters of the problems that are to be solved by each subsystem. A distributed-in-time feature is combined with a bundle method at the coordination layer that enables to enhance the performance and the real-time implementability of the proposed approach. The scheme performance is assessed using a real-life energy coordination problem in a building involving 20 zones that have to share a limited amount of total power.

9.1 Introduction

The principle of Distributed Model Predictive Control (DMPC) [1, 2] is to design local predictive controllers responsible of local decision making. In iterative schemes, these local controllers have to come up with an agreement throughout "negotiation" iterations in order to recover the solution (or to achieve a relevant suboptimal solution) of the original centralized problem. DMPC appears to be an interesting approach for large scale systems. Indeed, for such systems the centralized optimization problem is generally very hard (or even impossible) to solve given restrictions on the computational time and computational/communication resources. Moreover,

M. Y. Lamoudi (✉) · P. Béguery
Strategy & Innovation, Schneider-Electric Industries, Grenoble, France
e-mail: mohamed-yacine.lamoudi@fr.schneider-electric.com

P. Béguery
e-mail: patrick.beguery@fr.schneider-electric.com

M. Alamir
CNRS-University of Grenoble, Gipsa-lab, Saint Martin d'Hères, France
e-mail: mazen.alamir@gipsalab.grenoble-inp.fr

J. M. Maestre and R. R. Negenborn (eds.), *Distributed Model Predictive Control Made Easy*, Intelligent Systems, Control and Automation: Science and Engineering 69,
DOI: 10.1007/978-94-007-7006-5_9,

the non scalability of the centralized solution is not the only reason of making the distributed approach interesting, indeed centralizing the decision process on one physical controller is generally unsafe because any failure affects the whole system and leads to non-modular schemes which may be impractical. Several DMPC approaches have been proposed in the recent literature. They are generally based either on a primal or a dual decomposition of the centralized optimization problem [3]. As pointed out by [4], the convergence of the distributed solution to the centralized problem generally requires an important number of iterations even for simple case studies. This may be prohibitive for real-time implementability issues and for communication network saturation concern. Consequently, despite the apparent similarity of the many distributed MPC-based solutions, a relevant comparison should focus, among other indicators, on the specific issues of communication and computation time saving.

In order to deal with this issue, a general framework based on the combination of an efficient distributed optimization technique (*disaggregated bundle method*) and an original *distributed-in-time computation* feature is presented. The technique is based on a primal decomposition of the centralized optimization problem and leads to a hierarchical structure.

The original works leading to the scheme proposed here address the issue of energy management in buildings with a high number of zones and limited communication rate. As far as the building case is concerned, the results have been reported in [5, 6]. In the present chapter, a general setting of the underlying framework is proposed and some properties of the resulting algorithms are analyzed in deeper details.

The chapter is organized as follows: Sect. 9.2 presents succinctly the kind of problems that can be addressed by the proposed approach. Section 9.3 gives a step by step description of the framework. In Sect. 9.4, a brief discussion of the theoretical rationale that underlines the distributed-in-time iterations is given. In Sect. 9.5, a real-life example consisting of a 20 zones building submitted to global power restriction and time varying energy tariffs is analyzed using different algorithm parameters. Finally, Sect. 9.6 concludes the chapter and describes future research directions.

9.2 Boundary Conditions

In this section a general description of the class of problems targeted by the proposed scheme is given. Roughly speaking, a network of subsystems sharing limited resources is considered. This is detailed in the present section.

9.2.1 Subsystems

Consider a set of N_s subsystems that are dynamically uncoupled.[1] Each subsystem $\ell \in \mathcal{S} := \{1, \ldots, N_s\}$ is governed by the following general nonlinear dynamic:

[1] Potential coupling can be handled through dedicated observers as it is shown for instance in [5, 6] regarding the building energy management context.

$$x_{\ell,k+1} = f(x_{\ell,k}, u_{\ell,k}) \tag{9.1}$$

where:

- $x_{\ell,k} \in \mathbb{R}^{n^x_\ell}$ is the state vector of the subsystem ℓ at instant k,
- $u_{\ell,k} \in \mathbb{R}^{n^u_\ell}$ is the input vector of the subsystem ℓ at instant k,

In the sequel, the following notation is extensively used: given a vector quantity $v_\ell \in \mathbb{R}^{n_v}$ related to subsystem ℓ, the boldfaced vector $\mathbf{v}_{\ell,k}$ represents the future profiles v_ℓ over the prediction horizon of length N_p, namely:

$$\mathbf{v}_{\ell,k} := [\nu^T_{\ell,k}, \ldots, \nu^T_{\ell,k+N_p-1}]^T \tag{9.2}$$

when no ambiguity results, the time index k is dropped and the predicted profile $\mathbf{v}_{\ell,k}$ is simply denoted by $\mathbf{v}_\ell$. Moreover, all the quantities indexed by $\cdot_{\ell,k}$ refers to quantities related to subsystem ℓ at instant k.

Let us now define for each subsystem $\ell \in \mathcal{S}$ the resource vector $r_{\ell,k} \in \mathbb{R}^{n^s_\ell}$ and its future profile $\mathbf{r}_{\ell,k}$. The resource limitation for subsystems $\ell \in \mathcal{S}$ over the prediction horizon are expressed through the following local constraints inequality:

$$\forall \ell \in \mathcal{S}, \quad \mathbf{h}_\ell(\mathbf{x}_{\ell,k}, \mathbf{u}_{\ell,k}, \mathbf{r}_{\ell,k}) \leq 0 \tag{9.3}$$

It is assumed that each subsystem $\ell \in \mathcal{S}$ is locally controlled by a model predictive controller referred to hereafter by $\text{MPC}_\ell(\mathbf{r}_{\ell,k})$ where $\mathbf{r}_{\ell,k}$ is the available resource profile allocated to subsystem ℓ over the prediction horizon. Therefore, at each sampling time k and given a prescribed available resource profile $\mathbf{r}_{\ell,k}$ over the prediction horizon, the system ℓ has to solve the following optimization problem:

$$\text{MPC}_\ell(\mathbf{r}_{\ell,k}) : \underset{\mathbf{x}_{\ell,k} \in \mathcal{X}_{\ell,k}, \mathbf{u}_{\ell,k} \in \mathcal{U}_{\ell,k}}{\text{Minimize}} \quad L_\ell(\mathbf{x}_{\ell,k}, \mathbf{u}_{\ell,k}) \tag{9.4a}$$

$$\text{Subject to:} \quad \mathbf{h}_\ell(\mathbf{x}_{\ell,k}, \mathbf{u}_{\ell,k}, \mathbf{r}_{\ell,k}) \leq 0 \tag{9.4b}$$

where $L_\ell(\mathbf{x}_{\ell,k}, \mathbf{u}_{\ell,k}) \geq 0$ is the cost function related to the subsystem ℓ and the domains $\mathcal{X}_{\ell,k}$ and $\mathcal{U}_{\ell,k}$ denote respectively state and input constraints that are possibly time-varying. Once the problem (9.4) is solved at time k, the optimal predicted input and state trajectories $\mathbf{u}^\star_{\ell,k}$ and $\mathbf{x}^\star_{\ell,k}$ are obtained. The first component $u^\star_{\ell,k}$ of $\mathbf{u}^\star_{\ell,k}$ is applied to the subsystem during the time interval $[k, k+1]$. This operation is renewed at the next instant $k+1$ based on new measurement or estimation of the state $x_{\ell,k+1}$ and a new available resource profile $\mathbf{r}_{\ell,k+1}$ and so on. The next section details how the profile $\mathbf{r}_{\ell,k}$ is managed by a coordination level.

9.2.2 Resource Sharing

Consider now a global constraint on the available resource for the whole network of subsystems. Assume that it is expressed through the following inequality:

$$\mathbf{H}(\mathbf{r}_{1,k}, \ldots, \mathbf{r}_{N_s,k}) \leq 0 \tag{9.5}$$

The centralized formulation of the optimization problem becomes:

$$\underset{\{\mathbf{x}_{\ell,k} \in \mathcal{X}_{\ell,k}, \mathbf{u}_{\ell,k} \in \mathcal{U}_{\ell,k}\}_{\ell \in \mathcal{S}}}{\text{Minimize}} \quad \sum_{\ell \in \mathcal{S}} L_\ell(\mathbf{x}_{\ell,k}, \mathbf{u}_{\ell,k}) \tag{9.6a}$$

Subject to:

$$\mathbf{H}(\mathbf{r}_{1,k}, \ldots, \mathbf{r}_{N_s,k}) \leq 0 \tag{9.6b}$$

$$\mathbf{h}_\ell(\mathbf{x}_{\ell,k}, \mathbf{u}_{\ell,k}, \mathbf{r}_{\ell,k}) \leq 0, \quad \forall \ell \in \mathcal{S} \tag{9.6c}$$

When the number of subsystems N_s is high, the centralized optimization problem (9.6) becomes a large scale optimization problem which may become intractable. This is why a coordination layer is introduced in order to keep the global constraint (9.5) satisfied despite the fact that each subsystem still solves its own optimization problem. The coordinator is responsible of adjusting the available resources profiles $\{\mathbf{r}_{\ell,k}\}_{\ell \in \mathcal{S}}$ of each subsystem $\ell \in \mathcal{S}$. The communication is supposed take place between each subsystem and the coordinator but is not available between the subsystems, leading to a classical hierarchical scheme. The kind of information exchanged between these entities as well as an algorithmic description of the optimization taking place at the coordination layer are presented in the next section.

9.3 Description of the Approach

In order to coordinate the subsystems, an efficient iterative procedure is designed in this section. Throughout the negotiation iterations taking place between the coordinator and subsystems, the coordinator refines the optimal resource allocation in order to achieve a suitable repartition of the resources between the subsystems. To this end, let us first decompose the centralized problem (9.6) and introduce some notations and assumptions.

9.3.1 Problem Decomposition

Let $J_{\ell,k}(\mathbf{r}_{\ell,k})$ denotes the achieved optimal value for a given resource allocation $\mathbf{r}_{\ell,k}$:

$$J_{\ell,k}(\mathbf{r}_{\ell,k}) := L_\ell(\mathbf{x}^\star_{\ell,k}, \mathbf{u}^\star_{\ell,k}) \tag{9.7}$$

Moreover, consider that the (sub-)gradient $\mathbf{g}_{\ell,k}(\mathbf{r}_{\ell,k})$ of $J_{\ell,k}$ is available, namely:

$$\mathbf{g}_{\ell,k}(\mathbf{r}_{\ell,k}) \in \partial J_{\ell,k}(\mathbf{r}_{\ell,k}) \tag{9.8}$$

where $\partial J_{\ell,k}(\mathbf{r}_{\ell,k})$ is the subdifferential set of the function $J_{\ell,k}$ at $\mathbf{r}_{\ell,k}$.

Assumption 9.1 For all ℓ and all k, the function $J_{\ell,k}(\mathbf{r}_{\ell,k})$ is convex and its subgradient $\mathbf{g}_{\ell,k}(\mathbf{r}_{\ell,k}) \in \partial J_{\ell,k}(\mathbf{r}_{\ell,k})$ is available.

Note also that the local problems (9.4) are not always feasible. However, let us consider that $\mathbf{r}_{\ell,k} \in \mathcal{F}_{\ell,k}$ is a sufficient condition to ensure its feasibility and that $\mathcal{F}_{\ell,k}$ is available for all the subsystems $\ell \in \mathcal{S}$.

Assumption 9.2 The local sub-problems (9.4) are feasible $\forall \mathbf{r}_{\ell,k} \in \mathcal{F}_{\ell,k}$. Moreover, the domains $\mathcal{F}_{\ell,k}, \forall \ell \in \mathcal{S}$ are known by the coordinator.

Based on the previous definitions and assumptions, one can rewrite the optimization problem (9.6) in the following form:

$$\underset{\{\mathbf{r}_{\ell,k} \in \mathcal{F}_{\ell,k}\}_{\ell \in \mathcal{S}}}{\text{Minimize}} \quad J(\mathbf{r}_{\cdot,k}) := \sum_{\ell \in \mathcal{S}} J_{\ell,k}(\mathbf{r}_{\ell,k}) \quad \text{S.t.} \quad \mathbf{H}(\mathbf{r}_{1,k}, \ldots, \mathbf{r}_{N_s,k}) \le 0 \tag{9.9}$$

or more shortly

$$\underset{\mathbf{r}_{\cdot,k} \in \mathcal{D}}{\text{Minimize}} \quad J(\mathbf{r}_{\cdot,k}) \tag{9.10}$$

where $\mathbf{r}_{\cdot,k} := [\mathbf{r}^T_{1,k}, \ldots, \mathbf{r}^T_{N_s,k}]^T$. The notation $\mathbf{r}_{\cdot,k} \in \mathcal{D}$ means the fulfillment of both the feasibility conditions $\{\mathbf{r}_{\ell,k} \in \mathcal{F}_{\ell,k}\}_{\ell \in \mathcal{S}}$ and global resources constraints (9.5).

The optimization problem (9.10) is called the *master problem* and its resolution is performed at the coordination layer as explained in the following section.

9.3.2 Solving the Master Problem-disaggregated Bundle Method

In order to solve the master problem, the coordinator successively approximates the functions $\{J_{\ell,k}\}_{\ell \in \mathcal{S}}$. Indeed, let us remind that the coordinator does not have any information regarding the dynamics and current states of the subsystems. However, the coordinator is supposed to be able to affect any resource $\mathbf{r}_{\ell,k}$ to the subsystems $\ell \in \mathcal{S}$ and to request the values of the objective function of each subsystem ℓ as well as their corresponding sub-gradients (assumption 9.1). Bundle method for this kind of problems appears to be an interesting choice, it relies on iteratively approximating the objective function $J = \sum_{\ell \in \mathcal{S}} J_{\ell,k}$ by a so called *cutting plane model*. Since

Algorithm 9.1 Bundle update

1: **for** $i \leftarrow n_{\mathbf{B}} - 1, \ldots, 1$ **do** {Shift the bundle memory}
2: $\mathbf{s}_\ell^{(i+1)}, \epsilon_\ell^{(i+1)} \leftarrow \mathbf{s}_\ell^{(i)}, \epsilon_\ell^{(i)}$
3: **end for**
4: $\mathbf{s}_\ell^{(1)} \leftarrow \mathbf{g}_\ell^{(s)}$
5: $\epsilon_\ell^{(1)} \leftarrow J_\ell(\mathbf{r}_\ell^{(s)}) - \langle \mathbf{g}_\ell^{(s)}, \mathbf{r}_\ell^{(s)} \rangle$

the objective function J is separable, an individual *cutting plane model* $\check{J}_{\ell,k}$ of each function $J_{\ell,k}$ is built up. In this section, a presentation of the bundle method is given, the reader may refer to [7] for more details. For the sake of conciseness, the time index k is dropped in the notations (e.g : $\mathbf{r}_\ell \equiv \mathbf{r}_{\ell,k}$) since the procedure described here takes place at a given instant k.

In disaggregated bundle method, individual cutting plane approximations of the functions $\{J_\ell\}_{\ell \in \mathcal{S}}$ are built-up thanks to a memory $\mathbf{B}_\ell^{(s)}$ dedicated to each sub-function ℓ, updated at each iteration s and defined as follows:

$$\mathbf{B}_l^{(s)} := \{\mathrm{s}_l^{(i)}, \in_l^{(i)}\} i = 1, \ldots, n_{\mathbf{B}} \tag{9.11}$$

The bundle $\mathbf{B}_\ell^{(s)}$ keeps only the $n_{\mathbf{B}}$ last elements and behaves like a FIFO register in which the first element ($i = 1$) of the bundle is updated at the current iterate s after the whole bundle has been shifted and its last element dropped, namely:

$$\mathbf{B}_\ell^{(s)} = \text{Update}(\mathbf{B}_\ell^{(s-1)}, J_\ell(\mathbf{r}_\ell^{(s)}), \mathbf{g}_\ell^{(s)}, \mathbf{r}_\ell^{(s)}) \tag{9.12}$$

where the function update is defined by the algorithm (9.1).

The (sub-)gradients $\mathbf{s}_\ell^{(i)}$ and their corresponding linearization errors $\epsilon_\ell^{(i)}$ (:5) are computed thanks to the values of the function $J_\ell(\mathbf{r}_\ell^{(s)})$ and (sub-)gradients $\mathbf{g}_\ell^{(s)}$ returned by each the subsystem ℓ when the coordinator requests an evaluation at the current iterate $\mathbf{r}_\ell^{(s)}$. Each bundle of information $\mathbf{B}_\ell^{(s)}$ enables the so called *cutting plane approximation* $\check{J}_\ell^{(s)}(\cdot)$ to be defined according to:

$$\check{J}_\ell^{(s)}(\cdot) := \underset{i=1,\ldots,n_{\mathbf{B}}}{\text{Max}} \langle \mathbf{s}_\ell^{(i)}, \cdot \rangle + \epsilon_\ell^{(i)} \tag{9.13}$$

where each linear piece $\langle \mathbf{s}_\ell^{(i)}, \cdot \rangle + \epsilon_\ell^{(i)}, i = 1, \ldots, n_{\mathbf{B}}$ defines a half space as depicted on Fig. 9.1. Indeed, each *cutting plane* i is a supporting hyperplane of the epigraph $\mathbf{epi}(J_\ell)$ of the function J_ℓ and constitutes, since J_ℓ is supposed convex, a global under-estimator of J_ℓ.

Given the approximations $\check{J}_\ell^{(s)}, \ell \in \mathcal{S}$, an approximation $\check{J}^{(s)}$ of the centralized cost function can be obtained according to:

$$\check{J}^{(s)}(\mathbf{r}_\cdot) := \sum_{\ell \in \mathcal{S}} \check{J}_\ell^{(s)}(\mathbf{r}_\ell) \tag{9.14}$$

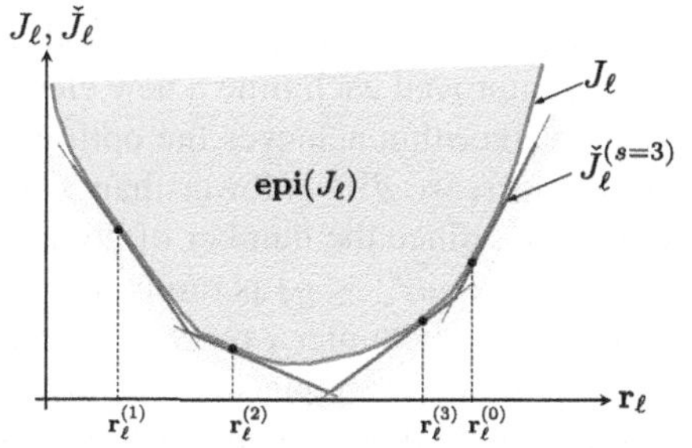

Fig. 9.1 Representation of J_ℓ and its piece-wise linear approximation $\check{J}_\ell^{(3)}$ after the third iteration. The approximate function is a global under-estimator of J_ℓ

Instead of minimizing $\check{J}^{(s)}(\cdot)$, the following regularized optimization problem, denoted Master$^{(s)}$, is considered:

$$\text{Master}^{(s)}: \quad \mathbf{r}_\cdot^{(s+1)} := \underset{\mathbf{r}_\cdot \in \mathcal{D}}{\text{Argmin}} \, [\check{J}^{(s)}(\mathbf{r}_\cdot) + D_{\gamma^{(s)}}(\mathbf{r}_\cdot - \overline{\mathbf{r}}_\cdot^{(s)})] \tag{9.15}$$

where the stabilization term $D_{\gamma^{(s)}}(\mathbf{r}_\cdot - \overline{\mathbf{r}}_\cdot^{(s)})$ is introduced in order to prevent any drastic movement from the current best candidate point $\overline{\mathbf{r}}_\cdot^{(s)}$, which is called the stability center (or central point). Quite weak assumptions on the properties of $D_{\gamma^{(s)}}(\cdot)$ are necessary in order to ensure the convergence of the algorithm [7], however the most common choice is the following (euclidean) measure:

$$D_{\gamma^{(s)}}(\mathbf{r}_\cdot - \overline{\mathbf{r}}_\cdot^{(s)}) := \frac{1}{2 \cdot \gamma^{(s)}} \|\mathbf{r}_\cdot - \overline{\mathbf{r}}_\cdot^{(s)}\|_{L_2} \tag{9.16}$$

The parameter $\gamma^{(s)} > 0$ is the proximity parameter, it is updated at each iteration s. The central point $\overline{\mathbf{r}}_\cdot^{(s)}$ plays a crucial role in the bundle method since it enables, contrarily to pure cutting planes techniques [8], to keep track of the best known solution until iteration s and to help avoiding eventual oscillations resulting from a potentially poor approximation of the functions $\{J_\ell\}_{\ell \in \mathcal{S}}$.

In order to derive updating rules for the central point $\overline{\mathbf{r}}_\cdot^{(s)}$ and the parameter $\gamma^{(s)}$ let us define on one hand the predicted decrease at iteration s:

$$\hat{d}^{(s)} := J(\overline{\mathbf{r}}_\cdot^{(s)}) - \check{J}^{(s)}(\mathbf{r}_\cdot^{(s+1)}) \geq 0 \tag{9.17}$$

and on the other hand the real decrease $d(s)$:

$$d^{(s)} := J(\overline{\mathbf{r}}_\cdot^{(s)}) - J(\mathbf{r}_\cdot^{(s+1)}) \tag{9.18}$$

If the real decrease is greater than a certain fraction $f \in [0, 1]$ of the predicted decrease then the current iterate s is called a *Serious Step*, in this case the coordinator enhanced the current solution : the central point $\bar{\mathbf{r}}^{(s)}$ is updated, and the proximity parameter $\gamma^{(s)}$ is increased. Otherwise ($d^{(s)} < f \cdot \hat{d}^{(s)}$), the step s is called a *Null Step*: the central point in kept unchanged and the proximity parameter $\gamma^{(s)}$ is decreased. Note however that in both situations the accuracy of the approximation $\check{J}^{(s)}$ is improved each time a new element is incorporated in the bundle (9.12).

The algorithm achieves the optimal solution with an accuracy η_J when the predicted decrease $\hat{d}^{(s)}$ is lower than a predefined accuracy on the objective function $\hat{d}^{(s)} \le \eta_J$. Since the number of iterations allowed in our framework is very limited the condition $\hat{d}^{(s)} \le \eta_J$ is rarely achieved. Thus, the algorithm is generally stopped if the iteration counter s reaches the maximum allowed number of iterations s_{max}.

Let us finally emphasize the fact that all iterates are feasible in the sense of respecting all global and also local constraints. This feature is very interesting since the algorithm can be stopped, if necessary, at any iteration. To find a feasible starting point, a projection of the best known solution at instant k, denoted $\mathbf{r}^{\sharp}_{\cdot}$, on the domain $\mathcal{D}$ is performed at the initialization phase. This operation is performed centrally, namely:

$$\mathbf{r}_{\cdot}^{(0)} := \text{Init}(\mathbf{r}^{\sharp}_{\cdot}) = \underset{\mathbf{r}_{\cdot} \in \mathcal{D}}{\text{Argmin}} \, \|\mathbf{r}_{\cdot} - \mathbf{r}^{\sharp}_{\cdot}\| \tag{9.19}$$

Some other technical details regarding, bundle compression techniques and update strategies of $\gamma^{(s)}$ have been omitted but are available in [7].

9.3.3 Distributing the Optimization Over Time

Ideally, all the iterations described in the previous section have to take place during the sampling period $[k-1, k]$. Moreover, many of these iterations have to be performed in order to achieve a sufficiently good approximation of cost functions $J_{\ell,k}$. This may be incompatible with the sampling period. The idea is then to still use a modified version of the past information contained in the previous bundles $\mathbf{B}_{\ell,k-j}$, $j = 1, 2, \ldots$. This operation relies on the assumption that the functions $J_{\ell,k}(\cdot)$ don't change drastically and therefore the approximations of these functions, stored in the bundle $\mathbf{B}_{\ell,k-1}$, can be used as an initialization at instant k of the bundle but "corrected" in the following way:

$$\mathbf{B}_{\ell,k}^{(0)} = \{m_{\ell,k} \cdot \overleftarrow{\mathbf{s}}_{\ell}^{(i)}, m_{\ell,k} \cdot \epsilon_{\ell}^{(i)}\}_{i=1,\ldots,n_{\mathbf{B}}} \tag{9.20}$$

where the approximated (sub-)gradient $\overleftarrow{\mathbf{s}}_{\ell}^{(i)}$ is obtained by conveniently shifting and completing the (sub-)gradient vectors obtained over the last instants. The operator shift $\overleftarrow{\mathbf{v}_k}$ of any profile $\mathbf{v}_k = [v_k^T, \ldots, v_{k+N-1}^T]^T$ is defined by:

$$\overleftarrow{\mathbf{v}_k} := [v_{k+1}^T, \ldots, v_{k+N-1}^T, v_{k+N-1}^T]^T \tag{9.21}$$

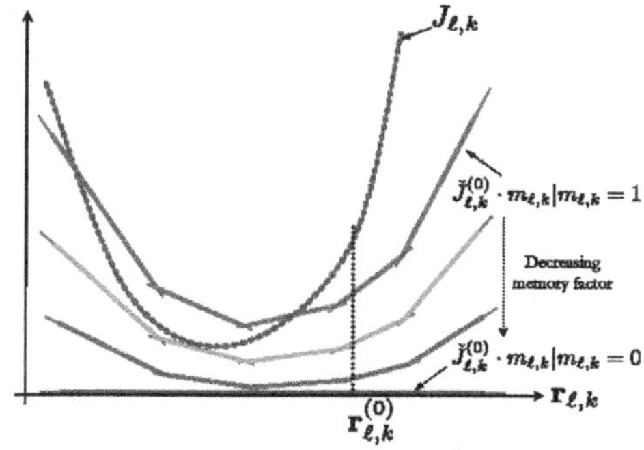

Fig. 9.2 Illustration of the role played by the memory factor $m_{\ell,k}$ enabling previously computed cutting planes to remain relevant as lower bound of the function to be approximated

The positive parameter $m_{\ell,k} \in [0\ ,1-\epsilon]$ $(0 < \epsilon \ll 1)$ is the memory factor and plays a central role in the scheme, it enables to "forget" a certain part of the information. The memory parameter expresses a trade-off between the quantity of information that one wants to keep and the fact the cuts used as initialization for the bundle at instant k should be under-estimators of the function $J_{\ell,k}$ in order to prevent the optimal solution from begin excluded from the search domain. Remark that when $m_{\ell,k} = 0$, no information of the previous instants is kept at the current instant and the whole bundle information is equivalent to $J_{\ell,k} > 0$. For these reasons, the memory factors $m_{\ell,k}$ are adjusted in accordance with the quality of the initial approximation $\check{J}^{\text{init}}_{\ell,k}$ computed at the initial point $\mathbf{r}^{(0)}_{\ell,k}$ and generalized over the whole bundle (Fig. 9.2).

$$m_{\ell,k} := 1 - \text{Sat}_{[0\ ,1-\epsilon]}\left(\frac{[\check{J}^{\text{init}}_{\ell,k}(\mathbf{r}^{(0)}_{\ell,k}) - J_{\ell,k}(\mathbf{r}^{(0)}_{\ell,k})]^2}{[J_{\ell,k}(\mathbf{r}^{(0)}_{\ell,k})]^2}\right) \tag{9.22}$$

where the initial approximation $\check{J}^{\text{init}}_{\ell,k}$ results from the bundle $\mathbf{B}^{(0)}_{\ell,k}|m_{\ell,k} = 1$ (with no forgetting factor). The function $\text{Sat}_{[a\ b]}(\cdot)$ is the saturation function.

9.3.4 The Correction Mechanism

It is essential to understand that the memory capability starts from the assumption that the functions $J_{\ell,k}$ from one decision instant to the next are supposed to have a slow change. In such conditions the initial rough approximation (9.20) is supposed to give a rough yet valuable starting approximation at instant k. However, one has to mention that this assumption is very hard to check. Moreover, note that the gradient approximation $\overleftarrow{\mathbf{s}}_{\ell}$ introduced previously introduces itself an intrinsic error and that no knowledge about the sub-systems respective states and/or disturbances is neither available nor taken into account.

Algorithm 9.2 Bundle correction mechanism

for $i \leftarrow 1, \ldots, n_{\mathbf{B}}$ **do**
  $\beta^{(i)} = J_{\ell,k}(\mathbf{r}_{\ell,k}^{(s)}) - (\langle \mathbf{s}_{\ell}^{(i)}, \mathbf{r}_{\ell,k}^{(s)} \rangle + \epsilon_{\ell}^{(i)})$
  if $\beta^{(i)} < 0$ **then**
    $\epsilon^{(i)} = \epsilon^{(i)} + 1.1 \cdot \beta^{(i)}$
  end if
  $\mathbf{B}_{\ell,k}^{(s)} = \{\mathbf{s}_{\ell}^{(i)}, \epsilon_{\ell}^{(i)}\}_{i=1,\ldots,n_{\mathbf{B}}}$ {Update the bundle information}
end for

This is why a correction mechanism has to be introduce in order to ensure, a priori, that the linear pieces $i = 1, \ldots, n_{\mathbf{B}}$ remain under-estimators of the functions $\{J_{\ell}\}_{\in \mathcal{S}}$ in order to prevent the optimal point to be excluded from the current search domain. Therefore, each time a new evaluation of the function $J_{\ell,k}(\mathbf{r}_{\ell,k}^{(s)})$ is performed, it is easy to check that each linear piece composing the current approximation $\check{J}_{\ell}^{(s)}$ under-estimates the function $J_{\ell,k}$ when compute at $(\mathbf{r}_{\ell,k}^{(s)})$, defining $\beta^{(i)}$ as the difference between the *true* value of the function at the current iterate s (returned by subsystems) and the value of the ith linear piece.

$$\beta^{(i)} := J_{\ell,k}(\mathbf{r}_{\ell,k}^{(s)}) - (\langle \mathbf{s}_{\ell}^{(i)}, \mathbf{r}_{\ell,k}^{(s)} \rangle + \epsilon_{\ell}^{(i)}) \quad i = 1, \ldots, n_{\mathbf{B}} \tag{9.23}$$

If $\beta^{(i)} \geq 0$ then the linear piece i is a priori a valid under-estimator of the function and it is kept in the bundle. Otherwise $\beta^{(i)} < 0$ and the linear piece i in the bundle has to be vertically translated in order to correct its position, namely:

$$\text{if} \quad \beta^{(i)} < 0 \quad \text{than} \quad \epsilon^{(i)} = \epsilon^{(i)} + 1.1 \cdot \beta^{(i)} \tag{9.24}$$

For the sake of clarity, the correction mechanism is defined as a function:

$$\text{Correct}(\mathbf{B}_{\ell,k}^{(s)}, J_{\ell,k}(\mathbf{r}_{\ell,k}^{(s)}), \mathbf{r}_{\ell,k}^{(s)}) \tag{9.25}$$

where the function Correct$(\cdot)$ is defined as in algorithm 9.2.

Finally, in order to make as clear as possible the presentation of the technique, the complete framework is presented on Fig. 9.3, in which the elements $\{\text{BM}_{\ell}\}_{\ell \in \mathcal{S}}$ are the bundle managers units, they perform the operations related to bundle storage, correction and memory factors described previously. The algorithm (9.3)[2] (see page 12) exhaustively summarizes the proposed framework.

[2] Contraction and dilatation parameters of $\gamma^{(s)}$, here 1.1 and 0.8 are provided as indications.

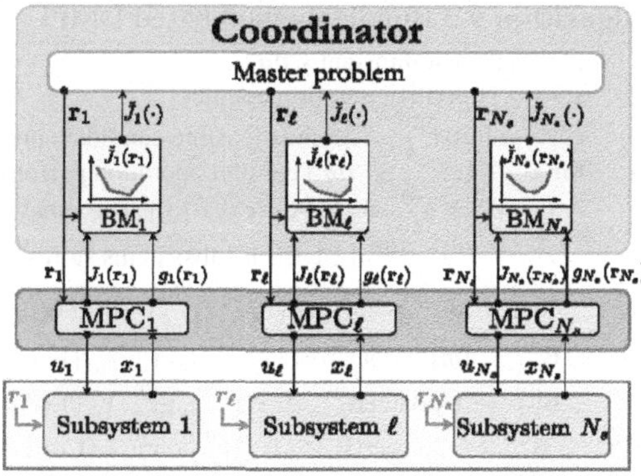

Fig. 9.3 Hierarchical Distributed Model Predictive Control scheme

9.4 Theoretical Results Availability

Convergence results for the classical bundle method (for an optimization problem defined once for all) can be found in [7] when the cost function is convex. The difficulties (in carrying the convergence analysis) associated to the dynamic character of the problem (that changes because of the state variation during a sampling period) can be overcome following the same guidelines recently used in [9]. Roughly speaking, it can be shown that the decrease in the cost function that is guaranteed in the ideal static case are disturbed by a term which is $O(\tau^2)$ where τ is the control updating

Algorithm 9.3 Bundle method based DMPC with memory

for each decision instant k **do**
  $s \leftarrow 0$ {Initialize iteration counter}
  $\mathbf{r}^{(0)}_{\cdot,k} \leftarrow \text{Init}(\overleftarrow{\mathbf{r}^{\star}_{\cdot,k-1}})$ {Sends feasible candidate profiles for subsystems}
  for $\ell \leftarrow 1, \ldots, N_s$ **do** {Parallel operation performed by the subsystems}
    $J_{\ell,k}(\mathbf{r}^{(0)}_{\ell,k}), \mathbf{g}^{(0)}_{\ell,k} \leftarrow \text{MPC}_{\ell}(\mathbf{r}^{(0)}_{\ell,k})$ {$\uparrow\uparrow$ The subsystems solve their optimization}
    $u^{\sharp}_{\ell,k} \leftarrow (u^{\star}_{\ell,k})^{(0)}$ {$\downarrow\downarrow$ Each subsystems stores the current optimal control solution}
    $m_{\ell,k} := 1 - \text{Sat}_{[0,1-\epsilon]}\left([\check{J}^{\text{init}}_{\ell}(\mathbf{r}^{(0)}_{\ell,k}) - J_{\ell,k}(\mathbf{r}^{(0)}_{\ell,k})]^2/[J_{\ell,k}(\mathbf{r}^{(0)}_{\ell,k})]^2\right)$ {Compute memory fact.}
    $\mathbf{B}^{(0)}_{\ell,k} = \{m_{\ell,k} \cdot \overleftarrow{\mathbf{s}}^{(i)}_{\ell}, m_{\ell,k} \cdot \epsilon^{(i)}_{\ell}\}_{i=1,\ldots,n_{\mathbf{B}}}$ {Forgetting operation}
    $\mathbf{B}^{(0)}_{\ell,k} \leftarrow \text{Update}(\mathbf{B}^{(0)}_{\ell,k}, J_{\ell,k}(\mathbf{r}^{(0)}_{\ell,k}), \mathbf{g}^{(0)}_{\ell,k}, \mathbf{r}^{(0)}_{\ell,k})$ {Add new cut to the bundle}
    $\mathbf{B}^{(0)}_{\ell,k} \leftarrow \text{Correct}(\mathbf{B}^{(0)}_{\ell,k}, J_{\ell,k}(\mathbf{r}^{(0)}_{\ell,k}), \mathbf{r}^{(0)}_{\ell,k})$ {Correct bundles information}
  end for
  $\bar{\mathbf{r}}^{(0)}_{\cdot} \leftarrow \mathbf{r}^{(0)}_{\cdot}$, $\hat{d}^{(0)} \leftarrow \infty$ {Initialize the central point}
  while $s \leq s_{max}$ **and** $\hat{d}^{(s)} \geq \eta_{\mathbf{J}}$ **do**
    $\mathbf{r}^{(s+1)}_{\cdot,k} \leftarrow \text{Master}^{(s)}$ { $\downarrow\downarrow$ Coordinator solves the master problem}
    $\hat{d}^{(s)} = J(\bar{\mathbf{r}}^{(s)}_{\cdot,k}) - \check{J}^{(s)}(\mathbf{r}^{(s+1)}_{\cdot,k})$ {Compute predicted decrease}
    for $\ell \leftarrow 1, \ldots, N_s$ **do**
      $J_{\ell,k}(\mathbf{r}^{(s+1)}_{\ell,k}), \mathbf{g}^{(s+1)}_{\ell,k} \leftarrow \text{MPC}_{\ell}(\mathbf{r}^{(s+1)}_{\ell,k})$ {$\uparrow\uparrow$ The subsystems solve their optimization}
      $d^{(s)} = J(\bar{\mathbf{r}}^{(s)}_{\cdot,k}) - \check{J}^{(s)}(\mathbf{r}^{(s+1)}_{\cdot,k})$ {Compute real decrease}
      if $\hat{d}^{(s)} > f \cdot d^{(s)}$ **then** {Serious step}
        $\gamma^{(s+1)} \leftarrow 1.1 \cdot \gamma^{(s)}$, $\bar{\mathbf{r}}^{(s+1)}_{\cdot,k} \leftarrow \mathbf{r}^{(s+1)}_{\cdot,k}$ {Increase γ, Update the central point}
        $u^{\sharp}_{\ell,k} \leftarrow (u^{\star}_{\ell,k})^{(s+1)}$ {$\downarrow\downarrow$ The subsystems update optimal solutions}
      else
        $\gamma^{(s+1)} \leftarrow 0.8 \cdot \gamma^{(s)}$, $\bar{\mathbf{r}}^{(s+1)}_{\cdot,k} \leftarrow \bar{\mathbf{r}}^{(s)}_{\cdot,k}$ {Decrease γ, central point is unchanged}
      end if
      $\mathbf{B}^{(s+1)}_{\ell,k} \leftarrow \text{Update}(\mathbf{B}^{(s)}_{\ell,k}, J_{\ell,k}(\mathbf{r}^{(s+1)}_{\ell,k}), \mathbf{g}^{(s+1)}_{\ell,k}, \mathbf{r}^{(s+1)}_{\ell,k})$ {Add new cut to the bundle}
      $\mathbf{B}^{(s+1)}_{\ell,k} \leftarrow \text{Correct}(\mathbf{B}^{(s+1)}_{\ell,k}, J_{\ell,k}(\mathbf{r}^{(s+1)}_{\ell,k}), \mathbf{r}^{(s+1)}_{\ell,k})$ {Correct bundles information}
    end for
    $s \leftarrow s + 1$ {Increment iteration counter}
  end while
  $\mathbf{r}^{\star}_{\cdot,k} \leftarrow \bar{\mathbf{r}}^{(s+1)}_{\cdot,k}$
  for $\ell \leftarrow 1, \ldots, N_s$ **do** {Parallel operation performed by the subsystems}
    Subsystem ℓ Applies $u^{\sharp}_{\ell,k}$
  end for
end for

period. As the latter is precisely reduced by the distributed in time scheme (requiring fewer iterations to be needed at each updating period) a sort of virtuous circle in favor of stability circle is favor of stability assessment takes place.

9.5 Application Results Availability

The proposed approach has been assessed in [5, 6]. The case study consists of a 20 zones building (20 subsystems). The building is able to store electrical energy in batteries in order to redistribute it to zones. This storage capability offers the benefit of shifting energy consumption of the building to periods in which electricity is cheaper since the final objective here is to minimize the energy invoice while respecting comfort constraints. The power grid can provide to the building a limited amount of power and constitute in this sense a hard constraint that must be respected. The local MPC controllers enable to handle local actuators at the least energetic cost besides ensuring occupants comfort [10, 11]. Therefore, the building should take into account this power limitation in advance in order to store energy in an electrical form or a thermal form in the zones in order to maintain occupants comfort within the prescribed level. Remind that the zones computation units (local MPCs) are much less powerful than the computation unit of the coordinator therefore the number iterations s_{max} is the most important criterion concerning the assessment of the scheme. Nevertheless, one has to mention that the introduction of memory feature should not result in an excessive increase of the computational burden of the coordinator. This is why, the computational time of the coordinator is examined in this section. In order to illustrate the benefit of the memory mechanism presented in this chapter a set of simulations on 12 h has been performed for different combinations of s_{max} (maximum number of iterations) and $n_{\mathbf{B}}$ (bundle memory size), the results are depicted on Fig. 9.5. Let however distinguish two main phases:

- **(12p.m–7a.m)**. During this phase the objective function is decreasing in order to enable the system to reach its steady state. Notice that the introduction of the memory feature enhances considerably the overall performance of the scheme. For instance, using only three iterations with memory enables to perform a fastest decrease of the objective function (excluding the very first decision instants) at a lower computational time compared to 10 iterations without memory. Note also that the computational time can be higher when using memory with four iterations than with five iterations, this results from the fact that the coordinator compensate the lack of new information by using a higher number of cuts previously stored in the bundle, this results in harder optimization problems to solve, implying a larger computational burden -which may be as high as in the case of 10 iterations without memory-.
- **(7a.m–12a.m)** during this phase the system reaches its steady state. The schemes exhibit quite similar performances. Nevertheless, the computational burden is still lower in the cases with memory -since less iterations are allowed.

9.6 Conclusions

In this chapter, a distributed-in-time algorithm for distributed model predictive control has been presented. The main advantages lie in decreasing drastically the number

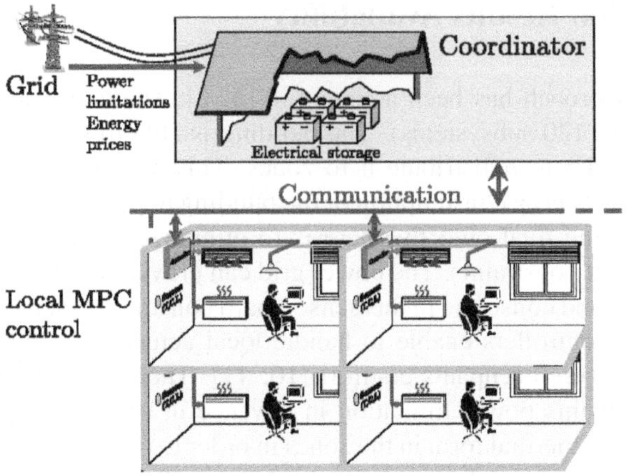

Fig. 9.4 Distributed model predictive in a multizone building. The coordinator gathers global information and ensures coordination of local controllers besides battery management

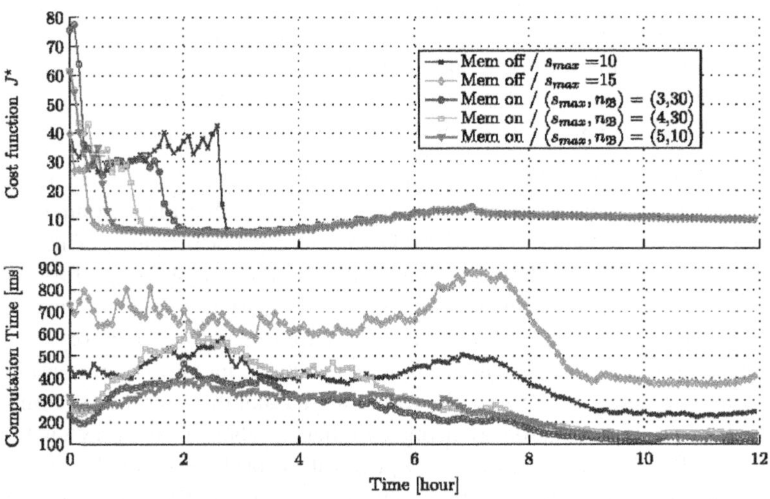

Fig. 9.5 Assessment of the distributed-in-time mechanism

negotiation iterations required in order to achieve relevant solution at each decision instant. The scheme is particularly adapted to situations in which the computation units at the subsystem layer are much more ineffective compared to the one at coordination layer. In the presented case study, this enabled to reduce the computational time by more than 60 % for quite similar performances. Future investigations concern the possible on-line tuning of the bundle size as well as the number of iterations. Moreover, a deeper and more rigorous analysis of the convergence is also undergoing.

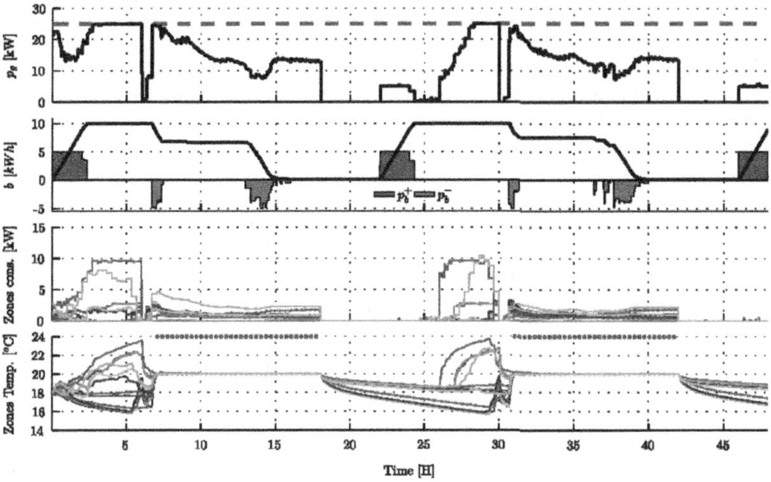

Fig. 9.6 Simulation results with $N_s = 20$, $s_{max} = 3$, $n_{\mathbf{B}} = 30$. Notice that the zones temperatures lie within their respective bounds during occupied hours of the building (CO_2 levels and lighting levels are also respected in all zones but not displayed for lack of space)

Finally, while the local subsystem MPC controller has been already implemented in the Schneider-Electric RoomBox device (see [11] for more details), implementation of the distributed scheme is currently investigated.

Acknowledgments This work is part of **HOMES** collaborative program. This program is funded by **OSEO**.

References

1. E. Camponogara, D. Jia, B.H. Krogh, S. Talukdar. Distributed model predictive control. IEEE Control Syst. Mag. **22**(1):44–52 (2002)
2. R.R. Negenborn, B. De Schutter, J. Hellendoorn, Multi-agent model predictive control: a survey. In Technical report 04–010 -Delft Center for Systems and Control -Delft University of Technology (2004)
3. H. Scheu, J.C. Calderon, D. Doan, J.F. Garcia, R.R. Negenborn, A. Tarau, F.V. Arroyave, B.D. Schutter, J.J. Espinosa, W. Marquardt, Report on assessment of existing coordination mechanisms for simple case studies, and on possible options for improving and extending these coordination mechanisms. Hierarchical and Distributed Model Predictive Control of Large-Scale Systems-Deliverable Number: 3.3.1/ Senventh Framework programme theme—ICT, 2009
4. M. Diehl, Report of literature survey, analysis, and comparison of on-line optimization methods for hierarchical and distributed MPC. Hierarchical and Distributed Model Predictive Control of Large-Scale Systems-Deliverable Number: 4.1.1/ Senventh Framework programme theme—ICT, 2009

5. M.Y. Lamoudi, M. Alamir, and P.Bèguery, Distributed constrained model predictive control based on bundle method for building energy management. in *50th IEEE conference on decision and control and european control conference*, Orlando, 12–15 December 2011
6. M.Y. Lamoudi, M. Alamir, P.Bèguery, Model predictive control for energy management in buildings- part 2: distributed model predictive control. in *IFAC workshop on nonlinear model predictive control* (The Netherlands, 2012)
7. A. Frangioni, Generalized bundle methods. SIAM J. Optim. **13**(1), 117–156 (2002)
8. O. Briant, C. Lemarèchal, Ph Meurdesoif, S. Michel, N. Perrot, F. Vanderbeck, Comparison of bundle and classical column generation. Math. Program. **113**(2), 299–344 (2008)
9. M. Alamir, A framework for real-time implementation of low dimensional parameterized NMPC. Automatica **48**, 198–204 (2011)
10. M.Y. Lamoudi, M. Alamir, and P.Bèguery, Unified NMPC for multi-variable control in smart buildings. in *IFAC 18th World Congress*, Milano, 28 August–2 September 2011
11. M.Y. Lamoudi, M. Alamir, P. Bèguery, Model predictive control for energy management in buildings- part 1: zone model predictive control. in *IFAC workshop on nonlinear model predictive control* (The Netherlands, 2012)

Chapter 10
Rate Analysis of Inexact Dual Fast Gradient Method for Distributed MPC

I. Necoara

Abstract In this chapter we propose a dual decomposition method based on inexact dual gradient information and constraint tightening for solving distributed model predictive control (MPC) problems for network systems with state-input constraints. The coupling constraints are tightened and moved in the cost using the Lagrange multipliers. The dual problem is solved by a fast gradient method based on approximate gradients for which we prove sublinear rate of convergence. We also provide estimates on the primal and dual suboptimality of the generated approximate primal and dual solutions and we show that primal feasibility is ensured by our method. Our analysis relies on the Lipschitz property of the dual MPC function and inexact dual gradients. We obtain a distributed control strategy that has the following features: state and input constraints are satisfied, stability of the plant is guaranteed, whilst the number of iterations for the suboptimal solution can be precisely determined.

10.1 Introduction

Different problems from control and estimation can be addressed within the framework of network systems [11]. In particular, model predictive control (MPC) has become a popular advanced control technology implemented in network systems due to its ability to handle hard input and state constraints. Network systems are complex and large in dimension, whose structure may be hierarchical, multistage or dynamical and they have multiple decision-makers.

Decomposition methods represent a very powerful tool for solving distributed control and estimation problems in network systems. The basic idea of these methods is to decompose the original large optimization problem into smaller subproblems.

I. Necoara(✉)
Automation and Systems Engineering Department, University Politehnica Bucharest, Bucharest, Romania
e-mail: ion.necoara@acse.pub.ro

J. M. Maestre and R. R. Negenborn (eds.), *Distributed Model Predictive Control Made Easy*, Intelligent Systems, Control and Automation: Science and Engineering 69, DOI: 10.1007/978-94-007-7006-5_10, © Springer Science+Business Media Dordrecht 2014

Decomposition methods can be divided in two main classes: primal decomposition and dual decomposition methods. In primal decomposition the optimization problem is solved using the original formulation and variables and the complicating constraints are handled via methods such as feasible directions, Gauss-Jacobi type and others [3, 6, 10, 11, 20]. In dual decomposition the original problem is rewritten using Lagrangian relaxation for the complicating constraints and the dual problem is solved with a Newton or (sub)gradient algorithm [1, 2, 5, 11, 12, 14, 18]. Dual decomposition methods based on subgradient iteration and averaging, that produce primal solutions in the limit, can be found in [7, 9]. Converge rate analysis for the dual subgradient method has been studied e.g. in [13], where the authors provide estimates of order $\mathcal{O}(1/\sqrt{k})$ for the approximate solutions and for feasibility violation. Thus, an important drawback of the dual decomposition methods consists of the fact that feasibility of the primal variables can be ensured only at optimality, which is usually impossible to attain in practice. However, in many applications, e.g. from control and estimation, the constraints can represent different requirements due to physical limitations of actuators, safety limits and operating conditions of the controlled plant. Therefore, any control or estimation scheme must ensure feasibility. Thus, we are interested in developing dual optimization algorithms which satisfy the requirement of feasibility.

One way to ensure feasibility of the primal variables in distributed MPC is through constraint tightening [5, 8, 17], or through distributed implementations of some classical methods, such as the method of feasible directions, Gauss-Jacobi type and others [3, 6, 20]. In [5] a dual distributed algorithm based on constraint tightening is presented for solving MPC problems for systems with coupled dynamics. The authors prove the convergence of the algorithm using the analysis of the dual subgradient method from [13], which has very slow convergence. In [8] the authors propose a decentralized MPC algorithm that uses the constraint tightening technique to achieve robustness while guaranteeing robust feasibility of the entire system. In [20] a cooperative based distributed MPC algorithm is proposed that converges to the centralized solution. In [3] a distributed algorithm based on the method of feasible directions is proposed that also converges to the centralized solution and guarantees primal feasibility. While most of the work cited above focuses on a primal approach, our work develops dual methods that ensure constraint feasibility, tackles more complex constraints and provides better estimates on suboptimality.

The main contribution in this chapter is to provide a novel distributed algorithm based on dual decomposition and constraint tightening for solving constrained MPC problems in network systems. The algorithm has better convergence rates than the algorithms in [5, 13] due to the fact that we exploit the Lipschitz property of the gradient of the dual MPC function. Further, we solve the inner problems only up to a certain accuracy by means of a parallel coordinate descent method that has linear rate of convergence. Even if we use inexact information of the gradient of the dual tightened function, after a finite number of iterations we are able to provide a primal feasible solution for our original MPC problem using averaging. We also derive estimates on dual and primal suboptimality of the generated approximate solutions for our MPC problem. Finally, we obtain a distributed MPC scheme that has the following features: state and input constraints are satisfied, stability of the plant is guaranteed, whilst the number of iterations for the suboptimal solution can be precisely determined.

The chapter is organized as follows. In Sect. 10.2 we introduce the MPC problem for network systems with state-input constraints. In Sect. 10.3 we develop an inexact dual fast gradient scheme for solving a tightened MPC problem and analyze its convergence. Section 10.4 shows how we can perform distributed computations in the MPC scheme. Its efficiency is demonstrated on a four-tank plant in Sect. 10.5.

Notation: For $u, v \in \mathbb{R}^n$ we denote the standard Euclidean inner product $\langle u, v\rangle = \sum_{i=1}^n u^i v^i$, the Euclidean norm $\|u\| = \sqrt{\langle u, u\rangle}$ and the projection onto $\mathbb{R}^n_+$ as $[u]^+$.

10.2 Distributed MPC for Network Systems

Many applications from control and estimation theory can be posed in the framework of optimization problem (10.6), e.g. the MPC problem for interconnected subsystems. We consider network systems which are comprised of M interconnected subsystems, whose dynamics are defined as:

$$x_i(k+1) = \sum_{j\in\mathcal{N}^i} A_{ij}x_j(k) + B_{ij}u_j(k), \tag{10.1}$$

where $x_i(k) \in \mathbb{R}^{n_{x_i}}$ and $u_i(k) \in \mathbb{R}^{n_{u_i}}$ represent the state and the input of ith subsystem at time k, $A_{ij} \in \mathbb{R}^{n_{x_i}\times n_{x_j}}$ and $B_{ij} \in \mathbb{R}^{n_{x_i}\times n_{u_j}}$ and $\mathcal{N}^i$ denotes the neighbors of the ith subsystem including i. In a particular case frequently found in literature [12, 20] the influence between neighboring subsystems is given only in terms of inputs:

$$x_i(k+1) = A_{ii}x_i(k) + \sum_{j\in\mathcal{N}^i} B_{ij}u_j(k). \tag{10.2}$$

We also impose local state and input constraints:

$$x_i(k) \in X_i, \quad u_i(k) \in U_i \quad \forall i = 1, \cdots, M, k \geq 0, \tag{10.3}$$

where $X_i \subseteq \mathbb{R}^{n_{x_i}}$ and $U^i \subseteq \mathbb{R}^{n_{u_i}}$ are simple convex sets. For the systems (10.1) or (10.2) and a prediction horizon of length N_p, we consider a convex cost function composed of a stage and a final cost for each subsystem i: $\sum_{l=0}^{N_\mathrm{p}-1} \ell_i(x_i(l), u_i(l)) + \ell_i^\mathrm{f}(x_i(N_\mathrm{p}))$. The centralized MPC problem for (10.1) for a given initial state x is formulated as:

$$J^*(x) = \min_{x_i(l) \in X_i,\ u_i(l) \in U_i} \sum_{i=1}^{M} \sum_{l=0}^{N_p - 1} \ell_i(x_i(l), u_i(l)) + \ell_i^f(x_i(N_p)) \tag{10.4}$$

$$\text{s.t: } x_i(l+1) = \sum_{j \in \mathcal{N}^i} A_{ij} x_j(l) + B_{ij} u_j(l),\ x_i(0) = x_i,\ x_i(N_p) \in X_i^f \ \ \forall i,$$

where X_i^f are terminal sets chosen under some appropriate conditions to ensure stability of the MPC scheme (see e.g. [19]). For the input trajectory of subsystem i and the overall input trajectory we use the notation (here $n_i = N_p n_{u_i}$ and $n = \sum_{i=1}^{M} n_i$):

$$\mathbf{u}_i = \left[u_i^T(0) \cdots u_i^T(N_p - 1)\right]^T \in \mathbb{R}^{n_i}, \quad \mathbf{u} = \left[\mathbf{u}_1^T \cdots \mathbf{u}_M^T\right]^T \in \mathbb{R}^n.$$

By eliminating the states from dynamics (10.1) and assuming that X_i and X_i^f are polyhedral sets for all i, problem (10.4) can be expressed as a large-scale convex problem:

$$J^*(x) = \min_{\mathbf{u}_1 \in \mathbf{U}_1, \cdots, \mathbf{u}_M \in \mathbf{U}_M} \{J(x, \mathbf{u}_1, \cdots, \mathbf{u}_M) : \ \mathbf{G}\mathbf{u} + \mathbf{E}x + \mathbf{g} \le 0\} \tag{10.5}$$

where the convex sets $\mathbf{U}_i$ are the Cartesian product of the sets U_i for N_p times and the inequalities $\mathbf{G}\mathbf{u} + \mathbf{E}x + \mathbf{g} \le 0$ are obtained by eliminating the states from the constraints $x_i(l) \in X_i$ and $x_i(N_p) \in X_i^f$ for all l and i.

10.3 Solving the Dual MPC Problem Using an Inexact Fast Gradient Method and Tightened Constraints

In this section we present briefly an optimization algorithm for solving distributively the MPC problem (10.5). For brevity, we remove the dependence of the cost and constraints in (10.5) on the initial state x and consider the smooth convex problem:

$$F^* = \min_{\mathbf{u} \in \mathbf{U}} \{F(\mathbf{u}) : \ \mathbf{G}\mathbf{u} + \mathbf{g} \le 0\}, \tag{10.6}$$

where $F : \mathbb{R}^n \to \mathbb{R}$ is a convex function, $\mathbf{U} \subseteq \mathbb{R}^n$ is a convex set and $\mathbf{G} \in \mathbb{R}^{m \times n}$, $\mathbf{g} \in \mathbb{R}^m$. We assume that projection on the set defined by the coupling constraints (called also *complicating constraints*) $\mathbf{G}\mathbf{u} + \mathbf{g} \le 0$ is hard to compute, but the local constraints are in the form of a Cartesian product $\mathbf{U} = \mathbf{U}_1 \times \cdots \times \mathbf{U}_M$, with $\mathbf{U}_i \subseteq \mathbb{R}^{n_i}$ being simple sets, i.e. the projection on these sets can be computed efficiently (e.g. box sets or entire space $\mathbb{R}^{n_i}$). We define the partition of identity matrix: $I_n = [E_1 \cdots E_M] \in \mathbb{R}^{n \times n}$, where $E_i \in \mathbb{R}^{n \times n_i}$ for all $i = 1, \cdots, M$. Thus, $\mathbf{u}$ can be represented as: $\mathbf{u} = \sum_{i=1}^{M} E^i \mathbf{u}_i$. We define the partial gradient of F at $\mathbf{u}$, denoted $\nabla_i F(\mathbf{u}) \in \mathbb{R}^{n_i}$, as: $\nabla_i F(\mathbf{u}) = (E^i)^T \nabla F(\mathbf{u})$ for all i. For simplicity we use the short notation: $h(\mathbf{u}) = \mathbf{G}\mathbf{u} + \mathbf{g}$. The following assumptions are considered in this paper:

Assumption 10.1 *(i) Function F is σ_F-strongly convex w.r.t $\|\cdot\|$.*

(ii) The gradient of F is coordinate-wise Lipschitz continuous with constants $L_i > 0$, i.e: $\|\nabla_i F(\mathbf{u} + E_i d_i) - \nabla_i F(\mathbf{u})\| \leq L_i \|d_i\|$ for all $\mathbf{u} \in \mathbb{R}^n, d_i \in \mathbb{R}^{n_i}$.

Assumption 10.2 *Slater constraint qualification holds for* (10.6)*, i.e. there exists vector $\tilde{\mathbf{u}} \in \mathbf{U}$ such that $\mathbf{G}\tilde{\mathbf{u}} + \mathbf{g} < 0$.*

In dual decomposition, instead of minimizing the primal function $F(\mathbf{u})$, one has to maximize the dual function $d(\lambda) = \min_{\mathbf{u}\in\mathbf{U}} L(\mathbf{u}, \lambda)$, where $L(\mathbf{u}, \lambda) = F(\mathbf{u}) + \langle \lambda, h(\mathbf{u})\rangle$ denotes the partial Lagrangian w.r.t. the complicating constraints $h(\mathbf{u}) \leq 0$. Since we assume the set $\mathbf{U}$ to be simple, any gradient or Newton based projection method for solving the inner subproblems, for a given λ, has cheap iterations. Moreover, based on Assumption 10.1 (i) the dual function $d(\lambda)$ has Lipschitz continuous gradient with constant $L_{\mathrm{d}} = \|\mathbf{G}\|^2/\sigma_{\mathrm{F}}$ (see e.g. [15]). As a consequence of Assumption 10.2, we also have that strong duality holds.

10.3.1 Tightening the Coupling Constraints

In many applications, e.g. the MPC problem from Sect. 10.2, the constraints may represent different requirements on physical limitations of actuators, safety limits and operating conditions of the controlled plant. Thus, ensuring the feasibility of the primal variables in (10.6), i.e. $\mathbf{u} \in \mathbf{U}$ and $\mathbf{G}\mathbf{u} + \mathbf{g} \leq 0$, becomes a prerequisite. However, dual decomposition methods can ensure these requirements only at optimality, which is usually impossible to attain in practice. Therefore, in our approach, instead of solving the original problem (10.6), we propose a first order optimization algorithm based on dual decomposition for a tightened problem (see also [5] for a similar approach where the tightened dual problem is solved using a subgradient algorithm). We introduce the following tightened problem associated with original problem (10.6):

$$F^*_{\epsilon_c} = \min_{\mathbf{u}\in\mathbf{U}} \{F(\mathbf{u}) : \quad \mathbf{G}\mathbf{u} + \mathbf{g} + \epsilon_c \mathbf{e} \leq 0\}, \tag{10.7}$$

where $\mathbf{e}$ denotes the vector with all entries 1 and

$$0 \leq \epsilon_{\mathrm{c}} < \min\{-(\mathbf{G}\tilde{\mathbf{u}} + \mathbf{g})_1, \cdots, -(\mathbf{G}\tilde{\mathbf{u}} + \mathbf{g})_m\}, \tag{10.8}$$

where $\tilde{\mathbf{u}}$ is a Slater vector for (10.6) as in Assumption 10.2. Note that for this choice of ϵ_{c}, we have that $\tilde{\mathbf{u}}$ is also a Slater vector for problem (10.7), so that Assumption 10.2 remains valid for this problem. Our goal is to solve the optimization problem (10.7) using dual decomposition. In order to update the dual variables we use an inexact dual fast gradient method. An important feature of our algorithm consists of the fact that even if we use inexact information of the gradient of the dual function, we are still able to compute a sequence of primal variables which, after a certain number of outer iterations, become feasible and ϵ_{out}-optimal for our original problem (10.6).

Recall that we use the short notation $h(\mathbf{u}) = \mathbf{G}\mathbf{u}+\mathbf{g}$. We assume that projection on $\mathbf{U}_i$ is simple but the projection on the set defined by the tightened coupling constraints $h(\mathbf{u}) + \epsilon_c\mathbf{e} \leq 0$ is hard to compute. Therefore, we remove the tightened constraints in the cost via Lagrange multipliers and define the dual function:

$$d_{\epsilon_c}(\lambda) = \min_{\mathbf{u}\in\mathbf{U}} L_{\epsilon_c}(\mathbf{u}, \lambda), \tag{10.9}$$

where $L_{\epsilon_c}(\mathbf{u}, \lambda) = F(\mathbf{u}) + \langle \lambda, h(\mathbf{u}) + \epsilon_c\mathbf{e}\rangle$. Further, the gradient of the dual function $d_{\epsilon_c}(\lambda)$ is Lipschitz continuous, with the same constant L_d as for the original dual function for (10.6), and is given by [2, 15]: $\nabla d_{\epsilon_c}(\lambda) = h(\mathbf{u}(\lambda)) + \epsilon_c\mathbf{e}$, where $\mathbf{u}(\lambda)$ is the optimal solution of the *inner problem*:

$$\mathbf{u}(\lambda) = \arg\min_{\mathbf{u}\in\mathbf{U}} L_{\epsilon_c}(\mathbf{u}, \lambda). \tag{10.10}$$

Under strong duality we have for the *outer problem*:

$$F^*_{\epsilon_c} = \max_{\lambda\geq 0} d_{\epsilon_c}(\lambda). \tag{10.11}$$

Remark 10.1 If Assumption 10.2 holds for (10.7) and $F^*_{\epsilon_c}$ is finite, then from [2] it follows that the set of optimal Lagrange multipliers for the inequalities $h(\mathbf{u}) + \mathbf{e} \leq 0$ is bounded. Therefore, problem (10.11) can be written equivalently as: $F^*_{\epsilon_c} = \max_{\lambda\in Q} d_{\epsilon_c}(\lambda)$, where

$$Q = \{\lambda \in \mathbb{R}^m : \lambda \geq 0, \ \|\lambda\| \leq R\}, \tag{10.12}$$

for some $R > 0$ such that $\lambda^* \in Q$, where λ^* denotes an optimal solution of (10.11). □

Since we cannot usually solve the inner optimization problem (10.10) exactly, but with some accuracy ϵ_{in} and obtaining an approximate solution $\bar{\mathbf{u}}(\lambda)$, we do not have available an exact gradient of the dual function in λ, but only an approximation:

$$\bar{\nabla} d_{\epsilon_c}(\lambda) = h(\bar{\mathbf{u}}(\lambda)) + \epsilon_c\mathbf{e}.$$

If we assume that $\bar{\mathbf{u}}(\lambda)$ is computed such that the following ϵ_{in}-optimality holds:

$$\bar{\mathbf{u}}(\lambda) \in \mathbf{U}, \quad L_{\epsilon_c}(\bar{\mathbf{u}}(\lambda), \lambda) - L_{\epsilon_c}(\mathbf{u}(\lambda), \lambda) \leq \epsilon_{\text{in}}/2, \tag{10.13}$$

then next lemma gives bounds for the dual function $d_{\epsilon_c}(\lambda)$ (see Sect. 3.1 in [4]):

Lemma 10.1 [4] *Let Assumptions 10.1 and 10.2 hold and for a given λ let $\bar{\mathbf{u}}(\lambda)$ be computed as in* (10.13). *Then, the following inequalities are valid for all $\tilde{\lambda} \in \mathbb{R}^m_+$:*

Algorithm 10.1 (IDFG)(λ^0)

Given $\lambda^0 \in \mathbb{R}^m_+$, for $p \geq 0$ compute:

1: $\bar{\mathbf{u}}^p \approx \arg\min_{\mathbf{u}\in\mathbf{U}} L_{\epsilon_c}(\mathbf{u}, \lambda^p)$ such that (10.13) holds

2: $\mu^p = \left[\lambda^p + \frac{1}{L_d}\bar{\nabla} d_{\epsilon_c}(\lambda^p)\right]^+$ and $\eta^p = \left[\lambda^0 + \frac{1}{L_d}\sum_{s=0}^p \frac{s+1}{2}\bar{\nabla} d_{\epsilon_c}(\lambda^s)\right]^+$

3: $\lambda^{p+1} = \frac{p+1}{p+3}\mu^p + \frac{2}{p+3}\eta^p$.

$$0 \geq d_{\epsilon_c}(\tilde{\lambda}) - [L_{\epsilon_c}(\bar{\mathbf{u}}(\lambda), \lambda) + \langle \bar{\nabla} d_{\epsilon_c}(\lambda), \tilde{\lambda} - \lambda\rangle] \geq -L_d\|\tilde{\lambda} - \lambda\|^2 - \epsilon_{in}. \quad (10.14)$$

Our goal is to solve the dual problem (10.11) using a fast gradient scheme with accuracy ϵ_{out}. We obtain such accuracy ϵ_{out} after p_{out} iterations and after which we construct a primal estimate $\hat{\mathbf{u}}^{p_{\text{out}}}$. We want to guarantee for this estimate primal feasibility and primal suboptimality of order ϵ_{out} for problem (10.6):

$$\hat{\mathbf{u}}^{p_{\text{out}}} \in \mathbf{U}, \quad \mathbf{G}\hat{\mathbf{u}}^{p_{\text{out}}} + \mathbf{g} \leq 0 \quad \text{and} \quad F(\hat{\mathbf{u}}^{p_{\text{out}}}) - F^* \leq \mathcal{O}(\epsilon_{\text{out}}).$$

10.3.2 Inexact Dual Fast Gradient Method for Solving Outer Problem

In this section we discuss an inexact dual fast gradient scheme for updating λ. The algorithm was proposed by Nesterov [15] and applied further in [12] for solving dual problems with exact gradient information. The scheme defines three sequences $(\lambda^p, \mu^p, \eta^p)_{p\geq 0}$ for the dual variables, see Algorithm 10.1.

Based on Theorem 1 in [4], which is an extension of the results in [12, 15] to the inexact case, we have the following result:

Lemma 10.2 [4, 12] *If Assumptions 10.1 and 10.2 hold and sequences* $(\bar{\mathbf{u}}^p, \lambda^p, \mu^p, \eta^p)_{p\geq 0}$ *are generated by Algorithm 10.1, then for all* $p \geq 0$ *we have:*

$$\frac{(p+1)(p+2)}{4} d_{\epsilon_c}(\mu^p) \geq \max_{\lambda\geq 0} -L_d\|\lambda-\lambda^0\|^2 + \sum_{s=0}^{p} \frac{s+1}{2}\left[L_{\epsilon_c}(\bar{\mathbf{u}}^s, \lambda^s) + \langle\bar{\nabla} d_{\epsilon_c}(\lambda^s), \lambda - \lambda^s\rangle\right]$$
$$- \frac{(p+1)(p+2)(p+3)}{12}\epsilon_{in} \ \forall \lambda \in \mathbb{R}^m_+. \quad (10.15)$$

The next theorem provides estimates on the dual suboptimality of the generated approximate dual solutions of **(IDFG)**:

Theorem 10.1 *Let Assumptions 10.1 and 10.2 hold and the sequences* $(\bar{\mathbf{u}}^p, \lambda^p, \mu^p, \eta^p)_{p\geq 0}$ *be generated by Algorithm 10.1 with* $\lambda^0 = 0$. *Then, an estimate on dual suboptimality for the original problem* (10.6) *is given by:*

$$F^* - d_{\epsilon_c}(\mu^p) \le \frac{4L_d R^2}{(p+1)^2} + \frac{p+3}{3}\epsilon_{in}. \tag{10.16}$$

Proof Using the first inequality from (10.14) in (10.15) we get:

$$\frac{(p+1)(p+2)}{4} d_{\epsilon_c}(\mu^p) \ge -\mathrm{L_d}\|\lambda^*\|^2 + \sum_{s=0}^{p} \frac{s+1}{2} d_{\epsilon_c}(\lambda^*) - \frac{(p+1)(p+2)(p+3)}{12}\epsilon_{\mathrm{in}}.$$

Dividing now both sides of the previous inequality by $\frac{(p+1)(p+2)}{4}$, using (10.12), rearranging the terms and taking into account that $d_{\epsilon_c}(\lambda^*) = F^*_{\epsilon_c} \ge F^*$ and $(p+1)^2 \le (p+1)(p+2)$ we obtain (10.16). □

We are interested now in finding estimates on primal suboptimality and primal infeasibility for our original problem (10.6). For this purpose we define the following average sequence for the primal variables:

$$\bar{\mathbf{u}}^p = \sum_{s=0}^{p} \frac{2(s+1)}{(p+1)(p+2)} \bar{\mathbf{u}}^s. \tag{10.17}$$

The next theorems give estimates on primal suboptimality and infeasibility, whose proofs can be found in Appendix (Sect. 10.7).

Theorem 10.2 *Assume the conditions from Theorem* 10.1 *hold and let* $\bar{\mathbf{u}}^p$ *be given by* (10.17). *Then, the following estimate on primal suboptimality for* (10.6) *can be derived:*

$$F(\bar{\mathbf{u}}^p) - F^* \le \sqrt{m} R \epsilon_c + \frac{p+3}{3}\epsilon_{in}. \tag{10.18}$$

Theorem 10.3 *Under the assumptions of Theorem 10.2, an estimate for primal infeasibility is given by: if we define* $v(p, \epsilon_{in}) = \frac{8L_d R\left(1+\sqrt{1+\frac{(p+1)^2(p+3)}{12 L_d R^2}\epsilon_{in}}\right)}{(p+1)^2}$, *then*

$$\|[\mathbf{G}\bar{\mathbf{u}}^p + \mathbf{g} + \epsilon_c \mathbf{e}]^+\| \le v(p, \epsilon_{in}). \tag{10.19}$$

Assume now that we fix the outer accuracy ϵ_{out} to a desired value such that:

$$\epsilon_{\text{out}} \le 2\sqrt{m} R \min_{j=1,\cdots,m} \{-(\mathbf{G}\tilde{\mathbf{u}} + \mathbf{g})_j\},$$

where $\tilde{\mathbf{u}}$ is a Slater vector for (10.6) (see Assumption 10.2). We are now interested in finding the number of outer iterations p_{out} and a relation between ϵ_{out}, ϵ_{in} and ϵ_c such that primal feasibility holds and primal suboptimality will be less than ϵ_{out}:

$$\hat{\mathbf{u}}^{p_{\text{out}}} \in \mathbf{U}, \quad \mathbf{G}\hat{\mathbf{u}}^{p_{\text{out}}} + \mathbf{g} \le 0 \quad \text{and} \quad F(\hat{\mathbf{u}}^{p_{\text{out}}}) - F^* \le \epsilon_{\text{out}}. \tag{10.20}$$

From (10.18), we can take for example:

$$\epsilon_{\mathrm{c}} = \frac{\epsilon_{\mathrm{out}}}{2\sqrt{m}R} \quad \text{and} \quad \epsilon_{\mathrm{in}} = \frac{3\epsilon_{\mathrm{out}}}{2(p_{\mathrm{out}}+3)}. \tag{10.21}$$

Using Theorem 10.2 we see that for this choice of ϵ_{c} and ϵ_{in} we ensure the second condition in (10.20), i.e. ϵ_{out}-optimality for original problem (10.6). To get primal feasibility we impose the condition $v(p_{\mathrm{out}}, \epsilon_{\mathrm{in}}) \le \epsilon_{\mathrm{c}}$, from which we obtain e.g. $(p_{\mathrm{out}}+1)^2 \ge \frac{16\sqrt{m}L_{\mathrm{d}}R^2}{\epsilon_{\mathrm{out}}}$ and thus we can take:

$$p_{\mathrm{out}} = \left\lfloor 4R\sqrt{\frac{\sqrt{m}L_{\mathrm{d}}}{\epsilon_{\mathrm{out}}}} \right\rfloor.$$

It follows immediately that for this choice of ϵ_{c}, ϵ_{in} and p_{out} the approximate primal solution produced for our original problem (10.6) is feasible (i.e. $\hat{\mathbf{u}}^{p_{\mathrm{out}}} \in \mathbf{U}$ and $\mathbf{G}\hat{\mathbf{u}}^{p_{\mathrm{out}}} + \mathbf{g} \le 0$) and ϵ_{out}-optimal (i.e. $F(\hat{\mathbf{u}}^{p_{\mathrm{out}}}) - F^* \le \epsilon_{\mathrm{out}}$) and thus (10.20) holds.

Note that all the results derived in this section hold also for $\lambda^0 \neq 0$, but in this case the formulas are more cumbersome. Further, our results hold also in the case when we solve the inner problems exactly, i.e. $\epsilon_{\mathrm{in}} = 0$ in (10.13), or when $\mathbf{U} = \mathbb{R}^n$, i.e. the inner problems are unconstrained.

10.3.3 Parallel Coordinate Descent Method for Solving Inner Problem

In this section, due to space limitations, we present briefly a block-coordinate descent algorithm which permits solving in parallel, for a fixed λ^p, the inner problem (10.9) approximately: $\min_{\mathbf{u}_1 \in \mathbf{U}_1, \cdots, \mathbf{u}_M \in \mathbf{U}_M} L_{\epsilon_{\mathrm{c}}}(\mathbf{u}, \lambda^p)$. Based on Assumption 10.1 it follows that $L_{\epsilon_{\mathrm{c}}}(\mathbf{u}, \lambda) = F(\mathbf{u}) + \langle \lambda, \mathbf{G}\mathbf{u} + \mathbf{g} + \epsilon_{\mathrm{c}}\mathbf{e}\rangle$ is strongly convex and with coordinate-wise Lipschitz continuous gradient in the first variable. Since the algorithm can be applied to a larger class of problems, which also includes our problem (10.9), we consider the more general problem:

$$f^* = \min_{\mathbf{u}_1 \in \mathbf{U}_1, \cdots, \mathbf{u}_M \in \mathbf{U}_1} f(\mathbf{u}), \tag{10.22}$$

where f is convex and satisfies Assumption 10.1 (i.e. it is σ_{F}-strongly convex and has L_i-coordinate-wise Lipschitz continuous gradient) and $\mathbf{U}_i \subseteq \mathbb{R}^{n_i}$ are simple, convex sets (e.g. box sets, entire space $\mathbb{R}^{n_i}$, etc). There exist many parallel algorithms in the literature for solving the optimization problem (10.22): e.g. Jacobi-type algorithms [2, 20]. However, the rate of convergence for these algorithms is guaranteed under more conservative assumptions than the ones required for the parallel coordinate descent method proposed in this section (see [10] for more details).

Algorithm 10.2 (PCD)($\mathbf{u}^0$)

Given $\mathbf{u}^0$, for $q \geq 0$ do:
1: compute in parallel $v_i(\mathbf{u}^q)$ for $i = 1, \dots, M$
2: update: $\mathbf{u}^{q+1} = \sum_{i=1}^{M} \frac{1}{M} w_i^+(\mathbf{u}^q)$.

Let us define the constrained coordinate update for our inner algorithm:

$$v_i(\mathbf{u}) = \arg\min_{v_i \in \mathbf{U}_i} \langle \nabla_i f(\mathbf{u}), v_i - \mathbf{u}_i \rangle + \frac{L_i}{2} \|v_i - \mathbf{u}_i\|^2$$
$$w_i^+(\mathbf{u}) = \mathbf{u} + E_i(v_i(\mathbf{u}) - \mathbf{u}_i) \ \forall i = 1, \cdots, M.$$

Algorithm 10.2 shows our *Parallel Coordinate Descent Method*, resembling the method in [20] but with a simpler iteration and guaranteed rate of convergence and which can also be viewed as a parallel version of the block-coordinate descent method from [16].

We can easily show that Algorithm 10.2 decreases the objective function at each iteration: $f(\mathbf{u}^{q+1}) \leq f(\mathbf{u}^q)$ for all $q \geq 0$ (see [10] for more properties of this algorithm). Note that if the sets $\mathbf{U}_i$ are simple (e.g. boxes) the projection on $\mathbf{U}_i$ can be easy (for boxes in $\mathcal{O}(n_i)$ operations) and if f has cheap coordinate derivatives (e.g. quadratic function with sparse Hessian), then the cost of computing $\nabla_i f(\mathbf{u})$ is less than $\mathcal{O}(n \cdot n_i)$. Thus, for quadratic problems the worst case complexity per iteration of our method is $\mathcal{O}(n^2)$. Note that the complexity per iteration of the method from [20] is at least $\mathcal{O}(n^2 + \sum_{i=1}^{M} n_i^3)$, provided that the local quadratic subproblems are solved with an interior point solver. The next theorem provides the convergence rate of Algorithm 10.2:

Theorem 10.4 [10] *If function f has L_i-coordinate-wise Lipschitz continuous gradient and is also σ_F-strongly convex, then the Algorithm 10.2 has linear rate of convergence.*

10.4 A Distributed MPC Scheme Based on Local Information

In this section we discuss some technical aspects for the implementation of our inexact dual decomposition method in the case of MPC problem (10.4) and its equivalent form (10.5) presented in Sect. 10.2. Usually, in the linear MPC framework, the local stage and final cost are taken of the following quadratic form:

$$\ell_i(x_i, u_i) = \|x_i\|_{Q_i}^2 + \|u_i\|_{R_i}^2, \quad \ell_i^{\mathrm{f}}(x_i) = \|x_i\|_{P_i}^2,$$

where $\|x\|_{P_i}^2 = x^T P_i x$, the matrix $Q_i, P_i \in \mathbb{R}^{n_{x_i} \times n_{x_i}}$ are positive semidefinite, whilst matrices $R_i \in \mathbb{R}^{n_{u_i} \times n_{u_i}}$ are positive definite. We also assume that the local constraints sets U_i, X_i and the terminal sets X_i^{f} are polyhedral. In this particular case, the objective function in (10.5) is quadratically strongly convex, having the form:

$$F(\mathbf{u}) = J(x,\mathbf{u}) = 0.5\,\mathbf{u}^T\mathbf{Q}\mathbf{u} + (\mathbf{W}x + \mathbf{w})^T\mathbf{u},$$

where $\mathbf{Q}$ is positive definite due to the assumption that all R_i are positive definite. Since we assume that the sets U_i, X_i and X_{f} are polyhedral, then after eliminating the dynamics (10.1) or (10.2) all the complicating constraints are gathered in:

$$h(\mathbf{u}) = \mathbf{G}\mathbf{u} + \mathbf{E}x + \mathbf{g} \leq 0.$$

If the projection on the set U_i is difficult we also move the input constraints in the complicating constraints defined above. In this case $\mathbf{U}_i = \mathbb{R}^{n_i}$. Otherwise (i.e. the local sets U_i are simple, e.g. boxes or the entire space $\mathbb{R}^{n_{u_i}}$) the convex sets $\mathbf{U}_i = \prod_{l=1}^{N_{\mathrm{p}}} U_i$. Usually, for the dynamics (10.1) the corresponding matrices $\mathbf{Q}$ and $\mathbf{G}$ obtained after eliminating the states are dense and despite the fact that both Algorithms 10.1 and 10.2 can perform parallel computations (i.e. each subsystem needs to solve small local problems) we need all to all communication between subsystems. However, for the dynamics (10.2) the corresponding matrices $\mathbf{Q}$ and $\mathbf{G}$ are sparse and in this case in our Algorithms 10.1 and 10.2 we can perform distributed computations (i.e. the subsystems solve small local problems in parallel and they need to communicate only with their neighbors). Indeed, if the dynamics of the system are given by (10.2), then $x_i(l) = A_{ii}^l x_i(0) + \sum_{s=1}^{l}\sum_{j\in\mathcal{N}^i} A_{ii}^{s-1} B_{ij} u_j(l-s)$ and thus the matrices $\mathbf{Q}$ and $\mathbf{G}$ have a sparse structure (see e.g. [3, 20]). In particular, the complicating constraints have the following structure: for matrix $\mathbf{G}$ the (i,j) block matrices of $\mathbf{G}$, denoted G_{ij}, are zero for all $j \notin \mathcal{N}^i$ for a given subsystem i, while the matrix $\mathbf{E}$ is block diagonal. Further, if we define the neighborhood subsystems of a certain subsystem i as $\hat{\mathcal{N}}^i = \mathcal{N}^i \cup \{l : l \in \mathcal{N}^j, j \in \bar{\mathcal{N}}^i\}$, where $\bar{\mathcal{N}}^i = \{j : i \in \mathcal{N}^j\}$, then the matrix $\mathbf{Q}$ has all the block matrices $\mathbf{Q}_{ij} = 0$ for all $j \notin \hat{\mathcal{N}}^i$ and the matrix $\mathbf{W}$ has all the block matrices $\mathbf{W}_{ij} = 0$ for all $j \notin \bar{\mathcal{N}}^i$, for any given subsystem i. Thus, the ith block components of both $\bar{\nabla} d_{\epsilon_{\mathrm{c}}}$ and $\nabla L_{\epsilon_{\mathrm{c}}}(\mathbf{u},\lambda)$ can be computed using local information:

$$\bar{\nabla}_i d_{\epsilon_{\mathrm{c}}}(\lambda) = \sum_{j\in\mathcal{N}^i} \mathbf{G}_{ij}\mathbf{u}_j + \mathbf{E}_{ii}x_i + \mathbf{g}_i + \epsilon_{\mathrm{c}}\mathbf{e}, \tag{10.23}$$

$$\nabla_i L_{\epsilon_{\mathrm{c}}}(\mathbf{u},\lambda) = \sum_{j\in\hat{\mathcal{N}}^i} \mathbf{Q}_{ij}\mathbf{u}_j + \sum_{j\in\bar{\mathcal{N}}^i}\left(\mathbf{W}_{ij}x_j + \mathbf{G}_{ij}^T\lambda_j\right) + \mathbf{w}_i. \tag{10.24}$$

Note that in the Algorithm 10.2 the only parameters that we need to compute are the Lipschitz constants L_i. However, in the MPC problem L_i does not depend on the initial state x and can be computed locally by each subsystem as: $L_i = \lambda_{\max}(\mathbf{Q}_{ii})$. From the previous discussion it follows immediately that the iterations of Algorithm 10.2 can be performed in parallel using distributed computations (see (10.24)).

Algorithm 10.3 Distributed MPC scheme $(x, \lambda^0, \mathbf{u}^0)$

Given x and accuracy ϵ_{out} choose $\lambda^0 \in \mathbb{R}^p_+$ and $\mathbf{u}^0 \in \mathbf{U}$.
Write MPC problem (10.4) into optimization problem (10.5).

1: *Outer Iteration:* for $0 \leq p \leq p_{\text{out}}$ do
Inner iteration: given $\mathbf{u}_i^0$ $\forall i$ and λ^p, for $q \geq 0$ repeat

- Each subsystem i computes in parallel $\nabla_i L_{\epsilon_c}(\mathbf{u}^q, \lambda^p)$ as in (10.24)
- Each subsystem i updates in parallel $\mathbf{u}_i^{q+1}$ using Algorithm 10.2 and $q \leftarrow q+1$.
- Until $\bar{\mathbf{u}}^q = [\mathbf{u}_1^q, \cdots \mathbf{u}_M^q] \approx \arg\min_{\mathbf{u} \in \mathbf{U}} L_{\epsilon_c}(\mathbf{u}, \lambda^k)$ such that (10.13) holds
- Define $\mathbf{u}^0 = \bar{\mathbf{u}}^q$ and compute $\hat{\mathbf{u}}^p$ as in (10.17)

2: Each subsystem i computes in parallel $\bar{\nabla}_i d_{\epsilon_c}(\lambda^p)$ as in (10.23).
3: Each subsystem i computes in parallel $(\mu_i^p, \eta_i^p, \lambda_i^{p+1})$ using Algorithm 10.1, $p \leftarrow p+1$.

Since the Algorithm 10.1 uses only first order information, we can observe that once $\bar{\nabla}_i d_{\epsilon_c}(\lambda)$ has been computed distributively, as proved in (10.23), all the computations for updating $(\lambda^p, \mu^p, \eta^p)$ can be done in parallel distributively, due to the fact that we have to do only vector operations. However, in the scheme 10.1 all subsystems need to know the global Lipschitz constant $L_{\text{d}} = \frac{\|\mathbf{G}\|^2}{\sigma_{\text{F}}} = \frac{\|\mathbf{G}\|^2}{\lambda_{\min}(\mathbf{Q})}$ that cannot be computed distributively. In practice, a good upper bound on $\frac{\|\mathbf{G}\|^2}{\lambda_{\min}(\mathbf{Q})}$ is sufficient. Note that we do not need to compute a Slater vector $\tilde{\mathbf{u}}$ for (10.4) as long as the MPC problem is well possed and the desired accuracy ϵ_{out} is sufficiently small such that $\epsilon_{\text{out}} \leq 2\sqrt{m}R \min_{j=1,\cdots,m}\{-(\mathbf{G}\tilde{\mathbf{u}} + \mathbf{E}x + \mathbf{g})_j\}$ for all x in some region of attraction $X_{N_{\text{p}}}$. In Algorithm 10.1 another global constant that has to be computed offline is an upper bound on the norm of the optimal multiplier, R_{d}, for any $x \in X_{N_{\text{p}}}$. Note that practical ways to compute an upper bound R_{d} can be found in [13, 18]. Once the desired accuracy was chosen and an upper bound R_{d} was computed offline, then we can take $\epsilon_{\text{c}} = \frac{\epsilon_{\text{out}}}{2\sqrt{m}R}$, as proven in previous sections.

From previous discussion we can conclude that the sequences $\left(\hat{\mathbf{u}}^p, \lambda^p, \mu^p, \eta^p\right)_{p\geq 0}$ generated by the Algorithms 10.1 and 10.2 can be computed in parallel and their updates can be performed distributively using local information, provided that dynamics (10.2) are considered and that good approximations for L_{d} and R_{d} can be computed offline. Closed loop stability for our MPC scheme as given as Algorithm 10.3 can be ensured by choosing the terminal costs ℓ_i^{f} and the terminal sets X_i^{f} for all i appropriately [19].

10.5 Numerical Example

To demonstrate the applicability of the new algorithms we consider the MPC problem for a 4-tank process, whose objective is to control the level of water in each tank. We obtain a continuous state-space model by linearizing the nonlinear model (see e.g. [1]) at an operating point given by h_i^e, γ_1^e, γ_2^e (where $h_1^e = 0.19\,\text{m}$, $h_2^e = 0.13\,\text{m}$, $h_3^e = 0.23\,\text{m}$, $h_4^e = 0.09\,\text{m}$ denote the water levels and $\gamma_1^e = 0.58$, $\gamma_2^e = 0.54$ represent the valve ratios) and the maximum inflows from the pumps $q_1^{max} = 0.39$ m^3/h, $q_2^{max} = 0.39$ m^3/h, with the deviation variables $x_i = h_i - h_i^e$, $u_1 = \gamma_1 - \gamma_1^e$, $u_2 = \gamma_2 - \gamma_2^e$:

$$\frac{dx}{dt} = \begin{bmatrix} -\frac{1}{\tau_1} & 0 & 0 & \frac{1}{\tau_4} \\ 0 & -\frac{1}{\tau_2} & \frac{1}{\tau_3} & 0 \\ 0 & 0 & -\frac{1}{\tau_3} & 0 \\ 0 & 0 & 0 & -\frac{1}{\tau_4} \end{bmatrix} x + \begin{bmatrix} \frac{q_1^{max}}{S} & 0 \\ 0 & \frac{q_2^{max}}{S} \\ \frac{-q_1^{max}}{S} & 0 \\ 0 & \frac{-q_2^{max}}{S} \end{bmatrix} \text{u},$$

where the state vector $x \in \mathbb{R}^4$, the input vector $u \in \mathbb{R}^2$, $\tau_i = \frac{S}{a_i}\sqrt{\frac{2h_e^i}{g}}$ denotes the time constant for tank i and $S = 0.02$, $a_i = 5.8 \cdot 10^{-5}$. Using zero-order hold method with a sampling time of $T = 5s$ we obtain the discrete time model of type (10.2):

$$\begin{aligned} x_1(k+1) &= A_{11}x_1(k) + B_{11}u_1(k) + B_{12}u_2(k), \\ x_2(k+1) &= A_{22}x_2(k) + B_{22}u_2(k) + B_{21}u_1(k), \end{aligned}$$

with the partition $x_1 \leftarrow [x_1\ x_4]^T$ and $x_2 \leftarrow [x_2\ x_3]^T$. For the input constraints we consider the practical constraint of the ratios of the three way valves for our plant, i.e $u \in [0.3, 0.8] - \gamma^e$. We also consider state constraints $0 \leq x \leq 0.31 - h^e$. For the stage cost we have taken the weighting matrices to be $Q_i = I_2$ and $R_i = 0.1$. Since the matrices A_{ii} are stable, in order to ensure stability of the MPC scheme we can compute matrices P_i for the final costs as the solutions of the discrete Lyapunov equations [19]: $A_{ii}^T P_i A_{ii} - P_i + Q_i = 0$.

For a prediction horizon $N = 20$ and starting from $h_0 = 0.3$m, the simulated closed-loop trajectories for the levels using Algorithm 10.1 are displayed in Fig. 10.1. Note that the closed-loop system is driven to the equilibrium point h^e.

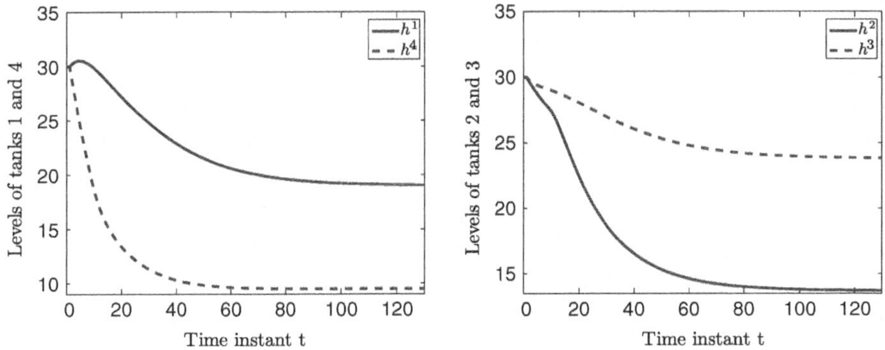

Fig. 10.1 Trajectories of the states (levels) for the four tanks

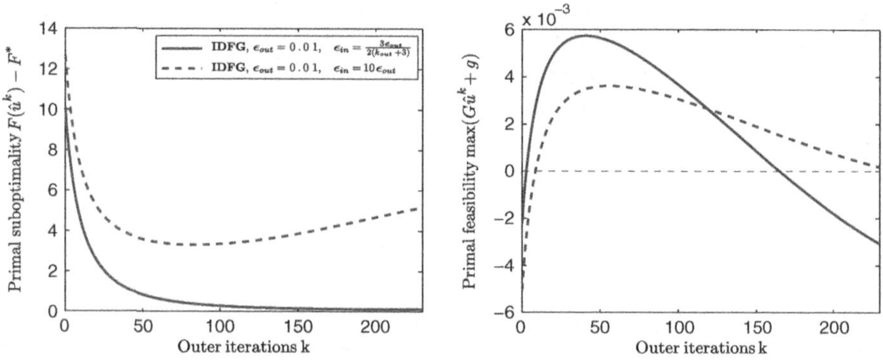

Fig. 10.2 Primal suboptimality and infeasibility for Algorithm 10.1

In Fig. 10.2 we represent the primal suboptimality and infeasibility for the MPC optimization problem solved at step $k = 10$ using Algorithm 10.1. We fix $\epsilon_{\text{out}} = 0.01$. The continuous line corresponds to ϵ_{in} and ϵ_{c} computed according to Sect. 10.3.2, while the dotted line corresponds to $\epsilon_{\text{in}} = 10\epsilon_{\text{out}}$. We can observe that Algorithm 10.1 is sensitive due to error accumulations.

10.6 Conclusions

Motivated by MPC problems for network systems, we have proposed a dual based method for solving constrained MPC problems in network systems. We moved the coupling constraints in the cost using duality theory and introduced a tightened version of the dual MPC problem. We solved the inner subproblems only up to a certain accuracy by means of a parallel coordinate descent method for which we have linear convergence. For solving the outer problems we developed an inexact dual fast gradient scheme. We proved primal feasibility for our original MPC problem and derived upper bounds on dual and primal suboptimality. We also discussed some implementation issues of the new algorithms for distributed MPC and tested them on a practical example.

For future research we intend to apply these newly developed algorithms to other practical problems and implement them in a distributed setting using MPI.

10.7 Appendix

Proof of Theorem 2: Using (10.15) and (10.17) and the convexity of F and h together with the fact that $Q \subseteq \mathbb{R}^m_+$, we obtain:

$$F(\hat{\mathbf{u}}^p) - d_{\epsilon_c}(\mu^p) \leq -\max_{\lambda \in Q} -\frac{4L_d}{(p+1)^2}\|\lambda\|^2 + \langle \lambda, h(\hat{\mathbf{u}}^p) + \epsilon_c \mathbf{e}\rangle + \frac{p+3}{3}\epsilon_{in}.$$

Since $\max_{\lambda \in Q} -\frac{4L_d}{(p+1)^2}\|\lambda\|^2 + \langle \lambda, h(\hat{\mathbf{u}}^p) + \epsilon_c \mathbf{e}\rangle \geq 0$ and taking into account that $d_{\epsilon_c}(\mu^p) \leq F^*_{\epsilon_c}$, we have: $F(\hat{\mathbf{u}}^p) - F^*_{\epsilon_c} \leq \frac{p+3}{3}\epsilon_{in}$. We can write further:

$$\begin{aligned} F^*_{\epsilon_c} &= \max_{\lambda \in Q} d_{\epsilon_c}(\lambda) = \max_{\lambda \in Q} \min_{\mathbf{u} \in \mathbf{U}} F(\mathbf{u}) + \langle \lambda, h(\mathbf{u}) + \epsilon_c \mathbf{e}\rangle \\ &\leq \max_{\lambda \in Q} \min_{\mathbf{u} \in \mathbf{U}} F(\mathbf{u}) + \langle \lambda, h(\mathbf{u})\rangle + \max_{\lambda \in Q}\langle \lambda, \epsilon_c \mathbf{e}\rangle = F^* + \sqrt{m} R \epsilon_c, \end{aligned}$$

which combined with the previous inequality proves (10.18). □

Proof of Theorem 3: Recall that $h(\mathbf{u}) = \mathbf{G}\mathbf{u} + \mathbf{g}$. Using (10.15) and convexity of F and h, we have:

$$\max_{\lambda \geq 0} -\frac{4L_d}{(p+1)^2}\|\lambda\|^2 + \langle \lambda, h(\hat{\mathbf{u}}^p) + \epsilon_c \mathbf{e}\rangle \leq \frac{p+3}{3}\epsilon_{in} + d_{\epsilon_c}(\mu^p) - F(\hat{\mathbf{u}}^p).$$

For the second term of the right hand-side we have:

$$\begin{aligned} &d_{\epsilon_c}(\mu^p) - F(\hat{\mathbf{u}}^p) \leq d_{\epsilon_c}(\lambda^*) - F(\hat{\mathbf{u}}^p) = \min_{\mathbf{u} \in \mathbf{U}} F(\mathbf{u}) + \langle \lambda^*, h(\mathbf{u}) + \epsilon_c \mathbf{e}\rangle - F(\hat{\mathbf{u}}^p) \\ &\leq F(\hat{\mathbf{u}}^p) + \langle \lambda^*, h(\hat{\mathbf{u}}^p) + \epsilon_c \mathbf{e}\rangle - F(\hat{\mathbf{u}}^p) = \langle \lambda^*, h(\hat{\mathbf{u}}^p) + \epsilon_c \mathbf{e}\rangle \leq \|\lambda^*\| \|[h(\hat{\mathbf{u}}^p) + \epsilon_c \mathbf{e}]^+\|, \end{aligned}$$

where in the last inequality we used that $\langle \lambda, y\rangle \leq \|\lambda\| \|[y]^+\|$ for any $y \in \mathbb{R}^m$ and $\lambda \geq 0$. Using now the fact that:

$$\max_{\lambda \geq 0} -\frac{4L_d}{(p+1)^2}\|\lambda\|^2 + \langle \lambda, h(\hat{\mathbf{u}}^p) + \epsilon_c \mathbf{e}\rangle = \frac{(p+1)^2}{16L_d}\|[h(\hat{\mathbf{u}}^p) + \epsilon_c \mathbf{e}]^+\|^2$$

and introducing the notation $\alpha = h(\hat{\mathbf{u}}^p) + \epsilon_c \mathbf{e}$, we obtain the following second order inequality in $\|[\alpha]^+\|$: $\frac{(p+1)^2}{16L_d}\|[\alpha]^+\|^2 - \|\lambda^*\| \|[\alpha]^+\| - \frac{p+3}{3}\epsilon_{in} \leq 0$. Therefore, $\|[\alpha]^+\|$ must be less or equal than the largest root of the corresponding second-order equation, from which, together with (10.12), we get the result. □

Acknowledgments The research leading to these results has received funding from: the European Union (FP7/2007–2013) under Grant agreement no 248940; CNCS (project TE-231, 19/11.08. 2010); ANCS (project PN II, 80EU/2010); POSDRU/89/1.5/S/62557.

References

1. I. Alvarado, D. Limon, D. Munoz de la Pena, J.M. Maestre, M.A. Ridao, H. Scheu, W. Marquardt, R.R. Negenborn, B. De Schutter, F. Valencia, J. Espinosa, A comparative analysis of distributed MPC techniques applied to the HD-MPC four-tank benchmark. J. Process Control **21**(5), 800–815 (2011)

2. D.P. Bertsekas, J.N. Tsitsiklis, *Parallel and Distributed Computation: Numerical Methods* (Prentice Hall, Englewood, 1989)
3. E. Camponogara, H.F. Scherer, Distributed optimization for model predictive control of linear dynamic networks with control-input and output constraints. IEEE Trans. Autom. Sci. Eng. **8**(1), 233–242 (2011)
4. O. Devolder, F. Glineur, Y. Nesterov. First-order methods of smooth convex optimization with inexact oracle. Technical report, CORE Discussion Paper 2011/02, UCL, Belgium (2011)
5. M.D. Doan, T. Keviczky, B. De Schutter, A distributed optimization-based approach for hierarchical model predictive control of large-scale systems with coupled dynamics and constraints, in *Proceedings of 50th IEEE Conference on Decision and, Control*, pp. 5236–5241 (2011)
6. M. Farina, R. Scattolini, Distributed predictive control: a non-cooperative algorithm with neighbor-to-neighbor communication for linear systems. Automatica **48**(6), 1088–1096 (2012)
7. K.C. Kiwiel, T. Larsson, P.O. Lindberg, Lagrangian relaxation via ballstep subgradient methods. Math. Oper. Res. **32**(3), 669–686 (2007)
8. Y. Kuwata, A. Richards, T. Schouwenaars, J.P. How, Distribted robust receding horizon control for multivariable guidance. IEEE Trans. Control Syst. Technol. **15**(4), 627–641 (2007)
9. T. Larsson, M. Patriksson, A. Stromberg, Ergodic convergence in subgradient optimization. Optim. Methods Softw. **9**, 93–120 (1998)
10. I. Necoara, D. Clipici. An efficient parallel coordinate descent algorithm for distributed MPC. J. Process Control **23**(3), 243–253 (2013)
11. I. Necoara, V. Nedelcu, I. Dumitrache, Parallel and distributed optimization methods for estimation and control in networks. J. Process Control **21**, 756–766 (2011)
12. I. Necoara, J. Suykens, Application of a smoothing technique to decomposition in convex optimization. IEEE Trans. Autom. Control **53**(11), 2674–2679 (2008)
13. A. Nedic, A. Ozdaglar, Approximate primal solutions and rate analysis for dual subgradient methods. SIAM J. Optim. **19**, 1757–1780 (2009)
14. R.R. Negenborn, B. De Schutter, J. Hellendoorn, Multi-agent model predictive control for transportation networks: serial versus parallel schemes. Eng. Appl. Artif. Intell. **21**(3), 353–366 (2008)
15. Y. Nesterov, Smooth minimization of non-smooth functions. Math. Program. **103**(1), 127–152 (2004)
16. Y. Nesterov, Efficiency of coordinate descent methods on huge-scale optimization problems. Technical report, CORE Discussion Paper 2010/2, UCL, Belgium (2010)
17. A. Richards, J.P. How, Robust distributed model predictive control. Int. J. Control **80**(9), 1517–1531 (2007)
18. S. Richter, M. Morari, C.N. Jones, Towards computational complexity certification for constrained MPC based on lagrange relaxation and the fast gradient method, in *Proceedings of 50th IEEE Conference on Decision and, Control*, pp. 5223–5229 (2011)
19. P.O.M. Scokaert, D.Q. Mayne, J.B. Rawlings, Suboptimal model predictive control (feasibility implies stability). IEEE Trans. Autom. Control **44**(3), 648–654 (1999)
20. B.T. Stewart, A.N. Venkat, J.B. Rawlings, S.J. Wright, G. Pannocchia, Cooperative distributed model predictive control. Syst. Control Lett. **59**, 460–469 (2010)

Chapter 11
Distributed MPC Via Dual Decomposition

B. Biegel, J. Stoustrup and P. Andersen

Abstract This chapter presents dual decomposition as a means to coordinate a number of subsystems coupled by state and input constraints. Each subsystem is equipped with a local model predictive controller while a centralized entity manages the subsystems via prices associated with the coupling constraints. This allows coordination of all the subsystems without the need of sharing local dynamics, objectives and constraints. To illustrate this, an example is included where dual decomposition is used to resolve power grid congestion in a distributed manner among a number of players coupled by distribution grid constraints.

11.1 Short Introduction

In this chapter we consider a number of dynamical subsystems; each subsystem has local inputs and states, a local objective function, and local state and input constraints. Moreover, global state and input constraints make the subsystems mutually dependent. The subsystems are not able (or willing) to share the local information; hence optimization of the operation of the subsystems cannot be performed centrally and a distributed approach is necessary.

We consider two small figurative examples to illustrate such global constraints causing coupling of the subsystems. As a first example, consider a number of subsystems that are dependent on the same shared limited resource: this could correspond

B. Biegel (✉) · J. Stoustrup · P. Andersen
Department of Electronic Systems, Automation and Control Aalborg University, Aalborg, Denmark
e-mail: bbi@es.aau.dk

J. Stoustrup
e-mail: jakob@es.aau.dk

P. Andersen
e-mail: pa@es.aau.dk

J. M. Maestre and R. R. Negenborn (eds.), *Distributed Model Predictive Control Made Easy*, Intelligent Systems, Control and Automation: Science and Engineering 69,
DOI: 10.1007/978-94-007-7006-5_11,

to coupling input inequality constraints. In the second example, consider a number of producing and consuming subsystems in a setup where balance must exist: this could correspond to coupling state equality constraints. In both cases the optimization problem is to minimize the total objective while honoring both local and global constraints—without sharing local information.

This chapter presents an approach to solve this problem via dual decomposition: by associating each coupling constraint with a price, the subsystems can be managed by a central entity to reach the solution. This allows coordination of the individual subsystems without sharing local dynamics, constraints or objectives. Further, the final prices of the coupling constraints, the so-called *shadow prices*, will reveal the marginal cost that each agent is willing to pay for the shared resources. This allows the shadow prices to be used for economical settlement purposes between the subsystems.

Dual decomposition is a huge area of research and there exists a large amount of literature on the topic. Dual decomposition appeared already in 1960s where it was used for solving large-scale optimization problems [6, 9]. Also within the area of coordination of dynamic systems via dual decomposition, which is the topic of this chapter, large amounts of literature exists; some references for this are [7, 10–12]. In this chapter, we show the basic idea in using dual decomposition in the coordination of coupled dynamic subsystems.

11.2 Boundary Conditions

We consider N subsystems each described by a discrete linear time-invariant state space model. The states and inputs of subsystem i are denoted $\mathbf{x}_i(k) \in \mathbb{R}^{n_{x,i}}$ and $\mathbf{u}_i(k) \in \mathbb{R}^{n_{u,i}}$, respectively. The state space model is formulated as

$$\mathbf{x}_i(k+1) = \mathbf{A}_i \mathbf{x}_i(k) + \mathbf{B}_i \mathbf{u}_i(k) \tag{11.1}$$

where $\mathbf{A}_i \in \mathbb{R}^{n_{x,i} \times n_{x,i}}$ is the state matrix and $\mathbf{B}_i \in \mathbb{R}^{n_{x,i} \times n_{u,i}}$ is the input matrix. Each subsystem is subject to state and input constraints:

$$\mathbf{x}_i(k) \in \mathcal{X}_i, \mathbf{u}_i(k) \in \mathcal{U}_i \tag{11.2}$$

where $\mathcal{X}_i$ and $\mathcal{U}_i$ are convex constraint sets with $0 \in \mathcal{X}_i, 0 \in \mathcal{U}_i$. The stage cost function of subsystem i is convex and denoted $\ell_i(\mathbf{x}_i(k), \mathbf{u}_i(k))$ and $\ell_i(0,0) = 0$.

Taking a receding horizon control approach with a finite control horizon of N_c time samples and a prediction horizon of $N_p = N_c$ time samples, a local control strategy at subsystem i can be formulated as follows. Let $\mathcal{K}$ be a set containing the current time sample k and the following $N_c - 1$ time samples: $\mathcal{K} = \{k, \dots, k+N_c-1\}$, and let $\mathcal{N}$ denote the set of all N subsystems: $\mathcal{N} = \{1, \dots, N\}$. Then we can formulate a decentralized model predictive control algorithm for subsystem i as in Algorithm 11.1.

Algorithm 11.1 Decentralized Model Predictive Control

1: Observe the current state $\mathbf{x}_i(k)$ and solve the optimization problem

$$\begin{aligned} \text{minimize} \; & \sum_{\kappa\in\mathcal{K}} \ell_i(\mathbf{x}_i(\kappa+1), \mathbf{u}_i(\kappa)) \\ \text{subject to} \; & \mathbf{x}_i(\kappa+1) = \mathbf{A}_i\mathbf{x}_i(\kappa) + \mathbf{B}_i\mathbf{u}_i(\kappa), \quad \forall\kappa\in\mathcal{K} \\ & \mathbf{x}_i(\kappa+1) \in \mathcal{X}_i, \; \mathbf{u}_i(\kappa) \in \mathcal{U}_i, \quad \forall\kappa\in\mathcal{K} \end{aligned} \tag{11.3}$$

where the variables are $\mathbf{x}_i(k+1:k+N_c)$, $\mathbf{u}_i(k:k+N_c-1)$ and $\mathbf{x}_i(k)$ is data. The solution is denoted $\mathbf{x}_i^\star(k+1:k+N_c)$, $\mathbf{u}_i^\star(k:k+N_c-1)$.

2: Apply the first control input solution $\mathbf{u}_i^\star(k)$ to subsystem i.

3: Increase k by one and repeat from 1.

Algorithm 11.1 is presented to illustrate the concept of receding horizon control as this control strategy forms the background for the method presented in this chapter. However, this algorithm is not applicable to the subsystems we have in scope: the N subsystems are not only subject to the local constraints 11.2, but also to global state and input constraints. Consider the following compact notation for inputs and states:

$$\mathbf{x}(k) = \left[\mathbf{x}_1(k)^{\mathrm{T}}, \ldots, \mathbf{x}_N(k)^{\mathrm{T}}\right]^{\mathrm{T}} \tag{11.4}$$

$$\mathbf{u}(k) = \left[\mathbf{u}_1(k)^{\mathrm{T}}, \ldots, \mathbf{u}_N(k)^{\mathrm{T}}\right]^{\mathrm{T}} \tag{11.5}$$

where $\mathbf{x}(k) \in \mathbb{R}^{n_x}$, $n_x = \sum_{i=1}^N n_{x,i}$, and $\mathbf{u}(k) \in \mathbb{R}^{n_u}$, $n_u = \sum_{i=1}^N n_{u,i}$. With this notation we can express the coupling constraints as

$$\mathbf{C}\mathbf{u}(k) \preceq \mathbf{c}, \qquad \mathbf{D}\mathbf{u}(k) = \mathbf{d}, \tag{11.6}$$

$$\mathbf{E}\mathbf{x}(k) \preceq \mathbf{e}, \qquad \mathbf{F}\mathbf{x}(k) = \mathbf{f}, \tag{11.7}$$

where $\preceq$ denotes componentwise inequality; $\mathbf{C} \in \mathbb{R}^{n_c \times n_u}$, $\mathbf{c} \in \mathbb{R}^{n_c}$, and $\mathbf{D} \in \mathbb{R}^{n_d \times n_u}$, $\mathbf{d} \in \mathbb{R}^{n_d}$ describe n_c input inequality constraints and n_d input equality constraints, respectively, while $\mathbf{E} \in \mathbb{R}^{n_e \times n_x}$, $\mathbf{e} \in \mathbb{R}^{n_e}$ and $\mathbf{F} \in \mathbb{R}^{n_f \times n_x}$, $\mathbf{f} \in \mathbb{R}^{n_f}$ describe n_e state inequality constraints and n_f state equality constraints, respectively. These types of constraints can for example express the previously described resource couplings or balancing couplings.

We illustrate this idea of coupled subsystems with a small figurative example. Consider $N = 4$ subsystems where subsystems 1, 2, and 3 share a limited resource while a production/consumption balance must exist between subsystems 3 and 4. This example can be visualized as in Fig. 11.1: subsystems 1, 2 and 3 are interconnected by a net of lines and subsystems 3 and 4 are interconnected by a single line representing the coupling constraints. In dual decomposition, each coupling constraint (each interconnection) will be associated with a price. These prices will be used to coordinate the subsystems to collectively honor the coupling constraints.

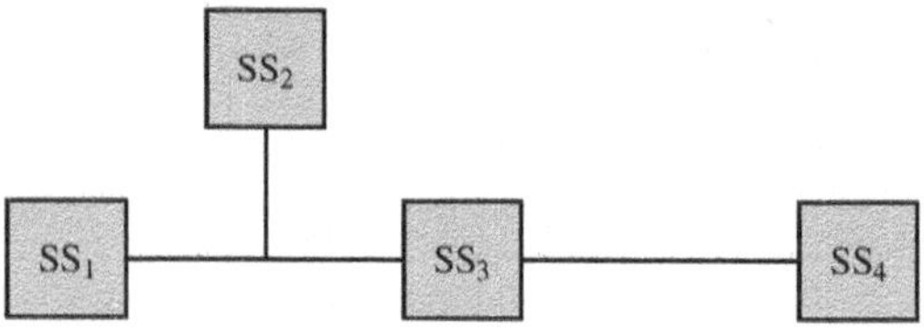

Fig. 11.1 Illustration of coupled subsystems: subsystems 1, 2, and 3 are coupled and subsystems 3 and 4 are coupled

Hereby the subsystems avoid sharing local information such as dynamics, objective and constraints. Two prices exist in the small example presented in Fig. 11.1: one for the coupling of subsystems 1, 2 and 3 and one for the coupling of subsystems 3 and 4.

Due to the coupling constraints 11.6 and 11.7, the subsystems depend on each other and must coordinate their actions to reach feasibility. In the following sections it will be shown that the subsystems can be coordinated via prices associated with the coupled resources by letting an external agent adjust these prices. It is therefore necessary to assume that each subsystems is able to establish a two-way communication link with such an external agent.

11.3 Description of the Approach

We only consider coupling constraints on the form $\mathbf{Cu}(k) \preceq \mathbf{c}$, $\mathbf{C} \in \mathbb{R}^{n_c \times n_x}$, $\mathbf{c} \in \mathbb{R}^{n_c}$ in the following and neglect the three other constraints presented in 11.6 and 11.7. This simplification is made to ease the notation. It is, however, straightforward to follow the method presented in the following to include all four of the presented constraints.

Let

$$\ell(\mathbf{x}(k), \mathbf{u}(k)) = \sum_{i \in \mathcal{N}} \ell_i(\mathbf{x}_i(k), \mathbf{u}_i(k)) \tag{11.8}$$

be the sum of the N convex objective functions of the subsystems and thereby itself a convex function. Based on this, we formulate a Algorithm 11.2 as a control algorithm using the receding horizon approach with a finite control and prediction horizon of $N_c = N_p$ time samples. This algorithm can be applied if all information is available centrally (which is not the case in our setup).

The centralized optimization problem 11.9 is completely separable except for the last coupling constraint $\mathbf{Cu}(\kappa) \preceq \mathbf{c}$. As the coupling constraints are affine, we are able to apply dual decomposition to eliminate the coupling (see, e.g., [5, 13]). This is exactly what we will do in the following.

First, we relax the coupling constraints by introducing the associated Lagrange multipliers; hereby the partial Lagrangian of problem 11.9 becomes

Algorithm 11.2 Centralized Model Predictive Control

1: Observe the current states $\mathbf{x}_i(k)$ for all subsystems $i \in \mathcal{N}$ and solve the centralized optimization problem

$$\begin{aligned} \text{minimize } & \sum_{\kappa \in \mathcal{K}} \ell(\mathbf{x}(\kappa+1), \mathbf{u}(\kappa)) \\ \text{subject to } & \mathbf{x}_i(\kappa+1) = \mathbf{A}_i \mathbf{x}_i(\kappa) + \mathbf{B}_i \mathbf{u}_i(\kappa), \quad \forall \kappa \in \mathcal{K},\ i \in \mathcal{N} \\ & \mathbf{x}_i(\kappa+1) \in \mathcal{X}_i,\ \mathbf{u}_i(\kappa) \in \mathcal{U}_i, \quad \forall \kappa \in \mathcal{K},\ i \in \mathcal{N} \\ & \mathbf{Cu}(\kappa) \preceq \mathbf{c}, \quad \forall \kappa \in \mathcal{K} \end{aligned} \tag{11.9}$$

where the variables are

$$\boldsymbol{\eta}(k) = \left[\mathbf{x}(k+1 : k+N_c)^{\mathrm{T}}, \mathbf{u}(k : k+N_c-1)^{\mathrm{T}}\right]^{\mathrm{T}}$$

and $\boldsymbol{\eta}(k) \in \mathbb{R}^{N_c(n_x+n_u)}$ is used as a compact representation of states and inputs in the following.

2: Apply the first control input solution $\mathbf{u}_i^{\star}(k)$, $\forall i \in \mathcal{N}$ to the N subsystems.

3: Increase k by one and repeat from 1.

$$L(\boldsymbol{\eta}(k), \boldsymbol{\Lambda}(k)) = \sum_{\kappa \in \mathcal{K}} \left(\ell(\mathbf{x}(\kappa+1), \mathbf{u}(\kappa)) + \boldsymbol{\lambda}(\kappa)^{\mathrm{T}} (\mathbf{Cu}(\kappa) - \mathbf{c})\right) \tag{11.10}$$

where $\boldsymbol{\lambda}(\kappa) \in \mathbb{R}^{n_c}$ are the Lagrange multipliers associated with the inequality constraint $\mathbf{Cu}(\kappa) \preceq \mathbf{c}$ and $\boldsymbol{\Lambda}(k) \in \mathbb{R}^{N_c n_c}$ is a compact representation of the Lagrange multipliers: $\boldsymbol{\Lambda}(k) = \boldsymbol{\lambda}(k : k+N_c-1)$.

Define $g(\boldsymbol{\Lambda}(k))$ as the optimal value of the problem

$$\begin{aligned} \text{minimize } & \sum_{\kappa \in \mathcal{K}} \left(\ell(\mathbf{x}(\kappa+1), \mathbf{u}(\kappa)) + \boldsymbol{\lambda}(\kappa)^{\mathrm{T}} (\mathbf{Cu}(\kappa) - \mathbf{c})\right) \\ \text{subject to } & \mathbf{x}_i(\kappa+1) = \mathbf{A}_i \mathbf{x}_i(\kappa) + \mathbf{B}_i \mathbf{u}_i(\kappa), \quad \kappa \in \mathcal{K},\ i \in \mathcal{N} \\ & \mathbf{x}_i(\kappa+1) \in \mathcal{X}_i,\ \mathbf{u}_i(\kappa) \in \mathcal{U}_i, \quad \kappa \in \mathcal{K},\ i \in \mathcal{N} \end{aligned} \tag{11.11}$$

where the variables are $\boldsymbol{\eta}(k)$. This problem is completely separable as both objective and constraints can be separated among the i subsystems. We see this clearly by separating the matrix $\mathbf{C}$ into blocks

$$\mathbf{C} = [\mathbf{C}_1, \ldots, \mathbf{C}_N] \tag{11.12}$$

where $\mathbf{C}_i \in \mathbb{R}^{n_c \times n_{u,i}}$ such that

$$\mathbf{Cu}(k) = \sum_{i \in \mathcal{N}} \mathbf{C}_i \mathbf{u}_i(k). \tag{11.13}$$

Evaluating a subgradient of $g(\boldsymbol{\Lambda}(k))$ can be done as follows. Solve problem 11.11 and let $\overline{\mathbf{u}}(\kappa)$ denote the optimal $\mathbf{u}(\kappa)$, $\forall \kappa \in \mathcal{K}$ for a given realization of $\boldsymbol{\Lambda}(k)$. By differentiation of the objective of problem 11.11 with respect to $\boldsymbol{\Lambda}(k)$ it is evident that a subgradient of $g(\boldsymbol{\Lambda}(k))$ can be described as

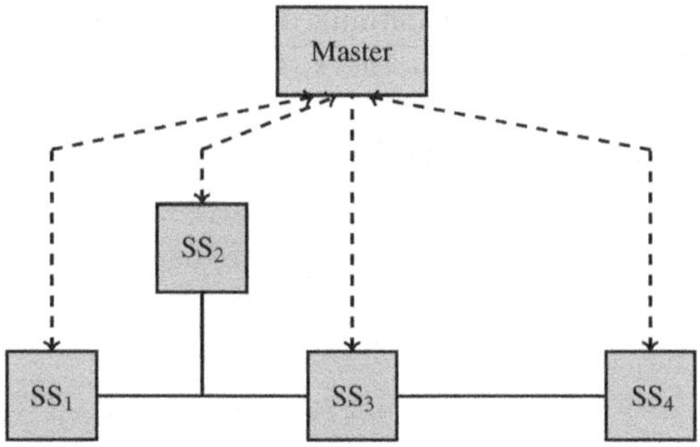

Fig. 11.2 Coupled subsystems interact with master: master broadcasts prices $\mathbf{\Lambda}(k)$ and subsystems respond by reporting how much they utilize the limited resources $\mathbf{C}_i\mathbf{x}_i(\kappa), \forall\kappa \in \mathcal{K}$. The dashed lines indicate the necessary two-way communication links between subsystems and master

$$\left[(\mathbf{C}\overline{\mathbf{u}}(k) - \mathbf{c})^{\mathrm{T}}, \dots, (\mathbf{C}\overline{\mathbf{u}}(k + N_c - 1) - \mathbf{c})^{\mathrm{T}}\right]^{\mathrm{T}} \in \partial(g)(\mathbf{\Lambda}(k)), \tag{11.14}$$

where $\partial(g)(\mathbf{\Lambda}(k))$ denotes the subdifferential of g at $\mathbf{\Lambda}(k)$.

We can formulate the dual of the original centralized problem 11.9 as

$$\begin{array}{ll} \text{maximize} & g(\mathbf{\Lambda}(k)) \\ \text{subject to} & \mathbf{\Lambda}(k) \succeq 0 \end{array} \tag{11.15}$$

with variables $\mathbf{\Lambda}(k)$. Based on the above, we are able to solve the original problem 11.9 in a distributed manner. The key idea is to solve the primal problem 11.9 by solving its dual problem 11.15 using a projected subgradient method. In the subgradient method, steps of appropriate length are taken in the direction of a subgradient of the dual problem which corresponds to iteratively updating the Lagrange multipliers $\mathbf{\Lambda}(k)$. We can do this in a distributed manner as a subgradient of the dual problem 11.15 is given by 11.14 which is separable among the subsystems as $\mathbf{C}\overline{\mathbf{u}}(\kappa) = \sum_{i\in\mathcal{N}} \mathbf{C}_i\overline{\mathbf{u}}_i(\kappa)$. Algorithm 11.3 illustrates this. Note that we use the term *Master* to denote a centralized entity able to perform two-way communication with all subsystems (an interpretation of this master entity is presented in the example in the end of this chapter).

Figure 11.2 illustrates Algorithm 11.3: each interconnection of solid lines represents a coupling constraint while the dashed lines illustrate the necessary communication. This shows that the master needs information form each subsystem in order to update the prices and communicate these prices to the subsystems. It is important to note that the master needs no information of local subsystem constraints, objectives or dynamics; it is sufficient that the master knows how much the limited resources will be used at each subsystem under a sequence of different price realizations. Finally we note that the resulting algorithm using dual decomposition has a straightforward interpretation: in step 5 the master observes if the shared resources $\mathbf{u}(k : k + N_c - 1)$ are overutilized or underutilized. If the subsystems overutilize a limited resource, the associated price is increased; if the subsystems underutilize a shared resource, the associated price is decreased (while keeping it non-negative).

Algorithm 11.3 Distributed Model Predictive Control

1: Master initializes the prices (Lagrange multipliers) $\mathbf{\Lambda}(k) \succeq 0$.
2: **repeat**
3: Master broadcasts the current prices $\mathbf{\Lambda}(k)$ to all subsystems.
4: Problem (11.1) is solved under the current $\mathbf{\Lambda}(k)$ distributedly by letting each subsystem $i \in \mathcal{N}$ *locally* solve the optimization problem

$$\begin{aligned} &\text{minimize} \sum_{\kappa\in\mathcal{K}} \left(\ell_i(\mathbf{x}_i(\kappa+1), \mathbf{u}_i(\kappa)) + \boldsymbol{\lambda}(\kappa)^{\mathrm{T}}\mathbf{C}_i\mathbf{u}_i(\kappa)\right) \\ &\text{subject to } \mathbf{x}_i(\kappa+1) = \mathbf{A}_i\mathbf{x}_i(\kappa) + \mathbf{B}_i\mathbf{u}_i(\kappa), \quad \forall\kappa \in \mathcal{K} \\ &\qquad\qquad \mathbf{x}_i(\kappa+1) \in \mathcal{X}_i, \; \mathbf{u}_i(\kappa) \in \mathcal{U}_i, \quad \forall\kappa \in \mathcal{K} \end{aligned} \tag{11.16}$$

where the variables are $\mathbf{x}_i(\kappa+1), \mathbf{u}_i(\kappa), \forall\kappa \in \mathcal{K}$. The solution is denoted $\overline{\mathbf{x}}_i(k+1 : k+N_c), \overline{\mathbf{u}}_i(k : k+N_c-1)$ and the vectors $\mathbf{C}_i\overline{\mathbf{u}}_i(\kappa) \in \mathbb{R}^{n_c}, \forall\kappa \in \mathcal{K}$ are determined locally at each subsystem and communicated to the master.
5: Master determines the violations $\mathbf{s}(\kappa) \in \mathbb{R}^{n_c}$ of the coupling inequality constraints: $\mathbf{s}(\kappa) = \sum_{i=1} \mathbf{C}_i\overline{\mathbf{u}}_i(\kappa) - \mathbf{c}, \forall\kappa \in \mathcal{K}$; $\mathbf{S}(k) = \mathbf{s}(k : k+N_c-1) \in \mathbb{R}^{N_c n_c}$ and assigns new prices via projection: $\mathbf{\Lambda}(k) := \max(0, \mathbf{\Lambda}(k) + \alpha\mathbf{S}(k))$.
6: **until** $\max(\mathbf{S}(k)) \leq \epsilon$ or maximum number of iterations reached.
7: Based on the final utilization of the input $\mathbf{u}(\kappa), \forall\kappa \in \mathcal{K}$, the master determines limits $\overline{\mathbf{c}}_i$ assuring feasibility of the overall problem and communicates the limits to all subsystems.
8: Each subsystem locally solves problem (11.3) with the additional constraint $\mathbf{C}_i\mathbf{u}_i(\kappa) \preceq \overline{\mathbf{c}}_i, \forall\kappa \in \mathcal{K}$ and applies the first control input solution.
9: Increase k by one and repeat from 1.

11.4 Theoretical Results Availability

In this section we briefly comment on the computational burden of the optimization algorithm and describe under what circumstances the algorithm will converge.

First, we note that the optimization problem of each subsystem in the distributed algorithm (Problem 11.16) is only slightly more complex than if the subsystem couplings were neglected problem (11.16). However, the complexity increases significantly as we are required to solve the distributed optimization problem (11.16) a number of times until convergence. Further we note that the update law of the master (Algorithm 11.3 step 5) requires only a single addition and multiplication operation. The computational burden of the master therefore scales well with the number of subsystems N.

A requirement for Algorithm 11.3 to converge is that we have no duality gap, i.e., the value of the primal and the dual solutions are identical. If the primal problem is convex, it often holds that the primal and dual solutions are identical but additional conditions are necessary to guarantee this. One such condition is Slater's condition [4, p. 226] which states that the primal and dual solutions are identical if the primal problem is convex and there exists a solution to the primal problem that is strictly feasible. In the formulation of problem (11.16), Slater's condition requires that a solution exists such that

$$\mathbf{x}_i(\kappa + 1) \in \mathbf{relint}\mathcal{X}_i, \quad \mathbf{u}_i(\kappa) \in \mathbf{relint}\mathcal{U}_i, \quad \forall \kappa \in \mathcal{K} \tag{11.17}$$

where $\mathbf{relint}\mathcal{X}_i$ denotes the relative interior of $\mathcal{X}_i$ and is a set that contains all points which are not on the edge of $\mathcal{X}_i$, relative to the smallest set in which $\mathcal{X}_i$ lies [1, p. 448]. Under this assumption, convergence can be guaranteed depending on the choice of step size in the subgradient method, which will be discussed in the following.

In the presented algorithm, a projected subgradient method is used to solve the constrained convex optimization problem 11.15. The subgradient method updates $\boldsymbol{\Lambda}(k)$ according to

$$\boldsymbol{\Lambda}(k) := \mathbf{P}(\boldsymbol{\Lambda}(k) - \alpha \mathbf{g}) \tag{11.18}$$

where $\mathbf{P}$ is a projection of $\boldsymbol{\Lambda}(k)$ onto the feasible set $\{\boldsymbol{\Lambda}(k) \in \mathbb{R}^{N_c n_c} | \boldsymbol{\Lambda}(k) \succeq 0\}$ and $\mathbf{g}$ is any subgradient to the dual problem and α is a (constant) step size. Using such constant step size assures that we will converge to a value that lies within some range of the optimum value. If the objective of problem 11.11 is differentiable, i.e., $\ell(\mathbf{x}(\kappa + 1), \mathbf{u}(\kappa))$, $\kappa \in \mathcal{K}$ is differentiable, the subgradient method will indeed converge to the optimum for sufficiently small α [3].

Another option is to let the step size vary with the iteration number j, hereby convergence to the optimal value can be guaranteed also for the case of a non-differentiable objective. One example is a non-summable diminishing step size

$$\lim_{j \to \infty} \alpha_j = 0, \qquad \sum_{j=1}^{\infty} \alpha_j = +\infty \tag{11.19}$$

where α_j is the step size at iteration i; this will guarantee that the subgradient method converges to the optimum [14, p. 215]. Other step size rules with same convergence result exist.

It is important to note that the subgradient method is chosen due to the fact that this allows us to decouple the problem. Other methods (such as second order methods) can provide much faster convergence than the subgradient method presented here. They are, however, not suitable for the type of decoupling presented in this chapter.

A final note concerns the convergence proofs of dual decomposition algorithms. Dual decomposition algorithms rely on subgradient methods as presented above. Generally, convergence proofs for gradient methods are based on the function value decreasing at each iteration; however, for subgradient methods this is not the case. In subgradient methods, the convergence proofs are generally based on the Euclidian distance to the optimal set by showing under what circumstances this distance will decrease at each iteration [3]. Therefore, the objective value can increase during the iterates in the subgradient method used in the algorithm; however, the distance to the optimal set will decrease at each iteration.

11.5 Application Results Availability

In this section, an application of distributed model predictive control via dual decomposition is presented. The example is taken from a *smart grid* setup where the basic idea is to increase the sustainability and stability of the electrical grid by utilizing the flexibility of the demand side (consumers) in the balancing efforts. Two main ideas of the smart grid concept are

- balancing of production and consumption by moving load temporally,
- avoiding distribution grid congestion by moving load temporally or spatially.

In this example, we address these two topics at an overall level.

Consider a number of balancing responsible parties (BRPs) each responsible for a number of consumers under their jurisdiction; each consumer belongs to exactly one BRP. The BRPs buy energy at the day-ahead electricity market on behalf of the consumers. In the following, we illustrate how BRPs can utilize the flexibility of the consumers under their jurisdiction to minimize the imbalance between the purchased energy and the consumed energy thereby avoiding trading compensating balancing energy at unfavorable prices. Further, we show how the BRPs can be coordinated such that distribution grid congestion is avoided. Due to the very competitive electricity market, the BRPs are not willing to share local information such as objectives and states; therefore we use the dual decomposition approach presented in this chapter to resolve grid congestion. In this way, congestion management can be achieved without information sharing between the BRPs. Finally, we show how the dual decomposition method can be interpreted as a *distribution grid capacity market*. Throughout the following, the notation from the previous section will be used to the extent possible.

Consider a star topology distribution grid (no loops) consisting of n_f distribution lines of limited capacity. A total of N BRPs are active in the distribution grid and BRP number i is responsible for $n_{x,i}$ consumers. The setup is illustrated in Fig. 11.3 and discussed in detail in the sequel.

The $n_{x,i}$ consumers under BRP i are characterized by hourly energy consumptions $\mathbf{u}_i(k) + \tilde{\mathbf{u}}_i(k)$ where $\mathbf{u}_i(k) \in \mathbb{R}^{n_{x,i}}$ is the controllable (flexible) part of the consumption and $\tilde{\mathbf{u}}_i(k) \in \mathbb{R}^{n_{x,i}}$ is an uncontrollable base consumption assuming a sampling time of 1 hour. Due to the flexible consumption, the devices are able to *store* energy. We denote the amount of stored energy $\mathbf{x}_i(k) \in \mathbb{R}^{n_{x,i}}$ for the consumers under BRP i; this may be energy stored as either heat, cold, energy in a battery, or similar. The stored energy depends on the controllable power consumption

$$\mathbf{x}_i(k+1) = \mathbf{A}_i \mathbf{x}_i(k) + \mathbf{B}_i \mathbf{u}_i(k), \tag{11.20}$$

where $\mathbf{A}_i, \mathbf{B}_i \in \mathbb{R}^{n_{x,i} \times n_{x,i}}$ are diagonal with diagonal elements describing drain losses of each energy storage. The consumers are limited by power and energy constraints

$$0 \preceq \mathbf{u}_i(k) + \tilde{\mathbf{u}}_i(k) \preceq \mathbf{u}_i^{\max}, \qquad \mathbf{x}_i^{\min} \preceq \mathbf{x}_i(k) \preceq \mathbf{x}_i^{\max} \tag{11.21}$$

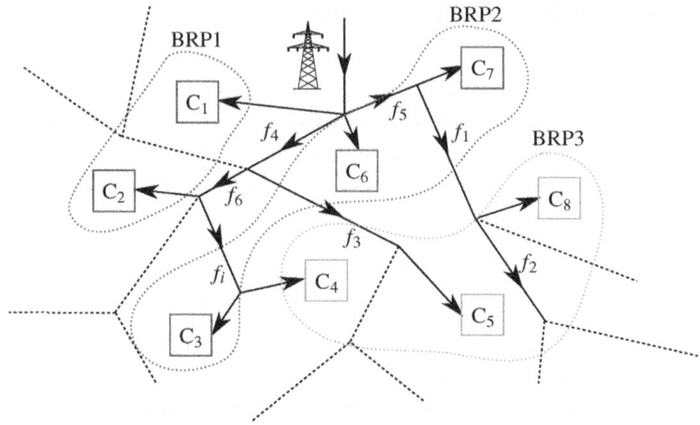

Fig. 11.3 Interconnected consumers under the jurisdiction of different BRPs sharing the same distribution grid (dotted lines indicate that only a small part of the total grid is shown)

where $\mathbf{u}_i^{\max}$, $\mathbf{x}_i^{\min}$, $\mathbf{x}_i^{\max} \in \mathbb{R}^{n_{x,i}}$ describe these limits. Consumer models described this way can be found for example in [8].

The consumers are powered through the distribution grid, as illustrated in Figure 11.3. Each BRP will contribute to the loading of the distribution lines. Let $\mathbf{r}_i(k) \in \mathbb{R}_+^{n_f}$ denote the partial flow caused by BRP i to the n_f distribution lines; these partial flows can by flow conversation be described as

$$\mathbf{r}_i(k) = \mathbf{R}_i\left(\mathbf{u}_i(k) + \tilde{\mathbf{u}}_i(k)\right) \tag{11.22}$$

where $\mathbf{R}_i \in \mathbb{R}^{n_f \times n_{x,i}}$ is given by

$$(\mathbf{R}_i)_{pq} = \begin{cases} 1 \text{ if consumer } q \text{ under BRP } i \text{ is supplied through link } p, \\ 0 \text{ otherwise.} \end{cases}$$

This simply states that the power to each consumer under BRP i must flow through a unique path of distribution lines; these paths are indicated in the $\mathbf{R}_i$ matrix.

The distribution grid is protected from overcurrents by electrical fuses; hence, the distribution lines are subject to constraints. The total flows $\mathbf{f}(k) \in \mathbb{R}_+^{n_f}$ over the distribution lines and associated fuse limits can be expressed as

$$\mathbf{f}(k) = \sum_{i \in \mathcal{N}} \mathbf{r}_i(k), \qquad \mathbf{f}(k) \preceq \mathbf{f}^{\max} \tag{11.23}$$

where $\mathbf{f}^{\max}(k) \in \mathbb{R}_+^{n_f}$ denotes the limits of the fuses and $\mathcal{N}$ is the set of all BRPs.

The BRPs buy energy at a day-ahead spot market for each hour of the following day. We denote the energy bought by BRP i at the day-ahead spot market $\mathbf{p}_i(k) \in \mathbb{R}$; this means that BRP i has bought the energy $\mathbf{p}_i(k)$ for the time interval from hour

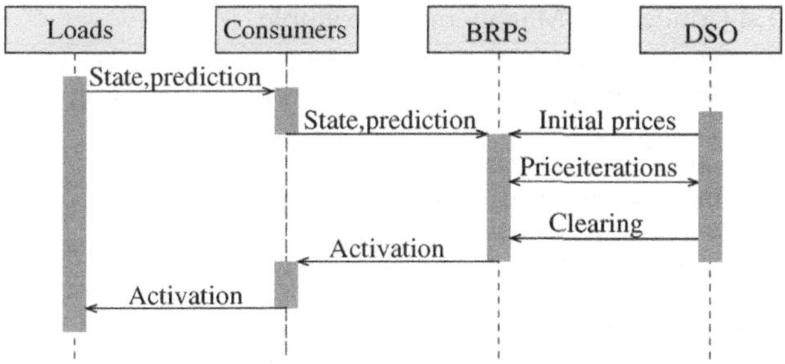

Fig. 11.4 Market interpretation of congestion alleviation via dual decomposition

k to $k+1$. The objective of each BRP is to minimize the imbalance between the consumed energy $\mathbf{1}^{\mathrm{T}}(\mathbf{u}_i(k)+\tilde{\mathbf{u}}_i(k))$ and the purchased energy $\mathbf{p}_i(k)$, i.e.,

$$\ell_i(\mathbf{u}_i(k)) = \|\mathbf{1}^{\mathrm{T}}(\mathbf{u}_i(k)+\tilde{\mathbf{u}}_i(k)) - \mathbf{p}_i(k)\|_2^2, \tag{11.24}$$

where it is chosen to minimize the imbalance in the two-norm sense and where $\mathbf{1}$ denotes a vector of appropriate dimension with all entries equal to one. By keeping this imbalance small, the BPR minimizes the energy imbalances and thereby avoids trading balancing energy possibly at very unfavorable price.

The modeling reveals that the optimization problem is completely separable among the BRPs except for the coupling via the distribution line capacity constraints 11.23. We apply Algorithm 11.3 to the presented application example and obtain the Algorithm 11.4 when performing receding horizon control with a control horizon N_c and prediction horizon of $N_p = N_c$ samples.

Algorithm 11.4 shows that the congestion management via dual decomposition can be interpreted as a new distribution grid market where each distribution line is associated with a time-varying cost per unit flow. If the lines are not congested, the BRPs are free to use the lines at no cost; however, if congestion occurs, the master will adjust the price on the lines until the congestion is resolved.

The sequence diagram in Fig. 11.4 illustrates how this market can be imagined in an electrical power system setup. First, the individual loads communicate their flexibility (via states and predictions) to the individual consumers. Following, the consumers communicate the flexibility of all their respective loads to the corresponding BRP. Further, the BRPs are provided with initial prices on the distribution grid from the distribution grid operator (DSO) which has the role of the master. Based on this, a price iteration follows where the DSO adjusts the prices until all grid congestions are resolved. When the iteration is completed, the DSO clears the market by communicating final prices and line capacity limits for each BRP. Here it is important to note that the prices at the moment of the market clearing are *real prices* that will determine the economical settlement between the BRPs. From the perspective of a BRP, the prices on the distribution lines reveal the cost that the BRP will have to pay (or be paid) for using more (or less) of the line capacity.

Algorithm 11.4 Congestion Management Example

1: Master initializes the prices $\mathbf{\Lambda}(k) \succeq 0$, $\mathbf{\Lambda}(k) = \boldsymbol{\lambda}(k : k+N_c-1)$, where $\boldsymbol{\lambda}(k) \in \mathbb{R}^{n_f}$ and $\boldsymbol{\lambda}_j(k)$ is the price associated with the capacity limit of distribution line j at time sample k.
2: **repeat**
3: Master broadcasts the current prices $\boldsymbol{\lambda}(\kappa)$, $\forall \kappa \in \mathcal{K}$ to the subsystems.
4: Each BRP locally solves the price dependent problem

$$\begin{array}{ll} \text{minimize} & \sum_{\kappa \in \mathcal{K}} \left(\|\mathbf{1}^{\mathrm{T}}(\mathbf{u}_i(\kappa) + \tilde{\mathbf{u}}_i(\kappa)) - \mathbf{p}_i(\kappa)\|_2^2 + \boldsymbol{\lambda}(\kappa)^{\mathrm{T}} \mathbf{r}_i(\kappa) \right) \\ \text{subject to} & \mathbf{x}_i(\kappa+1) = \mathbf{A}_i \mathbf{x}_i(\kappa) + \mathbf{B}_i \mathbf{u}_i(\kappa), \quad \forall \kappa \in \mathcal{K} \\ & 0 \preceq \mathbf{u}_i(\kappa) + \tilde{\mathbf{u}}_i(\kappa) \preceq \mathbf{u}_i^{\max}, \quad \forall \in \mathcal{K} \\ & \mathbf{x}_i^{\min} \preceq \mathbf{x}_i(\kappa) \preceq \mathbf{x}_i^{\max}, \quad \forall \kappa \in \mathcal{K} \\ & \mathbf{r}_i(\kappa) = \mathbf{R}_i \left(\mathbf{u}_i(\kappa) + \tilde{\mathbf{u}}_i(\kappa) \right), \quad \forall \kappa \in \mathcal{K} \end{array} \tag{11.25}$$

where the variables are $\mathbf{x}_i(k+1 : k+N_c)$, $\mathbf{u}_i(k : k+N_c-1)$, $\mathbf{r}_i(k : k+N_c-1)$. The solution is denoted $\overline{\mathbf{x}}_i(k+1 : k+N_c)$, $\overline{\mathbf{u}}_i(k : k+N_c-1)$, $\overline{\mathbf{r}}_i(k : k+N_c-1)$.
5: Each BRP reports local partial flows $\overline{\mathbf{r}}_i(\kappa)$ to the master. The master centrally determines line capacity violations $\mathbf{s}(\kappa) = \sum_{i \in \mathcal{N}} \mathbf{r}_i(\kappa) - \mathbf{f}^{\max} \in \mathbb{R}^{n_f}$, $\forall \kappa \in \mathcal{K}$ where $\mathbf{s}_j$ is the capacity violation of line j and $\mathbf{S}(k) = \mathbf{s}(k : k+N_c-1) \in \mathbb{R}^{N_c n_f}$.
6: Master updates prices $\mathbf{\Lambda}(k)$ via projection: $\mathbf{\Lambda}(k) := \max\left(0, \mathbf{\Lambda}(k) + \alpha \mathbf{S}(k)\right)$. Again notice that this corresponds to increasing the cost on congested lines and reducing the price on lines where there is free capacity; however, always assuring non-negative line prices.
7: **until** $\max(\mathbf{S}(k)) \leq \epsilon$ or maximum number of iterations reached.
8: Master determines limits $\overline{\mathbf{c}}_i \in \mathbb{R}^{n_f}$ and communicates limits and final prices (shadow prices) to the BRPs.
9: Each subsystem locally solves problem (11.25) with the additional constraint $\mathbf{r}_i(\kappa) \preceq \overline{\mathbf{c}}_i$ and applies the first control input of the solution.
10: Increase k by one and repeat from 1.

Finally, we consider a small numerical example to illustrate the price iteration. The example is kept at a conceptual level to clearly illustrate the concept. The details of the simulation are not presented here but can be found in [2]. Consider two BRPs responsible for one and two consumers, respectively, as presented in Fig. 11.5. The example is constructed with dynamics and objectives fitting the structure of Algorithm 11.4; we assume we are at time sample $k = 1$ and use a control horizon and prediction horizon of $N_c = N_p = 10$. Both BRP 1 and 2 desire to increase the controllable consumption in the first hours, and decrease the consumption in the later hours. If no action is taken, this will violate the capacity constraint on line 2: $\mathbf{f}_2 \leq \mathbf{f}_2^{\max}$. To remedy the problem without information sharing, Algorithm 11.4 is used. The DSO starts by publishing the initial prices $\mathbf{\Lambda}(1) = 0$ where after the two BRPs report back to the DSO how they will utilize the distribution grid under this price, by respectively sending $\overline{\mathbf{r}}_1$ and $\overline{\mathbf{r}}_2$ to the DSO. The DSO discovers that congestion will occur under the initial prices and therefore updates the prices according to $\mathbf{\Lambda}(1) := \mathbf{\Lambda}(1) + \alpha \mathbf{S}(1)$. The top plot of Fig. 11.6 shows the price adjustments, converging to the shadow prices $\mathbf{\Lambda}^{\star}(1)$, optimally resolving the congestion (within the given horizon). The solid line shows the primal objective when using feasible flows, the dashed line is the dual objective, and the dotted line is the optimal value within the control horizon. The lower plot shows the iteration of the prices associated

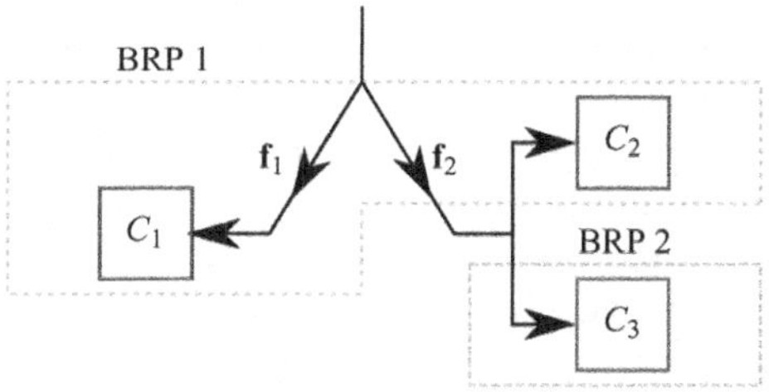

Fig. 11.5 Three consumers under the jurisdiction of two different BRPs sharing the same distribution grid

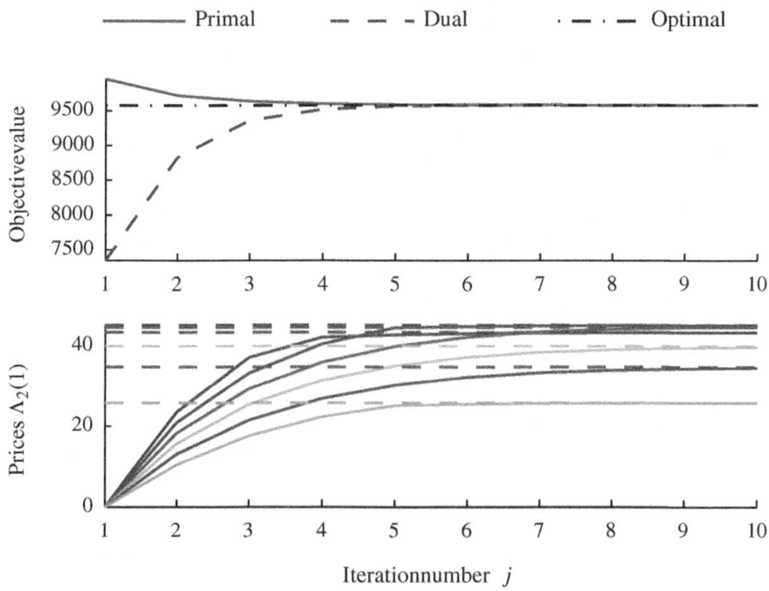

Fig. 11.6 Top: objective value progress. Bottom: convergence of $\lambda_2(1), \ldots, \lambda_2(6)$ (solid lines) towards the shadow prices (dashed lines)

with capacity constraint at line 2 from time sample $k = 1$ to $k = 6$; the prices at time $k = 7$ to $k = 10$ remain at zero as there is no congestion at these hours.

The large benefit of resolving congestion management by prices is that the global economical optimum is reached within the control horizon N_c without the need of a centralized optimization. In the presented example, consumer 3 under PRB 2 is a storage of high quality (low drainage) while consumer 2 under BRP 1 is a storage of low quality (high drainage). In this market approach, this results in consumer 3 being the main user of the shared distribution line because BRP 2 is willing to pay a higher price for the line utilization due to the fact that he can profit much from this high quality storage. BRP 1, on the other hand, is willing to decrease the use of his low quality storage by receiving a payment from BRP 2 as he is not able to profit much from his poor storage.

To illustrate the benefit of using the distribution grid market approach to resolve grid congestion, consider an alternative very simple strategy: congestion is simply resolved by splitting the capacity of the shared line equally among the players sharing the line. In this case, the high quality storage would be used less and the low quality storage would be used more. As a result, a larger amount of energy would be lost due to the higher utilization of the low quality storage; hence, we would not have reached the global economical optimum.

Acknowledgments The work is completed as a part of the project *iPower* and supported by the Danish government via the DSR-SPIR program 10-095378.

References

1. D.P. Bertsekas, *Nonlinear Programming* (Athena Scientific, Belmont, 1999)
2. B. Biegel, P. Andersen, J. Stoustrup, J. Bendtsen, *Congestion management in a smart grid via shadow prices*, in IFAC Power Plant and Power Systems Control, Toulouse, 2012
3. S. Boyd, L. Xiao, A. Mutapic, Notes on subgradient, methods (2007)
4. S. Boyd, L. Vandenberghe, *Convex Optimization* (Cambridge University Press, New York, 2004)
5. S. Boyd, L. Xiao, A. Mutapcic, J. Mattingley, Notes on decomposition methods (2008)
6. G.B. Danzig, P. Wolfe, The decomposition algorithm for linear programming. Econometrica **29**, 767–778 (1961)
7. M.D. Doan, T. Keviczky, B. De Schutter, A dual decomposition-based optimization method with guaranteed primal feasibility for hierarchical MPC problems, in *Proceedings of the 18th IFAC World Congress*, pp. 392–397, Milan, Italy, August–September (2011)
8. K. Heussen, S. Koch, A. Ulbig, G. Andersson, Unified system-level modeling of intermittent renewable energy sources and energy storage for power system operation. IEEE Syst. J. **99**, 1 (2011)
9. H Everett III, Generalized Lagrange multiplier method for solving problems of optimum allocation of resources. Oper. Res. **11**(3), 399–417 (1962)
10. F.P. Kelly, A.K. Maulloo, D.K.H. Tan, Rate control in communication networks: shadow prices, proportional fairness and stability. J. Oper. Res. Soc. **49**, 237–252 (1998)
11. A. Rantzer, Dynamic dual decomposition for distributed control. In American Control Conference, 2009. ACC '09, pp. 884–888, June 2009
12. Anders Rantzer, On prize mechanisms in linear quadratic team theory, in *Proceedings of 46th IEEE Conference on Decision and Control*, New Orleans, LA, December 2007
13. S. Samar, S. Boyd, D. Gorinevsky, Distributed estimation via dual decomposition, in *IEEE Multi-Conference on Systems and Control*, Denver, CO., USA, Kos, Greece, July 2008
14. N.Z. Shor, *Minimization Methods for Non-Differentiable Functions* (Springer, New York, 1979)

Chapter 12
Distributed Optimization for MPC of Linear Dynamic Networks

E. Camponogara

Abstract This chapter presents existing models and distributed optimization algorithms for model predictive control (MPC) of linear dynamic networks (LDNs). The models consist of networks of subsystems with deterministic and uncertain dynamics subject to local and coupling constraints on the control and output signals. The distributed optimization algorithms are based on gradient-projection, subgradient, interior-point, and dual strategies that depend on the nature of the couplings and constraints of the underlying networks. The focus will be on a class of LDNs in which the dynamics of the subsystems are influenced by the control signals of the upstream subsystems with constraints on state and control variables. A distributed gradient-based algorithm is presented for implementing an interior-point method distributively with a network of agents, one for each subsystem.

12.1 Introduction

Dynamic networks are systems of subsystems that can model large, geographically distributed systems such as urban traffic networks, sewage systems, and petrochemical plants. A dynamic network (DN) consists of a graph whose nodes represent dynamic subsystems, each characterized by a local state that evolves dynamically depending on its control signals and the signals from other subsystems. When the differential equations governing the dynamics of the subsystems are linear the network is called linear dynamic network (LDN).

In [3] a distributed model predictive control (DMPC) framework was developed for operating LDNs in which the dynamics of the subsystems depend on their local control signals and state, but only on the control inputs of the upstream subsystems

E. Camponogara(✉)
Department of Automation and Systems Engineering, Federal University of Santa Catarina, Florianópolis, Brazil
e-mail: camponog@das.ufsc.br

J. M. Maestre and R. R. Negenborn (eds.), *Distributed Model Predictive Control Made Easy*, Intelligent Systems, Control and Automation: Science and Engineering 69,
DOI: 10.1007/978-94-007-7006-5_12, © Springer Science+Business Media Dordrecht 2014

while being subject to constraints on local controls. This work was extended by accounting for uncertainties in the dynamic models and by proposing a distributed subgradient optimization algorithm for implementing DMPC [4]. In [5] constraints on the state variables are admitted which induce algebraic couplings between subsystems in addition to the dynamic interconnections. Distributed algorithms based on the dual strategy [10] and the interior-point method [5] have been proposed in the literature for handling such algebraic couplings. The focus of this chapter is on the latter strategy.

A number of other distributed optimization strategies have also appeared in the literature. In [11] the solution of optimization problems within a distributed model predictive control framework is addressed, whereby Lagrangian duality is applied to handle coupling variables among neighboring agents. In [14] a cooperation-based iterative process was developed for optimization and control of linear systems with constrains on local control signals. In [15] distributed primal-dual subgradient algorithms are developed for general convex optimization with multi-agent systems, considering global objectives and constraints.

This chapter is organized as follows. Section 12.2 introduces three classes of linear dynamic networks with focus on LDNs with coupling state constraints. Section 12.3 presents a distributed optimization algorithm for solving quadratic programs that arise from the model predictive control (MPC) of LDNs with state constraints. Section 12.4 briefly discusses some applications of distributed MPC on LDNs.

12.2 Linear Dynamic Networks

A dynamic network consists of a directed graph with nodes modeling dynamic subsystems and arcs representing the direct influence between the subsystems. Dynamic networks can serve as models for geographically distributed systems such as urban traffic networks, electric power systems, and water pipelines. The structure of a DN is represented by a coupling graph $\mathcal{G} = (\mathcal{N}, \mathcal{E})$ with $\mathcal{N} = \{1, \ldots, n\}$ being the set of nodes and $\mathcal{E} \subseteq \mathcal{N} \times \mathcal{N}$ being the set of arcs.

Each node $i \in \mathcal{N}$ is characterized by a local state vector $\mathbf{x}_i$ of dimension $n_{\mathbf{x}_i}$ and a local control-input vector $\mathbf{u}_i$ of dimension $n_{\mathbf{u}_i}$. The state $\mathbf{x}_i$ of subsystem i evolves dynamically depending on the local state and controls and also on the state and controls of its *input neighborhood* $\mathcal{I}(i) = \{j : (j, i) \in \mathcal{E}\} \cup \{i\}$.

To illustrate these concepts consider the small urban traffic network and the corresponding coupling graph depicted in Fig. 12.1. In this dynamic system the subsystems correspond to intersections, the state of an intersection consists of the number of vehicles in the roads leading to it, and the control signals are the green times assigned to each stage which is assumed to be one for each state for simplicity. For instance, the state of subsystem 3 is $\mathbf{x}_3 = [x_{3,1}\ x_{3,4}]^{\mathrm{T}}$ and its control vector is $\mathbf{u}_3 = [u_{3,1}\ u_{3,4}]^{\mathrm{T}}$. The input neighborhood of this subsystem is $\mathcal{I}(3) = \{1, 3, 4\}$. In such traffic systems, the discharge of queues at the upstream subsystems becomes the arrival of vehicles in the roads that correspond to the state of subsystem 3.

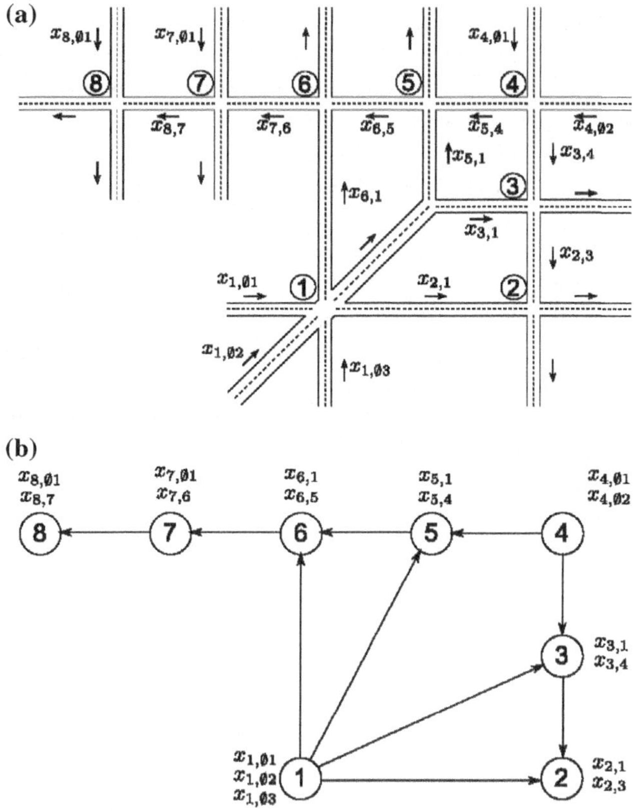

Fig. 12.1 Dynamic network of an urban traffic system. **a** Urban traffic system with 8 intersections. **b** Graph of the linear dynamic network

The potential to represent actual systems and the complexity of the control strategy will depend greatly on the nature of the differential equations governing the dynamics of the nodes, which in general are given by:

$$\mathbf{x}_i(k+1) = \mathbf{F}_i(\mathbf{x}_i^I(k), \mathbf{u}_i^I(k)). \tag{12.1}$$

where $\mathbf{x}_i^I = [\mathbf{x}_j^{\mathrm{T}} : j \in \mathcal{I}(i)]^{\mathrm{T}}$ and $\mathbf{u}_i^I = [\mathbf{u}_j^{\mathrm{T}} : j \in \mathcal{I}(i)]^{\mathrm{T}}$ are vectors collecting all of the states and control inputs influencing the state of subsystem i, respectively. Previous works have developed optimization and control frameworks assuming specialized structures for the dynamic equation (12.1) and constraints. Frameworks for three general classes of dynamic networks are presented below.

1. *Dynamic dependency on upstream control signals, subject to constraints on local control signals*

A first work on dynamic networks was inspired by the store-and-forward modeling of urban traffic flow [9]. Traffic flow can be modeled with linear dynamic equations depending on the control signals from the input neighborhood and local state, being expressed for a subsystem $i \in \mathcal{N}$ as:

$$\mathbf{x}_i(k+1) = \mathbf{A}_i \mathbf{x}_i(k) + \sum_{j \in \mathcal{I}(i)} \mathbf{B}_{i,j} \mathbf{u}_j(k) \tag{12.2}$$

where $\mathbf{A}_i \in \mathbb{R}^{n_{\mathbf{x}_i} \times n_{\mathbf{x}_i}}$ and $\mathbf{B}_{i,j} \in \mathbb{R}^{n_{\mathbf{x}_i} \times n_{\mathbf{u}_j}}$ are matrices of suitable dimensions. For the particular case of traffic flow modeling with store-and-forward, the matrix $\mathbf{A}_i$ is the identity while the $\mathbf{B}_{i,j}$ matrices model the discharges from and arrivals in the vehicle queues defining the state of intersection i. Considering intersection 3 the store-and-forward approach defines these matrices in terms of physical characteristics of the traffic network as:

$$\mathbf{B}_{3,1} = T_3 \begin{bmatrix} \rho_{3,1,\emptyset 1} \cdot \frac{S_{1,\emptyset 1}}{C_1} & \rho_{3,1,\emptyset 2} \cdot \frac{S_{1,\emptyset 2}}{C_1} & \rho_{3,1,\emptyset 3} \cdot \frac{S_{1,\emptyset 3}}{C_1} \\ 0 & 0 & 0 \end{bmatrix},$$

$$\mathbf{B}_{3,3} = T_3 \begin{bmatrix} -\frac{S_{3,1}}{C_3} & 0 \\ 0 & -\frac{S_{3,4}}{C_3} \end{bmatrix},$$

$$\mathbf{B}_{3,4} = T_3 \begin{bmatrix} 0 & 0 \\ \rho_{3,4,\emptyset 1} \cdot \frac{S_{4,\emptyset 1}}{C_4} & \rho_{3,4,\emptyset 2} \cdot \frac{S_{4,\emptyset 2}}{C_4} \end{bmatrix}$$

where T_3 is the sample time (in seconds), $S_{i,j}$ is the saturation flow on the link/road from intersection j to i (in vehicles per second), $\rho_{m,i,j}$ is the rate at which vehicles from link j to i enter link i to m, and C_i (in seconds) is the cycle time of junction i with a cycle being composed by a cyclic sequence of stages which are specific traffic light configurations.

This class of systems is referred to as linear dynamic networks (LDNs) in view of the nature of linear dynamics. The controls of any subsystem i can be subject to linear constraints in the form:

$$\mathbf{D}_i \mathbf{u}_i(k) \leq \mathbf{d}_i, \tag{12.3a}$$

$$\mathbf{E}_i \mathbf{u}_i(k) = \mathbf{e}_i. \tag{12.3b}$$

In the case of the traffic system these constraints are used to force the green times to add up to cycle time and impose bounds on these control signals.

A distributed MPC framework based on gradient projection [1] was developed for controlling the LDN with a network of agents, one for each subsystem, which minimizes a quadratic cost function on states and control inputs:

$$J = \min \sum_{i \in \mathcal{N}} \sum_{k=1}^{N_p} \frac{1}{2} \mathbf{x}_i(k)^T \mathbf{Q}_i \mathbf{x}_i(k) + \sum_{i \in \mathcal{N}} \sum_{k=0}^{N_c - 1} \frac{1}{2} \mathbf{u}_i(k)^T \mathbf{R}_i \mathbf{u}_i(k). \tag{12.4}$$

Following a synchronization protocol that allows only uncoupled agents to iterate simultaneously, the gradient-projection algorithm was shown to converge to an optimal solution to the quadratic program of centralized MPC [3].

2 *Dynamic dependency on upstream control signals with uncertainty, subject to constraints on local control signals*
The previous class of LDNs was recently generalized to consider uncertainty on the dynamic equations:

$$\mathbf{x}_i(k+1) = \mathbf{A}_i(k)\mathbf{x}_i(k) + \sum_{j \in \mathcal{I}(i)} \mathbf{B}_{i,j}(k)\mathbf{u}_j(k) \tag{12.5}$$

where $(\mathbf{A}_i(k), \mathbf{B}_{i,j}(k) : j \in \mathcal{I}(i)) \in \Omega_i$ with Ω_i being a set of possible dynamic realizations of subsystem i. For an urban traffic network, Ω_i can model uncertainties related to the conversion rates and traffic patterns. The LDN is also subject to linear constraints of the form of Eq. (12.3) imposed on the control signals.
A robust MPC strategy would hedge against the worst-case scenario by optimizing the objective:

$$J = \min \sum_{i \in \mathcal{N}} \max \left\{ \sum_{k=1}^{N_p} \frac{1}{2} \mathbf{x}_i(k|\omega)^T \mathbf{Q}_i \mathbf{x}_i(k|\omega) + \sum_{k=0}^{N_c-1} \frac{1}{2} \mathbf{u}_i(k)^T \mathbf{R}_i \mathbf{u}_i(k) : \omega \in \Omega_i(N_p) \right\} \tag{12.6}$$

where $\Omega_i(N_p)$ is the set of all possible dynamic trajectories for subsystem i during a horizon of length N_p, ω is a particular trajectory, and $\mathbf{x}_i(k|\omega)$ is the predicted state at time t_k assuming a dynamic trajectory ω.
Because the objective function to be minimized in (12.6) is nondifferentiable, derivative-based approaches such as gradient projection could not be directly applied. Thus a distributed subgradient algorithm was developed for solving the optimization problem with a network of distributed agents [4]. The algorithm and synchronization protocol are similar to the ones developed for the differentiable case, differing in the iterative process followed by the agents which implement projections onto the feasible space of iterates produced by taking steps along the subgradient. This distributed algorithm was also shown to produce a sequence of iterates converging to the optimum.

3. *Dynamic dependency on upstream states and control signals, subject to constraints on local states and control signals*
The above LDN models were extended to allow constraints on state variables. These constraints introduce hard couplings among the subsystems and the control

agents, thereby rendering distributed MPC more challenging. The gradient and subgradient projection strategies will fail to converge to the optimal solution in the presence of such constraints. In this generalized class of LDNs the state of a subsystem i depends on the state of its upstream subsystem, leading to the following dynamic equation:

$$\mathbf{x}_i(k+1) = \sum_{j \in \mathcal{I}(i)} \left(\mathbf{A}_{i,j}\mathbf{x}_j(k) + \mathbf{B}_{i,j}\mathbf{u}_j(k)\right). \tag{12.7}$$

Each subsystem i is also subject to constraints on state and control variables:

$$\mathbf{C}_i\mathbf{x}_i(k) \leq \mathbf{c}_i, \tag{12.8a}$$

$$\mathbf{D}_i\mathbf{u}_i(k) \leq \mathbf{d}_i, \tag{12.8b}$$

$$\mathbf{E}_i\mathbf{u}_i(k) = \mathbf{e}_i. \tag{12.8c}$$

In the application to urban traffic networks the constraint (12.8a) can be used to limit the number of vehicles in each road link, while the remaining ones enforce bounds on the green-time signals and cycle time.

Because $\mathbf{x}_i$ is now a function of $\mathbf{x}_i^I$, the state constraint (12.8a) is directly affected by all the agents that belong to the input neighborhood of subsystem i. The objective function for this LDN is identical to objective (12.4). In [5], a distributed MPC framework was proposed to control such LDNs whereby a network of agents implements an interior-point method, solving a sequence of unconstrained approximation problems with a distributed gradient-descent algorithm.

From now on the focus of this chapter is on this class of LDNs for being a generalization of the others. Actually, the nondifferentiable objective function (12.6) can be represented by a system of inequalities and thereby handled by a straightforward extension of the inequalities (12.8). The following section presents a distributed optimization framework for solving the MPC optimization problem with a network of agents, thereby implementing a distributed MPC strategy for this class of LDNs.

12.3 Distributed Optimization Framework

This section presents a distributed optimization algorithm for the class of LDNs in which constraints are imposed on the state of the subsystems. To keep the presentation simple but without losing generality, the states are assumed to be independent of the state of the upstream subsystems, namely:

$$\mathbf{x}_i(k+1) = \mathbf{A}_i\mathbf{x}_i(k) + \sum_{j \in \mathcal{I}(i)} \mathbf{B}_{i,j}\mathbf{u}_j(k). \tag{12.9}$$

Given the state of subsystem i at time t_k and the future control signals from the input neighborhood, the state at time t_{k+l} can be anticipated as

$$\mathbf{x}_i(k+l) = \mathbf{A}_i^l \mathbf{x}_i(k) + \sum_{t=1}^{l} \sum_{j \in \mathcal{I}(i)} \mathbf{A}_i^{l-1} \mathbf{B}_{i,j} \mathbf{u}_j(k+l-t). \tag{12.10}$$

This means that the future states of subsystem i depend on the input neighborhood regardless of the length of the prediction horizon. However, the predictions for the state of subsystem i will depend on subsystems that are farther away from the input neighborhood, extending outwards as the prediction horizon increases, when the subsystem state evolves according to Eq. (12.7) rather (12.9). For further details the reader can refer to [5].

The concern here is on the distributed solution at time t_k of the following MPC optimization problem for linear dynamic networks:

$$P(k): \min \sum_{i \in \mathcal{N}} \sum_{l=k+1}^{k+N_\mathrm{p}} \frac{1}{2} \mathbf{x}_i(l)^\mathrm{T} \mathbf{Q}_i \mathbf{x}_i(l) + \sum_{i \in \mathcal{N}} \sum_{l=k}^{k+N_\mathrm{c}-1} \mathbf{u}_i(l)^\mathrm{T} \mathbf{R}_i \mathbf{u}_i(l) \tag{12.11a}$$

$$\text{s.t.}: \quad \text{for } i \in \mathcal{N},\ l = k, \ldots, k+N_\mathrm{p}-1:$$

$$\mathbf{x}_i(l+1) = \mathbf{A}_i \mathbf{x}_i(l) + \sum_{j \in \mathcal{I}(i)} \mathbf{B}_{i,j} \mathbf{u}_j(l), \tag{12.11b}$$

$$\mathbf{C}_i \mathbf{x}_i(l+1) \leq \mathbf{c}_i, \tag{12.11c}$$

$$\mathbf{D}_i \mathbf{u}_i(l) \leq \mathbf{d}_i, \tag{12.11d}$$

$$\mathbf{E}_i \mathbf{u}_i(l) = \mathbf{e}_i \tag{12.11e}$$

where the prediction and control horizons have the same length for convenience of mathematical development, i.e. $N_\mathrm{p} = N_\mathrm{c}$. The MPC strategy solves $P(k)$ at each sample time t_k, implementing only the control signals $\mathbf{u}_i(k)$ for the time interval $[t_k, t_{k+1}]$. Then the horizon is rolled forward at the next sample time, $P(k+1)$ is instantiated from time t_{k+1} until $t_{k+N_\mathrm{p}+1}$, and the process is repeated.

Let $\widetilde{\mathbf{u}}_i(k) = [\mathbf{u}_i(k+l)^\mathrm{T} : l = 0, \ldots, N_\mathrm{c}-1]^\mathrm{T}$ be the vector with the control predictions for subsystem i. Using Eq. (12.10) to express subsystem states as a function of control signals, the MPC optimization problem can be recast in terms of the control signals and the current state as follows:

$$P(k): \min \frac{1}{2} \sum_{i \in \mathcal{N}} \sum_{j \in \mathcal{I}(i)} \sum_{l \in \mathcal{I}(i)} \widetilde{\mathbf{u}}_j(k)^\mathrm{T} \widetilde{\mathbf{H}}_{i,j,l} \widetilde{\mathbf{u}}_l(k)$$
$$+ \sum_{i \in \mathcal{N}} \sum_{j \in \mathcal{I}(i)} \widetilde{\mathbf{g}}_{i,j}(k)^\mathrm{T} \widetilde{\mathbf{u}}_j(k) + \sum_{i \in \mathcal{N}} \widetilde{c}_i(k) \tag{12.12a}$$

$$\text{s.t.}: \quad \text{for all } i \in \mathcal{N}:$$

$$\widetilde{\mathbf{C}}_i \left(\widetilde{\mathbf{A}}_i \mathbf{x}_i(k) + \sum_{j \in \mathcal{I}(i)} \widetilde{\mathbf{B}}_{i,j} \widetilde{\mathbf{u}}_j(k) \right) \leq \widetilde{\mathbf{c}}_i, \tag{12.12b}$$

$$\widetilde{\mathbf{D}}_i \widetilde{\mathbf{u}}_i(k) \leq \widetilde{\mathbf{d}}_i, \tag{12.12c}$$

$$\widetilde{\mathbf{E}}_i \widetilde{\mathbf{u}}_i(k) = \widetilde{\mathbf{e}}_i. \tag{12.12d}$$

where $\widetilde{\mathbf{H}}_{i,j,l}$, $\widetilde{\mathbf{A}}_i$, and $\widetilde{\mathbf{B}}_{i,j}$ are constant matrices, $\widetilde{\mathbf{g}}_{i,j}(k)$ is a constant vector, and $\widetilde{c}_i(k)$ is a constant obtained from the structure of the LDN and problem $P(k)$ defined in Eq. (12.11). The terms $\widetilde{\mathbf{g}}_{i,j}(k)$ and $\widetilde{c}_i(k)$ are functions of k because they depend on the initial state $\mathbf{x}_i(k)$ of the subsystems. Detailed procedures for computing these parameters are found in [3, 5]. Further, $\widetilde{\mathbf{C}}_i$, $\widetilde{\mathbf{D}}_i$, and $\widetilde{\mathbf{E}}_i$ are block diagonal matrices whose blocks are $\mathbf{C}_i$, $\mathbf{D}_i$, and $\mathbf{E}_i$ respectively, and $\widetilde{\mathbf{c}}_i$, $\widetilde{\mathbf{d}}_i$, and $\widetilde{\mathbf{e}}_i$ are vectors with stacked copies of $\mathbf{c}_i$, $\mathbf{d}_i$, and $\mathbf{e}_i$ respectively.

Problem $P(k)$ is further simplified by explicitly removing the linear dependencies induced by Eq. (12.12d). Let $\widetilde{\mathbf{u}}_i^{\dagger}$ be any solution for the system of linear equations (12.12d) and let $\Phi(\widetilde{\mathbf{E}}_i)$ be a basis for the null space of $\widetilde{\mathbf{E}}_i$. Then, replacing $\widetilde{\mathbf{u}}_i(k)$ with $\widetilde{\mathbf{u}}_i^{\dagger} + \Phi(\widetilde{\mathbf{E}}_i)\widehat{\mathbf{u}}_i(k)$ produces the following equivalent form of the problem:

$$P(k): \ \min \ \frac{1}{2} \sum_{i \in \mathcal{N}} \sum_{j \in \mathcal{I}(i)} \sum_{l \in \mathcal{I}(i)} \widehat{\mathbf{u}}_j(k)^{\mathrm{T}} \widehat{\mathbf{H}}_{i,j,l} \widehat{\mathbf{u}}_l(k) + \sum_{i \in \mathcal{N}} \sum_{j \in \mathcal{I}(i)} \widehat{\mathbf{g}}_{i,j}(k)^{\mathrm{T}} \widehat{\mathbf{u}}_j(k) + \sum_{i \in \mathcal{N}} \widehat{c}_i(k) \tag{12.13a}$$

s.t. : for all $i \in \mathcal{N}$:

$$\sum_{j \in \mathcal{I}(i)} \widehat{\mathbf{B}}_{i,j} \widehat{\mathbf{u}}_j(k) \leq \widehat{\mathbf{c}}_i(k), \tag{12.13b}$$

$$\widehat{\mathbf{D}}_i \widehat{\mathbf{u}}_i(k) \leq \widehat{\mathbf{d}}_i. \tag{12.13c}$$

where the matrices $\widehat{\mathbf{H}}_{i,j,l}$, $\widehat{\mathbf{B}}_{i,j}$, and $\widehat{\mathbf{D}}_i$, vectors $\widehat{\mathbf{g}}_{i,j}(k)$, $\widehat{\mathbf{c}}_i(k)$, and $\widehat{\mathbf{d}}_i$, and constants $\widehat{c}_i(k)$ are easily obtained. In particular, $\widehat{\mathbf{H}}_{i,j,l} = \Phi(\widetilde{\mathbf{E}}_i)^{\mathrm{T}} \widetilde{\mathbf{H}}_{i,j,l} \Phi(\widetilde{\mathbf{E}}_i)$. The design of algorithms will be simplified by casting $P(k)$ in the form:

$$P(k): \ \min \ f(\widehat{\mathbf{u}}(k)) \tag{12.14a}$$

$$\text{s.t.}: \ h_i(\widehat{\mathbf{u}}(k)) \leq 0, \ i \in \mathcal{H}, \tag{12.14b}$$

where $\widehat{\mathbf{u}}(k) = [\widehat{\mathbf{u}}_i(k)^{\mathrm{T}} : i \in \mathcal{N}]^{\mathrm{T}}$ is the vector with the control variables, $f : \mathbb{R}^{\widehat{n}} \to \mathbb{R}$ defines the objective, and $h_i : \mathbb{R}^{\widehat{n}} \to \mathbb{R}$, $i \in \mathcal{H}$, define the constraints given by (12.13b)–(12.13c) with $\widehat{n}$ being the dimension of $\widehat{\mathbf{u}}$. Actually, $f(\widehat{\mathbf{u}}) = \frac{1}{2}\widehat{\mathbf{u}}(k)^{\mathrm{T}} \widehat{\mathbf{H}} \widehat{\mathbf{u}}(k) + \widehat{\mathbf{g}}^{\mathrm{T}} \widehat{\mathbf{u}}(k) + \widehat{c}$ for suitable $\widehat{\mathbf{H}}$, $\widehat{\mathbf{g}}$, and $\widehat{c}$, and $h_i = \widehat{\mathbf{a}}_i^{\mathrm{T}} \widehat{\mathbf{u}}(k) - \widehat{b}_i$ for suitable $\widehat{\mathbf{a}}_i$ and $\widehat{b}_i$, $i \in \mathcal{H}$. Notice that $\widehat{\mathbf{H}} \succ 0$. The feasible set is $\Omega = \{\widehat{\mathbf{u}}(k) : h_i(\widehat{\mathbf{u}}(k)) \leq 0, \ i \in \mathcal{H}\}$, whereas the interior of the feasible set is $\bar{\Omega} = \{\widehat{\mathbf{u}}(k) : h_i(\widehat{\mathbf{u}}(k)) < 0, \ i \in \mathcal{H}\}$.

12.3.1 Distributed Modeling

The distributed solution of $P(k)$ starts with the decomposition into a set $\{P_i(k) : i \in \mathcal{N}\}$ of subproblems, $\{P_i(k)\}$ for short. The proposed decomposition is carried out based on the following definitions of relationships of agent i with agents controlling other subsystems:

- $\mathcal{O}(i) = \{j : i \in \mathcal{I}(j)\}$ is the *output neighborhood* which consists of the subsystems that are affected by the decisions at subsystem i;
- $\mathcal{C}(i) = \{j : \exists l \neq i, j, \text{ such that } i, j \in \mathcal{I}(l)\} \setminus (\mathcal{I}(i) \cup \mathcal{O}(i))$ is the *indirect neighborhood* which encompasses the subsystems j that do not influence nor are influenced by subsystem i, but both affect some other subsystem l;
- $\mathcal{N}(i) = (\mathcal{I}(i) \cup \mathcal{O}(i) \cup \mathcal{C}(i)) \setminus \{i\}$ is the *neighborhood* which comprises the subsystems that are coupled with subsystem i.

In the linear network of Fig. 12.2, subsystem 4 has $\mathcal{I}(4) = \{4\}$, $\mathcal{O}(4) = \{3, 4, 5\}$, $\mathcal{C}(4) = \{1\}$, and $\mathcal{N}(4) = \{1, 3, 5\}$.

A decomposition $\{P_i(k)\}$ is said to be *perfect* if each $P_i(k)$ is obtained from $P(k)$ by dropping from the objective all of the terms, and discarding all of the constraints, that do not depend on $\widehat{\mathbf{u}}_i(k)$. Models and algorithms for obtaining approximate decompositions are found in [6]. For a perfect decomposition, agent i's view of the dynamic network is divided in:

- *local variables*: the variables $\widehat{\mathbf{u}}_i(k)$ whose values are set by agent i;
- *neighborhood variables*: the vector $\widehat{\mathbf{w}}_i(k) = (\widehat{\mathbf{u}}_j(k) : j \in \mathcal{N}(i))$ of variables set by the agents in the neighborhood;
- *remote variables*: the vector $\widehat{\mathbf{r}}_i(k) = (\widehat{\mathbf{u}}_j(k) : j \notin \mathcal{N}(i) \cup \{i\})$ with all of the other variables.

Perfect decomposition allows for the decision variables of any agent i to be arranged as $\widehat{\mathbf{u}}(k) = [\widehat{\mathbf{u}}_i(k)^{\mathrm{T}}\ \widehat{\mathbf{w}}_i(k)^{\mathrm{T}}\ \widehat{\mathbf{r}}_i(k)^{\mathrm{T}}]^{\mathrm{T}}$ and for subproblem $P_i(k)$ to be cast in the form:

$$P_i(k, \widehat{\mathbf{w}}_i(k)) : \min_{\widehat{\mathbf{u}}_i(k)} \ f_i(\widehat{\mathbf{u}}_i(k), \widehat{\mathbf{w}}_i(k)) = \frac{1}{2}\widehat{\mathbf{u}}_i(k)^{\mathrm{T}}\widehat{\mathbf{H}}_i\widehat{\mathbf{u}}_i(k) + \widehat{\mathbf{g}}_i(k)^{\mathrm{T}}\widehat{\mathbf{u}}_i(k) + \hat{c}_i(k) \tag{12.15a}$$

$$\text{s.t.} : \sum_{l \in \mathcal{I}(j)} \widehat{\mathbf{B}}_{j,l}\widehat{\mathbf{u}}_l(k) \leq \widehat{\mathbf{c}}_j(k), \ j \in \mathcal{O}(i), \tag{12.15b}$$

$$\widehat{\mathbf{D}}_i\widehat{\mathbf{u}}_i(k) \leq \widehat{\mathbf{d}}_i, \tag{12.15c}$$

where:

$$\widehat{\mathbf{g}}_i(k) = \sum_{j \in \mathcal{O}(i)} \widehat{\mathbf{g}}_{j,i}(k) + \frac{1}{2} \sum_{j \in \mathcal{O}(i)} \sum_{l \in \mathcal{I}(j) \setminus \{i\}} (\widehat{\mathbf{H}}_{j,l,i}^{\mathrm{T}} + \widehat{\mathbf{H}}_{j,i,l})\widehat{\mathbf{u}}_l(k), \tag{12.16a}$$

$$\widehat{\mathbf{H}}_i = \sum_{j \in \mathcal{O}(i)} \widehat{\mathbf{H}}_{j,i,i}. \tag{12.16b}$$

The development of a distributed algorithm for solving $\{P_i(k)\}$ is simplified by recasting $P_i(k)$ in the following form:

$$P_i(k, \widehat{\mathbf{w}}_i(k)): \min_{\widehat{\mathbf{u}}_i(k)} \quad f_i(\widehat{\mathbf{u}}_i(k), \widehat{\mathbf{w}}_i(k)) \tag{12.17a}$$

$$\text{s.t.}: \quad h_j(\widehat{\mathbf{u}}_i(k), \widehat{\mathbf{w}}_i(k)) \leq 0, \; j \in \mathcal{H}_i, \tag{12.17b}$$

where $\mathcal{H}_i = \{j \in \mathcal{H} : h_j$ is a function of $\widehat{\mathbf{u}}_i(k)\}$. Notice that h_j does not depend on $\widehat{\mathbf{r}}_i(k)$ under a perfect problem decomposition for all $j \in \mathcal{H}_i$. Further, h_j is not a function of $\widehat{\mathbf{u}}_i(k)$ for all $j \in \mathcal{H} \setminus \mathcal{H}_i$.

12.3.2 Distributed Algorithm

Iterative strategies in which an agent i reacts with a solution $\widehat{\mathbf{u}}_i^{(k+1)}$ for $P_i(k, \widehat{\mathbf{w}}_i^{(k)})$ to the decisions of its neighbors at iteration p, $\widehat{\mathbf{w}}_i^{(k)}$, may lead to undesirable behavior due to the coupling constraints given by Eq. (12.11c), as illustrated below.

Let $\mathbf{h}(k, \cdot) = [h_1 \cdots h_p]^{\mathrm{T}}$ denote the vector function with all of the constraints in $P(k)$. Then, $\Omega = \{\widehat{\mathbf{u}}(k) : \mathbf{h}(k, \widehat{\mathbf{u}}(k)) \leq \mathbf{0}\}$ is the feasible set. A vector $\widehat{\mathbf{u}}(k)^\star \in \Omega$ is a fixed point for the subproblem set $\{P_i(k)\}$ if and only if $\widehat{\mathbf{u}}_i(k)^\star$ is an optimal solution to $P_i(k, \widehat{\mathbf{w}}_i(k)^\star)$ for each $i \in \mathcal{N}$. Without constraints coupling the subsystems, $\widehat{\mathbf{u}}(k)^\star$ is a fixed point for $\{P_i(k)\}$ if and only if $\widehat{\mathbf{u}}(k)^\star$ satisfies first-order optimality conditions for $P(k)$, which in turn imply that $\widehat{\mathbf{u}}(k)^\star$ is globally optimal for $P(k)$ due to convexity [3]. The solving process of $\{P_i(k)\}$ can be thought of as a dynamic game with agents reacting to one another's decisions so as to improve their payoff [7].

However, the equivalence between a fixed point of $\{P_i(k)\}$ and an optimal solution to $P(k)$ does not hold in the presence of coupling constraints. As an illustration, consider the quadratic program:

$$P: \min \; f(u_1, u_2) = \frac{1}{2}u_1^2 - u_1 u_2 + u_2^2 - 4u_2$$

$$\text{s.t.}: -u_1 + 2u_2 \leq 2, \quad 1 \leq u_1 \leq 7, \quad u_2 \geq \frac{1}{2}.$$

The solution of P_i over the decision variable u_i while holding $u_{(i \bmod 2)+1}$ fixed is a process defined by the reaction function r_i of the agents as follows:

$$r_1(u_2) = \arg\min_{u_1}\{\frac{1}{2}u_1^2 - u_1 u_2 : -u_1 + 2u_2 \leq 2, \; 1 \leq u_1 \leq 7\}$$

$$r_2(u_1) = \arg\min_{u_2}\{u_2^2 - u_1 u_2 - 4u_2 : -u_1 + 2u_2 \leq 2, \; u_2 \geq \frac{1}{2}\}$$

Starting from an initial point $\mathbf{u}^{(0)}$ the iterative process following the serial protocol yields a sequence of iterates $\{\mathbf{u}^{(p)}\}$ defined by:

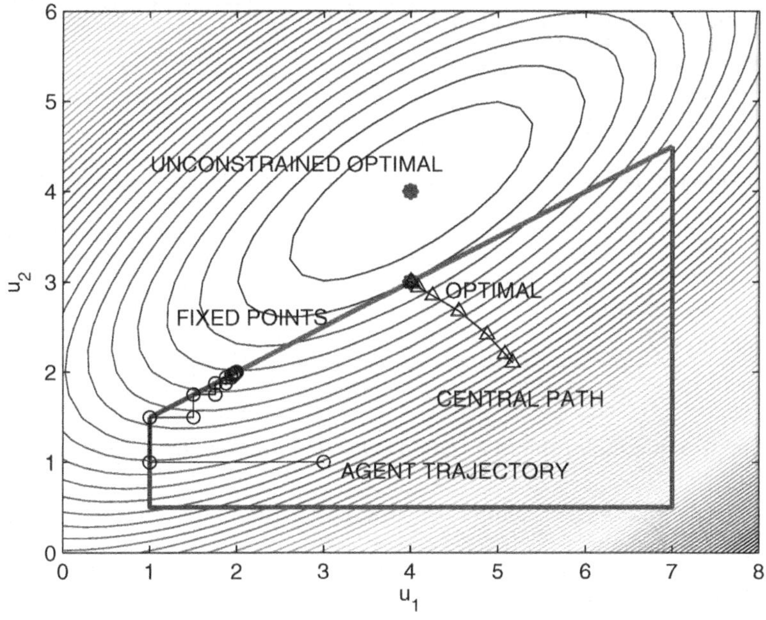

Fig. 12.2 Illustration of iterative processes based on the reactive functions and the interior-point method (extracted from [5]).

$$u_1^{(p)} = \begin{cases} r_1(u_2^{(p-1)}) & \text{if } p \text{ is odd} \\ u_1^{(p-1)} & \text{if } p \text{ is even} \end{cases} \qquad u_2^{(p)} = \begin{cases} u_2^{(p-1)} & \text{if } p \text{ is odd} \\ r_2(u_1^{(p-1)}) & \text{if } p \text{ is even} \end{cases}$$

Figure 12.2 shows the contour curves of f, the feasible region, and the set of fixed points. The trajectory $\mathbf{u}^{(p)} = [u_1^{(p)}\ u_2^{(p)}]^{\mathrm{T}}$ traced by the agents when they start from $[u_1^{(0)}\ u_2^{(0)}]^{\mathrm{T}} = [3\ 1]^{\mathrm{T}}$ is also shown, which converges to the fixed point $\mathbf{u} = [2\ 2]^{\mathrm{T}}$. Actually, all the points in the line segment $\{(u_1, (2+u_1)/2) : 2 \leq u_1 \leq 7\}$ are fixed points for $P_1(u_2)$ and $P_2(u_1)$. The figure also depicts the trajectory followed by agents that implement an interior-point strategy converging to the optimal solution $\mathbf{u}^\star = [4\ 3]^{\mathrm{T}}$. This trajectory is known as *central path*.

The algorithmic solution presented here relies on an interior-point method to approximate problem $P(k)$ given in Eq. (12.14) with an unconstrained problem to which gradient descent is applied [2]. The approximation accounts for the constraints with a *logarithmic barrier* function:

$$\phi(\widehat{\mathbf{u}}(k)) = -\sum_{i \in \mathcal{H}} \log(-h_i(\widehat{\mathbf{u}}(k))) \tag{12.18}$$

which is real-valued within the interior set $\bar{\Omega}$ but tends to infinity as the solution $\widehat{\mathbf{u}}(k)$ is drawn towards the boundary of any constraint.

Algorithm 12.1 Barrier method for solving $P(k)$

1: **Input:** strictly feasible $\widehat{\mathbf{u}}(k)^{\mathrm{s}}$, initial $\epsilon(k)^{(0)}$, decrease rate $\mu < 1$, tolerance τ;
2: **Initialize:** $l := -1$;
3: **Repeat**

a. $l := l + 1$;
b. **Centering step:** obtain $\widehat{\mathbf{u}}(k)^{(l)}$ by solving $P(k, \epsilon(k)^{(l)})$ with initial solution $\widehat{\mathbf{u}}(k)^{\mathrm{s}}$;
c. **If** $\epsilon(k)^{(l)} > \tau/|\mathcal{H}|$ **then**

i. $\widehat{\mathbf{u}}(k)^{\mathrm{s}} := \widehat{\mathbf{u}}(k)^{(l)}$;
ii. $\epsilon(k)^{(l+1)} := \mu\epsilon(k)^{(l)}$;

4: **Until** $\epsilon(k)^{(l)} \leq \tau/|\mathcal{H}|$;
5: **Output:** $\widehat{\mathbf{u}}(k)^{(l)}$;

The approximation problem is called *centering problem* being defined as:

$$P(k, \epsilon) : \quad \min_{\widehat{\mathbf{u}}(k) \in \operatorname{dom} \theta} \quad \theta(\widehat{\mathbf{u}}(k)) = f(\widehat{\mathbf{u}}(k)) + \epsilon\phi(\widehat{\mathbf{u}}(k)) \tag{12.19}$$

where $\operatorname{dom} \theta = \bar{\Omega}$ and the parameter $\epsilon > 0$ sets the accuracy of the approximation. The barrier method solves $P(k, \epsilon(k)^{(l)})$ for a decreasing sequence $\{\epsilon(k)^{(l)}\}_{l=0}^{\infty}$. The solution $\widehat{\mathbf{u}}(k, \epsilon(k)^{(l)})$ to the centering problem $P(k, \epsilon(k)^{(l)})$ is drawn towards the optimal solution $\widehat{\mathbf{u}}(k)^{\star}$ to $P(k)$ as $\epsilon(k)^{(l)}$ tends to 0. The barrier method is described in Algorithm 12.1.

A strictly feasible solution $\widehat{\mathbf{u}}(k)^{\mathrm{s}}$ can be obtained by solving the auxiliary convex program [2]:

$$P(k)^{\mathrm{S}} : \min \quad \sum_{i \in \mathcal{H}} s_i \tag{12.20a}$$

$$\text{s.t.} : \quad h_i(\widehat{\mathbf{u}}(k)^{\mathrm{S}}) \leq s_i, \ i \in \mathcal{H}, \tag{12.20b}$$

$$s_i \geq 0, \ i \in \mathcal{H}. \tag{12.20c}$$

The distributed optimization strategy aims at solving the centering problem $P(k, \epsilon)$ with the agent network. The agent of a subsystem i will sense the local state variables $\mathbf{x}_i$ and decide upon the control signals $\mathbf{u}_i$. Communication among agents will ensure a perfect decomposition of the centering problem whereby each agent i solves, given the neighborhood variables $\widehat{\mathbf{w}}_i(k)$, the centering subproblem:

$$P_i(k, \epsilon, \widehat{\mathbf{w}}_i(k)) : \min_{\widehat{\mathbf{u}}_i(k) \in \operatorname{dom} \theta_i} \theta_i(\widehat{\mathbf{u}}_i(k)) = f_i(\widehat{\mathbf{u}}_i(k), \widehat{\mathbf{w}}_i(k)) + \epsilon\phi_i(\widehat{\mathbf{u}}_i(k), \widehat{\mathbf{w}}_i(k)) \tag{12.21}$$

where:

$$\phi_i(\widehat{\mathbf{u}}_i(k), \widehat{\mathbf{w}}_i(k)) = -\sum_{j \in \mathcal{H}_i} \log(-h_j(\widehat{\mathbf{u}}_i(k), \widehat{\mathbf{w}}_i(k))) \tag{12.22}$$

is the logarithmic barrier for the constraints depending on $\widehat{\mathbf{u}}_i$. Assuming fixed neighborhood variables any converging algorithm could be applied by agent i to solve $P_i(k, \epsilon, \widehat{\mathbf{w}}_i(k))$. Instead, a simple gradient-descent strategy is suggested to yield the optimal solution to $P(k, \epsilon)$ provided that the agents coordinate their iterations. This descent strategy does not require any agent i to optimally solve $P_i(k, \epsilon, \widehat{\mathbf{w}}_i(k))$, only a sufficient decrease on the objective is needed.

Given $(\widehat{\mathbf{u}}_i(k)^{(l,p)}, \widehat{\mathbf{w}}_i(k)^{(l,p)})$, agent i yields the next iterate $\widehat{\mathbf{u}}_i(k)^{(l,p+1)}$ by taking a step $s_i(k)^{(l,p)} > 0$ in the gradient descent direction $-\nabla\theta_i(\widehat{\mathbf{u}}_i(k)^{(l,p)})$:

$$\widehat{\mathbf{u}}_i(k)^{(l,p+1)} = \widehat{\mathbf{u}}_i(k)^{(l,p)} - s_i(k)^{(l,p)} \nabla\theta_i(\widehat{\mathbf{u}}_i(k)^{(l,p)}). \tag{12.23}$$

The iterations of the barrier method given by counter l are called *outer iterations*, whereas the iterations of the distributed gradient-descent method given by counter p are called *inner iterations*. Notice that for each outer iteration l there is a series of inner iterations l being denoted by (l, p). The distributed descent strategy is detailed in Algorithm 12.2 which is followed by the agent network to solve $\{P_i(k, \epsilon(k)^{(l)})\}$ in place of $P(k, \epsilon(k)^{(l)})$. The algorithm requires that at least one agent of maximum descent works in each iteration, with the set of agents of maximum descent being:

$$\mathcal{N}(k)^{(l,p)} = \{i \in \mathcal{N} : \|\nabla\theta_i(\widehat{\mathbf{u}}_i(k)^{(l,p)})\| = \max_{j \in \mathcal{N}} \|\nabla\theta_j(\widehat{\mathbf{u}}_j(k)^{(l,p)})\|\}.$$

The use of Newton's direction $-\nabla^2\theta_i(\widehat{\mathbf{u}}_i(k)^{(l,p)})^{-1}\nabla\theta_i(\widehat{\mathbf{u}}_i(k)^{(l,p)})$, multiple backtracking iterations, and off-the-shelf solvers are discussed in [5]. The distributed identification of a set $\mathcal{V}(k)^{(l,p)}$ containing at least one agent of maximum descent can be performed by defining an agent cycle $\mathcal{A} = \langle \mathcal{A}_0, \mathcal{A}_1, \ldots, \mathcal{A}_{K-1} \rangle$ such that $\cup_{k=0}^{K-1} \mathcal{A}_k = \mathcal{N}$, and i and j are nonneighbors for all $i, j \in \mathcal{A}_k$.

Then, a message-exchange protocol can be implemented to follow the sequence $\langle \mathcal{A}_0, \mathcal{A}_1, \ldots, \mathcal{A}_{K-1}, \mathcal{A}_0, \mathcal{A}_1, \ldots, \mathcal{A}_{K-1}, \ldots \rangle$, skipping sets until finding $\mathcal{V}(k)^{(l,p)} = \mathcal{A}_t$ such that $\mathcal{A}_t \cap \mathcal{N}(k)^{(l,p)} \neq \emptyset$. In essence, each agent $i \in \mathcal{A}_l$ would send a message with a token to all the agents in $\mathcal{A}_{(l+1) \bmod K}$, which would be blocked waiting for the tokens.

Distributed detection of global convergence can be achieved by keeping track of the number of past iterations that all of the preceding agents have converged with respect to their problems. Besides state and control information, each agent i receives from a neighbor $j \in \mathcal{N}(i)$ a parameter t_j such that: $t_j = 0$ if agent j changed its control decisions in the latest iteration; otherwise $t_j = \min\{t_l : l \in \mathcal{N}(j)\} + 1$. So, if agent j changes its decisions, it will define $t_j = 0$ and pass this value to its neighbors when they request its state and control variables, otherwise $t_j = \min\{t_l : l \in \mathcal{N}(j)\} + 1$. If in any iteration (l, p), an agent $i \in \mathcal{V}(k)^{(l,p)}$ detects

Algorithm 12.2 Distributed gradient descent algorithm for solving $P(k, \epsilon(k)^{(l)})$

1: **Input:** strictly feasible $\widehat{\mathbf{u}}(k)^{(l)}$, barrier parameter $\epsilon(k)^{(l)}$, backtracking parameters $\alpha \in (0, 1/2)$ and $\beta \in (0, 1)$, and tolerance τ;
2: **Initialize:** $p := 0$; $\widehat{\mathbf{u}}(k)^{(l,0)} := \widehat{\mathbf{u}}(k)^{(l)}$;
3: **While** $\|\nabla\theta(\widehat{\mathbf{u}}(k)^{(l,p)})\| > \tau$ **do**

a. Let $\mathcal{V}(k)^{(l,p)} \subseteq \mathcal{N}$ be a subset of non-neighboring agents such that $\mathcal{V}(k)^{(l,p)} \cap \mathcal{N}(k)^{(l,p)} \neq \emptyset$;
b. **For** each $i \in \mathcal{V}(k)^{(l,p)}$ in parallel **do**

i. Obtain $\widehat{\mathbf{w}}_i(k)^{(l,p)}$ from the neighborhood $\mathcal{N}(i)$;
ii. $s_i(k)^{(l,p)} := 1$;
iii. **While** $\theta_i(\widehat{\mathbf{u}}_i(k)^{(l,p)} - s_i(k)^{(l,p)}\nabla\theta_i(\widehat{\mathbf{u}}_i(k)^{(l,p)})) > \theta_i(\widehat{\mathbf{u}}_i(k)^{(l,p)}) - \alpha s_i(k)^{(l,p)}\|\nabla\theta_i(\widehat{\mathbf{u}}_i(k)^{(l,p)})\|^2$ **do**

$$s_i(k)^{(l,p)} := \beta s_i(k)^{(l,p)};$$

iv. $\widehat{\mathbf{u}}_i(k)^{(l,p+1)} := \widehat{\mathbf{u}}_i(k)^{(l,p)} - s_i(k)^{(l,p)}\nabla\theta_i(\widehat{\mathbf{u}}_i(k)^{(l,p)})$;

c. **For** each $i \in \mathcal{N} \setminus \mathcal{V}(k)^{(l,p)}$ in parallel **do**

$$\widehat{\mathbf{u}}_i(k)^{(l,p+1)} := \widehat{\mathbf{u}}_i(k)^{(l,p)};$$

d. $p := p + 1$;

4: **Output:** $\widehat{\mathbf{u}}(k)^{(l,p)}$;

that $\min\{t_l : l \in \mathcal{N}(i)\} \geq K$, then global convergence has been achieved assuming that the LDN is a connected graph. This agent i can broadcast a message to all the others or let them detect convergence on their own.

12.3.3 Theoretical Results

In [5], it was shown that the iterates $\{\widehat{\mathbf{u}}(k)^{(l,p)}\}_{p=0}^{\infty}$ produced by the distributed gradient-descent algorithm converge to the solution $\widehat{\mathbf{u}}(k)^{(l)}$ to the centering problem $P(k, \epsilon(k)^{(l)})$. The convergence is a consequence of the assumptions stated below which are satisfied by Algorithm 12.2.

Assumption 12.1 (Synchronous Work) If agent i updates its variables in iteration p, then:

1. agent i uses $\widehat{\mathbf{w}}_i(k) = \widehat{\mathbf{w}}_i(k)^{(l,p)}$ and follows the descent method to obtain an approximate solution $\widehat{\mathbf{u}}_i(k)^{(l,p+1)}$ to $P_i(k, \epsilon(k)^{(l)}, \widehat{\mathbf{w}}_i(k))$;
2. $\widehat{\mathbf{u}}_i(k)^{(l,p)}$ is not an optimal solution to $P_i(k, \epsilon(k)^{(l)}, \widehat{\mathbf{w}}_i(k))$ since otherwise the agent does not yield any improvement; and
3. each neighbor of agent i does not iterate, meaning that $\widehat{\mathbf{u}}_j(k)^{(l,p+1)} = \widehat{\mathbf{u}}_j(k)^{(l,p)}$ for all $j \in \mathcal{N}(i)$.

Notice that condition 1 of the Assumption 12.1 is met by the algorithm since the agents $i \in \mathcal{V}(k)^{(l,p)}$ request $\widehat{\mathbf{w}}_i(k)^{(l,p)}$ from their neighbors, condition 2 can be satisfied by agents that are not locally optimal, and condition 3 is satisfied by the agents $i \in \mathcal{N} \setminus \mathcal{V}(k)^{(l,p)}$ which do not iterate.

Assumption 12.2 (Continuous and Maximum Work) If $\widehat{\mathbf{u}}(k)^{(l,p)}$ is not optimal to $P(k, \epsilon(k)^{(l)})$ at outer iteration l, then any agent $i(k, l, p) \in \mathcal{N}(k)^{(l,p)}$ performs a backtracking line search, starting at $\widehat{\mathbf{u}}_{i(k,l,p)}(k)^{(l,p)}$, to produce its next solution $\widehat{\mathbf{u}}_{i(k,l,p)}(k)^{(l,p+1)}$.

This assumption is ensured by Algorithm 12.2 at the step that defines the set $\mathcal{V}(k)^{(l,p)}$ of the agents that iterate.

Assumption 12.3 The objective function $\theta(\widehat{\mathbf{u}}(k))$ of the centering problem $P(k, \epsilon)$ is strongly convex.

With strong convexity of $\theta(\widehat{\mathbf{u}}(k))$ the convergence of Algorithm 12.2 is established as stated below. Actually the rate of convergence can be determined in terms of bounds induced by strong convexity.

Theorem 12.1 *Under Assumptions 12.1, 12.2, and 12.3, the distributed gradient-descent algorithm yields a sequence $\{\widehat{\mathbf{u}}(k)^{(l,p)}\}_{p=0}^{\infty}$ of iterates converging to the optimal solution $\widehat{\mathbf{u}}(k)^{(l)}$ to the centering problem $P(k, \epsilon(k)^{(l)})$ using exact or back-tracking line search.*

12.4 Applications

Although distributed MPC has been applied to operate and control several distributed dynamic systems, here a reference is made only to the ones closely related to the control of linear dynamic networks. In [8] the green-time control of urban traffic networks was modeled as the MPC of an LDN with constraints on the control signals. Distributed optimization was carried out using a gradient-projection strategy. In [5] this work was extended to incorporate road capacity by introducing constraints on state variables. A simulation study was performed in the network of the city of Macaé (north of Rio de Janeiro) consisting of 15 junctions and 28 road links [13]. The quadratic programs for MPC were solved using the interior-point strategy and the distributed gradient-descent algorithm. In [12] the dynamic models for LDNs were expressed using transfer functions and the constraints were imposed on the control inputs and output signals. The resulting distributed transfer-function MPC approach was applied to the control of a distillation column.

Acknowledgments This work was funded in part by Conselho Nacional de Desenvolvimento Científico e Tecnológico (CNPq), Brazil.

References

1. D.P. Bertsekas, *Nonlinear Programming* (Athena Scientific, Belmont, MA, 1995)
2. S. Boyd, L. Vandenberghe, *Convex Optimization* (Cambridge University Press, Cambridge, 2004)

3. E. Camponogara, L.B. de Oliveira, Distributed optimization for model predictive control of linear dynamic networks. IEEE Trans. Syst. Man Cybern. Part A **39**(6), 1331–1338 (2009)
4. E. Camponogara, M.L. Lima, Distributed optimization for MPC of linear networks with uncertain dynamics. IEEE Trans. Autom. Control **57**(3), 804–809 (2012)
5. E. Camponogara, H.F. Scherer, Distributed optimization for model predictive control of linear dynamic networks with control-input and output constraints. IEEE Trans. Autom. Sci. Eng. **8**(1), 233–242 (2011)
6. E. Camponogara, S.N. Talukdar, Designing communication networks to decompose network control problems. INFORMS J. Comput. **17**(2), 207–223 (2005)
7. E. Camponogara, H. Zhou, S.N. Talukdar, Altruistic agents in uncertain, dynamic games. J. Comput. Syst. Sci. Int. **45**(4), 536–552 (2006)
8. L.B. de Oliveira, E. Camponogara, Multi-agent model predictive control of signaling split in urban traffic networks. Transp. Res. Part C **18**(1), 120–139 (2010)
9. C. Diakaki, M. Papageorgiou, K. Aboudolas, A multivariable regulator approach to traffic-responsive network-wide signal control. Control Eng. Pract. **10**(2), 183–195 (2002)
10. I. Necoara, V. Nedelcu, I. Dumitrache, Parallel and distributed optimization methods for estimation and control in networks. J. Process Control **21**, 756–766 (2011)
11. R.R. Negenborn, B. De Schutter, J. Hellendoorn, Multi-agent model predictive control for transportation networks: serial versus parallel schemes. Eng. Appl. Artif. Intell. **21**(3), 353–366 (2007)
12. H.F. Scherer, E. Camponogara, A. Codas, Transfer function modeling of linear dynamic networks for distributed MPC, in *Proceedings of the 7th IEEE Conference on Automation Science and Engineering*, pp. 613–618, 2011
13. F.A. Souza, V.B. Peccin, E. Camponogara, Distributed model predictive control applied to urban traffic networks: implementation, experimentation, and analysis, in *Proceedings of the 6th IEEE Conference on Automation Science and Engineering*, pp. 399–405, 2010
14. A.N. Venkat, I.A. Hiskens, J.B. Rawlings, S.J. Wright, Distributed MPC strategies with application to power system automatic generation control. IEEE Trans. Control Syst. Technol. **16**(6), 1192–1206 (2008)
15. M. Zhu, S. Martínez, On distributed convex optimization under inequality and equality constraints. IEEE Trans. Autom. Control **57**(1), 151–164 (2012)

Chapter 13
Adaptive Quasi-Decentralized MPC of Networked Process Systems

Y. Hu and N. H. El-Farra

Abstract This work presents a framework for quasi-decentralized model predictive control (MPC) design with an adaptive communication strategy. In this framework, each unit of the networked process system is controlled by a local control system for which the measurements of the local process state are available at each sampling instant. And we aim to minimize the cross communication between each local control system and the sensors of the other units via the communication network while preserving stability and certain level of control system performance. The quasi-decentralized MPC scheme is designed on the basis of distributed Lyapunov-based bounded control with sampled measurements and then the stability properties of each closed-loop subsystem are characterized. By using this obtained characterization, an adaptive communication strategy is proposed that forecasts the future evolution of the local process state within each local control system. Whenever the forecast shows signs of instability of the local process state, the measurements of the entire process state are transmitted to update the model within this particular control system to ensure stability; otherwise, the local control system will continue to rely on the model within the local MPC controller. The implementation of this theoretical framework is demonstrated using a simulated networked chemical process.

13.1 Introduction

Driven by the need to maximize economic efficiency as well as the demand of the market, operations, the size and complexity of modern industrial and commercial processes have been continuing to increase. As an example, chemical plants are

Y. Hu (✉) · N. H. El-Farra
University of California, Davis, California
e-mail: yehhu@ucdavis.edu

N. H. El-Farra
e-mail: nhelfarra@ucdavis.edu

J. M. Maestre and R. R. Negenborn (eds.), *Distributed Model Predictive Control Made Easy*, Intelligent Systems, Control and Automation: Science and Engineering 69,
DOI: 10.1007/978-94-007-7006-5_13,

typically large-scale processes involving many units that have complex dynamical behavior; tight interconnections usually exist among the various units and thus the dynamics of each unit are strongly coupled with the dynamics of the rest of the plant through the exchange of material and energy. The traditional solution for control of plants with interconnected and distributed units usually falls within either the centralized or the decentralized framework (e.g., see [1, 5, 11, 13] and the references therein). The centralized framework offers satisfactory performance for small- to medium-scale plants as the single control agent that controls the entire plant is able to account for the interconnections of all the units; however, for large-scale plants, the single agent has to maintain control for a large number of subsystems which may have different objectives, and this poses significant problems to the computation and coordination of the plant-wide control. While for decentralized control where each subsystem is controlled independently by an agent, computational burden and coordination of all the subsystems are no longer a problem, but it may lead to poor and unreliable performance when the various subsystems have strong interactions with each other. These issues associated with the traditional solutions for control of large-scale plants have motivated the development of distributed control framework (e.g., see [2, 7, 15] for some results and references in this area) which, on the one hand, avoids the complexity and inflexibility that are typical of centralized approaches by adopting a decentralized topology, and on the other hand, addresses the stability and performance issues associated with decentralized approaches by handling the inter-subsystem interactions in a more systematic way.

The increase in size and complexity of the industrial processes is also accompanied by the increasing use of communication networks in place of the traditional point-to-point dedicated link because the use of networks can significantly reduce the cost of installation and maintenance and can also provide more flexibility for diagnosis and reconfiguration of the control systems. Nevertheless, control over networks introduces some new issues that challenge the traditional process control theory due to the inherent limitations on the capabilities of the networked devices and communication medium. These issues, for example, resource constraints, data losses and communication delays, limit the overall achievable control quality if they are not explicitly handled in the control system design.

In this work, we propose an adaptive quasi-decentralized MPC framework for networked process systems. Specifically, each unit of the interconnected system is controlled locally by an MPC controller and the local process state is available to the local control system at each sampling instant. Our objective is to minimize the cross communication between each local control system and the sensors of the other units within this interconnected system while maintaining stability and some degree of performance. The quasi-decentralized MPC controllers are designed based on distributed Lyapunov-based bounded control with sampled measurements and the stability properties of each closed-loop subsystem are analyzed. An adaptive communication strategy is then proposed to minimize the cross communication. The key idea of this strategy is to use the stability properties of each closed-loop subsystem to forecast the evolution of the local process state in the future. If the forecast indicates possible instability, then the cross communication is established to curb the growth of the local model estimation error that could lead to instability; otherwise, the communication is kept dormant for as long as the stability is not at risk.

The rest of the chapter is organized as follows. In Sect. 13.2, the class of systems is described together with the problem formulation and solution overview. We then proceed to the step-by-step description of the adaptive quasi-decentralized MPC scheme in Sect. 13.3 and the stability properties of this scheme are also investigated. The proposed framework is then implemented on a simulated networked chemical process in Sect. 13.4 and some concluding remarks are given in Sect. 13.5.

13.2 Preliminaries

13.2.1 System Description

We consider the following large-scale nonlinear process system:

$$\dot{\mathbf{x}} = \mathbf{f}(\mathbf{x}) + \mathbf{G}(\mathbf{x})\mathbf{u} \tag{13.1}$$

which is comprised of a number of interconnected subsystems with each subsystem i represented by the following state-space description:

$$\dot{\mathbf{x}}_i = \mathbf{f}_i(\mathbf{x}) + \mathbf{G}_i(\mathbf{x})\mathbf{u}_i, \; i \in \mathcal{N} \triangleq [1, |\mathcal{N}|] \cap \mathbb{Z} \tag{13.2}$$

where $|\mathcal{N}|$ represents the cardinality of the set $\mathcal{N}$, $\mathbf{x}_i = [x_{i,1} \; x_{i,2} \; \cdots \; x_{i,n_{\mathbf{x}_i}}]^{\mathrm{T}} \in \mathbb{R}^{n_{\mathbf{x}_i}}$ denotes the state vector for the i-th subsystem, $\mathbf{x} = [\mathbf{x}_1^{\mathrm{T}} \; \mathbf{x}_2^{\mathrm{T}} \; \cdots \; \mathbf{x}_{|\mathcal{N}|}^{\mathrm{T}}]^{\mathrm{T}} \in \mathbb{R}^{n_{\mathbf{x}}}$ denotes the state vector of the entire system where $n_{\mathbf{x}} \triangleq \sum_{i=1}^{|\mathcal{N}|} n_{\mathbf{x}_i}$, $\mathbf{u}_i = [u_{i,1} \; u_{i,2} \; \cdots \; u_{i,n_{\mathbf{u}_i}}]^{\mathrm{T}} \in \mathcal{U}_i \subset \mathbb{R}^{n_{\mathbf{u}_i}}$ denotes the vector of constrained manipulated input variables associated with the i-th subsystem which takes values in a nonempty compact convex set $\mathcal{U}_i \triangleq \{\mathbf{u}_i \in \mathbb{R}^{n_{\mathbf{u}_i}} : \|\mathbf{u}_i\| \le u_{i,\max}\}$ where $u_{i,\max} > 0$ and $\|\cdot\|$ denotes the Euclidean norm of a vector or matrix, $\mathbf{u} = [\mathbf{u}_1^{\mathrm{T}} \; \mathbf{u}_2^{\mathrm{T}} \; \cdots \; \mathbf{u}_{|\mathcal{N}|}^{\mathrm{T}}]^{\mathrm{T}} \in \mathbb{R}^{n_{\mathbf{u}}}$ where $n_{\mathbf{u}} \triangleq \sum_{i=1}^{|\mathcal{N}|} n_{\mathbf{u}_i}$. For the sake of generality, the specific interconnection topology within the system (13.1) is not considered; therefore, $\mathbf{f}_i(\cdot)$ and $\mathbf{G}_i(\cdot) = [\mathbf{g}_{i,1}(\cdot) \; \mathbf{g}_{i,2} \; \cdots \; \mathbf{g}_{i,n_{\mathbf{u}_i}}]$ are assumed to be smooth vector- and matrix-valued functions respectively of the state vector of the entire system $\mathbf{x}$, which implies that interactions between any two subsystems are allowed.

An approximate dynamic model is assumed to be available for the system (13.1):

$$\dot{\hat{\mathbf{x}}} = \hat{\mathbf{f}}(\hat{\mathbf{x}}) + \hat{\mathbf{G}}(\hat{\mathbf{x}})\mathbf{u} \tag{13.3}$$

with the model for each subsystem being

$$\dot{\hat{\mathbf{x}}}_i = \hat{\mathbf{f}}_i(\hat{\mathbf{x}}) + \hat{\mathbf{G}}_i(\hat{\mathbf{x}})\mathbf{u}_i, \; i \in \mathcal{N} \tag{13.4}$$

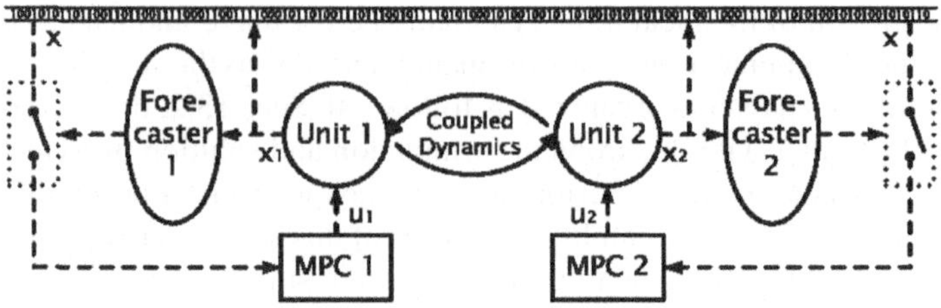

Fig. 13.1 A schematic representation of the quasi-decentralized MPC architecture for a networked process system with forecast-triggered communication strategy implemented

where $\hat{\mathbf{x}}_i = [\hat{x}_{i,1}\ \hat{x}_{i,2}\ \cdots\ \hat{x}_{i,n_{\mathbf{x}_i}}]^{\mathrm{T}} \in \mathbb{R}^{n_{\mathbf{x}_i}}$ denotes the model state for the i-th subsystem and $\hat{\mathbf{x}} = [\hat{\mathbf{x}}_1^{\mathrm{T}}\ \hat{\mathbf{x}}_2^{\mathrm{T}}\ \cdots\ \hat{\mathbf{x}}_{|\mathcal{N}|}^{\mathrm{T}}]^{\mathrm{T}} \in \mathbb{R}^{n_{\mathbf{x}}}$ denotes the model state for the entire system, $\hat{\mathbf{f}}_i(\cdot)$ and $\hat{\mathbf{G}}_i(\cdot) = [\hat{\mathbf{g}}_{i,1}(\cdot)\ \hat{\mathbf{g}}_{i,2}\ \cdots\ \hat{\mathbf{g}}_{i,n_{\mathbf{u}_i}}]$ are approximate models of $\mathbf{f}_i(\cdot)$ and $\mathbf{G}_i(\cdot)$ and are also assumed to be sufficiently smooth on the domain of interest which contains the origin $\hat{\mathbf{x}} = \mathbf{0}$ in its interior, $\hat{\mathbf{f}}(\cdot) = [\hat{\mathbf{f}}_1^{\mathrm{T}}(\cdot)\ \hat{\mathbf{f}}_2^{\mathrm{T}}(\cdot)\ \cdots\ \hat{\mathbf{f}}_{|\mathcal{N}|}^{\mathrm{T}}(\cdot)]^{\mathrm{T}}$, $\hat{\mathbf{G}}(\cdot) = \mathrm{diag}(\hat{\mathbf{G}}_1(\cdot), \hat{\mathbf{G}}_2(\cdot), \cdots, \hat{\mathbf{G}}_{|\mathcal{N}|}(\cdot))$ is a block matrix whose main diagonal blocks are $\hat{\mathbf{G}}_i(\cdot)$ for all $i \in \mathcal{N}$. Without loss of generality, it is assumed that the origin is the open-loop equilibrium point of the model (13.3) (i.e., $\hat{\mathbf{f}}(\mathbf{0}) = \mathbf{0}$).

13.2.2 Problem Formulation and Solution Overview

The communication architecture we consider in this work is depicted in Fig. 13.1. Each subsystem that is controlled by a local MPC controller transmits its local state information to the plant-wide communication network at each sampling time, and those data can be accessed by the local MPC controllers in order to update their model states at the sampling instants. The problems under consideration in this work include how to practically stabilize the interconnected system (13.1) using quasi-decentralized MPC controllers and how to minimize the information transfer from the network to the local MPC controllers without sacrificing practical stability. To address these problems, we propose the following approach:

1. Construct model-based distributed Lyapunov-based bounded controllers for (13.3) and characterize the closed-loop stability properties.
2. Analyze the stability properties when the model-based distributed Lyapunov-based bounded controllers are implemented with sampled measurements (i.e., sample-and-hold control implementation).
3. Synthesize the quasi-decentralized MPC controllers on the basis of the model-based distributed Lyapunov-based bounded control, and characterize the stability properties when the quasi-decentralized MPC controllers are implemented on the plant (13.1) with model states updated at every sampling instant.

4. Propose an adaptive communication strategy that suspends/restores communication from the network to the local MPC controller based on the forecast of the future dynamical behavior of each subsystem which is provided by the corresponding forecaster that uses only the local state information.

13.3 Adaptive Quasi-Decentralized MPC

In this section, step-by-step description of the adaptive quasi-decentralized MPC approach is given together with the stability analysis of the closed-loop system.

13.3.1 Step 1: Model-Based Distributed Lyapunov-Based Bounded Control with Continuous Measurements

Referring to the subsystem model (13.4), we assume that, for $i \in \mathcal{N}$, there exists a locally Lipschitz nonlinear feedback control law $\mathbf{u}_i = \mathbf{k}_i(\hat{\mathbf{x}})$ that satisfies the constraint on the magnitude of $\mathbf{u}_i$ for all the states $\hat{\mathbf{x}}$ within a stability region $\mathcal{W}$ and asymptotically stabilizes the origin of each subsystem with $\dot{V}_i(\hat{\mathbf{x}}_i) \leq -\lambda_i V_i(\hat{\mathbf{x}}_i)$ for $\hat{\mathbf{x}} \in \mathcal{W}$ where $\lambda_i > 0$ and the control Lyapunov function [6, 8, 12, 14] $V_i(\hat{\mathbf{x}}_i)$ is a continuously differentiable positive definite function that also satisfies

$$\alpha_{1i}(\|\hat{\mathbf{x}}_i\|) \leq V_i(\hat{\mathbf{x}}_i) \leq \alpha_{2i}(\|\hat{\mathbf{x}}_i\|) \tag{13.5a}$$

$$\left\| \frac{\partial V_i(\hat{\mathbf{x}}_i)}{\partial \hat{\mathbf{x}}_i} \right\| \leq \alpha_{3i}(\|\hat{\mathbf{x}}_i\|) \tag{13.5b}$$

where $\alpha_{1i}(\cdot)$, $\alpha_{2i}(\cdot)$, and $\alpha_{3i}(\cdot)$ are class $\mathbf{K}$ functions.[1] By using the control Lyapunov functions, the stability region of the closed-loop model (13.3) can be characterised: $\mathcal{W} \triangleq \mathcal{W}_1 \times \mathcal{W}_2 \times \cdots \times \mathcal{W}_{|\mathcal{N}|}$ where $\mathcal{W}_i \triangleq \{\hat{\mathbf{x}}_i \in \mathbb{R}^{n_{\hat{\mathbf{x}}_i}} : V_i(\hat{\mathbf{x}}_i) \leq c_i\}$ with $c_i > 0$. One example of the controller designs that satisfy these requirements is constrained Sontag-type controller reported in [3].

13.3.2 Step 2: Model-Based Distributed Lyapunov-Based Bounded Control with Sampled Measurements

When the measurements are only available at the sampling instants, the bounded controllers are implemented in a discrete (sample-and-hold) fashion:

---

[1] A function $\alpha(\cdot)$ is said to be of class $\mathbf{K}$ if it is strictly increasing and $\alpha(0) = 0$.

$$\dot{\hat{\mathbf{x}}}_i(t) = \hat{\mathbf{f}}_i(\hat{\mathbf{x}}(t)) + \hat{\mathbf{G}}_i(\hat{\mathbf{x}}(t))\mathbf{u}_i(t) \tag{13.6a}$$

$$\mathbf{u}_i(t) = \mathbf{k}_i(\hat{\mathbf{x}}(t_k)), \ t \in [t_k, t_{k+1}), \ k \in \mathbb{Z} \tag{13.6b}$$

where $t_{k+1} - t_k \triangleq \Delta$ is the sampling period. Due to the presence of the measurement errors, only practical stability can be achieved and the model state $\hat{\mathbf{x}}$ will converge to a small terminal neighborhood of the nominal equilibrium point. Specifically, for all i, let $c_i' \in (0, c_i)$, $\Delta > 0$ and $\sigma_i > 0$ satisfy

$$-\lambda_i c_i' + (\Phi_{L_{\hat{\mathbf{f}}_i} V_i} + \Phi_{L_{\hat{\mathbf{G}}_i} V_i} u_{i,\max})\gamma\Delta \leq -\sigma_i \tag{13.7a}$$

$$c_i' + (\Phi_{L_{\hat{\mathbf{f}}_i} V_i} + \Phi_{L_{\hat{\mathbf{G}}_i} V_i} u_{i,\max})\gamma\Delta^2 \triangleq \epsilon_i < c_i \tag{13.7b}$$

where $\gamma \triangleq \Theta_{\hat{\mathbf{f}}} + \Theta_{\hat{\mathbf{G}}} M_{\mathbf{u}}$, $M_{\mathbf{u}} \triangleq \left\| [u_{1,\max} \ \ u_{2,\max} \ \ \cdots \ \ u_{|\mathcal{N}|,\max}] \right\|$, $\Theta_{\hat{\mathbf{f}}} \triangleq \max_{\hat{\mathbf{x}} \in \mathcal{W}} \|\hat{\mathbf{f}}(\hat{\mathbf{x}})\|$, $\Theta_{\hat{\mathbf{G}}} \triangleq \max_{\hat{\mathbf{x}} \in \mathcal{W}} \|\hat{\mathbf{G}}(\hat{\mathbf{x}})\|$, $\Phi_{L_{\hat{\mathbf{f}}_i} V_i}$ and $\Phi_{L_{\hat{\mathbf{G}}_i} V_i}$ satisfy, for all $\hat{\mathbf{x}}, \hat{\mathbf{x}}' \in \mathcal{W}$,

$$\left\| \frac{\partial V_i(\hat{\mathbf{x}}_i)}{\partial \hat{\mathbf{x}}_i} \hat{\mathbf{f}}_i(\hat{\mathbf{x}}) - \frac{\partial V_i(\hat{\mathbf{x}}_i')}{\partial \hat{\mathbf{x}}_i'} \hat{\mathbf{f}}_i(\hat{\mathbf{x}}') \right\| \triangleq \left\| L_{\hat{\mathbf{f}}_i} V_i(\hat{\mathbf{x}}) - L_{\hat{\mathbf{f}}_i} V_i(\hat{\mathbf{x}}') \right\| \leq \Phi_{L_{\hat{\mathbf{f}}_i} V_i} \|\hat{\mathbf{x}} - \hat{\mathbf{x}}'\|$$

$$\left\| \frac{\partial V_i(\hat{\mathbf{x}}_i)}{\partial \hat{\mathbf{x}}_i} \hat{\mathbf{G}}_i(\hat{\mathbf{x}}) - \frac{\partial V_i(\hat{\mathbf{x}}_i')}{\partial \hat{\mathbf{x}}_i'} \hat{\mathbf{G}}_i(\hat{\mathbf{x}}') \right\| \triangleq \left\| L_{\hat{\mathbf{G}}_i} V_i(\hat{\mathbf{x}}) - L_{\hat{\mathbf{G}}_i} V_i(\hat{\mathbf{x}}') \right\| \leq \Phi_{L_{\hat{\mathbf{G}}_i} V_i} \|\hat{\mathbf{x}} - \hat{\mathbf{x}}'\|$$

Then, given any $\hat{\mathbf{x}}(t_0) \in \mathcal{W}$, $\hat{\mathbf{x}}(t) \in \mathcal{W}$ for all $t \in [t_0, \infty)$ and

$$\limsup_{t \to \infty} V_i(\hat{\mathbf{x}}_i(t)) \leq \epsilon_i \tag{13.8}$$

furthermore, when $\hat{\mathbf{x}}_i(t_k) \in \mathcal{W}_i \backslash \mathcal{W}_i'$ where $\mathcal{W}_i' \triangleq \{\hat{\mathbf{x}}_i \in \mathbb{R}^{n_{\hat{\mathbf{x}}_i}} : V_i(\hat{\mathbf{x}}_i) \leq c_i'\}$, the following inequality holds:

$$\dot{V}_i(\hat{\mathbf{x}}_i(t)) \leq -\sigma_i, \ \forall t \in [t_k, t_{k+1}) \tag{13.9}$$

The proof of this result is omitted here for brevity and interested readers are referred to [9] and [10] for proofs of some conceptually similar results. The above characterization of the closed-loop stability properties shows that the controller possesses a robustness property against the measurement errors; however, the controller can only bring the model state within a terminal set $\mathcal{M} \triangleq \mathcal{M}_1 \times \mathcal{M}_2 \times \cdots \times \mathcal{M}_{|\mathcal{N}|} \subset \mathcal{W}$ where $\mathcal{M}_i \triangleq \{\hat{\mathbf{x}}_i \in \mathbb{R}^{n_{\hat{\mathbf{x}}_i}} : V_i(\hat{\mathbf{x}}_i) \leq \epsilon_i\} \subset \mathcal{W}_i$. It is also important to note that the size of the sampling period Δ has great influence on the stability as can be seen from (13.7a) and (13.7b). It has to be sufficiently small so that practical stability can be preserved with the discrete implementation of the bounded controller; an excessively large sampling period could lead to instability.

13.3.3 Step 3: Quasi-Decentralized MPC

13.3.3.1 Quasi-Decentralized MPC Formulation

The quasi-decentralized MPC scheme is based on the distributed Lyapunov-based bounded control with sampled measurements and the optimization problem for each subsystem (13.2) is formulated as follows:

$$\min_{\mathbf{u}_i \in \mathcal{U}_i} \int_{t_k}^{t_{k+N_p}} [\|\hat{\mathbf{x}}_i^i(s)\|_{Q_i}^2 + \|\mathbf{u}_i(s)\|_{R_i}^2]\, \mathrm{d}s \tag{13.10a}$$

Subject to

$$\dot{\hat{\mathbf{x}}}_i^i(t) = \hat{\mathbf{f}}_i(\hat{\mathbf{x}}^i(t)) + \hat{\mathbf{G}}_i(\hat{\mathbf{x}}^i(t))\mathbf{u}_i(t) \tag{13.10b}$$

$$\dot{\hat{\mathbf{x}}}_j^i(t) = \hat{\mathbf{f}}_j(\hat{\mathbf{x}}^i(t)) + \hat{\mathbf{G}}_j(\hat{\mathbf{x}}^i(t))\mathbf{k}_j(\hat{\mathbf{x}}^i(t_p)), \ \forall j \in \mathcal{N}\backslash\{i\}, \ p \in [k, k+N_p-1] \cap \mathbb{Z} \tag{13.10c}$$

$$\hat{\mathbf{x}}^i(t_k) = \mathbf{x}(t_k) \tag{13.10d}$$

$$\dot{V}_i(\hat{\mathbf{x}}_i^i(t)) \le -\sigma_i, \ \forall t \in [t_k, t_{k+1}), \ \text{if } V_i(\hat{\mathbf{x}}_i^i(t_k)) > c_i' \tag{13.10e}$$

$$V_i(\hat{\mathbf{x}}_i^i(t)) \le \epsilon_i, \ \forall t \in [t_k, t_{k+1}), \ \text{if } V_i(\hat{\mathbf{x}}_i^i(t_k)) \le c_i' \tag{13.10f}$$

where $\hat{\mathbf{x}}^i = [\hat{\mathbf{x}}_1^{i\mathrm{T}} \ \hat{\mathbf{x}}_2^{i\mathrm{T}} \ \cdots \ \hat{\mathbf{x}}_{|\mathcal{N}|}^{i\mathrm{T}}]^\mathrm{T} \in \mathbb{R}^{n_\mathbf{x}}$ denotes the state vector of the copy of the plant model that is embedded in the MPC controller for the i-th subsystem, N_p is the prediction horizon and the control horizon N_c takes the same value, Q_i and R_i are positive definite weighting matrices. The initial process state and model state must satisfy $\hat{\mathbf{x}}^i(t_0) = \mathbf{x}(t_0) \in \mathcal{W}$ for all i.

Remark 13.1 Since the evolution of each local model state $\hat{\mathbf{x}}_i^i$ is affected by the entire model state, the models that govern the evolution of the model states $\hat{\mathbf{x}}_j^i$ for all $j \in \mathcal{N}\backslash\{i\}$ must also be embedded within the i-th MPC controller, which naturally involves the solution of the j-th optimization problem $\mathbf{u}_j$. However, all the optimization problems are computed simultaneously and $\mathbf{u}_j$ cannot be made available for the computation of the i-th optimization problem (unless we allow real-time iterative communication of the solutions of the optimization problems between the controllers). Therefore, we use $\mathbf{k}_j(\hat{\mathbf{x}}^i(t_p))$ as an assumed value for $\mathbf{u}_j$ in (13.10c). Moreover, as long as $\hat{\mathbf{x}}^i(t_0) = \mathbf{x}(t_0) \in \mathcal{W}$, the use of $\mathbf{k}_j(\hat{\mathbf{x}}^i(t_p))$ in place of $\mathbf{u}_j$ in (13.10c) guarantees the feasibility of the optimization problem for the first sampling interval. Furthermore, if the plant state $\mathbf{x}(t)$ is bounded within $\mathcal{W}$ for all t (which will be shown next), the optimization problem will continue to be feasible.

13.3.3.2 Stability Analysis

The stability analysis of the system (13.1) in closed-loop with the quasi-decentralized MPC scheme consists of two parts. In the first part, the dynamics of the model estimation error are analyzed during a sampling interval assuming that $\mathbf{x}$ stays within $\mathcal{W}$ at all times. In the second part, the bound on the estimation error obtained in the first part is used to show the stability properties of the closed-loop subsystem through Lyapunov techniques and the assumption made in the first part is verified.

Part 1: We define $\hat{\mathbf{u}}^i(t) \triangleq [\mathbf{k}_1^{\mathrm{T}}(\hat{\mathbf{x}}^i(t_k)) \ \cdots \ \mathbf{u}_i^{\mathrm{T}} \ \cdots \ \mathbf{k}_{|\mathcal{N}|}^{\mathrm{T}}(\hat{\mathbf{x}}^i(t_k))]^{\mathrm{T}}$ and $\mathbf{e}^i(t) \triangleq \mathbf{x}(t) - \hat{\mathbf{x}}^i(t)$ for $t \in [t_k, t_{k+1})$. Calculating the derivative of $\mathbf{e}^i(t)$ on this time interval gives

$$\begin{aligned}\dot{\mathbf{e}}^i(t) &= \mathbf{f}(\mathbf{x}) - \hat{\mathbf{f}}(\mathbf{x}) + \hat{\mathbf{f}}(\mathbf{x}) - \hat{\mathbf{f}}(\hat{\mathbf{x}}^i) \\ &\quad + [\mathbf{G}(\mathbf{x}) - \hat{\mathbf{G}}(\mathbf{x}) + \hat{\mathbf{G}}(\mathbf{x}) - \hat{\mathbf{G}}(\hat{\mathbf{x}}^i)]\mathbf{u} + \hat{\mathbf{G}}(\hat{\mathbf{x}}^i)(\mathbf{u} - \hat{\mathbf{u}}^i)\end{aligned}$$

Given $\hat{\mathbf{x}}^i(t_0) = \mathbf{x}(t_0) \in \mathcal{W}$, $\hat{\mathbf{x}}^i(t)$ will remain within $\mathcal{W}$ for all $t \in [t_0, t_1)$ because of the enforced stability constraints (which ensures the boundedness of $\hat{\mathbf{x}}_i^i$) and the properties of the bounded controller (which ensures the boundedness of $\hat{\mathbf{x}}_j^i$). If $\mathbf{x}(t)$ also remains within $\mathcal{W}$ during this interval, then the bound on $\mathbf{e}^i(t)$ satisfies

$$\|\mathbf{e}^i(t)\| \le \|\mathbf{e}^i(t_0)\| + (\Theta_{\mathbf{d_f}} + \Theta_{\mathbf{d_G}} M_{\mathbf{u}} + 2\Theta_{\hat{\mathbf{G}}} M_{\mathbf{u}})(t - t_0) + \int_{t_0}^{t} (\Phi_{\hat{\mathbf{f}}} + \Phi_{\hat{\mathbf{G}}} M_{\mathbf{u}})\|\mathbf{e}^i(s)\| \, \mathrm{d}s$$

where

$$\begin{gathered}\Theta_{\mathbf{d_f}} \triangleq \max_{\mathbf{x}\in\mathcal{W}} \|\mathbf{d_f}(\mathbf{x})\|, \quad \Theta_{\mathbf{d_G}} \triangleq \max_{\mathbf{x}\in\mathcal{W}} \|\mathbf{d_G}(\mathbf{x})\| \\ \mathbf{d_f}(\cdot) \triangleq \mathbf{f}(\cdot) - \hat{\mathbf{f}}(\cdot), \quad \mathbf{d_G}(\cdot) \triangleq \mathbf{G}(\cdot) - \hat{\mathbf{G}}(\cdot) \\ \Phi_{\hat{\mathbf{G}}} = \|\Phi_{\hat{\mathbf{G}}_1} \ \Phi_{\hat{\mathbf{G}}_2} \ \cdots \ \Phi_{\hat{\mathbf{G}}_{|\mathcal{N}|}}\|, \quad \Phi_{\hat{\mathbf{G}}_i} = \|\Phi_{\hat{\mathbf{g}}_{i,1}} \ \Phi_{\hat{\mathbf{g}}_{i,2}} \ \cdots \ \Phi_{\hat{\mathbf{g}}_{i,n_{\mathbf{u}_i}}}\|\end{gathered}$$

$\Phi_{\hat{\mathbf{f}}}$ and $\Phi_{\hat{\mathbf{g}}_{i,q_i}}$ (where $q_i \in [1, n_{\mathbf{u}_i}] \cap \mathbb{Z}$) are the Lipschitz constants of $\hat{\mathbf{f}}(\cdot)$ and $\hat{\mathbf{g}}_{i,q_i}(\cdot)$ on $\mathcal{W}$. Applying Gronwall-Bellman inequality with $\mathbf{e}^i(t_0) = \mathbf{0}$ yields

$$\|\mathbf{e}^i(t)\| \le \frac{\Theta_{\mathbf{d_f}} + \Theta_{\mathbf{d_G}} M_{\mathbf{u}} + 2\Theta_{\hat{\mathbf{G}}} M_{\mathbf{u}}}{\Phi_{\hat{\mathbf{f}}} + \Phi_{\hat{\mathbf{G}}} M_{\mathbf{u}}} [e^{(\Phi_{\hat{\mathbf{f}}} + \Phi_{\hat{\mathbf{G}}} M_{\mathbf{u}})(t - t_0)} - 1] \triangleq \zeta(t - t_0) \tag{13.12}$$

Following a similar approach and using (13.12), the following bound on $\mathbf{e}_i^i(t) \triangleq \mathbf{x}_i(t) - \hat{\mathbf{x}}_i^i(t)$ can also be obtained:

$$\begin{aligned}\|\mathbf{e}_i^i(t)\| &\le \|\mathbf{e}_i^i(t_0)\| + \int_{t_0}^{t} \left\|[\mathbf{G}_i(\mathbf{x}(s)) - \hat{\mathbf{G}}_i(\mathbf{x}(s)) + \hat{\mathbf{G}}_i(\mathbf{x}(s)) - \hat{\mathbf{G}}_i(\hat{\mathbf{x}}^i(s))]\mathbf{u}_i(s)\right\| \mathrm{d}s \\ &\quad + \int_{t_0}^{t} \left\|[\mathbf{f}_i(\mathbf{x}(s)) - \hat{\mathbf{f}}_i(\mathbf{x}(s)) + \hat{\mathbf{f}}_i(\mathbf{x}(s)) - \hat{\mathbf{f}}_i(\hat{\mathbf{x}}^i(s))]\right\| \mathrm{d}s \\ &\le \|\mathbf{e}_i^i(t_0)\| + [(\Phi_{\hat{\mathbf{f}}_i} + \Phi_{\hat{\mathbf{G}}_i} u_{i,\max})\zeta(t - t_0) + (\Theta_{\mathbf{d}_{\mathbf{f}_i}} + \Theta_{\mathbf{d}_{\mathbf{G}_i}} u_{i,\max})](t - t_0) \\ &\triangleq \beta_i(\|\mathbf{e}_i^i(t_0)\|, \zeta(t - t_0), t - t_0) = \beta_i(0, \zeta(t - t_0), t - t_0)\end{aligned} \tag{13.13}$$

where

$$\Theta_{\mathbf{d}_{\mathbf{f}_i}} \triangleq \max_{\mathbf{x}\in\mathcal{W}} \|\mathbf{d}_{\mathbf{f}_i}(\mathbf{x})\|, \quad \Theta_{\mathbf{d}_{\mathbf{G}_i}} \triangleq \max_{\mathbf{x}\in\mathcal{W}} \|\mathbf{d}_{\mathbf{G}_i}(\mathbf{x})\| \tag{13.14a}$$

$$\mathbf{d}_{\mathbf{f}_i}(\cdot) \triangleq \mathbf{f}_i(\cdot) - \hat{\mathbf{f}}_i(\cdot), \quad \mathbf{d}_{\mathbf{G}_i}(\cdot) \triangleq \mathbf{G}_i(\cdot) - \hat{\mathbf{G}}_i(\cdot) \tag{13.14b}$$

and $\Phi_{\hat{\mathbf{f}}_i}$ is the Lipschitz constant of $\hat{\mathbf{f}}_i(\cdot)$ on $\mathcal{W}$.

Part 2: For the subsystems whose initial state $\hat{\mathbf{x}}_i^i(t_0) = \mathbf{x}_i(t_0) \in \mathcal{W}_i \backslash \mathcal{W}_i'$, the derivative of the Lyapunov function along the trajectories of the subsystem (13.2) satisfies

$$\dot{V}_i(\mathbf{x}_i) = \frac{\partial V_i}{\partial \hat{\mathbf{x}}_i^i}[\hat{\mathbf{f}}_i(\hat{\mathbf{x}}^i)+\hat{\mathbf{G}}_i(\hat{\mathbf{x}}^i)\mathbf{u}_i]+\frac{\partial V_i}{\partial \mathbf{x}_i}\mathbf{f}_i(\mathbf{x})-\frac{\partial V_i}{\partial \hat{\mathbf{x}}_i^i}\hat{\mathbf{f}}_i(\hat{\mathbf{x}}^i)+\frac{\partial V_i}{\partial \mathbf{x}_i}\mathbf{G}_i(\mathbf{x})\mathbf{u}_i-\frac{\partial V_i}{\partial \hat{\mathbf{x}}_i^i}\hat{\mathbf{G}}_i(\hat{\mathbf{x}}^i)\mathbf{u}_i$$

Then by using (13.5), (13.10e), and (13.12), we can obtain that

$$\dot{V}_i(\mathbf{x}_i(t)) \le -\sigma_i + (\Phi_{L_{\hat{\mathbf{f}}_i}V_i} + \Phi_{L_{\hat{\mathbf{G}}_i}V_i}u_{i,\max})\zeta(t-t_0) + \alpha_{3i}(\alpha_{1i}^{-1}(c_i))(\Theta_{\mathbf{d}_{\mathbf{f}_i}} + \Theta_{\mathbf{d}_{\mathbf{G}_i}}u_{i,\max}) \tag{13.15}$$

If the following inequality holds:

$$-\sigma_i + (\Phi_{L_{\hat{\mathbf{f}}_i}V_i} + \Phi_{L_{\hat{\mathbf{G}}_i}V_i}u_{i,\max})\zeta(\Delta) + \alpha_{3i}(\alpha_{1i}^{-1}(c_i))(\Theta_{\mathbf{d}_{\mathbf{f}_i}} + \Theta_{\mathbf{d}_{\mathbf{G}_i}}u_{i,\max}) \le -\omega_i \tag{13.16}$$

then we have

$$\dot{V}_i(\mathbf{x}_i(t)) \le -\omega_i, \ \forall t \in [t_0, t_1) \tag{13.17}$$

For the subsystems whose initial state $\hat{\mathbf{x}}_i^i(t_0) = \mathbf{x}_i(t_0) \in \mathcal{W}_i'$, we use the following inequality derived from the Taylor expansion of $V_i(\mathbf{x}_i)$ which holds for all $\mathbf{x}_i, \hat{\mathbf{x}}_i^i \in \mathcal{W}_i$:

$$V_i(\mathbf{x}_i) \le V_i(\hat{\mathbf{x}}_i^i) + \frac{\partial V_i(\hat{\mathbf{x}}_i^i)}{\partial \hat{\mathbf{x}}_i^i}\|\mathbf{x}_i - \hat{\mathbf{x}}_i^i\| + \eta_i\|\mathbf{x}_i - \hat{\mathbf{x}}_i^i\|^2$$

where η_i is a positive number and the term $\eta_i\|\mathbf{x}_i - \hat{\mathbf{x}}_i^i\|^2$ bounds the second and higher order terms of the Taylor expansion. Together with (13.5), (13.10f), and (13.13), it can be shown that

$$V_i(\mathbf{x}_i(t)) \le \epsilon_i + \alpha_{3i}(\alpha_{1i}^{-1}(c_i))\beta_i(0, \zeta(t-t_0), t-t_0) + \eta_i[\beta_i(0, \zeta(t-t_0), t-t_0)]^2 \tag{13.18}$$

If the following condition is satisfied:

$$\epsilon_i + \alpha_{3i}(\alpha_{1i}^{-1}(c_i))\beta_i(0, \zeta(\Delta), \Delta) + \eta_i[\beta_i(0, \zeta(\Delta), \Delta)]^2 \le \mu_i < c_i \tag{13.19}$$

then the state of the subsystem will be bounded within $\mathcal{M}_i' \triangleq \{\mathbf{x}_i \in \mathbb{R}^{n_{x_i}} : V_i(\mathbf{x}_i) \le \mu_i\} \subset \mathcal{W}_i$ for $t \in [t_0, t_1)$.

The above analysis for the initial interval can be performed recursively for all the following intervals, which shows that the state $\mathbf{x}(t)$ is bounded within $\mathcal{W}$ for all $t \in [t_0, \infty)$ and in turn validates the assumption made on the boundedness of $\mathbf{x}$ previously. Therefore, if (13.7a), (13.7b), (13.16), and (13.19) are satisfied, we can conclude that given any $\hat{\mathbf{x}}^i(t_0) = \mathbf{x}(t_0) \in \mathcal{W}$ for all i, $\mathbf{x}(t) \in \mathcal{W}$ for all $t \in [t_0, \infty)$, and

$$\limsup_{t\to\infty} V_i(\mathbf{x}_i(t)) \leq \mu_i \tag{13.20}$$

furthermore, when $\hat{\mathbf{x}}_i^i(t_k) = \mathbf{x}_i(t_k) \in \mathcal{W}_i \backslash \mathcal{W}_i'$, the following inequality holds:

$$\dot{V}_i(\mathbf{x}_i(t)) \leq -\omega_i, \ \forall t \in [t_k, t_{k+1}) \tag{13.21}$$

Remark 13.2 (13.15) and (13.18) show that, in addition to the size of the sampling period Δ, the model estimation errors greatly affect the stability properties of the closed-loop system as well. The stability of a particular subsystem is not only susceptible to the estimation error of the local state, but actually also influenced by the estimation errors of the states of the rest of the plant. Therefore, the embedded models must be sufficiently accurate and the sampling period must be sufficiently small to ensure practical stability. Also note that the errors caused by the plant-model mismatch will in general slow down the decay of the value of the Lyapunov function when $\hat{\mathbf{x}}_i^i(t_k) = \mathbf{x}_i(t_k) \in \mathcal{W}_i \backslash \mathcal{W}_i'$ and enlarge the size of the ultimate bound of the process state $\mathcal{M}' \triangleq \mathcal{M}_1' \times \mathcal{M}_2' \times \cdots \times \mathcal{M}_{|\mathcal{N}|}'$ compared with the size of $\mathcal{M}$ associated with the distributed bounded control using sampled measurements.

13.3.4 Step 4: Adaptive Forecast-Triggered Communication Strategy

In order to reduce sensor-controller information transfer (required by (13.10d)), it is ideal if we can suspend the feedback so that the optimization problems are solved by relying on the model estimates that are not updated by the actual process state. However, the expressions of the bounds on $\mathbf{e}^i$ (13.12) and $\mathbf{e}_i^i$ (13.13) show that both $\zeta(\tau)$ and $\beta_i(\kappa, \zeta, \tau)$ are class $\mathbf{K}$ functions with respect to τ; therefore, if the model state is allowed to evolve without being updated by the actual state, the model estimation error will continue to grow and, as can be seen from (13.15) and (13.18), eventually lead to instability. Once the process state leaves the stability region, the MPC controllers can no longer stabilize the process even if the sensor-controller communication is restored. As a result, our objective is to minimize the communication between the sensors and the controllers (via the communication network) while maintaining stability and some degree of performance.

In order to achieve this objective, the local process state $\mathbf{x}_i$ which is available at each sampling instant is monitored by the i-th local control system for all i, and then the estimation error of the local state $\mathbf{e}_i^i$ will be available at each sampling instant

Algorithm 13.1 Forecast-triggered communication strategy

1: Initialize $\hat{\mathbf{x}}^i(t_0) = \mathbf{x}(t_0) \in \mathcal{W}$ and set $k = 0, \theta = 0$
2: Solve (13.10) for $[t_0, t_1)$ and implement the first step of the control sequence
3: If $\hat{\mathbf{x}}_i^i(t_{k+1}) \in \mathcal{W}_i \backslash \mathcal{W}_i'$ then
4: Calculate $\bar{V}_i(\mathbf{x}_i(t_{k+2}))$ (estimate of $V_i(\mathbf{x}_i(t_{k+2}))$) using (13.22) and $V_i(\mathbf{x}_i(t_{k+1}))$
5: Else
6: Calculate $\bar{V}_i(\mathbf{x}_i(t_{k+2}))$ (estimate of $V_i(\mathbf{x}_i(t_{k+2}))$) using (13.23) and $\mathbf{e}_i^i(t_{k+1})$
7: End if
8: If $\bar{V}_i(\mathbf{x}_i(t_{k+2})) < V_i(\mathbf{x}_i(t_{k+1}))$ then
9: Solve (13.10) without (13.10d) for $[t_{k+1}, t_{k+2})$
10: Else if $\bar{V}_i(\mathbf{x}_i(t_{k+2})) \ge V_i(\mathbf{x}_i(t_{k+1}))$ and $\bar{V}_i(\mathbf{x}_i(t_{k+2})) \le \mu_i$ then
11: Solve (13.10) without (13.10d) for $[t_{k+1}, t_{k+2})$
12: Else
13: Solve (13.10) for $[t_{k+1}, t_{k+2})$ and set $\theta = k + 1$
14: End if
15: Implement the first step of the control sequence on $[t_{k+1}, t_{k+2})$
16: Set $k = k + 1$ and go to step 3

for the i-th subsystem. For $t \in [t_k, t_{k+1})$, (13.15) and (13.18) can be modified as follows:

$$\dot{V}_i(\mathbf{x}_i(t)) \le -\sigma_i + (\Phi_{L_{\hat{\mathbf{f}}_i} V_i} + \Phi_{L_{\hat{\mathbf{G}}_i} V_i} u_{i,\max})\zeta(t - t_\theta) + \alpha_{3i}(\alpha_{1i}^{-1}(c_i))(\Theta_{\mathbf{d}_{\mathbf{f}_i}} + \Theta_{\mathbf{d}_{\mathbf{G}_i}} u_{i,\max}) \tag{13.22}$$

$$\begin{aligned} V_i(\mathbf{x}_i(t)) \le{} & \epsilon_i + \alpha_{3i}(\alpha_{1i}^{-1}(c_i))\beta_i(\|\mathbf{e}_i^i(t_k)\|, \zeta(t - t_\theta), t - t_k) \\ & + \eta_i[\beta_i(\|\mathbf{e}_i^i(t_k)\|, \zeta(t - t_\theta), t - t_k)]^2 \end{aligned} \tag{13.23}$$

where t_θ denotes the time that the last update prior to t_k takes place.

Algorithm 13.1 summarizes the adaptive communication strategy and it should be executed by the control system of every subsystem.

Remark 13.3 The forecaster as shown in Fig. 13.1 receives the measurements of the local process state and performs steps 3–7 in Algorithm 13.1 and then decides whether or not to update the model state using the state of the entire process based on the criteria described in steps 8–14. Specifically, at $t = t_{k+1}$, the evolution of the local process state within the next sampling interval is forecast using the actual value of $V_i(\mathbf{x}_i(t_{k+1}))$ or $\mathbf{e}_i^i(t_{k+1})$ as well as the constraint on the Lyapunov function that will become active within the next sampling interval $[t_{k+1}, t_{k+2})$ which is dictated by the location of $\hat{\mathbf{x}}_i^i(t_{k+1})$. If the estimated local process state $\mathbf{x}_i$ enters a smaller level set of V_i or lies within $\mathcal{M}_i'$, no updates of the local model state need to be performed at t_{k+1} because practical stability is still guaranteed within the next sampling interval; otherwise, $\hat{\mathbf{x}}^i(t_{k+1})$ must be updated by $\mathbf{x}(t_{k+1})$ to suppress the destabilizing tendency of the local process state. Note that the decision of whether or not to update the model state is triggered by the prediction of a future event instead of a current event (i.e., a simple comparison of the situations at the current and previous sampling instants). And the reason that a "current-event–triggered" algorithm is not adopted here is that, even if the local process state currently appears to be stable, it may have already become unstable or can become unstable within the next sampling interval if an indeed necessary update is skipped at the current sampling time. It is only by forecasting the worst-case plant behavior within the next sampling interval that one can safely determine whether or not the current update should be skipped.

Remark 13.4 In step 8 of Algorithm 13.1, $\bar{V}_i(\mathbf{x}_i(t_{k+2})) < V_i(\mathbf{x}_i(t_{k+1}))$ is used as a criterion for skipping an update at t_{k+1} because this is sufficient to ensure stability and can minimize the possibility of performing noncritical updates that merely improves the performance of the control system. However, when minimization of communication is not crucial or when better control system performance is equally important, one can impose a more stringent constraint on the decay rate of V_i which could enhance the performance at the cost of increased sensor-controller communication.

Remark 13.5 In this communication strategy, the prediction of how the system evolves in the future is based on the characterized stability properties of the closed-loop subsystem as well as the available information about its current behavior, which allows us to adaptively adjust the communication frequency. It is worth noting that when an update is necessary, say, for subsystem i, the entire model state $\hat{\mathbf{x}}^i$ should be updated by the state of the entire process system $\mathbf{x}$. The reason is that the evolution of the i-th subsystem is coupled with other subsystems and this update could be caused by the need of an update for $\hat{\mathbf{x}}^i_j$ in the first place.

In Algorithm 13.1, if an update is found to be necessary at a sampling instant, this update is immediately implemented and then the optimization problem is solved with updated model state. For certain applications, however, it may be impractical to perform an "immediate" update and the communication strategy depicted as Algorithm 13.2, which is a variation of Algorithm 13.1, is proposed to account for this issue.

Remark 13.6 It is important to note that the estimation of the value of $V_i(\mathbf{x}_i(t_{k+2}))$ performed during the time interval $[t_k, t_{k+1})$ which uses either (13.22) or (13.23) implicitly requires that the constraints on the value of the Lyapunov function in (13.10e) and (13.10f) must be valid for the first two sampling intervals of the prediction horizon. Therefore, (13.10e) and (13.10f) must be modified to allow the estimation of the value of $V_i(\mathbf{x}_i(t_{k+2}))$ using (13.22) and/or (13.23).

Remark 13.7 In steps 8–10 of Algorithm 13.2, an estimate of the value of $V_i(\mathbf{x}_i(t_{k+2}))$ is obtained by calculating estimates using both (13.22) and (13.23) and then taking the maximum of the two. A calculated upper bound of $V_i(\hat{\mathbf{x}}^i_i(t_{k+1}))$ doesn't provide much insight into whether (13.22) or (13.23) should used in the estimation of $V_i(\mathbf{x}_i(t_{k+2}))$ (there won't be confusion, however, when the estimate of $\hat{\mathbf{x}}^i_i(t_{k+1})$ is inside $\mathcal{W}^{\prime}_i$) because the actual value of $V_i(\hat{\mathbf{x}}^i_i(t_{k+1}))$ can be smaller than its estimate and thus the chosen inequality for estimation of $V_i(\mathbf{x}_i(t_{k+2}))$ based on the estimate of $V_i(\hat{\mathbf{x}}^i_i(t_{k+1}))$ can be different from the inequality that should be used based on the actual value of

Algorithm 13.2 Two-step ahead forecast-triggered communication strategy

```
1: Initialize x̂^i(t_0) = x(t_0) ∈ W and set k = 0, θ = 0
2: Solve (13.10) for [t_0, t_1) and implement the first step of the control sequence
3: If x̂_i^i(t_k) ∈ W_i \ W_i' then
4:   Calculate V̄_i(x_i(t_{k+1})) (estimate of V_i(x_i(t_{k+1}))) using (13.22) and V_i(x_i(t_k))
5: Else
6:   Calculate V̄_i(x_i(t_{k+1})) (estimate of V_i(x_i(t_{k+1}))) using (13.23) and e_i^i(t_k)
7: End if
8: Calculate Ṽ_i(x_i(t_{k+2})) (estimate of V_i(x_i(t_{k+2}))) using (13.22) and V̄_i(x_i(t_{k+1}))
9: Calculate V̌_i(x_i(t_{k+2})) (estimate of V_i(x_i(t_{k+2}))) using (13.23) and e_i^i(t_k)
10: Set V̄_i(x_i(t_{k+2})) = max{Ṽ_i(x_i(t_{k+2})), V̌_i(x_i(t_{k+2}))}
11: If V̄_i(x_i(t_{k+2})) < V̄_i(x_i(t_{k+1})) then
12:   Solve (13.10) without (13.10d) for [t_{k+1}, t_{k+2})
13: Else if V̄_i(x_i(t_{k+2})) ≥ V̄_i(x_i(t_{k+1})) and V̄_i(x_i(t_{k+2})) ≤ μ_i then
14:   Solve (13.10) without (13.10d) for [t_{k+1}, t_{k+2})
15: Else
16:   Solve (13.10) for [t_{k+1}, t_{k+2}) and set θ = k + 1
17: End if
18: Implement the first step of the control sequence on [t_{k+1}, t_{k+2})
19: Set k = k + 1 and go to 3
```

$V_i(\hat{\mathbf{x}}_i^i(t_{k+1}))$. In light of this, it is only safe to obtain both estimates of $V_i(\mathbf{x}_i(t_{k+2}))$ and use the larger one as the basis for suspending/restoring communication.

13.4 Application to a Simulated Networked Chemical Process

To illustrate the implementation of the proposed methodology, here we consider a process with two well-mixed continuous stirred tank reactors (CSTR) and a flash tank. The description and the dynamics of this process can be found in Sect. 6.6 of [4]. The plant-model mismatch is simulated by $\Delta H_1 - \Delta\hat{H}_1 = 0.01|\Delta\hat{H}_1|\sin(\pi t)$ and $\Delta H_2 - \Delta\hat{H}_2 = 0.01|\Delta\hat{H}_2|\sin(\pi t)$ with $\Delta\hat{H}_1$ and $\Delta\hat{H}_2$ being the nominal values used in the model of the process. The rates of heat transfer from the jackets Q_1, Q_2, and Q_3 are constrained manipulated inputs: $|Q_1| \le 5 \times 10^4$ kJ/h, $|Q_2| \le 1.5 \times 10^5$ kJ/h, and $|Q_3| \le 2 \times 10^5$ kJ/h. In order to stabilize the local process state of each unit around the nominal equilibrium point, quasi-decentralized MPC controllers are designed and the communication strategy described in Algorithm 13.1 is implemented where $V_i(\mathbf{x}_i) = \mathbf{x}_i^{\mathrm{T}} P_i \mathbf{x}_i$, $P_i = \mathrm{diag}(10, 1000, 1000, 1000)$, $Q_i = \mathrm{diag}(10, 1000, 1000, 1000)$, $R_i = 10^{-7}$, $N_p = N_c = 5$, and $\Delta = 0.01$ h.

As can be seen from Figs. 13.2a–13.2c, T_1, T_2, and T_3 are successfully stabilized around their corresponding nominal steady-state values (the other state variables are stabilized as well and their closed-loop profiles are omitted here). By using the adaptive communication strategy in Algorithm 13.1, the update rates (ratio between the number of updates on the local model state and the number of sampling instants) for the local model states of CSTR 1, CSTR 2, and the separator are 15 %, 42 %, and

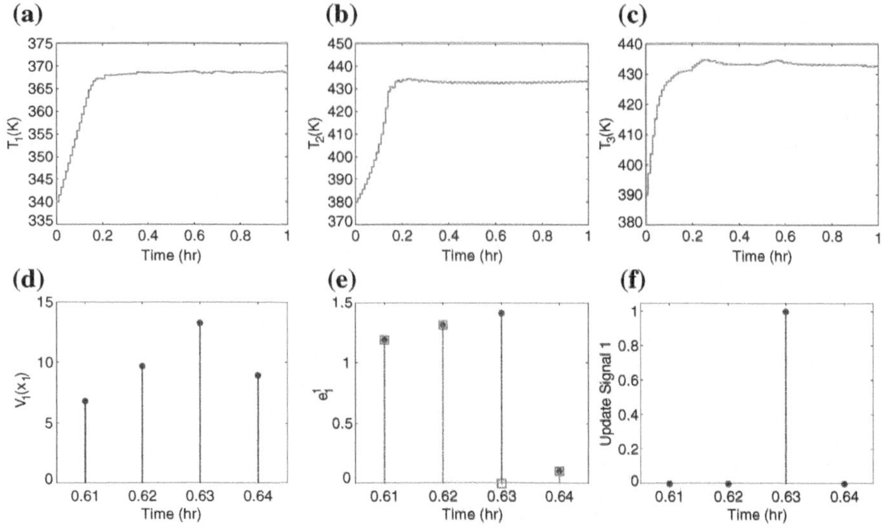

Fig. 13.2 Evolution of the temperature profiles of CSTR 1 (**a**), CSTR 2 (**b**), and the separator (**c**), and the profiles of $V_1(\mathbf{x}_1)$ (**d**), $\mathbf{e}_1^1$ (**e**) (where blue circle markers represent $\mathbf{e}_1^1(t_k^-)$ before potential updates and red square markers represent $\mathbf{e}_1^1(t_k)$ at the time of potential updates), and the update signal for CSTR 1 (**f**) on the time interval of [0.605 h, 0.645 h)

33 %, respectively. As an example to illustrate the implementation of Algorithm 13.1, the profiles of $V_1(\mathbf{x}_1)$, $\mathbf{e}_1^1$, and update signal 1 for $t \in [0.605\,\text{h}, 0.655\,\text{h})$ are shown in Figs. 13.2d-13.2f. At $t = 0.61$ h and 0.62 h, the measurements of the local state $\mathbf{x}_1$ are sampled and transmitted to the control system for CSTR 1 so that the values of $V_1(\mathbf{x}_1)$ and $\mathbf{e}_1^1$ can be calculated. Together with the estimates of the values of $\mathbf{e}^1$, the estimates of $V_1(\mathbf{x}_1(t))$ at $t = 0.62$ h and $t = 0.63$ h are calculated which indicate that the local process state will not escape from the terminal set associated with CSTR 1 ($\mathcal{M}_1' \triangleq \{\mathbf{x}_1 \in \mathbb{R}^4 : V_1(\mathbf{x}_1) \leq 15\}$) during [0.61 h, 0.62 h) and [0.62 h, 0.63 h). At $t = 0.63$ h, the same procedures are carried out but the estimate of the value of $V_1(\mathbf{x}_1(t))$ at $t = 0.64$ h shows that $\mathbf{x}_1$ may escape from $\mathcal{M}_1'$ and therefore an update on the local model state is performed that resets the model estimation errors and maintains practical stability which can be verified from the actual value of $V_1(\mathbf{x}_1(t))$ at $t = 0.64$ h in Fig. 13.2d.

13.5 Conclusions

In this chapter, we presented an adaptive quasi-decentralized MPC strategy for networked process systems. Based on the results of distributed Lyapunov-based bounded control, the quasi-decentralized MPC control scheme was developed and its stability properties were investigated. In order to minimize the sensor-controller information transfer via the communication network, a forecast-triggered communication strategy was proposed that forecasts the future evolution of each subsystem and generates a worst-case estimate by using the closed-loop stability properties as well as the information about the current operating status of each subsystem. Whenever the forecast indicates possible instability of a particular subsystem in the future, its model state will be updated by the actual process state; otherwise, the local control system will continue to rely on the model. Finally, simulation studies were presented to demonstrate the implementation and evaluate the effectiveness of the proposed methodology.

References

1. L. Acar, Ü. Özgüner, A completely decentralized suboptimal control strategy for moderately coupled interconnected systems, *in Proc. American Control Conf.*, pp. 1521–1524, Atlanta (1988)
2. E. Camponogara, D. Jia, B. Krogh, S. Talukdar, Distributed model predictive control. *IEEE Control Syst. Mag.* **22**(1), 44–52 (2002)
3. P.D. Christofides, N.H. El-Farra, *Control of Nonlinear and Hybrid Process Systems: Designs for Uncertainty, Constraints and Time-Delays* (Springer, Berlin/Heidelberg, 2005)
4. P.D. Christofides, J. Liu, D. Muñoz de la Peña, *Networked and Distributed Predictive Control: Methods and Nonlinear Process Network Applications* (Springer, London, 2011)
5. H. Cui, E.W. Jacobsen, Performance limitations in decentralized control. *J. Process Control* **12**(4), 485–494 (2002)
6. R.A. Freeman, P.V. Kokotović, *Robust Nonlinear Control Design: State-Space and Lyapunov Techniques* (Birkhäuser, Boston, 1996)
7. M.R. Katebi, M.A. Johnson, Predictive control design for large-scale systems. *Automatica* **33**(3), 421–425 (1997)
8. M. Krstić, H. Deng, *Stabilization of Nonlinear Uncertain Systems*, 1st edn. (Springer, Berlin, 1998)
9. P. Mhaskar, N.H. El-Farra, P.D. Christofides, Predictive control of switched nonlinear systems with scheduled mode transitions. *IEEE Trans. Autom. Control* **50**(11), 1670–1680 (2005)
10. D. Muñoz de la Peña, P.D. Christofides, Lyapunov-based model predictive control of nonlinear systems subject to data losses. *IEEE Trans. Autom. Control* **53**(9), 2076–2089 (2008)
11. N.R. Sandell, P. Varaiya, M. Athans, M.G. Safonov, Survey of decentralized control methods for large-scale systems. *IEEE Trans. Autom. Control* **23**(2), 108–128 (1978)
12. R. Sepulchre, M. Janković, P. Kokotović, *Constructive Nonlinear Control* (Springer, Berlin/Heidelberg, 1997)
13. D.D. Šiljak, *Decentralized Control of Complex Systems* (Academic Press, London, 1991)
14. E.D. Sontag, Smooth stabilization implies coprime factorization. *IEEE Trans. Autom. Control* **34**(4), 435–443 (1989)
15. Y. Sun, N.H. El-Farra, A quasi-decentralized approach for networked state estimation and control of process systems. *Ind. Eng. Chem. Res.* **49**(17), 7957–7971 (2010)

Chapter 14
Distributed Lyapunov-Based MPC

R. Hermans, M. Lazar and A. Jokić

Abstract We provide an almost decentralized solution to the problem of stabilizing a network of discrete-time nonlinear systems with coupled dynamics that are subject to local state/input constraints. By "almost decentralized" we mean that each local controller is allowed to use the states of neighboring systems for feedback, whereas it is not permitted to employ iterations between the systems in the network to compute the control action. The controller synthesis method used in this work is Lyapunov-based model predictive control. The closed-loop stability conditions are decentralized via a set of structured control Lyapunov functions (CLFs) for which the maximum over all the functions in the set is a CLF for the global network of systems. However, this does not necessarily imply that each function is a CLF for its corresponding subsystem. Additionally, a solution is provided for relaxing the temporal monotonicity of the network-wide CLF. For infinity-norm based structured CLFs and input-affine dynamics, we show that the decentralized MPC algorithm can be implemented by solving a single linear program in each network node. Two application examples are provided to illustrate the effectiveness of the developed theory and to show that the proposed method can perform as well as more complex distributed, iteration-based MPC algorithms.

R. Hermans (✉) · M. Lazar
Department of Electrical Engineering, Eindhoven University of Technology, Eindhoven, The Netherlands
e-mail: r.m.hermans@tue.nl

M. Lazar
e-mail: m.lazar@tue.nl

A. Jokić
Faculty of Mechanical Engineering and Naval Architecture, University of Zagreb, Zagreb, Croatia
e-mail: andrej.jokic@fsb.hr

J. M. Maestre and R. R. Negenborn (eds.), *Distributed Model Predictive Control Made Easy*, Intelligent Systems, Control and Automation: Science and Engineering 69, DOI: 10.1007/978-94-007-7006-5_14,

14.1 Introduction

The basic requirement of any control system is that the corresponding closed-loop dynamics are stable. In standard model predictive control schemes, this is typically guaranteed via monotonic convergence of the subsequent optimal performance cost values, see, e.g., [11]. Hence, in these cost-based approaches, attaining *globally optimal performance* is a key prerequisite for stability. Unfortunately, when non-centralized MPC for large-scale networks of interconnected dynamical systems (NDS) is the main focus, the demand for optimization of a system-wide performance cost function inherently comes with a need for intensive iterative exchange of information or global coordination among the agents that control the various subsystems in the network. Such coordination may be hampered by limitations of the communication infrastructure that is available in practice, or may be undesired in competitive environments such as the deregulated electrical power market [3].

A successful alternative to cost based stabilization is *Lyapunov-based MPC (L-MPC)*, see, e.g., [1]. L-MPC makes use of an explicit control Lyapunov function (CLF) to characterize the set of stabilizing control actions, and therefore decouples the need for globally optimal performance from the desire for closed-loop stability. Although L-MPC has already been successfully applied to networked control systems in, for example, [10], therein the focus is more on communication network effects such as time delays and packet dropouts, rather than on non-centralized stabilization of large-scale systems. However, as the large size and complexity of networked dynamical systems generally impede the application of centralized control laws, the search for non-centralized formulations of CLF-based stability conditions has become a major concern.

In this chapter we propose a non-centralized Lyapunov-based model predictive control scheme for discrete-time nonlinear NDS that are subject to coupled local dynamics and separable constraints. The key ingredient of the proposed approach is a set of *structured CLFs* with a particular type of convergence conditions. While these conditions do not force each of the structured functions to decrease monotonously, as typically required for a CLF [14], they provide a standard CLF for the overall network. Still, the conservatism associated with a demand for monotonous convergence of the overall CLF might be restrictive in practice. Therefore, we provide a solution for relaxing the temporal monotonicity of the global CLF based on an adaptation of the Lyapunov-Razumikhin technique [9], which was originally developed for systems with time delays. The proposed L-MPC scheme needs no global coordination and can be implemented in an *almost decentralized fashion*. By this we mean that the controller only requires one run of information exchange between direct neighbors per sampling instant. This is in contrast to many of the existing non-centralized MPC schemes, which either require iterative computations or global information, see e.g., [2, 15], or, employ contractive constraints or small gain conditions, see e.g., [12, 13], to guarantee closed-loop stability.

For systems that are affine in the control input, we show that by employing infinity-norm based structured CLFs, the proposed L-MPC setup can be implemented by solving a single linear problem per sampling instant and node. The effectiveness and computational complexity of the proposed scheme is assessed on a non-trivial nonlinear example and a benchmark model of an electricity network that is subject to competitive optimization of local performance objectives.

This chapter is organized as follows. We start in Sect. 14.2 with a brief overview of the stability notions that are instrumental in our control setup. In Sect. 14.3, we define the class of systems and the control problem considered in this work. The proposed L-MPC framework is described and analyzed in Sect. 14.4, and simulation results are provided in Sect. 14.5. We finish with conclusions in Sect. 14.6.

14.1.1 Nomenclature

Let $\mathbb{R}$, $\mathbb{R}_+$, $\mathbb{Z}$ and $\mathbb{Z}_+$ denote the field of real numbers, the set of non-negative reals, the set of integer numbers and the set of non-negative integers, respectively. For every $c \in \mathbb{R}$ and $\Pi \subseteq \mathbb{R}$ we define $\Pi_{\geq c(\leq c)} := \{k \in \Pi \mid k \geq c\,(k \leq c)\}$ and $\mathbb{Z}_\Pi := \{k \in \mathbb{Z} \mid k \in \Pi\}$. For a finite set of vectors $\{\mathbf{X}_i\}_{i\in\mathbb{Z}_{[1,N]}}$, $\mathbf{x}_i \in \mathbb{R}^{n_i}$, $N \in \mathbb{Z}_+$, we use $\text{col}(\{\mathbf{X}_i\}_{i\in\mathbb{Z}_{[1,N]}})$, and equivalently $\text{col}(\mathbf{x}_1, \ldots, \mathbf{x}_N)$, to denote the column vector $\left(\mathbf{x}_1^{\mathrm{T}}, \ldots, \mathbf{x}_n^{\mathrm{T}}\right)^{\mathrm{T}}$. Let $\mathbf{0}_n$ denote the zero vector in $\mathbb{R}^n$. For a set $\mathcal{S} \subseteq \mathbb{R}^n$, we denote by $\text{int}(\mathcal{S})$ the interior of $\mathcal{S}$. For a vector $\mathbf{x} \in \mathbb{R}^n$, let $\|\mathbf{x}\|$ denote an arbitrary p-norm and let $[\mathbf{x}]_i$, $i \in \mathbb{Z}_{[1,n]}$ be the i-th scalar component of $\mathbf{x}$. The ∞-norm of a vector $\mathbf{x} \in \mathbb{R}^n$ is defined as $\|\mathbf{x}\|_\infty := \max_{i=1,\ldots,n} |[\mathbf{x}]_i|$, where $|\cdot|$ denotes the absolute value. For a matrix $\mathbf{M} \in \mathbb{R}^{m\times n}$, let $\|\mathbf{M}\| := \max_{\mathbf{x}\neq\mathbf{0}_n} \frac{\|\mathbf{Mx}\|}{\|\mathbf{x}\|}$ denote the corresponding induced matrix norm. Let $\underline{\mathbf{z}} := \{\mathbf{z}(l)\}_{l\in\mathbb{Z}_+}$ with $\mathbf{z}(l) \in \mathbb{R}^n$ for all $l \in \mathbb{Z}_+$ denote an arbitrary sequence. Define $\|\underline{\mathbf{z}}\| := \sup\{\|\mathbf{z}(l)\| \mid l \in \mathbb{Z}_+\}$ and $\underline{\mathbf{z}}_{[0,k]} := \{\mathbf{z}(l)\}_{l\in\mathbb{Z}_{[0,k]}}$. For some $s \in \mathbb{R}$, let $\lfloor s \rfloor := \max\{n \in \mathbb{Z} \mid n \leq s\}$ be the floor function. A function $\varphi : \mathbb{R}_+ \to \mathbb{R}_+$ belongs to class $\mathcal{K}$ if it is continuous, strictly increasing and $\varphi(0) = 0$. A function $\varphi : \mathbb{R}_+ \to \mathbb{R}_+$ belongs to class $\mathcal{K}_\infty$ if $\varphi \in \mathcal{K}$ and $\lim_{s\to\infty} \varphi(s) = \infty$.

14.2 Preliminary Stability Notions

Consider the discrete-time, time-invariant nonlinear system

$$\mathbf{x}(k+1) \in \mathbf{\Phi}\left(\mathbf{x}(k)\right), \qquad k \in \mathbb{Z}_+, \tag{14.1}$$

where $\mathbf{x}(k) \in \mathbb{X} \subseteq \mathbb{R}^n$ is the state at the discrete-time instant $k \in \mathbb{Z}_+$. The set-valued mapping $\mathbf{\Phi} : \mathbb{R}^n \rightrightarrows \mathbb{R}^n$ is such that $\mathbf{\Phi}(\mathbf{x})$ is compact and nonempty for all $\mathbf{x} \in \mathbb{X}$. We assume that the origin is an equilibrium of (14.1), i.e., $\mathbf{\Phi}\left(\mathbf{0}_n\right) = \{\mathbf{0}_n\}$.

Definition 14.1 A set $\mathcal{P} \subseteq \mathbb{R}^n$ is *Positively Invariant (PI)* for system (14.1) if $\forall \mathbf{x} \in \mathcal{P}$ it holds that $\mathbf{\Phi}(\mathbf{x}) \subseteq \mathcal{P}$.

Definition 14.2 (i) System (14.1) is *Lyapunov stable* if $\forall \varepsilon > 0$, $\exists \delta(\varepsilon) > 0$ such that for all state trajectories of (14.1) it holds that $\|\mathbf{x}(0)\| \leq \delta(\varepsilon) \Rightarrow \|\mathbf{x}(k)\| \leq \varepsilon$ for all $k \in \mathbb{Z}_+$. (ii) Let $\mathbb{X} \subseteq \mathbb{R}^n$ and $\mathbf{0}_n \in \text{int}(\mathbb{X})$. The origin of (14.1) is *attractive in* $\mathbb{X}$ if for any $\mathbf{x}(0) \in \mathbb{X}$ it holds that all corresponding trajectories of (14.1) satisfy $\lim_{k\to\infty} \|\mathbf{x}(k)\| = 0$. (iii) System (14.1) is *asymptotically stable in* $\mathbb{X}$ if it is Lyapunov stable and attractive in $\mathbb{X}$.

Theorem 14.1 *Let* $\mathbb{X}$ *be a PI set for system* (14.1) *and let* $\mathbf{0}_n \in \text{int}(\mathbb{X})$. *Furthermore, let* $\alpha_1, \alpha_2 \in \mathcal{K}_\infty$, $\rho \in \mathbb{R}_{[0,1)}$ *and let* $V : \mathbb{R}^n \to \mathbb{R}_+$ *be a function such that*

$$\alpha_1(\|\mathbf{x}\|) \leq V(\mathbf{x}) \leq \alpha_2(\|\mathbf{x}\|) \tag{14.2a}$$

$$V(\mathbf{x}^+) \leq \rho V(\mathbf{x}) \tag{14.2b}$$

for all $\mathbf{x} \in \mathbb{X}$ *and all* $\mathbf{x}^+ \in \mathbf{\Phi}(x)$. *Then system* (14.1) *is asymptotically stable in* $\mathbb{X}$.

A function V that satisfies the conditions of Theorem 14.1 is called a *Lyapunov function*. The proof of Theorem 14.1 can be obtained from [7], Theorem 2.8. Note that in [7] continuity of the function V on $\mathbb{X}$, i.e., not solely at the origin as specified by Theorem 14.1, is required only to show certain robustness properties.

Next, consider the discrete-time constrained system

$$\mathbf{x}(k+1) = \phi(\mathbf{x}(k), \mathbf{u}(k)), \quad k \in \mathbb{Z}_+, \tag{14.3}$$

where $\mathbf{x}(k) \in \mathbb{X} \subseteq \mathbb{R}^n$ is the state and $\mathbf{u}(k) \in \mathbb{U} \subseteq \mathbb{R}^m$ is the control input at the discrete-time instant $k \in \mathbb{Z}_+$. The function $\phi : \mathbb{R}^n \times \mathbb{R}^m \to \mathbb{R}^n$ is nonlinear with $\phi(\mathbf{0}_n, \mathbf{0}_m) = \mathbf{0}_n$. We assume that $\mathbb{X}$ and $\mathbb{U}$ are bounded sets with $\mathbf{0}_n \in \text{int}(\mathbb{X})$ and $\mathbf{0}_m \in \text{int}(\mathbb{U})$. Next, let $\alpha_1, \alpha_2 \in \mathcal{K}_\infty$ and let $\rho \in \mathbb{R}_{[0,1)}$.

Definition 14.3 A function $V : \mathbb{R}^n \to \mathbb{R}_+$ that satisfies

$$\alpha_1(\|\mathbf{x}\|) \leq V(\mathbf{x}) \leq \alpha_2(\|\mathbf{x}\|), \quad \forall \mathbf{x} \in \mathbb{R}^n, \tag{14.4}$$

and for which there exists a possibly set-valued control law $\pi : \mathbb{R}^n \rightrightarrows \mathbb{U}$ such that

$$V(\phi(\mathbf{x}, \mathbf{u})) \leq \rho V(\mathbf{x}), \quad \forall \mathbf{x} \in \mathbb{X}, \forall \mathbf{u} \in \pi(\mathbf{x}), \tag{14.5}$$

is called a control Lyapunov function in $\mathbb{X}$ for (14.3).

14.3 Modeling Framework and Problem Definition

In order to set-up the control algorithm, we first introduce a framework for defining a network of interconnected dynamical systems. Consider a *directed connected graph* $\mathcal{G} = (\mathcal{S}, \mathcal{E})$ with a finite number of vertices $\mathcal{S} = \{\varsigma_1, \ldots, \varsigma_N\}$ and a set of directed edges $\mathcal{E} \subseteq \{(\varsigma_i, \varsigma_j) \in \mathcal{S} \times \mathcal{S} \mid i \neq j\}$. In an NDS, a dynamical system is assigned to each vertex $\varsigma_i \in \mathcal{S}$, whose dynamics are governed by the difference equation

$$\mathbf{x}_i(k+1) = \phi_i(\mathbf{x}_i(k), \mathbf{u}_i(k), \mathbf{v}_i(\mathbf{x}_{\mathcal{B}_i}(k))), \quad k \in \mathbb{Z}_+, \tag{14.6}$$

for vertex indices $i \in \mathcal{N} := \mathbb{Z}_{[1,N]}$, $N := |\mathcal{N}|$. In (14.6), $\mathbf{x}_i \in \mathbb{X}_i \subseteq \mathbb{R}^{n_i}$ denotes the state and $\mathbf{u}_i \in \mathbb{U}_i \subseteq \mathbb{R}^{m_i}$ represents the control input of the i-th system, i.e., the system assigned to vertex ς_i. With each directed edge $(\varsigma_j, \varsigma_i) \in \mathcal{E}$ we associate a function $\mathbf{v}_{ij} : \mathbb{R}^{n_j} \to \mathbb{R}^{n_{v_{ij}}}$ that defines the interconnection signal $\mathbf{v}_{ij}(\mathbf{x}_j(k)) \in \mathbb{R}^{n_{v_{ij}}}$, $k \in \mathbb{Z}_+$, between system j and system i, i.e., $\mathbf{v}_{ij}(\mathbf{x}_j(k))$ characterizes how the states of system j influence the dynamics of system i. We use $\mathcal{B}_i := \{j \mid (\varsigma_j, \varsigma_i) \in \mathcal{E}\}$ to denote the set of indices corresponding to the *direct neighbors* of system i. A direct neighbor of system i is any system in the network whose dynamics (e.g., states or outputs) appear explicitly (via the function $\mathbf{v}_{ij}(\cdot)$) in the state equations that govern the dynamics of system i. Clearly, if system j is a direct neighbor of system i, this does not necessarily imply the reverse. Let $\overline{\mathcal{B}}_i := \mathcal{B}_i \cup \{i\}$. We define $\mathbf{x}_{\mathcal{B}_i}(k) := \mathrm{col}(\{\mathbf{x}_j(k)\}_{j \in \mathcal{B}_i})$ as the vector that collects all the state vectors of the direct neighbors of system i and $\mathbf{v}_i(\mathbf{x}_{\mathcal{B}_i}(k)) := \mathrm{col}(\{\mathbf{v}_{ij}(\mathbf{x}_j(k))\}_{j \in \mathcal{B}_i}) \in \mathbb{R}^{n_{v_i}}$ as the vector that collects all the vector-valued interconnection signals that "enter" system i. The functions $\phi_i(\cdot,\cdot,\cdot)$ and $\mathbf{v}_{ij}(\cdot)$ $\phi_i(\mathbf{0}_{n_i}, \mathbf{0}_{m_i}, \mathbf{0}_{n_{v_i}}) = \mathbf{0}_{n_i}$ for all $i \in \mathcal{N}$ and $\mathbf{v}_{ij}(\mathbf{0}_{n_j}) = \mathbf{0}_{n_{v_{ij}}}$ for all $(i, j) \in \mathcal{N} \times \mathcal{B}_i$. For all $i \in \mathcal{N}$ we assume that $\mathbf{0}_{n_i} \in \mathrm{int}\,(\mathbb{X}_i)$ and $\mathbf{0}_{m_i} \in \mathrm{int}\,(\mathbb{U}_i)$.

The following standing assumption is instrumental for obtaining the results presented in this chapter.

Assumption 14.1 The values of all the interconnection signals $\{\mathbf{v}_{ij}(\mathbf{x}_j(k))\}_{j \in \mathcal{B}_i}$ are known at each discrete-time instant $k \in \mathbb{Z}_+$, for any system $i \in \mathcal{N}$.

Notice that Assumption 14.1 does not require knowledge of any of the interconnection signals at future time instants. From a technical point of view, Assumption 14.1 is satisfied, e.g., if all interconnection signals $\mathbf{v}_{ij}(\mathbf{x}_j(k))$ are directly measurable[1] at all $k \in \mathbb{Z}_+$. Alternatively, Assumption 14.1 is satisfied if all directly neighboring systems $j \in \mathcal{B}_i$ are able to communicate their local measured state $\mathbf{x}_j(k)$ to system $i \in \mathcal{N}$.

[1] For example, in electrical power systems, where a dynamical system is a power generator, the interconnection signal is the generator bus voltage and line power (or current) flow in the corresponding power line, which can be directly measured.

Finally, let

$$\mathbf{x}(k+1) = \boldsymbol{\phi}(\mathbf{x}(k), \mathbf{u}(k)), \qquad k \in \mathbb{Z}_+, \tag{14.7}$$

denote the dynamics of the overall network of interconnected systems (14.6), written in a compact form. In (14.7), $\mathbf{x} = \text{col}(\{\mathbf{x}_i\}_{i \in \mathcal{N}}) \in \mathbb{R}^n$, $n = \sum_{i \in \mathcal{N}} n_i$, and $\mathbf{u} = \text{col}(\{\mathbf{u}_i\}_{i \in \mathcal{N}}) \in \mathbb{R}^m$, $m = \sum_{i \in \mathcal{N}} m_i$, are vectors that collect all local states and inputs, respectively.

Now, our objective is to find a tractable solution to the following control problem.

Problem 14.1 Given NDS (14.6), design a set of control laws $\boldsymbol{\mu}_i(\mathbf{x}_i, \{\mathbf{x}_j\}_{j \in \mathcal{B}_i})$, $\boldsymbol{\mu}_i : \mathbb{X}_i \times \left(\prod_{j \in \mathcal{B}_i} \mathbb{X}_j\right) \to \mathbb{U}_i$, $i \in \mathcal{N}$, such that the corresponding closed-loop dynamics

$$\mathbf{x}(k+1) = \boldsymbol{\phi}(\mathbf{x}(k), \boldsymbol{\mu}(\mathbf{x}(k))), \quad k \in \mathbb{Z}_+, \tag{14.8}$$

with $\boldsymbol{\mu}(x) := \text{col}(\{\boldsymbol{\mu}_i(\mathbf{x}_i, \{\mathbf{x}_j\}_{j \in \mathcal{B}_i})\}_{i \in \mathcal{N}})$, are asymptotically stable and $\mathbf{x}_i(k) \in \mathbb{X}_i$ for all $k \in \mathbb{Z}_+$, $i \in \mathcal{N}$. ■

Problem 14.1 belongs to the class of *structured control problems*. This is reflected by the feedback laws $\boldsymbol{\mu}_i(\mathbf{x}_i, \{\mathbf{x}_j\}_{j \in \mathcal{B}_i})$, which define local control actions based on local and neighboring state information only.

14.4 Main Result: Structured Max-CLFs

In order to solve Problem 14.1, we introduce the notion of a set of "structured max-CLFs", which provides an alternative to the structured CLFs defined recently in [6].

Definition 14.4 Let $\alpha_1^i, \alpha_2^i \in \mathcal{K}_\infty$ for $i \in \mathcal{N}$ and let $\{V_i\}_{i \in \mathcal{N}}$ be a set of functions $V_i : \mathbb{R}^{n_i} \to \mathbb{R}_+$ that satisfy

$$\alpha_1^i(\|\mathbf{x}_i\|) \leq V_i(\mathbf{x}_i) \leq \alpha_2^i(\|\mathbf{x}_i\|), \ \forall \mathbf{x}_i \in \mathbb{R}^{n_i}, \ \forall i \in \mathcal{N}. \tag{14.9a}$$

Then, given $\rho_i \in \mathbb{R}_{[0,1)}$ for $i \in \mathcal{N}$, if there exists a non-empty set of possibly set-valued control laws $\boldsymbol{\pi}_i : \mathbb{R}^{n_i} \times \mathbb{R}^{n_{v_i}} \rightrightarrows \mathbb{U}_i$ such that and

$$V_i(\boldsymbol{\phi}_i(\mathbf{x}_i, \mathbf{u}_i, \mathbf{v}_i(\mathbf{x}_{\mathcal{B}_i}))) \leq \rho_i \max_{j \in \overline{\mathcal{B}}_i} V_j(\mathbf{x}_j), \tag{14.9b}$$

$\forall \mathbf{x}_i \in \mathbb{X}_i$, $\forall \mathbf{u}_i \in \boldsymbol{\pi}_i(\mathbf{x}_i, \mathbf{v}_i(\mathbf{x}_{\mathcal{B}_i}))$, the set of functions $\{V_i\}_{i \in \mathcal{N}}$ is called a set of "structured max control Lyapunov functions" in $\mathbb{X} := \{\text{col}(\{\mathbf{x}_i\}_{i \in \mathcal{N}}) \mid \mathbf{x}_i \in \mathbb{X}_i\}$ for (14.7).

In the above definition the term *structured* emphasizes the fact that each V_i is a function of $\mathbf{x}_i$ only, i.e., the structural decomposition of the dynamics of the overall interconnected system (14.6) is reflected in the functions $\{V_i\}_{i \in \mathcal{N}}$. Moreover, the term *max* originates from the corresponding convergence condition, i.e., (14.9b). Next, based on Definition 14.4, we formulate the following feasibility problem.

Problem 14.2 Let $\rho_i \in \mathbb{R}_{[0,1)}$, $i \in \mathcal{N}$ and a set of structured max-CLFs $\{V_i\}_{i\in\mathcal{N}}$ be given. At time $k \in \mathbb{Z}_+$, let the state vector $\{\mathbf{x}_i(k)\}_{i\in\mathcal{N}}$, the set of interconnection signals $\{\mathbf{v}_i(\mathbf{x}_{\mathcal{B}_i}(k))\}_{i\in\mathcal{N}}$ and the values $\{V_i(\mathbf{x}_i(k))\}_{i\in\mathcal{N}}$ be known, and calculate a set of control actions $\{\mathbf{u}_i(k)\}_{i\in\mathcal{N}}$, such that

$$\mathbf{u}_i(k) \in \mathbb{U}_i,\ \ \phi_i(\mathbf{x}_i(k), \mathbf{u}_i(k), \mathbf{v}_i(\mathbf{x}_{\mathcal{B}_i}(k))) \in \mathbb{X}_i, \tag{14.10a}$$

$$V_i(\phi_i(\mathbf{x}_i(k), \mathbf{u}_i(k), \mathbf{v}_i(\mathbf{x}_{\mathcal{B}_i}(k)))) \leq \rho_i \max{}_{j\in\overline{\mathcal{B}}_i} V_j(\mathbf{x}_j(k)), \tag{14.10b}$$

for all $i \in \mathcal{N}$. ■

Let $\pi(\mathbf{x}(k)) := \{\mathrm{col}(\{\mathbf{u}_i(k)\}_{i\in\mathcal{N}}) \mid (13.10) \text{ holds}\}$ and let

$$\mathbf{x}(k+1) \in \phi_{\mathrm{CL}}(\mathbf{x}(k), \pi(\mathbf{x}(k))) := \{\phi(\mathbf{x}(k), \mathbf{u}(k)) \mid \mathbf{u}(k) \in \pi(\mathbf{x}(k))\} \tag{14.11}$$

denote the difference inclusion corresponding to system (14.7) in "closed loop" with the set of feasible solutions obtained by solving Problem 14.2 at each instant $k \in \mathbb{Z}_+$.

Theorem 14.2 *Let $\alpha_1^i, \alpha_2^i \in \mathcal{K}_\infty$ and $\rho_i \in \mathbb{R}_{[0,1)}$, $\forall i \in \mathcal{N}$ be given and choose a set of structured max-CLFs $\{V_i\}_{i\in\mathcal{N}}$ in $\mathbb{X} = \{\mathrm{col}(\{\mathbf{x}_i\}_{i\in\mathcal{N}}) \mid \mathbf{x}_i \in \mathbb{X}_i\}$ for system* (14.7). *Suppose that Problem 14.2 is feasible for all $\mathbf{x}(k) \in \mathbb{X}$ and the corresponding signals $\{\mathbf{v}_i(\mathbf{x}_{\mathcal{B}_i}(k))\}_{i\in\mathcal{N}}$. Then* (14.11) *is asymptotically stable in $\mathbb{X}$.*

Proof Let $\mathbf{x}(k) \in \mathbb{X}$ for some $k \in \mathbb{Z}_+$. Then, feasibility of Problem 14.2 ensures that $\mathbf{x}(k+1) \in \phi_{\mathrm{CL}}(\mathbf{x}(k), \pi(\mathbf{x}(k))) \subseteq \mathbb{X}$ due to constraint (14.10a). Hence, Problem 14.2 remains feasible and thus, $\mathbb{X}$ is a PI set for system (14.11). Now consider the function $V(\mathbf{x}) := \max_{i\in\mathcal{N}} V_i(\mathbf{x}_i)$. Together with condition (14.10b) this yields

$$\begin{aligned} V(\mathbf{x}(k+1)) &= \max_{i\in\mathcal{N}} V_i(\mathbf{x}_i(k+1)) \\ &\leq \rho \max_{i\in\mathcal{N}} \max_{j\in\overline{\mathcal{B}}_i} V_j(\mathbf{x}_j(k)) = \rho \max_{i\in\mathcal{N}} V_i(\mathbf{x}_i(k)) = \rho V(\mathbf{x}(k)), \end{aligned} \tag{14.12}$$

for all $\mathbf{x}(k) \in \mathbb{X}$, where $\rho := \max_{i\in\mathcal{N}} \rho_i \in \mathbb{R}_{[0,1)}$.

Next, we derive a lower bound for $V(\mathbf{x})$. Observing that the maximum element of a set always equals or exceeds the average value of the elements and using (14.9a) yields

$$V(\mathbf{x}) := \max_{i\in\mathcal{N}} V_i(\mathbf{x}_i) \geq \frac{1}{N} \sum_{i\in\mathcal{N}} V_i(\mathbf{x}_i) \geq \frac{1}{N} \sum_{i\in\mathcal{N}} \alpha_1^i(\|\mathbf{x}_i\|). \tag{14.13}$$

Next, note that

$$\sum_{i\in\mathcal{N}} \alpha_1^i(\|\mathbf{x}_i\|) \geq \sum_{i\in\mathcal{N}} \tilde{\alpha}_1(\|\mathbf{x}_i\|) \geq \tilde{\alpha}_1(\max_{i\in\mathcal{N}} \|\mathbf{x}_i\|) \geq \tilde{\alpha}_1\left(\frac{1}{N}\sum_{i\in\mathcal{N}} \|\mathbf{x}_i\|\right), \tag{14.14}$$

where $\tilde{\alpha}_1(s) := \min_{i\in\mathcal{N}} \alpha_1^i(s) \in \mathcal{K}_\infty$. With $\hat{\mathbf{x}}_i := \text{col}(\mathbf{0}_{n_1}, \ldots, \mathbf{0}_{n_{i-1}}, \mathbf{x}_i, \mathbf{0}_{n_{i+1}}, \ldots, \mathbf{0}_{n_N})$ we have that

$$\sum_{i\in\mathcal{N}} \|\mathbf{x}_i\| = \sum_{i\in\mathcal{N}} \|\hat{\mathbf{x}}_i\| \geq \left\|\sum_{i\in\mathcal{N}} \hat{\mathbf{x}}_i\right\| = \|\mathbf{x}\|. \tag{14.15}$$

Using this property, the fact that $\tilde{\alpha}_1 \in \mathcal{K}_\infty$ is strictly increasing and (14.13) gives the desired lower bound, i.e.,

$$V(\mathbf{x}) \geq \frac{1}{N}\sum_{i\in\mathcal{N}} \alpha_1^i(\|\mathbf{x}_i\|) \geq \frac{1}{N}\tilde{\alpha}_1\left(\frac{1}{N}\|\mathbf{x}\|\right) =: \alpha_1(\|\mathbf{x}\|), \tag{14.16}$$

for all $\mathbf{x} \in \mathbb{R}^n$ and where $\alpha_1 \in \mathcal{K}_\infty$.

Next, we search for an upper bound on $V(\mathbf{x})$. For this, we first prove that $\|\mathbf{x}_i\| \leq \|\mathbf{x}\|$, $\forall \mathbf{x} = \text{col}(\{\mathbf{x}_i\}_{i\in\mathcal{N}}) \in \mathbb{R}^n$, $\forall i \in \mathcal{N}$, and any p-norm. For $1 \leq p < \infty$, the inequality follows from the definition of the p-norm:

$$\|\mathbf{x}\|_p^p := \sum_{j=1}^{n} |[\mathbf{x}]_j|^p = \sum_{l\in\mathcal{N}}\sum_{j=1}^{n_l} |[\mathbf{x}_l]_j|^p = \sum_{l\in\mathcal{N}} \|\mathbf{x}_l\|_p^p. \tag{14.17}$$

Hence

$$\|\mathrm{x}_i\|_p^p = \|\mathrm{x}\|_p^p - \sum_{l\in\{N\setminus\{i\}\}} \|\mathrm{X}\|_p^p \leq \|\mathrm{X}\|_p^p, \forall i \in N. \tag{14.18}$$

From this and the observation that $f(s) : \mathbb{R}_+ \to \mathbb{R}_+$, $f(s) := s^{\frac{1}{p}}$ and $p \geq 1$ is strictly increasing it follows that $\|\mathbf{x}_i\|_p \leq \|\mathbf{x}\|_p$ for $1 \leq p < \infty$. $f(s) : \mathbb{R}_+ \to \mathbb{R}_+$ with $f(s) := s^{\frac{1}{p}}$, for $p \geq 1$ is strictly increasing. It is straightforward to see that the inequality holds for the ∞-norm as well:

$$\|\mathbf{x}\|_\infty = \max_{j\in\mathbb{Z}_{[1,n]}} |[\mathbf{x}]_j| = \max_{l\in\mathcal{N}} \max_{j\in\mathbb{Z}_{[1,n_l]}} |[\mathbf{x}_l]_j| = \max_{l\in\mathcal{N}} \|\mathbf{x}_l\|_\infty \geq \|\mathbf{x}_i\|_\infty, \quad \forall i \in \mathcal{N}. \tag{14.19}$$

Next, using (14.9a), the fact that α_2^i is strictly increasing for all $i \in \mathcal{N}$ and (14.18), (14.19), we obtain the desired upper bound, i.e.,

$$V(\mathbf{x}) := \max_{i\in\mathcal{N}} V_i(\mathbf{x}_i) \leq \max_{i\in\mathcal{N}} \alpha_2^i(\|\mathbf{x}_i\|) \leq \max_{i\in\mathcal{N}} \alpha_2^i(\|\mathbf{x}\|) =: \alpha_2(\|\mathbf{x}\|), \tag{14.20}$$

for all $\mathbf{x} \in \mathbb{R}^n$ and where $\alpha_2 \in \mathcal{K}_\infty$.

The result now directly follows from Theorem 14.1, with $V(\mathbf{x}) := \max_{i\in\mathcal{N}} V_i(\mathbf{x}_i)$ as a CLF for the overall system. □

Notice that in Problem 14.2 the functions V_i do not need to be CLFs (in conformity with Definition 14.3) in $\mathbb{X}_i$ for each system $i \in \mathcal{N}$, respectively. Condition (14.10b) permits a *spatially* non-monotonous evolution of V_i. More precisely, the local functions are allowed to increase, as long as for each system the value of its function V_i at the next time instant is less than ρ_i times the maximum over the current values of its own function and those of its direct neighbors.

Moreover, observe that Problem 14.2 is separable in $\{\mathbf{u}_i\}_{i\in\mathcal{N}}$. The set of feasible control inputs is defined by (14.10a), which only contains inequalities at a local level. Therefore, it is possible to solve Problem 14.2 by solving N feasibility problems independently, with each problem assigned to one local controller, corresponding to one system $i \in \mathcal{N}$. In order to compute $\mathbf{u}_i(k)$, each controller needs to measure or estimate the current state $\mathbf{x}_i(k)$ of its system, and have knowledge of the interconnection signals $\{\mathbf{v}_i(\mathbf{x}_{\mathcal{B}_i})\}_{i\in\mathcal{B}_i}$ and the values $\{V_i(\mathbf{x}_i(k))\}_{i\in\mathcal{B}_i}$. Clearly, a single run of information exchange among direct neighbors per sampling instant is sufficient to acquire this knowledge. Therefore, an attractive feature of the control scheme considered in this chapter is that it can be implemented in an almost decentralized fashion.

14.4.1 Temporal Non-Monotonicity

In general, it may be difficult to find functions $\{V_i\}_{i\in\mathcal{N}}$ that satisfy (14.9a) for all $\mathbf{x}_i \in \mathbb{X}_i$. Systematic methods for synthesizing CLFs for an arbitrary nonlinear system do not exist, although candidate CLFs can often be generated using linearized system dynamics. However, the region of validity for these CLFs is often limited to a neighborhood of the origin. Supposing that we have a set of structured max-CLFs in $\widetilde{\mathbb{X}} \subset \mathbb{X}$, we propose a method to relax the conditions on the candidate CLFs, based on an adaptation of the Lyapunov-Razumikhin (LR) technique for time-delay systems [9]. The LR method allows the Lyapunov function to be non-monotonous in order to compensate for the effects of the delay. Next, we show how the LR technique can be applied to discrete-time systems as well, to permit a *temporal* non-monotonous evolution of the candidate CLF for the full network.

Problem 14.3 Let $N_\tau \in \mathbb{Z}_{\geq 1}$ be given. Consider Problem 14.2 for a set of "structured max-CLFs" $\{V_i\}_{i\in\mathcal{N}}$ in $\widetilde{\mathbb{X}} \subset \mathbb{X}$, with (14.10b) replaced by

$$V_i(\phi_i(\mathbf{x}_i(k), \mathbf{u}_i(k), \mathbf{v}_i(\mathbf{x}_{\mathcal{B}_i}(k)))) \leq \rho_i \max_{\tau\in\mathbb{Z}_{[0,N_\tau-1]}} \max_{j\in\overline{\mathcal{B}}_i} V_j(\mathbf{x}_j(k-\tau)), \tag{14.21}$$

for all $k \in \mathbb{Z}_{\geq} N_\tau - 1$ and $i \in \mathcal{N}$. ■

Let $\bar{\pi}(\mathbf{x}(k)) := \{\, \text{col}(\{\mathbf{u}_i(k)\}_{i\in\mathcal{N}}) \mid (13.10\text{a}) \text{ and } (13.21) \text{ hold} \,\}$ and let

$$\mathbf{x}(k+1) \in \bar{\phi}_{\mathrm{CL}}(\mathbf{x}(k), \bar{\pi}(\mathbf{x}(k))) := \{\phi(\mathbf{x}(k), \mathbf{u}(k)) \mid \mathbf{u}(k) \in \bar{\pi}(\mathbf{x}(k))\} \tag{14.22}$$

denote the difference inclusion corresponding to system (14.7) in "closed loop" with the set of feasible solutions obtained by solving Problem 14.3 at each instant $k \in \mathbb{Z}_+$.

Theorem 14.3 *Let $\alpha_1^i, \alpha_2^i \in \mathcal{K}_\infty$, $N_\tau \in \mathbb{Z}_{\geq 1}$ and $\rho_i \in \mathbb{R}_{[0,1)}$, $\forall i \in \mathcal{N}$ be given and choose a set of structured max-CLFs $\{V_i\}_{i\in\mathcal{N}}$ in $\widetilde{\mathbb{X}} \subseteq \mathbb{X} = \{\mathrm{col}(\{\mathbf{x}_i\}_{i\in\mathcal{N}}) \mid \mathbf{x}_i \in \mathbb{X}_i\}$ for system* (14.7)*. Suppose that Problem 14.3 is feasible for all $\mathbf{x}(k) \in \mathbb{X}$, all $k \in \mathbb{Z}_+$ and the corresponding signals $\{\mathbf{v}_i(\mathbf{x}_{\mathcal{B}_i}(k))\}_{i\in\mathcal{N}}$. Then* (14.22) *is asymptotically stable in $\mathbb{X}$.*

Proof Let $\mathbf{x}(k) \in \mathbb{X}$ for some $k \in \mathbb{Z}_+$. Positive invariance of $\mathbb{X}$ follows from feasibility of (14.10a), as shown in the proof of Theorem 14.2. Now consider the function $V(\mathbf{x}) := \max_{i\in\mathcal{N}} V_i(\mathbf{x}_i)$. Condition (14.21) implies that

$$\begin{aligned} V(\mathbf{x}(k+1)) &= \max_{i\in\mathcal{N}} V_i(\mathbf{x}_i(k+1)) \\ &\leq \rho \max_{i\in\mathcal{N}} \max_{\tau\in\mathbb{Z}_{[0,N_\tau-1]}} \max_{j\in\overline{\mathcal{B}}_i} V_j(\mathbf{x}_j(k-\tau)) \\ &= \rho \max_{\tau\in\mathbb{Z}_{[0,N_\tau-1]}} \max_{i\in\mathcal{N}} V_i(\mathbf{x}_i(k-\tau)) = \rho \max_{\tau\in\mathbb{Z}_{[0,N_\tau-1]}} V(\mathbf{x}(k-\tau)), \end{aligned} \tag{14.23}$$

for all $k \in \mathbb{Z}_{\geq N_\tau - 1}$ and $\rho := \max_{i\in\mathcal{N}} \rho_i \in \mathbb{R}_{[0,1)}$. Recursive application of (14.23) gives

$$V(\mathbf{x}(k)) \leq \rho^{\left\lfloor \frac{k}{N_\tau} \right\rfloor} \max_{l\in\mathbb{Z}_{[0,N_\tau-1]}} V(\mathbf{x}(l)), \quad \forall k \in \mathbb{Z}_+. \tag{14.24}$$

Furthermore, we derived in the proof of Theorem 14.2 that

$$\alpha_1(\|\mathbf{x}\|) \leq V(\mathbf{x}) \leq \alpha_2(\|\mathbf{x}\|), \quad \forall \mathbf{x} \in \mathbb{R}^n, \tag{14.25}$$

with $\alpha_1(s) := \frac{1}{N} \min_{i\in\mathcal{N}} \alpha_1^i(\frac{1}{N}s) \in \mathcal{K}_\infty$ and $\alpha_2(s) := \max_{i\in\mathcal{N}} \alpha_2^i(s) \in \mathcal{K}_\infty$. As $\mathcal{K}_\infty$-functions are strictly increasing, we know that

$$\max_{l\in\mathbb{Z}_{[0,N_\tau-1]}} V(\mathbf{x}(l)) \leq \max_{l\in\mathbb{Z}_{[0,N_\tau-1]}} \alpha_2(\|\mathbf{x}(l)\|) = \alpha_2(\|\underline{\mathbf{x}}_{[0,N_\tau-1]}\|). \tag{14.26}$$

Combining this bound with (14.24), (14.25) gives $\|\mathbf{x}(k)\| \leq \alpha_1^{-1}(\rho^{\left\lfloor \frac{k}{N_\tau} \right\rfloor} \alpha_2(\|\underline{\mathbf{x}}_{[0,N_\tau-1]}\|))$. The fact that $\lim_{k\to\infty} \|\mathbf{x}(k)\| \leq \lim_{k\to\infty} \alpha_1^{-1}(\rho^{\left\lfloor \frac{k}{N_\tau} \right\rfloor} \alpha_2(\|\underline{\mathbf{x}}_{[0,N_\tau-1]}\|)) = 0$ proves attractivity of the closed-loop system (14.22). Lyapunov stability follows, as for every $\varepsilon > 0$ we can find a $\delta(\varepsilon) := \alpha_2^{-1}(\alpha_1(\varepsilon)) > 0$, such that $\|\underline{\mathbf{x}}_{[0,N_\tau-1]}\| < \delta$ implies that $\|\mathbf{x}(k)\| < \alpha_1^{-1}(\rho^{\left\lfloor \frac{k}{N_\tau} \right\rfloor} \alpha_2(\delta)) \leq \varepsilon$ for all $k \in \mathbb{Z}_+$. This proves asymptotic stability of (14.22) in $\mathbb{X}$. with $V(x) := \max_{i\in\mathcal{N}} V_i(\mathbf{x}_i)$ as candidate CLF. □

Algorithm 14.1 Almost-decentralized MPC

At each instant $k \in \mathbb{Z}_+$ and node $i \in \mathcal{N}$:

1: Measure or estimate the current local state $\mathbf{x}_i(k)$ and transmit $\mathbf{v}_{ji}(\mathbf{x}_i(k))$ and $V_i(\mathbf{x}_i(k))$ to nodes $\{j \in \mathcal{N} \mid i \in \mathcal{B}_j\}$.

2: Specify the set of feasible local control actions $\bar{\pi}_i(\mathbf{x}_i(k), \mathbf{v}_i(\mathbf{x}_{\mathcal{B}_i}(k))) := \{\mathbf{u}_i(k) \mid$ (14.10a) and (14.21) hold$\}$. Minimize the cost $J_i(\mathbf{u}_i(k), \mathbf{x}_i(k))$ over the set of actions $\bar{\pi}_i(\mathbf{x}_i(k), \mathbf{v}_i(\mathbf{x}_{\mathcal{B}_i}(k)))$ and denote the optimizer by $\mathbf{u}_i^*(k)$;

3: Use $\mathbf{u}_i(k) = \mathbf{u}_i^*(k)$ as control action.

For more details on Theorem 14.3, the interested reader is referred to [4]. The distinctive feature of Problem 14.3 is that it allows the trajectories of the local functions $V_i(\mathbf{x}_i(k))$ to be non-monotonous, and relaxes the convergence condition on the candidate CLF for the overall network, i.e., $V(\mathbf{x}(k))$, as well. The evolution of $V(\mathbf{x}(k))$ can be arbitrary, as long as it remains within the asymptotically converging envelope generated by (14.21) and the first N_τ values of $V(\mathbf{x}(k))$.

Note that if we combine Problem 14.2 or Problem 14.3 with the optimization of a set of local cost functions, the feasibility-based stability guarantee and the possibility of an almost decentralized implementation still hold. This enables the formulation of a one-step-ahead predictive control algorithm in which stabilization is decoupled from performance, and in which the controllers do not need to attain the global optimum at each sampling instant, as typically required for stability in classical MPC [11].

For the remainder of the article we therefore consider the almost-decentralized MPC algorithm given as Algorithm 14.1, supposing that a set of local objective functions $\{J_i(\mathbf{u}_i(k), \mathbf{x}_i(k))\}_{i \in \mathcal{N}}$ is known.

Remark 14.1 In Algorithm 14.1, each controller optimizes its own local performance. However, many distributed MPC schemes (see, for instance, [2, 15]) optimize a global cost function (e.g., some convex combination of local objectives) and aim for optimization of global performance by employing network-wide information or iterations. Therein, stability is attained by assuming optimality (for example, in [15]) or by imposing a contractive constraint on the norm of the local states (e.g., in [2]). The L-MPC conditions proposed in this chapter can be used in those implementations as well, as an alternative way to achieve stability that is less conservative than contractive constraints, while time-consuming iterations would only be required for improving global performance.

Remark 14.2 The dependence of conditions (14.10b) and (14.21) on $V_j(\cdot)$ for j in the set $\overline{\mathcal{B}}_i$ is convenient, because the corresponding communication graph coincides with the graph that characterizes the exchange of $\mathbf{v}_{ij}(\mathbf{x}_j)$ as required by the local state predictions, i.e., (14.6). However, in fact, stability also follows if (14.10a) and

$$V_i(\phi_i(\mathbf{x}_i(k), \mathbf{u}_i(k), \mathbf{v}_i(\mathbf{x}_{\mathcal{B}_i}(k)))) \leq \rho_i \max_{\tau \in \mathbb{Z}_{[0,N_\tau-1]}} \max_{j \in \mathcal{P}_i} V_j(\mathbf{x}_j(k-\tau)), \forall k \in \mathbb{Z}_{\geq N_\tau - 1}, \tag{14.27}$$

hold, where $\{\mathcal{P}_i\}_{i\in\mathcal{N}}$ is an *arbitrary* collection of non-empty sets such that

$$\bigcup_{i\in\mathcal{N}}\mathcal{P}_i = \mathcal{N}. \tag{14.28}$$

As $\max_{j\in\overline{\mathcal{B}}_i} V_j \leq \max_{j\in\mathcal{P}_i} V_j$ for $\overline{\mathcal{B}}_i \subseteq \mathcal{P}_i$, enlarging the set over which the maximum V_j value is evaluated is an attractive method of relaxing the local CLF convergence conditions if this is supported by the available communication infrastructure.

Analogously, it can be shown that Problems 14.2 and 14.3 provide some robustness against communication failures such as packet loss. As long as the perturbed communication graph is characterized by a collection of sets $\{\mathcal{P}_i\}_{i\in\mathcal{N}}$ that satisfies (14.28), the asymptotic stability guarantee continues to hold.

14.4.2 Implementation Issues

In what follows, we consider nonlinear NDS that are affine in the control input, i.e.,

$$\begin{aligned}\mathbf{x}_i(k+1) &= \boldsymbol{\phi}_i(\mathbf{x}_i(k), \mathbf{u}_i(k), \mathbf{v}_i(\mathbf{x}_{\mathcal{B}_i}(k)))\\ &= \mathbf{f}_i(\mathbf{x}_i(k), \mathbf{v}_i(\mathbf{x}_{\mathcal{B}_i}(k))) + \mathbf{g}_i(\mathbf{x}_i(k), \mathbf{v}_i(\mathbf{x}_{\mathcal{B}_i}(k)))\mathbf{u}_i(k),\end{aligned} \tag{14.29}$$

with $\mathbf{f}_i : \mathbb{R}^{n_i} \times \mathbb{R}^{n_{v_i}} \to \mathbb{R}^{n_i}$, $\mathbf{f}_i(\mathbf{0}_{n_i}, \mathbf{0}_{n_{v_i}}) = \mathbf{0}_{n_i}$, $\mathbf{g}_i : \mathbb{R}^{n_i} \times \mathbb{R}^{n_{v_i}} \to \mathbb{R}^{n_i\times m_i}$ and $\mathbf{g}_i(\mathbf{0}_{n_i}, \mathbf{0}_{n_{v_i}}) = 0$ for all $i \in \mathcal{N}$. For these systems and polytopic state and input sets $\mathbb{X}_i$ and $\mathbb{U}_i$, respectively, it is possible to implement step 2 of Algorithm 14.1 by solving a single linear program, without introducing conservatism.

For this, we restrict our attention to infinity-norm based structured CLFs, i.e.,

$$V_i(\mathbf{x}_i) = \|\mathbf{P}_i\mathbf{x}_i\|_\infty, \tag{14.30}$$

where $\mathbf{P}_i \in \mathbb{R}^{p_i\times n_i}$ is a full-column rank matrix. Note that this type of structured max-CLFs satisfies (14.9a), for $\alpha_1^i(s) := \frac{\sigma_{\mathbf{P}_i}}{\sqrt{p_i}}s$, where $\sigma_{\mathbf{P}_i} > 0$ is the smallest singular value of $\mathbf{P}_i$, and $\alpha_2^i(s) := \|\mathbf{P}_i\|_\infty s$.

By definition of the infinity norm, for $\|\mathbf{x}\|_\infty \leq c$ to be satisfied for some vector $\mathbf{x} \in \mathbb{R}^n$ and constant $c \in \mathbb{R}$, it is necessary and sufficient to require that $\pm[\mathbf{x}]_j \leq c$ for all $j \in \mathbb{Z}_{[1,n]}$.

So, for (14.21) to be satisfied, it is necessary and sufficient to require that

$$\pm[\mathbf{P}_i\{\mathbf{g}_i(\mathbf{x}_i(k), \mathbf{v}_i(\mathbf{x}_{\mathcal{B}i}(k)))\mathbf{u}_i(k)\}]_j \leq \zeta_i(k) \mp [\mathbf{P}_i\{\mathbf{f}_i(\mathbf{x}_i(k), \mathbf{v}_i(\mathbf{x}_{\mathcal{B}i}(k)))\}]_j, \tag{14.31}$$

for $j \in \mathbb{Z}_{[1,p_i]}$ and $k \in \mathbb{Z}_{\geq N_\tau - 1}$, and where

$$\zeta_i(k) := \rho_i \max_{\tau\in\mathbb{Z}_{[0,N_\tau-1]}} \max_{j\in\overline{\mathcal{B}}_i} V_j(\mathbf{x}_j(k-\tau)) \in \mathbb{R}_+ \tag{14.32}$$

is constant at any $k \in \mathbb{Z}_{\geq N_\tau - 1}$. This yields a total of $2p_i$ linear inequalities in $\mathbf{u}_i$.

Then, by choosing infinity-norm based local cost functions of the form

$$J_i(\mathbf{x}_i, \mathbf{u}_i) := \|\mathbf{Q}_1^i \phi_i(\mathbf{x}_i, \mathbf{u}_i, \mathbf{v}_i(\mathbf{x}_{\mathcal{B}_i}))\|_\infty + \|\mathbf{Q}_0^i \mathbf{x}_i\|_\infty + \|\mathbf{R}^i \mathbf{u}_i\|_\infty, \tag{14.33}$$

with full-rank matrices $\mathbf{Q}_1^i \in \mathbb{R}^{n_{q1,i} \times n_i}$, $\mathbf{Q}_0^i \in \mathbb{R}^{n_{q0,i} \times n_i}$ and $\mathbf{R}^i \in \mathbb{R}^{n_{r_i} \times m_i}$, we can reformulate step 2 of Algorithm 14.1 as the linear program

$$\min_{\mathbf{u}_i(k), \varepsilon_1, \varepsilon_2} \quad \varepsilon_1 + \varepsilon_2 \tag{14.34}$$

subject to (14.10a), (14.31) and

$$\pm \left[\mathbf{Q}_1^i \phi_i(\mathbf{x}_i(k), \mathbf{u}_i(k), \mathbf{v}_i(\mathbf{x}_{\mathcal{B}_i}(k)))\right]_j + \|\mathbf{Q}_0^i \mathbf{x}_i(k)\|_\infty \leq \varepsilon_1$$
$$\pm [\mathbf{R}^i \mathbf{u}_i(k)]_l \leq \varepsilon_2,$$

for $j \in \mathbb{Z}_{[1, n_{q_{1,i}}]}$ and $l \in \mathbb{Z}_{[1, n_{r_i}]}$.

14.5 Illustrative Examples

Consider NDS (14.6) with $\mathcal{S} = \{\varsigma_1, \varsigma_2\}$, $\mathcal{B}_1 = \{2\}$, $\mathcal{B}_2 = \{1\}$, $\mathbb{X}_1 = \mathbb{X}_2 = \{\mathbf{x} \in \mathbb{R}^2 \mid \|\mathbf{x}\|_\infty \leq 5\}$ and $\mathbb{U}_1 = \mathbb{U}_2 = \{u \in \mathbb{R} \mid |u| \leq 2\}$. Its dynamics are given by:

$$\phi_1(\mathbf{x}_1, u_1, \mathbf{v}_1(\mathbf{x}_{\mathcal{B}_1})) := \begin{bmatrix} 1 & 0.7 \\ 0 & 1 \end{bmatrix} \mathbf{x}_1 + \begin{bmatrix} \sin([\mathbf{x}_1]_2) \\ 0 \end{bmatrix} + \begin{bmatrix} 0.245 \\ 0.7 \end{bmatrix} u_1 + \begin{bmatrix} 0 \\ ([\mathbf{x}_2]_1)^2 \end{bmatrix}, \tag{14.35a}$$

$$\phi_2(\mathbf{x}_2, u_2, \mathbf{v}_2(\mathbf{x}_{\mathcal{B}_2})) := \begin{bmatrix} 1 & 0.5 \\ 0 & 1 \end{bmatrix} \mathbf{x}_2 + \begin{bmatrix} \sin([\mathbf{x}_2]_2) \\ 0 \end{bmatrix} + \begin{bmatrix} 0.125 \\ 0.5 \end{bmatrix} u_2 + \begin{bmatrix} 0 \\ [\mathbf{x}_1]_1 \end{bmatrix}. \tag{14.35b}$$

The method of [8] was used to compute the weights $\mathbf{P}_1, \mathbf{P}_2 \in \mathbb{R}^{2\times 2}$ of two local infinity-norm based candidate CLFs, i.e., $V_1(\mathbf{x}_1) = \|\mathbf{P}_1 \mathbf{x}_1\|_\infty$ and $V_2(\mathbf{x}_2) = \|\mathbf{P}_2 \mathbf{x}_2\|_\infty$ with $\rho = \rho_1 = \rho_2 = 0.8$ and linearizations of (14.35a), (14.35b), respectively, around the origin, in closed-loop with the local state-feedback laws $u_1(k) := \mathbf{K}_1 \mathbf{x}_1(k)$, $u_2(k) := \mathbf{K}_2 \mathbf{x}_2(k)$, $\mathbf{K}_1, \mathbf{K}_2 \in \mathbb{R}^{1\times 2}$, yielding

$$\mathbf{P}_1 = \begin{bmatrix} 2.56 & 0.33 \\ 1.86 & 5.02 \end{bmatrix}, \mathbf{K}_1 = \begin{bmatrix} -0.97 \\ -2.12 \end{bmatrix}^{\mathrm{T}}, \mathbf{P}_2 = \begin{bmatrix} -0.39 & -0.38 \\ 0.27 & 0.98 \end{bmatrix}, \mathbf{K}_2 = \begin{bmatrix} -0.39 \\ -2.78 \end{bmatrix}^{\mathrm{T}}.$$

Note that the control laws $u_1(k) = \mathbf{K}_1 \mathbf{x}_1(k)$ and $u_2(k) = \mathbf{K}_2 \mathbf{x}_2(k)$ are only employed off-line, to calculate the weight matrices $\mathbf{P}_1, \mathbf{P}_2$ and they are not used for controlling

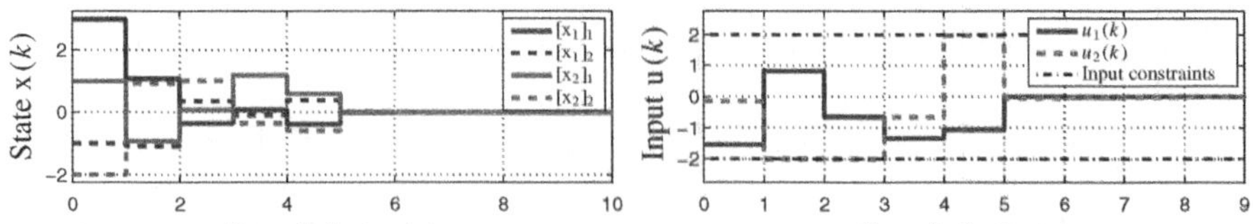

Fig. 14.1 State and input trajectories for $\rho = 0.8$ and $N_{\mathcal{T}} = 3$

the system. For each system $i \in \mathcal{N}$, we employed cost function (14.33), where $\mathbf{Q}_1^1 = \mathbf{Q}_1^2 = 4\mathbf{I}_2$, $\mathbf{Q}_0^1 = \mathbf{Q}_0^2 = 0.1\mathbf{I}_2$ and $\mathbf{R}^1 = \mathbf{R}^2 = 0.4$.

In the simulation scenario we tested the system response in closed-loop with Algorithm 14.1 for $\mathbf{x}_1(0) = [3, \ -1]^{\mathrm{T}}$, $\mathbf{x}_2(0) = [1, \ -2]^{\mathrm{T}}$ and $N_{\mathcal{T}} = 3$. Figure 14.1 shows the state and control inputs for systems 1 and 2, along with the input constraints that are represented by the dash-dotted lines. Note that these constraints are not violated, although they are active at some time instants. The corresponding evolution of $V_1(\mathbf{x}_1(k))$, $V_2(\mathbf{x}_2(k))$, $V(\mathbf{x}(k))$ and the upper bounds generated by (14.21) are shown in Fig. 14.2. The simulation illustrates that $V(\mathbf{x}(k))$ is allowed to vary arbitrarily within the asymptotically converging envelope defined by (14.21), thus inducing state convergence. Moreover, the proposed L-MPC algorithm allows a spatially non-monotonous evolution of the structured max-CLFs (at time instant $k = 2$, $V_2(\mathbf{x}_2(k))$ increases although $V(\mathbf{x}(k))$ does not), whereas the candidate CLF itself can be non-decreasing as well (which is the case for $k = 4$). The closed-loop convergence rate is similar to the one attained by the method in [6] for the same example and initial conditions. However, the technique in [6] requires global coordination and iterative optimization to guarantee closed-loop stability, whereas the method proposed in this work does not.

Next, we briefly consider almost-decentralized L-MPC as a possible solution to the *frequency-control problem*, i.e., the problem of asymptotically stabilizing the frequency or *supply-demand balance* in large-scale electrical power systems [3, 15]. This problem is of particular interest due to the liberalization of the electrical energy market and the increasing share of intermittent, distributed generation. As a result, there is a growing need for advanced, non-centralized control methods that are able to guarantee network stability even under competitive market operation.

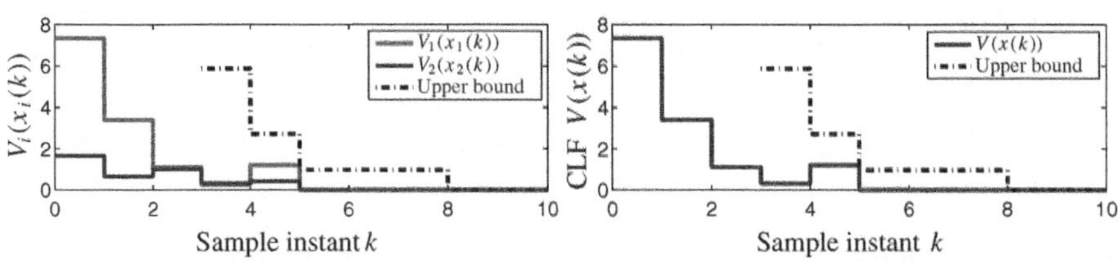

Fig. 14.2 $V_1(\mathbf{x}_1(k))$, $V_2(\mathbf{x}_2(k))$ and $V(\mathbf{x}(k))$ trajectories for $\rho = 0.8$ and $N_{\mathcal{T}} = 3$

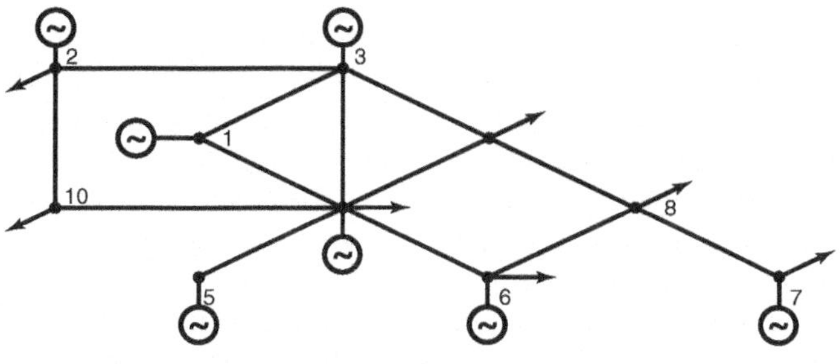

Fig. 14.3 Single-line representation of the seven-machine CIGRÉ test system

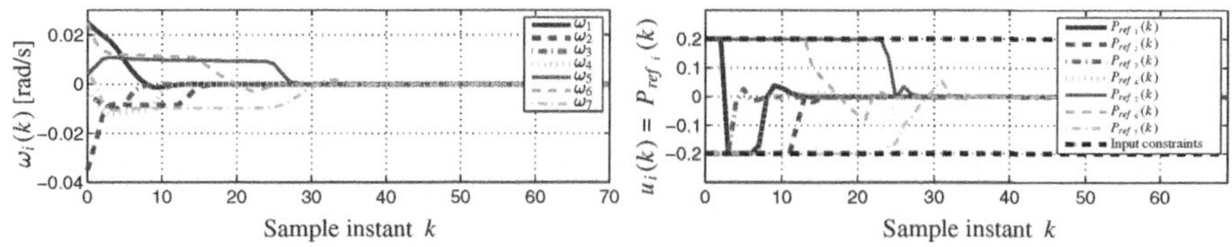

Fig. 14.4 Frequency and control action trajectories

To evaluate the suitability of L-MPC for frequency control, Algorithm 14.1 was simulated in closed-loop with the 7-machine CIGRÉ benchmark system [3], see Fig. 14.3. Each of the $N = 7$ generators was modeled by a linear state space model with state $x_i := [\delta_i\ \omega_i\ P_{M_i}\ P_{G_i}]^T$ (rotor/voltage phase angle and frequency, turbine and governor states). The corresponding discrete-time prediction model is given by

$$\mathbf{x}_i(k+1) = \phi_i(\mathbf{x}_i(k), u_i(k), \mathbf{v}_i(\mathbf{x}_{\mathcal{B}_i}(k))) := \mathbf{A}_{ii}\mathbf{x}_i(k) + \mathbf{B}_{ii}u_i(k) + \mathbf{v}_i(\mathbf{x}_{\mathcal{B}_i}(k)), \tag{14.36}$$

for $i \in \mathcal{N} := \mathbb{Z}_{[1,7]}$, $\mathbf{x}_i \in \mathbb{R}^4$, $u_i \in \mathbb{R}$, $\mathbf{v}_i(\mathbf{x}_{\mathcal{B}_i}) := \sum_{j \in \mathcal{B}_i} \mathbf{A}_{ij}\mathbf{x}_j$, $\mathbf{A}_{ij} \in \mathbb{R}^{n_i \times n_j}$, $\mathbf{B}_{ii} \in \mathbb{R}^{n_i}$.

Closed-loop performance was evaluated when recovering from a state perturbation/imbalance $\mathbf{x}(0) \neq \mathbf{0}_n$. The control actions were constrained as $-0.2 \leq u_i \leq 0.2$, $i \in \mathcal{N}$. Figure 14.4 shows the generator frequencies ω_i, along with the corresponding control inputs u_i, $i \in \mathcal{N}$. Clearly, all the ω_i converge to 0, and the input constraints are respected at all time instants. This is achieved despite the non-cooperative nature of the generators, while requiring local exchange of information only.

For more details on the above example, the interested reader is referred to [5].

14.6 Conclusions

This chapter proposed an almost decentralized solution to the problem of stabilizing a network of discrete-time nonlinear systems with coupled dynamics that are subject to local state/input constraints. By "almost decentralized" we mean that each local controller is allowed to use the states of neighboring systems for feedback, whereas it is not permitted to employ iterations between the systems in the network to compute the control action. The stability conditions were decentralized via a set of structured control Lyapunov functions for which the maximum over all the functions in the set is a CLF for the overall network of systems. However, this does not necessarily imply that each function is a local CLF for its corresponding system. Additionally, we provided a solution for relaxing the temporal monotonicity of the CLF for the overall network. Two non-trivial examples illustrated the effectiveness of the developed scheme and demonstrated that the proposed L-MPC technique can perform as well as more complex distributed, iteration-based MPC algorithms.

Acknowledgments This research is supported by the Veni grant "Flexible Lyapunov Functions for Real-time Control", grant number 10230, awarded by STW (Dutch Science Foundation) and NWO (The Netherlands Organization for Scientific Research), and is part of the EOS-LT Regelduurzaam project (funded by the Dutch Ministry of Economic Affairs) and the European Commission Research Project FP7–ICT–249096 "Price-based Control of Electrical Power Systems" (E-Price).

References

1. A. Bemporad, A predictive controller with artificial Lyapunov function for linear systems with input/state constraints. Automatica **34**(10), 1255–1260 (1998)
2. E. Camponogara, D. Jia, B.H. Krogh, S. Talukdar, Distributed model predictive control. IEEE Control Syst. Mag. **22**(1), 44–52 (2002)
3. R.M. Hermans, A. Jokić, M. Lazar, A. Alessio, P.P.J. van den Bosch, I.A. Hiskens, A. Bemporad, Assessment of non-centralised model predictive control techniques for electrical power networks. Int. J. Control **85**(8), 1162–1177 (2012)
4. R.M. Hermans, M. Lazar, A. Jokić, Almost decentralized Lyapunov-based nonlinear model predictive control, in *29th American Control Conference* (Baltimore, MD, 2010), pp. 3932–3938
5. R.M. Hermans, M. Lazar, A. Jokić, Distributed predictive control of the 7-machine CIGRÉ power system, in *30th American Control Conference*(San Francisco, CA, 2011), pp. 5225–5230
6. A. Jokić, M. Lazar, On decentralized stabilisation of discrete-time nonlinear systems. In *28th American Control Conference* (St. Louis, MO, 2009), pp. 5777–5782
7. C.M. Kellett, A.R. Teel, On the robustness of $\mathcal{KL}$-stability for difference inclusions: smooth discrete-time Lyapunov functions. SIAM J. Control Optim. **44**(3), 777–800 (2005)
8. M. Lazar, W.P.M.H. Heemels, S. Weiland, A. Bemporad, Stabilizing model predictive control of hybrid systems. IEEE Trans. Autom. Control **51**(11), 1813–1818 (2006)
9. B. Liu, H.J. Marquez, Uniform stability of discrete delay systems and synchronization of discrete delay dynamical networks via Razumikhin technique. IEEE Trans. Circuits Syst. I: Regul. Pap. **55**(9), 2795–2805 (2008)
10. J. Liu, D. Muñoz de la Peña, P.D. Christofides, J.F. Davis, Lyapunov-based model predictive control of nonlinear systems subject to time-varying measurement delays. Int. J. Adapt. Control Signal Process. **23**(8), 788–807 (2009)
11. D.Q. Mayne, J.B. Rawlings, C.V. Rao, P.O.M. Scokaert, Constrained model predictive control: stability and optimality. Automatica **36**(6), 789–814 (2000)
12. D.M. Raimondo, P. Hokayem, J. Lygeros, M. Morari, An iterative decentralized MPC algorithm for large-scale nonlinear systems, in *Proceedings of the 1st IFAC Workshop on Estimation and Control of Networked Systems* (Venice, Italy, 2009), pp. 162–167

13. A. Richards, J.P. How, Robust distributed model predictive control. Int. J. Control **80**(9), 1517–1531 (2007)
14. E.D. Sontag, A Lyapunov-like characterization of asymptotic controllability. SIAM J. Control Optim. **21**, 462–471 (1983)
15. A.N. Venkat, I.A. Hiskens, J.B. Rawlings, S.J. Wright, Distributed MPC strategies with application to power system automatic generation control. IEEE Trans. Control Syst. Technol. **16**(6), 1192–1206 (2008)

Chapter 15
A Distributed Reference Management Scheme in Presence of Non-Convex Constraints: An MPC Based Approach

F. Tedesco, D. M. Raimondo and A. Casavola

Abstract This chapter deals with distributed coordination problems which include the fulfillment of non-convex constraints. A Distributed Command Governor (D-CG) strategy is here proposed to coordinate a set of dynamically decoupled subsystems. The approach results in a receding horizon strategy that requires the computation of mixed-integer optimization programs.

15.1 Short Introduction

In this chapter we consider the problem of satisfying a set of coordination tasks over a collection of dynamically decoupled subsystems. As a solution we propose a distributed coordination strategy based on *Command Governor* (CG). Consider Fig. 15.1.

Each agent is in charge of supervising one specific subsystem. Variables, $\mathbf{r}_i$, $\mathbf{g}_i$, $\mathbf{x}_i$, $\mathbf{y}_i$ and $\mathbf{c}_i$ represent respectively: the nominal references, the applied references, the states, the performance-related and the coordination-related outputs of the slave systems. Moreover, the agents are able to exchange the state $\mathbf{x}_i$ and the applied references $\mathbf{g}_i$ with the other agents via a data-network.

In such a context, the coordination task can be expressed as the requirement of satisfying some tracking performance, viz. $\mathbf{y}_i \approx \mathbf{r}_i$, while at the same time enforcing some pointwise-in-time constraints on the coordination-related outputs $\mathbf{c}_i$ of each

F. Tedesco · A. Casavola (✉)
Università degli Studi della Calabria, DEIS, Rende (CS), Calabria, Italy
e-mail: casavola@deis.unical.it

F. Tedesco
e-mail: tedesco@deis.unical.it

D. M. Raimondo
Università degli Studi di Pavia, Dipartimento di Informatica e Sistemistica, Pavia, Italy
e-mail: davide.raimondo@unipv.it

J. M. Maestre and R. R. Negenborn (eds.), *Distributed Model Predictive Control Made Easy*, Intelligent Systems, Control and Automation: Science and Engineering 69,
DOI: 10.1007/978-94-007-7006-5_15, © Springer Science+Business Media Dordrecht 2014

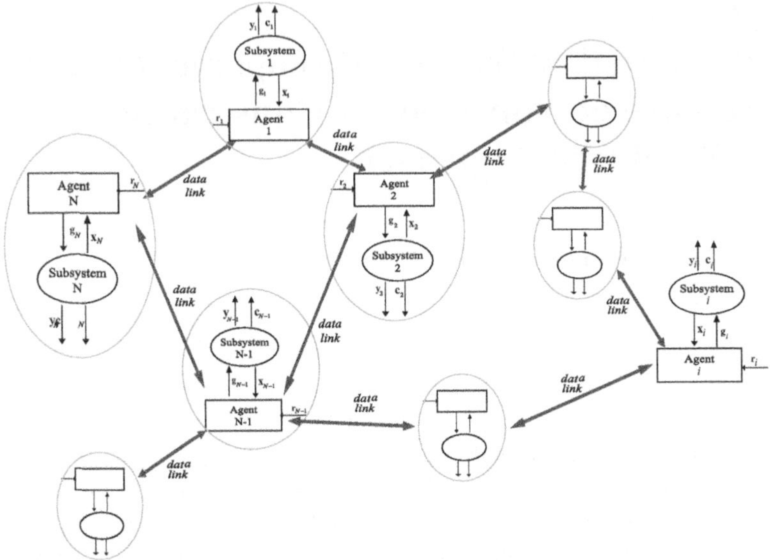

Fig. 15.1 Multi-agent distributed architecture

slave system and/or of the overall network. Each subsystem is provided with a primary controller in charge of stabilizing the system and ensuring good tracking performance as depicted in Fig. 15.2. The agents are responsible of modifying the nominal local references $\mathbf{r}_i$ into feasible ones $\mathbf{g}_i$, when the tracking of the nominal references would produce constraint violations.

The CG unit is a nonlinear device which is added to a plant, regulated by a primal controller that is separately designed to guarantee stability and tracking performance requirements. Its main objective is that of modifying, whenever necessary, the reference signal to be supplied to the pre-compensated system in order to avoid constraint violations. This modification is typically achieved according to a receding horizon philosophy consisting in solving on-line at each time instant a constrained optimization problem. As in MPC schemes, only the first sample of such a sequence is applied and the entire procedure is repeated at the next time instant.

It is worth remarking that the CG just solves a reference management problem and it does not deal with stability issues already demanded to the primal controller. Hence the CG can be seen as a simplified MPC that does not have to stabilize the plant. In this respect, Fig. 15.2 illustrates a CG scheme in the distributed context depicted in Fig. 15.1 seen from an MPC perspective.

Actually the joint action of CG and the local primal controller is "equivalent" to a particular MPC that computes inputs $\mathbf{u}_i$ and references $\mathbf{g}_i$ on the basis of states $\mathbf{x}_i^p$ and system constraints. A CG-based approach could be preferable than a MPC-based strategy when a primal controller pre-compensating the system already exists and one is not allowed to modify or replace it.

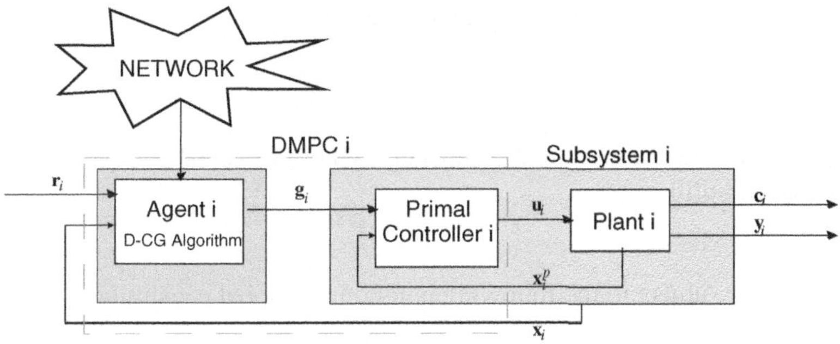

Fig. 15.2 CG and MPC schemes equivalence in a distributed context

In such a context, Command Governor represents an interesting (even though suboptimal because of the fixed primal controller) option and a distributed CG-framework allows one to deal with large-scale networked systems with a relatively low computational load. To this end, in this chapter a distributed sequential non-iterative CG-based strategy (D-CG) is presented which is capable of handling non-convex constraints. Such a problem involves a mixture of continuous and integer variables and requires the solution of mixed-integer programs. The method proposed in this chapter can be applied to distributed vehicle coordination problems where collision avoidance is required.

15.2 Boundary Conditions

15.2.1 Subsystems Description and Assumptions

Let us consider a set $\mathcal{N} = \{1, \ldots, |\mathcal{N}|\}$ of closed-loop dynamical subsystems. Each subsystem is regulated by a local controller which ensures stability and tracking performance when the constraints are supposedly not active (small-signal regime). Let the generic i-th subsystem be described by the following discrete-time model

$$\begin{cases} \mathbf{x}_i(k+1) = \mathbf{A}_{ii}\mathbf{x}_i(k)+\mathbf{B}_i\mathbf{g}_i(k) \\ \mathbf{y}_i(k) \quad = \mathbf{C}_i^y\mathbf{x}_i(k) \\ \mathbf{z}_i(k) \quad = \mathbf{C}_i^z\mathbf{x}_i(k) \\ \mathbf{c}_i(k) \quad = \mathbf{C}_i^c\mathbf{x}_i(k) + \mathbf{L}_i\mathbf{g}_i(k) \end{cases} \tag{15.1}$$

where: $k \in \mathbb{Z}_+$, $\mathbf{x}_i$ is the state vector (which includes the controller states under dynamic regulation), $\mathbf{g}_i(k)$ the manipulable input that represents a reference rather a control signal and $\mathbf{y}_i(k) \in \mathbb{R}^{n_{\mathbf{y}_i}} \equiv \mathbb{R}^{n_{\mathbf{g}_i}}$ is a performance related vector. Please notice that the apparent restriction $\mathbb{R}^{n_{\mathbf{y}_i}} \equiv \mathbb{R}^{n_{\mathbf{g}_i}}$ is not dictated by any technical reason and it could be removed without consequence. It is usually adopted in the CG literature for practical reasons because reasonable in many cases. Finally, $\mathbf{c}_i$

represents a constrained vector which has to fulfill constraints of the form

$$\mathbf{c}_i(k) \in \mathcal{C}_i, \ \forall k \in \mathbb{Z}_+, \tag{15.2}$$

with $\mathcal{C}_i$ representing a convex and compact polytopic set. Finally, the output $\mathbf{z}_i(k)$ is constrained to fulfill

$$\mathbf{z}_i(k) \in \Omega \subset \mathbb{R}^{n_{\mathbf{z}_i}} \tag{15.3}$$

and it is further subject to the following non-convex global constraints

$$||\mathbf{z}_i(k) - \mathbf{z}_j(k)||_\infty \geq d \ \ \forall i, j (i \neq j) \ \text{ and } \ k \in \mathbb{Z}_+, \tag{15.4}$$

which explicitly involve the outputs of all other subsystems. More formally, depending on the variable $\mathbf{z}_j(k)$, $j \in \mathcal{N} \setminus \{i\}$ related to other subsystems, the i-th output $\mathbf{z}_i(k)$ has to satisfy

$$\mathbf{z}_i(k) \in \mathcal{Z}_i(d|\{\mathbf{z}_j(k)\}_{j \in \mathcal{N}\setminus\{i\}}), \forall k \in \mathbb{Z}_+ \tag{15.5}$$

where

$$\mathcal{Z}_i(d|\mathbf{z}_{j\in\mathcal{N}\setminus\{i\}}) := \{\mathbf{z}_i \in \Omega : ||\mathbf{z}_i - \mathbf{z}_j||_\infty \geq d \ \ \forall j \in \mathcal{N} \setminus \{i\}\} \tag{15.6}$$

From now on, we will make use of the more compact notation $\{\mathbf{z}_j\}_i$ in place of $\{\mathbf{z}_j\}_{j\in\mathcal{N}\setminus\{i\}}$, in order to indicate a set of variables provided to agent i and related to the other agents. Notice also that, for the sake of clarity we prefer to separate convex from not-convex constraints, i.e. to separate (15.5) from (15.2). Constraints of type (15.5) could be of interest in a distributed context because they can model for example collision avoidance requirements amongst autonomous vehicles operating in the same area. It is further assumed that

A1. Each subsystem (15.1) is asymptotically stable.
A2. Each subsystem (15.1) is off-set free, which means that $\lim_{t\to\infty} \mathbf{y}_i(k) = \bar{\mathbf{g}}_i$ if a constant signal $\mathbf{g}_i(k) = \bar{\mathbf{g}}_i$ is applied.

15.2.2 Available Communication Architecture

We assume that the agents are connected via a communication network. Such a network is modeled by a *communication graph*: an undirected graph (see Fig. 15.3) $\Gamma = (\mathcal{N}, \mathcal{E})$, where $\mathcal{E} \subseteq \mathcal{N} \times \mathcal{N}$ denotes the set of edges representing the existence of direct communication links. More precisely, the edge (i, j) belongs to $\mathcal{E}$ if and only if the agents governing the i-th and the j-th subsystems are able to directly share data within one sampling step. The communication graph is assumed to be connected, i.e. for each couple of agents $i \in \mathcal{N}$, $j \in \mathcal{N}$ there exists at least one

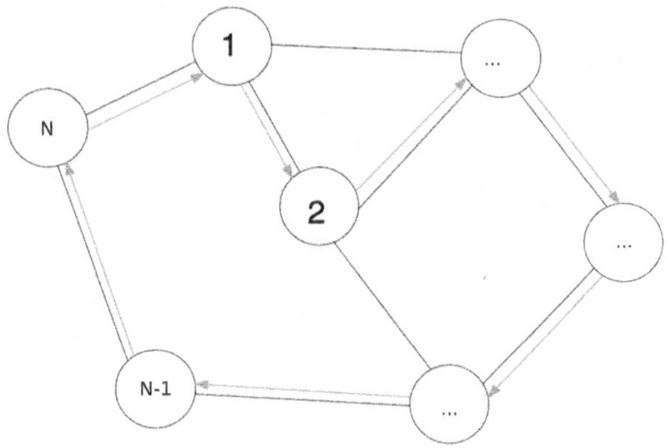

Fig. 15.3 Graph Γ and Hamiltonian cycle $\mathcal{H}$ (*yellow arrows*)

sequence of edges connecting i to j, with the minimum number of edges connecting the two agents denoted by $d_{i,j}$.

The set of all agents with a direct connection with the i-th agent will be referred to as the *neighborhood of the i-th agent* $\mathcal{N}_i = \{j \in \mathcal{N} : d_{i,j} = 1\}$. We will also assume that each agent acts as a gateway in redistributing data to the agents that are not directly connected. It is finally assumed that each agent is aware of the model describing the dynamics of the other systems.

15.2.3 Coordination Problem to be Solved

In words, the coordination problem we want to face is that of determining in a distributed fashion, at each time step k and for each agent $i \in \mathcal{N}$ associated to each subsystem, a suitable reference signal $\mathbf{g}_i(k)$ which is the best approximation, compatible with the constraints, of the given reference $\mathbf{r}_i(k) \in \mathbb{R}^{n_{g_i}}$ such that its application never produces violation of constraints (15.2) and (15.3).

15.3 Description of the Approach

Here we introduce a distributed solution for the above stated coordination problem. The idea behind such an approach is that only one agent at a time is allowed to manipulate its local command signal $\mathbf{g}_i(k)$ while the others keep applying their current commands. As a consequence, each agent applies the same command $\mathbf{g}_i$ for at least $|\mathcal{N}| - 1$ steps. The approach clearly requires agents to know when to change their local commands $\mathbf{g}_i$. In this respect, let us assume, without loss of generality, that the sequence $\mathcal{H} = \{1, 2, \ldots, |\mathcal{N}| - 1, |\mathcal{N}|\}$ is an Hamiltonian cycle defined on Γ. Roughly speaking we want the agents to carry out the following basic actions at each time k:

1. receive data from neighbors;
2. for the non-updating agent, apply the old local command $\mathbf{g}_i(k-i)$ and, for the updating agent, update $\mathbf{g}_i(k)$ according to a proper optimization procedure that guarantees the feasibility of $\mathbf{g}_i(k)$ for future $|\mathcal{N}|-1$ steps. Apply $\mathbf{g}_i(k)$;
3. transmit data to neighbors; the non-updating agents behave as simple gateways in redistributing data amongst neighbors.

At step $k+1$ the above procedure is repeated with the next updating agent in $\mathcal{H}$.

In the following, we will detail the procedure above by describing the network communication policy, the usage of received data by the updating agent and the optimization problem that needs to be solved.

15.3.1 Communication Policy

In this section we explain **how communication takes place** at each time k. Because in the communication graph Γ each agent acts as gateway in redistributing data (local and remote $\mathbf{g}_i$, $\mathbf{x}_i$), then, at each time instant k, the generic i-th agent is aware of the overall status of the network via the following vectors:

$$\begin{aligned}\mathbf{p}_i(k) =& [\mathbf{g}_1^T(k-d_{i,1}),\ldots,\mathbf{g}_i^T(k-1),\ldots,\mathbf{g}_{|\mathcal{N}|}^T(k-d_{i,|\mathcal{N}|})]^T\\ \mathbf{s}_i(k) =& [\mathbf{x}_1^T(k-d_{i,1}),\ldots,\mathbf{x}_i^T(k),\ldots,\mathbf{x}_{|\mathcal{N}|}^T(k-d_{i,|\mathcal{N}|})]^T\end{aligned}$$

The i-th agent updates its reference and transmits the updated vector

$$\mathbf{p}_i(k) = [\mathbf{g}_1^T(k-d_{i,1}),\ldots,\mathbf{g}_i^T(k),\ldots,\mathbf{g}_{|\mathcal{N}|}^T(k-d_{i,|\mathcal{N}|})]^T$$

and $\mathbf{s}_i(k)$ to its neighbors.

15.3.2 Updating the Local Reference

Here we describe **how the received vectors $\mathbf{p}_i(k)$ and $\mathbf{s}_i(k)$ are exploited**. Taking into account that agents freeze their commands for $|\mathcal{N}|-1$ steps, the updating agent is always provided with the last applied aggregate command $\mathbf{g}(k-1) = [\mathbf{g}_1(k-1)^T,\ldots,\mathbf{g}_i(k-1)^T,\ldots,\mathbf{g}_{|\mathcal{N}|}(k-1)^T]$. Therefore, after a preliminary initialization cycle, given $\mathbf{p}_i(k-1)$ and $\mathbf{s}_i(k-1)$, the generic i-th "agent in charge" is able to compute $\mathbf{x}_j(k)$, $j\in\mathcal{N}\setminus\{i\}$:

$$\mathbf{x}_j(l+1) = \mathbf{A}_{jj}\mathbf{x}_j(l) + \mathbf{B}_j\mathbf{g}_j(l),\ l = k-d_{i,|\mathcal{N}|},\ldots,k \qquad (15.7)$$

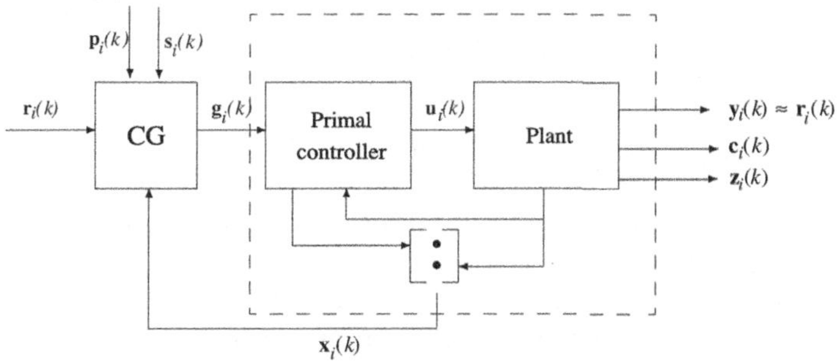

Fig. 15.4 Local command governor structure. The vector $\mathbf{x_i(k)}$ collects the state of the local controller and of the local plant

Hence, the updating agent i can estimate the outputs $\mathbf{c}_j(k)$, $\mathbf{z}_j(k)$ and, in turn, as described in the next section, locally solve an optimization problem involving the constraints (15.2–15.5).

In Fig. 15.4 the local supervision scheme for the generic i-th agent is depicted.

At each time k, the local CG action $\mathbf{g}_i(k)$ is computed on the basis of the current reference $\mathbf{r}_i(k)$, the locally measured state $\mathbf{x}_i(k)$ and the available states $\mathbf{x}_j(k)$ evaluated by means of (15.7). It is worth commenting that, the knowledge of the overall initial conditions and the communication of references $\mathbf{g}_i(k)$ would be sufficient to solve the above stated coordination problem. The sharing of local states amongst agents would be mandatory in the presence of disturbances though. The proposed scheme can be easily modified to take that into account them but, for the sake of clearness, we do not consider explicitly such a case.

15.3.3 Building the Underlying Optimization Problem

Here we present the **optimization problem** carried out by the updating agent at time k for fulfilling constraints (15.2–15.5). In order to introduce such a problem first we have to reformulate the non-convex constraints (15.5) into linear constraints depending on continuous and integer variables [11]. For the sake of notational clarity, the dimension of vectors $\mathbf{z}_i$ is assumed to be $n_{\mathbf{z}_i} = 2$, i.e. $\mathbf{z}_i = [\mathbf{z}_i^1, \mathbf{z}_i^2]^T$. Nevertheless, an extension to the more general case where $n_{\mathbf{z}_i} > 2$ is straightforward.

The first step towards a mixed-integer formulation relies upon rewriting the constraints (15.3), according to [9], as

$$|z_i^1 - z_j^1| \geq d \text{ or } |z_i^2 - z_j^2| \geq d, \forall j \in \mathcal{N} \setminus \{i\} \tag{15.8}$$

By taking the absolute values, constraints (15.8) modify into

$$\begin{aligned} & z_i^1 - z_j^1 \geq d \\ \text{or } & z_i^2 - z_j^2 \geq d \\ \text{or } & z_j^1 - z_i^1 \geq d \\ \text{or } & z_j^2 - z_i^2 \geq d, \forall j \in \mathcal{N} \setminus \{i\} \end{aligned} \tag{15.9}$$

A standard way to transform the *or* conditions into more useful *and* conditions is that of introducing binary slack variables. To this end, let $b_{ij}^p \in \{0, 1\}$ be a binary variable and μ a sufficiently large positive number ([9]). Then, constraints (15.9) can be replaced by the following mixed-integer linear constraints

$$\begin{aligned} & z_i^1 - z_j^1 \geq d - \mu b_{ij}^1 \\ \text{and } & z_i^2 - z_j^2 \geq d - \mu b_{ij}^2 \\ \text{and } & z_j^1 - z_i^1 \geq d - \mu b_{ij}^3 \\ \text{and } & z_j^2 - z_i^2 \geq d - \mu b_{ij}^4 \\ \text{and } & \textstyle\sum_{p=1}^4 b_{ij}^p \leq 3, \qquad \forall j \in \mathcal{N} \setminus \{i\}| \end{aligned} \tag{15.10}$$

If the p-th *or* condition in (15.9) is not satisfied, the corresponding binary variable b_{ij}^p is equal to 1. Finally, observe that the last condition in (15.10) ensures that at least one of the conditions in (15.9) is satisfied. Hence, in order to belong to the set $\mathcal{Z}_i(d|\{\mathbf{z}_j\}_{j\in\mathcal{N}\setminus\{i\}})$, the vector $\mathbf{z}_i$ has to be contained into the $\mathbf{z}$-projection of the following set

$$\bar{\mathcal{Z}}_i(d|\{\mathbf{z}_j\}_i) := \left\{ (\mathbf{z}_i, \mathbf{b}_i) \in \mathbb{R}^{n_{\mathbf{z}_i}} \times \{0,1\}^{2n_{\mathbf{z}_i}} \left| \begin{aligned} & z_i^1 - z_j^1 \geq d - \mu b_i j^1 \\ \text{and } & z_i^2 - z_j^2 \geq d - \mu b_{ij}^2 \\ \text{and } & z_j^1 - z_i^1 \geq d - \mu b_{ij}^3 \\ \text{and } & z_j^2 - z_i^2 \geq d - \mu b_{ij}^4 \\ \text{and } & \textstyle\sum_p = 1^4 b_j^p \leq 3, \qquad \forall j \in \mathcal{N} \setminus \{i\} \end{aligned} \right. \right\} \tag{15.11}$$

where $\mathbf{b}_i := [b_{i1}^1, \ldots, b_{i1}^4 \ldots b_{ij}^1, \ldots, b_{ij}^4 \ldots, b_{i|\mathcal{N}|}^1, \ldots, b_{i|\mathcal{N}|}^4] \in \{0,1\}^{2n_{\mathbf{z}_i}(|\mathcal{N}|-1)}, i \neq j$. More formally, the relation between $\mathcal{Z}_i(d|\{\mathbf{z}_j\}_i)$ and $\bar{\ominus}\left(d|\{\mathbf{z}_j\}_i\right)$ can be expressed as

$$\mathbf{z}_i \in \mathcal{Z}_i(d|\{\mathbf{z}_j\}_i) \Leftrightarrow \exists \mathbf{b}_i \in \{0,1\}^{2n_{\mathbf{z}_i}(|\mathcal{N}|-1)} \text{ such that } (\mathbf{z}_i, \mathbf{b}_i) \in \bar{\mathcal{Z}}_i(d) \tag{15.12}$$

Now we have all the elements to build the local optimization procedure related to i-th agent. The objective, as usual in predictive control, lies in the computation, at time k, of an open-loop feasible command sequence $\mathbf{g}_i(\cdot)$ defined on a certain virtual time horizon starting from k. In particular, we choose the following family of constant virtual command sequences to be computed

$$\mathbf{g}_i(\cdot) = \left\{\mathbf{g}_i(l|k) \equiv \mathbf{w}_i \in \mathbb{R}^{n_{\mathbf{g}_i}}\right\}_{l=k}^{\infty} \tag{15.13}$$

where $\mathbf{w}_i$ will be the decision variable of the local optimization problem. Obviously, according to the receding horizon approach, such a sequence will be replaced with a new one when the agent i will be again in charge of modifying its local reference (after $|\mathcal{N}| - 1$ steps).

Moreover, we assume that in the virtual interval $[k, \infty)$, considered in the optimization, other agents do not modify their local references, i.e.,

$$\mathbf{g}_j(\cdot) = \left\{\mathbf{g}_j(l|k) \equiv \mathbf{g}_j(k) = \mathbf{g}_j(k-1) \in \mathbb{R}^{n_{\mathbf{g}_j}}\right\}_{l=k}^{\infty}. \tag{15.14}$$

Please notice that, because other agents $j \in \mathcal{N} \setminus \{i\}$ do not modify their reference at time k, we can be ensured that $g_j(k) = g_j(k-1), \forall j \in \mathcal{N} \setminus \{i\}$.

The first set of constraints, to be satisfied in the local optimization, is related to the equilibrium solution of (15.1)

$$\begin{aligned}
\mathbf{x}_i^{\mathbf{w}_i} &:= (\mathbf{I}_{n_{\mathbf{x}_i}} - \mathbf{A}_{ii})^{-1}\mathbf{B}_i\mathbf{w}_i \\
\mathbf{z}_i^{\mathbf{w}_i} &:= \mathbf{C}_i^z(\mathbf{I}_{n_{\mathbf{x}_i}} - \mathbf{A}_{ii})^{-1}\mathbf{B}_i\mathbf{w}_i \\
\mathbf{c}_i^{\mathbf{w}_i} &:= \mathbf{C}_i^c(\mathbf{I}_{n_{\mathbf{x}_i}} - \mathbf{A}_{ii})^{-1}\mathbf{B}_i\mathbf{w}_i + \mathbf{L}_i\mathbf{w}_i
\end{aligned} \tag{15.15}$$

In particular, we impose that the steady-state equilibria related to $\mathbf{w}_i$ satisfy the constraints (15.2) and the constraints (15.3–15.5) with margin $\delta > 0$, i.e.,

$$\mathbf{c}_i^{\mathbf{w}_i} \in \mathcal{C}_i \sim \mathcal{B}^{\delta} \tag{15.16}$$

$$\mathbf{z}_i^{\mathbf{w}_i} \in \Omega \sim \mathcal{B}^{\delta} \tag{15.17}$$

$$\begin{aligned}
&(\mathbf{z}_i^{\mathbf{w}_i}, \mathbf{b}_i) \in \bar{\mathcal{Z}}_i\left(d + \delta|\{\mathbf{z}_j^{g_j(k)}\}_i\right), \forall j \in \mathcal{N} \setminus \{i\}, \\
&\quad \text{for any } \mathbf{b}_i \in \{0, 1\}^{2n_{\mathbf{z}_i}(|\mathcal{N}|-1)}
\end{aligned} \tag{15.18}$$

where $\mathcal{B}^{\delta}$ is a ball of radius δ centered at the origin and, for given sets $\mathcal{A}$ and $\mathcal{P} \subset \mathbf{R}^n$, $\mathcal{A} \sim \mathcal{P}$ denotes the Pontryagin set difference $\mathcal{A} \sim \mathcal{P} := \{a : a + e \in \mathcal{A}, \forall e \in \mathcal{P}\}$. $\mathbf{z}_j^{\mathbf{g}_j(k-1)}$ is the $\mathbf{z}_j$ steady-state solution related to $\mathbf{g}_j(k-1)$, i.e.,

$$\mathbf{z}_j^{\mathbf{g}_j(k)} := \mathbf{C}_i^z(\mathbf{I}_{n_{\mathbf{x}_j}} - \mathbf{A}_{jj})^{-1}\mathbf{B}_i\mathbf{g}_j(k-1)$$

The introduction of the above constraints and restricted sets is motivated by feasibility needs as explained in [1, 14]. Notice that the choice of $\mathbf{w}_i$, admissible for (15.16–15.18), only ensures constraints fulfillment in steady-state conditions, and nothing is said about the constraints fulfillment along the transient from $\mathbf{x}_i(k)$ to $\mathbf{x}_i^{\mathbf{w}_i}$.

For this reason, we need to define a further set of constraints that, given $\mathbf{x}_i(k)$ and $\mathbf{x}_j(k)$, guarantee the selection of a local CG action $\mathbf{g}_i(\cdot) = \mathbf{w}_i$, admissible for (15.16–15.18), such that if constantly applied from the time instant k onwards, gives rise to system evolutions that do not produce constraint violations along the transient from $\mathbf{x}_i(k)$ to $\mathbf{x}_i^{\mathbf{w}_i}$.

In this respect, future predictions of the state, $\mathbf{c}_i$-vector and $\mathbf{z}_i$-vector from $\mathbf{x}_i(k)$ under a constant command $\mathbf{w}_i$ have to be considered

$$\begin{aligned} \mathbf{x}_i(l, \mathbf{x}_i(k), \mathbf{w}_i) &:= \mathbf{A}_{ii}^l \mathbf{x}_i(k) + \sum_{j=0}^{l-1} \mathbf{A}_{ii}^{l-j-1} \mathbf{B}_i \mathbf{w}_i \\ \mathbf{c}_i(l, \mathbf{x}_i(k), \mathbf{w}_i) &:= \mathbf{C}_i^{\mathbf{c}} \mathbf{x}_i(k, \mathbf{x}_i(k), \mathbf{w}_i) + \mathbf{L}_i \mathbf{w}_i \\ \mathbf{z}_i(l, \mathbf{x}_i(k), \mathbf{w}_i) &:= \mathbf{C}_i^{\mathbf{z}} \mathbf{x}_i(l, \mathbf{x}_i(k), \mathbf{w}_i). \end{aligned} \tag{15.19}$$

for $l \geq 0$. Finally, we define the constraints

$$\mathbf{c}_i(l, \mathbf{x}_i(k), \mathbf{w}_i) \in \mathcal{C}_i, \forall l \geq 0 \tag{15.20}$$

$$(\mathbf{z}_i(l, \mathbf{x}_i(k), \mathbf{w}_i), \tilde{\mathbf{b}}_i(l)) \in \bar{\mathcal{Z}}_i(d|\{\mathbf{z}_j(l, \mathbf{x}_j(k), \mathbf{g}_j(k))\}_i), \forall l \geq 0 \tag{15.21}$$

where, $\forall j \in \mathcal{N} \setminus \{i\}$ and

$$\mathbf{z}_j(l, \mathbf{x}_j(k), \mathbf{g}_j(k)) := \mathbf{C}_i^{\mathbf{z}} \left(\mathbf{A}_{jj}^l \mathbf{x}_j(k) + \sum_{h=0}^{l-1} \mathbf{A}_{jj}^{l-h-1} \mathbf{B}_j \mathbf{g}_j(k) \right).$$

The above constraints allow us to formulate a constrained optimization in order to find, according to a performance index, the "best" feasible approximation $\mathbf{w}_i$ (constant virtual command) to $\mathbf{r}_i(k)$ for which the evolutions of $\mathbf{c}_i$ and $\mathbf{z}_i$, satisfy the constraints for all $l \geq 0$ along the predictions. Therefore, the local CG action can be chosen according to the solution of the following constrained optimization problem

$$\begin{aligned} &\mathbf{g}_i(k) = \arg\min_{(\mathbf{w}_i, \mathbf{b}_i, \tilde{\mathbf{b}}_i(\cdot))} \|\mathbf{w}_i - \mathbf{r}_i(k)\|_{\mathbf{R}_i}^2 \\ &\text{subject to} : (15.16\text{–}15.18), (15.20\text{–}15.21) \end{aligned} \tag{15.22}$$

where $\mathbf{R}_i > 0$ is a weighting matrix.

15.3.4 Resulting Distributed Algorithm and Related Properties

Finally, we can formulate the distributed D-CG algorithm as in Algorithm 15.1.

It can be proved (see [15]) that the following properties hold true for the above described distributed CG.

Theorem 15.1 *Let assumptions **A1-A2** be fulfilled. Consider subsystems (15.1) with the local D-CG selection rules (15.22) and let assume the existence of admissible solutions* $g_i(0)$, $\forall i \in \mathcal{N}$ *for constraints (15.16–15.18, 15.20–15.21). Then:*

Algorithm 15.1: Sequential-CG Algorithm (D-CG) - Agent i

Initialization

1: choose δ for constraints (15.16–15.18), and scalars μ and d in (15.11)
2: compute l_0 for constraints (15.20–15.21) (see [15])
3: provide each agent with $\mathbf{x}_j(0)$ and $\mathbf{g}_j(0), \forall j \in \mathcal{N} \setminus \{i\}$

On-line phase

1: repeat at each time k

1.1 receive $\mathbf{p}_j(k-1)$ and $\mathbf{s}_j(k-1)$ from $\mathcal{N}_i$
1.2 evaluate $\mathbf{p}_i(k)$ and $\mathbf{s}_j(k)$
1.1 if(k mod$|\mathcal{N}|$) == i

1.2.1 compute $\mathbf{x}_j(k)$, $j \in \mathcal{N} \setminus \{i\}$ by means of (15.7)
1.2.2 solve (15.22)
1.2.3 apply $\mathbf{g}_i(k)$

1.3 else

1.3.1 apply $\mathbf{g}_i(k) = \mathbf{g}_i(k-1)$

1.4 update $\mathbf{p}_i(k) = [\mathbf{g}_1^T(k - d_{i,1}), \ldots, \mathbf{g}_i^T(k), \ldots, \mathbf{g}_{|\mathcal{N}|}^T(k - d_{i,|\mathcal{N}|})]^T$
1.5 transmit $\mathbf{p}_i(k)$ and $\mathbf{s}_j(k)$ to $\mathcal{N}_i$

where k mod$|\mathcal{N}|$ is the remainder of the integer division $k/|\mathcal{N}|$.

1. (***Feasibility***) the existence of an admissible aggregate vector $g(0)$ for constraints (15.16–15.18,15.20–15.21) implies the existence of an admissible aggregate vector $g(k)$ for constraints (15.16–15.18,15.20–15.21) evaluated along the trajectories generated by the local D-CG commands (15.22) for all $i \in \mathcal{N}$.
2. (***Computability***) Concerning constraints (15.20–15.21), there exists an integer l_0 such that if $\mathbf{c}_i(l, \mathbf{x}_i, \mathbf{w}_i) \in \mathcal{C}_i$, $\mathbf{z}_i(l, \mathbf{x}_i, \mathbf{w}_i) \in \bar{\mathcal{Z}}_i(d|\{\mathbf{z}_j(l, \mathbf{x}_j, \mathbf{g}_j)\}_i)$, $l \in \{0, 1, \ldots l_0\}$, then $\mathbf{c}_i(l, \mathbf{x}_i, \mathbf{w}_i) \in \mathcal{C}_i$, $\mathbf{z}_i(l, \mathbf{x}_i, \mathbf{w}_i) \in \bar{\mathcal{Z}}_i(d|\{\mathbf{z}_j(l, \mathbf{x}_j, \mathbf{g}_j)\}_i)$, $\forall l \in \mathbb{Z}_+$. Such a *constraint horizon* l_0 can be determined off-line (see [15]).
3. (***Stability***) The overall system is BIBO stable; in particular, whenever $\mathbf{r}_i(k) = [\mathbf{r}_1^T(k), \ldots, \mathbf{r}_i^T(k), \ldots, \mathbf{r}_{|\mathcal{N}|}^T(k)] \equiv \mathbf{r}$, $\mathbf{g}_i(k) = [\mathbf{g}_1^T(k), \ldots, \mathbf{g}_i^T(k), \ldots, \mathbf{g}_{|\mathcal{N}|}^T(k)]$ in general, asymptotically converges either to $\mathbf{r}$, if it is feasible, or to a set of Pareto Optimal solutions.

15.3.5 Computational Details

In this section we will present all computational details for the design and the implementation of the above distributed D-CG strategy. In particular, we will present an explicit formulation of the optimization local problem (15.22).

If linear constraints for $\mathbf{c}_i(k) \in \mathcal{C}_i$ (15.2) are assumed, they can be represented as

$$\mathbf{c}_i(k) \in \mathcal{C}_i \Longleftrightarrow \mathcal{C}_i := \{\mathbf{c}_i \in \mathbb{R}^{n_{c_i}} : \mathbf{T}_i \mathbf{c}_i \le \mathbf{q}_i\}, \quad (15.23)$$

$$\mathbf{T}_i \in \mathbb{R}^{p \times n_{\mathbf{c}_i}} \quad (15.24)$$

$$\text{rank}\,(\mathbf{T}_i) = n_{\mathbf{c}_i} \quad (15.25)$$

with $\mathbf{T}_{i,h}^T$ and $q_{i,h}$ denoting respectively the h-th rows of $\mathbf{T}_i$ and $\mathbf{q}_i$. Notice that the number p of rows of $\mathbf{T}_i$ is in general larger than the number of columns and the rank condition on $\mathbf{T}_i$ ensures that $\mathcal{C}_i$ is a bounded set. The set $\mathcal{C} \sim \mathcal{B}^\delta$ is given by

$$\mathcal{C}_i \sim \mathcal{B}^\delta = \{\mathbf{c}_i \in \mathbb{R}^{n_{\mathbf{c}_i}} \;:\; \mathbf{T}_{i,h}^T \mathbf{c}_i \le \mathbf{q}_i - \delta\sqrt{\mathbf{T}_{i,h}^T \mathbf{T}_{i,h}},\ h = 1, \ldots, p\}$$

where the terms $\delta\sqrt{\mathbf{T}_{i,h}^T \mathbf{T}_{i,h}}$ is the support function of the ball $\mathcal{B}^\delta$ (see [8]). Then, constraints (15.16) can be reformulated as

$$\mathbf{T}_i \left(\mathbf{C}_i^c (\mathbf{I}_{n_{\mathbf{x}_i}} - \mathbf{A}_{ii})^{-1} \mathbf{B}_i + \mathbf{L}_i\right) \mathbf{w}_i \le \mathbf{q}_i - \delta[\sqrt{\mathbf{T}_{i,h}^T \mathbf{T}_{i,h}}] \quad (15.26)$$

where

$$[\sqrt{\mathbf{X}_{i,h}^T \mathbf{X}_{i,h}}] := \begin{bmatrix} \sqrt{\mathbf{X}_{i,1}^T \mathbf{X}_{i,1}} \\ \sqrt{\mathbf{X}_{i,2}^T \mathbf{X}_{i,2}} \\ \cdots\cdot \\ \sqrt{\mathbf{X}_{i,p}^T \mathbf{X}_{i,p}} \end{bmatrix} \quad (15.27)$$

In order to evaluate constraints (15.20), it is easy to understand that

$$\mathbf{x}_i(l, \mathbf{x}_i, \mathbf{w}_i) = \mathbf{A}^l \mathbf{x}_i + \left(\sum_{h=0}^{l-1} \mathbf{A}_{ii}^h \mathbf{B}_i\right) \mathbf{w}_i = \mathbf{A}^l \mathbf{x}_i + \mathbf{P}_l^{\mathbf{c}_i} \mathbf{w}_i \quad (15.28)$$

$$\begin{aligned} \mathbf{c}_i(l, \mathbf{x}_i, \mathbf{w}_i) &= \mathbf{C}_i^c \mathbf{x}_i(l, \mathbf{x}_i, \mathbf{w}_i) + \mathbf{L}_i \mathbf{w}_i = \mathbf{C}_i^c \mathbf{A}^l \mathbf{x}_i + \left(\mathbf{C}^{\mathbf{c}_i} \mathbf{P}_l^{\mathbf{c}_i} + \mathbf{L}_i\right) \mathbf{w}_i \\ &= \mathbf{C}_i^c \mathbf{A}^l \mathbf{x}_i + \mathbf{P}_l^{\mathbf{c}_i} \mathbf{w}_i \end{aligned} \quad (15.29)$$

and hence (15.20) becomes

$$\mathbf{T}_i \mathbf{C}_i^c \mathbf{A}^l \mathbf{x}_i + \mathbf{T}_i \mathbf{P}_l^{\mathbf{c}_i} \mathbf{w}_i \le \mathbf{q}_i,\ l = 0, \ldots, l_0 \quad (15.30)$$

Concerning constraints (15.17–15.18), an explicit reformulation can be straightforwardly achieved by looking at the $\bar{\mathcal{Z}}_i(d|\{\mathbf{z}_j\}_i)$ structure (15.11) and by defining

$$\mathbf{Q}_i^z = \left(\mathbf{C}_i^c (\mathbf{I}_{n_{\mathbf{x}_i}} - \mathbf{A}_{ii})^{-1} \mathbf{B}_i + \mathbf{L}_i\right) \quad (15.31)$$

Then, (15.17–15.18) are recast into

$$\mathbf{E}_i\mathbf{Q}_i^{\mathbf{z}}\mathbf{w}_i \le \mathbf{q}_i^z - \delta_2[\sqrt{E_{i,h}^T E_{i,h}}]$$

$$\begin{cases} \mathbf{E}_{i,1}\mathbf{Q}_i^z\mathbf{w}_i - \mathbf{z}_j^1 \ge d - \mu b_{ij}^1 \\ \mathbf{E}_{i,2}\mathbf{Q}_i^z\mathbf{w}_i - \mathbf{z}_j^2 \ge d - \mu b_{ij}^2 \\ \mathbf{z}_j^1 - \mathbf{E}_{i,1}\mathbf{Q}_i^z\mathbf{w}_i \ge d - \mu b_{ij}^3 \\ \mathbf{z}_j^2 - \mathbf{E}_{i,2}\mathbf{Q}_i^z\mathbf{w}_i \ge d - \mu b_{ij}^4 \\ \sum_{p=1}^4 b_{ij}^p \le 3,\, b_{ij}^p \in \{0,\,1\},\ \forall j \in \mathcal{N} \setminus \{i\} \end{cases} \tag{15.32}$$

Please note that the constraint $\mathbf{z}_i \in \Omega$ in (15.3) has been assumed to be boxed, in the sense that the vector $\mathbf{z}_i$ is component-wise limited by two fixed vectors

$$\underline{\mathbf{z}}_i \le \mathbf{z}_i \le \bar{\mathbf{z}}_i \Leftrightarrow \mathbf{E}_i\mathbf{z}_i \le \mathbf{q}_i^z := \begin{bmatrix} \bar{\mathbf{z}}_i \\ -\underline{\mathbf{z}}_i \end{bmatrix} = \tag{15.33}$$

where

$$\mathbf{E}_i = [\mathbf{E}_{i,1}^T, \mathbf{E}_{i,2}^T, -\mathbf{E}_{i,1}^T - \mathbf{E}_{i,2}^T]^T = \begin{bmatrix} 1 & 0 \\ 0 & 1 \\ -1 & 0 \\ 0 & -1 \end{bmatrix}. \tag{15.34}$$

It remains to characterize constraints (15.21). In a similar way w.r.t. (15.28)–(15.29) we can explicit the structure of predictions $\mathbf{z}_i(l, \mathbf{x}_i, \mathbf{w}_i)$ with the following expression

$$\begin{aligned} \mathbf{z}_i(l, \mathbf{x}_i, \mathbf{w}_i) &= \mathbf{C}_i^z\mathbf{x}_i(l, \mathbf{x}_i, \mathbf{w}_i) + \mathbf{L}_i\mathbf{w}_i = \mathbf{C}_i^z\mathbf{A}^l\mathbf{x}_i + \left(\mathbf{C}_i^z\mathbf{P}_l^{\mathbf{x}_i} + \mathbf{L}_i\right)\mathbf{w}_i \\ &= \mathbf{C}_i^z\mathbf{A}^l\mathbf{x}_i + \mathbf{P}_l^{\mathbf{x}_i}\mathbf{w}_i \end{aligned} \tag{15.35}$$

and hence constraints (15.21) become

$$\mathbf{E}_i\left(\mathbf{C}_i^z\mathbf{A}^l\mathbf{x}_i + \mathbf{P}_l^{\mathbf{z}_i}\mathbf{w}_i\right) \le \mathbf{q}_i^z$$

$$\begin{cases} \mathbf{E}_{i,1}\left(\mathbf{C}_i^z\mathbf{A}^l\mathbf{x}_i + \mathbf{P}_l^{\mathbf{z}_i}\mathbf{w}_i\right) - \mathbf{E}_{i,1}\left(\mathbf{C}_i^z\mathbf{A}^l\mathbf{x}_i + \mathbf{R}_l^{\mathbf{z}_i}\mathbf{g}_j(k)\right) \ge d - \mu\tilde{b}_{ij}^1(l) \\ \mathbf{E}_{i,2}\left(\mathbf{C}_i^z\mathbf{A}^l\mathbf{x}_i + \mathbf{P}_l^{\mathbf{z}_i}\mathbf{w}_i\right) - \mathbf{E}_{i,2}\left(\mathbf{C}_i^z\mathbf{A}^l\mathbf{x}_i + \mathbf{P}_l^{\mathbf{z}_i}\mathbf{g}_j(k)\right) \ge d - \mu\tilde{b}_{ij}^2(l) \\ \mathbf{E}_{i,1}\left(\mathbf{C}_i^z\mathbf{A}^l\mathbf{x}_i + \mathbf{P}_l^{\mathbf{z}_i}\mathbf{g}_j(k)\right) - \mathbf{E}_{i,1}\left(\mathbf{C}_i^z\mathbf{A}^l\mathbf{x}_i + \mathbf{R}_l^{\mathbf{z}_i}\mathbf{w}_i\right) \ge d - \mu\tilde{b}_{ij}^3(l) \\ \mathbf{E}_{i,2}\left(\mathbf{C}_i^z\mathbf{A}^l\mathbf{x}_i + \mathbf{P}_l^{\mathbf{z}_i}\mathbf{g}_j(k)\right) - \mathbf{E}_{i,2}\left(\mathbf{C}_i^z\mathbf{A}^l\mathbf{x}_i + \mathbf{P}_l^{\mathbf{z}_i}\mathbf{w}_i\right) \ge d - \mu\tilde{b}_{ij}^4(l) \\ \sum_{p=1}^4 \tilde{b}_{ij}^p(l) \le 3,\, \tilde{b}_{ij}^p(l) \in \{0,\,1\}, \forall j \in \mathcal{N} \setminus \{i\}, l = 0, \ldots, l_0 \\ \mathbf{g}_i(k) = \min\limits_{\mathbf{w}_i, \mathbf{b}_i, \tilde{\mathbf{v}}_i} (\mathbf{w}_i - \mathbf{r}_i(k))'\mathbf{R}_i(\mathbf{w}_i - \mathbf{r}_i(k)) \end{cases} \tag{15.36}$$

Then, Problem (15.22) is rephrased as

$$
s.t.\ \begin{cases}
\mathbf{T}_i\left(\mathbf{C}_i^c(\mathbf{I}_{n_{x_i}}-\mathbf{A}_{ii})^{-1}\mathbf{B}_i+\mathbf{L}_i\right)\mathbf{w}_i \le \mathbf{q}_i-\delta\lfloor\sqrt{\mathbf{T}_{i,h}^T\mathbf{T}_{i,h}}\rfloor \\
\mathbf{T}_i\mathbf{R}_l^{\mathbf{c}_i}\mathbf{w}_i \le \mathbf{q}_i-\mathbf{T}_i\mathbf{C}_i^c\mathbf{A}^l\mathbf{x}_i(k),\ l=0,\dots,l_0 \\
\mathbf{E}_i\mathbf{Q}_i^z\mathbf{w}_i \le \mathbf{q}_i^z-\delta\lfloor\sqrt{\mathbf{E}_{i,h}^T\mathbf{E}_{i,h}}\rfloor \\
\mathbf{E}_i\left(\mathbf{C}_i^z\mathbf{A}^l\mathbf{x}_i+\mathbf{R}_l^{\mathbf{z}_i}\mathbf{w}_i\right) \le \mathbf{q}_i^z \\
\begin{cases}
-\mathbf{E}_{i,1}Q_i^z\mathbf{w}_i \le -d+\mu b_{ij}^1-\mathbf{E}_{i,1}\mathbf{Q}_j^z\mathbf{g}_j(k) \\
-\mathbf{E}_{i,2}Q_i^z\mathbf{w}_i \le -d+\mu b_{ij}^2-\mathbf{E}_{i,2}\mathbf{Q}_j^z\mathbf{g}_j(k) \\
\mathbf{E}_{i,1}\mathbf{Q}_i^z\mathbf{w}_i \le -d+\mu b_{ij}^3+\mathbf{E}_{i,1}\mathbf{Q}_j^z\mathbf{g}_j(k) \\
\mathbf{E}_{i,2}\mathbf{Q}_i^z\mathbf{w}_i \le -d+\mu b_{ij}^4+\mathbf{E}_{i,2}\mathbf{Q}_j^z\mathbf{g}_j(k) \\
\sum_{p=1}^4 b_{ij}^p \le 3,\ b_{ij}^p \in \{0,1\},\ \forall j\in\mathcal{N}\setminus\{i\}
\end{cases} \\
\begin{cases}
-\mathbf{E}_{i,1}\mathbf{P}_l^{\mathbf{z}_i}\mathbf{w}_i \le -d+\mu\tilde{b}_{ij}^1(l)-\mathbf{E}_{i,1}\left(\mathbf{C}_j^z\mathbf{A}^l\mathbf{x}_j(k)+\mathbf{P}_l^{\mathbf{z}_i}\mathbf{g}_j(k)+\mathbf{C}_i^z\mathbf{A}^l\mathbf{x}_i(k)\right) \\
-\mathbf{E}_{i,2}\mathbf{P}_l^{\mathbf{z}_i}\mathbf{w}_i \le -d+\mu\tilde{b}_{ij}^2(l)-\mathbf{E}_{i,2}\left(\mathbf{C}_j^z\mathbf{A}^l\mathbf{x}_j(k)+\mathbf{P}_l^{\mathbf{z}_i}\mathbf{g}_j(k)+\mathbf{C}_i^z\mathbf{A}^l\mathbf{x}_i(k)\right) \\
\mathbf{E}_{i,1}\mathbf{P}_l^{\mathbf{z}_i}\mathbf{w}_i \le -d+\mu\tilde{b}_{ij}^3(l)+\mathbf{E}_{i,1}\left(\mathbf{C}_j^z\mathbf{A}^l\mathbf{x}_j(k)+\mathbf{P}_l^{\mathbf{z}_i}\mathbf{g}_j(k)-\mathbf{C}_i^z\mathbf{A}^l\mathbf{x}_i(k)\right) \\
\mathbf{E}_{i,2}\mathbf{P}_l^{\mathbf{z}_i}\mathbf{w}_i \le -d+\mu\tilde{b}_{ij}^4(l)+\mathbf{E}_{i,2}\left(\mathbf{C}_j^z\mathbf{A}^l\mathbf{x}_j(k)+\mathbf{P}_l^{\mathbf{z}_i}\mathbf{g}_j(k)-\mathbf{C}_i^z\mathbf{A}^l\mathbf{x}_i(k)\right) \\
\sum_{p=1}^4 \tilde{b}_{ij}^p(l) \le 3,\ \tilde{b}_{ij}^p(l) \in \{0,1\},\ \forall j\in\mathcal{N}\setminus\{i\},\ l=0,\dots,l_0
\end{cases}
\end{cases}
$$

15.4 Theoretical Results Availability

Centralized solutions of the above stated CG design problem, without no convex constraints of the form (15.4), have been proposed in [1, 4–7]. In particular, in [7] there is a detailed discussion about the computation of the scalar i_0 (there referred to as *constraint horizon*), whereas in [6] an exhaustive analysis of the existence and evaluation of l_0 (there referred to as *Generalized Settling Time*) is presented. Distributed CG schemes for convex constraints were presented in [2, 3, 12, 13].

A mixed-integer based approach was exploited for the development of distributed and centralized CG schemes in the presence of non-convex constraints in [15] and, in more details, in [14]. In particular, in the latter a proof of Theorem 15.1 is provided. Nowadays mixed-integer linear programs (MILP) can be efficiently solved with commercial solvers, as for example CPLEX.

The interested reader can find more information on the comparison between CG and MPC schemes in [10], where the two strategies are compared and their equivalence formally proved.

15.5 Application Results Availability

In [14, 15] the above described scheme is used to solve collision avoidance problems arising in formations of autonomous vehicles acting in the same area. In particular, detailed comparisons between centralized and distributed solutions have been presented by means of several numerical experiments.

References

1. A. Bemporad, A. Casavola, E. Mosca, Nonlinear control of constrained linear systems via predictive reference management. IEEE Trans. Automat. Control **42**, 340–349 (1997)
2. A. Casavola, E. Garone, F. Tedesco, Distributed coordination-by-constraint strategies for multi-agent networked systems. in *Decision and Control and European Control Conference (CDC-ECC), 2011 50th IEEE Conference on Decision and Control* (Orlando, USA, 2011), pp. 6888–6893
3. A. Casavola, E. Garone, F. Tedesco, A liveliness analysis of a distributed constrained coordination strategy for multi-agent linear systems. in *Decision and Control and European Control Conference (CDC-ECC), 2011 50th IEEE Conference on Decision and Control* (Orlando, USA, 2011), pp. 8133–8138
4. A. Casavola, E. Mosca, D. Angeli, Robust command governors for constrained linear systems. IEEE Trans. Automat. Control **45**, 2071–2077 (2000)
5. A. Casavola, M. Papini, G. Franzè, Supervision of networked dynamical systems under coordination constraints. IEEE Trans. Automat. Control **51**(3), 421–437 (2006)
6. E. Garone, F. Tedesco, A. Casavola, Sensorless supervision of linear dynamical systems: The feed-forward command governor approach. Automatica **47**(7), 1294–1303 (2011)
7. E.G. Gilbert, I.V. Kolmanovsky, K.T. Tan, Discrete-time reference governors and the nonlinear control of systems with state and control constraints. Int. J. Robust Nonlinear Control **5**, 487–504 (1995)
8. A. Kurzhanski, I. Valyi, *Ellipsoidal Calculus for Estimation and Control* (Birkhauser, Germany, 1997)
9. A. Richards, P.G. How, *Aircraft trajectory planning with collision avoidance using mixed integer linear programming*. in *Proceeding of American Control Conference* (Anchorage, USA, 2002)
10. J.A. Rossiter, B. Kouvaritakis, Reference governors and predictive control. in *Proceeding of IEEE American Control Conference* (Philadelphia, USA, 1998)
11. T. Schouwenaars, B. DeMoor, E. Feron, J. How, Mixed integer programming for safe multivehicle cooperative path planning. in *Proceeding of IEEE European Control Conference* (Porto, Portugal, 2001)
12. F. Tedesco, *Distributed Command Governor Strategies for Multi-agent Dynamical Systems*, Ph.D. thesis, (Università della Calabria, Rende (CS), Italy, 2012)
13. F. Tedesco, A. Casavola, E. Garone, Distributed command governor strategies for constrained coordination of multi-agent networked systems. in *Proceding of American Control Conference* (Montreal, Canada, 2012)
14. F. Tedesco, D.M. Raimondo, A. Casavola. A command governor approach for multiple vehicles collision avoidance problems. (submitted) http://tedescof.wordpress.com/publication/ (2012)
15. F. Tedesco, D.M. Raimondo, A. Casavola, J. Lygeros, Distributed collision avoidance for interacting vehicles: a command governor approach. in *Proceding of Distributed Estimation and Control in Networked Systems (NecSys'10)* (Annecy, France, 2010)

Chapter 16
The Distributed Command Governor Approach in a Nutshell

A. Casavola, E. Garone and F. Tedesco

Abstract The term Command Governor (CG) refers to a particular class of Model Predictive Control (MPC) strategies designed to manage the reference of a pre-compensated system in such a way that set-membership constraints on relevant system variables are not violated. More precisely, a CG unit is added to a primal compensated plant, which has supposedly been designed so as to exhibit stability and good tracking performance in the absence of constraints, and is in charge of modifying the prescribed reference signal whenever its direct application would lead to constraint violations. This chapter describes a distributed CG strategy for the supervision of a network of interconnected, possibly dynamically coupled, subsystems. Such an approach could be useful in situations where the use of a centralized coordination unit may be impracticable because requiring unrealistic or unavailable communication infrastructures.

16.1 Short Introduction

The problem of interest here is the design of distributed supervision strategies for large-scale multi-agent systems in situations where the use of centralized solutions might be impracticable. To this end, a distributed Command Governor (CG) approach will be presented.

A. Casavola (✉) · F. Tedesco
Università degli Studi della Calabria, DEIS, Rende (CS), Calabria, Italy
e-mail: casavola@deis.unical.it

F. Tedesco
e-mail: ftedesco@deis.unical.it

E. Garone
Universite Libre de Bruxelles, Bruxelles, Belgium
e-mail: egarone@ulb.ac.be

J. M. Maestre and R. R. Negenborn (eds.), *Distributed Model Predictive Control Made Easy*, Intelligent Systems, Control and Automation: Science and Engineering 69,
DOI: 10.1007/978-94-007-7006-5_16, © Springer Science+Business Media Dordrecht 2014

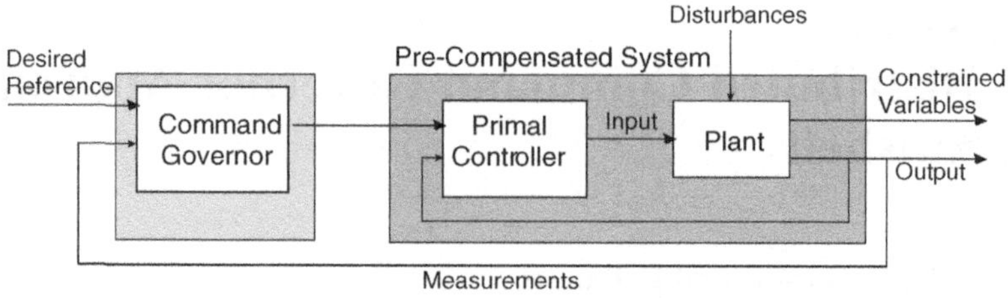

Fig. 16.1 Command governor control scheme

A CG is a non-linear device that outputs a reference trajectory rather than a control input and that it is superimposed to a primal compensated control system, as depicted in Fig. 16.1. The main requirement is the presence of a primal controller, (e.g. PID) designed to stabilize the plant and provide good tracking and disturbance attenuation properties regardless of any constraints fulfillment. This latter task is in fact accomplished by CG that, whenever necessary, modifies the reference supplied to the primal control system in order to avoid constraint violations. As in MPC schemes, a modified reference sequence is computed at each sampling instant and, according to a receding horizon control (RHC) policy, only the first sample of such a sequence is applied. The entire procedure is repeated at the next time instant.

One of the advantages of CG usage over the more general MPC approach lies in its simplicity in addressing constrained tracking problems. For such reason, the CG strategy seems to be very suitable in the supervision of large-scale networked systems where constraints fulfillment is a crucial task. Indeed, in such a case a centralized control architecture could not be an adequate choice because the corresponding computational burdens could be prohibitive and the required communication infrastructures not available. All these reasons call for distributed CG approaches, as the one described in this chapter.

The distributed context under consideration is depicted in Fig. 16.2, where the supervisory task is distributed amongst many agents which are assumed to be able to communicate each other through a communication network. There, each agent is in charge of supervising and coordinating one specific subsystem. In particular, $\mathbf{r}_i$, $\mathbf{g}_i$, $\mathbf{x}_i$, $\mathbf{y}_i$ and $\mathbf{c}_i$ represent respectively: the nominal reference, the applied reference, the state, the performance-related and the coordination-related output vectors of the i−th subsystem.

In such a context, supervision can be expressed as the task of satisfying some tracking performance, viz. $\mathbf{y}_i \approx \mathbf{r}_i$ while ensuring set-membership constraints $\mathbf{c}_i \in \mathcal{C}_i$ and/or $f(\mathbf{c}_1, \mathbf{c}_2, \ldots, \mathbf{c}_{|\mathcal{N}|}) \in \mathcal{C}$ on each subsystem and/or on the overall network evolutions. To do so, each i−th supervising node is in charge of modifying its local desired reference $\mathbf{r}_i$ into the applied one $\mathbf{g}_i$, ensuring that their joint application would not produce constraint violations and hence loss of coordination.

In this chapter a *parallel* distributed approach able to locally compute applied reference while ensuring global constraints satisfaction is presented. The main technical expedient behind such a scheme is the on-line determination of a suitable Cartesian

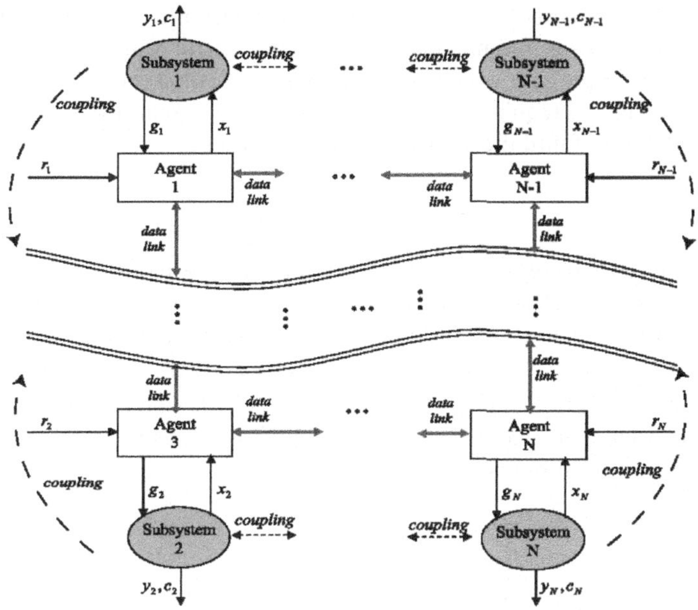

Fig. 16.2 Multi-agent architectures

Inner Approximation of the global constraint set, which allows the agents to optimize independently and simultaneously their reference signals while ensuring the fulfillment of the global constraints.

16.2 Boundary Conditions

16.2.1 System Model

Consider a set $\mathcal{N}$ of subsystems, each one being a LTI closed-loop dynamical system regulated by a local controller which ensures stability and good closed-loop properties within linear regimes (when the constraints are not active). Let the ith closed-loop subsystem be described by the following discrete-time model

$$\begin{cases} \mathbf{x}_i(k+1) = \mathbf{A}_{ii}\mathbf{x}_i(k)+\mathbf{B}_i\mathbf{g}_i(k)+\sum\limits_{j\in\mathcal{N}-\{i\}}\mathbf{A}_{ij}\mathbf{x}_j(k) \\ \mathbf{y}_i(k) \quad = \mathbf{C}_i\mathbf{x}_i(k) \\ \mathbf{c}_i(k) \quad = \mathbf{H}_i^c\mathbf{x}(k) + \mathbf{L}_i\mathbf{g}(k) \end{cases} \tag{16.1}$$

where: $k \in \mathbb{Z}_+$, $\mathbf{x}_i$ is the state vector (which includes the controller states under dynamic regulation), $\mathbf{g}_i$ the manipulable reference vector which, if no constraints (and no CG) were present, would coincide with the desired reference $\mathbf{r}_i$ and $\mathbf{y}_i$ the output vector which is required to track $\mathbf{r}_i$. Finally, $\mathbf{c}_i$ represents the local constrained vector which has to fulfill the set-membership constraint

$$\mathbf{c}_i(k) \in \mathcal{C}_i, \ \forall k \in \mathbb{Z}_+, \tag{16.2}$$

$\mathcal{C}_i$ being a convex and compact polytopic set. It is worth pointing out that, in order to possibly characterize global (coupling) constraints amongst states of different subsystems, the vector $\mathbf{c}_i$ in (16.1) is allowed to depend on the aggregate state and manipulable reference vectors $\mathbf{x} = [\mathbf{x}_1^T, \dots, \mathbf{x}_{|\mathcal{N}|}^T]^T$, and $\mathbf{g} = [\mathbf{g}_1^T, \dots, \mathbf{g}_{|\mathcal{N}|}^T]^T$. Moreover, we denote by $\mathbf{r} = [\mathbf{r}_1^T, \dots, \mathbf{r}_{|\mathcal{N}|}^T]^T$, $\mathbf{y} = [\mathbf{y}_1^T, \dots, \mathbf{y}_{|\mathcal{N}|}^T]^T$ and $\mathbf{c} = [\mathbf{c}_1^T, \dots, \mathbf{c}_{|\mathcal{N}|}^T]^T$, the other relevant aggregate vectors. The overall system arising by the composition of the above $|\mathcal{N}|$ subsystems can be described as

$$\begin{cases} \mathbf{x}(k+1) = \mathbf{A}\mathbf{x}(k) + \mathbf{B}\mathbf{g}(k) \\ \mathbf{y}(k) \quad\quad = \mathbf{C}\mathbf{x}(k) \\ \mathbf{c}(k) \quad\quad = \mathbf{H}^c\mathbf{x}(k) + \mathbf{L}\mathbf{g}(k) \end{cases} \tag{16.3}$$

where

$$\mathbf{A} = \begin{pmatrix} \mathbf{A}_{11} & \dots & \mathbf{A}_{1|\mathcal{N}|} \\ \vdots & \ddots & \vdots \\ \mathbf{A}_{|\mathcal{N}|1} & \dots & \mathbf{A}_{|\mathcal{N}||\mathcal{N}|} \end{pmatrix}, \mathbf{B} = \begin{pmatrix} \mathbf{B}_1 & \dots & 0 \\ \vdots & \ddots & \vdots \\ 0 & \dots & \mathbf{B}_{|\mathcal{N}|} \end{pmatrix}$$

$$\mathbf{C} = \begin{pmatrix} \mathbf{C}_1^y & \dots & 0 \\ \vdots & \ddots & \vdots \\ 0 & \dots & \mathbf{C}_{|\mathcal{N}|}^y \end{pmatrix}, \mathbf{H}^c = \begin{pmatrix} \mathbf{H}_1^c \\ \dots \\ \mathbf{H}_{|\mathcal{N}|}^c \end{pmatrix}, L = \begin{pmatrix} L_1 \\ \dots \\ L_{|\mathcal{N}|} \end{pmatrix},$$

and subject to

$$\mathbf{c}(k) \in \mathcal{C}, \ \forall k \in \mathbb{Z}_+, \tag{16.4}$$

where $\mathcal{C} = \mathcal{C}_1 \times \dots \times \mathcal{C}_{|\mathcal{N}|}$. It is further assumed that system (16.3) satisfies the following Assumptions.

A1. The overall system (16.3) is asymptotically stable, i.e., eigenvalues of $\mathbf{A}$ have a magnitude strictly smaller than 1.

A2. System (16.3) is offset free, that is $\lim_{k\to\infty} \mathbf{y}(k) = \bar{\mathbf{g}}$ if a constant $\mathbf{g}(k) = \bar{\mathbf{g}}$ is applied.

The first assumption captures the fact that the system is pre-compensated. The second assumption, which is dictated only by practical reasons, represents the fact that this controller is built in such a way that there is not static tracking error for a constant reference.

16.2.2 Communication Assumptions

It is assumed that the agents are connected via a communication network. Such a network may be modeled by a *communication graph* defined as follows:

Definition 16.1 ***(Communication Graph)*** Let a set of $|\mathcal{N}|$ dynamically interconnected subsystems be given. Then, a Communication Graph is an undirected graph $\Gamma = (\mathcal{N}, \mathcal{B})$, where

(i) $\mathcal{N}$ denotes the set of the $|\mathcal{N}|$ subsystems;
(ii) $\mathcal{B} \subseteq \mathcal{N} \times \mathcal{N}$ the set of edges representing the existing communication links amongst agents in the Communication Graph, i.e., the edge (i, j) belongs to $\mathcal{B}$ if and only if the agents governing the ith and the jth subsystems are able to directly share information within one sampling time;
(iii) Γ is assumed to be connected, i.e., for each couple of agents $i \in \mathcal{N}$, $j \in \mathcal{N}$ there exists at least one sequence of edges connecting i and j, with the minimum number of edges connecting the two agents denoted by $d_{i,j}$.
(iv) The set of all agents with a direct connection with the ith agent represents the **Neighborhood of the** i**th agent** $|\mathcal{N}|_i = \{j \in \mathcal{N} : d_{i,j} = 1\}$.

It is hereafter assumed that Γ is a Hamiltonian graph and, without loss of generality, that the sequence $\mathcal{H} = \{1, 2, \dots, |\mathcal{N}| - 1, |\mathcal{N}|\}$ is a Hamiltonian cycle, i.e, a graph cycle through Γ that visits each node exactly once. Moreover, it is also assumed that each agent communicates with its neighbors once per sampling time and that each of them may act as a gateway in redistributing data amongst the other, no directly connected, agents. As a consequence, at each time instant k, the most recent information on the local state measurements and on the applied commands available to the ith agent is represented by the following **Local Information** vectors:

$$\begin{aligned} \mathbf{p}_i(k) &= [\mathbf{g}_1^T(k - d_{i,1}), \dots, \mathbf{g}_i^T(k-1), \dots, \mathbf{g}_{|\mathcal{N}|}^T(k - d_{i,|\mathcal{N}|})]^T \\ \mathbf{s}_i(k) &= [\mathbf{x}_1^T(k - d_{i,1}), \dots, \mathbf{x}_i^T(k), \dots, \mathbf{x}_{|\mathcal{N}|}^T(k - d_{i,|\mathcal{N}|})]^T \end{aligned} \tag{16.5}$$

while the **Common Information** shared by all agents at each time k is given by

$$\begin{aligned} \mathbf{p}(k) &= [\mathbf{g}_1^T(k - d_1), \dots, \mathbf{g}_i^T(k-1), \dots, \mathbf{g}_{|\mathcal{N}|}^T(k - d_{|\mathcal{N}|})]^T \\ \mathbf{s}(k) &= [\mathbf{x}_1^T(k - d_1), \dots, \mathbf{x}_i^T(k), \dots, \mathbf{x}_{|\mathcal{N}|}^T(k - d_{|\mathcal{N}|})]^T \end{aligned} \tag{16.6}$$

with d_i the maximum amongst all distances $d_{i,j}$ from the ith agent to any other in the graph, i.e., $d_i = \max_{j \in |\mathcal{N}|} d_{i,j}$. Finally, the latest complete state measurement available to all agents at time k is

$$\mathbf{x}(k - d_{\max}) = \left[\mathbf{x}_1^T(k - d_{\max}), \dots, \mathbf{x}_{|\mathcal{N}|}^T(k - d_{\max})\right]^T \tag{16.7}$$

where $d_{\max} = \max_j d_j$.

16.2.3 Problem Statement

Problem 16.1 Consider a network of $|\mathcal{N}|$ pre-compensated, possibly coupled, subsystems $\mathcal{N} = \{1, \ldots, |\mathcal{N}|\}$ where each subsystem (16.1) is subject to local and global constraints (16.2). Moreover, let the overall aggregated system (16.3) fulfill assumptions **A1** and **A2** and denote with $\mathbf{r}_i(k)$ the desired reference signal. Then at each time step k, the CG design problem is that of determining, for each agent $i \in \mathcal{N}$ and on the basis of the available information, a suitable reference signal $\mathbf{g}_i(k)$ such that:

- its application never produces constraints violation, i.e., $\mathbf{c}_i(k) \in \mathcal{C}_i$, $\forall k \in \mathbb{Z}_+$, $i \in \mathcal{N}$.
- $\mathbf{g}_i(k)$ approximates as close as possible $\mathbf{r}_i(k)$.

16.3 Description of the Approach

16.3.1 A Recall of the Centralized CG Design Approach

Before detailing the decentralized CG approach, let us briefly recall how the centralized CG works. The main idea behind the CG approach is that of applying at each time step a command $\mathbf{g}(k) = \mathbf{w}$ chosen in such a way that, if constantly applied over a semi-infinite horizon $l \in [0, \infty)$, it would never produce constraints violation. Two main technical expedients are used:

1. The applied command $\mathbf{w}$ is chosen in such a way that the corresponding equilibrium point $\mathbf{x_w} := (\mathbf{I}_n - \mathbf{A})^{-1}\mathbf{B}\mathbf{w}$ satisfies constraints with margin $\delta > 0$, i.e.,
$$\mathbf{g} \in \mathcal{W}^{\delta} := \{\mathbf{w} : \mathbf{c_w} \in \mathcal{C} \sim \mathcal{B}\}$$
where $\mathbf{c_w} := \mathbf{H}_c\mathbf{x_w} + \mathbf{L}\mathbf{w}$, $\mathcal{B}_\delta$ is the ball of radius δ centered at the origin, δ is an arbitrarily small scalar and $\sim$ denotes the Pontryagin set difference defined as $\mathcal{A} \sim \mathcal{E} := \{a : a + e \in \mathcal{A}, \forall e \in \mathcal{E}\}$. Being $\mathcal{C}$ closed and convex, $\mathcal{W}^{\delta}$ is a closed and convex set as well. Hereafter $\mathcal{W}^{\delta}$ is assumed to be non-empty.
2. The applied command $\mathbf{w}$ is chosen in such a way that the future predictions (*virtual evolutions*) of the $\mathbf{c}$-variable along the *virtual time* k under a constant *virtual command* $\mathbf{g}(l) \equiv \mathbf{w}$ from the initial state $\mathbf{x}$ (at virtual time $l = 0$)
$$\bar{\mathbf{c}}(l, \mathbf{x}, \mathbf{w}) = \mathbf{H}_c\left(\mathbf{A}^l\mathbf{x} + \sum_{i=0}^{l-1}\mathbf{A}^{l-i-1}\mathbf{B}\mathbf{w}\right) + \mathbf{L}\mathbf{w} \tag{16.8}$$
do not violate constraints. This is obtained by selecting the applied command as follows
$$\mathbf{w} \in \mathcal{V}(\mathbf{x}) := \left\{\mathbf{w} \in \mathcal{W}^{\delta} : \bar{\mathbf{c}}(l, \mathbf{x}, \mathbf{w}) \in \mathcal{C},\ \forall l \in \mathbb{Z}_+\right\} \tag{16.9}$$

It is worth remarking that the set $\mathcal{V}(\mathbf{x})$, $\forall \mathbf{x} \in \mathbb{R}^n_{\mathbf{x}}$, is convex and finitely determined, viz. there exists an *a priori* known integer l_0 (see [10]) such that if $\mathbf{c}(l, \mathbf{x}, \mathbf{g}) \in \mathcal{C}$, $l \in \{0, 1, \dots l_0\}$, then $\mathbf{c}(l, \mathbf{x}, \mathbf{w}) \in \mathcal{C}, \;\; \forall l \in \mathbb{Z}_+$.

Finally, the centralized CG problem is solved by choosing at each time instant k a command $\mathbf{g}(k)$ which is the solution of the following convex optimization problem

$$\mathbf{g}(k) = \arg \min_{\mathbf{w} \in \mathcal{V}(\mathbf{x}(k))} \|\mathbf{w} - \mathbf{r}(k)\|^2_{\mathbf{R}} \tag{16.10}$$

The following Theorem, proved in [1] and [7], summarizes the properties of the described CG Algorithm:

Theorem 16.1 *Let assumptions* ***A1-A2*** *be fulfilled. Consider system (16.3)–(16.4) along with the* ***CG*** *selection rule (16.10) and let* $\mathcal{V}(\boldsymbol{x}(0))$ *be non-empty. Then:*

1. *Problem (16.10) is feasible and at each decision time k, viz. the minimizer of (16.10) always uniquely exists;*
2. *The system never violates the constraints, i.e.,* $\mathbf{c}(k) \in \mathcal{C}$, $\forall k \in \mathbb{Z}_+$*;*
3. *The overall system is asymptotically stable. Moreover, whenever* $\mathbf{r}(k) \equiv \mathbf{r}$*, with* $\mathbf{r}$ *a constant set-point, the sequence of* $\mathbf{g}(k)$ *converges in finite time either to* $\mathbf{r}$ *or to its best admissible steady-state approximation* $\hat{\mathbf{r}}$*:*

$$\exists k_s > 0 \;\; k.c. \;\; g(k) = \hat{r} := \arg \min_{\mathbf{w} \in \mathcal{W}_\delta} \|\mathbf{w} - r\|^2_{\mathbf{R}}, \forall k \geq k_s \tag{16.11}$$

and $\lim_{k\to\infty} \hat{\boldsymbol{x}}(k) = \boldsymbol{x}_{\hat{\mathbf{r}}}, \quad \lim_{k\to\infty} \hat{\boldsymbol{y}}(k) = \boldsymbol{y}_{\hat{\mathbf{r}}} = \hat{\mathbf{r}}, \quad \lim_{k\to\infty} \hat{\boldsymbol{c}}(k) = \boldsymbol{c}_{\mathbf{r}}.$

16.3.2 Distributed Command Governor: The Philosophy

The approach adopted to develop the proposed distributed CG strategy is based on the following two key ideas

- On the basis of the information available to all agents (the **Common Information**), they locally compute the same family of ***local artificial constraints*** $\{\Delta\mathcal{V}_1(k), \Delta\mathcal{V}_2(k), \dots, \Delta\mathcal{V}_{|\mathcal{N}|}(k)\}$. This family of sets is computed to ensure that for any choice of the local reference such that

$$\mathbf{g}_i(k) - \mathbf{g}_i(k-1) \in \Delta\mathcal{V}_i(k), i = 1, \dots, |\mathcal{N}| \tag{16.12}$$

  the global constraints (16.9) are not violated. In order to ensure feasibility and computability $\Delta\mathcal{V}_i(k) \subseteq \mathbb{R}^{n_{g_i}}$, $i \in \mathcal{N}$ have to be convex and compact sets containing 0_{m_i} as an internal point.
- On the basis of the **Local Information** each agent computes the command to be applied minimizing its local cost function $\|\mathbf{w}_i - \mathbf{r}_i\|^2_{\mathbf{R}_i}$, with $\mathbf{R}_i > 0$, and complying with the above mentioned local artificial constraint.

16.3.3 State and Command Predictions

In order to be able to compute the collection of sets $\{\Delta\mathcal{V}_1(k), \Delta\mathcal{V}_2(k), \ldots, \Delta\mathcal{V}_{|\mathcal{N}|}(k)\}$, the first step is the determination of the set of all possible commands and states compatible with the common information available to all agents of the networks. To this end, we assume that each agent, at time k, is provided with a collection $\{\Delta\mathcal{V}_i(k-l)\}_{l=1}^{d_{\max}}$, $\forall i \in \mathcal{N}$ of sets computed at the previous step. Such information, together with the common information vector $\mathbf{p}(k)$, can be exploited to define the set of all possible feasible values which $\mathbf{g}(k)$ could have been assumed in the last $d_{\max}$ time steps,

$$\begin{array}{ll}\hat{\mathbf{g}}_i(k-l|k) = \mathbf{g}_i(k-l) & \text{if } l \geq d_i \\ \hat{\mathbf{g}}_i(k-l|k) = \mathbf{g}_i(k-d_i) \oplus \Delta\mathcal{V}_i(k-d_i+1) \oplus \ldots \oplus \Delta\mathcal{V}_i(k-l) & \text{if } l < d_i.\end{array} \tag{16.13}$$

where $\oplus$ denotes *Pontryagin—Minkowski set sum, i.e.,* $\mathcal{A} \oplus \mathcal{B} := \{a + b | a \in \mathcal{A}, b \in \mathcal{B}\}$. The set of all possible values for $\mathbf{g}(k-l)$ is then given by

$$\hat{\mathcal{G}}(k-l|k) = \hat{\mathbf{g}}_1(k-l|k) \times, \ldots, \times \hat{\mathbf{g}}_{|\mathcal{N}|}(k-l|k). \tag{16.14}$$

Moreover, the set of all possible state predictions at time k, computed on the basis of the measured state $\mathbf{x}(k-d_{\max})$ available to all agents of the network is given by

$$\hat{\mathcal{X}}(k) := \mathbf{A}^{d_{\max}}\mathbf{x}(k-d_{\max}) + \bigoplus_{j=0}^{d_{\max}-1} \mathbf{A}^{d_{\max}-1-j}\mathbf{B}\hat{\mathcal{G}}(k-d_{\max}+j|k) \tag{16.15}$$

By following the same lines, for each ith agent one can define the predictions of the commands computed on the basis of the information available

$$\begin{array}{ll}\hat{\mathbf{g}}_j(k-l|k,i) = \mathbf{g}_j(k-l) & \text{if } l \geq d_{i,j} \\ \hat{\mathbf{g}}_j(k-l|k,i) = \mathbf{g}_j(k-d_{i,j}) \oplus \Delta\mathcal{V}_j(k-d_{i,j}+1) \oplus \ldots \oplus \Delta\mathcal{V}_j(k-l) & \text{if } l < d_{i,j}\end{array} \tag{16.16}$$

and the corresponding prediction sets:

$$\hat{\mathcal{G}}_i(k-l|k) = \hat{\mathbf{g}}_1(k-l|k,i) \times, \ldots, \times \hat{\mathbf{g}}_{|\mathcal{N}|}(k-l|k,i), \tag{16.17}$$

$$\hat{\mathcal{X}}_i(k) := \mathbf{A}^{d_i}\mathbf{x}(k-d_i) + \bigoplus_{k=0}^{d_i-1} \mathbf{A}^{d_i-1-l}\mathbf{B}\hat{\mathcal{G}}_i(k-d_i+l|k). \tag{16.18}$$

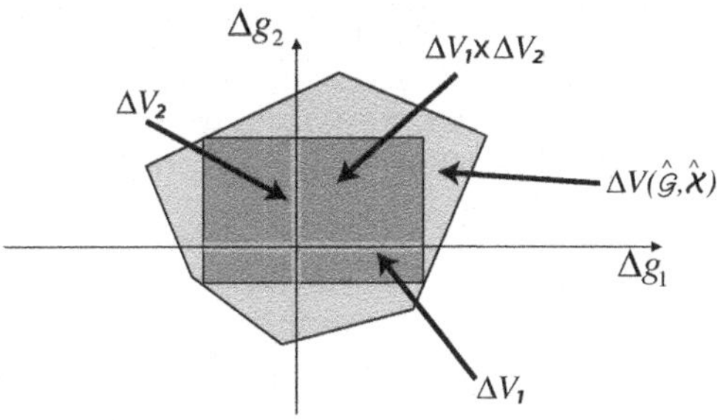

Fig. 16.3 Cartesian decomposition of the set $\Delta\mathcal{V}(\hat{\mathcal{G}}, \hat{\mathcal{X}})$ in the case where two agents are operating in a bi-dimensional decision space

16.3.4 Computation of $\{\Delta\mathcal{V}_1(k), \Delta\mathcal{V}_2(k), \ldots, \Delta\mathcal{V}_{|\mathcal{N}|}(k)\}$

The core of the distributed CG approach is the computation of a set of orthogonal artificial constraints $\{\Delta\mathcal{V}_1(k), \Delta\mathcal{V}_2(k), \ldots, \Delta\mathcal{V}_{|\mathcal{N}|}(k)\}$ such that for any choice of the local reference variation complying with them, global constraints (16.9) are not violated. In principle, there are several ways to build such kind of sets. The two main approaches are

- Maximal Volume Set Cartesian Decomposition (see Fig. 16.3);
- Decomposition along the ith direction;

which are detailed hereafter.

16.3.4.1 Maximal Volume Set Cartesian Decomposition

Based on the sets introduced in Sect. 16.3.3 we may compute the set of admissible aggregate command variations computed on the basis of the common information as follows:

$$\Delta\mathcal{V}(\hat{\mathcal{G}}, \hat{\mathcal{X}}) := \left\{\Delta\mathbf{w} : (\mathbf{w} + \Delta\mathbf{w}) \in \mathcal{V}(\mathbf{x}), \forall\mathbf{w} \in \hat{\mathcal{G}}, \forall\mathbf{x} \in \hat{\mathcal{X}}\right\} \tag{16.19}$$

The **Maximal Volume Set Cartesian Decomposition** determines the maximum volume inner box approximation of (16.19). It can be computed independently and locally by each agent by solving the following optimization problem:

$$\begin{array}{l} \max\limits_{\Delta\mathcal{V}_i(k), i=1,\ldots,|\mathcal{N}|} \quad Volume(\Delta\mathcal{V}_1(k) \times \ldots \times \Delta\mathcal{V}_{|\mathcal{N}|}(k)) \\ subject\ to \\ \Delta\mathcal{V}_1(k) \times \ldots \times \Delta\mathcal{V}_{|\mathcal{N}|}(k) \subseteq \Delta\mathcal{V}(\hat{\mathcal{G}}(k-1|k), \hat{\mathcal{X}}(k)), \\ 0_1 \in In(\Delta\mathcal{V}_1(k)), \ldots, 0_{|\mathcal{N}|} \in In(\Delta\mathcal{V}_{|\mathcal{N}|}(k)) \end{array} \tag{16.20}$$

Efficient exact solutions for the case $m_i = 1,\ i = 1, \ldots, |\mathcal{N}|$ and approximated solutions for the other cases can be found in [2]. For notational com-

Algorithm 16.1 Distributed-CG Algorithm (D-CG), for Time k, Agent ith

1: Receive $\mathbf{p}_j(k-1), \mathbf{s}_j(k-1)$ from the neighbors
2: Compute $\{\Delta\mathcal{V}_1(k), \Delta\mathcal{V}_2(k), \ldots, \Delta\mathcal{V}_{|\mathcal{N}|}(k)\}$ using a predefined policy
3: Solve $\mathbf{g}_i(k) = \arg\min_{\mathbf{w}_i \in \Delta\mathcal{V}_i} \|\mathbf{w}_i - \mathbf{r}_i\|^2_{\mathbf{R}_i}$
4: Apply $\mathbf{g}_i(k)$
5: Send $\mathbf{p}_i(k), \mathbf{s}_i(k), \mathbf{g}_i(k)$ to the neighbors

pactness we will denote this decomposition as : $\{\Delta\mathcal{V}_1(k) \times \ldots \times \Delta\mathcal{V}_{|\mathcal{N}|}(k)\} = \text{maxVol}(\Delta\mathcal{V}(\hat{\mathcal{G}}(k-1|k), \hat{\mathcal{X}}(k)))$.

16.3.4.2 Decomposition Along the ith Direction

The idea is to compute a decomposition where all agents but the ith one are forced to hold their previously applied references. The latter corresponds to considering the following local artificial constraints:

$$\begin{aligned} &\Delta\mathcal{V}_i(k) = \left\{\Delta\mathbf{w}_i | [0_1^T, \ldots, \Delta\mathbf{w}_i^T, \ldots, 0_{|\mathcal{N}|}^T]^T \in \mathcal{V}(\hat{\mathcal{G}}_i(k-1|k), \hat{\mathcal{X}}_i(k))\right\}, \\ &\Delta\mathcal{V}_j(k) = \{0_j\}, \forall j \neq i. \end{aligned} \tag{16.21}$$

For notational compactness we will denote this decomposition as : $\{\Delta\mathcal{V}_1(k) \times \ldots \times \Delta\mathcal{V}_{|\mathcal{N}|}(k)\} = \text{maxDir}(i, \Delta\mathcal{V}(\hat{\mathcal{G}}(k-1|k), \hat{\mathcal{X}}(k)))$.

In practical uses, the direction of the decomposition is a time-varying function of the time chosen so as to ensure that in a certain period all the directions have been chosen. Hereafter, we will always make use of $i = k \bmod |\mathcal{N}|$ where mod is the remainder of the integer division of k by $|\mathcal{N}|$.

16.3.5 Distributed Command Governor: The Algorithm

Once that a policy to determine the sets $\{\Delta\mathcal{V}_1(k), \Delta\mathcal{V}_2(k), \ldots, \Delta\mathcal{V}_{|\mathcal{N}|}(k)\}$ is selected, the distributed CG algorithm can be defined as in Algorithm 16.1.

In the previous Section, two decomposition policies have been defined. Preliminary works [8, 9] have shown that each one of them has pros and cons and in particular:

- **Maximal Volume Set Cartesian Decomposition** works well when the possible applied commands $\hat{\mathcal{G}}(k-1|k)$ are far from the borders of $\mathcal{W}^\delta$ while it becomes very slow close to them.
- **Decomposition along the** $i = k \bmod |\mathcal{N}|$**-th direction** is slow when $\hat{\mathcal{G}}(k-1|k)$ is far from the borders of $\mathcal{W}^\delta$ while works well close to them.

In order to overcome the above limitations, it is convenient to use the Maximal Volume Set Cartesian Decomposition when the last command is far from the borders

Algorithm 16.2 Improved Distributed-CG Algorithm (D-CG), for Agent ith

1: Initialization

1.1 compute $\mathcal{W}^\delta$ after choosing δ
1.2 compute l_0 for set $\mathcal{V}(\mathbf{x})$ (see [10])
1.3 provide each agent with $\mathbf{x}_j(0)$ and $\mathbf{g}_j(0), \forall j \in \mathcal{N} \setminus \{i\}$

2: On-line phase, repeated at each time k

2.1. receive $\mathbf{p}_j(k-1), \mathbf{s}_j(k-1)$ from all neighbors $j \in \mathcal{N}_i$
2.2. solve

$$\mu = \min_{\mathbf{w} \in \hat{\mathcal{G}}(k-1|k), \mathbf{w}' \in \partial(\mathcal{W}^\delta)} \|\mathbf{w} - \mathbf{w}'\| \tag{16.22}$$

2.3. if $(\mu \leq \epsilon_S)$ OR $(0 < counter \leq |\mathcal{N}|_S)$

2.3.1. $(\Delta\mathcal{V}_1(k) \times \ldots \times \Delta\mathcal{V}_{|\mathcal{N}|}(k)) = \text{maxDir}(k \text{mod} |\mathcal{N}|, \Delta\mathcal{V}(\hat{\mathcal{G}}_i(k-1|k), \hat{\mathcal{X}}_i(k)))$
2.3.2. $counter++;$

2.4. else

2.4.1. $(\Delta\mathcal{V}_1(k) \times \ldots \times \Delta\mathcal{V}_{|\mathcal{N}|}(k)) = \text{maxVol}(\Delta\mathcal{V}(\hat{\mathcal{G}}(k-1|k), \hat{\mathcal{X}}(k)))$
2.4.2. $counter = 0;$

2.5. Solve $\mathbf{g}_i(k) = \arg\min_{\mathbf{w}_i \in \Delta\mathcal{V}_i} \|\mathbf{w}_i - \mathbf{r}_i\|^2_{\mathbf{R}_i}$
2.6. apply $\mathbf{g}_i(k)$
2.7. transmit $\mathbf{p}_i(k), \mathbf{s}_i(k), \mathbf{g}_i(k)$ to neighborhood $\mathcal{N}_i$

of $\mathcal{W}^\delta$ and the Decomposition along the ith direction when, on the contrary, is near to them. This leads to the definition of the improved scheme Algorithm 16.2, presented for the first time in [8]. In Algorithm 16.2, μ computed in (16.22) represents the distance between $\hat{\mathcal{G}}(k-1|k)$ and the border of $\mathcal{W}^\delta$. When μ is smaller than the threshold maxDir will be used instead of maxVol. Notice that $counter$ is an integer initialized at zero and used to force the system to use the decomposition along the $i = k \bmod |\mathcal{N}|$-th direction for at least $|\mathcal{N}|_S \geq |\mathcal{N}|$ consecutive steps before switching back to maxVol. The threshold $\epsilon_S > 0$ and the minimum number of sequential actions $|\mathcal{N}|_S \geq |\mathcal{N}|$ are design parameters that can be tuned accordingly to the specific plant.

16.4 Theoretical Results Availability

The described distributed CG scheme has been analyzed in several works. In [5] and [4] a complete **liveliness** analysis was presented. There it has been shown that, unlike the centralized case, Algorithm 16.2 could remain stuck in certain pathological configurations. To avoid that, a geometrical Constraints Qualification (CQ) condition on the prescribed constrained set is proposed whose fulfillment avoids the occurrence of deadlock situations and always ensures viable solutions. Such a condition can be easily checked from the outset via a numerical procedure provided in [4] where an algorithm to compute Constrained Qualified (CD) arbitrarily accurate inner approximations of the original constraint set is also presented for systems where each local reference is scalar, i.e., $g_i \in \mathbb{R}, i = 1, \ldots, |\mathcal{N}|$.

Under CQ, the properties of the **D-CG** scheme can be summarized in the following Theorem whose proof is detailed in [13],

Theorem 16.2 *Let assumptions* **A1-A2** *be fulfilled for the system arising from the composition of* $|\mathcal{N}|$ *subsystems (16.1) and let CQ condition hold true. Consider the distributed* **D-CG** *Algorithm 16.1 and let* $\mathcal{V}(\boldsymbol{x}(k))$ *be non empty at time* $k = 0$. *Then*

1. *for each agent* $i \in |\mathcal{N}|$, *at each time* k, *the minimizer related the optimization problem in Algorithms 16.1-16.2, to be solved* $\forall i \in \mathcal{N}$, *uniquely exists and can be obtained by solving a convex constrained optimization problem;*
2. *The overall system acted by agents implementing the D-CG supervisory policy never violates the constraints, i.e.,* $\boldsymbol{c}(k) \in \mathcal{C}$ *for all* $k \in \mathbb{Z}_+$.
3. *Whenever* $\boldsymbol{r}_i(k) \equiv \boldsymbol{r}_i, \forall i \in \mathcal{N}$, *with* $\boldsymbol{r}_i$ *a constant set-point the sequence of* $\boldsymbol{g}(k) = [\boldsymbol{g}_1^T(k), \ldots, \boldsymbol{g}_{|\mathcal{N}|}^T(k)]^T$ *asymptotically converges either to* $r = [r_1^T(k), \ldots, r_{|\mathcal{N}|}^T(k)]^T$ *if* $r \in \mathcal{W}^\delta$ *or to a point* $\hat{r}$ *that is Pareto-Optimal.*

For the details concerning the above stated properties and Constraints Qualification algorithms, please refer to [12].

16.5 Application Results Availability

In recent years, a few interesting application examples making use of the distributed CG scheme here presented have been published. Amongst them it is worth mentioning that:

- Sequential and parallel non-iterative FF-CG schemes with application to networks of autonomous vehicles have been presented in [8, 9].
- Non-iterative CG schemes with application to networks of pumps and water tanks appeared in [6, 14, 13].
- A sequential non-iterative CG scheme applied to Load/Frequency control problems in multi-area power systems appeared in [3].
- A mixed-integer distributed CG formulation for collision avoidance in autonomous vehicle supervision problems appeared in [15].

16.6 Some Hints for the Implementation

In what follows we present the computational details related to the implementation of the strategy presented in this chapter. The main difficulty in implementing Algorithms 16.1-16.2 relies in the computation of the set $\Delta\mathcal{V}(\hat{\mathcal{G}}, \hat{\mathcal{X}})$ in (16.19) that has to be

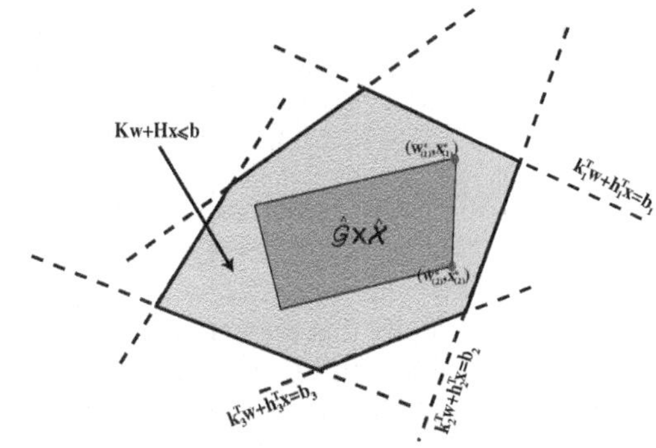

Fig. 16.4 Some points $(\mathbf{w}^*_{(j)}, \mathbf{x}^*_{(j)})$ computed by means of (16.25)

evaluated for an infinite number of points $(w, x) \in \hat{\mathcal{G}} \times \hat{\mathcal{X}}$. Nevertheless, it can be proved that sets $\hat{\mathcal{G}}$ and $\hat{\mathcal{X}}$ are characterized by a polytopic structure. Then, if the set $\mathcal{C}$ is assumed to be a polytope as well, the set $\Delta\mathcal{V}(\hat{\mathcal{G}}, \hat{\mathcal{X}})$ can be defined as follows

$$\Delta\mathcal{V}(\hat{\mathcal{G}}, \hat{\mathcal{X}}) = \{\Delta\mathbf{w} | \mathbf{K}(\mathbf{w} + \Delta\mathbf{w}) + \mathbf{H}\mathbf{x} \le \mathbf{b}, \forall \mathbf{w} \in \hat{\mathcal{G}}, \forall \mathbf{x} \in \hat{\mathcal{X}}\} \tag{16.23}$$

where $\mathbf{K} \in \mathbb{R}^{p \times n_\mathbf{w}}$, $\mathbf{H} \in \mathbb{R}^{p \times n_\mathbf{x}}$ and $\mathbf{b} \in \mathbb{R}^p$.

In [12] it has been proved that it is not necessary to evaluate $\Delta\mathcal{V}(\hat{\mathcal{G}}, \hat{\mathcal{X}})$ for all points belonging to $\hat{\mathcal{G}} \times \hat{\mathcal{X}}$. On the contrary, it can be done, as depicted in Fig. 16.4, by separately considering each of the p hyper-planes described by matrices $\mathbf{H}, \mathbf{K}$ and $\mathbf{b}$ and computing the related sets

$$\Delta\mathcal{V}^j(\mathbf{w}^*_j, \mathbf{x}^*_j) := \{\Delta\mathbf{w} | \mathbf{k}_j^T(\mathbf{w}^*_j + \Delta\mathbf{w}) + \mathbf{h}_j^T\mathbf{x}^*_j \le \mathbf{b}_j\}, \; j = 1, \ldots, p \tag{16.24}$$

where $\mathbf{w}^*_j$ and $\mathbf{x}^*_j$ are obtainable by solving the following Linear Programming optimization problem

$$(\mathbf{w}^*_j, \mathbf{x}^*_j) = \arg \min_{\mathbf{g} \in \hat{\mathcal{G}}, \mathbf{x} \in \hat{\mathcal{X}}} \left(\mathbf{b}_j - \mathbf{k}_j^T(\mathbf{w}^*_j) - \mathbf{h}_j^T\mathbf{x}^*_j\right) \tag{16.25}$$

In this way one is able to replace $\Delta\mathcal{V}(\hat{\mathcal{G}}, \hat{\mathcal{X}})$ with its inner-approximation

$$\begin{aligned} \bigcap_{j=1}^{p} \Delta\mathcal{V}^j(\mathbf{w}^*_j, \mathbf{x}^*_j) &:= \left\{\Delta\mathbf{w} | \mathbf{k}_j^T(\mathbf{w}^*_j + \Delta\mathbf{w}) + \mathbf{h}_j^T\mathbf{x}^*_j \le \mathbf{b}_j, \; j = 1, \ldots, p\right\} \\ &:= \left\{\Delta\mathbf{w} | \mathbf{K}\Delta\mathbf{w} \le \tilde{\mathbf{b}}\right\} \end{aligned} \tag{16.26}$$

where

$$\tilde{\mathbf{b}} = \begin{bmatrix} \mathbf{b}_1 - \mathbf{k}_1^T\mathbf{w}_1^* + \mathbf{h}_1^T\mathbf{x}_1^* \\ \vdots \quad\quad \vdots \\ \mathbf{b}_j - \mathbf{k}_j^T\mathbf{w}_j^* + \mathbf{h}_j^T\mathbf{x}_j^* \\ \vdots \quad\quad \vdots \\ \mathbf{b}_p - \mathbf{k}_p^T\mathbf{w}_p^* + \mathbf{h}_p^T\mathbf{x}_p^* \end{bmatrix} \tag{16.27}$$

Another difficulty that may arise in the implementation is how to solve problems in the form (16.20). In [2] several polynomially-in-time algorithms are presented for computing approximated Cartesian Decomposition with maximum volume contained in a given convex polytope as $\bigcap_{j=1}^{p} \Delta\mathcal{V}^j(\mathbf{w}_j^*, \mathbf{x}_j^*)$. One of these methods has been extended to the case where the set $\bigcap_{j=1}^{p} \Delta\mathcal{V}^j(\mathbf{w}_j^*, \mathbf{x}_j^*)$ describes a polyhedron. The idea is to maximize the volume of the box set

$$\mathcal{M}(\underline{\Delta\mathbf{w}}, \overline{\Delta\mathbf{w}}) = \left\{ \Delta\mathbf{w} \in \bigcap_{j=1}^{p} \Delta\mathcal{V}^j(\mathbf{w}_j^*, \mathbf{x}_j^*) : \underline{\Delta\mathbf{w}} \leq \Delta\mathbf{w} \leq \overline{\Delta\mathbf{w}} \right\} \tag{16.28}$$

Hence, an algorithm that maximizes the volume of $\mathcal{M}(\underline{\Delta\mathbf{w}}, \overline{\Delta\mathbf{w}})$ can be formulated as

$$\max_{\underline{\Delta\mathbf{w}}, \Delta\mathbf{w}^+} \quad \textstyle\sum_{i=1}^{m} \ln \Delta\mathbf{w}_i^+ \tag{16.29}$$

$$\text{s.t. } \mathbf{K}\underline{\Delta\mathbf{w}} + \mathbf{K}^+\Delta\mathbf{w}^+ \leq \tilde{\mathbf{b}} \tag{16.30}$$

$$\underline{\Delta\mathbf{w}} \leq -\varepsilon \tag{16.31}$$

$$\underline{\Delta\mathbf{w}} + \Delta\mathbf{w}^+ \geq \varepsilon \tag{16.32}$$

where $\Delta\mathbf{w}^+ := \overline{\Delta\mathbf{w}} - \underline{\Delta\mathbf{w}}$ and $\tilde{\mathbf{K}}^+$ is the positive part of $\tilde{\mathbf{K}}$, namely $\mathbf{k}_{i,j}^+ := \max(0, \mathbf{w}_{i,j})$. More details can be found in [2].

As far as the computation of sets $\Delta\mathcal{V}_i(k)$ in (16.21) is concerned, they can be evaluated by taking into account the inner-approximation (16.26), which leads

$$\begin{aligned} \Delta\mathcal{V}_i(k) &:= \left\{ \Delta\mathbf{w}_i | [0_1^T, \ldots, \Delta\mathbf{w}_i^T, \ldots, 0_{|\mathcal{N}|}^T]^T \in \textstyle\bigcap_{j=1}^{p} \Delta\mathcal{V}^j(\mathbf{w}_j^*(k), \mathbf{x}_j^*(k)), \, j = 1, \ldots, p \right\} \\ &:= \left\{ \Delta\mathbf{w}_i | \mathbf{k}_j^T(\mathbf{w}_j^*(k) + [0_1^T, \ldots, \Delta\mathbf{w}_i^T, \ldots, 0_{|\mathcal{N}|}^T]^T) + \mathbf{h}_j^T\mathbf{x}_j^*(k) \leq \mathbf{b}_j, \, j = 1, \ldots, p \right\} \\ &:= \left\{ \Delta\mathbf{w}_i | \mathbf{K}_i \Delta\mathbf{w}_i \leq \tilde{\mathbf{b}}(k) \right\} \end{aligned} \tag{16.33}$$

with $\mathbf{K}_i$ representing the ith column of $\mathbf{K}$.

Finally, a reasoning path similar to that followed for the evaluation of the set $\Delta\mathcal{V}(\hat{\mathcal{G}}, \hat{\mathcal{X}})$ can be used to solve Problem (16.22). In fact, because $\mathcal{C}$ is assumed to be a polytope (intersection of z hyper-planes), $\mathcal{W}_\delta$ takes a polytopic structure as well

$$\Delta\mathcal{W}_\delta = \left\{\mathbf{w} : \mathbf{K}_{\mathcal{W}_\delta}\mathbf{w} \leq \mathbf{b}_{\mathcal{W}_\delta}\right\} \tag{16.34}$$

As a consequence, $\mu(t)$ can be easily computed as

$$\mu(t) = \min_j \left(\min_{g \in \hat{\mathcal{G}}} \frac{\mathbf{b}_{j,\mathcal{W}_\delta} - \mathbf{k}^T_{j,\mathcal{W}_\delta}}{|\mathbf{k}^T_{j,\mathcal{W}_\delta}|} \right), j = 1, \ldots, z \tag{16.35}$$

Please note that, because $\hat{\mathcal{G}}$ is a polytope (a box indeed), the resolution of the latter expression requires the optimization of a finite number of LP problems. All the above presented computational details have been rigorously detailed in [12].

References

1. A. Bemporad, A. Casavola, E. Mosca, Nonlinear control of constrained linear systems via predictive reference management. IEEE Trans. Automat. Control **42**, 340–349 (1997)
2. A. Bemporad, C. Filippi, F.D. Torrisi, Inner and outer approximations of polytopes using boxes. Comput. Geom. **27**, 151–178 (2004)
3. A. Casavola, G. Franzè, E. Garone, F. Tedesco, Distributed coordination-by-constraint strategies in networked multi-area power systems, in *Proceeding of 21st IEEE International Symposium on Industrial Electronics (ISIE2012)*, Gdansk, 2011, http://tedescof.wordpress.com/publication/
4. A. Casavola, E. Garone, and F. Tedesco, A liveliness analysis of a distributed constrained coordination strategy for multi-agent linear systems. in *50th IEEE Decision and Control and European Control Conference (CDC-ECC)*, Orlando, 2011, pp. 8133–8138
5. A. Casavola, E. Garone, and F. Tedesco, Distributed coordination-by-constraint strategies for multi-agent networked systems, in *50th IEEE Decision and Control and European Control Conference (CDC-ECC)*, Orlando, 2011, pp. 6888–6893
6. A. Casavola, E. Garone, F. Tedesco, Distributed reference management strategies for a networked water distribution system, in *Proceeding of 18th IFAC World Congress*, Milan, 2011, pp. 8951–8956, http://tedescof.wordpress.com/publication/
7. A. Casavola, E. Mosca, D. Angeli, Robust command governors for constrained linear systems. IEEE Trans. Automat. Control **45**, 2071–2077 (2000)
8. E. Garone, F. Tedesco, A. Casavola, Distributed coordination strategies for interconnected multi-agent systems, in *Proceeding of 8th IFAC Symposium on Nonlinear Control Systems (NOLCOS'10)*, Bologna, 2010, http://tedescof.wordpress.com/publication/
9. E. Garone, F. Tedesco, A. Casavola, Distributed coordination-by-constraints strategies for networked control systems, in *Proceeding of Distributed Estimation and Control in Networked Systems (NecSys'09)*, Venice, 2009, http://tedescof.wordpress.com/publication/
10. E.G. Gilbert, I.V. Kolmanovsky, K.T. Tan, Discrete-time reference governors and the nonlinear control of systems with state and control constraints. Int. J. Robust Nonlinear Control **5**, 487–504 (1995)
11. A. Kurzhanski, I. Valyi, *Ellipsoidal Calculus for Estimation and Control* (Birkhauser, Berlin, 1997)
12. F. Tedesco, Distributed Command Governor Strategies for Multi-agent Dynamical Systems. PhD thesis, Università della Calabria, Rende , March 2012

13. F. Tedesco, A. Casavola, E. Garone, A distributed parallel command governor strategy for the coordination of multi-agent networked systems, in *Proceeding of 4th IFAC Nonlinear Model Predictive Control conference (NMPC'12)*, Noordwijkerhout, 2012, http://tedescof.wordpress.com/publication/
14. F. Tedesco, A. Casavola, E. Garone, Distributed command governor strategies for constrained coordination of multi-agent networked systems, in *Proceeding of American Control Conference*, Montreal, 2012, http://tedescof.wordpress.com/publication/
15. F. Tedesco, D.M. Raimondo, A. Casavola, J. Lygeros, Distributed collision avoidance for interacting vehicles: a command governor approach, in *Proceeding of Distributed Estimation and Control in Networked Systems (NecSys'10)*, Annecy, 2010, http://tedescof.wordpress.com/publication/

Chapter 17
Mixed-Integer Programming Techniques in Distributed MPC Problems

I. Prodan, F. Stoican, S. Olaru, C. Stoica and S.-I. Niculescu

Abstract This chapter proposes a distributed approach for the resolution of a multi-agent problem under collision and obstacle avoidance conditions. Using hyperplane arrangements and mixed integer programming, we provide an efficient description of the feasible region verifying the avoidance constraints. We exploit geometric properties of hyperplane arrangements and adapt this description to the distributed scheme in order to provide an efficient Model Predictive Control (MPC) solution. Furthermore, we prove constraint validation for a hierarchical ordering of the agents.

17.1 Introduction

Distributed control usually means a decomposition of a large scale system into a set of several smaller subsystems ("neighborhoods"). The rationale of this approach is to provide subsystems which have fewer decision variables and are affected by fewer

I. Prodan (✉)· S. Olaru · C. Stoica
Automatic Control Department, SUPELEC Systems Sciences (E3S), Gif Sur Yvette, France
e-mail: ionela.prodan@supelec.fr

S. Olaru
e-mail: sorin.olaru@supelec.fr

C. Stoica
e-mail: cristina.stoica@supelec.fr

F. Stoican
Department of Engineering Cybernetics, Norwegian University of Science and Technology (NTNU), Trondheim, Norway
e-mail: florin.stoican@itk.ntnu.no

I. Prodan · S.-I. Niculescu
Signals and Systems Laboratory, SUPELEC Systems Sciences (E3S), Gif Sur Yvette, France
e-mail: ionela.prodan@lss.supelec.fr

S.-I. Niculescu
e-mail: silviu.niculescu@lss.supelec.fr

J. M. Maestre and R. R. Negenborn (eds.), *Distributed Model Predictive Control Made Easy*, Intelligent Systems, Control and Automation: Science and Engineering 69,
DOI: 10.1007/978-94-007-7006-5_17,

constraints (thus making the optimization problems easier to solve). This design requirement means that we aim for subsystems which are loosely inter-coupled, i.e., a given subsystem is affected by only a few other subsystems [11].

A classical area in which distributed control can be applied is the control of *multi-agent systems* [1]. Here, the subsystems affect each other mainly through obstacle and collision avoidance requirements. The primary challenge in applying these requirements is that they model a non-convex feasible region, i.e., the agent state trajectory has to avoid a convex region representing an obstacle (static constraints) or another agent (dynamic constraints—leading to a parametrization of the set of constraints with respect to the current state).

An extensively used method for formulating this type of problems is represented by *Mixed-Integer Programming* (MIP) [2] which provides the advantage of *explicitly* including non-convex constraints and discrete decisions in the optimization problem. These techniques have proven their usefulness in various applications. Among them, we cite [8] for task assignment with coordinated control of multiple agents, subject to dynamics and collision avoidance constraints, and [12] for fault detection and isolation.

A sensitive aspect of MIP techniques is the *computational complexity* which can increase exponentially with the number of binary variables used in the problem formulation. There are some works where the original decision problems are reformulated in a simplified MIP form [14], but the complexity still remains significant. Other works try to reduce the number of binary variables, e.g., through a logarithmic formulation, as recalled in [13]. A similar technique with geometric insights into the description of the feasible region is discussed in [9] and provides a notable improvement of the overall MIP formulation. The main improvement is the use of hyperplane arrangements which give a formal way of describing the usually *non-connected and non-convex* feasible region.

With all these improvements considered, the problem at hand is still difficult to solve, even more so when a Model Predictive Control (MPC) scheme is applied (both the dimension of the solution space and the difficulty of describing the feasible region become larger with an increase in the length of the prediction horizon). MPC and, naturally, Distributed Model Predictive Control (DMPC) can be used to tackle this problem and the present chapter aims to offer a MIP solution in the context of predictive control. Here we propose to present a two-pronged approach. On one side, we consider the geometric description of the feasible region and on the other side we introduce a distributed MPC approach, upon the principles in [11]. In both cases we exploit the topology of the problem: we consider neighborhoods which partition the agents into groups and give a MIP description of the feasible region in which they stand. With an adequate communication between the resulting groups of agents, we are able to respect performance and stability constraints while in the same time we greatly reduce the computational load.

The rest of the chapter is organized as follows. The preliminaries are given in Sect. 17.2. Improvements in the geometric representation are shown in Sect. 17.3 and a hierarchical distributed MPC is discussed in Sect. 17.4. The conclusions are drawn in Sect. 17.5.

The following notation will be used throughout the paper. The closure of a set S, $cl(S)$, is the intersection of all closed sets containing S. The collection of all possible combinations of N binary variables will be noted $\{0, 1\}^N := \{(b_1, \ldots, b_N) : b_i \in \{0, 1\}, \forall i = 1 : N\}$. For a scalar $x \in \mathbb{R}$, we denote by $\lceil x \rceil$ the upper integer part. A finite intersection of inequalities which describes a non-empty region is called a polyhedral set. A polytope is a bounded polyhedral set. Minkowski's addition of two sets $\mathcal{X}$ and $\mathcal{Y}$ is defined as $\mathcal{X} \oplus \mathcal{Y} = \{x + y : x \in \mathcal{X},\ y \in \mathcal{Y}\}$. We write $\mathbf{R} \succ 0$ to denote that $\mathbf{R}$ is a positive definite matrix and $\mathbf{Q} \succeq 0$ a positive semidefinite matrix.

17.2 Preliminaries and Prerequisites

In this section, we recall results from [9] and the appropriate framework which permit to describe a non-convex and non compact feasible region using the MIP formalism. In the second part, we use this codification to construct a typical centralized optimization problem for a multi-agent system [8]. Lastly, we shed light on the shortcomings afflicting this construction (most importantly the numerical difficulties) and provide a sketch of the approaches we consider in the rest of the chapter for improving the solution.

17.2.1 Mixed Integer Representation of a Non-Convex Feasible Region

Let us consider a collection of $N > 0$ hyperplanes

$$\mathcal{H}_i = \{x \in \mathbb{R}^n : h_i x = k_i\},\ i = 1 : N, \tag{17.1}$$

with $(h_i, k_i) \in \mathbb{R}^{1\times n} \times \mathbb{R}$, each of them partitioning the space $\mathbb{R}^n$ into two disjoint regions (up to their common boundary—the hyperplane $\mathcal{H}_i$):

$$\mathcal{R}^+(\mathcal{H}_i) = \{x \in \mathbb{R}^n : h_i x \le k_i\},\quad \mathcal{R}^-(\mathcal{H}_i) = \{x \in \mathbb{R}^n : -h_i x \le -k_i\}. \tag{17.2}$$

Assuming their intersection non-empty and bounded, we define the polytopic set $P \subset \mathbb{R}^n$ through its implicit half-space description[1]:

$$P = \bigcap_{i=1:N} \mathcal{R}^+(\mathcal{H}_i), \tag{17.3}$$

[1] We have made the simplifying convention that all the half-spaces appearing in (17.3) are of form $\mathcal{R}^+(\cdot)$.

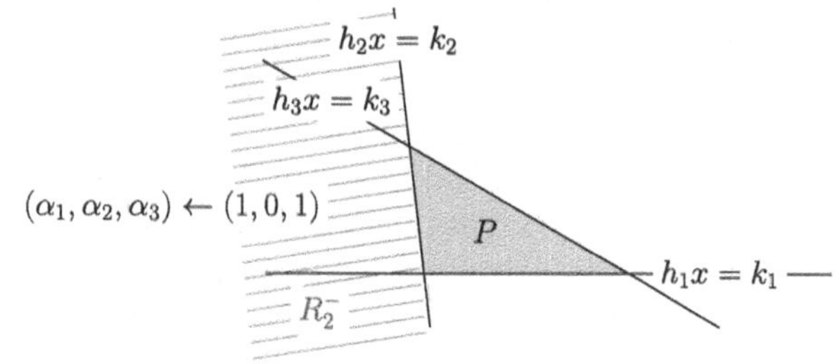

Fig. 17.1 Exemplification of mixed integer codification

and its complement as:

$$\mathcal{C}_X(\mathbb{P}) = cl(X \setminus P) = \bigcup_{i=1:N} \mathcal{R}^+(\mathcal{H}_i). \tag{17.4}$$

In order to obtain a manageable formulation for (17.4), one has to use mixed integer techniques with the aim of defining a polyhedra in the extended space of *state* and *auxiliary binary variables* of the form:

$$-h_i x \leq -k_i + M\alpha_i, \; i = 1 : N \tag{17.5a}$$

$$\sum_{i=1:N} \alpha_i \leq N - 1, \tag{17.5b}$$

with a positive scalar M chosen appropriately (that is, significantly larger than the rest of the variables in the right hand side of the inequalities).[2]

Remark 17.1 Inequality (17.5a) becomes active when its associated binary variable α_i is "0" and redundant when the binary variable is "1" (due to term M, the right hand side is much larger than the left hand side and variable "x" can have any value). E.g., region $\mathcal{R}^-(\mathcal{H}_i)$ can be obtained from (17.5a)–(17.5b) by projecting along:

$$\alpha_i := (1, \ldots, 1, \underbrace{0}_{i}, 1, \ldots, 1). \tag{17.6}$$

Condition (17.5b) forces at least one binary variable "0" and thus, makes at least one inequality active. ♦

Illustrative Example

Consider the following illustrative example depicted in Fig. 17.1, where the complement of the triangle $P = R^+(\mathcal{H}_1) \cap R^+(\mathcal{H}_2) \cap R^+(\mathcal{H}_3)$ represents the feasible region.

---

[2] Sometimes this construction is called in the literature the "big M" formulation.

In formulation (17.5a)–(17.5b) there corresponds a unique binary variable to each inequality. Then, by choosing adequate combinations of binary variables, each region $\mathcal{R}^-(\mathcal{H}_i)$ can be recovered from the extended formulation (17.5a)–(17.5b). For example, by taking $(\alpha_1, \alpha_2, \alpha_3) \leftarrow (1, 0, 1)$ we can recover region $\mathcal{R}^-(\mathcal{H}_2)$, as depicted in Fig. 17.1.

17.2.2 Mixed Integer Representation of a Non-Convex and Non-Connected Feasible Region

The results shown in Sect. 17.2.1 have several limitations. Firstly, the feasible region is the complement of a polytope and thus has a restricted feasibility and secondly, the used MIP formulation requires a large number of binary variables (one for each inequality). Here we address both these shortcomings by considering the feasible region as the complement of a union of bounded polyhedral sets and by employing a logarithmic formulation for the MIP description.

Let us define a non-convex and possibly non compact feasible region as the complement of a union of bounded polyhedral sets $\mathbb{P} = \bigcup_l P_l$, with $P_l = \bigcap_{i \in \mathcal{K}_l} \mathcal{R}^+(\mathcal{H}_i)$ where $\mathcal{K}_l$ denotes the set of indices of the hyperplanes defining polytope P_l:

$$\mathcal{C}_X(\mathbb{P}) = cl(X \setminus \mathbb{P}). \tag{17.7}$$

We denote the collection of hyperplanes which define $\mathbb{P}$ as $\mathbb{H}$ and their number as $N \triangleq \sum_l |\mathcal{K}_l|$. The reduced notation $\mathcal{C}(\mathbb{P})$ is used whenever $X \subseteq \mathbb{R}^n$ is presumed known or is considered to be the entire space $\mathbb{R}^n$.

To describe (17.7) we introduce the next combinatorial notion.

Definition 17.1 (Hyperplane arrangements—[15]**)** A collection of hyperplanes $\mathbb{H} = \{\mathcal{H}_i\}_{i=1:N}$ will partition the space into a union of disjoint cells $\mathcal{A}_l$ defined as follows:

$$\mathcal{A}(\mathbb{H}) = \bigcup_{l=1,\dots,\gamma(N)} \left(\underbrace{\bigcap_{i=1}^{N} \mathcal{R}^{\sigma_l(i)}(\mathcal{H}_i)}_{\mathcal{A}_l} \right) \tag{17.8}$$

where $\sigma_l \in \{-, +\}^N$ denotes feasible combinations of regions (17.2) that are obtained for the hyperplanes in $\mathbb{H}$. □

We note that there exists a subset $\{\mathcal{B}_j\}$ of *feasible* cells from (17.8) which describes region (17.7):

$$\mathcal{C}_X(\mathbb{P}) = \bigcup_{j=1,\ldots,\gamma^b(N)} \mathcal{B}_j, \text{ with } \mathcal{B}_j \in \left\{\mathcal{A}_i \in \mathcal{A}(\mathbb{H}) : \mathcal{A}_i \bigcap \mathcal{P} = \emptyset\right\} \tag{17.9}$$

where $\gamma^b(N) \leq \gamma(N)$ denotes the number of cells $\mathcal{B}_j$.

Recall that any of the cells of (17.9) is described by a unique sign tuple ($\mathcal{B}_j \leftrightarrow \sigma_j$). As such, we obtain that the cells are disjoint and cover the entire feasible space (17.7). For our purposes we are satisfied with any collection of regions not necessarily disjoint which covers the feasible space.

We can formally represent the problem by requiring that (17.7) is described by a union of regions $\{C_k\}_{k=1:\gamma^c(N)}$ (where $\gamma^c(N)$ denotes the number of cells),

$$\mathcal{C}_X(\mathbb{P}) = \bigcup_{k=1,\ldots,\gamma^c(N)} C_k, \tag{17.10}$$

which verifies the next conditions:

- the new polyhedra are formed as unions of the old polyhedra (i.e., for any k there exists a set $\mathcal{I}_k$ which selects indices from $1 : \gamma^b(N)$ such that $C_k = \bigcup_{j \in \mathcal{I}_k} \mathcal{B}_j$);
- the union is minimal, i.e., the number $\gamma^c(N)$ of regions is minimal.

Existing merging algorithms are usually computationally expensive but here we can simplify the problem by noting that the sign tuples σ_l describe an adjacency graph since any two cells whose sign tuples differ at only one position are neighbors and that the union of any two adjacent cells is a polyhedron. In particular, in [9], it is shown that boolean algebra methods can be applied for the generation of the merged cells.

We can now state the mixed integer formulation of the non-convex region (17.10) which allows to express it as a polyhedra in the extended space $X \times \{0, 1\}^q$ of *state and auxiliary binary variables* as follows:

$$\begin{array}{c} \vdots \\ \left.\begin{array}{c} \sigma_k(i_1)h_{i_1}x \leq \sigma_k(i_1)k_{i_1} + M\alpha_k(\lambda) \\ \vdots \\ \sigma_k(i_s)h_{i_s}x \leq \sigma_k(i_s)k_{i_s} + M\alpha_k(\lambda) \end{array}\right\} C_k \\ \vdots \end{array} \tag{17.11}$$

with functions $\alpha_l(\cdot)$ with binary arguments which link (17.10) with (17.11). The sequence $1 \leq i_1 < i_2 \cdots < i_s \leq N$ denotes the hyperplanes appearing in the definition of cell C_k.

In order to provide an explicit representation of the binary part of (17.11) we recall a slightly modified form of Proposition 3.1 from [9].

Proposition 17.1 *For each cell C_k we associate a unique combination of binary variables $\lambda^k \in \{0, 1\}^q$ where $q = \lceil log_2 \gamma^c(N) \rceil$. Then, we can construct the functions*

$\alpha_k : \{0, 1\}^q \rightarrow \{0\} \cup [1, \infty)$ *such that they are affine in* λ *and verify the relations:*

$$\alpha_k(\lambda) = \begin{cases} \geq 0, \; for\; \lambda = \lambda^k \\ \geq 1, \; for\; \lambda \neq \lambda^k \end{cases} \tag{17.12}$$

and are defined as follows:

$$\alpha_k(\lambda) = \sum_{i=0}^{q} \left(\lambda_i^k + (1 - 2\lambda_i^k) \cdot \lambda_i\right). \tag{17.13}$$

Index 'i' denotes the ith *variable and* λ_i^k *its value for the tuple* λ^k*, associated to cell* C_k*. A similar construction can be found in [13].* □

Sketch of proof. See the proof of Proposition 3.1 of [9].

Remark 17.2 The validation of (17.12) for functions (17.13) assures that projecting (17.11) along the direction $\lambda = \lambda^k$ results in the space $\mathbb{R}^n$ in the cell C_k (since only $\alpha_k(\lambda^k)$ will be zero). Note also that the converse is false: if a binary variable is "1", the corresponding inequality degenerates such that it covers any point $x \in \mathbb{R}^n$ (this represents the limit case for $M \rightarrow \infty$). The last issue means that all the combinations of binary variables not associated with cells from (17.10) need to be made infeasible by additional inequalities. See Corollary 3.1 and Proposition 3.2 of [9] for constructive details. □

Illustrative Example

Consider the following illustrative example depicted in Fig. 17.2, where the complement of the union of two triangles ($\mathbb{P} = P_1 \cup P_2$) represents the feasible region. We take $\mathbb{H} = \{\mathcal{H}_i\}_{i=1:4}$ the collection of $N = 4$ hyperplanes (given as in (17.1)) which define P_1 and P_2 as follows: $P_1 = R^+(\mathcal{H}_1) \cap R^+(\mathcal{H}_2) \cap R^+(\mathcal{H}_3)$, $P_2 = R^-(\mathcal{H}_1) \cap R^-(\mathcal{H}_2) \cap R^+(\mathcal{H}_4)$.

We observe that we have 11 cells obtained as in the arrangement (17.8). From them, a total of 9, $\mathcal{B}_1, \ldots, \mathcal{B}_9$, describe the non-convex region (17.7). To each of them corresponds a unique tuple of signs, e.g., $B_1 \leftrightarrow (-, -, -, -)$ means that B_1 is the result of intersecting half-spaces $\mathcal{R}^-(\mathcal{H}_1)$, $R^-(\mathcal{H}_2)$, $R^-(\mathcal{H}_3)$ and $R^-(\mathcal{H}_4)$.

Applying the merging methods of [9] we obtain 4 overlapping regions: $C_1 = B_1 \cup B_2 \cup B_3 \cup B_4$, $C_2 = B_4 \cup B_5 \cup B_6$, $C_3 = B_6 \cup B_7 \cup B_8 \cup B_1$ and $C_4 = B_8 \cup B_9 \cup B_1 \cup B_2$, depicted in Fig. 17.2 through the dotted lines. Consequently, we note that $q = 2$ binary variables suffice in coding these regions. As for (17.11) and (17.13), we are now able to describe cells $\{C_i\}$ in the MIP formalism attaching to each of the regions a tuple in lexicographical order as seen in the left side of Fig. 17.2.

Note that, in addition to reducing the number of regions from 9 to 4, we also have reduced the number of hyperplanes appearing in the region's half-space represen-

$$-h_3 x \leq -k_3 + M(\quad \lambda_1 + \lambda_2)\} \quad C_1$$

$$\left.\begin{aligned} -h_1 x &\leq -k_1 \\ -h_4 x &\leq -k_4 \end{aligned}\right. + M(1 + \lambda_1 - \lambda_2)\Big\} \quad C_2$$

$$-h_4 x \leq -k_4 + M(1 - \lambda_1 + \lambda_2)\} \quad C_3$$

$$\left.\begin{aligned} h_1 x &\leq k_1 \\ -h_2 x &\leq -k_2 \end{aligned}\right. + M(2 - \lambda_1 - \lambda_2)\Big\} \quad C_4$$

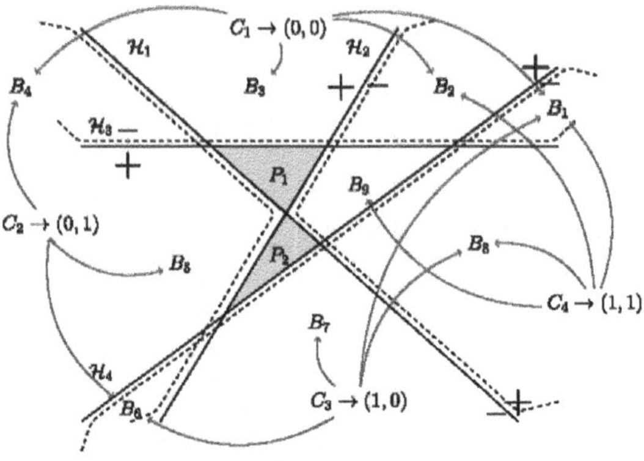

Fig. 17.2 Exemplification of hyperplane arrangement and cell merging

tation. In this particular case no tuple remains unallocated and thus, no additional constraints need to be added (see Remark 17.2).

17.2.3 Numerical Issues

Although functional, the representation of a feasible region defined as in Sect. 17.2.1 does not scale favorably with increases in space dimension and number of hyperplanes. More precisely, the number of feasible cells of an arrangement of form (17.8), denoted as $\gamma(N)$, is bounded by Buck's formula (see for more details [9]):

$$\gamma(N) \leq \sum_{i=0}^{d} \binom{N}{i} \tag{17.14}$$

with equality satisfied if the hyperplanes are in general position and $X = \mathbb{R}^n$ (in relation with the space dimension—d and the number of hyperplanes—N).

The increase in the number of cells makes their enumeration more difficult and increases (even after applying merging algorithms) the number of necessary binary variables. Therein lies the second issue which plagues the centralized formulation: the mixed programming algorithms increase in worst case situations exponentially with respect to the number of binary auxiliary variables and the computation becomes fragile for high dimensions.

To alleviate these issues, we propose two enhancements:

- to reduce the complexity of the problem and the dimension of the space in which the problem is solved, we consider a distributed MPC approach. We partition the agent collection into neighborhoods and compute the solutions locally while using information provided by the other neighborhoods. The stability of the overall formation and a reasonable performance of the agents also needs to be verified;

- to reduce the complexity of the feasible region representation we decompose only the "visible" part of the feasible region (i.e., the region which is reachable along the prediction horizon of the agents from a given neighborhood).

In the forthcoming sections we will detail these improvements and show that the computational load is reduced significantly while the performance of the scheme remains within acceptable bounds.

17.3 Feasible Region Decomposition Improvements

Consider a set of N_a linear systems which model the behavior of individual agents.

Let us define:

$$\mathrm{i} \triangleq \{1, \ldots, N_a\}, \tag{17.15}$$

as the collection of all agents indices. In the following, a basic multi-agent problem is considered. Besides normal constraints upon input/output magnitude and rate of variance, we need to consider collision and obstacle avoidance constraints. Let us for the moment ignore the dynamics governing the agents trajectory and discuss only the non-convex constraints coming from the obstacle and collision avoidance. To this end, let $x_i \in \mathbb{R}^d$ be the state and S_i the invariant with respect to time safety region (see [7]) associated with the ith agent. The collection of fixed obstacles is described through a union of polyhedra $\{\mathcal{O}\}_{o \in \mathcal{I}_o}$ with $\mathcal{I}_o \triangleq \{1, \ldots, N_o\}$ and where N_o denotes the number of polyhedra $\mathcal{O}_o$.

This allows us to write the collision/obstacle avoidance conditions as follows:

$$(\{x_i\} \oplus \mathcal{S}_i) \cap \big(\{x_j\} \oplus \mathcal{S}_j\big) = \emptyset, i, j \in \mathrm{i},\ i \neq j, \tag{17.16a}$$
$$(\{x_i\} \oplus \mathcal{S}_i) \cap \mathcal{O}_o = \emptyset, i \in \mathcal{I},\ o \in \mathcal{I}_o. \tag{17.16b}$$

Concatenating the variables x_i into the extended variable $\mathbf{x} \triangleq \left[x_1^T \ldots x_{N_a}^T\right]^T$ and using the fact that $[(\{a\} \oplus A) \cap (\{b\} \oplus B) = \emptyset] \leftrightarrow [a - b \notin B \oplus \{-A\}]$ we reach the equivalent formulation

$$\big(P_i - P_j\big)\mathbf{x} \notin \big(\{-S_i\} \oplus S_j\big), i, j \in \mathrm{i},\ i \neq j \tag{17.17a}$$
$$P_i\mathbf{x} \notin (\{-S_i\} \oplus \mathcal{O}_o), i = 1 : N_a,\ o \in \mathcal{I}_o, \tag{17.17b}$$

where matrix $P_i \triangleq [0 \ \ldots \underbrace{I}_{i} \ldots \ 0]$ "extends" the state x_i into the extended state $\mathbf{x}$ (i.e., $x(k)_i = P_i\mathbf{x}(k)$). Clearly, conditions (17.17) describe a non-convex region of form (17.7) which ultimately results in a description of form (17.10). We abuse the notation and thereafter we use the same notation for describing region (17.17):

$$\mathbf{x} \notin \mathbb{P}, \quad \mathcal{C}(\mathbb{P}) = \bigcup_{k=1,\ldots,\gamma^c(N)} C_k. \tag{17.18}$$

The above construction corresponds to a centralized approach where all the agents are considered simultaneously. In contrast, the distributed approach partitions the set of agent indices into subsets ("neighborhoods") and solves a series of local optimization problems.

For further use, we introduce the following definition of a neighborhood of agents.

Definition 17.2 Let $\mathbb{N} \triangleq \bigcup_i \mathcal{N}_i$ be the collection of neighborhoods $\mathcal{N}_i \subseteq \mathrm{i}$ which are considered to be *disjoint* (for any $i \neq j, \mathcal{N}_i \cap \mathcal{N}_j = \emptyset$) and to cover i (for any $i \in \mathrm{i}$ there exists j such that $i \in \mathcal{N}_j$), with i defined as in (17.15). ♦

For each optimization involving the agents of a given neighborhood, the feasible region can be obtained from conditions (17.17) by selecting only the constraints involving the agents of the current neighborhood and taking the other agents as additional obstacles. This approach has the drawback of necessitating a continuous recalculation of decomposition (17.10). An alternative solution which uses the previously calculated $\mathcal{C}(\mathbb{P})$ to compute the feasible region corresponding to $\mathcal{N}_i$ is proposed in the following statement.

Proposition 17.2 *Let the variable* $\mathbf{x}_{\mathcal{N}_i} \triangleq \left[x_{i_1}^T \; x_{i_2}^T \; \ldots\right]^T$ *denote the agents of the neighborhood* $\mathcal{N}_i$ *and* $\mathbf{x}_{\mathrm{i}\setminus\mathcal{N}_i}$ *denote the remaining agents. With the notation of* (17.17) *and considering that* $\mathbf{x}_{\mathrm{i}\setminus\mathcal{N}_i} \in \mathcal{X}_{\mathrm{i}\setminus\mathcal{N}_i}$ *we have that the feasible region characterizing* $\mathbf{x}_{\mathcal{N}_i}$ *is defined as:*

$$\mathcal{C}(\mathbb{P})|\mathcal{X}_{\mathrm{i}\setminus\mathcal{N}_i} = \bigcup_{k=1,\ldots,\gamma^c(N)} C_k|\mathcal{X}_{\mathrm{i}\setminus\mathcal{N}_i}, \tag{17.19}$$

with

$$C_k|\mathcal{X}_{\mathrm{i}\setminus\mathcal{N}_i} \triangleq \left\{\mathbf{x}_{\mathcal{N}_i} : \begin{bmatrix} \mathbf{x}_{\mathcal{N}_i} \\ \mathbf{x}_{\mathrm{i}\setminus\mathcal{N}_i} \end{bmatrix} \in C_k, \; \forall x_{\mathrm{i}\setminus\mathcal{N}_i} \in \mathcal{X}_{\mathrm{i}\setminus\mathcal{N}_i}\right\}. \tag{17.20}$$

□

Proof The proof is constructive. By intersecting every cell C_k with $\mathbb{R}^{n\cdot|\mathcal{N}_i|} \times \mathcal{X}_{\mathrm{i}\setminus\mathcal{N}_i}$ and then projecting along the $\mathbf{x}_{\mathcal{N}_i}$ subspace we obtain the restrictions $C_k|\mathcal{X}_{\mathrm{i}\setminus\mathcal{N}_i}$ of the from (17.20), allowing us to conclude the proof. □

17.4 Distributed MPC Optimization Problem

A number of commonly found situations in the control related to multi-agent systems imply a cost function that has to be minimized, while in the same time, the agent avoids collision with obstacles and other agents. To solve this problem, there exists various methods in the literature. Arguably, they can be gathered in methods which penalize through the cost function as the violation of the constraints (e.g. Potential Field Method [5] and Navigation Functions [10], and methods which impose hard constraints which cannot be broken. The latter group usually employs receding horizon techniques as they naturally take into account constraints.

We start by describing in Sect. 17.4.1 the centralized optimization problem and make use of it in Sect. 17.4.2 to construct and analyze a particular distributed approach.

17.4.1 Centralized Multi-Agent Problem

Let us recall the N_a agents considered in Sect. 17.3 and provide for them the LTI dynamics:

$$x(k+1)_i = A_i x(k)_i + B_i u(k)_i, \quad i \in \mathrm{i}, \tag{17.21}$$

where $x(k)_i \in \mathbb{R}^n$ and $u(k)_i \in \mathbb{R}^m$ represent the ith agent state and input, respectively, at time[3] k. The matrices $A_i \in \mathbb{R}^{n\times n}$ and $B_i \in \mathbb{R}^{n\times m}$ describe the dynamics and the pair (A_i, B_i) is controllable.

Furthermore, let us consider the centralized optimization problem formulation. We take a cost function $V(\mathbf{x}, \mathbf{u}) : \mathbb{R}^{N_a\cdot n} \times \mathbb{R}^{N_a\cdot m} \to \mathbb{R}$ which aims at maintaining a formation, following a reference path or simply to gather the agents towards the origin and the constraints defined as in (17.18). The centralized optimization problem using a finite receding horizon technique is formulated as follows:

$$\mathbf{u}^* = \arg \min_{\mathbf{u}(k|k),\ldots,\mathbf{u}(k+N_p-1)|k)} \sum_{l=0}^{N_p-1} V(\mathbf{x}(k+l|k), \mathbf{u}(k+l|k)) \tag{17.22a}$$

$$\text{s.t.:} \quad \mathbf{x}(k+l+1|k) = \mathbf{A}\mathbf{x}(k+l|k) + \mathbf{B}\mathbf{u}(k+l|k), \quad l = 0 : N_p - 1, \tag{17.22b}$$

$$\mathbf{x}(k+l|k) \in \mathcal{C}(\mathbb{P}), \quad l = 1 : N_p. \tag{17.22c}$$

The optimization problem (17.22) requires the minimization of the cost function over a finite prediction horizon N_p. From the optimal sequence of inputs $\mathbf{u}(k)^*, \ldots, \mathbf{u}(k+N_p-1|k)^*$ the first control input, $\mathbf{u}(k)^*$, is selected and applied to the centralized system, thus closing the loop.

Remark 17.3 The problem can be further simplified by noting that the agents' dynamics are usually subject to operational constraints (e.g., magnitude or rate of variation constraints) which limit their actual range. This means that we need to consider in the description of $\mathcal{C}(\mathbb{P})$ only these cells (17.20) which are intersecting the *reachable set* of the agents. The simplification of the scheme can be significant,

---

[3] Whenever the time instant is clear we abuse the notation and denote the current state, $x(k)_i$, as x_i. The same simplified notation is applied to the input.

e.g., if no obstacle is in the "line of sight" of the agents and the agents themselves are far away from each other, then the resulting feasible domain will be convex. □

17.4.2 Distributed Multi-Agent Problem

As detailed in the introduction, the distributed approach has obvious computational benefits. Consequently, we reformulate problem (17.22) into a distributed form. For the agents of neighborhood $\mathcal{N}_i$, the local optimization problem becomes:

$$\mathbf{u}^*_{\mathcal{N}_i} = \arg \min_{\mathbf{u}_{\mathcal{N}_i}(k|k),\dots,\mathbf{u}_{\mathcal{N}_i}(k+N_p-1|k)} \sum_{l=0}^{N_p-1} V_{\mathrm{i}\backslash\mathcal{N}_i}(\mathbf{x}_{\mathcal{N}_i}(k+l|k), \mathbf{u}_{\mathcal{N}_i}(k+l|k)) \tag{17.23a}$$

$$\text{s.t.:}\quad \mathbf{x}_{\mathcal{N}_i}(k+l+1|k) = \mathbf{A}_{\mathcal{N}_i}\mathbf{x}_{\mathcal{N}_i}(k+l|k) + \mathbf{B}_{\mathcal{N}_i}\mathbf{u}_{\mathcal{N}_i}(k+l|k), \quad l = 0 : N_p - 1, \tag{17.23b}$$

$$\mathbf{x}_{\mathcal{N}_i}(k+l|k) \in \mathcal{C}(\mathbb{P})|_{\mathcal{X}_{\mathrm{i}\backslash\mathcal{N}_i}}. \tag{17.23c}$$

The use of indexing "$\mathcal{N}_i$" in (17.23) is to be understood as in Preposition 17.2, e.g., $\mathbf{A}_{\mathcal{N}_i}$ denotes the concatenation (block-diagonal in this case) of state matrices A_j where the indices j are found in the neighborhood $\mathcal{N}_i$.

The local cost function $V_{\mathrm{i}\backslash\mathcal{N}_i}(\mathbf{x}_{\mathcal{N}_i}, \mathbf{u}_{\mathcal{N}_i}) : \mathbb{R}^{|\mathcal{N}_i|\cdot n} \times \mathbb{R}^{|\mathcal{N}_i|\cdot m} \to \mathbb{R}$ is defined as

$$V_{\mathrm{i}\backslash\mathcal{N}_i}(\mathbf{x}_{\mathcal{N}_i}, \mathbf{u}_{\mathcal{N}_i}) = \min_{\substack{\mathbf{u}_{\mathrm{i}\backslash\mathcal{N}_i} \in \mathcal{U}_{\mathrm{i}\backslash\mathcal{N}_i} \\ \mathbf{x}_{\mathrm{i}\backslash\mathcal{N}_i} \in \mathcal{X}_{\mathrm{i}\backslash\mathcal{N}_i}}} V(\mathbf{x}, \mathbf{u}), \tag{17.24}$$

where $\mathcal{U}_{\mathrm{i}\backslash\mathcal{N}_i}$ denotes the values taken by the inputs $u_{\mathrm{i}\backslash\mathcal{N}_i}$ (similarly with the definition of set $\mathcal{X}_{\mathrm{i}\backslash\mathcal{N}_i}$).

Remark 17.4 The use of operator "min" assumes a cooperative approach (the agents exterior to the current neighborhood will try to accommodate the inputs/states suggested by the optimization problem). If replacing with the "max" operator we assume instead an adversarial or indifferent approach where the worst combination of inputs/states of the exterior agents has to be taken into account. □

Both the cost function (17.23a) and the feasible domain (17.23b)–(17.23c) depend explicitly upon the values found in the sets $\mathcal{U}_{\mathrm{i}\backslash\mathcal{N}_i}$ and $\mathcal{X}_{\mathrm{i}\backslash\mathcal{N}_i}$ which characterize the behavior (input and state) of the agents exterior to the current neighborhood. Consequently, the content of these sets can accommodate a large span of distributed control strategies. If no information is forthcoming from the exterior, then these sets are defined using reachable analysis, thus resulting in a decentralized control. At the other extreme, when the exact state of the exterior agents is communicated we have a distributed cooperative approach.

Usually, distributed approaches necessitate several iterations in-between consecutive discretization steps. That is, at discretization step k for a state $x(k)_i$ we may have $\bar{p}$ iterations with the intermediate values x_i^p where $p = 0 : \bar{p}$ and $x_i^0 \leftarrow x(k)$ and $x(k+1)_i \leftarrow x_i^{\bar{p}}$. The computations stop after some predefined number of iterations or when a consensus is reached.

Here we propose a hierarchical implementation which avoids by construction consensus verification and requires only one iteration. To this end we consider a hierarchical ordering of the neighborhoods.[4] For neighborhood $\mathcal{N}_i$, the remaining indices, represented by $\mathrm{i} \setminus \mathcal{N}_i$, are partitioned into $\mathcal{N}_i^-$ and $\mathcal{N}_i^+$ which denote the neighborhoods with lower, respectively higher, priority. Then, the sets $\mathcal{U}_{\mathrm{i}\setminus\mathcal{N}_i}$ and $\mathcal{X}_{\mathrm{i}\setminus\mathcal{N}_i}$ are constructed as follows:

$$\mathcal{U}_{\mathrm{i}\setminus\mathcal{N}_i} = \left(\bigcup_{j\in\mathcal{N}_i^-}\{u_j^0\}\right) \cup \left(\bigcup_{j\in\mathcal{N}_i^+}\{u_j^1\}\right), \tag{17.25a}$$

$$\mathcal{X}_{\mathrm{i}\setminus\mathcal{N}_i} = \left(\bigcup_{j\in\mathcal{N}_i^-}\{x_j^0\}\right) \cup \left(\bigcup_{j\in\mathcal{N}_i^+}\{x_j^1\}\right), \tag{17.25b}$$

where superscript “0” and “1” denote the current iteration for a certain variable (e.g., u_j^0 means that the state is not yet updated and the initial value $u(k)_j$ will be used, whereas u_j^1 means that we use the updated value, the one which will define the next discretization step, $u(k+1)_j$).

Under the aforementioned constructive assumptions we state the following lemma dealing with constraints verification.

Lemma 17.1 *Let the agents be in a feasible agent formation at discretization step k:*

$$\mathbf{x}(k) \in \mathcal{C}(\mathbb{P}). \tag{17.26}$$

By applying optimization problems (17.23) *with sets* (17.25) *for each of the neighborhoods $\mathcal{N}_i$ which partition* i*, we preserve the formation feasibility at the next discretization step $k+1$:*

$$\mathbf{x}(k+1) \in \mathcal{C}(\mathbb{P}). \tag{17.27}$$

□

Proof The proof is inductive. Consider neighborhood $\mathcal{N}_i$ and its attached optimization problem (17.23). Due to the construction of sets $\mathcal{U}_{\mathrm{i}\setminus\mathcal{N}_i}$ and $\mathcal{X}_{\mathrm{i}\setminus\mathcal{N}_i}$ as in (17.25), the optimization problem “sees” the agents of higher order in their updated positions and the ones of lower order in their initial positions. Then, the resulting control

[4] Note that in the hierarchical implementation the neighborhoods are disjoint, see also the definition of neighborhoods $\mathcal{N}_i$ in Sect. 17.3.

$u_{\mathcal{N}_i}(k+1)$ will lead to a state $x_{\mathcal{N}_i}(k+1)$ which respects the new states of the higher order agents (since they are explicitly included in the constraint set) and the un-updated states of the lower-order agents. On the other hand, the lower-order agents will not break the constraints involving agents with index in $\mathcal{N}_i$ because from their point of view, this neighborhood has a superior position in the hierarchy.

The presence of the constraints associated to the lower order agents guarantees the feasibility of $\mathbf{x}(k+1)$. At the ith optimization problem, the initial state of the lower order agents is respected, thus, if no movement is possible for them, they can at lest keep the same state. This final argument completes the global recursive feasibility proof of the control scheme. □

Remark 17.5 As mentioned in the proof of Lemma 17.1 we consider for lower order agents the un-updated state and thus we guarantee the existence of a solution (at worse, the lower order agent will be able to keep the same state[5]). This approach can be generalized by assuming not a single point, but rather all the points that can be reached by those agents. In other words, this means that agents situated lower in the hierarchy could be "pushed-around" within acceptable bounds. □

Remark 17.6 Note that the partitioning between neighborhoods needs not to be time invariant. If we consider that the neighborhoods have a geometric meaning (i.e., take the indices of agents which are physically close) it may be necessary to change their content at every (few) discretization steps. In this sense, we note the k-means clustering algorithms, which permit partitioning a collection of agents into a predefined number of groupings and partitions the state space into Voronoi cells [3]. □

To clarify the exposition we provide in Algorithm 17.1 a sketch of the distributed control problem.

17.4.3 Illustrative Example

For illustrative purposes let us consider the following example. Consider 3 agents, each one of them its own neighborhood: $\mathcal{N}_i = \{i\}$, $i = 1:3$. We order these neighborhoods lexicographically, that is, $\mathcal{N}_1 < \mathcal{N}_2 < \mathcal{N}_3$ in the hierarchical point of view and apply the optimization procedure (17.23). For clarity of the exposition we keep a one-step MPC problem (i.e., $N_p = 1$), such that only one step-ahead has to be considered in the constraints.

In Fig. 17.3 we consider the 3 agents and show their evolution. Note that the first and last frames represent the current (k) and next discretization step ($k+1$) respectively and that the 3 intermediate frames represent the single iteration executed in-between the discretization steps. Further, we detail the execution of this iteration.

[5] Not necessarily true when the dynamics describe systems which have a minimal velocity—unmanned aerial vehicles (UAVs) for example.

Algorithm 17.1 Distributed MPC scheme

Require: obstacles $\{\mathcal{O}_o\}$, safety regions S_i, neighborhoods $\mathbb{N}$, initial inputs and states $\mathcal{U}_{i\setminus\mathcal{N}_i}$, $\mathcal{X}_{i\setminus\mathcal{N}_i}$, agent dynamics (A_i, B_i) and global cost function $V(\cdot,\cdot)$

describe $\mathbb{P}$ based on collision and obstacle avoidance conditions (17.17) and extract the collection of hyperplanes $\mathbb{H}$
obtain the cell arrangement as in (17.8) for $\mathbb{H}$
obtain the feasible cells (17.9) and merge them in representation (17.10)
for $k = 1, 2, \ldots$ **do**
  for each $\mathcal{N}_i \in \mathbb{N}$ **do**
    for neighborhood $\mathcal{N}_i$ calculate $\mathcal{U}_{i\setminus\mathcal{N}_i}$ and $\mathcal{X}_{i\setminus\mathcal{N}_i}$ as in (17.25)
    construct the feasible region $\mathcal{C}(\mathbb{P})|_{\mathcal{X}_{i\setminus\mathcal{N}_i}}$ as in Proposition 17.2
    write $\mathcal{C}(\mathbb{P})|_{\mathcal{X}_{i\setminus\mathcal{N}_i}}$ in MI formulation as in (17.11) with Proposition 17.1
    solve optimization problem (17.23)
  end for
  $k = k + 1$
end for

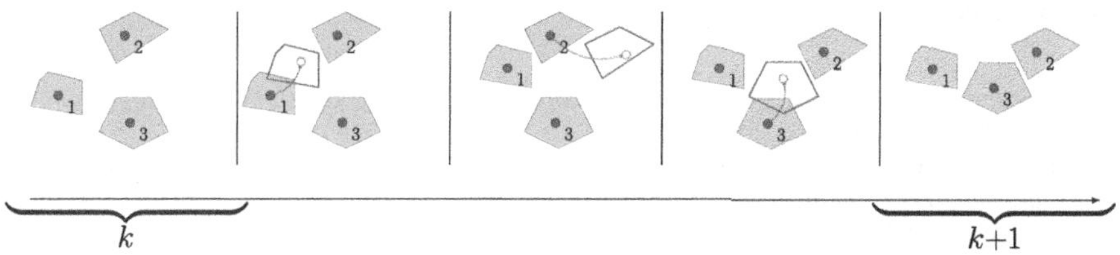

Fig. 17.3 Exemplification of the hierarchical distributed approach

In the second frame, we solve optimization (17.23) for $\mathcal{N}_1$. Since this neighborhood is the highest in the ordering ($\mathcal{N}_1^+ = \emptyset$ and $\mathcal{N}_1^- = \{2, 3\}$) all the other agents are kept in their initial position and agent 1 positions itself as depicted by the dashed red contour. At the next frame, the third, we solve the optimization problem for $\mathcal{N}_2$ to which correspond $\mathcal{N}_2^+ = \{1\}$ and $\mathcal{N}_2^- = \{3\}$. In this case, $\mathcal{N}_2^+$ is not empty and thus the agent's 1 updated state is used and agent 3 has its initial state. It can be seen that agent 2 finds a better state which respects both the updated and the initial constraints. The same procedure is repeated at the next frame for $\mathcal{N}_3$ to which correspond $\mathcal{N}_3^+ = \{1, 2\}$ and $\mathcal{N}_3^- = \emptyset$. In the last frame it can be seen that each of the agents has changed its position and that the constraints were respected.

17.5 Conclusions

In this chapter we revisited a technique which transforms a non-convex and possibly non-connected region into a polyhedra in an augmented space (state and auxiliary binary variables) through the use of hyperplane arrangements. This enables the geometric description of the feasible regions and use them for describing a typical multi-agent collision and obstacle avoidance problem. Furthermore, the agents can be portioned into neighborhoods and a distributed Model Predictive problem can be solved through a hierarchical ordering of the neighborhoods in order to guarantee constraint validation and avoid consensus seeking. "Proof of concept" illustrations are provided. The interested reader is encouraged to seek further details on the mathematical concepts related to hyperplane arrangements in [6] and Mixed Integer Programming (MIP) based Model Predictive Control (MPC) in [4]. For the distributed MPC version, the implementation of these concepts leads to challenging open questions, as for example how to make use of reachable set calculation for MIP enhancement or how to efficiently decompose the feasible regions.

Acknowledgments The research of Ionela Prodan is financially supported by the EADS Corporate Foundation (091-AO09-1006).

References

1. D. Grundel, R. Murphey, P.M. Pardalos, *Cooperative Systems, Control and Optimization*, vol. 588 (Springer, New York, 2007)
2. M. Jünger, M. Junger, T.M. Liebling, D. Naddef, G. Nemhauser, W.R. Pulleyblank, *50 Years of Integer Programming 1958–2008: From the Early Years to the State-of-the-Art* (Springer, New York, 2009)
3. T. Kanungo, D.M. Mount, N.S. Netanyahu, C.D. Piatko, R. Silverman, A.Y. Wu, An efficient k-means clustering algorithm: analysis and implementation. IEEE Trans. Pattern Anal. Mach. Intell. **24**(7), 881–892 (2002)
4. S. Lin, B. De Schutter, Y. Xi, H. Hellendoorn, Model predictive control for urban traffic networks via milp, in *Proceedings of the 29th American Control Conference*, pp. 2272–2277, Baltimore, 30 June–2 July 2010
5. R. Olfati-Saber, R.M. Murray, Distributed cooperative control of multiple vehicle formations using structural potential functions, in *Proceedings of the 15th IFAC World Congress*, pp. 346–352, Barcelona, 21–26 July 2002
6. P.M. Pardalos, Hyperplane arrangements in optimization, in *Encyclopedia of Optimization* (Springer, New York, 2009), pp. 1547–1548
7. I. Prodan, S. Olaru, C. Stoica, S.-I. Niculescu, Predictive control for tight group formation of multi-agent systems, in *Proceedings of the 18th IFAC World Congress*, pp. 138–143, Milano, 28 Aug–2 Sept 2011
8. I. Prodan, S. Olaru, C. Stoica, S.-I. Niculescu, On the tight formation for multi-agent dynamical systems, in *KES—Agents and Multi-agent Systems, Technologies and Applications*, LNAI 7327. (Springer, New York, 2012), pp. 554–565
9. I. Prodan, F. Stoican, S. Olaru, S.-I. Niculescu, Enhancements on the hyperplanes arrangements in mixed-integer techniques. J. Optim. Theory Appl. **154**(2):549–572 (2012). doi:10.1007/s10957-012-0022-9
10. E. Rimon, D.E. Koditschek, Exact robot navigation using artificial potential functions. IEEE Trans. Robot. Autom. **8**(5), 501–518 (1992)
11. R. Scattolini, Architectures for distributed and hierarchical model predictive control—a review. J. Process Control **19**(5), 723–731 (2009)
12. F. Stoican, S. Olaru, M.M. Seron, J.A. De Doná, Reference governor design for tracking problems with fault detection guarantees. J. Process Control **22**(5), 829–836 (2012)
13. J.P. Vielma, G.L. Nemhauser, Modeling disjunctive constraints with a logarithmic number of binary variables and constraints. Math. Program. **128**(1), 49–72 (2011)

14. M.P. Vitus, V. Pradeep, G. Hoffmann, S.L. Waslander, C.J. Tomlin, Tunnel-milp: path planning with sequential convex polytopes, in *Proceedings of the 26th AIAA Guidance, Navigation, and Control Conference*, Honolulu, 18–21 Aug 2008
15. G.M. Ziegler, *Lectures on Polytopes*, vol. 152. (Springer, New York, 1995)

Chapter 18
Distributed MPC of Interconnected Nonlinear Systems by Dynamic Dual Decomposition

A. Grancharova and T. A. Johansen

Abstract A suboptimal approach to distributed Nonlinear Model Predictive Control (NMPC) for systems consisting of nonlinear subsystems with nonlinearly coupled dynamics subject to both state and input constraints is proposed. The approach applies a dynamic dual decomposition method to reformulate the original centralized NMPC problem into a distributed quasi-NMPC problem by linearization of the nonlinear system dynamics and taking into account the couplings between the subsystems. The developed approach is based entirely on distributed on-line optimization (by gradient iterations) and can be applied to large-scale nonlinear systems. The theoretical results related to the application of the distributed MPC approach to both linear and nonlinear systems are outlined and some simulation results are provided.

18.1 Introduction

Nonlinear Model Predictive Control (NMPC) involves the solution at each sampling instant of a finite horizon optimal control problem subject to the system dynamics, and state and input constraints. However, solving in a centralized way NMPC problems for large-scale systems may be impractical due to the complexity of the Nonlinear Programming (NLP) problem, the topology of the plant and data communication, and the large number of decision variables. Therefore, there is a strong motivation for

A. Grancharova (✉)
Institute of System Engineering and Robotics, Bulgarian Academy of Sciences, Sofia, Bulgaria
e-mail: alexandra.grancharova@abv.bg

A. Grancharova
Department of Industrial Automation, University of Chemical Technology and Metallurgy, Sofia, Bulgaria

T. A. Johansen
Department of Engineering Cybernetics, Norwegian University of Science and Technology, Trondheim, Norway
e-mail: Tor.Arne.Johansen@itk.ntnu.no

J. M. Maestre and R. R. Negenborn (eds.), *Distributed Model Predictive Control Made Easy*, Intelligent Systems, Control and Automation: Science and Engineering
DOI: 10.1007/978-94-007-7006-5_18,

development of methods for distributed solution of NMPC problems. At the same time, the multi-core computer architectures available nowadays would encourage parallel and distributed NMPC computations [4]. Recently, several approaches for decentralized implementation of MPC algorithms have been developed [18]. As it is pointed out in [14], the possibility to use MPC in a decentralized fashion has the advantage to reduce the original, large size, optimization problem into a number of smaller and more tractable ones. In [7, 12, 19, 20], approaches for distributed MPC for systems consisting of *linear* interconnected subsystems have been developed.

Approaches for distributed MPC for systems composed of several *nonlinear* subsystems have been proposed in [6, 9, 11, 14, 15]. In [14], a stabilizing decentralized MPC algorithm for nonlinear systems consisting of several interconnected local subsystems is developed. It is derived under the main assumptions that no information can be exchanged between local control laws. In [6], it is supposed that the dynamics and constraints of the nonlinear subsystems are decoupled, but their state vectors are coupled in a single cost function of a finite horizon optimal control problem. In [11], an optimal control problem for a set of dynamically decoupled nonlinear systems, where the cost function and constraints couple the dynamical behavior of the systems, is solved. In [15], novel control techniques for the control of transportation networks are proposed, which are based on a combination of ideas from the fields of multi-agent systems and MPC.

The distributed MPC approach in [7] is based on the *dual decomposition* methods [3, 5], whose rationale is to reformulate the large-scale optimization problems by using Lagrange multipliers for relaxing the couplings between the sub-problems. It is applied to systems composed by a set of *linear* subsystems, subject to both state and input constraints. Since the approach applies to the complex case when the subsystems are coupled through their dynamics, it is referred to as *dynamic dual decomposition* [7, 17]. The fact that it takes into account the dynamic couplings between the subsystems when optimizing the overall system behavior and at the same time allows for decentralized MPC solution makes it very attractive. In [9], the suboptimal approach from [7] is extended to an approach to distributed NMPC for a more general class of systems consisting of *nonlinear* subsystems with *linearly coupled* dynamics subject to both state and input constraints. Thus, the approach in [9] applies the dynamic dual decomposition method [3, 7, 17] and reformulates the original centralized NMPC problem into a distributed quasi-NMPC problem by linearization of the nonlinear system dynamics. It is based entirely on distributed on-line optimization (by gradient iterations) and can be applied to large-scale nonlinear systems. In [10], the suboptimal quasi-NMPC approach from [9] where linear couplings between the subsystems have been assumed, is further extended to solve distributed NMPC problems for interconnected *nonlinear* systems with separable *nonlinearly coupled* dynamics.

The purpose of this chapter is to summarize the main ideas of the distributed quasi-NMPC approach in [9] with the extensions in [10], as well as to review the theoretical and the simulation results associated to both linear and nonlinear interconnected systems.

18.2 Boundary Conditions on the System, Control Objectives, and Constraints

Consider a system composed by the interconnection of a set of subsystems $\mathcal{N} = \{S_1, S_2, \ldots, S_{|\mathcal{N}|}\}$ (Fig. 18.1), which is described by the following nonlinear discrete-time models [14]:

$$\mathbf{x}_i(k+1) = \mathbf{f}_i(\mathbf{x}_i(k), \mathbf{u}_i(k)) + \mathbf{g}_i(\mathbf{x}(k)), \quad i = 1, 2, \ldots, |\mathcal{N}| \tag{18.1}$$

where $\mathbf{x}_i(k) \in \mathbb{R}^{n_{x_i}}$ and $\mathbf{u}_i(k) \in \mathbb{R}^{n_{u_i}}$ are the state and control input vectors, related to the ith subsystem, and $\mathbf{f}_i : \mathbb{R}^{n_{x_i}} \times \mathbb{R}^{n_{u_i}} \to \mathbb{R}^{n_{x_i}}$ and $\mathbf{g}_i : \mathbb{R}^{n_x} \to \mathbb{R}^{n_{x_i}}$ are nonlinear functions. In (18.1), the mutual influence of the $|\mathcal{N}|$ subsystems is described by the functions $\mathbf{g}_i$, which depend on the overall state:

$$\mathbf{x}(k) = [\mathbf{x}_1(k), \mathbf{x}_2(k), \ldots, \mathbf{x}_{|\mathcal{N}|}(k)] \in \mathbb{R}^{n_x}, \quad n_x = \sum_{i=1}^{|\mathcal{N}|} n_{x_i} \tag{18.2}$$

Similarly, the overall control input is denoted:

$$\mathbf{u}(k) = [\mathbf{u}_1(k), \mathbf{u}_2(k), \ldots, \mathbf{u}_{|\mathcal{N}|}(k)] \in \mathbb{R}^{n_u}, \quad n_u = \sum_{i=1}^{|\mathcal{N}|} n_{u_i} \tag{18.3}$$

The following control input and state constraints are imposed on the subsystems:

$$\mathbf{u}_{\min,i} \le \mathbf{u}_i(k) \le \mathbf{u}_{\max,i}, \ \mathbf{x}_{\min,i} \le \mathbf{x}_i(k) \le \mathbf{x}_{\max,i}, \ i = 1, \ldots, |\mathcal{N}| \tag{18.4}$$

and the following assumptions are made:

A1. The functions $\mathbf{f}_i$ and $\mathbf{g}_i$, $i = 1, \ldots, |\mathcal{N}|$ are C^1 functions with $\mathbf{f}_i(0,0) = 0$, $\mathbf{g}_i(0) = 0$.

A2. $\mathbf{x}_{\min,i} < 0 < x_{\max,i}$, $\mathbf{u}_{\min,i} < 0 < \mathbf{u}_{\max,i}$, $i = 1, \ldots, |\mathcal{N}|$.

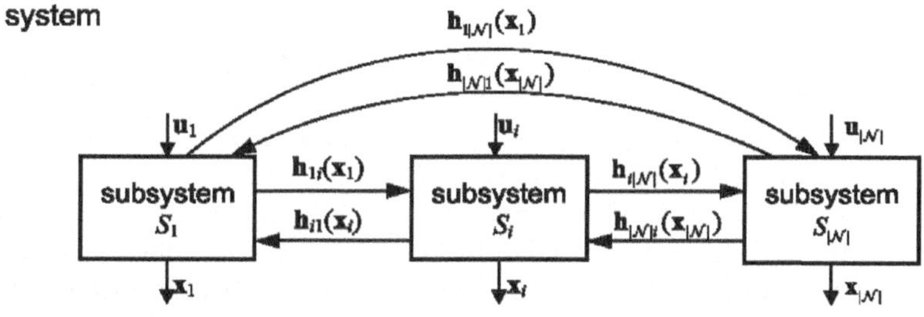

Fig. 18.1 System composed by the interconnection of $|\mathcal{N}|$ subsystems

A3. The nonlinear functions $\mathbf{g}_i(\mathbf{x}(k))$ of the couplings are separable, i.e., they have the form:

$$\mathbf{g}_i(\mathbf{x}(k)) = \sum_{j=1,\, j\neq i}^{|\mathcal{N}|} \mathbf{h}_{ij}(\mathbf{x}_j(k)), \quad i = 1, \ldots, |\mathcal{N}| \tag{18.5}$$

where $\mathbf{h}_{ij} : \mathbb{R}^{n_{x_j}} \to \mathbb{R}^{n_{x_i}}$ are nonlinear functions with $\mathbf{h}_{ij}(0) = 0$.

It is also assumed that the distributed NMPC controllers for the interconnected subsystems can communicate to each other the computed optimal state trajectories of the subsystems and the trajectories of the Lagrange multipliers, associated to the application of the dual decomposition method. A detailed communication scheme corresponding to the proposed distributed NMPC method is given in Sect. 18.3.

It is supposed that a full measurement $\bar{\mathbf{x}} = [\bar{\mathbf{x}}_1, \bar{\mathbf{x}}_2, \ldots, \bar{\mathbf{x}}_{|\mathcal{N}|}]$ of the overall initial state is available, i.e., $\mathbf{x}(0) = \bar{\mathbf{x}}$. The optimal regulation problem is considered where the goal is to steer the overall state of the system (18.1) to the origin. This leads to the infinite horizon optimal control problem:

Problem P0 (Infinite Horizon Optimization):

$$J^{\infty}(\bar{\mathbf{x}}) = \min_{\mathbf{u}} \sum_{k=0}^{\infty} \sum_{i=1}^{|\mathcal{N}|} s_i(\mathbf{x}_i(k), \mathbf{u}_i(k)) \tag{18.6}$$

subject to (18.1) and:

$$\mathbf{x}_i(k) \in \mathcal{X}_i,\ i = 1, \ldots, |\mathcal{N}| \quad \text{for all } k \tag{18.7}$$

$$\mathbf{u}_i(k) \in \mathcal{U}_i,\ i = 1, \ldots, |\mathcal{N}| \quad \text{for all } k \tag{18.8}$$

Here $s_i(\mathbf{x}_i(k), \mathbf{u}_i(k)) = \|\mathbf{x}_i(k)\|^2_{\mathbf{Q}_i} + \|\mathbf{u}_i(k)\|^2_{\mathbf{R}_i}$ is the stage cost for the ith subsystem with weighting matrices $\mathbf{Q}_i, \mathbf{R}_i \succ 0$, and the sets $\mathcal{X}_i$ and $\mathcal{U}_i$ are defined by:

$$\mathcal{X}_i = \left\{\boldsymbol{\lambda}_i \in \mathbb{R}^{n_{x_i}} \mid \mathbf{x}_{\min,i} \leq \boldsymbol{\lambda}_i \leq \mathbf{x}_{\max,i}\right\} \tag{18.9}$$

$$\mathcal{U}_i = \left\{\boldsymbol{\eta}_i \in \mathbb{R}^{n_{u_i}} \mid \mathbf{u}_{\min,i} \leq \boldsymbol{\eta}_i \leq \mathbf{u}_{\max,i}\right\} \tag{18.10}$$

It follows from (18.9)–(18.10) that $\mathcal{X}_i$ and $\mathcal{U}_i$ are convex (polyhedral) sets, which include the origin in their interior (due to assumption A2).

Similar to [7], the infinite horizon optimal control problem P0 is approximated with an NMPC problem without a terminal cost and terminal constraint, where the objective function is obtained by truncating the infinite horizon objective. Let the measured state at time k be $\mathbf{x} = [\mathbf{x}_1, \mathbf{x}_2, \ldots, \mathbf{x}_{|\mathcal{N}|}]$. For the current $\mathbf{x}$, the regulation NMPC solves the problem:

Problem P1 (Centralized NMPC):

$$J^{\mathrm{opt}}(\mathbf{x}) = \min_{\mathbf{u}(k:k+N_\mathrm{p})} J(\mathbf{u}(k:k+N_\mathrm{p}), \mathbf{x}) \tag{18.11}$$

subject to $\mathbf{x}(k) = \mathbf{x}$ and:

$$\mathbf{x}_i(k+l) \in \mathcal{X}_i,\ i = 1, \ldots, |\mathcal{N}|,\ l = 1, \ldots, N_\mathrm{p} \tag{18.12}$$

$$\mathbf{u}_i(k+l) \in \mathcal{U}_i,\ i = 1, \ldots, |\mathcal{N}|,\ l = 0, 1, \ldots, N_\mathrm{p} - 1 \tag{18.13}$$

$$\mathbf{x}_i(k+l+1) = \mathbf{f}_i(\mathbf{x}_i(k+l), \mathbf{u}_i(k+l)) + \mathbf{g}_i(\mathbf{x}(k+l))$$
$$i = 1, \ldots, |\mathcal{N}|,\ l = 0, 1, \ldots, N_\mathrm{p} - 1 \tag{18.14}$$

$$\mathbf{x}(k+l) = [\mathbf{x}_1(k+l), \ldots, \mathbf{x}_{|\mathcal{N}|}(k+l)],\ l = 0, 1, \ldots, N_\mathrm{p} \tag{18.15}$$

$$\mathbf{u}(k+l) = [\mathbf{u}_1(k+l), \ldots, \mathbf{u}_{|\mathcal{N}|}(k+l)],\ l = 0, 1, \ldots, N_\mathrm{p} - 1 \tag{18.16}$$

where the cost function is given by:

$$J(\mathbf{u}(k:k+N_\mathrm{p}), \mathbf{x}) = \sum_{l=0}^{N_\mathrm{p}} \sum_{i=1}^{|\mathcal{N}|} s_i(\mathbf{x}_i(k+l), \mathbf{u}_i(k+l)) \tag{18.17}$$

Here, N_p is a finite horizon.

18.3 Description of the Distributed NMPC Approach

18.3.1 Distributed NMPC by Dynamic Dual Decomposition

Problem P1 can be decomposed by using the dynamic dual decomposition approach [3, 17]. The following *decoupled* state equations can be formulated:

$$\mathbf{x}_i(k+1) = \mathbf{f}_i(\mathbf{x}_i(k), \mathbf{u}_i(k)) + \mathbf{v}_i(k), \quad i = 1, \ldots, |\mathcal{N}| \tag{18.18}$$

with the additional constraints that:

$$\mathbf{v}_i(k) = \sum_{j=1,\, j\neq i}^{|\mathcal{N}|} \mathbf{h}_{ij}(\mathbf{x}_j(k)), \quad i = 1, \ldots, |\mathcal{N}| \text{ for all } k \tag{18.19}$$

The variable $\mathbf{v}_i \in \mathbb{R}^{n_{x_i}}$ can be interpreted as the influence of the other subsystems in the update of $\mathbf{x}_i$.

Then, the constraints (18.19) are relaxed by introducing the corresponding Lagrange multipliers $\mathbf{p}_i \in \mathbb{R}^{n_{x_i}}$ in the cost function (18.17) and the following distributed NMPC problem is formulated:

Problem P2 (Distributed NMPC):

$$
\begin{aligned}
\tilde{J}^{\mathrm{opt}}(\mathbf{x}) &= \max_{\mathbf{p}(k:k+N_\mathrm{p})} \min_{\substack{\mathbf{u}(k:k+N_\mathrm{p})\\ \mathbf{x}(k:k+N_\mathrm{p})\\ \mathbf{v}(k:k+N_\mathrm{p})}} \sum_{l=0}^{N_\mathrm{p}} \sum_{i=1}^{|\mathcal{N}|} \Bigg[s_i(\mathbf{x}_i(k+l), \mathbf{u}_i(k+l)) \\
&\quad + \mathbf{p}_i(k+l)^\mathrm{T}(\mathbf{v}_i(k+l) - \sum_{j=1,\, j\neq i}^{|\mathcal{N}|} \mathbf{h}_{ij}(\mathbf{x}_j(k+l))) \Bigg] \\
&= \max_{\mathbf{p}(k:k+N_\mathrm{p})} \sum_{i=1}^{|\mathcal{N}|} \left(\min_{\substack{\mathbf{u}_i(k:k+N_\mathrm{p})\\ \mathbf{x}_i(k:k+N_\mathrm{p})\\ \mathbf{v}_i(k:k+N_\mathrm{p})}} \sum_{l=0}^{N_\mathrm{p}} \Bigg[s_i(\mathbf{x}_i(k+l), \mathbf{u}_i(k+l)) \right. \\
&\quad \left. + \mathbf{p}_i(k+l)^\mathrm{T}\mathbf{v}_i(k+l) - \sum_{j=1,\, j\neq i}^{|\mathcal{N}|} \mathbf{h}_{ji}^\mathrm{T}(\mathbf{x}_i(k+l))\mathbf{p}_j(k+l) \Bigg] \right)
\end{aligned}
\tag{18.20}
$$

subject to $\mathbf{x}(k) = \mathbf{x}$, constraints (18.12)–(18.13) and:

$$
\begin{aligned}
&\mathbf{x}_i(k+l+1) = \mathbf{f}_i(\mathbf{x}_i(k+l), \mathbf{u}_i(k+l)) + \mathbf{v}_i(k+l) \\
&i = 1, \ldots, |\mathcal{N}|,\ l = 0,\ 1, \ldots, N_\mathrm{p} - 1 && (18.21) \\
&\mathbf{p}(k+N_\mathrm{p}) = 0 && (18.22)
\end{aligned}
$$

The requirement (18.22) follows from the optimality conditions of Pontryagin's principle for discrete-time systems [2] and the fact that the state is not specified at the terminal time $k + N_\mathrm{p}$. Here the following notation is used: $\mathbf{p}(k+l) = [\mathbf{p}_1(k+l), \ldots, \mathbf{p}_{|\mathcal{N}|}(k+l)]$, $\mathbf{v}(k+l) = [\mathbf{v}_1(k+l), \ldots, \mathbf{v}_{|\mathcal{N}|}(k+l)]$, $l = 0,\ 1, \ldots, N_\mathrm{p}$.

The stage cost in the second equality in (18.20) is denoted:

$$
\begin{aligned}
&s_i^p(\mathbf{x}_i(k+l), \mathbf{u}_i(k+l), \mathbf{v}_i(k+l), \mathbf{p}(k:k+N_\mathrm{p})) \\
&= s_i(\mathbf{x}_i(k+l), \mathbf{u}_i(k+l)) + \mathbf{p}_i(k+l)^\mathrm{T}\mathbf{v}_i(k+l) - \sum_{j=1,\, j\neq i}^{|\mathcal{N}|} \mathbf{h}_{ji}^\mathrm{T}(\mathbf{x}_i(k+l))\mathbf{p}_j(k+l)
\end{aligned}
\tag{18.23}
$$

The Lagrange multipliers $\mathbf{p}$ are also referred to as prices [17]. In the special case when the problem P1 is convex and the Slater's condition holds for the inequality constraints (18.12)–(18.13), then $J^{\text{opt}}(\mathbf{x}) = \widetilde{J}^{\text{opt}}(\mathbf{x})$ (there will be no duality gap [1]).

The inner *decoupled* optimization problems in problem P2 are Nonlinear Programming (NLP) sub-problems, since the constraints (18.21) and the stage cost function $s_i^p(\mathbf{x}_i(k+l), \mathbf{u}_i(k+l), \mathbf{v}_i(k+l), \mathbf{p}(k:k+N_\text{p}))$ are nonlinear:

Problem P3$^\mathbf{i}$ (ith NLP Sub-problem):

$$\widetilde{J}_i^{\text{opt}}(\mathbf{p}(k:k+N_\text{p}), \mathbf{x}_i) = \min_{\substack{\mathbf{u}_i(k:k+N_\text{p})\\ \mathbf{x}_i(k:k+N_\text{p})\\ \mathbf{v}_i(k:k+N_\text{p})}} \sum_{l=0}^{N_\text{p}} s_i^p(\mathbf{x}_i(k+l), \mathbf{u}_i(k+l), \mathbf{v}_i(k+l), \mathbf{p}(k:k+N_\text{p})) \tag{18.24}$$

subject to $\mathbf{x}_i(k) = \mathbf{x}_i$ and:

$$\mathbf{x}_i(k+l) \in \mathcal{X}_i\,,\ l = 1, \ldots, N_\text{p} \tag{18.25}$$

$$\mathbf{u}_i(k+l) \in \mathcal{U}_i\,,\ l = 0, 1, \ldots, N_\text{p} - 1 \tag{18.26}$$

$$\mathbf{x}_i(k+l+1) = \mathbf{f}_i(\mathbf{x}_i(k+l), \mathbf{u}_i(k+l)) + \mathbf{v}_i(k+l),\ l = 0, 1, \ldots, N_\text{p} - 1 \tag{18.27}$$

Denote with $\mathbf{u}_i^{\text{opt}}(k:k+N_\text{p})$, $\mathbf{x}_i^{\text{opt}}(k:k+N_\text{p})$ and $\mathbf{v}_i^{\text{opt}}(k:k+N_\text{p})$ the optimal solution of problem P3^i. Note that the optimizers depend on the prices $\mathbf{p}$.

18.3.2 Local Quadratic Programming Approximations

The constraints (18.27) in the problems P3^i, $i = 1, 2, \ldots, |\mathcal{N}|$ may be non-convex in the general case (due to the nonlinearity of the functions $\mathbf{h}_{ji}(\cdot)$ and $\mathbf{f}_i(\cdot,\cdot)$). Below, the functions $\mathbf{h}_{ji}(\cdot)$ and $\mathbf{f}_i(\cdot,\cdot)$ are locally approximated by linear functions, leading to a convex *quasi-nonlinear* approach. An advantage of this approach is that strong duality can be easily claimed, i.e., there will be no duality gap between the solution obtained by solving the centralized quasi-NMPC and the solution corresponding to the distributed quasi-NMPC (see Proposition 1 later in this section). Let $\mathbf{x}_i(k) = \bar{\mathbf{x}}_i \in \mathcal{X}_i$ be arbitrary and denote the corresponding optimal solution to the sub-problem P3^i with:

$$\mathbf{u}_i^0 = \mathbf{u}_i^{\text{opt}}(\bar{\mathbf{x}}_i, k:k+N_\text{p}),\ \mathbf{x}_i^0 = \mathbf{x}_i^{\text{opt}}(\bar{\mathbf{x}}_i, k:k+N_\text{p}),\ \mathbf{v}_i^0 = \mathbf{v}_i^{\text{opt}}(\bar{\mathbf{x}}_i, k:K+N_\text{p}) \tag{18.28}$$

The optimal solution (18.28) depends on the values of the prices $\mathbf{p}$. Therefore, it would be necessary for $\mathbf{p}$ and the solution (18.28) to be updated iteratively. In order to simplify the computations while iterating, the system model (18.1) with the coupling functions given by (18.5) is linearized around the point $(\mathbf{u}_i^0, \mathbf{x}_i^0, \mathbf{v}_i^0, \bar{\mathbf{x}}_i)$ and the predicted states are obtained with the linearized model:

$$\mathbf{x}_i(k+l+1) = \mathbf{S}_{\mathbf{x}_i}(k+l)\mathbf{x}_i(k+l) + \mathbf{S}_{\mathbf{u}_i}(k+l)\mathbf{u}_i(k+l) + \sum_{\substack{j=1 \\ j \neq i}}^{|\mathcal{N}|} \mathbf{A}_{ij}(k+l)\mathbf{x}_j(k+l) + \mathbf{e}_{0i}(k+l),\ i = 1, \ldots, |\mathcal{N}|,\ l = 0, 1, \ldots, N_\mathrm{p} - 1 \tag{18.29}$$

where the matrices $\mathbf{S}_{\mathbf{x}_i}(k+l)$, $\mathbf{S}_{\mathbf{u}_i}(k+l)$, $\mathbf{A}_{ij}(k+l)$, and the vector $\mathbf{e}_{0i}(k+l)$ are given by:

$$\mathbf{S}_{\mathbf{x}_i}(k+l) = \nabla_{\mathbf{x}_i}\mathbf{f}_i(\mathbf{x}_i^0(k+l), \mathbf{u}_i^0(k+l)) \tag{18.30}$$

$$\mathbf{S}_{\mathbf{u}_i}(k+l) = \nabla_{\mathbf{u}_i}\mathbf{f}_i(\mathbf{x}_i^0(k+l), \mathbf{u}_i^0(k+l)) \tag{18.31}$$

$$\mathbf{A}_{ij}(k+l) = \nabla_{\mathbf{x}_j}\mathbf{h}_{ij}(\mathbf{x}_j^0(k+l)),\ j = 1, 2, \ldots, |\mathcal{N}|,\ j \neq i \tag{18.32}$$

$$\mathbf{e}_{0i}(k+l) = \mathbf{f}_i(\mathbf{x}_i^0(k+l), \mathbf{u}_i^0(k+l)) + \sum_{j=1,\, j\neq i}^{|\mathcal{N}|} \mathbf{h}_{ij}(\mathbf{x}_j^0(k+l)) - \mathbf{S}_{\mathbf{x}_i}(k+l)\mathbf{x}_i^0(k+l) - \mathbf{S}_{\mathbf{u}_i}(k+l)\mathbf{u}_i^0(k+l) - \sum_{j=1,\, j\neq i}^{|\mathcal{N}|} \mathbf{A}_{ij}(k+l)\mathbf{x}_j^0(k+l) \tag{18.33}$$

Then similar to (18.18), the following *decoupled* linear state equations are formulated:

$$\mathbf{x}_i(k+l+1) = \mathbf{S}_{\mathbf{x}_i}(k+l)\mathbf{x}_i(k+l) + \mathbf{S}_{\mathbf{u}_i}(k+l)\mathbf{u}_i(k+l) + \mathbf{w}_i(k+l) + \mathbf{e}_{0i}(k+l)$$
$$i = 1, \ldots, |\mathcal{N}|,\ l = 0, 1, \ldots, N_\mathrm{p} - 1 \tag{18.34}$$

with the additional constraints that:

$$\mathbf{w}_i(k+l) = \sum_{j=1,\, j\neq i}^{|\mathcal{N}|} \mathbf{A}_{ij}(k+l)\mathbf{x}_j(k+l), \quad i = 1, \ldots, |\mathcal{N}| \tag{18.35}$$

Then, the NLP sub-problems $\mathrm{P3}^i$ are approximated with the QP sub-problems:
Problem P4$^{\mathbf{i}}$ (ith QP Sub-problem):

$$J_i^*(\mathbf{p}(k:k+N_\mathrm{p}), \mathbf{x}_i) = \min_{\substack{\mathbf{u}_i(k:k+N_\mathrm{p}) \\ \mathbf{x}_i(k:k+N_\mathrm{p}) \\ \mathbf{w}_i(k:k+N_\mathrm{p})}} \sum_{l=0}^{N_\mathrm{p}} s_{\mathrm{QP},i}^{p}(\mathbf{x}_i(k+l), \mathbf{u}_i(k+l), \mathbf{w}_i(k+l), \mathbf{p}(k:k+N_\mathrm{p})) \tag{18.36}$$

subject to $\mathbf{x}_i(k) = \mathbf{x}_i$, (18.25), (18.26), and:

$$\mathbf{x}_i(k+l+1) = \mathbf{S}_{\mathbf{x}_i}(k+l)\mathbf{x}_i(k+l) + \mathbf{S}_{\mathbf{u}_i}(k+l)\mathbf{u}_i(k+l) + \mathbf{w}_i(k+l) + \mathbf{e}_{0i}(k+l)$$
$$l = 0,\ 1, \ldots,\ N_\mathrm{p} - 1 \tag{18.37}$$

In (18.36), the function $s^p_{\mathrm{QP},i}$ is given by:

$$s^p_{\mathrm{QP},i}(\mathbf{x}_i(k+l), \mathbf{u}_i(k+l), \mathbf{w}_i(k+l), \mathbf{p}(k:k+N_\mathrm{p})) = s_i(\mathbf{x}_i(k+l), \mathbf{u}_i(k+l))$$
$$+ \mathbf{p}_i(k+l)^\mathrm{T}\mathbf{w}_i(k+l) - \mathbf{x}_i(k+l)^\mathrm{T} \sum_{j=1,\, j\neq i}^{|\mathcal{N}|} \mathbf{A}_{ji}(k+l)^\mathrm{T}\mathbf{p}_j(k+l) \tag{18.38}$$

Denote with $\mathbf{u}_i^*(k:k+N_\mathrm{p})$, $\mathbf{x}_i^*(k:k+N_\mathrm{p})$ and $\mathbf{w}_i^*(k:k+N_\mathrm{p})$ the optimal solution of $\mathrm{P4}^i$. The following centralized quasi-NMPC problem with quadratic cost and linear constraints is formulated:

Problem P5 (Centralized Quasi-NMPC):

$$J^*(\mathbf{x}) = \min_{\mathbf{u}(k:k+N_\mathrm{p})} J(\mathbf{u}(k:k+N_\mathrm{p}), \mathbf{x}) \tag{18.39}$$

subject to $\mathbf{x}(k) = \mathbf{x}$, constraints (18.12), (18.13), and (18.29).

The cost function $J(\mathbf{u}(k:k+N_\mathrm{p}), \mathbf{x})$ is given by (18.17).

Then, the distributed quasi-NMPC problem is as follows:

Problem P6 (Distributed Quasi-NMPC):

$$\max_{\mathbf{p}(k:k+N_\mathrm{p})} \sum_{i=1}^{|\mathcal{N}|} J_i^*(\mathbf{p}(k:k+N_\mathrm{p}), \mathbf{x}_i) = \tag{18.40}$$

$$\max_{\mathbf{p}(k:k+N_\mathrm{p})} \sum_{i=1}^{|\mathcal{N}|} \min_{\substack{\mathbf{u}_i(k:k+N_\mathrm{p})\\ \mathbf{x}_i(k:k+N_\mathrm{p})\\ \mathbf{w}_i(k:k+N_\mathrm{p})}} \sum_{l=0}^{N_\mathrm{p}} s^p_{\mathrm{QP},i}(\mathbf{x}_i(k+l), \mathbf{u}_i(k+l), \mathbf{w}_i(k+l), \mathbf{p}(k:k+N_\mathrm{p})) \tag{18.41}$$

subject to $\mathbf{x}(k) = \mathbf{x}$, constraints (18.12), (18.13), (18.34), and:

$$\mathbf{p}(k+N_\mathrm{p}) = 0 \tag{18.42}$$

Then, the decomposition of the optimization problem P5 is given by the following proposition:

Proposition 18.1 *Suppose that* $\mathbf{x} = [\mathbf{x}_1,\ \mathbf{x}_2, \ldots,\ \mathbf{x}_{|\mathcal{N}|}]$ *is a feasible point for problem P5. Then:*

$$J^*(\mathbf{x}) = \max_{\mathbf{p}(k:k+N_\mathrm{p})} \sum_{i=1}^{|\mathcal{N}|} J_i^*(\mathbf{p}(k:k+N_\mathrm{p}), \mathbf{x}_i) \tag{18.43}$$

where maximization is subject to $\mathbf{p}(k+N_\mathrm{p}) = 0$.

The proof follows similar arguments as in [7, 9] and the fact that the Slater's condition always holds for a feasible convex QP [1, pp. 226–227].

Proposition 18.1 shows that the computation of $\mathbf{u}_i^*(k : k+N_\mathrm{p})$, $\mathbf{x}_i^*(k : k+N_\mathrm{p})$ and $\mathbf{w}_i^*(k : k+N_\mathrm{p})$ for given prices $\mathbf{p}(k : k+N_\mathrm{p})$ is completely decentralized. However, finding the optimal prices requires coordination [7]. Given a price prediction sequence $\mathbf{p}_i^r(k : k+N_\mathrm{p})$ for the r-th iteration, the corresponding sequences $\mathbf{u}_i^{*r}(k : k+N_\mathrm{p})$, $\mathbf{x}_i^{*r}(k : k+N_\mathrm{p})$ and $\mathbf{w}_i^{*r}(k : k+N_\mathrm{p})$ are computed locally by solving $\mathrm{P4}^i$. Then, the prices are updated distributedly by a gradient step:

$$\mathbf{p}_i^{r+1}(k+l) = \mathbf{p}_i^r(k+l) + \gamma_i^r[\mathbf{w}_i^{*r}(k+l) - \sum_{j=1,\, j\neq i}^{|\mathcal{N}|} \mathbf{A}_{ij}(k+l)\mathbf{x}_j^{*r}(k+l)]$$
$$l = 0, 1, \ldots, N_\mathrm{p}-1 \text{ with } \mathbf{p}_i^{r+1}(k+N_\mathrm{p}) = \mathbf{p}_i^r(k+N_\mathrm{p}) = 0 \tag{18.44}$$

where γ_i^r is the step size. In a neighborhood of the solution $\mathbf{u}_i^0$, $\mathbf{x}_i^0$, $\mathbf{v}_i^0$ to the NLP sub-problems $\mathrm{P3}^i$, $i = 1, \ldots, |\mathcal{N}|$, the linearized model (18.29) can sufficiently accurately approximate the nonlinear model (18.1). Therefore, to approximate the NMPC solution it would be necessary to periodically update the linearized model (18.29) and to apply formula (18.44) for a number of steps. This is done by a slight modification of the suboptimal algorithm for distributed quasi-NMPC in [9].

18.3.3 An Algorithm for Distributed Quasi-NMPC

In [7], an approach to distributed MPC for *linear* systems has been suggested, where the prices are updated according to (18.44). In [9], a suboptimal algorithm for distributed quasi-NMPC for *nonlinear* systems with *linear* couplings has been proposed. Here, the algorithm in [9] is slightly modified to consider a more general class of systems consisting of *nonlinear* subsystems with separable *nonlinear* couplings (see Sect. 18.2). The algorithm includes two loops. In the outer loop, the NLP sub-problems $\mathrm{P3}^i$, $i = 1, 2, \ldots, |\mathcal{N}|$, are solved and the matrices of the linearized model are computed, which are used in the approximating QP sub-problems $\mathrm{P4}^i$, $i = 1, 2, \ldots, |\mathcal{N}|$. Then, in the inner loop, the price sequences and solution are updated based on Proposition 18.1 and applying formula (18.44) for a given number of steps. The algorithm is presented as Algorithm 18.1.

The steps 4 to 11 in Algorithm 18.1 include an iterative solution of the NLP sub-problems $\mathrm{P3}^i$, approximating them with the QP sub-problems $\mathrm{P4}^i$, and then updating the prices by utilizing Proposition 18.1.

Algorithm 18.1 Distributed quasi-NMPC by on-line optimization

1: Given integer numbers Q and L, step sizes γ_i, $i = 1, 2, \ldots, |\mathcal{N}|$ and arbitrary guesses $\mathbf{p}_i^0(k : k + N_\mathrm{p})$, $i = 1, 2, \ldots, |\mathcal{N}|$ for the price sequences. Let $k = 0$.
2: Let the state at time k be $\mathbf{x}(k) = \mathbf{x} = [\mathbf{x}_1, \ldots, \mathbf{x}_{|\mathcal{N}|}]$.
3: **for** $q = 1, 2, \ldots, Q$ **do**
4: For $\mathbf{x}_i(k) = \mathbf{x}_i$ compute distributedly the optimal solutions $\mathbf{u}_i^0 = \mathbf{u}_i^{\mathrm{opt}}(\mathbf{x}_i, k : k + N_\mathrm{p})$, $\mathbf{x}_i^0 = \mathbf{x}_i^{\mathrm{opt}}(\mathbf{x}_i, k : k+N_\mathrm{p})$, $\mathbf{v}_i^0 = \mathbf{v}_i^{\mathrm{opt}}(\mathbf{x}_i, k : k+N_\mathrm{p})$ to the NLP sub-problems P3^i, $i = 1, 2, \ldots, |\mathcal{N}|$, corresponding to the price sequences $\mathbf{p}_i^0(k : k+N_\mathrm{p})$. Compute the matrices $\mathbf{S}_{\mathbf{x}_i}(k+l)$, $\mathbf{S}_{\mathbf{u}_i}(k+l)$, $\mathbf{A}_{ij}(k+l)$, $j = 1, \ldots, |\mathcal{N}|$, $j \neq i$, and $\mathbf{e}_{0i}(k+l)$, $l = 0, 1, \ldots, N_\mathrm{p} - 1$, associated to the approximating QP sub-problems P4^i, $i = 1, 2, \ldots, |\mathcal{N}|$.
5: **for** $r = 0, 1, \ldots, L - 1$ **do**
6: For i-th subsystem, $i = 1, 2, \ldots, |\mathcal{N}|$, communicate the price sequences $\mathbf{p}_j^r(k : k + N_\mathrm{p})$, $j = 1, \ldots, |\mathcal{N}|$, $j \neq i$ of the interconnected subsystems.
7: Compute the sequences $\mathbf{u}_i^{*r}(k : k + N_\mathrm{p})$, $\mathbf{x}_i^{*r}(k : k + N_\mathrm{p})$ and $\mathbf{w}_i^{*r}(k : k + N_\mathrm{p})$ corresponding to the price sequence $\mathbf{p}^r(k : k + N_\mathrm{p})$ by solving distributedly the QP sub-problems P4^i, $i = 1, 2, \ldots, |\mathcal{N}|$.
8: For i-th subsystem, $i = 1, 2, \ldots, |\mathcal{N}|$, communicate the state trajectories $\mathbf{x}_j^{*r}(k : k + N_\mathrm{p})$, $j = 1, \ldots, |\mathcal{N}|$, $j \neq i$ of the interconnected subsystems.
9: Compute distributedly the updates $\mathbf{p}_i^{r+1}(k : k + N_\mathrm{p})$, $i = 1, 2, \ldots, |\mathcal{N}|$ of the price sequences by applying (18.44) for $\gamma_i^r = \gamma_i$, $i = 1, 2, \ldots, |\mathcal{N}|$.
10: **end**
11: Let $\mathbf{p}_i^0(k : k + N_\mathrm{p}) = \mathbf{p}_i^L(k : k + N_\mathrm{p})$, $i = 1, 2, \ldots, |\mathcal{N}|$.
12: **end**
13: Apply to the overall system the control inputs $\mathbf{u}_i(k) = \mathbf{u}_i(k)^{*L-1}$, $i = 1, 2, \ldots, |\mathcal{N}|$.
14: Let $k = k + 1$ and go to step 2.

It should be noted that alternatively, an approach similar to [13, 16] can be applied, where the idea would be to avoid solving the NLP sub-problems P3^i in step 4 and to formulate the approximating QP sub-problems P4^i by using the optimal sequences $\mathbf{u}_i^*(k : k + N_\mathrm{p})$, $\mathbf{x}_i^*(k : k + N_\mathrm{p})$ and $\mathbf{w}_i^*(k : k + N_\mathrm{p})$, computed in the previous time instant.

The communication between the distributed MPC controllers is illustrated in Fig. 18.2 for three subsystems (only the inner loop in Algorithm 18.1 is depicted). The solution of the QP sub-problems P4^i and the update of prices is distributed, but requires for the neighboring subsystems to communicate their price sequences and optimal state trajectories.

18.4 Theoretical Results

The dynamic dual decomposition approach is applied for the first time to distributed MPC in [7], where a system composed by the interconnection of a set of *linear* subsystems $\mathcal{N} = \{S_1, S_2, \ldots, S_{|\mathcal{N}|}\}$ is considered. The interconnected subsystems are described by the following linear discrete-time models:

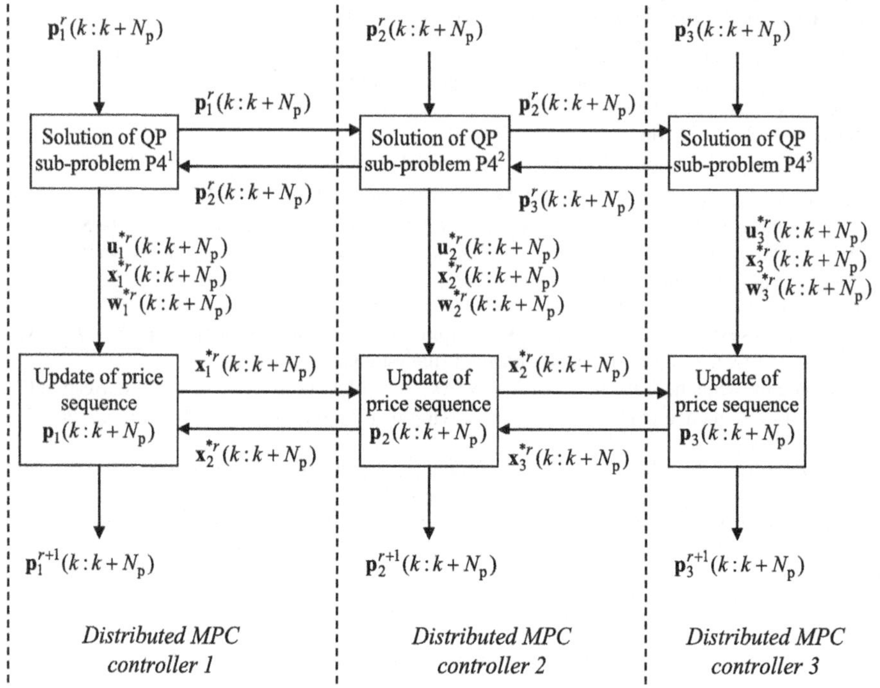

Fig. 18.2 The communication between the distributed MPC controllers

$$\mathbf{x}_i(k+1) = \sum_{j=1}^{|\mathcal{N}|} \mathbf{A}_{ij}\mathbf{x}_j(k) + \mathbf{B}_i\mathbf{u}_i(k),\ i = 1,\ 2, \ldots,\ |\mathcal{N}| \tag{18.45}$$

where $\mathbf{x}_i(k) \in \mathbb{R}^{n_{x_i}}$ and $\mathbf{u}_i(k) \in \mathbb{R}^{n_{u_i}}$ are the state and control input vectors, related to the ith subsystem, and $\mathbf{A}_{ij} \in \mathbb{R}^{n_{x_i} \times n_{x_j}}$ and $\mathbf{B}_i \in \mathbb{R}^{n_{x_i} \times n_{u_i}}$ are constant matrices.

For the system (18.45), the infinite horizon optimal control problem P0 is considered, where the nonlinear equality constraints (18.1) are replaced by the linear equality constraints (18.45). The problem P0 is approximated with a linear centralized MPC problem P1, where the nonlinear constraints (18.14) are replaced by linear ones, associated to the model (18.45).

In [7], a suboptimal algorithm for distributed *linear* MPC is developed, which includes an iterative solution of the QP sub-problems for the linear subsystems and updating the price sequences (illustrated in Fig. 18.2). Further in [7], a stopping criterion for the iterative distributed MPC scheme is proposed that can be locally verified at each subsystem and that guarantees closed loop suboptimality above a pre-specified level $\alpha \in (0,\ 1)$ and asymptotic stability of the overall linear system. With this criterion, at each time instant the minimal number of iterations is obtained such that the following holds (cf. Theorem 4 in [7]):

$$\alpha J^{\infty}_{\mathrm{MPC}}(\bar{\mathbf{x}}) \leq J^{\infty}(\bar{\mathbf{x}}) \text{ and } \|\mathbf{x}(k)\|^2 \to 0 \text{ as } k \to \infty \tag{18.46}$$

where $\bar{\mathbf{x}}$ is the initial state of the overall system (18.45), $\mathbf{x}(k)$ is the response of the system (18.45) in closed-loop with the distributed MPC law, $J^{\infty}_{\mathrm{MPC}}(\bar{\mathbf{x}})$ is the cost associated to infinite horizon performance of the distributed MPC law, and $J^{\infty}(\bar{\mathbf{x}})$ is the cost obtained by solving the infinite horizon optimal control problem P0 for the system (18.45).

It is shown in [10] that under certain assumptions the infinite horizon cost, associated with the suboptimal distributed quasi-NMPC control (designed by applying the approach in Sect. 18.3) in closed-loop with the *nonlinear* system (18.1), is bounded and an estimate of the degree of suboptimality is derived.

The relation of the distributed quasi-NMPC approach [9, 10] to other distributed MPC approaches and their comparison is a subject of a future research.

18.5 Application Results

In [7], the performance of the developed distributed MPC approach for *linear* interconnected systems is evaluated by applying it to a simulation example with equally sized water containers, connected in series.

In [8], the distributed *linear* MPC approach is applied to the MPC-based path planning for a UAV (Unmanned Aerial Vehicles) communication chain under radio path loss constraints. The centralized path planning optimization problem with linear model of the UAVs and coupled inequality constraints is reformulated into an equivalent distributed path planning optimization problem, where the resulting QP sub-problems are completely decoupled. The MPC-based optimization sub-problems are computed autonomously within each UAV, using convex quadratic programming and gradient iterations, with the requirement that each UAV communicates its current measured position and the computed optimal velocity trajectory to its neighboring UAVs.

In [9], the distributed quasi-NMPC approach is applied to a system composed by two *nonlinear* subsystems with *linear* couplings in the presence of state and input constraints and bounded disturbances.

Here, the distributed quasi-NMPC approach described in Sect. 18.3 is applied to a system consisting of two *nonlinear* subsystems with *nonlinear* couplings. The two subsystems S_1 and S_2 are described by:

$$S_1 : x_1(k+1) = \sqrt{3x_1(k)^2 + 1} + u_1(k) - 1 + \mu_1 x_2^3(k) \tag{18.47}$$

$$S_2 : x_2(k+1) = e^{-\sin(x_2(k))} + u_2(k) - 1 + \mu_2 x_1^3(k) \tag{18.48}$$

The functions f_1, f_2, h_{12}, h_{21} in the description (18.1) with the assumption A3 are:

$$f_1(x_1(k), u_1(k)) = \sqrt{3x_1(k)^2 + 1} + u_1(k) - 1 \tag{18.49}$$

$$f_2(x_2(k), u_2(k)) = e^{-\sin(x_2(k))} + u_2(k) - 1 \tag{18.50}$$

$$h_{12}(x_2(k)) = \mu_1 x_2^3(k)\,, \quad h_{21}(x_1(k)) = \mu_2 x_1^3(k) \tag{18.51}$$

The functions h_{12} and h_{21} describe the mutual influence of the two subsystems. The following constraints are imposed on the system (18.47)–(18.48):

$$-0.3 \le u_i(k) \le 0.5\,, \quad i = 1, 2 \tag{18.52}$$

$$-2 \le x_1(k) \le 2\,, \quad -1.5 \le x_2(k) \le 1.5 \tag{18.53}$$

The coefficients related to the couplings between the two subsystems are $\mu_1 = \mu_2 = 0.2$. The prediction horizon in the centralized NMPC problem P1 is $N_p = 10$ and the weighting matrices are $Q_i = R_i = 1$, $i = 1, 2$.

The centralized NMPC problem is represented as a distributed NMPC problem (problem P6) by applying the dual decomposition approach. Then, Algorithm 18.1 with parameters $Q = 5$, $L = 3$, $\gamma_i = 0.3$, $i = 1, 2$ is used to generate the two control inputs for an initial state $x(0) = [1.2 \;\; 1.2]$. The corresponding trajectories of the control inputs u_1, u_2 and the states x_1, x_2, associated to the two subsystems are depicted in Figs. 18.3 and 18.4. The trajectories obtained with the following approaches are compared:

- The suboptimal distributed quasi-NMPC approach with linearization (described in Sect. 18.3);
- A suboptimal distributed NMPC approach without linearization. In this case, a modification of Algorithm 18.1 is used for the on-line computation of the control inputs. It has only one loop, where the optimal solutions of the NLP sub-problems $P3^i$, $i = 1, 2, \ldots, |\mathcal{N}|$ are computed distributedly, and then the price sequences

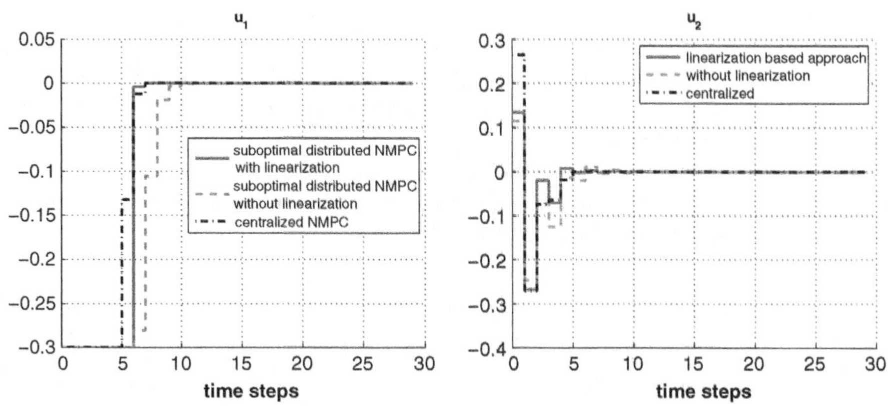

Fig. 18.3 The control inputs u_1 and u_2 for subsystems S_1 and S_2

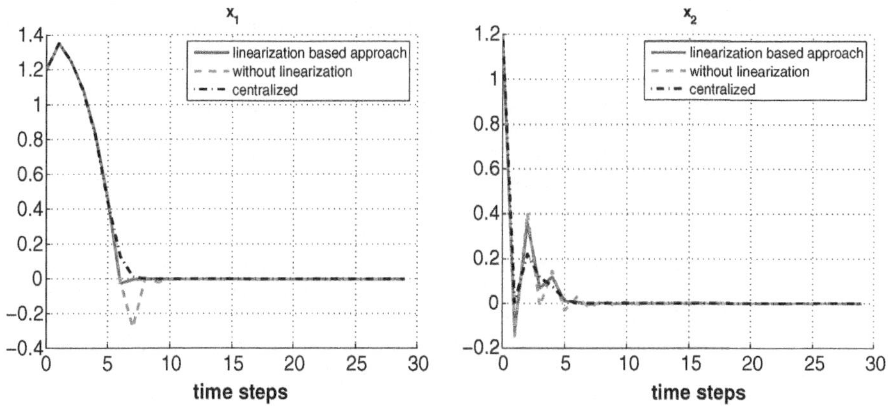

Fig. 18.4 The states x_1 and x_2 of subsystems S_1 and S_2

are updated by applying (18.44) by using the computed optimal solutions. The loop is repeated $Q = 5$ times and the step size in (18.44) is $\gamma_i = 0.3,\ i = 1,\ 2$.

- The centralized NMPC approach, which solves problem P1 at each time instant.

18.6 Conclusions

In this chapter, the distributed quasi-NMPC approach is described in details and an algorithm based on on-line optimization is presented. The theoretical and the application results associated to both linear and nonlinear interconnected systems are reviewed. The relation of the suggested distributed NMPC approach to other distributed approaches is a subject of a future research.

References

1. S. Boyd, L. Vandenberghe, in *Convex Optimization* (Cambridge University Press, Cambridge, 2004)
2. A. Bryson, Y. Ho, *Applied Optimal Control: Optimization Estimation and Control* (Blaisdell, Waltham, 1969)
3. G. Cohen, B. Miara, Optimization with an auxiliary constraint and decomposition. SIAM J. Control Optim. **28**, 137–157 (1990)
4. G.A. Constantinides. Parallel architectures for model predictive control, in *Proceedings of the European Control Conference*, pp 138–143, Budapest, Hungary, 2009
5. G.B. Dantzig, P. Wolfe, The decomposition algorithm for linear programs. Econometrica **29**, 767–778 (1961)
6. W.B. Dunbar, R.M. Murray, Distributed receding horizon control for multi-vehicle formation stabilization. Automatica **42**, 549–558 (2006)

7. P. Giselsson, A. Rantzer, Distributed model predictive control with suboptimality and stability guarantees, in *Proceedings of the IEEE Conference on Decision and Control*, pp. 7272–7277, Atlanta, GA, (2010)
8. A. Grancharova, E.I. Grøtli, T.A. Johansen, Distributed path planning for a UAV communication chain by dual decomposition, in *Proceedings of the 2nd IFAC Workshop on Multivehicle Systems*, Espoo, Finland, (2012)
9. A. Grancharova, T.A. Johansen, Distributed quasi-nonlinear model predictive control by dual decomposition, in *Proceedings of the 18-th IFAC World Congress*, pp. 1429–1434, Milano, Italy, (2011)
10. A. Grancharova, T.A. Johansen, Suboptimality analysis of distributed quasi-nonlinear model predictive control, in *Proceedings of the IFAC Conference on Nonlinear Model Predictive Control, NMPC'12* (Noordwijkerhout, Netherlands, 2012)
11. T. Keviczky, F. Borrelli, G.J. Balas, Decentralized receding horizon control for large scale dynamically decoupled systems. Automatica **42**, 2105–2115 (2006)
12. T.H. Kim, T. Sugie, Robust decentralized MPC algorithm for a class of dynamically interconnected constrained systems, in *Proceedings of the IEEE Conference on Decision and Control*, pp. 290–295, Seville, Spain, (2005)
13. W.C. Li, L.T. Biegler, Multistep, Newton-type control strategies for constrained nonlinear processes. Chem. Eng. Res. Des. **67**, 562–577 (1989)
14. L. Magni, R. Scattolini, Stabilizing decentralized model predictive control of nonlinear systems. Automatica **42**, 1231–1236 (2006)
15. R.R. Negenborn, Multi-Agent Model Predictive Control with Applications to Power Networks. PhD thesis, Delft University of Technology, Delft, Netherlands, December 2007
16. N.M.C. Oliveira, L.T. Biegler, An extension of Newton-type algorithms for nonlinear process control. Automatica **31**, 281–286 (1995)
17. A. Rantzer, Dynamic dual decomposition for distributed control, in *Proceedings of the American Control Conference*, pp. 884–888, St. Louis, MO, USA, (2009)
18. R. Scattolini, Architectures for distributed and hierarchical model predictive control: a review. J. Process Control **19**, 723–731 (2009)
19. A.N. Venkat, J.B. Rawlings, S.J. Wright, Plant-wide optimal control with decentralized MPC, in *Proceedings of the IFAC Conference on Dynamics and Control of Process Systems*, Boston, MA, (2004)
20. Y. Zhang, S. Li, Networked model predictive control based on neighbourhood optimization for serially connected large-scale processes. J. Process Control **17**, 37–50 (2007)

Chapter 19
Generalized Accelerated Gradient Methods for Distributed MPC Based on Dual Decomposition

P. Giselsson and A. Rantzer

Abstract We consider distributed model predictive control (DMPC) where a sparse centralized optimization problem without a terminal cost or a terminal constraint set is solved in distributed fashion. Distribution of the optimization algorithm is enabled by dual decomposition. Gradient methods are usually used to solve the dual problem resulting from dual decomposition. However, gradient methods are known for their slow convergence rate, especially for ill-conditioned problems. This is not desirable in DMPC where the amount of communication should be kept as low as possible. In this chapter, we present a distributed optimization algorithm applied to solve optimization problems arising in DMPC that has significantly better convergence rate than the classical gradient method. This improved convergence rate is achieved by using accelerated gradient methods instead of standard gradient methods and by in a well-defined manner, incorporating Hessian information into the gradient-iterations. We also present a stopping condition to the distributed optimization algorithm that ensures feasibility, stability and closed loop performance of the DMPC-scheme, without using a stabilizing terminal cost or terminal constraint set.

19.1 Introduction

Dual decomposition techniques have often been applied to solve distributed optimization problems with a strongly convex objective arising in distributed model predictive control. Traditionally, the dual problem resulting from dual decomposition is solved using gradient methods. These methods are known to have bad convergence rate properties $O(1/p)$ with p the iteration number [12], especially for ill-conditioned

P. Giselsson (✉) · A. Rantzer
Department of Automatic Control, Lund University, Lund, Sweden
e-mail: pontusg@control.lth.se

A. Rantzer
e-mail: rantzer@control.lth.se

J. M. Maestre and R. R. Negenborn (eds.), *Distributed Model Predictive Control Made Easy*, Intelligent Systems, Control and Automation: Science and Engineering 69,
DOI: 10.1007/978-94-007-7006-5_19,

problems. Thus, to implement a DMPC controller based on dual decomposition and a gradient method would require extensive communication. In this chapter we propose a method that reduces the communication requirement significantly when solving the dual problem obtained from dual decomposition. The communication reduction is made possible by two main improvements compared to the classical gradient method. The first improvement is to use accelerated gradient methods, which converges as $O(1/p^2)$, instead of gradient methods to solve the dual problem. For more on accelerated gradient methods, the reader is referred to [1, 12, 14, 15]. Also, the optimal step size for the algorithm is provided which is important for performance reasons. The other improvement is on the gradient step. In every iteration in gradient or accelerated gradient methods, a quadratic upper bound with the same curvature in every direction is minimized to compute the new iterate. If this quadratic upper bound does not well approximate the cost function, the number of iterations can be significant. By allowing for different curvature in different directions of the quadratic upper bound, a much closer fit between the cost function and the upper bound can be obtained, especially for ill-conditioned problems. This can reduce significantly the number of iterations, hence the amount of communication, needed to achieve the desired accuracy of the solution. The distributed optimization algorithm with improved convergence rate was published in [3].

We also present a stopping condition for the presented distributed optimization algorithm. The stopping condition is developed to guarantee feasibility, stability, and a prespecified performance of the closed loop system. Traditionally, a terminal cost and a terminal constraint set is used to prove stability in MPC. These are usually dense, i.e., involve all state variables. In dual decomposition, the cost function should be separable and the constraints should be sparse. Hence, we cannot rely on a terminal cost or a terminal constraint set prove stability in DMPC based on dual decomposition. However, stability and performance of the closed loop system can be established by choosing the control horizon such that the optimal value function is decreasing with a certain amount in each time step. Further, feasibility in duality-based optimization can only be guaranteed in the limit. To guarantee feasibility with a finite number of iterations, an adaptive constraint tightening approach is presented. The constraint tightening enables a feasible solution within finite number of iterations and the adaptation adapts the amount of constraint tightening to guarantee that the optimal value function is decreasing in each time step. This implies that feasibility, stability, and a prespecified performance of the closed loop system are guaranteed using the presented stopping condition. The stopping condition was published in [8].

19.2 Problem Formulation

We consider control of linear dynamical systems of the form

$$\mathbf{x}(k+1) = \mathbf{A}\mathbf{x}(k) + \mathbf{B}\mathbf{u}(k), \qquad \mathbf{x}(0) = \bar{\mathbf{x}}$$

where $\mathbf{x} \in \mathbb{R}^{n_x}$, $\mathbf{u} \in \mathbb{R}^{n_u}$, $\mathbf{A} \in \mathbb{R}^{n_x \times n_x}$, and $\mathbf{B} \in \mathbb{R}^{n_x \times n_u}$. We assume that the pair $(\mathbf{A}, \mathbf{B})$ is controllable. The state and control vectors are partitioned according to

$$\mathbf{x} = [\mathbf{x}_1^T, \ldots, \mathbf{x}_{|\mathcal{N}|}^T]^T, \mathbf{u} = [\mathbf{u}_1^T, \ldots, \mathbf{u}_{|\mathcal{N}|}^T]^T$$

where $\mathbf{x}_i \in \mathbb{R}^{n_{x_i}}$, $\mathbf{u}_i \in \mathbb{R}^{n_{u_i}}$, for all $i \in \mathcal{N}$ are referred to as local variables, $\mathcal{N} = \{1, \ldots, |\mathcal{N}|\}$ is the set of subsystems, and $|\mathcal{N}|$ is the number of subsystems. The dynamics matrices are partitioned according to the state and control variable partitions

$$\mathbf{A} = \begin{Bmatrix} \mathbf{A}_{11} & \cdots & \mathbf{A}_{1|\mathcal{N}|} \\ \vdots & \ddots & \vdots \\ \mathbf{A}_{|\mathcal{N}|1} & \cdots & \mathbf{A}_{|\mathcal{N}||\mathcal{N}|} \end{Bmatrix}, \qquad \mathbf{B} = \begin{Bmatrix} \mathbf{B}_{11} & \cdots & \mathbf{B}_{1|\mathcal{N}|} \\ \vdots & \ddots & \vdots \\ \mathbf{B}_{|\mathcal{N}|1} & \cdots & \mathbf{B}_{|\mathcal{N}||\mathcal{N}|} \end{Bmatrix}$$

where $\mathbf{A}_{ij} = \mathbb{R}^{n_{x_i} \times n_{x_j}}$ and $\mathbf{B}_{ij} = \mathbb{R}^{n_{x_i} \times n_{u_j}}$. We assume that the matrices have a sparse structure, i.e., that some $\mathbf{A}_{ij} = 0$ and $\mathbf{B}_{ij} = 0$. The neighboring interaction is defined by the following sets

$$\mathcal{N}_i = \{j \in \mathcal{N} \mid \mathbf{A}_{ij} \neq 0 \text{ or } \mathbf{B}_{ij} \neq 0\},$$
$$\mathcal{M}_i = \{j \in \mathcal{N} \mid \mathbf{A}_{ji} \neq 0 \text{ or } \mathbf{B}_{ji} \neq 0\}.$$

Using the introduced local variables and the neighborhood sets, the local dynamics is described by

$$\mathbf{x}_i(k+1) = \sum_{j \in \mathcal{N}_i} \left(\mathbf{A}_{ij}\mathbf{x}_j(k) + \mathbf{B}_{ij}\mathbf{u}_j(k)\right), \qquad \mathbf{x}_i(0) = \bar{\mathbf{x}}_i.$$

There are constraints on the local state and control variables, they should satisfy $\mathbf{x}_i \in \mathcal{X}_i$, and $\mathbf{u}_i \in \mathcal{U}_i$ where

$$\mathcal{X}_i = \{\mathbf{x}_i \in \mathbb{R}^{n_{x_i}} \mid \mathbf{C}_{x,i}\mathbf{x}_i \leq \mathbf{d}_{x,i}\}, \qquad \mathcal{U}_i = \{\mathbf{u}_i \in \mathbb{R}^{n_{u_i}} \mid \mathbf{C}_{u,i}\mathbf{u}_i \leq \mathbf{d}_{u,i}\}$$

and $\mathbf{C}_{x,i}$, $\mathbf{C}_{u,i}$, $\mathbf{d}_{x,i} > 0$, and $\mathbf{d}_{u,i} > 0$ are real matrices/vectors with appropriate dimensions. The assumption that $\mathbf{d}_{x,i} > 0$ and $\mathbf{d}_{u,i} > 0$ implies that $0 \in \text{int}(\mathcal{X}_i)$ and $0 \in \text{int}(\mathcal{U}_i)$. The global constraint sets, $\mathcal{X}$ and $\mathcal{U}$, are products of local sets, i.e.,

$$\mathcal{X} = \mathcal{X}_1 \times \ldots \times \mathcal{X}_{|\mathcal{N}|}, \qquad \mathcal{U} = \mathcal{U}_1 \times \ldots \times \mathcal{U}_{|\mathcal{N}|}.$$

The total number of inequalities describing the sets $\mathcal{X}$ and $\mathcal{U}$ is denoted by n_c. We assume quadratic local stage-cost functions, i.e., local stage-cost functions of the form

$$\ell_i(\mathbf{x}_i, \mathbf{u}_i) = \frac{1}{2}\left(\mathbf{x}_i^T\mathbf{Q}_i\mathbf{x}_i + \mathbf{u}_i^T\mathbf{R}_i\mathbf{u}_i\right)$$

where cost matrices $\mathbf{Q}_i \in \mathbb{R}^{n_{x_i} \times n_{x_i}}$ and $\mathbf{R}_i \in \mathbb{R}^{n_{u_i} \times n_{u_i}}$ for all $i \in \mathcal{N}$ are assumed symmetric and positive definite. This gives the following stage-cost for the full system

$$\ell(\mathbf{x}, \mathbf{u}) = \sum_{i=1}^{|\mathcal{N}|} \ell_i(\mathbf{x}_i, \mathbf{u}_i) = \frac{1}{2} \sum_{i=1}^{|\mathcal{N}|} \left(\mathbf{x}_i^T \mathbf{Q}_i \mathbf{x}_i + \mathbf{u}_i^T \mathbf{R}_i \mathbf{u}_i \right) = \frac{1}{2} \left(\mathbf{x}^T \mathbf{Q} \mathbf{x} + \mathbf{u}^T \mathbf{R} \mathbf{u} \right)$$

where $\mathbf{Q} = \text{blkdiag}(\mathbf{Q}_1, \ldots, \mathbf{Q}_{|\mathcal{N}|})$ and $\mathbf{R} = \text{blkdiag}(\mathbf{R}_1, \ldots, \mathbf{R}_{|\mathcal{N}|})$. For future reference we also introduce

$$\ell^*(\mathbf{x}) := \min_{\mathbf{u} \in \mathcal{U}} \ell(\mathbf{x}, \mathbf{u}) = \ell(\mathbf{x}, \mathbf{0}) = \frac{1}{2} \mathbf{x}^T \mathbf{Q} \mathbf{x}.$$

We use prediction horizon $N_p = N$ and control horizon $N_c = N$. Also, neither a terminal cost nor a terminal constraint set is used in the DMPC optimization problem formulation. Hence, the optimization problem to be solved for initial condition $\bar{\mathbf{x}}$ in the DMPC scheme is:

$$\begin{aligned} V_N(\bar{\mathbf{x}}) := \min_{\hat{\mathbf{x}}, \hat{\mathbf{u}}} & \sum_{l=0}^{N-1} \frac{1}{2} (\hat{\mathbf{x}}^T(l) \mathbf{Q} \hat{\mathbf{x}}(l) + \hat{\mathbf{u}}^T(l) \mathbf{R} \hat{\mathbf{u}}(l)) \\ \text{s.t. } & (\hat{\mathbf{x}}(l), \hat{\mathbf{u}}(l)) \in \mathcal{X} \times \mathcal{U}, \qquad l = 0, \ldots, N-1 \\ & \hat{\mathbf{x}}(l+1) = A\hat{\mathbf{x}}(l) + B\hat{\mathbf{u}}(l), \qquad l = 0, \ldots, N-2 \\ & \hat{\mathbf{x}}(0) = \bar{\mathbf{x}} \end{aligned} \tag{19.1}$$

where $\hat{\mathbf{x}}(l)$ and $\hat{\mathbf{u}}(l)$ denote the predicted state and control variables l steps ahead. To describe the optimization problem in a more compact form, the following stacked vectors are introduced

$$\mathbf{z}_i = [\hat{\mathbf{x}}_i^T(0), \ldots, \hat{\mathbf{x}}_i^T(N-1), \hat{\mathbf{u}}_i^T(0), \ldots, \hat{\mathbf{u}}_i^T(N-1)]^T$$

for all $i \in \mathcal{N}$ and $\mathbf{z} = [\mathbf{z}_1^T, \ldots, \mathbf{z}_{|\mathcal{N}|}^T]^T$. This implies that the optimization problem (19.1) more compactly can be written as

$$\begin{aligned} V_N(\bar{\mathbf{x}}) := \min_{\mathbf{z}} & \frac{1}{2} \mathbf{z}^T \mathbf{H} \mathbf{z} \\ \text{s.t. } & \mathbf{F}\mathbf{z} = \mathbf{g}\bar{\mathbf{x}} \\ & \mathbf{C}\mathbf{z} \leq \mathbf{d} \end{aligned} \tag{19.2}$$

where

$$\begin{aligned} \mathbf{H} &= \text{blkdiag}(\mathbf{H}_1, \ldots, \mathbf{H}_{|\mathcal{N}|}), & \bar{\mathbf{x}} &= [\bar{\mathbf{x}}_1^T, \ldots, \bar{\mathbf{x}}_{|\mathcal{N}|}^T]^T, \\ \mathbf{F} &= [\mathbf{F}_1^T, \ldots, \mathbf{F}_{|\mathcal{N}|}^T]^T, & \mathbf{g} &= [\mathbf{g}_1^T, \ldots, \mathbf{g}_{|\mathcal{N}|}^T]^T, \\ \mathbf{C} &= \text{blkdiag}(\mathbf{C}_1, \ldots, \mathbf{C}_{|\mathcal{N}|}), & \mathbf{d} &= [\mathbf{d}_1^T, \ldots, \mathbf{d}_{|\mathcal{N}|}^T]^T \end{aligned}$$

and

$$
\begin{aligned}
\mathbf{H}_i &= \text{blkdiag}(\mathbf{Q}_i, \dots, \mathbf{Q}_i, \mathbf{R}_i, \dots, \mathbf{R}_i),\\
\mathbf{F}_i &= [\mathbf{F}_{i1}, \dots, \mathbf{F}_{i|\mathcal{N}|}],\\
\mathbf{F}_{ij} &= \begin{cases}
\begin{bmatrix}
\mathbf{0} & & & \mathbf{B}_{ij} & & \\
\mathbf{A}_{ij} & \ddots & & & \ddots & \\
& \ddots & \ddots & & & \ddots \\
& & \mathbf{A}_{ij} & \mathbf{0} & & \mathbf{B}_{ij}
\end{bmatrix}, & j \in \mathcal{N}_i \backslash \{i\}\\
\begin{bmatrix}
-\mathbf{I} & & & \mathbf{B}_{ii} & & \\
\mathbf{A}_{ii} & \ddots & & & \ddots & \\
& \ddots & \ddots & & & \ddots \\
& & \mathbf{A}_{ii} & -\mathbf{I} & & \mathbf{B}_{ii}
\end{bmatrix}, & j = i\\
\mathbf{0}, & j \notin \mathcal{N}_i
\end{cases}\\
\mathbf{g}_i &= [\mathbf{g}_{i1}, \dots, \mathbf{g}_{i|\mathcal{N}|}],\\
\mathbf{g}_{ij} &= \begin{cases} [-\mathbf{A}_{ij}^T, \mathbf{0}^T, \dots, \mathbf{0}^T]^T, & j \in \mathcal{N}_i\\ \mathbf{0}, & j \notin \mathcal{N}_i \end{cases}\\
\mathbf{C}_i &= \text{blkdiag}(\mathbf{C}_{x,i}, \dots, \mathbf{C}_{x,i}, \mathbf{C}_{u,i}, \dots, \mathbf{C}_{u,i}),\\
\mathbf{d}_i &= [\mathbf{d}_{x,i}^T, \dots, \mathbf{d}_{x,i}^T, \mathbf{d}_{u,i}^T, \dots, \mathbf{d}_{u,i}^T]^T.
\end{aligned}
$$

The optimization problem (19.2) is solved in every time instant in the DMPC controller with the latest measurement as initial condition to the state predictions. Communication between subsystems i and j is allowed if $j \in \mathcal{N}_i \cup \mathcal{M}_i$.

19.3 Description of the DMPC Method

In this section the proposed DMPC methodology is presented. We present a distributed algorithm based dual decomposition and a generalized accelerated gradient method. We also present a stopping condition that can be used to guarantee feasibility, stability, and a prespecified performance of the closed loop system.

19.3.1 Dual Problem Formulation

We introduce dual variables $\lambda \in \mathbb{R}^{n_\lambda}$ for the equality constraints and dual variables $\mu \in \mathbb{R}_{\geq 0}^{n_\mu}$ for the inequality constraints in (19.2), where $n_\lambda = (N-1)(n_x + n_u)$ and $n_\mu = Nn_c$. This gives the following dual problem

$$\max_{\mu \geq 0, \lambda} \min_{\mathbf{z}} \frac{1}{2}\mathbf{z}^T\mathbf{H}\mathbf{z} + \lambda^T(\mathbf{F}\mathbf{z} - \mathbf{g}\bar{\mathbf{x}}) + \mu^T(\mathbf{C}\mathbf{z} - \mathbf{d}).$$

By solving the inner minimization problem explicitly, the dual problem becomes

$$\max_{\mu \geq 0, \lambda} -\frac{1}{2}(\mathbf{F}^T\lambda + \mathbf{C}^T\mu)^T\mathbf{H}^{-1}(\mathbf{F}^T\lambda + \mathbf{C}^T\mu) - \lambda^T\mathbf{g}\bar{\mathbf{x}} - \mu^T\mathbf{d}. \tag{19.3}$$

The dual function for initial condition $\bar{\mathbf{x}}$ is defined as

$$D_N(\bar{\mathbf{x}}, \lambda, \mu) := -\frac{1}{2}(\mathbf{F}^T\lambda + \mathbf{C}^T\mu)^T\mathbf{H}^{-1}(\mathbf{F}^T\lambda + \mathbf{C}^T\mu) - \lambda^T\mathbf{g}\bar{\mathbf{x}} - \mu^T\mathbf{d} \tag{19.4}$$

which is quadratic, concave, and differentiable with gradient

$$\nabla D_N(\bar{\mathbf{x}}, \lambda, \mu) = -\begin{bmatrix}\mathbf{F}\\\mathbf{C}\end{bmatrix}\mathbf{H}^{-1}(\mathbf{F}^T\lambda + \mathbf{C}^T\mu) - \begin{bmatrix}\mathbf{g}\bar{\mathbf{x}}\\\mathbf{d}\end{bmatrix}. \tag{19.5}$$

We partition the dual variables into local dual variables $\lambda_i \in \mathbb{R}^{n_{\lambda_i}}$ and $\mu_i \in \mathbb{R}^{n_{\mu_i}}_{\geq 0}$ according to $\lambda = [\lambda_1^T, \ldots, \lambda_{|\mathcal{N}|}^T]^T$ and $\mu = [\mu_1^T, \ldots, \mu_{|\mathcal{N}|}^T]^T$, where the partitions are introduced according to matrices $\mathbf{F}$ and $\mathbf{C}$ respectively. This gives that the dual function gradients w.r.t. to local dual variables are given by

$$\nabla_{\lambda_i} D_N(\bar{\mathbf{x}}, \lambda, \mu) = \sum_{j \in \mathcal{N}_i}\left[-\mathbf{F}_{ij}\mathbf{H}_j^{-1}\left(\sum_{l \in \mathcal{M}_j}\mathbf{F}_{lj}^T\lambda_l + \mathbf{C}_j^T\mu_j\right) - \mathbf{g}_{ij}\bar{\mathbf{x}}_j\right]$$

$$\nabla_{\mu_i} D_N(\bar{\mathbf{x}}, \lambda, \mu) = -\mathbf{C}_i\mathbf{H}_i^{-1}\left(\sum_{j \in \mathcal{M}_i}\mathbf{F}_{ji}^T\lambda_j + \mathbf{C}_i^T\mu_i\right) - \mathbf{d}_i.$$

By setting the primal variable

$$\mathbf{z}_i = -\mathbf{H}_i^{-1}\left(\sum_{j \in \mathcal{M}_i}\mathbf{F}_{ji}^T\lambda_j + \mathbf{C}_i^T\mu_i\right) \tag{19.6}$$

we get that the local gradients are described by

$$\nabla_{\lambda_i} D_N(\bar{\mathbf{x}}, \lambda, \mu) = \sum_{j \in \mathcal{N}_i}(\mathbf{F}_{ij}\mathbf{z}_j - \mathbf{g}_{ij}\bar{\mathbf{x}}_j), \; \nabla_{\mu_i} D_N(\bar{\mathbf{x}}, \lambda, \mu) = \mathbf{C}_i\mathbf{z}_i - \mathbf{d}_i. \tag{19.7}$$

Further, the gradient of the dual function ∇D_N is Lipschitz continuous with constant

$$L = \|[\mathbf{F}^T \; \mathbf{C}^T]^T\mathbf{H}^{-1}[\mathbf{F}^T \; \mathbf{C}^T]\|. \tag{19.8}$$

19.3.2 Gradient-Based Methods

Optimization problems with a differentiable and convex cost function $f : \mathbb{R}^{n_v} \to \mathbb{R}$ and closed and convex constraint set $\mathcal{V}$, can be solved using projected gradient methods. If the gradient to f is Lipschitz continuous with constant L, then the optimal step size for the gradient-step can be shown to be $1/L$. The projected gradient method is described by the following iteration

$$\mathbf{v}^{p+1} = \arg\min_{\mathbf{v}\in\mathcal{V}} \left(\left\| \mathbf{v} - \mathbf{v}^p + \frac{1}{L}\nabla f(\mathbf{v}^p) \right\| \right)$$

where p is the iteration number. The iteration is a step in the gradient direction with step length $1/L$. The resulting point is projected using Euclidean projection onto the feasible set $\mathcal{V}$. By introducing the notation $\langle x, y\rangle = x^T y$, we get the following equivalent formulation of the projected gradient algorithm

$$\mathbf{v}^{p+1} = \arg\min_{\mathbf{v}\in\mathcal{V}} \left[f(\mathbf{v}^p) + \langle \nabla f(\mathbf{v}^p), \mathbf{v} - \mathbf{v}^p \rangle + \frac{L}{2}\|\mathbf{v} - \mathbf{v}^p\|^2 \right].$$

Hence, in a gradient method, a quadratic function with the same curvature in every direction is minimized in every iteration of the algorithm. The quadratic function is actually an upper bound to f since the Lipschitz continuity assumption implies that for every $\mathbf{v}_1, \mathbf{v}_2 \in \mathbb{R}^{n_v}$ the following holds

$$f(\mathbf{v}_1) \le f(\mathbf{v}_2) + \langle \nabla f(\mathbf{v}_2), \mathbf{v}_1 - \mathbf{v}_2 \rangle + \frac{L}{2}\|\mathbf{v}_1 - \mathbf{v}_2\|^2. \tag{19.9}$$

The classical gradient method is known to have bad convergence rate properties, $O(1/p)$. However, this convergence rate can be improved significantly by instead using accelerated gradient methods that have a convergence rate of $O(1/p^2)$. A simple accelerated projected gradient algorithm is given by the following iterations

$$\begin{aligned} \bar{\mathbf{v}}^p &= \mathbf{v}^p + \frac{p-1}{p+2}(\mathbf{v}^p - \mathbf{v}^{p-1}) \\ \mathbf{v}^{p+1} &= \arg\min_{\mathbf{v}\in\mathcal{V}} \left[f(\bar{\mathbf{v}}^p) + \langle \nabla f(\bar{\mathbf{v}}^p), \mathbf{v} - \bar{\mathbf{v}}^p \rangle + \frac{L}{2}\|\mathbf{v} - \bar{\mathbf{v}}^p\|^2 \right]. \end{aligned}$$

The increase in algorithm complexity compared to the classical gradient method is minor, but the improvement in convergence rate is vast. However, this improvement in convergence rate is not always enough to achieve satisfactory accuracy within a small number of iterations. Another improvement to the convergence rate of the algorithm can be obtained by letting the quadratic upper bound that is minimized in every iteration, have different curvatures in different directions. The resulting generalized accelerated gradient algorithm is described by the following iterations

$$\bar{\mathbf{v}}^p = \mathbf{v}^p + \frac{p-1}{p+2}(\mathbf{v}^p - \mathbf{v}^{p-1})$$
$$\mathbf{v}^{p+1} = \arg\min_{\mathbf{v}\in\mathcal{V}} \left[f(\bar{\mathbf{v}}^p) + \langle \nabla f(\bar{\mathbf{v}}^p), \mathbf{v} - \bar{\mathbf{v}}^p \rangle + \frac{1}{2}\|\mathbf{v} - \bar{\mathbf{v}}^p\|_{\mathbf{L}}^2 \right]$$

where $\mathbf{L} \in \mathbb{R}^{n_v \times n_v}$ is a symmetric positive definite matrix that must be chosen such that for every $\mathbf{v}_1, \mathbf{v}_2 \in \mathbb{R}^{n_v}$ the following holds

$$f(\mathbf{v}_1) \leq f(\mathbf{v}_2) + \langle \nabla f(\mathbf{v}_2), \mathbf{v}_1 - \mathbf{v}_2 \rangle + \frac{1}{2}\|\mathbf{v}_1 - \mathbf{v}_2\|_{\mathbf{L}}^2. \tag{19.10}$$

This requirement is very similar to the requirement for Lipschitz continuity (19.9). The only difference is in the quadratic term. The generalized accelerated gradient method can equivalently be written as

$$\bar{\mathbf{v}}^p = \mathbf{v}^p + \frac{p-1}{p+2}(\mathbf{v}^p - \mathbf{v}^{p-1}) \tag{19.11}$$
$$\mathbf{v}^{p+1} = \arg\min_{\mathbf{v}\in\mathcal{V}} \left(\left\| \mathbf{v} - \bar{\mathbf{v}}^p + \mathbf{L}^{-1}\nabla f(\bar{\mathbf{v}}^p) \right\|_{\mathbf{L}} \right). \tag{19.12}$$

where $\mathbf{L}^{-1}$ serves as a step matrix for the gradient-step. By choosing the $\mathbf{L}$-matrix wisely, Hessian-information can be introduced to the gradient algorithm, which further improves the convergence rate. The generalized accelerated gradient method differs from the accelerated gradient method only in the weight used in the quadratic penalty. It also shares the same theoretical convergence rate $O(1/p^2)$.

19.3.3 The Distributed Algorithm

By choosing the $\mathbf{L}$-matrix appropriately, the generalized accelerated gradient method can be applied to solve the dual problem (19.3). The $\mathbf{L}$-matrix must satisfy

$$\mathbf{L} \succeq [\mathbf{F}^T\ \mathbf{C}^T]^T \mathbf{H}^{-1} [\mathbf{F}^T\ \mathbf{C}^T]$$

which implies that (19.10) holds for $f = -D_N$ where D_N is the dual function defined in (19.4). The objective of the algorithm is to enable for a distributed implementation, hence the $\mathbf{L}$-matrix should be chosen to accompany this requirement. We introduce a structural constraint on the $\mathbf{L}$-matrix described by set

$$\mathcal{L} = \{v\mathbf{L}\ \in \mathbb{R}^{(n_\lambda+n_\mu)\times(n_\lambda+n_\mu)} \mid \mathbf{L} = \text{blkdiag}(\mathbf{L}_{\lambda_1}, \ldots, \mathbf{L}_{\lambda_M}, \mathbf{L}_{\mu_1}, \ldots, \mathbf{L}_{\mu_M})v\}$$

where $\mathbf{L}_{\lambda_i} \in \mathbb{R}^{n_{\lambda_i} \times n_{\lambda_i}}$ and $\mathbf{L}_{\mu_i} \in \mathbb{R}^{n_{\mu_i} \times n_{\mu_i}}$ for all $i \in \mathcal{N}$. We also introduce $\mathbf{L}_\lambda = \text{blkdiag}(\mathbf{L}_{\lambda_1}, \ldots, \mathbf{L}_{\lambda_M})$ and $\mathbf{L}_\mu = \text{blkdiag}(\mathbf{L}_{\mu_1}, \ldots, \mathbf{L}_{\mu_M})$ for future reference. An $\mathbf{L}$-matrix satisfying the mentioned requests is computed by solving:

$$\begin{aligned}
&\min_{\mathbf{L}\in\mathcal{L}} \operatorname{trace}(\mathbf{L}) \\
&\text{s.t. } \mathbf{L} \succeq [\mathbf{F}^T\ \mathbf{C}^T]^T\mathbf{H}^{-1}[\mathbf{F}^T\ \mathbf{C}^T] \\
&\qquad \mathbf{L} \succ 0
\end{aligned} \tag{19.13}$$

The idea behind solving a convex optimization problem to compute the step matrix is similar to the idea used in explicit MPC; by increasing the amount of offline computational burden, the online execution time or amount of communication can be reduced significantly. Using the algorithm description in (19.11)–(19.12) and the gradient computations in (19.5) it is straightforward to verify that the generalized accelerated gradient algorithm when applied to the dual problem becomes

$$\begin{aligned}
\mathbf{z}^p &= -\mathbf{H}^{-1}(\mathbf{F}^T\boldsymbol{\lambda}^p + \mathbf{C}^T\boldsymbol{\mu}^p) \\
\bar{\mathbf{z}}^p &= \mathbf{z}^p + \frac{p-1}{p+2}(\mathbf{z}^p - \mathbf{z}^{p-1}) \\
\bar{\boldsymbol{\lambda}}^p &= \boldsymbol{\lambda}^p + \frac{p-1}{p+2}(\boldsymbol{\lambda}^p - \boldsymbol{\lambda}^{p-1}) \\
\boldsymbol{\lambda}^{p+1} &= \bar{\boldsymbol{\lambda}}^p + \mathbf{L}_\lambda^{-1}(\mathbf{F}\bar{\mathbf{z}}^p - \mathbf{g}\bar{x}) \\
\bar{\boldsymbol{\mu}}^p &= \boldsymbol{\mu}^p + \frac{p-1}{p+2}(\boldsymbol{\mu}^p - \boldsymbol{\mu}^{p-1}) \\
\boldsymbol{\mu}^{p+1} &= \arg\min_{\mu\geq 0}\left(\left\|\mu - \bar{\boldsymbol{\mu}}^p - \mathbf{L}_\mu^{-1}(\mathbf{C}\bar{\mathbf{z}}^p + \mathbf{d})\right\|^2_{\mathbf{L}_\mu}\right)
\end{aligned}$$

These computations can be distributed by using the local gradients defined in (19.6) and (19.7). The resulting distributed algorithm is presented in Algorithm 19.1.

If all $\mathbf{L}_{\mu_i}$, $i \in \mathcal{N}$ are chosen diagonal, the minimization to find $\boldsymbol{\mu}_i^{p+1}$ in Algorithm 19.1 can be replaced by $\max(0, \cdot)$ which is computed uncostly. Using non-diagonal $\mathbf{L}_{\mu_i}$ gives more elaborate iterations, but the number of iterations to achieve satisfactory accuracy of the solution may decrease significantly. This is advantageous in DMPC, where the amount of communication should be kept as small as possible, without compromising the global closed loop performance.

The algorithm converges in both primal variables $\mathbf{z}_i^p$ and dual function value at the rate $O(1/p^2)$. In [3] methods to compute iteration complexity bounds to achieve a prespecified accuracy of the solution are presented. In [7] similar methods are presented for the case where $\mathbf{L}$ is a multiple of the identity matrix. Also, in [7] it was shown how to precondition the matrices describing the inequality constraints and equality constraints optimally. The optimal preconditioning refers to the preconditioning that minimizes the iteration complexity bound that guarantees a dual function value accuracy.

Algorithm 19.1: Distributed optimization algorithm

Initialize $\boldsymbol{\lambda}_i^0 = \boldsymbol{\lambda}_i^{-1}$, $\boldsymbol{\mu}_i^0 = \boldsymbol{\mu}_i^{-1}$ and $\mathbf{z}_i^0 = \mathbf{z}_i^{-1}$
for every node i **do**
  for $p \geq 0$ **do**
    Update primal variables according to:

$$\mathbf{z}_i^p = -\mathbf{H}_i^{-1}\left(\left(\sum_{j\in\mathcal{M}_i} \mathbf{F}_{ji}^T \boldsymbol{\lambda}_j^p\right) + \mathbf{C}_i^T \boldsymbol{\mu}_i^p\right)$$

$$\bar{\mathbf{z}}_i^p = \mathbf{z}_i^p + \frac{p-1}{p+2}(\mathbf{z}_i^p - \mathbf{z}_i^{p-1})$$

    Send $\bar{\mathbf{z}}_i^p$ to each $j \in \mathcal{M}_i$, receive $\bar{\mathbf{z}}_j^p$ from each $j \in \mathcal{N}_i$
    Update dual variables according to:

$$\bar{\boldsymbol{\lambda}}_i^p = \boldsymbol{\lambda}_i^p + \frac{p-1}{p+2}(\boldsymbol{\lambda}_i^p - \boldsymbol{\lambda}_i^{p-1})$$

$$\boldsymbol{\lambda}_i^{p+1} = \bar{\boldsymbol{\lambda}}_i^p + \mathbf{L}_{\lambda_i}^{-1}\left(\sum_{j\in\mathcal{N}_i} (\mathbf{F}_{ij}\bar{\mathbf{z}}_j^p - \mathbf{g}_{ij}\bar{x}_j)\right)$$

$$\bar{\boldsymbol{\mu}}_i^p = \boldsymbol{\mu}_i^p + \frac{p-1}{p+2}(\boldsymbol{\mu}_i^p - \boldsymbol{\mu}_i^{p-1})$$

$$\boldsymbol{\mu}_i^{p+1} = \arg\min_{\mu\geq 0}\left(\left\|\mu - \bar{\boldsymbol{\mu}}_i^p - \mathbf{L}_{\mu_i}^{-1}(\mathbf{C}_i\bar{\mathbf{z}}_i^p - \mathbf{d}_i)\right\|_{L_\mu^i}\right)$$

    Send $\boldsymbol{\lambda}_i^{p+1}$ to each $j \in \mathcal{N}_i$, receive $\boldsymbol{\lambda}_j^{p+1}$ from each $j \in \mathcal{M}_i$
  end for
end for

19.3.4 Stopping Condition

One drawback of using duality-based optimization in DMPC is that feasibility of the primal problem can be guaranteed only in the limit of iterations. Also, the optimization problem used in Algorithm 19.1 has neither a terminal cost nor a terminal constraint set. These are usually required to prove stability in MPC. In this section we will briefly present a stopping condition for the duality-based optimization algorithm in Algorithm 19.1. The stopping condition guarantees feasibility, stability and performance of the closed loop system and it reduces the amount of communication needed since it enables for early termination of the optimization algorithm. The stopping condition is based on relaxed dynamic programming for stability and performance, and adaptive constraint tightening for feasibility. We start by describing relaxed dynamic programming when applied to MPC with optimization problems without a terminal cost or a terminal constraint set.

Relaxed dynamic programming applied to MPC states that if the control horizon N is such that for every $\mathbf{x} \in \mathcal{X}$ the following holds

$$V_N(\mathbf{x}) \geq V_N(\mathbf{A}\mathbf{x} + \mathbf{B}\nu_N(\mathbf{x})) + \alpha\ell(\mathbf{x}, \nu_N(\mathbf{x})) \tag{19.14}$$

where $\alpha \in (0, 1)$, V_N is the optimal value function to the optimization problem (19.2) without terminal constraint set of terminal cost, and ν_N is the MPC feedback control law obtained by solving (19.2) and applying the first control action in optimal control trajectory in every sample. Then we get asymptotic stability and closed loop performance as specified by

$$\alpha \sum_{k=0}^{\infty} \ell(\mathbf{x}(k), \nu_N(\mathbf{x}(k))) \leq V_\infty(\mathbf{x}(0)) \tag{19.15}$$

where $\mathbf{x}(k+1) = \mathbf{A}\mathbf{x}(k) + \mathbf{B}\nu_N(\mathbf{x}(k))$. A method to find control horizon N such that (19.14) holds is presented in [10]. This was further extended in [9] where it was shown how to compute the minimal control horizon for systems that satisfies a certain controllability assumption on the stage costs. Another, more explicit, characterization of the relation between the control horizon N and the performance parameter α was presented in [8]. Once a control horizon N is known such that (19.14) holds for every $\mathbf{x} \in \mathcal{X}$, performance and stability is guaranteed by computing the optimal solution to (19.2). However, stability and performance can be guaranteed also for suboptimal solutions, which indicates a potential benefit of using a stopping condition that ensures this.

The issue that feasibility can be guaranteed only in the limit of iterations in duality-based optimization, is addressed by an adaptive constraint tightening approach. We use the following optimization problem with tightened constraint sets

$$\begin{aligned} V_N^\delta(\bar{\mathbf{x}}) := \min_{\mathbf{z}} \; & \frac{1}{2}\mathbf{z}^T\mathbf{H}\mathbf{z} \\ \text{s.t. } & \mathbf{F}\mathbf{z} = \mathbf{g}\bar{\mathbf{x}} \\ & \mathbf{C}\mathbf{z} \leq (1-\delta)\mathbf{d} \end{aligned} \tag{19.16}$$

where $\delta \in (0, 1)$ specifies the relative constraint tightening used and the matrices in (19.16) are specified at the end of Sect. 19.2. The objective is to choose constraint tightening δ such that it can be guaranteed that (19.14) holds and that feasibility can be guaranteed with finite number of iterations. The problem (19.16) is solved through the dual problem

$$\max_{\boldsymbol{\lambda}, \boldsymbol{\mu} \geq 0} -\frac{1}{2}(\mathbf{F}^T\boldsymbol{\lambda} + \mathbf{C}^T\boldsymbol{\mu})^T\mathbf{H}^{-1}(\mathbf{F}^T\boldsymbol{\lambda} + \mathbf{C}^T\boldsymbol{\mu}) - \boldsymbol{\lambda}^T\mathbf{g}\bar{\mathbf{x}} - \boldsymbol{\mu}^T\mathbf{d}(1-\delta)$$

using Algorithm 19.1. The dual function for initial condition $\bar{\mathbf{x}}$ and with constraint tightening δ is defined as

$$D_N^\delta(\bar{\mathbf{x}}, \boldsymbol{\lambda}, \boldsymbol{\mu}) := -\frac{1}{2}(\mathbf{F}^T\boldsymbol{\lambda} + \mathbf{C}^T\boldsymbol{\mu})^T\mathbf{H}^{-1}(\mathbf{F}^T\boldsymbol{\lambda} + \mathbf{C}^T\boldsymbol{\mu}) - \boldsymbol{\lambda}^T\mathbf{g}\bar{\mathbf{x}} - \boldsymbol{\mu}^T\mathbf{d}(1-\delta)$$

To online guarantee that the condition (19.14) holds, a lower bound to the l.h.s. is needed and an upper bound to the r.h.s. is needed. A lower bound to the l.h.s. of (19.14) is obtained by using the following lemma which is proven in [8, Lemma 1].

Lemma 19.1 *For every* $\bar{\mathbf{x}} \in \mathbb{R}^{n_x}$, $\boldsymbol{\lambda} \in \mathbb{R}^{n_\lambda}$ *and* $\boldsymbol{\mu} \in \mathbb{R}^{n_\mu}_{\geq 0}$ *we have that*

$$V_N^0(\bar{x}) \geq D_N^\delta(\bar{x}, \boldsymbol{\lambda}, \boldsymbol{\mu}) - \delta\boldsymbol{\mu}^T\mathbf{d}.$$

Hence, a lower bound to V_N^0 is readily available from the optimization algorithm data in each iteration p. To compute an upper bound to the r.h.s. of (19.14) we define the following primal cost for initial condition $\bar{\mathbf{x}}$ using control trajectory $\mathbf{u}(0 : N-1)$

$$P_N(\bar{\mathbf{x}}, \mathbf{u}(0:N-1)) = \begin{cases} \sum_{l=0}^{N-1} \ell(\boldsymbol{\xi}(l), \mathbf{u}(l)), & \text{if } \boldsymbol{\xi}(l+1) = \mathbf{A}\boldsymbol{\xi}(l) + \mathbf{B}\mathbf{u}(l), \\ & \quad \boldsymbol{\xi}(0) = \bar{\mathbf{x}}, \\ & \quad \boldsymbol{\xi}(l) \in \mathcal{X}, \mathbf{u}(l) \in \mathcal{U}, \\ & \quad l = 0, \dots, N-1 \\ \infty, & \text{else.} \end{cases}$$

The cost P_N is finite if the state and control trajectories are feasible and ∞ if they are not. Hence

$$P_N(\bar{\mathbf{x}}, \mathbf{u}(0:N-1)) \geq V_N(\bar{\mathbf{x}})$$

for any control trajectory $\mathbf{u}(0 : N-1)$. The cost at iteration p in the algorithm is $P_N(\bar{\mathbf{x}}, \hat{\mathbf{u}}^p(0 : N-1))$ where $\hat{\mathbf{u}}^p$ is extracted from $\mathbf{z}^p$ in Algorithm 19.1. By introducing the shifted control trajectory

$$\hat{\mathbf{u}}_s^p(0:N-1) = [\hat{\mathbf{u}}^p(1)^T, \dots, \hat{\mathbf{u}}^p(N-1)^T \; 0^T]^T$$

a cost for the next time step at iteration p is

$$P_N(\mathbf{A}\bar{\mathbf{x}} + \mathbf{B}\hat{\mathbf{u}}^p(0), \hat{\mathbf{u}}_s^p(0:N-1)).$$

We have that

$$P_N(\mathbf{A}\bar{\mathbf{x}}+\mathbf{B}\hat{\mathbf{u}}^p(0), \hat{\mathbf{u}}_s^p(0:N-1)) + \ell(\bar{\mathbf{x}}, \hat{\mathbf{u}}^p(0)) \geq V_N(\mathbf{A}\bar{\mathbf{x}}+\mathbf{B}\hat{\mathbf{u}}^p(0)) + \ell(\bar{\mathbf{x}}, \hat{\mathbf{u}}^p(0)) \tag{19.17}$$

which implies that an upper bound to the r.h.s. of (19.14) can readily be computed in any iteration p. The objective of the stopping condition is to adapt the constraint tightening δ and stop at iteration p such that

$$D_N^\delta(\bar{\mathbf{x}}, \boldsymbol{\lambda}^p, \boldsymbol{\mu}^p) - \delta(\boldsymbol{\mu}^p)^T\mathbf{d} \geq P_N(\mathbf{A}\bar{\mathbf{x}} + \mathbf{B}\hat{\mathbf{u}}^p(0), \hat{\mathbf{u}}_s^p(0:N-1)) + \alpha\ell(\bar{\mathbf{x}}, \hat{\mathbf{u}}^p(0)).$$

This implies feasibility of the next step due to the definition of P_N and that

$$V_N^0(\bar{\mathbf{x}}) \geq V_N^0(\mathbf{A}\bar{\mathbf{x}} + \mathbf{B}\hat{\mathbf{u}}^p(0)) + \alpha\ell(\bar{\mathbf{x}}, \hat{\mathbf{u}}^p(0))$$

holds due to Lemma 19.1 and (19.17). This, in turn, implies asymptotic stability and performance as specified by (19.15). In the stopping condition, there are three parameters that need to be set. These are the desired performance $\alpha \in (0, 1)$, the initial constraint tightening $\delta_{\text{init}} \in (0, 1)$, and an optimality tolerance $\epsilon \in [0.05, 0.001]$. The stopping condition is presented in Algorithm 19.2.

To guarantee that the stopping condition will terminate with finite number of iterations and that the feasibility, stability, and performance results hold, there are restrictions on how the control horizon N must be chosen for a given $\alpha \in (0, 1)$. The reader is referred to [8] for details of this choice.

19.3.5 Evaluation of Algorithm Efficiency

The efficiency of the algorithms presented in this chapter is evaluated by applying them to a randomly generated system. The cost function, constraints, and dynamics of the randomly generated system are specified in [4, Supplement A.1]. The system consists of three interconnected subsystems with five states and one input each, i.e., 15 states and 3 inputs in total. The magnitude of the largest eigenvalue is 1.1 and the upper and lower bounds on the states are chosen randomly from the intervals $[0.5, 1.5]$ and $[-0.15, 0.05]$ respectively. The upper and lower bounds on the inputs are chosen randomly from the interval $[0.5, 1.5]$ and $[-1.5, -0.5]$ respectively. The cost matrices $\mathbf{Q}$ and $\mathbf{R}$ are diagonal with diagonal elements chosen randomly from the interval $[1, 100]$. Using the method in [8] it is shown that using $\alpha = 0.01$, (19.14) is satisfied with control horizon $N = 6$.

In Table 19.1 the algorithm evaluation is presented. We compare the iteration complexity when using Algorithm 19.1 with a block-diagonal $\mathbf{L}$-matrix computed as in (19.13). This distributed generalized accelerated gradient algorithm is referred to as Algorithm 19.1, GAG. This is compared to when using Algorithm 19.1 with the optimal step size, i.e., with $\mathbf{L} = LI$ where L is the (smallest) Lipschitz constant to the dual function (19.8). This distributed accelerated gradient algorithm is referred to as Algorithm 19.1, AG. We also compare these algorithms to when using a standard gradient method with the optimal step size, which is the traditional way of solving the dual problem in distributed fashion. This gradient method is obtained by setting $\bar{\mathbf{z}}_i^p = \mathbf{z}_i^p$, $\bar{\boldsymbol{\lambda}}_i^p = \boldsymbol{\lambda}_i^p$, and $\bar{\boldsymbol{\mu}}_i^p = \boldsymbol{\mu}_i^p$ in Algorithm 19.1. This distributed gradient method is referred to as Algorithm 19.1, G. The methods are compared with and without the preconditioning presented in [7] and with and without the stopping condition in Algorithm 19.2. The comparison is made on 1000 randomly generated points that are chosen from a uniform distribution over the state constraint set. Also, an estimate of the region of attraction is presented and compared to the estimated region of attraction in centralized MPC where the terminal constraint set is chosen as the maximal output admissible set. The centralized MPC formulation is referred to as CMPC. All data in Table 19.1 that comes from Algorithm 19.1 are obtained by cold-starting the algorithm.

Algorithm 19.2: Stopping condition

Input: $\bar{\mathbf{x}}$
Set $p = 0, l = 0, \delta = \delta_{\text{init}}$
Run Δp iterations of Algorithm 19.1 based on optimization problem (19.16)
repeat
  if $D_N^\delta(\bar{\mathbf{x}}, \boldsymbol{\lambda}^p, \boldsymbol{\mu}^p) \geq P_N(\bar{\mathbf{x}}, \hat{\mathbf{u}}^p(0:N-1)) - \frac{\epsilon}{l+1}\ell^*(\bar{\mathbf{x}})$ **or** $\delta \mathbf{d}^T \boldsymbol{\mu}^p > \epsilon \ell^*(\bar{\mathbf{x}})$ **then**
    Set $\delta \leftarrow \delta/2$ {reduce constraint tightening}
    Set $l \leftarrow l + 1$ {reduce optimality tolerance}
    Set $p = 0$ {reset step size and iteration counter}
  end if
  Run Δp iterations of Algorithm 19.1 based on optimization problem (19.16)
  Set $p \leftarrow p + \Delta p$
until $D_N^\delta(\bar{\mathbf{x}}, \boldsymbol{\lambda}^p, \boldsymbol{\mu}^p) \geq P_N(\mathbf{A}\bar{\mathbf{x}} + \mathbf{B}\hat{\mathbf{u}}^p(0), \hat{\mathbf{u}}_s^p(0:N-1)) + \alpha\ell(\bar{\mathbf{x}}, \hat{\mathbf{u}}_0^p)$ **and** $\delta \mathbf{d}^T \boldsymbol{\mu}^p \leq \epsilon$ $\ell^*(\bar{\mathbf{x}})$
Output: $\hat{\mathbf{u}}^p(0)$

The first column in Table 19.1 specifies the algorithm used. The second and third columns specify the optimality tolerance ϵ and the initial constraint tightening δ_{init} in the stopping condition in Algorithm 19.2. If Algorithm 19.2 is not used, the parameters specify the following optimality condition

$$D_N^\delta(\bar{\mathbf{x}}, \boldsymbol{\lambda}^p, \boldsymbol{\mu}^p) \geq P_N(\bar{\mathbf{x}}, \hat{\mathbf{u}}^p(0:N-1)) - \epsilon\ell^*(\bar{\mathbf{x}}).$$

Columns four and five specify if preconditioning and the stopping condition in Algorithm 19.2 are used respectively. Columns six and seven specify the average and max number of iterations needed. The seventh column specifies the average final constraint tightening and the final columns specifies the estimated region of attraction, i.e., the percentage of initial conditions steered to the origin.

Table 19.1 reveals that by using accelerated gradient method instead of gradient methods, the algorithm is improved by a factor 20. Further, by using preconditioning an additional improvement by a factor 1-4 is achieved. The stopping condition, besides guaranteeing stability, improves the convergence by another factor 1-3. By allowing for **L**-matrices instead of only scalar step sizes, we get another improvement by a factor around 10. Hence, using the presented methods, the number of iterations is, for this example, decreased by almost three orders of magnitude compared to the traditional and straightforward application of a gradient method, although the gradient method here is equipped with the optimal step size. Further, the comparison in estimated region of attraction for the presented method without a terminal constraint set and CMPC with a terminal constraint set reveals that the region of attraction can be increased significantly by not using a terminal constraint set.

Table 19.1 Algorithm comparison for Algorithm 19.1 based on the generalized accelerated gradient method (GAG), the accelerated gradient method (AG), and the gradient method (G)

Algorithm	ϵ	δ_{init}	Precond	Stop cond	Avg. # iters	Max # iters	Avg. δ	R.o.A. (%)
Algorithm 19.1, GAG	0.005	0.1	y	y	8.86	41	0.052	57.6
Algorithm 19.1, GAG	0.005	0.01	y	n	19.36	47	0.01	57.6
Algorithm 19.1, GAG	0.005	0.1	n	y	9.65	48	0.053	57.6
Algorithm 19.1, GAG	0.005	0.01	n	n	25.00	53	0.01	57.6
Algorithm 19.1, AG	0.005	0.1	y	y	64.92	122	0.054	57.6
Algorithm 19.1, AG	0.005	0.01	y	n	146.68	400	0.01	57.6
Algorithm 19.1, AG	0.005	0.1	n	y	126.68	455	0.053	57.6
Algorithm 19.1, AG	0.005	0.01	n	n	515.30	1161	0.01	57.6
Algorithm 19.1, G	0.005	0.1	y	y	1185.8	2906	0.055	57.6
Algorithm 19.1, G	0.005	0.01	y	n	2458.9	10194	0.01	57.6
Algorithm 19.1, G	0.005	0.1	n	y	3807.7	13991	0.089	57.6
Algorithm 19.1, G	0.005	0.01	n	n	9954.9	21225	0.01	57.6
CMPC	-	-	-	-	-	-	-	0.7

The algorithms are compared with and without preconditioning and the stopping condition in Algorithm 19.2. Average and max number of iterations as well as average constraint tightening is presented. Also, the region of attraction (R.o.A.) is estimated and compared to centralized MPC with a terminal constraint set

19.4 Theoretical Results

Application of accelerated gradient methods to DMPC based on dual decomposition and with an additional 1-norm term in the objective is presented in [6]. Iteration complexity bounds for the algorithm in [6] is presented in [7]. Based on the iteration complexity bounds, it is also shown how to precondition the optimization problem data optimally in [7]. The algorithm in [6] is in [3] extended to allow for step matrices instead of scalar step sizes. The algorithm in [3] is presented in Algorithm 19.1 in this chapter. Also, iteration complexity bounds to achieve a prespecified accuracy of the solution for Algorithm 19.1 are provided in [3]. Compared to existing methods that use dual decomposition to solve the optimization problem arising in distributed model predictive control, Algorithm 19.1 gives considerably lower iteration complexity.

In [8] it is shown for which initial conditions the stopping condition in Algorithm 19.2 is guaranteed to terminate. Also, a result guaranteeing feasibility, closed loop stability, and a prespecified performance when using the stopping condition in Algorithm 19.2 is presented. The region of attraction using the presented stopping condition compared to using a centralized MPC formulation with a terminal cost and a terminal constraint set, where the terminal constraint set is chosen as the maximal output admissible set, is evaluated in [5, 8]. For some systems, the region of attraction when using the presented stopping condition is significantly larger. Robustness to bounded errors when using the stopping condition in Algorithm 19.2 is shown in [5]. In [5] also an output feedback DMPC scheme based on the stopping condition in Algorithm 19.2 is presented.

19.5 Applications

The presented methods have been applied in a simulated hydro power valley benchmark problem [13]. The hydro power valley benchmark consists of six dams that are placed along a river and three lakes that are interconnected to the river. The dams are equipped with turbines to generate power and interconnections between the lakes and the river are equipped with pumps and turbines such that water can flow in any direction. The objective of the control problem is to follow a time-varying power reference with the power production, while keeping flows and water levels within allowed limits. The model contains nonlinearities and binary constraints that are addressed to enable for an efficient and well performing implementation of the methods presented in this chapter. The case study is presented in [2].

Another application where these methods have been used is for disturbance management in the process industry. Chemicals in the process industry are often manufactured in several production areas within a production site. The production areas are interconnected by the product flow and each interconnection usually has a buffer-tank. If problems occur in one of the production areas, the other production areas are affected. The objective of the control is to set the production rates for the individual production areas to maximize the total throughput while avoiding shut-down of any production area due to lack of product inflow, i.e., to avoid empty buffer-tanks. This case study is presented in [11].

Acknowledgments The authors were supported by the Swedish Research Council through the Linneaus center (LCCC).

References

1. A. Beck, M. Teboulle, A fast iterative shrinkage-thresholding algorithm for linear inverse problems. SIAM J. Imag. Sci. **2**(1), 183–202 (2009)
2. M.D. Doan, P. Giselsson, T. Keviczky, B. De Schutter, A. Rantzer, A distributed accelerated gradient algorithm for DMPC of a hydro power valley. Submitted to Control Eng. Pract. (2012)
3. P. Giselsson, A generalized distributed accelerated gradient method for DMPC with iteration complexity bounds. Submitted to *2013 American Control Conference*, 2012
4. P. Giselsson, Gradient-based distributed model predictive control. Ph.D. thesis, Department of Automatic Control, Lund University, Sweden, November 2012
5. P. Giselsson, Output feedback distributed model predictive control with inherent robustness properties. Submitted to *2013 American Control Conference*, 2012
6. P. Giselsson, M.D. Doan, T. Keviczky, B. De Schutter, A. Rantzer, Accelerated gradient methods and dual decomposition in distributed model predictive control. Automatica 2012 (To appear)
7. P. Giselsson, Optimal preconditioning and iteration complexity bounds for gradient-based optimization in model predictive control. Submitted to *2013 American Control Conference*, 2012
8. P. Giselsson, A. Rantzer, On feasibility, stability and performance in distributed model predictive control. Submitted to IEEE Trans. Autom. Control (2012)
9. L. Grüne, Analysis and design of unconstrained nonlinear MPC schemes for finite and infinite dimensional systems. SIAM J. Control Optim. **48**, 1206–1228 (2009)

10. L. Grüne, A. Rantzer, On the infinite horizon performance of receding horizon controllers. IEEE Trans. Autom. Control **53**, 2100–2111 (2008)
11. A. Lindholm, P. Giselsson, Formulating an optimization problem for minimization of losses due to utilities, in *8th IFAC International Symposium on Advanced Control of Chemical Processes*, Singapore, 2012
12. Y. Nesterov, *Introductory Lectures on Convex Optimization: A Basic Course*, 1st edn. (Springer, Netherlands, 2003)
13. C. Savorgnan, M. Diehl, Control benchmark of a hydro-power plant. Technical report, Hierarchical and Distributed Model Predictive Control for Large-Scale Systems (HD-MPC) (2011)
14. P. Tseng, On accelerated proximal gradient methods for convex-concave optimization. Submitted to SIAM J. Optim. (2008)
15. W. Zuo, Z. Lin, A generalized accelerated proximal gradient approach for total-variation-based image restoration. IEEE Trans. Image Process. **20**(10), 2748–2759 (2011)

Chapter 20
Distributed Multiple Shooting for Large Scale Nonlinear Systems

A. Kozma, C. Savorgnan and M. Diehl

Abstract The distributed multiple shooting method is tailored for large scale optimal control problems with decoupled structure. It can be used as a fast and distributed solver for model predictive control subproblems. The algorithm may be regarded as a generalization of the standard multiple shooting method that decomposes the original large scale optimal control problem in both the time and spatial domain to obtain high parallelizability. In each iteration, the linearization of the original problem is calculated in a parallel manner, which is then solved by a centralized structure-exploiting optimizer. We demonstrate the approach on a simple mechanical example of two coupled pendula.

20.1 Introduction

In this chapter, we deal with optimal control problems of large scale systems employing shooting methods. Since model predictive control (MPC) necessitates the solution of optimal control problems repeatedly, we pay special attention to how to do this efficiently. Typically, throughout the solution of optimal control problems with shooting methods the majority of the computation time is spent by solving initial value problems (IVP) and particularly in the sensitivity generation. The employed IVP solvers are treated as functions, which are evaluated and linearized repeatedly. The parallel evaluation of these functions may thus lead to significant speed-up. The

A. Kozma (✉) · C. Savorgnan · M. Diehl
Department Of Electrical Engineering (ESAT/SCD) and Optimization in Engineering Center (OPTEC), Leuven, Belgium
e-mail: Attila.Kozma@esat.kuleuven.be

C. Savorgnan
e-mail: Carlo.Savorgnan@esat.kuleuven.be

M. Diehl
e-mail: Moritz.Diehl@esat.kuleuven.be

J. M. Maestre and R. R. Negenborn (eds.), *Distributed Model Predictive Control Made Easy*, Intelligent Systems, Control and Automation: Science and Engineering 69,
DOI: 10.1007/978-94-007-7006-5_20,

distributed multiple shooting (DMS) method from [9] can speed up the solution time of nonlinear optimal control problems compared to standard methods by introducing parallelization in the simulation process, and thus makes problems with longer horizons or with greater number of variables feasible. The proposed approach, which is based on the direct multiple shooting method [1], may be regarded as a method that introduces parallel simulation both in the time and spatial domain of the problem.

20.2 Considered Problem Class

We consider an optimal control problem of a large scale nonlinear system that is composed of a set of subsystems $\mathcal{N}$ having $|\mathcal{N}|$ agents in the form of

$$\min_{\substack{\mathbf{x},\mathbf{u},\\ \mathbf{z},\mathbf{y}}} \sum_{i=1}^{|\mathcal{N}|} \int_0^{N_p} \ell_i(\mathbf{x}_i(t), \mathbf{u}_i(t), \mathbf{z}_i(t))dt \tag{20.1a}$$

$$\text{s.t.}\ \dot{\mathbf{x}}_i(t) = \mathbf{f}_i(\mathbf{x}_i(t), \mathbf{u}_i(t), \mathbf{z}_i(t)) \tag{20.1b}$$

$$\mathbf{y}_i(t) = \mathbf{g}_i(\mathbf{x}_i(t), \mathbf{u}_i(t), \mathbf{z}_i(t)) \tag{20.1c}$$

$$\mathbf{x}_i(0) = \bar{\mathbf{x}}_i^0 \tag{20.1d}$$

$$\mathbf{z}_i(t) = \sum_{j=1}^{|\mathcal{N}|} \mathbf{A}_{ij}\mathbf{y}_j(t) \tag{20.1e}$$

$$\mathbf{p}_i(\mathbf{x}_i(t), \mathbf{u}_i(t)) \geq 0, \quad t \in [0, N_p], \tag{20.1f}$$

where i denotes the subsystem index. The variables $\mathbf{x}_i(t) \in \mathbb{R}^{n_{x,i}}$, $\mathbf{u}_i(t) \in \mathbb{R}^{n_{u,i}}$ and $\mathbf{z}_i(t) \in \mathbb{R}^{n_{z,i}}$ are the differential state, the control input and the coupling input variables, respectively. The objective function (20.1a) is the sum of the local performance functions $l_i(\cdot)$ of each subsystem. For each subsystem i, the function $\mathbf{f}_i(\cdot)$ determines the nonlinear dynamic behaviour enforced by the constraint (20.1b), while the function $\mathbf{g}_i(\cdot)$ determines the output imposed by the constraint (20.1c). The coupling variable $\mathbf{z}_i(t)$, defined by (20.1e), provides information about the other subsystems. The coupling matrices $\mathbf{A}_{ij}$ characterize the nature of the coupling between subsystems. For example, in the simplest case subsystem $i+1$ might depend on the states of subsystem i, which could be formulated as $\mathbf{z}_{i+1}(t) = \mathbf{y}_i(t)$ with $\mathbf{y}_i(t) = \mathbf{x}_i(t)$. The formulation summarizes state and control input constraints via (20.1f).

The remainder of this chapter proceeds as follows. We discuss the building blocks of the optimal control problem (20.1) on a mechanical example of coupled pendula in Sect. 20.3. In Sect. 20.4, the direct multiple shooting method is discussed first, which can exploit parallel simulation in the time domain. In the second part, we extend this to obtain the distributed multiple shooting method, which introduces parallel simulation of different subsystems.We describe and analyze numerical results in Sect. 20.5.

20.3 A Guiding Example: Coupled Pendula

In order to make the discussion easily followable and tutorial, we derive a special case of (20.1) using the simple example of two coupled pendula. However, it should be emphasized that our algorithm is tailored for large scale optimal control problems.

The system consists of two pendula (see Fig. 20.1) hanging at position $x_{f,1}$ and $x_{f,2}$ with lengths L_1 and L_2 and masses m_1 and m_2 that are connected by a spring with spring constant k, thus $\mathcal{N} = \{1, 2\}$. The objective is to bring the pendula into their equilibrium position. Throughout our discussion each pendulum will be treated as a seperate subsystem. For each pendulum, we introduce states $\mathbf{x}_i(t) := (\theta_i(t), \omega_i(t))^T$ denoting the angle and the angular velocity and input $u_i(t)$ denoting the controlled force acting on the head of pendulum i. We define the angular acceleration $\alpha_i(.)$ as

$$\begin{aligned}\alpha_1(\theta_1(t), u_1(t), \theta_2(t)) :=& \frac{u_1(t)\cos(\theta_1(t)) - 2k\cos(\theta_1(t))(x_{f,1} - x_{f,2})}{L_1 m_1} \\ &- \frac{2L_2 k \sin(\theta_1(t) - \theta_2(t))}{L_1 m_1} - \frac{g\sin(\theta_1(t))}{L_1^2},\end{aligned} \tag{20.2a}$$

$$\begin{aligned}\alpha_2(\theta_2(t), u_2(t), \theta_1(t)) :=& \frac{u_2(t)\cos(\theta_2(t)) + 2k\cos(\theta_2(t))(x_{f,1} - x_{f,2})}{L_2 m_2} \\ &+ \frac{2L_1 k \sin(\theta_1(t) - \theta_2(t))}{L_2 m_2} - \frac{g\sin(\theta_2(t))}{L_2^2}.\end{aligned} \tag{20.2b}$$

The differential equations of the first pendulum read as

$$\dot{\theta}_1(t) = \omega_1(t), \tag{20.3a}$$

$$\dot{\omega}_1(t) = \alpha_1(\theta_1(t), u_1(t), \theta_2(t)). \tag{20.3b}$$

We note that the angle of the second pendulum $\theta_2(t)$ appears in dynamic equation (20.3b), thus, in order to decouple the systems, we introduce coupling input $z_1(t)$ to replace $\theta_2(t)$. Now we can define $\mathbf{f}_1(\mathbf{x}_1(t), u_1(t), z_1(t))$ as

$$\mathbf{f}_1(\mathbf{x}_1(t), u_1(t), z_1(t)) := \begin{bmatrix} \omega_1(t) \\ \alpha_1(\theta_1(t), u_1(t), \theta_2(t)) \end{bmatrix} \tag{20.4a}$$

The second pendulum is governed by similar dynamics:

$$\dot{\theta}_2(t) = \omega_2(t), \tag{20.5a}$$

$$\dot{\omega}_2(t) = \alpha_2(\theta_2(t), u_2(t), \theta_1(t)). \tag{20.5b}$$

Here, we similarly introduce $z_2(t)$ to replace $\theta_1(t)$ and define $\mathbf{f}_2(\mathbf{x}_2(t), u_2(t), z_2(t))$ as

$$\mathbf{f}_2(\mathbf{x}_2(t), u_2(t), z_2(t)) := \begin{bmatrix} \omega_2(t) \\ \alpha_2(\theta_2(t), u_2(t), \theta_1(t)) \end{bmatrix} \tag{20.6a}$$

In the next step, we define the system outputs. Each subsystem gives its angle directly as system output, thus

$$g_1(\mathbf{x}_1(t)) := \theta_1(t) \tag{20.7a}$$
$$g_2(\mathbf{x}_2(t)) := \theta_2(t). \tag{20.7b}$$

Now we define the coupling constraint that is now a simple input-output relation.

$$z_1(t) = y_2(t) \tag{20.8a}$$
$$z_2(t) = y_1(t) \tag{20.8b}$$

We define the local cost functions as well that penalize state deviations from the equilibrium state, i.e.

$$\ell_1(\mathbf{x}_1(t), u_1(t)) := \frac{1}{2}\mathbf{x}_1(t)^T \mathbf{Q}_1 \mathbf{x}_1(t) + \frac{\gamma_1}{2} u_1(t)^T u_1(t) \tag{20.9a}$$
$$\ell_2(\mathbf{x}_2(t), u_2(t)) := \frac{1}{2}\mathbf{x}_2(t)^T \mathbf{Q}_2 \mathbf{x}_2(t) + \frac{\gamma_2}{2} u_2(t)^T u_2(t). \tag{20.9b}$$

Here, $\mathbf{Q}_1, \mathbf{Q}_2$ are symmetric positive definite matrices and $\gamma_1, \gamma_2 > 0$ are regularization parameters that make the optimization problem non-singular. Since the actuators normally have physical limits, this should be reflected in control constraints on the inputs as

$$\underline{u}_1 \le u_1(t) \le \overline{u}_1 \tag{20.10a}$$
$$\underline{u}_2 \le u_2(t) \le \overline{u}_2, \tag{20.10b}$$

where $\underline{u}_1, \overline{u}_1, \underline{u}_2, \overline{u}_2 \in \mathbb{R}$. We can summarize our optimal control problem as

$$\min_{\substack{\mathbf{x},\mathbf{u},\\ \mathbf{z},\mathbf{y}}} \sum_{i=1}^{2} \int_0^{N_p} \ell_i(\mathbf{x}_i(t), u_i(t))dt \tag{20.11a}$$
$$\text{s.t. } \dot{\mathbf{x}}_i(t) = \mathbf{f}_i(\mathbf{x}_i(t), u_i(t), z_i(t)) \qquad i = 1, 2 \tag{20.11b}$$
$$y_i(t) = g_i(\mathbf{x}_i(t)) \qquad i = 1, 2 \tag{20.11c}$$
$$\mathbf{x}_i(0) = \bar{\mathbf{x}}_i^0 \qquad i = 1, 2 \tag{20.11d}$$
$$z_1(t) = y_2(t) \tag{20.11e}$$
$$z_2(t) = y_1(t) \tag{20.11f}$$
$$\underline{u}_i \le u_i(t) \le \overline{u}_i, \quad t \in [0, N_p] \qquad i = 1, 2 \tag{20.11g}$$

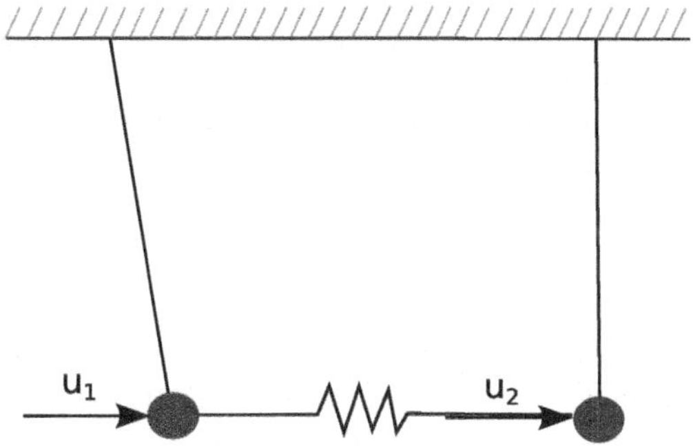

Fig. 20.1 Sketch of two coupled pendula with a spring

Note that (20.11) is a special case of (20.1). In the following, we will use the example to explain distributed multiple shooting in details.

20.4 The Distributed Multiple Shooting Method

In this section, we discuss and extend the *direct multiple shooting* method [1]. Our aim is to simulate and calculate sensitivities in parallel not only in the time domain, as it can be done by direct multiple shooting, but also in the spatial or subsystem domain. In particular, our aim is to simulate all subsystems in each time interval simultaneously.

20.4.1 Direct Multiple Shooting in Optimal Control

Shooting methods typically treat one system and thus we regard subsystem i and assume that $\mathbf{z}_i(t)$ is known. An important subproblem of finding the solution of (20.1) is to solve an initial value problem (IVP) of form

$$\dot{\mathbf{x}}_i(t) = \mathbf{f}_i(\mathbf{x}_i(t), \mathbf{u}_i(t), \mathbf{z}_i(t)) \tag{20.12a}$$

$$\mathbf{x}_i(0) = \overline{\mathbf{x}}_i^0, \tag{20.12b}$$

where $\mathbf{x}_i(N_p)$ is sought for. The direct multiple shooting method introduces a time grid on the control horizon $[0, N_p]$ as

$$0 = t_0 \leq \cdots \leq t_N = N_p. \tag{20.13}$$

For each time point $t_0, \ldots, t_{N-1}$, initial state values $\mathbf{x}_{i,j}$, $j = 0, \ldots, N-1$ are introduced, which become optimization variables. The control input $\mathbf{u}_i(t)$ is discretized in each time interval $[t_j, t_{j+1}]$ with piecewise constant values $\mathbf{u}_{i,j}$, which are also degrees of freedom.

On each time interval $[t_j, t_{j+1}]$, we solve the initial value problem

$$\dot{\mathbf{x}}_i(t) = \mathbf{f}_i(\mathbf{x}_i(t), \mathbf{u}_{i,j}, \mathbf{z}_i(t)) \tag{20.14a}$$

$$\mathbf{x}_i(t_j) = \mathbf{x}_{i,j}, \tag{20.14b}$$

i.e. we propagate the state trajectory with initial value $\mathbf{x}_{i,j}$, control inputs $\mathbf{u}_{i,j}$ using an integration rule, e.g. Runge-Kutta or backward differentiation formula (BDF) schemes, resulting in the end-state $\mathbf{F}_{i,j}(\mathbf{x}_{i,j}, \mathbf{u}_{i,j}, \mathbf{z}_i(t)) := \mathbf{x}_i(t_{j+1})$. However, by this construction, we may introduce a non-zero residual between $\mathbf{F}_{i,j}(\mathbf{x}_{i,j}, \mathbf{u}_{i,j}, \mathbf{z}_i(t))$ and $\mathbf{x}_{i,j+1}$. In other words, the state trajectories are propagated, but they are not continuous. This property can be recovered in the optimization routine by imposing a continuity constraint of form

$$\mathbf{F}_{i,j}(\mathbf{x}_{i,j}, \mathbf{u}_{i,j}, \mathbf{z}_i(t)) = \mathbf{x}_{i,j+1}. \tag{20.15}$$

The very first initial value is enforced by imposing $\mathbf{x}_{i,0} = \overline{\mathbf{x}}_i^0$. The benefit of direct multiple shooting is that it can better treat unstable systems. Moreover, the simulation of the system on the different time intervals can be executed in parallel, i.e. evaluating and linearizing $\mathbf{F}_{i,j}(\cdot)$ may take place in parallel for each $j = 0, \ldots, N-1$. For example, if we have N time intervals, the parallized simulation and sensitivity calculation can lead to a speed-up of up to N.

20.4.2 The Distributed Multiple Shooting Method

The *distributed multiple shooting* algorithm uses two different decomposition methods to tackle with problem (20.1). First, we apply the state discretization of direct multiple shooting on each subsystem. Second, the spatial domain of the whole system is decomposed such that on the boundary of the spatial subdomains a finite approximation is introduced. The resulting subsystems can be simulated and linearized independently from each other and on each time interval.

In the previous subsection, we have assumed that for subsystem i the coupling input variable $\mathbf{z}_i(t)$ is known, although this is determined by the state and control variables of the other subsystems. Our aim is to obtain a finite dimensional approximation of $\mathbf{z}_i(t)$ in order to be able to simulate subsystems independently. It should be understood that $\mathbf{z}_i(t)$ and $\mathbf{y}_i(t)$ must have the same discretization so that the coupling equations are compatible. While other choices are possible, we propose to use orthogonal polynomials to obtain a finite dimensional description of $\mathbf{z}_i(t)$ and $\mathbf{y}_i(t)$.

We define *normalized Legendre time-polynomials* $\gamma_p(t)$ of order p and $\gamma_q(t)$ of order q as

$$\int_{-1}^{1} \gamma_p(t)\gamma_q(t)dt = \begin{cases} 1 & \text{if } p \neq q \\ 0 & \text{otherwise} \end{cases}. \tag{20.16}$$

Also, we define the *normalized Legendre basis* of order S as

$$\Gamma_S(t) = [\gamma_0(t), \gamma_1(t), \ldots, \gamma_S(t)]^T. \tag{20.17}$$

Note that $\Gamma_s(t)$ is a column vector containing Legendre time-polynomials up to order S whose coefficients can be precomputed offline.

Let us detail how the approximation based on Legendre polynomials can be calculated. We assume that we want to approximate a variable $\mathbf{w}_i(t) \in \mathbb{R}^n$ of subsystem i by computing the corresponding coefficient matrix $\mathbf{W}_{i,j} \in \mathbb{R}^{(S+1)\times n}$ on time interval $[t_j, t_{j+1}]$, then it can be shown that the best approximation in the L_2-norm is given by the coefficients

$$\mathbf{W}_{i,j} = \frac{2}{t_{j+1} - t_j} \int_{t_j}^{t_{j+1}} \Gamma_S(\hat{t})\mathbf{w}_i^T(t)dt, \tag{20.18}$$

where $\hat{t}$ is the translated and scaled time variable defined by

$$\hat{t} := \frac{2t - t_j - t_{j+1}}{t_{j+1} - t_j}. \tag{20.19}$$

It should be noticed that the integrand on the right-hand side of (20.18) is a matrix and the quadrature should be computed element-wise.

With the given coefficient matrix $\mathbf{W}_{i,j} \in \mathbb{R}^{(S+1)\times n}$, valid on $[t_j, t_{j+1}]$, we can compute the original variable $\mathbf{w}_i(t) \in \mathbb{R}^n$, up to approximation error, by

$$\mathbf{w}_i(t) = \Gamma_S(\hat{t})^T \mathbf{W}_{i,j}, \quad t \in [t_j, t_{j+1}] \tag{20.20}$$

If we apply the previously indtroduced discretization methods to the general optimal control problem (20.1) we result in the NLP of form

$$\min_{\substack{\mathbf{x},\mathbf{y},\\ \mathbf{z},\mathbf{y}}} \sum_{i=1}^{|\mathcal{N}|} \sum_{j=0}^{N} L_{i,j}(\mathbf{x}_{i,j}, \mathbf{u}_{i,j}, \mathbf{Z}_{i,j}) \tag{20.21a}$$

$$\text{s.t.}\ \mathbf{x}_{i,j+1} = \mathbf{F}_{i,j}(\mathbf{x}_{i,j}, \mathbf{u}_{i,j}, \mathbf{Z}_{i,j}) \tag{20.21b}$$

$$\mathbf{Y}_{i,j} = \mathbf{G}_{i,j}(\mathbf{x}_{i,j}, \mathbf{u}_{i,j}, \mathbf{Z}_{i,j}) \tag{20.21c}$$

$$\mathbf{x}_{i,0} = \overline{\mathbf{x}}_i^0 \tag{20.21d}$$

$$\mathbf{Z}_{i,j} = \sum_{j=1}^{|\mathcal{N}|} \mathbf{A}_{i,j}\mathbf{Y}_{i,j} \tag{20.21e}$$

$$\mathbf{p}_i(\mathbf{x}_{i,j}, \mathbf{u}_{i,j}) \geq 0, \tag{20.21f}$$

where $i = 1, \ldots, |\mathcal{N}|,\ j = 0, \ldots, N-1$. This problem can be solved by a nonlinear programming algorithm such as the ones published in [3, 5, 11] with the modification that the linearizations of functions $L_{i,j}(\cdot)$, $\mathbf{F}_{i,j}(\cdot)$ and $\mathbf{G}_{i,j}(\cdot)$ are computed in parallel. In the following, we detail the steps how to obtain this NLP on the coupled pendula.

20.4.3 DMS on the Coupled Pendula

In the special case of the pendula, the coupling output $\mathbf{Y}_{1,j} \in \mathbb{R}^{(S+1)\times 1}$ approximates the angle of the first pendulum, which is used by the second pendulum as a finite dimensional coupling input $\mathbf{Z}_{2,j} \in \mathbb{R}^{(S+1)\times 1}$. And conversely, the second pendulum generates $\mathbf{Y}_{2,j} \in \mathbb{R}^{(S+1)\times 1}$ by calculating quadratures, which is used by the first pendulum in a form of input $\mathbf{Z}_{1,j} \in \mathbb{R}^{(S+1)\times 1}$.

In subsystem i and on shooting interval $[t_j, t_{j+1}]$, we have

$$\dot{\mathbf{x}}_i(t) = \mathbf{f}_i(\mathbf{x}_i(t), u_{i,j}, \Gamma_S(\hat{t})^T \mathbf{Z}_{i,j}) \tag{20.22a}$$

$$\mathbf{x}_i(t_j) = \mathbf{x}_{i,j} \tag{20.22b}$$

$$y_i(t) = g_i(\mathbf{x}_i(t)). \tag{20.22c}$$

We denote the solution of the ODE (20.22a) and (20.22b) with $\mathbf{F}_{i,j}(\mathbf{x}_{i,j}, u_{i,j}, \mathbf{Z}_{i,j})$. The function $\mathbf{F}_{i,j}(\cdot)$ may be regarded as a simulator function that simulates subsystem i on time interval $[t_j, t_{j+1}]$ with initial state $\mathbf{x}_{i,j}$, control input $u_{i,j}$ and coupling input approximation $\mathbf{Z}_{i,j}$. We will see later that the differentiability of $\mathbf{F}_{i,j}(\cdot)$ is needed. We define the $\mathbf{G}_{i,j}(\cdot)$ and $L_{i,j}(\cdot)$ as

$$\mathbf{G}_{i,j}(\mathbf{x}_{i,j}) := \frac{2}{t_{j+1} - t_j} \int_{t_j}^{t_{j+1}} \Gamma_S(\hat{t}) g_i(\mathbf{x}_i(t)) dt \tag{20.23}$$

$$L_{i,j}(\mathbf{x}_{i,j}, u_{i,j}) := \int_{t_j}^{t_{j+1}} \ell_i(\mathbf{x}_i(t), u_{i,j}) dt \tag{20.24}$$

where $\mathbf{x}_i(t)$ respects the ODE defined in (20.22a) and (20.22b). Note that $\mathbf{F}_{i,j}(\cdot)$ and $\mathbf{G}_{i,j}(\cdot)$ are simultaneously computable if we extend (20.22a) and (20.22a) with $S+1$ extra dynamic variables with zero initial value. Now we can build up an NLP that reads as

$$\min_{\substack{\mathbf{x},\mathbf{u},\\ \mathbf{z},\mathbf{y}}} \sum_{i=1}^{2} \sum_{j=0}^{N} L_{i,j}(\mathbf{x}_{i,j}, u_{i,j}) \tag{20.25a}$$

$$\text{s.t. } \mathbf{x}_{i,j+1} = \mathbf{F}_{i,j}(\mathbf{x}_{i,j}, u_{i,j}, \mathbf{Z}_{i,j}) \tag{20.25b}$$

$$\mathbf{Y}_{i,j} = \mathbf{G}_{i,j}(\mathbf{x}_{i,j}) \tag{20.25c}$$

$$\mathbf{x}_{i,0} = \overline{\mathbf{x}}_i^0 \tag{20.25d}$$

$$\mathbf{Z}_{1,j} = \mathbf{Y}_{2,j} \tag{20.25e}$$
$$\mathbf{Z}_{2,j} = \mathbf{Y}_{1,j} \tag{20.25f}$$
$$\underline{u}_i \leq u_{i,j} \leq \overline{u}_i, \tag{20.25g}$$

where $i = 1, 2$ and $j = 0, \ldots, N-1$. Let us discuss the meaning of the different parts of this optimization problem. First, (20.25a) penalizes non-zero states and control inputs of each pendulum on each shooting interval. The constraint (20.25b) ensures that the state profile is continuous along the time domain and within each pendulum. With the constraint (20.25c) the finite approximation of the angle of each pendulum on each shooting interval is calculated. The constraint (20.25d) fixes the initial angle and angular speed of both pendula. The constraints (20.25e) and (20.25f) impose that the polynomial coefficients of the two pendula are equal in each time interval.

20.4.4 DMS: Central Optimization, Parallel Linearization

Throughout the iterations of the optimization routine that solves (20.25) the evaluation of $L_{i,j}(\cdot)$, $\mathbf{F}_{i,j}(\cdot)$ and $\mathbf{G}_{i,j}(\cdot)$ along with their derivatives are required, which are computationally expensive operations. The most powerful feature of DMS is that it parallelizes these computations in both the time and the subsystem domain. In other words, the evaluation and linearization of $L_{i,j}(\cdot)$, $\mathbf{F}_{i,j}(\cdot)$ and $\mathbf{G}_{i,j}(\cdot)$ may be calculated for all i, $i = 1, 2$ and j, $j = 0, \ldots, N-1$ simultaneously. This requires $|\mathcal{N}|N$ parallel processes where process (i, j) calculates linearization of subsystem i on time interval $[t_j, t_{j+1})$ and a dedicated process, which collects the calculated linearizations and finds the next linearization point by the optimization subroutine.

Note that the computation time of both the linearization and the calculation of the next linearization point depends on the order of the approximation S. Once we employ higher order polyonomials, on one hand the linearization of functions $\mathbf{F}_{i,j}(\cdot)$ and $\mathbf{G}_{i,j}(\cdot)$ becomes more involved. On the other hand, the central optimization problem becomes higher dimensional. It is important to choose the value of S properly. Since this is typically application specific, a generic method does not exist how to compute it. The distributed multiple shooting method is summarized in Algorithm 20.1.

It has been shown in [10] that in the case of large systems one can save time in the sensitivity generation already by a serial implementation of DMS. This is due to the fact that the coupling between subsystems can be approximated by a low-dimensional basis, which results in fewer directional derivatives.

20.4.5 Theoretical Results

The distributed multiple shooting method is essentially a tailored *sequential quadratic programming* (SQP) method. In each iteration, the linearization of the original NLP is

---

Algorithm 20.1 Distributed multiple shooting method

1: *Problem preparation*

1.1 Introduce time mesh $0 = t_0 \leq \cdots \leq t_N = N_p$ and optimization variables $\mathbf{x}_{i,j}$ and $\mathbf{u}_{i,j}$.
1.2 Choose S and introduce optimization variables $\mathbf{Z}_{i,j}$ and $\mathbf{Y}_{i,j}$.
1.3 Reformulate local dynamic equations $\mathbf{f}_i$ for each time interval $[t_j, t_{j+1})$: plug $\Gamma_S(\hat{t})^T \mathbf{Z}_{i,j}$ into $\mathbf{z}_i(t)$.
1.4 Extend local dynamic equations with quadratures that calculate local output approximations and the objective function.

2: *Solution method*

2.1 Evaluate and linearize $L_{i,j}(\cdot)$, $\mathbf{F}_{i,j}(\cdot)$ and $\mathbf{G}_{i,j}(\cdot)$ in the actual linearization point using $|\mathcal{N}| \cdot N$ parallel processes.
2.2 Collect linearizations at the dedicated optimizer process.
2.3 Determine the next linearization point using an NLP solver.
2.4 Communicate the new linearization point to the $|\mathcal{N}| \cdot N$ processes.
2.5 If convergence is achieved then quit, otherwise go to Step 2.1.

---

calculated in parallel and the optimization variables are updated centrally by solving a quadratic programming problem. The first SQP method was proposed by Powell [7]. Later on, it was shown that with quasi-Newton Hessian approximation locally superlinear convergence can be obtained. By using globalization techniques (e.g. line-search, trust-region methods) one can ensure the global convergence for SQP methods. In other words, the optimization procedure may be initialized arbitrarily, the convergence is guaranteed. For a very good survey of SQP methods and their convergence analysis we refer the reader to the article [2] and the textbook [6].

20.5 Applications and Numerical Illustration

In this section, we carry out simulations in two different scenarios. First, we solve an optimal control problem with different order of approximations and would like to analyze the quality of the solution given by DMS. It is also of our interest how the different approximations compare to the centralized solution given by direct multiple shooting. Second, we apply DMS inside a model predictive controller loop. One may find further applications of DMS in [4, 9].

20.5.1 Nonlinear Optimal Control of Coupled Pendula

We have solved the optimal control problem of (20.11) with initial values

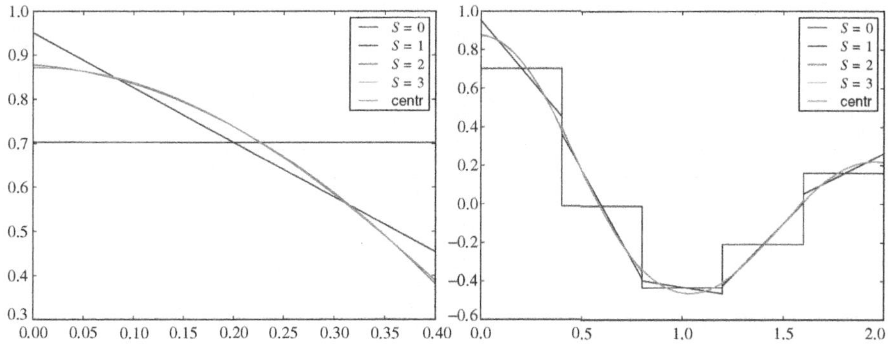

Fig. 20.2 Polynomial approximation of $\theta_1(t)$ with normalized Legendre polynomials of order $S = 0, \ldots, 3$ together with the centralized optimal state trajectory on the control interval of [0, 0.4] (*left*) and on the whole prediction horizon of [0, 2.0] (*right*)

$$\bar{\mathbf{x}}_1^0 = (0.87, 0.0)^T \tag{20.26a}$$

$$\bar{\mathbf{x}}_2^0 = (-0.17, 0.0)^T \tag{20.26b}$$

which correspond to an inital angle of 50° and −10°, respectively. Note that with this initialization the system is not stabilizable with a linear controller. We introduce 5 shooting intervals of 0.4 s, with a prediction horizon of 2 s.

We have seen that the system output approximation has a tuning parameter that is the order of the basis function S. We would like to see how well the approximation behaves with different values of S and whether it is reasonable to use higher order approximations. We have plotted the optimal trajectory of $\theta_1(t)$ given by direct multiple shooting together with its polynomial approximation of order $S = 0, \ldots, 3$ by DMS in Fig. 20.2. The piece-wise constant approximation, i.e. $S = 0$, gives exactly the mean value of the approximated variable over the time interval considered. If we increase S we obtain better and better approximations. For our particular example, an approximation of order $S = 2$ may be regarded as sufficiently good. Higher order approximations result in negligable approximation error.

It is interesting to see how the optimal solution of the optimal control problem is affected by the polynomial approximation order. For this reason, we have compared the optimal state trajectories and control plan given by DMS with different approximations of order $S = 0, \ldots, 3$ against the optimal centralized state prediction and control profile in Fig. 20.3. We can conclude that the piece-wise constant approximation yields slightly different control actions from the centralized solution, although the state trajectories are similar. Once we have increased the approximation order, both the optimal state trajectories and control actions are practically the same.

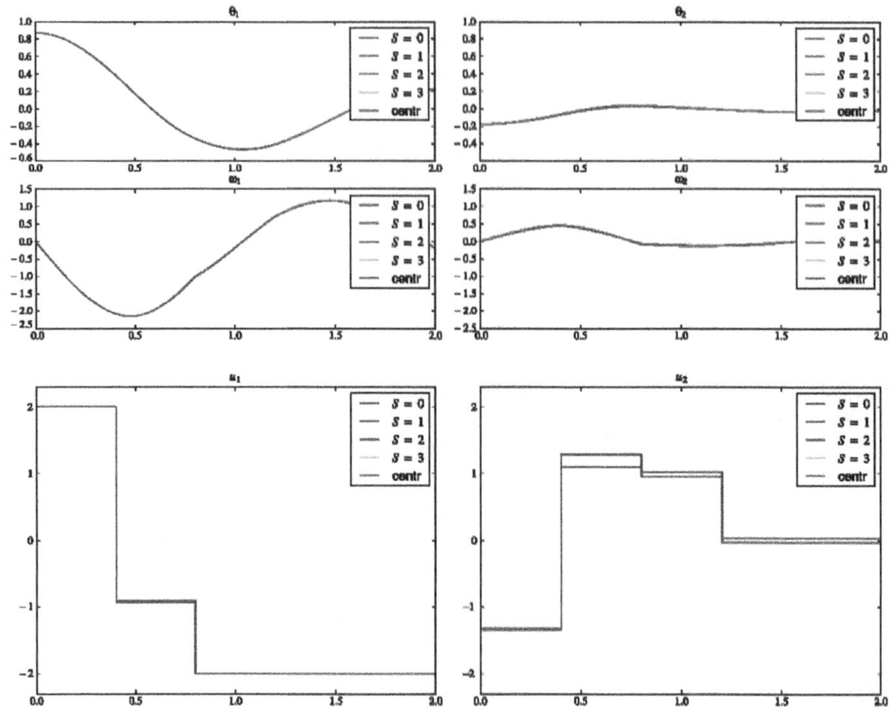

Fig. 20.3 Solution of optimal control problems: state trajectories (*top-left* and *top-right*) using normalized Legendre polynomial approximation of order $S = 0, \ldots, 3$ and control input plan (*bottom-left* and *bottom-right*) together with the centralized solution

20.5.2 Model Predictive Control of Coupled Pendula

In the following, we introduce a model predictive controller (MPC) [8] that uses DMS as a solver. In essence, MPC solves an optimal control problem in each time instance using a prediction model that is initialized by the real state values of the controlled plant. In our context, the centralized model of the two pendula with added absolute Gaussian noise $\Delta \mathbf{x} = 0.05$ was used to generate the plant response, while the model discretized by DMS was used as a prediction model. In each MPC step, a prediction horizon of 2 s and shooting interval length of 0.4 s was considered. In Fig. 20.4, we have plotted the evolution of the real system states along with the state values that were fed to the prediction model. We have also depicted the injected control inputs. In the beginning, the control bounds are hit at the first pendulum, which is harder to stabilize. After the sixth MPC step both pendula are approaching to their equilibrium states, although the added noise makes the tracking imperfect.

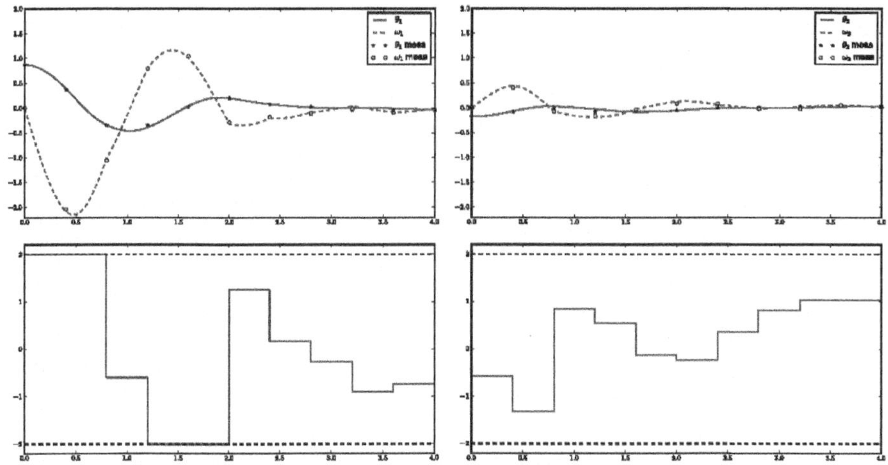

Fig. 20.4 State evolution of the first (*top-left*) and the second (*top-right*) pendulum and the applied control input of the first (*bottom-left*) and the second (*bottom-right*) pendulum

20.6 Conclusions

In this chapter, we have introduced the distributed multiple shooting algorithm, which is designed to solve large scale optimal control problems of distributed systems. The discussed method can be used as a fast solver for MPC subproblems by introducing parallel simulation and sensitivity generation. The local nonlinear dynamics of each subsystem is simulated and linearized simultaneously in each shooting interval. The linearizations are collected on a dedicated node, where the new linearization point is calculated, which is then broadcasted to the simulator nodes. We have pointed out that with this approach one can reach fast local convergence to the centralized solution and that the procedure can be extended to reach global convergence.

We have demonstrated the behaviour of DMS on a system of coupled pendula. We have experimented with approximations of different order and concluded that for this particular example a low-order representation is sufficiently good. DMS was used an MPC loop as an OCP solver that managed to drive the pendula into their equilibrium position.

Acknowledgments I am very grateful and would like to thank to Sebastien (Grandpapa) for giving help with the model equations and to Marco, who helped with the numerical experiments.
This research was supported by Research Council KUL: PFV/10/002 Optimization in Engineering Center OPTEC, GOA/10/09 MaNet and GOA/10/11 Global real- time optimal control of autonomous robots and mechatronic systems. Flemish Government: IOF/KP/SCORES4CHEM, FWO: Ph.D./postdoc grants and projects: G.0320.08 (convex MPC), G.0377.09 (Mechatronics MPC); IWT: Ph.D. Grants, projects: SBO LeCoPro; Belgian Federal Science Policy Office: IUAP P7 (DYSCO, Dynamical systems, control and optimization, 2012-2017); EU: FP7-EMBOCON (ICT-248940), FP7-SADCO (MC ITN-264735), ERC ST HIGHWIND (259 166), Eurostars SMART, ACCM.

References

1. H.G. Bock, K.J. Plitt, A multiple shooting algorithm for direct solution of optimal control problems, in *Proceedings 9th IFAC World Congress Budapest* (Pergamon Press, 1984), pp. 243–247
2. P.T. Boggs, J.W. Tolle, Sequential quadratic programming. Acta Numeric, pp. 1–51 (1995)
3. R.H. Byrd, J. Nocedal, R.A. Waltz, KNITRO: An integrated package for nonlinear optimization, in G. Pillo, M. Roma (eds.), *Large Scale Nonlinear Optimization* (Springer Verlag, 2006), pp. 35–59
4. A. Kozma, J. Andersson, C. Savorgnan, M. Diehl, Distributed multiple shooting for optimal control of large interconnected systems, in *Proceedings of the International Symposium on Advanced Control of Chemical Processes*, 2012
5. T. Nikolayzik, C. Büskens, M. Gerdts, Nonlinear large-scale optimization with WORHP, in *Proceedings of the 13th AIAA/ISSMO Multidisciplinary Analysis Optimization Conference*, 2010
6. J. Nocedal, S.J. Wright, *Numerical optimization*. Springer Series in Operations Research and Financial Engineering, 2nd edn. (Springer, 2006)
7. M.J.D. Powell, Algorithms for nonlinear constraints that use Lagrangian functions. Math. Program. **14**(3), 224–248 (1978)
8. J.B. Rawlings, D.Q. Mayne, *Model predictive control: theory and design* (Nob Hill, 2009)
9. C. Savorgnan, C. Romani, A. Kozma, M. Diehl, Multiple shooting for distributed systems with applications in hydro electricity production. J. Process Control **21**, 738–745 (2011)
10. C. Savorgnan, A. Kozma, J. Andersson, M. Diehl, Adjoint-based distributed multiple shooting for large-scale systems, in *Proceedings of the 18th IFAC World Congress*, 2011
11. A. Wächter, L. Biegler, IPOPT - an interior point OPTimizer. https://projects.coin-or.org/Ipopt. Accessed 2009

Chapter 21
Nash-Based Distributed MPC for Multi-Rate Systems

S. Roshany-Yamchi, R. R. Negenborn and A. A. Cornelio

Abstract In this chapter, a new Nash-based distributed MPC method is proposed to control large-scale multi-rate systems with linear dynamics that are coupled via inputs. These systems are multi-rate systems in the sense that either output measurements or input updates are not available at certain sampling times. Such systems can arise when the number of sensors is less than the number of variables to be controlled or when measurements of outputs cannot be completed simultaneously because of applicational limitations. The multi-rate nature gives rise to a lack of information which will cause uncertainty in the system's performance. To compensate for the information loss due to the multi-rate nature of the systems under study, a distributed Kalman filter is proposed to provide an optimal estimate of the missing information.

21.1 Introduction

Many practical control problems involve measurements sampled at multiple rates. One of the common cases arises when the process outputs or controlled variables are measurable only with large sampling delays and intervals and secondary measurements must be utilized in order to design an effective control system [1]. The secondary measurements are usually sampled at much higher rates and consequently,

---

S. Roshany-Yamchi (✉)
NIMBUS Center for Embedded Research, Cork Institute of Technology, Cork, Ireland
e-mail: samira.roshany@cit.ie

R. R. Negenborn
Department of Marine and Transport Technology, Delft University of Technology, Delft, The Netherlands
e-mail: r.r.negenborn@tudelft.nl

A. A. Cornelio
Department of Electrical Engineering, Mercedes Benz R&D, Bangalore, India
e-mail: avi_anthony.cornelio@daimler.com

J. M. Maestre and R. R. Negenborn (eds.), *Distributed Model Predictive Control Made Easy*, Intelligent Systems, Control and Automation: Science and Engineering 69,
DOI: 10.1007/978-94-007-7006-5_21,

a multi-rate system results. In multi-rate systems, either output measurements or input updates are not available at certain sampling times. Such systems can arise, e.g. when the number of sensors is less than the number of variables to be controlled or when measurements of outputs cannot be completed simultaneously because of applicational limitations. The multi-rate nature gives rise to a lack of information which will cause uncertainty in the system's performance. In order to deal with the lack of information at the faster rates, state estimation techniques in combination with control methods are typically used.

In multi-rate plants, either the measurements are available less frequently or the control actions are made at a lower rate. Such systems can be encountered in many industrial applications [2–5]. In the process industry, for instance, quality variables such as product concentration or average molecular weight distribution in a polymerization process, can be evaluated/updated at much slower rates compared to other process measurements. Moreover, in certain biomedical applications [3] the input injection rate is inherently slower than the output measurement, e.g. in drug infusion systems, the drug injection to the patient occurs less frequently than the body symptom's measurements such as blood pressure, body temperature, etc. Multi-rate systems with large number of inputs and outputs can be considered as distributed multi-rate systems.

On the other hand, economical and technological demands motivate the development of large-scale plants such as process plants, manufacturing systems and satellite orbit formations with low complexity and high performance accuracy. Large-scale systems consist of many subsystems that interact and have to be controlled. Large-scale distributed systems are present in many engineering application domains including process plants, road traffic networks, water and sewer networks, power distribution systems, wind farms and wireless sensor/actuator networks [6–8].

Research on Decentralized MPC dates back to the work of Wang et al. [9] and since then, the interest has grown significantly due to attempts to reduce the complexity of the centralized problem [10]. Since in the implementation of decentralized MPC only local process data are required, it has become an attractive method to practitioners [8]. However, there is a well-known defect with decentralized techniques and that is their performances are usully far from optimal, especially when the involved subsystems interact strongly. On the other hand, in centralized MPC, the optimal nominal control solution is achievable for any system [8, 11]. However, implementation of centralized MPC for large-scale systems is neither feasible nor practical. Controlling large-scale networked systems will be even more challenging when the systems are multi-rate. In multi-rate systems neither output measurements nor input updates are available at certain sampling times. Controlling multi-rate systems with centralized MPC and also decentralized MPC have been studied before in [5] and [1].

In this chapter we discuss a distributed predictive control and estimation framework with guaranteed performance properties, an assured performance improvement over decentralized schemes and the capability of approaching centralized MPC performance for multi-agent multi-rate systems, [2, 3, 5].

21.2 A Distributed Multi-Rate System

Consider a linear dynamic plant consisting of m subsystems where the model of subsystem i, $\{i \in 1, \ldots, m\}$ can be written as:

$$\mathbf{x}_{\mathrm{p}_i}(k+1) = \mathbf{A}_{\mathrm{p}_i}\mathbf{x}_{\mathrm{p}_i}(k) + \mathbf{B}_{\mathrm{p}_i}\mathbf{u}_i(k) + \mathbf{B}_{\mathrm{d}_i}\mathbf{d}_i(k) + \sum_{\substack{j=1\\ j\neq i}}^{m} \mathbf{B}_{\mathrm{p}i,j}\mathbf{u}_j(k) \tag{21.1}$$

$$\mathbf{y}_{\mathrm{p}_i}(k) = \mathbf{C}_{\mathrm{p}_i}\mathbf{x}_{\mathrm{p}_i}(k) + \mathbf{w}_i(k), \tag{21.2}$$

where k is the discrete time index of the system under control. Also, $\mathbf{x}_{\mathrm{p}_i} \in \mathbb{R}^p$, $\mathbf{u}_i \in \mathbb{R}^l$, $\mathbf{y}_{\mathrm{p}_i} \in \mathbb{R}^q$ and $\mathbf{d}_i$, $\mathbf{w}_i$ are the process and measurement noise vectors, respectively. Let:

$$\Delta\mathbf{x}_{\mathrm{p}_i}(k) = \mathbf{x}_{\mathrm{p}_i}(k) - \mathbf{x}_{\mathrm{p}_i}(k-1), \tag{21.3}$$

$$\Delta\mathbf{u}_i(k) = \mathbf{u}_i(k) - \mathbf{u}_i(k-1). \tag{21.4}$$

The state space equation for $\Delta\mathbf{x}_{\mathrm{p}_i}$ can be obtained from (21.2) resulting in:

$$\Delta\mathbf{x}_{\mathrm{p}_i}(k+1) = \mathbf{A}_{\mathrm{p}_i}\Delta\mathbf{x}_{\mathrm{p}_i}(k) + \mathbf{B}_{\mathrm{p}_i}\Delta\mathbf{u}_i(k) + \mathbf{B}_{\mathrm{d}_i}\mathbf{w}_{\mathrm{p}_i}(k) + \sum_{\substack{j=1\\ j\neq i}}^{m} \mathbf{B}_{\mathrm{p}i,j}\Delta\mathbf{u}_j(k), \tag{21.5}$$

where

$$\mathbf{w}_{\mathrm{p}_i}(k) = \mathbf{d}_i(k) - \mathbf{d}_i(k-1). \tag{21.6}$$

Similarly, for the output we obtain:

$$\mathbf{y}_{\mathrm{p}_i}(k+1) = \mathbf{C}_{\mathrm{p}_i}\Delta\mathbf{x}_{\mathrm{p}_i}(k+1) + \mathbf{y}_{\mathrm{p}_i}(k) + \mathbf{w}_{\mathrm{m}_i}(k), \tag{21.7}$$

where

$$\mathbf{w}_{\mathrm{m}_i}(k) = \mathbf{w}_i(k+1) - \mathbf{w}_i(k). \tag{21.8}$$

Variables $\mathbf{w}_{\mathrm{m}_i}(k)$ and $\mathbf{w}_{\mathrm{p}_i}(k)$ are assumed to be a Wiener process. The Wiener process can be used to represent the integral of a Gaussian white noise process, and so is useful as a model of noise and instrument errors. Substituting (21.5) into (21.7) we have:

$$\begin{aligned}\mathbf{y}_{\mathrm{p}_i}(k+1) &= \mathbf{C}_{\mathrm{p}_i}\Delta\mathbf{x}_{\mathrm{p}_i}(k+1) + \mathbf{y}_{\mathrm{p}_i}(k) + \mathbf{w}_{\mathrm{m}_i}(k) = \mathbf{C}_{\mathrm{p}_i}\mathbf{A}_{\mathrm{p}_i}\Delta\mathbf{x}_{\mathrm{p}_i}(k) + \mathbf{C}_{\mathrm{p}_i}\mathbf{B}_{\mathrm{p}_i}\Delta\mathbf{u}_i(k)\\ &\quad + \mathbf{C}_{\mathrm{p}_i}\mathbf{B}_{\mathrm{d}_i}\mathbf{w}_{\mathrm{p}_i}(k) + \mathbf{y}_{\mathrm{p}_i}(k) + \mathbf{w}_{\mathrm{m}_i}(k) + \sum_{\substack{j=1\\ j\neq i}}^{m} \mathbf{C}_{\mathrm{p}_i}\mathbf{B}_{\mathrm{p}i,j}\Delta\mathbf{u}_j(k).\end{aligned} \tag{21.9}$$

Now, considering (21.1), (21.2), (21.5) and (21.9), we define new state and control input vectors as $\mathbf{x}_i(k)$ and $\Delta\mathbf{u}_i(k)$, respectively, where

$$\mathbf{x}_i(k) = \begin{bmatrix} \Delta\mathbf{x}_{\mathrm{p}_i}(k) \\ \mathbf{y}_{\mathrm{p}_i}(k) \end{bmatrix} \tag{21.10}$$

The following model represents a distributed multi-rate system with input coupling:

$$\mathbf{x}_i(k+1) = \mathbf{A}_i\mathbf{x}_i(k) + \mathbf{B}_{ii}\Delta\mathbf{u}_i(k) + \mathbf{D}_i\mathbf{v}_i(k) + \sum_{\substack{j=1\\ j\neq i}}^{m} \mathbf{B}_{ij}\Delta\mathbf{u}_j(k) \tag{21.11}$$

$$\mathbf{y}_i(k) = \mathbf{C}_i\mathbf{x}_i(k) + \mathbf{z}_i(k), \tag{21.12}$$

where for each subsystem i, $\mathbf{x}_i$ and $\mathbf{y}_i$ denote the state and output variables, respectively, $\mathbf{v}_i$, $\mathbf{z}_i$ are the process and measurement noise sequences, respectively, and $\mathbf{A}_i$, $\mathbf{B}_{ii}$, $\mathbf{B}_{ij}$, $\mathbf{D}_i$ and $\mathbf{C}_i$ are matrices defined as:

$$\mathbf{A}_i = \begin{bmatrix} \mathbf{A}_{\mathrm{p}_i} & \mathbf{O}_{l\times q} \\ \mathbf{C}_{\mathrm{p}_i}\mathbf{A}_{\mathrm{p}_i} & \mathbf{I}_{q\times q} \end{bmatrix}, \mathbf{B}_{ii} = \begin{bmatrix} \mathbf{B}_{\mathrm{p}_i} \\ \mathbf{C}_{\mathrm{p}_i}\mathbf{B}_{\mathrm{p}_i} \end{bmatrix},$$

$$\mathbf{B}_{ij} = \begin{bmatrix} \mathbf{B}_{\mathrm{p}_{i,j}} \\ \mathbf{C}_{\mathrm{p}_i}\mathbf{B}_{\mathrm{p}_{i,j}} \end{bmatrix}, \mathbf{D}_i = \begin{bmatrix} \mathbf{B}_{\mathrm{d}_i} & \mathbf{O}_{l\times q} \\ \mathbf{C}_{\mathrm{p}_i}\mathbf{B}_{\mathrm{d}_i} & \mathbf{I}_{q\times q} \end{bmatrix},$$

$$\mathbf{v}_i(k) = \begin{bmatrix} \mathbf{w}_{\mathrm{p}_i}(k) \\ \mathbf{w}_{\mathrm{m}_i}(k) \end{bmatrix}, \mathbf{C}_i = \begin{bmatrix} \mathbf{O}_{l\times q} & \mathbf{I}_{q\times q} \end{bmatrix}.$$

Matrix $\mathbf{I}_{q\times q}$ is the q by q identity matrix and $\mathbf{O}_{l\times q}$ is the l by q zero matrix. Variable $\Delta\mathbf{u}_i(k)$ is the multi-rate input signal (the input signal that includes the *input holding mechanism* described later in this section) which is injected into subsystem i at sampling time t. As can be seen from (21.11), subsystems are coupled through inputs only ($\Delta\mathbf{u}_j(k)$).

Assume that the sampling instants for the system vary as $t = 1, 2, \dots, T_\mathrm{f}$ where T_f is the final sampling time. A strategy similar to [5] is followed in implementing multi-rate measurement or input updating mechanisms for subsystem i. In a multi-rate output setting, the output vector $\mathbf{y}_i(k)$ of subsystem i can be measured every T_{y_i} sample steps, where $T_{\mathrm{y}_i} > 0$. Define the output switching function for subsystem i, γ_{i_j}, for $j = 1, 2, \dots, q_i$ with q_i being the number of outputs of subsystem i as follows:

$$\gamma_{i_j}(k) = \begin{cases} 1 & \text{if } t = \tau T_{\mathrm{y}_j}, \text{ for some integer } \tau \\ 0 & \text{otherwise,} \end{cases} \tag{21.13}$$

The following multi-rate output vector $\varphi_i(k)$ can now be defined:

$$\varphi_i(k) = \boldsymbol{\Upsilon}_i(k)\mathbf{y}_i(k), \tag{21.14}$$

where

$$\boldsymbol{\Upsilon}_i(k) = \text{diag}[\gamma_{i_1}(k) \;\; \gamma_{i_2}(k) \;\; \ldots \;\; \gamma_{i_{qi}}(k)]. \tag{21.15}$$

The multi-rate output vector $\varphi_i(k)$ represents the measured outputs of the multi-rate system at sampling time k. In a multi-rate input setting, the input vector $\Delta u_i(k)$ of subsystem i is updated every T_{u_i} sample steps, where $T_{\mathrm{u}_i} > 0$. Introduce the input switching function μ_{i_j} for $j = 1, 2, \ldots, l_i$ with l_i being the number of inputs of subsystem i. Define the inputs holding mechanism as:

$$\mu_{i_j}(k) = \begin{cases} 1 & \text{if } k = \tau T_{\mathrm{u}_j}, \text{ for some integer } \tau \\ 0 & \text{otherwise,} \end{cases} \tag{21.16}$$

The following input matrix $\boldsymbol{\Psi}_i(k)$ for subsystem i can be defined.

$$\boldsymbol{\Psi}_i(k) = \text{diag}\left[\mu_{i_1}(k) \;\; \mu_{i_2}(k) \;\; \ldots \;\; \mu_{i_{l_i}}(k)\right]. \tag{21.17}$$

A new control variable $\boldsymbol{\theta}_i(k)$ is introduced to implement the input administering mechanism:

$$\Delta \boldsymbol{u}_i(k) = \boldsymbol{\Psi}_i(k)\boldsymbol{\theta}_i(k). \tag{21.18}$$

The control variable $\boldsymbol{\theta}_i(k)$ represents the computed inputs using the proposed distributed MPC scheme at sampling time k and the multi-rate control variable $\Delta \boldsymbol{u}_i(k)$ represents the injected inputs into the multi-rate system at sampling time k. After substituting (21.18) into (21.11) we get:

$$\mathbf{x}_i(k+1) = \mathbf{A}_i\mathbf{x}_i(k) + \mathbf{B}_{ii}\boldsymbol{\Psi}_i(k)\boldsymbol{\theta}_i(k) + \mathbf{D}_i\mathbf{v}_i(k) + \sum_{\substack{j=1 \\ j \neq i}}^{m} \mathbf{B}_{ij}\boldsymbol{\Psi}_j(k)\boldsymbol{\theta}_j(k). \tag{21.19}$$

As in multi-rate systems, output measurements are made at specific sampling times and the output sampling mechanism needs to be included in the system model. To do that, consider (21.12) and (21.14), therefore:

$$\varphi_i(k) = \boldsymbol{\Upsilon}_i(k)\mathbf{C}_i\mathbf{x}_i(k) + \boldsymbol{\Upsilon}_i(k)\mathbf{z}_i(k). \tag{21.20}$$

Equations (21.19) and (21.20) give the linear state-space representation of the distributed multi-rate system for $i = 1, 2, \ldots, m$. Next, the Nash-based solution to the distributed MPC problem will be formulated for such a system.

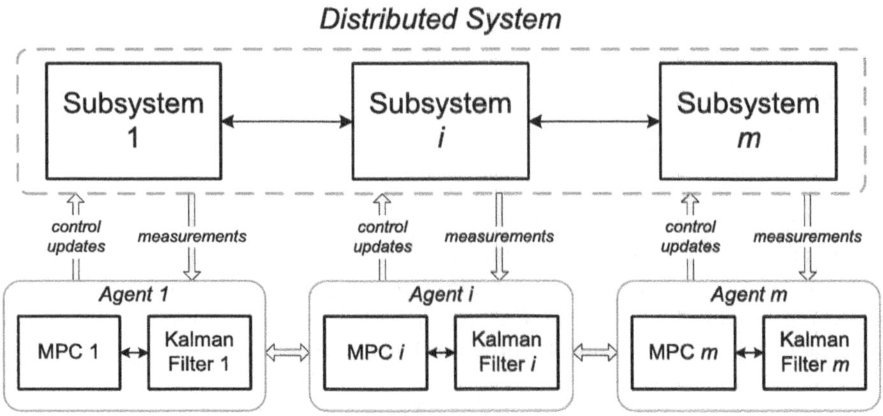

Fig. 21.1 Distributed control and estimation architecture

21.3 Nash-Based Distributed Multi-Rate MPC

In the distributed control structure, input coupling among subsystems is given by (21.19). These subsystems communicate with one another to accomplish a global objective (see Fig. 21.1). One type of Distributed MPC based on Nash optimality has been investigated by [12, 13]. In this approach, the agents communicate but they do not take a cooperative decision. The agents iterate to resolve their local optimization problem simultaneously and obtain their optimal solution [13, 14]. An initial guess for each agent is first given based on the solution found at the last sampling time. Then each agent checks if its terminal iteration condition satisfies a user-defined threshold. This implies that the agents do not share information about the utility of each decision; the agreement (Nash equilibrium) between the agents is reached when neither of their solutions can be improved. The main advantage of this scheme is that the on-line optimization of a large-scale problem can be converted into several small-scale subproblems, thus reducing the computational complexity significantly while keeping satisfactory performance.

Consider a linear multi-rate system consisting of m subsystems (21.19), (21.20) and m control agents. In Nash-based distributed MPC each control agent calculates the manipulated variable $\boldsymbol{\theta}_i(k)$ by minimizing its local cost function as follows:

$$\min_{\boldsymbol{\theta}_i(k),\ldots,\boldsymbol{\theta}_i(k+N_\mathrm{c}-1)} J_i(k) = \sum_{n=1}^{N_\mathrm{p}} \|\mathbf{y}_i(k+n) - \mathbf{y}_{\mathrm{ref,i}}(k+n)\|^2_{\mathbf{Q}_i} + \sum_{n=0}^{N_\mathrm{c}-1} \|\boldsymbol{\theta}_i(k+n)\|^2_{\mathbf{R}_i}, \tag{21.21}$$

subject to

$$\boldsymbol{\theta}_{i,\min} \leq \boldsymbol{\theta}_i(k+n) \leq \boldsymbol{\theta}_{i,\max}, \quad n = 0, 1, \ldots, N_{\mathrm{c}-1}, \tag{21.22}$$

$$\mathbf{x}_i(k+n+1) = \mathbf{A}_i\mathbf{x}_i(k+n) + \mathbf{B}_{ii}\Delta\mathbf{u}_i(k+n) + \mathbf{D}_i\mathbf{v}_i(k+n) + \sum_{\substack{j=1 \\ j\neq i}}^{m} \mathbf{B}_{ij}\Delta\mathbf{u}_j(k+n), \tag{21.23}$$

$$n = 0, 1, \ldots, N_{p-1}.$$

where $\boldsymbol{\theta}_{i,\min}$ and $\boldsymbol{\theta}_{i,\max}$ are the lower and upper limits for the inputs, respectively. Variables $\mathbf{y}_{\text{ref},i}(k+n)$ are the reference values to be tracked by the predicted outputs, $\mathbf{Q}_i \geq 0$ and $\mathbf{R}_i > 0$ denote the weighting matrices, and N_p and N_c are the prediction and control horizons, respectively.

In order to solve the problem (21.21)–(21.23), (21.19) is first substituted into (21.20). Based on (21.19), (21.20), the future state variables are calculated using the sequence of future control inputs. The matrices obtained can be written in a compact form as

$$\mathbf{Y}_i(k) = \mathbf{F}_i\mathbf{x}_i(k) + \tilde{\phi}_{ii}(k)\tilde{\boldsymbol{\theta}}_i(k) + \boldsymbol{\Gamma}_i\tilde{\zeta}_i(k) + \sum_{\substack{j=1 \\ j\neq i}}^{m} \tilde{\phi}_{ij}(k)\tilde{\boldsymbol{\theta}}_j(k), \tag{21.24}$$

with

$$\mathbf{Y}_i(k) = [\mathbf{y}_i^\text{T}(k+1)\;\; \mathbf{y}_i^\text{T}(k+2), \ldots, \mathbf{y}_i^\text{T}(k+N_\text{p})]^\text{T}, \tag{21.25}$$

$$\tilde{\boldsymbol{\theta}}_i(k) = [\boldsymbol{\theta}_i^\text{T}(k)\;\; \boldsymbol{\theta}_i^\text{T}(k+1), \ldots, \boldsymbol{\theta}_i^\text{T}(k+N_\text{c}-1)]^\text{T}, \tag{21.26}$$

$$\tilde{\zeta}_i(k) = [\mathbf{v}_i^\text{T}(k)\;\; \mathbf{v}_i^\text{T}(k+1), \ldots, \mathbf{v}_i^\text{T}(k+N_\text{c}-1)]^\text{T}, \tag{21.27}$$

$$\mathbf{F}_i = [(\mathbf{C}_i\mathbf{A}_i)^\text{T}\;\; (\mathbf{C}_i\mathbf{A}_i^2)^\text{T}, \ldots, (\mathbf{C}_i\mathbf{A}_i^{N_\text{p}})^\text{T}]^\text{T}, \tag{21.28}$$

$$\tilde{\phi}_{ii}(k) = \begin{bmatrix} \mathbf{C}_i\mathbf{B}_{ii}\boldsymbol{\Psi}_i(k) & 0 & \ldots & 0 \\ \mathbf{C}_i\mathbf{A}_i\mathbf{B}_{ii}\boldsymbol{\Psi}_i(k) & \vdots & \ddots & \vdots \\ \vdots & \vdots & \vdots & 0 \\ \mathbf{C}_i\mathbf{A}_i^{N_\text{p}-1}\mathbf{B}_{ii}\boldsymbol{\Psi}_i(k) & \ldots & \ldots & \mathbf{C}_i\mathbf{A}_i^{N_\text{p}-N_\text{c}}\mathbf{B}_{ii}\boldsymbol{\Psi}_i(k+N_\text{c}-1) \end{bmatrix} \tag{21.29}$$

$$\tilde{\phi}_{ij}(k) = \begin{bmatrix} \mathbf{C}_i\mathbf{B}_{ij}\boldsymbol{\Psi}_j(k) & 0 & \ldots & 0 \\ \mathbf{C}_i\mathbf{A}_i\mathbf{B}_{ij}\boldsymbol{\Psi}_j(k) & \vdots & \ddots & \vdots \\ \vdots & \vdots & \vdots & 0 \\ \mathbf{C}_i\mathbf{A}_i^{N_\text{p}-1}\mathbf{B}_{ij}\boldsymbol{\Psi}_j(k) & \ldots & \ldots & \mathbf{C}_i\mathbf{A}_i^{N_\text{p}-N_\text{c}}\mathbf{B}_{ij}\boldsymbol{\Psi}_j(k+N_\text{c}-1) \end{bmatrix} \tag{21.30}$$

$$\boldsymbol{\Gamma}_i = \begin{bmatrix} \mathbf{C}_i\mathbf{D}_i & 0 & \cdots & 0 \\ \mathbf{C}_i\mathbf{A}_i\mathbf{D}_i & \mathbf{C}_i\mathbf{D}_i & \ddots & \vdots \\ \vdots & \vdots & \vdots & 0 \\ \mathbf{C}_i\mathbf{A}_i^{N_\mathrm{p}-1}\mathbf{D}_i & \mathbf{C}_i\mathbf{A}_i^{N_\mathrm{p}-2}\mathbf{D}_i & \cdots & \mathbf{C}_i\mathbf{A}_i^{N_\mathrm{p}-N_\mathrm{c}}\mathbf{D}_i \end{bmatrix}, \tag{21.31}$$

where, in (21.27), we assume that the process noise is zero from $t + N_\mathrm{c} - 1$ on: $\mathbf{v}_i(k+n) = 0$ for $n = N_\mathrm{c}, \ldots, N_\mathrm{p} - 1$.[1]
Note that in (21.24), $\tilde{\boldsymbol{\theta}}_i(k)$ represents the control inputs over the control horizon and the input sampling matrix $\boldsymbol{\Psi}_j(k)$ is embedded in (21.30). Also, note that the control input computed by the local MPC controller is $\theta_i(k)$ and not $\Delta\mathbf{u}_i(k)$. However, in the multi-rate system the manipulated variables that are injected into the system are defined as in (21.18) which includes both the computed inputs and the input updating matrix $\boldsymbol{\Psi}_i(k)$.

21.3.1 Computation

In order to solve the problem in (21.21)–(21.23) consider the predicted output vector in (21.24). In practice, the current state $\mathbf{x}_i(k)$ is usually not available from measurements and a state observer needs to be used to reconstruct the full state vector. In this case, we replace $\mathbf{x}_i(k)$ by its estimate $\widehat{\mathbf{x}}_i(k)$, hence

$$\widehat{\mathbf{Y}}_i(k) = \mathbf{F}_i\widehat{\mathbf{x}}_i(k) + \tilde{\phi}_{ii}(k)\tilde{\boldsymbol{\theta}}_i(k) + \boldsymbol{\Gamma}_i\zeta_i(k) + \sum_{\substack{j=1\\ j\neq i}}^{m} \tilde{\phi}_{ij}(k)\tilde{\boldsymbol{\theta}}_j(k). \tag{21.32}$$

If $\mathbf{Y}_{\mathrm{ref,i}}(k) = [\mathbf{y}_{\mathrm{ref,i}}(k+1)\ \mathbf{y}_{\mathrm{ref,i}}(k+2), \ldots, \mathbf{y}_{\mathrm{ref,i}}(k+N_p)]^T$, the local optimization problem for agent i can be reformulated as:

$$\min_{\tilde{\boldsymbol{\theta}}_i(k)} \ J_i(k) = \|\widehat{\mathbf{Y}}_i(k) - \mathbf{Y}_{\mathrm{ref,i}}(k)\|^2_{\mathbf{Q}_i} + \|\tilde{\boldsymbol{\theta}}_i(k)\|^2_{\mathbf{R}_i}, \tag{21.33}$$

$$\text{subject to } \ \tilde{\boldsymbol{\theta}}_{i,\min} \leq \tilde{\boldsymbol{\theta}}_i(k) \leq \tilde{\boldsymbol{\theta}}_{i,\max}, \tag{21.34}$$

where $\tilde{\boldsymbol{\theta}}_{i,\min}$ and $\tilde{\boldsymbol{\theta}}_{i,\max}$ are the lower and upper bounds for the inputs, respectively. It can be shown that problem (21.33), (21.34) is equivalent to a quadratic programming problem which can be solved efficiently and reliably using standard off-the-shelf solvers. Notice that although the global objective function can be split into several local objective functions, the output of each subsystem is still related to all the input variables due to the input coupling.

[1] Since the control horizon constraint only holds for $\Delta\mathbf{u}_i$ via $\tilde{\boldsymbol{\theta}}_i$ and not for $\mathbf{v}_i$, the assumption is made that the process noise is zero from $t + N_\mathrm{c} - 1$ on.

The Nash-based MPC algorithm proceeds by allowing each subsystem/agent to optimise its objective function using its own control decision $\tilde{\vartheta}_i(k)$ assuming that other subsystem's solutions $\tilde{\vartheta}_j(k)$ are known. Let $\tilde{\vartheta}_i^p(k)$ define the computed control input for subsystem i at iteration p, $(p \geq 0)$. At sampling time k, in the first iteration $(p = 0)$, each agent makes initial estimation of the input variables $\tilde{\vartheta}_i^p(k)$ and announces it to the other agents. Then, each agent solves its optimization problem (21.33), (21.34) simultaneously and gets its optimal solution $\tilde{\vartheta}_i^{p+1}(k)$. Next, all the agent compares the new solution with that obtained in the previous iteration and checks the convergence condition:

$$||\tilde{\vartheta}_i^{p+1}(k) - \tilde{\vartheta}_i^p(k)|| \leq \epsilon_i, \tag{21.35}$$

in which ϵ_i, (for $i = 1, 2, \ldots, m$) is the error accuracy If the Nash optimal solution is achieved, each subsystem does not change its decision $\tilde{\vartheta}_i^p(k)$ because it has achieved an equilibrium point of the coupling decision process [14]; otherwise the local cost function $J_i(k)$ will degrade.

In the following section, a novel distributed Kalman Filter algorithm is proposed to provide optimal estimation $\widehat{\mathbf{x}}_i(k)$ of the state vector $\mathbf{x}_i(k)$ while compensating for the inter-sampling information loss due to the multi-rate nature of the systems under study.

21.4 Distributed Multi-Rate Kalman Filter

Consider the linear model in (21.19), (21.20). It is required to use the available measurements $\mathbf{y}$ to estimate the state of the system $\mathbf{x}$. To understand the distributed Kalman Filter (DMKF) equations, let us consider the process noise $\mathbf{v}_i(k)$ to be a discrete-time white noise for each subsystem i. The following covariance matrix for each agent can hence be defined:

$$\mathbb{E}\{\mathbf{v}_i(k)\mathbf{v}_i^T(k)\} = S_{p_i}(k) \tag{21.36}$$

where $\mathbb{E}[\cdot]$ denotes the expectation of the argument and $S_{p_i}(k)$ represents the covariance matrix of the process noise. Consider the measurement noise $z_i(k)$ in (21.20) to be a discrete-time white noise. The following covariance matrix for the measurement noise, $S_{m_i}(k)$, can be similarly defined:

$$\mathbb{E}\{z_i(k)z_i^T(k)\} = S_{m_i}(k). \tag{21.37}$$

Let the estimated states by the distributed Kalman Filter for a multi-rate system be given by:

$$\widehat{\mathbf{x}}_i(k+1|k) = \mathbf{A}_i\widehat{\mathbf{x}}_i(k|k-1) + \mathbf{B}_{ii}\Delta\mathbf{u}_i(k) + \mathbf{L}_i(k)[\varphi_i(k) - \boldsymbol{\Upsilon}_i(k)\mathbf{C}_i\widehat{\mathbf{x}}_i(k|k-1)]$$
$$+ \sum_{\substack{j=1\\ j\neq i}}^{m}\big[\mathbf{B}_{ij}\Delta\mathbf{u}_j(k) + \mathbf{L}_j(k)[\varphi_j(k) - \boldsymbol{\Upsilon}_j(k)\mathbf{C}_j\widehat{\mathbf{x}}_j(k|k-1)]\big], \tag{21.38}$$

where the terms $\mathbf{L}_i(k)$ and $\mathbf{L}_j(k)$ are referred as the *Kalman Gains*. From (21.38), it is clear that local estimators share their gains and also estimated states to accomplish their estimation task. Substituting (21.20) into (21.38) and combining with (21.19) yields the next step in obtaining the estimation error $\mathbf{e}_i(k+1|k) = \mathbf{x}_i(k+1|k) - \widehat{\mathbf{x}}_i(k+1|k)$ at sampling time k. The index $(k|k-1)$ refers to the information at sampling time k given knowledge of the process prior to sampling time k. Therefore,

$$\mathbf{e}_i(k+1|k) = \mathbf{x}_i(k+1|k) - \widehat{\mathbf{x}}_i(k+1|k)$$
$$= [\mathbf{A}_i - \mathbf{L}_i(k)\boldsymbol{\Upsilon}_i(k)\mathbf{C}_i]\mathbf{e}_i(k|k-1) + \mathbf{D}_i\mathbf{v}_i(k) - \mathbf{L}_i(k)\boldsymbol{\Upsilon}_i(k)\mathbf{z}_i(k)$$
$$- \sum_{\substack{j=1\\ j\neq i}}^{m}\big(\mathbf{L}_j(k)\boldsymbol{\Upsilon}_j(k)\mathbf{C}_j\mathbf{e}_j(k|k-1) + \mathbf{L}_j(k)\boldsymbol{\Upsilon}_j(k)\mathbf{z}_j(k)\big). \tag{21.39}$$

To initialize the estimator algorithm, consider $\mathbb{E}[\mathbf{x}_i(0|-1)] = \widehat{\mathbf{x}}_i(0|-1)$ then $\mathbb{E}[\mathbf{e}_i(k|k-1)] = \mathbf{0},\ \forall k$. It is assumed that the mean of the estimates should be equal to the mean of the expected value in Kalman Filter design. In order to develop the Kalman Filter for the multi-rate and distributed case a covariance matrix, $\mathbf{S}_i(k)$, is defined where,

$$\mathbf{S}_i(k+1) = \mathbb{E}\{\mathbf{e}_i(k+1|k)\mathbf{e}_i^T(k+1|k)\}$$
$$= \mathbf{A}_i\mathbf{S}_i(k)\mathbf{A}_i^T + \mathbf{D}_i\,\mathbf{S}_{p_i}(k)\,\mathbf{D}_i^T - \mathbf{A}_i\mathbf{S}_i(k)\mathbf{C}_i^T\boldsymbol{\Upsilon}_i(k)\Omega_i^{-1}(k)\boldsymbol{\Upsilon}_i(k)\mathbf{C}_i\mathbf{S}_i(k)\mathbf{A}_i^T$$
$$+ \sum_{\substack{j=1\\ j\neq i}}^{m}\mathbf{A}_j\mathbf{S}_i(k)\mathbf{C}_j^T\boldsymbol{\Upsilon}_j(k)\Omega_j^{-1}(k)\,\boldsymbol{\Upsilon}_j(k)\mathbf{C}_j\mathbf{S}_j(k)\mathbf{A}_j^T, \tag{21.40}$$

with $\Omega_i(k)$ and $\Omega_j(k)$ positive definite and defined as:

$$\Omega_i(k) = \boldsymbol{\Upsilon}_i(k)\mathbf{C}_i\mathbf{S}_i(k)\mathbf{C}_i^T\boldsymbol{\Upsilon}_i(k) + \boldsymbol{\Upsilon}_i(k)\mathbf{S}_{m_i}(k)\boldsymbol{\Upsilon}_i(k) + [\mathbf{I}_{q\times q} - \boldsymbol{\Upsilon}_i(k)] \tag{21.41}$$

$$\Omega_j(k) = \boldsymbol{\Upsilon}_j(k)\mathbf{C}_j\mathbf{S}_j(k)\mathbf{C}_j^T\boldsymbol{\Upsilon}_j(k) + \boldsymbol{\Upsilon}_j(k)\mathbf{S}_{m_j}(k)\boldsymbol{\Upsilon}_j(k) + [\mathbf{I}_{q\times q} - \boldsymbol{\Upsilon}_j(k)]. \tag{21.42}$$

It should be noted that (21.40) is the algebraic Riccati equation. The solution of the Ricatti equation is found iteratively backwards in time by using (21.41) and (21.42). Then, the Kalman gains are computed as:

Algorithm 21.1 Nash-based distributed MPC and distributed Kalman filter algorithm for multi-rate systems

1: $k \leftarrow 0, p \leftarrow 0, f \leftarrow 0, flag \leftarrow 0$
2: $\theta_i^{\mathrm{p}}(k) = [\theta_i^{\mathrm{p}}(k)\theta_i^{\mathrm{p}}(k+1)\ldots\theta_i^{\mathrm{p}}(k+N_{\mathrm{c}}-1)]^{\mathrm{T}}$
3: while $flag = 0$ do
  while $f = 0$ do
  $\theta_i^*(k) \leftarrow \min_{\theta_i(k),\ldots,\theta_i(k+N_{\mathrm{c}}-1)}$ $J_i(k) = \sum_{n=1}^{N_{\mathrm{p}}} \|\mathbf{y}_i(k+n) - \mathbf{y}_{\mathrm{ref,i}}(k+n)\|_{\mathbf{Q}_i}^2 + \sum_{n=0}^{N_{\mathrm{c}}-1}$ $\|\theta_i(k+n)\|_{\mathbf{R}_i}^2$
4: if $\left\|\theta_i^{\mathrm{p}+1}(k) - \theta_i^{\mathrm{p}}(k)\right\| \le \epsilon$ then
  $f \leftarrow 1$
  end if
5: $\theta_i^{\mathrm{p}}(k) \leftarrow \theta_i^*(k)$
6: $p \leftarrow p+1$
  end while
7: $\tilde{\boldsymbol{\theta}}_i(k) \leftarrow [I0\cdots0]\,\theta_i^*(k)$
8: Compute $\varphi_i(k)$
9: $\widehat{\mathbf{x}}_i(k+1) \leftarrow DMKF(\mathbf{S}_i(k))$
10: $k \leftarrow k+1$
11: if $k = k_{\mathrm{end}}$
  $flag \leftarrow 1$
  end if
  end while

$$\mathbf{L}_i(k) = \mathbf{A}_i\mathbf{S}_i(k)\mathbf{C}_i^T\boldsymbol{\Upsilon}_i(k)\Omega_i^{-1}(k), \tag{21.43}$$

$$\mathbf{L}_j(k) = \mathbf{A}_j\mathbf{S}_i(k)\mathbf{C}_j^T\boldsymbol{\Upsilon}_j(k)\Omega_j^{-1}(k). \tag{21.44}$$

In order to guarantee the non-singularity of $\Omega_i(k)$ and $\Omega_j(k)$ at any time instant the extra terms $\left[\mathbf{I}_{q\times q} - \boldsymbol{\Upsilon}_i(k)\right]$ and $\left[\mathbf{I}_{q\times q} - \boldsymbol{\Upsilon}_j(k)\right]$ have been added to (21.41) and (21.42), respectively, in which $\mathbf{I}_{q\times q}$ is the identity matrix of size q, [5]. The matrix $\left[\mathbf{I}_{q\times q} - \boldsymbol{\Upsilon}(k)\right]$ only adds non-zero terms to the scalar diagonal elements of $\Omega_i(k)$ and $\Omega_j(k)$ during the output sampling mechanism and in no way affects the predictor equation.

The proposed Nash-based distributed MPC and distributed Kalman filter algorithm is summarized in Algorithm 21.1.

21.5 Theoretical Properties

In the distributed multi-rate system, considered in this chapter, the subsystems are coupled via inputs. The presented method can also be applied to a distributed multi-rate system, in which subsystems are coupled through their states [15].

The proposed framework consists of two main parts, control and estimation. In the control part, a distributed MPC via a Nash game has been studied for multi-rate sampled-data systems and in the estimation part a distributed KF has been proposed to provide the state values for inter-sampling times. The algorithm provides a reliable control and estimation and compensation mechanism for the information loss due to the multi-rate nature of the systems using the proposed distributed KF.

The proposed method has been applied to different case studies inorder to demonstrate the effectiveness of the method (see [15, 16]). Moreover, several simulation scenarios including slow and fast input sampling as well as slow and fast output sampling in both synchronous and asynchronous arrangements have been considered showing feasibility and high effectiveness. By analysing the simulation results, it can be observed that the proposed scheme offers efficient tracking for constrained problems including both process and measurement noise. For further theoretical properties, such as convergence, computation cost comparison and optimality see [15] and [17, 18].

In the presented method, each agent has knowledge of its own dynamics and also is aware of the neighboring agents' computed inputs. The presented method uses a communication-based optimization based on a Nash Equilibrium, which is non-cooperative and could be unstable [8]. The best achievable performance is characterized by a Pareto set, which represents the set of optimal trade-offs among the competing controller objectives. To cover this drawback more research needs to be done on cooperative MPC algorithms for multi-rate systems.

21.6 Conclusions

Design and synthesis of multi-agents in large scale plants with multi-rateness in a distributed MPC framework has been the focus of this chapter. In a multi-rate system, the output sampling and input updating mechanisms are non-uniform and/or non-synchronous. A distributed method was proposed to control such distributed multi-rate systems based on communicative Nash game theory. In the presented method, each agent has knowledge of its own transient and coupled to inputs of the neighboring agent. These linear dynamic plants with their agents reach a co-operative consensus for their control moves.

Further research is also needed to develop an efficient distributed multi-rate control and estimation approaches that are robust to system parameter variation and model uncertainty.

Acknowledgments This research is supported by the Irish Programme for Research in Third Level Institutions (Cycle 4) (funded under the National Development Plan 2007-2013 with assistance from the European Regional Development Fund) and the VENI project "Intelligent multi-agent control for flexible coordination of transport hubs" (project 11210) of the Dutch Technology Foundation STW, a subdivision of The Netherlands Organisation for Scientific Research (NWO).

References

1. J.H. Lee, M.S. Gelormino, M. Morari, Model predictive control of multi-rate sampled-data systems: a state-space approach. Int. J. Control **55**(1), 153–191 (1992)
2. M. Embiruçu, C. Fontes, Multirate multivariable generalized predictive control and its application to a slurry reactor for ethylene polymerization. Chem. Eng. Sci **61**(17), 5754–5767 (2006)
3. R.S. Gopinath, B.W. Bequettelt, R.J. Roy, H. Kaufman, Multirate MPC design for a nonlinear drug infusion system, in *Proceedings of the American Control Conference*, pp. 102–106 (Baltimore, Maryland, June–July 1994)
4. M. Ohshima, I. Hashimoto, H. Ohno, M. Takeda, T. Yoneyama, F. Gotoh, Multirate multivariable model predictive control and its application to a polymerization reactor. Int. J. Control **59**(3), 731–742 (1994)
5. R. Scattolini, N. Schiavoni, A multi-rate model based predictive control. IEEE Trans. Autom. Control **40**(6), 1093–1097 (June 1995)
6. R.R. Negenborn, B. De Schutter, J. Hellendoorn, Multi-agent model predictive control for transportation networks: Serial versus parallel schemes. Eng. Appl. Artif. Intell. **21**(3), 353–366 (2008)
7. R.R. Negenborn, Z. Lukszo, H. Hellendoorn, *Intelligent Infrastructures* (Springer, Dordrecht, 2010)
8. A.N. Venkat, I.A. Hiskens, J.B. Rawlings, S.J. Wright, Distributed MPC strategies with application to power system automatic generation control. IEEE Trans. Control Syst. Technol. **16**(6), 1192–1206 (2008)
9. S. Wang, E.J. Davison, On the stabilization of decentralized control systems. IEEE Trans. Autom. Control **18**(5):473–478 (1973)
10. V.D. Blondel, J.N. Tsitsiklis, A survey of computational complexity results in systems and control. Automatica **36**(9):1249–1274 (2000)
11. A.N. Venkat, J.B. Rawlings, S.J. Wright, Stability and optimality of distributed model predictive control, in *44th IEEE Conference on Decision and Control, and the European Control Conference* (Seville, Spain, 12–15 December 2005)
12. W. Al-Gherwi, H. Budman, A. Elkamel, Selection of control structures for distributed model predictive control in the presence of model errors. J. Process Control **20**(3), 270–284 (2010)
13. S. Li, Y. Zhang, Q. Zhu, Nash-optimization enhanced distributed model predictive control applied to the Shell benchmark problem. Inf. Sci. **170**(2–4), 329–349 (2005)
14. L. Giovanini, J. Balderud, Game approach to distributed model predictive control, in *Proceedings of the International Control Conference* (Glasgow, UK, 2006)
15. S. Roshany-Yamchi, M. Cychowski, R.R. Negenborn, B. De Schutter, K. Delaney, J. Connell, Kalman filter-based distributed predictive control of large-scale multi-rate systems: Application to power networks. IEEE Trans. Control Syst. Technol. **99**:1–13 (2011)
16. S. Roshany-Yamchi, R. R. Negenborn, M. Cychowski, B. De Schutter, J. Connell, K. Delaney, Distributed model predictive control and estimation of large-scale multi-rate systems, in *Proceedings of the 18th IFAC World Congress* (Milano, Italy, August 28–September 2 2011)
17. S. Roshany-Yamchi, A Distributed Model Predictive Control For Multi-Agent Multi-Rate Systems, Ph.D. thesis, Cork Institute of Technology, Cork, Ireland, 2012
18. S. Roshany-Yamchi, A.A. Cornelio, K. Delaney, J. Connell, An application of nash game to distributed multi-rate predictive control, in *Proceedings of the 14th IASTED International Conference on Control and Applications* (Crete, Greece, 18–20 June 2012)

Part II
From Large-Scale to Small-Scale: The Decomposed Monolithic System Perspective

Chapter 22
Cooperative Dynamic MPC for Networked Control Systems

I. Jurado, D. E. Quevedo, K. H. Johansson and A. Ahlén

Abstract This work studies cooperative MPC for Networked Control Systems with multiple wireless nodes. Communication between nodes is affected by random packet dropouts. An algorithm is presented to decide at each time instant which nodes will calculate the control input and which will only relay data. The nodes chosen to calculate the control values solve a cooperative MPC by communicating with their neighbors. This algorithm makes the control architecture flexible by adapting it to the possible changes in the network conditions.

22.1 Introduction

Networked Control Systems (NCSs) are systems in which practical communication links are used to exchange system information and control signals between various components of the system that may be physically distributed. Major advantages of

I. Jurado(✉)
Departamento de Ingeniería de Sistemas y Automática, Escuela Superior de Ingenieros, Universidad de Sevilla, Sevilla, Spain
e-mail: ijurado@cartuja.us.es

D. E. Quevedo
School of Electrical Engineering and Computer Science, The University of Newcastle, Newcastle, Australia
e-mail: dquevedo@ieee.org

K. H. Johansson
ACCESS Linnaeus Centre, School of Electrical Engineering, Royal Institute of Technology, Stockholm, Sweden
e-mail: kallej@ee.kth.se

A. Ahlén
Department of Engineering Sciences, Signals and Systems, Uppsala University, Uppsala, Sweden
e-mail: Anders.Ahlen@signal.uu.se

J. M. Maestre and R. R. Negenborn (eds.), *Distributed Model Predictive Control Made Easy*, Intelligent Systems, Control and Automation: Science and Engineering 69, DOI: 10.1007/978-94-007-7006-5_22,

NCSs include low cost, reduced weight and power requirements, simple installation and maintenance, and high reliability. Nonetheless, closing a control loop on a shared communication network introduces additional dynamics and constraints in the control problem. In addition to being bit-rate limited [1, 2], practical communitacion channels are commonly affected by packet dropouts and time delays, mainly due to transmission errors and waiting times to access the medium; see, e.g., [3–8].

This chapter studies NCS in which the transmissions are affected by random packet dropouts. The network is composed of a certain number of nodes forming a matrix strucure. These nodes follow an algorithm, that decides which node will calculate the control input. This node will solve a cooperative MPC communicating with its neighbors. Each node knows a part of the whole system model and it shares its information with a group of neighbor nodes, so they cooperate in order to exchange their information about the system. At each sampling time, we have a different group of nodes chosen to calculate the control signal. This group of nodes will be chosen depending on the particular network outcomes for that sampling time. The present work extends our recent conference contribution [9] to encompass NCSs with parallel links and the use of cooperative MPC. The idea is motivated by the fact that the link transmission outcomes may change at each sampling instant, so one particular node is not always the best suited to perform the control calculation.

In the network under consideration in this work, the only node that receives the state of the plant without any dropouts is the sensor node, which is located next to the plant. The actuator node is directly connected to the plant input, therefore this data is received without problems. The actuator node is also the only node that provides transmission acknowledgments. We assume as well that there is an array

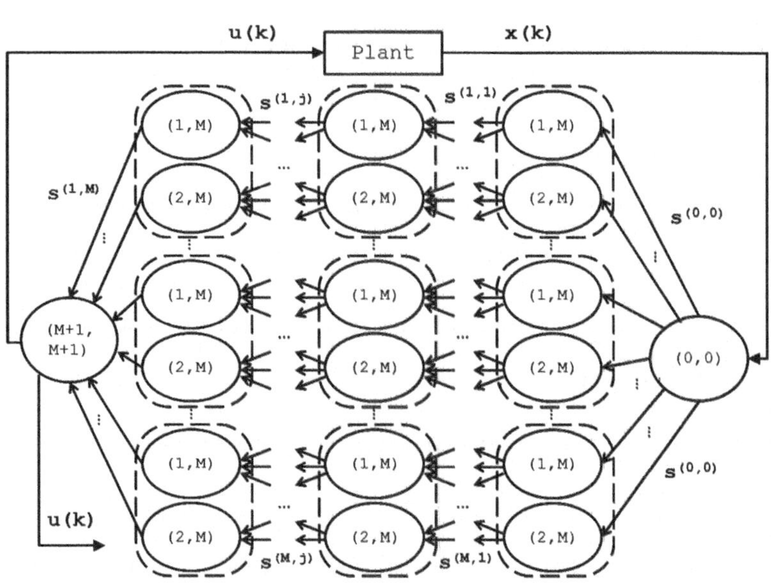

Fig. 22.1 Control over a graph with dropouts and unreliable acknowledgments of actuator values

of nodes between the sensor and the actuator nodes, as shown in Fig. 22.1. Each node in a column sends the information to the three closest nodes in the following column. We will assume that the nodes, except the sensor and the actuator nodes, can communicate with some of its neighbors in the same column, and may thereby cooperate and exchange information. The sensor and actuator nodes can't calculate control values; they can only transmit information. The communication between nodes is limited to a maximum number of iterations, and subject to dropouts.

We are supposing that the model of the plant is divided into a certain number of incomplete subsystems. Each node will know only a part of the model of the system, that is why it has to collaborate with its neighbors, which know the other parts of the system. Therefore, each node will estimate just a part of the state.

The control policy to be used will be a cooperative MPC. Within this context, we present a flexible NCS architecture where the role played by cooperative nodes depends upon transmission outcomes and their acknowledgments. With the algorithm proposed, transmission outcomes and their acknowledgments will determine, at each time instant, whether the control input will be calculated at the actuator node, or closer to the sensor node.

Therefore, the distinguishing feature of this approach is the dynamic architecture of the controlled system, plus the fact that we are using a matrix of nodes which do not know the whole information of the plant model. The algorithm seeks to find the best group of nodes to calculate the control, depending on the network transmission outcomes.

The remainder of this chapter is organized as follows. Section 22.2 describes the NCS topology of interest. In Sect. 22.3.1 we present the control law. Section 22.3.2 provides the state estimation. Section 22.3.3 presents the algorithm that chooses the best node to calculate the control action. In Sect. 22.4 an example illustrating the control algorithm is presented and in Sect. 22.5 conclusions are drawn.

Notation: We write $\mathbb{N}$ for $\{1, 2, \ldots\}$, and $\mathbb{N}_0$ for $\{0, 1, 2, \ldots\}$; $\mathbb{R}$ are the real numbers, whereas $\mathbb{R}_{\geq 0} \triangleq [0, \infty)$. The trace of a matrix $\mathbf{A}$ is denoted by $\mathbf{tr}(\mathbf{A})$. If a matrix $\mathbf{Q}$ is positive definite, then we write $\mathbf{Q} > 0$. We adopt the convention $\sum_{j=1}^{0} a_j = 0$, for all $a_0, a_1 \in \mathbb{R}$. The expectation of a random variable μ is denoted by $\mathbf{E}\{\mu\}$. A real random variable μ, which is zero-mean Gaussian with covariance Γ is denoted by $\mu \sim \mathcal{N}(0, \Gamma)$.

22.2 Considerations About the System Setup Control Objectives and the Network Constraints

We consider MIMO LTI plant models of the form

$$\mathbf{x}(k+1) = \mathbf{A}\mathbf{x}(k) + \mathbf{B}\mathbf{u}(k) + \mathbf{d}(k) \tag{22.1}$$

where $\mathbf{x}(0) \sim \mathcal{N}(0, \mathbf{P}_0), \mathbf{P}_0 > 0$. In (22.1), $\mathbf{u}(k) \in \mathbb{R}^{n_u}$ is the plant input, $\mathbf{x}(k) \in \mathbb{R}^{n_x}$ is the state, and $\mathbf{d}(k) \sim \mathcal{N}(0, \mathbf{D}), \mathbf{D} > 0$ is driving noise.

The model described in (22.1) represents the whole plant. But, as foreshadowed in the introduction, individual nodes do not have knowlegde of this whole model. Thus, nodes have to interact with their neighbors to get all the information about the plant. We are considering that, between a certain number of nodes, they have all the information about the plant model. Figure 22.1 shows a particular situation in which the whole information is shared by two nodes.

22.2.1 The Composite Model

For each node, the *composite model (CM)* [10], is the combination of the decentralized model and all the interaction models. In order to make the problem formulation simpler, we will consider the particular case shown in Fig. 22.1, in wich the cooperation is done between two nodes in the same column, i.e., we have one interation model. The decentralized state vector in node (i, j), $\mathbf{x}_{(i,j)}$, is augmented with the state from the neighbor node (i^*, j).

Therefore, the augmented state $\mathbf{x}_{(i,j)} = [\mathbf{x}^T_{(i,j)(i,j)}, \mathbf{x}^T_{(i,j)(i^*,j)}]^T$ represents the CM states for the node (i, j), (i^*, j) being the neighbor node interacting, pairwise, with (i, j), which is in the same column and $i^* \in \{i-1, i+1\}$. In this augmented state, $x_{(i,j)(i^*,j)}$ is the influence of the node (i^*, j) on the node (i, j), and $x_{(i,j)(i,j)}$ is the part of the state that take into account just the part of the model that the node (i, j) knows, so it is a decentralized state. In this case, the CM for the node (i, j) is written as

$$\mathbf{x}_{(i,j)}(k+1) = \mathbf{A}_{(i,j)}\mathbf{x}_{(i,j)}(k) + \mathbf{B}_{(i,j)}\mathbf{u}_{(i,j)}(k) + \mathbf{W}_{(i,j)(i^*,j)}\mathbf{u}_{(i^*,j)}(k)$$

where

$$\mathbf{A}_{(i,j)} = \begin{bmatrix} \mathbf{A}_{(i,j)(i,j)} & \\ & \mathbf{A}_{(i,j)(i^*,j)} \end{bmatrix}, \quad \mathbf{B}_{(i,j)} = \begin{bmatrix} \mathbf{B}_{(i,j)(i,j)} \\ \mathbf{0} \end{bmatrix}, \quad \mathbf{W}_{(i,j)(i^*,j)} = \begin{bmatrix} \mathbf{0} \\ \mathbf{B}_{(i,j)(i^*,j)} \end{bmatrix}.$$

Since

$$\mathbf{A}_{(i,j)(i-1,j)} = \mathbf{A}_{(i,j)(i+1,j)} \quad and \quad \mathbf{B}_{(i,j)(i-1,j)} = \mathbf{B}_{(i,j)(i+1,j)},$$

$\mathbf{A}_{(i,j)}, \mathbf{B}_{(i,j)}$ and $\mathbf{W}_{(i,j)(i^*,j)}$ do not depend on the value of i^*. Therefore, two neighbor cooperating nodes have available the entire information about the system model. Therefore, the node (i, j) has one part of the model, and the neighbor nodes $(i-1, j)$ and $(i+1, j)$ have the other part of the model.

22.2.2 Control Problem

The augmented control signal to calculate is denoted as

$$\mathbf{u}_{(i,j)}^{(i^*,j)}(k) = [\mathbf{u}_{(i,j)}(k)^T, \mathbf{u}_{(i^*,j)}(k)^T]^T, \quad k \in \mathbb{N}_0, \tag{22.2}$$

and will be calculated with MPC techniques employing pair-wise cooperation among the neighboring nodes. Thus, the control problem solved for this kind of systems is

$$\mathbf{u}_{(i,j)}^{(i^*,j)}(k) = \text{Cooperative MPC}\left(\mathbf{x}_{(i,j)}(k), \mathbf{x}_{(i^*,j)}(k)\right), \quad k \in \mathbb{N}_0, \tag{22.3}$$

where $\mathbf{x}_{(i,j)}(k)$ and $\mathbf{x}_{(i^*,j)}(k)$ represent the CM states for the nodes (i, j) and (i^*, j), respectively.

22.2.3 Network Issues

Sensor and actuator nodes are connected via a wireless network, characterised via a graph having $M \times M + 2$ nodes, see Fig. 22.1. Control values cannot be calculated by the sensor nor the actuator nodes, they are just used to measure the plant state and apply the control signal, respectively. Therefore, according to Fig. 22.1, the network has $M \times M$ nodes that could act as the controller. Transmissions are done in a sequential manner as shown in Fig. 22.2. More precisely, the packet $\mathbf{s}^{(i,j)}(k)$ is transmitted from node (i, j) to its closest neighbors[1] at times $kT + j\tau$, where T

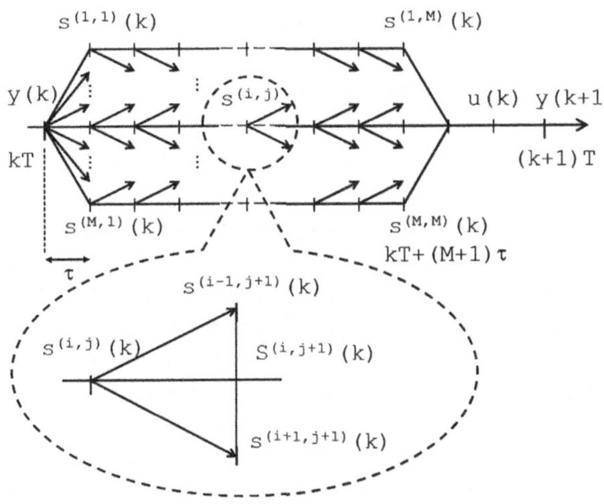

Fig. 22.2 Transmission schedule; $t \in \mathbb{R}_{\geq 0}$ is actual time

[1] These are $(i + 1, j + 1)$, $(i, j + 1)$ and $(i - 1, j + 1)$.

is the sampling period of (22.1) and $\tau \ll T/(M+1)$ refers to the times between transmissions of packets. The plant input $\mathbf{u}(k)$ is applied at time $kT + (M+1)\tau$. We, thus assume that in-network processing is much faster than the plant dynamics (22.1) and, as in, e.g., [11], neglects delays introduced by the network.

A distinguishing characteristic of the situation at hand is that (due to channel fading) the network introduces packet dropouts. To study the situation, we adopt an analog erasure channel model and introduce the binary success random processes

$$\gamma_{(i,j)}^{(i,j-1)}(k) \in \{0, 1\}, \quad k \in \mathbb{N}_0, \ i \in \{0, 1, 2, \dots, M+1\}, j \in \{0, 1, 2, \dots, M+1\}$$

where $\gamma_{(i,j)}^{(i,j-1)}(k) = 1$ indicates that transmission of the packet $\mathbf{s}^{(i,j-1)}(k)$ from node $(i, j-1)$ to node (i, j) at time $kT + (j+1)\tau$, is successful, i.e., error-free; $\gamma_{(i,j)}^{(i,j-1)}(k) = 0$ refers to a packet-dropout. Throughout this work we assume that the sensor node $(0, 0)$ has direct access to plant output measurements. For notational convenience, we write $\gamma^{(0,0)}(k) = 1$, for all $k \in \mathbb{N}_0$.

To save energy, in our formulation the wireless nodes (i, j), where

$$i, j \in \{0, 1, 2, \dots, M\},$$

do not provide acknowledgments of receipt of the packets. However, the actuator node, $(M+1, M+1)$, will in general have less stringent energy constraints, so at time $kT + (M+1)\tau$ the control signal is received, at $kT + (M+2)\tau$ this control value is applied and at time $kT + (M+3)\tau$, the actuator broadcasts the control value applied, namely $\mathbf{u}(k) = [\mathbf{u}_{(i,j)}^A(k)^T, \mathbf{u}_{(i^*,j)}^A(k)^T]^T$, back to the wireless nodes (i, j), see Fig. 22.1. This acknowledgment-like signal is unreliable and affected by dropouts with associated success processes

$$\delta^{(i,j)}(k) \in \{0, 1\}, \quad k \in \mathbb{N}_0, \ i, j \in \{0, 1, 2, \dots, M\}.$$

More precisely, if $\mathbf{u}(k)$ is successfully received at node (i, j), then we set $\delta^{(i,j)}(k) = 1$; see also [12] and [13] for studies on the importance of acknowledgments in closed loop control. We assume that the actuator node has perfect knowledge of plant inputs, and thus, write $\delta^{(M+1,M+1)}(k) = 1, \forall k \in \mathbb{N}_0$.

Due to packet dropouts, plant state measurements are not always available at the actuator node. On the other hand, the sensor node will, in general, not have perfect information of previous plant inputs. This makes the implementation of (22.3) via estimated state feedback a challenging task. The main purpose of the present work is to decide which nodes of the network (with the exception of the sensor and actuator nodes) should use their local state estimates to implement the control law (22.3), that is, which node will play the role of the controller and which ones only relay the received information. We foresee that our approach will lead to a dynamic assignment of the role played by the individual network nodes. Which tasks are carried out by each node at each time instant, will depend upon transmission outcomes, i.e., on $\gamma_{(i,j)}^{(i,j-1)}(k)$ and $\delta^{(i,j)}(k)$.

22.2.4 Dynamic Ccontroller Placement

The packets transmitted by each node (i, j) have three fields, namely, state measurements, tentative plant inputs (if available) and the value of the objective function under consideration:

$$\mathbf{s}^{(i,j)}(k) = \big(\mathbf{x}(k), \mathbf{u}^{(\alpha^*,\beta)}_{(\alpha,\beta)}(k), J(k)\big),$$
$$\alpha \in \{1, \ldots, i\}, \alpha^* \in \{\alpha - 1, \alpha + 1\}, \beta \in \{1, \ldots, j\}. \qquad (22.4)$$

The plant states $\mathbf{x}(k)$ includes the two components corresponding to the cooperation nodes, that is $\mathbf{x}(k) = [\mathbf{x}_{(i,j)}(k)^T, \mathbf{x}_{(i^*,j)}(k)^T]$, with $i^* \in \{i - 1, i + 1\}$.

The control signal $\mathbf{u}(k)^{(\alpha^*,\beta)}_{(\alpha,\beta)}$ in (22.4) with the structure shown in (22.2), is the plant input which is applied at the plant provided the packet $\mathbf{s}^{(i,j)}(k)$ is delivered at the actuator node. If $\mathbf{s}^{(i,j)}(k)$ is lost, then following Algorithm 22.2, which will be described in Sect. 22.3.3, the plant input will be provided by one of the nodes in subsequent columns, see Fig. 22.1, which thereby takes on the controller role at time k. For further reference, we will refer to the node which calculates the plant input at time k as

$$c(k) \in \{1, 2, \ldots, M\}^2.$$

22.2.5 Example

Consider the network in Fig. 22.3. Some nodes in the network only have one part of the plant model and the other nodes will have the other part. Therefore, nodes (1, 1), (3, 1), (1, 2), (3, 2), (1, 3) and (3, 3) will have the same information about the plant, that is $\mathbf{A}_{(i,j)} = \mathbf{A}_{(1,1)}$, $\mathbf{B}_{(i,j)} = \mathbf{B}_{(1,1)}$ for all of them, and $\mathbf{W}_{(1,1)(2,1)} = \mathbf{W}_{(3,1)(2,1)} = \mathbf{W}_{(1,2)(2,2)} = \mathbf{W}_{(3,2)(2,2)} = \mathbf{W}_{(1,3)(2,3)} = \mathbf{W}_{(3,3)(2,3)}$.

On the other hand, the rest of the nodes will have the other part of the information about the plant: $\mathbf{A}_{(2,1)} = \mathbf{A}_{(2,2)} = \mathbf{A}_{(2,3)}$, $\mathbf{B}_{(2,1)} = \mathbf{B}_{(2,2)} = \mathbf{B}_{(2,3)}$ and $\mathbf{W}_{(2,1)(1,1)} = \mathbf{W}_{(2,1)(3,1)} = \mathbf{W}_{(2,2)(1,2)} = \mathbf{W}_{(2,2)(3,2)} = \mathbf{W}_{(2,3)(1,3)} = \mathbf{W}_{(2,3)(3,3)}$.

Moreover, the cooperating couples are: $(1, 1) \leftrightarrow (2, 1)$, $(1, 2) \leftrightarrow (2, 2)$, $(1, 3) \leftrightarrow (2, 3)$, $(3, 1) \leftrightarrow (2, 1)$, $(3, 2) \leftrightarrow (2, 2)$, and $(3, 3) \leftrightarrow (2, 3)$.

So, it is easy to see that:

$$\mathbf{A}_{(1,1)(1,1)} = \mathbf{A}_{(1,2)(1,2)} = \mathbf{A}_{(1,3)(1,3)} = \mathbf{A}_{(3,1)(3,1)} = \mathbf{A}_{(3,2)(3,2)} = \mathbf{A}_{(3,3)(3,3)},$$

$$\mathbf{B}_{(1,1)(1,1)} = \mathbf{B}_{(1,2)(1,2)} = \mathbf{B}_{(1,3)(1,3)} = \mathbf{B}_{(3,1)(3,1)} = \mathbf{B}_{(3,2)(3,2)} = \mathbf{B}_{(3,3)(3,3)},$$

$$\mathbf{A}_{(2,1)(2,1)} = \mathbf{A}_{(2,2)(2,2)} = \mathbf{A}_{(2,3)(2,3)}, \quad \mathbf{B}_{(2,1)(2,1)} = \mathbf{B}_{(2,2)(2,2)} = \mathbf{B}_{(2,3)(2,3)},$$

$$\mathbf{A}_{(1,1)(2,1)} = \mathbf{A}_{(3,1)(2,1)} = \mathbf{A}_{(1,2)(2,2)} = \mathbf{A}_{(3,2)(2,2)} = \mathbf{A}_{(1,3)(2,3)} = \mathbf{A}_{(3,3)(2,3)},$$

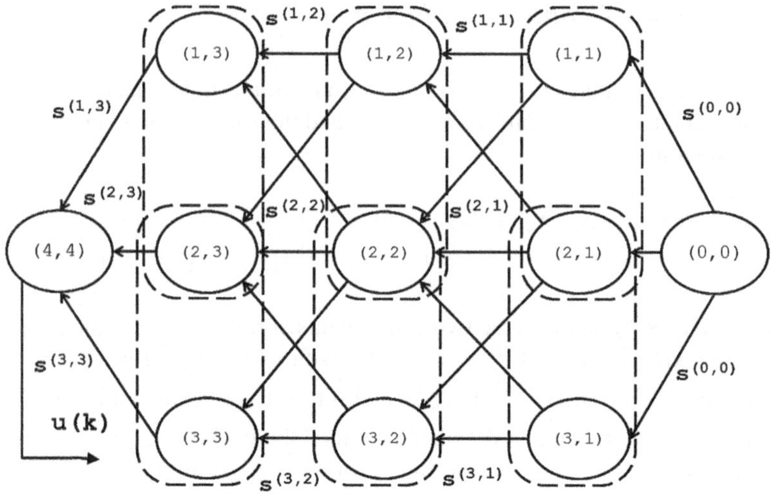

Fig. 22.3 Graph with $3 \times 3 + 2$ nodes

$$\mathbf{B}_{(1,1)(2,1)} = \mathbf{B}_{(3,1)(2,1)} = \mathbf{B}_{(1,2)(2,2)} = \mathbf{B}_{(3,2)(2,2)} = \mathbf{B}_{(1,3)(2,3)} = \mathbf{B}_{(3,3)(2,3)},$$

$$\mathbf{A}_{(2,1)(1,1)} = \mathbf{A}_{(2,1)(3,1)} = \mathbf{A}_{(2,2)(1,2)} = \mathbf{A}_{(2,2)(3,2)} = \mathbf{A}_{(2,3)(1,3)} = \mathbf{A}_{(2,3)(3,3)}$$

and

$$\mathbf{B}_{(2,1)(1,1)} = \mathbf{B}_{(2,1)(3,1)} = \mathbf{B}_{(2,2)(1,2)} = \mathbf{B}_{(2,2)(3,2)} = \mathbf{B}_{(2,3)(1,3)} = \mathbf{B}_{(2,3)(3,3)}.$$

For example, if

$$\mathbf{A}_{(1,1)} = \begin{bmatrix} \begin{bmatrix} 1 & 1 \\ 0 & 1 \end{bmatrix} & \begin{bmatrix} 0 & 0 \\ 0 & 0 \end{bmatrix} \\ \begin{bmatrix} 0 & 0 \\ 0 & 0 \end{bmatrix} & \begin{bmatrix} 0.1 & 0 \\ 0 & 0.2 \end{bmatrix} \end{bmatrix}, \quad \mathbf{A}_{(2,1)} = \begin{bmatrix} \begin{bmatrix} 1 & 1 \\ 0 & 1 \end{bmatrix} & \begin{bmatrix} 0 & 0 \\ 0 & 0 \end{bmatrix} \\ \begin{bmatrix} 0 & 0 \\ 0 & 0 \end{bmatrix} & \begin{bmatrix} 0 & 0 \\ 0 & 0.1 \end{bmatrix} \end{bmatrix},$$

$$\mathbf{B}_{(1,1)} = \begin{bmatrix} \begin{bmatrix} 0 \\ 1 \end{bmatrix} \\ \begin{bmatrix} 0 \\ 0 \end{bmatrix} \end{bmatrix}, \quad \mathbf{B}_{(2,1)} = \begin{bmatrix} \begin{bmatrix} 0 \\ 1 \end{bmatrix} \\ \begin{bmatrix} 0 \\ 0 \end{bmatrix} \end{bmatrix},$$

$$\mathbf{W}_{(1,1)(2,1)} = \begin{bmatrix} \begin{bmatrix} 0 \\ 0 \end{bmatrix} \\ \begin{bmatrix} 0 \\ 0.4 \end{bmatrix} \end{bmatrix}, \quad \mathbf{W}_{(2,1)(1,1)} = \begin{bmatrix} \begin{bmatrix} 0 \\ 0 \end{bmatrix} \\ \begin{bmatrix} 0 \\ 1 \end{bmatrix} \end{bmatrix}.$$

then, the decentralized models are

$$\mathbf{A}_{(1,1)(1,1)} = \mathbf{A}_{(2,1)(2,1)} = \begin{bmatrix} 1 & 1 \\ 0 & 1 \end{bmatrix}, \mathbf{B}_{(1,1)(1,1)} = \mathbf{B}_{(2,1)(2,1)} = \begin{bmatrix} 0 \\ 1 \end{bmatrix},$$

whereas the interacting models are given by:

$$\mathbf{A}_{(1,1)(2,1)} = \begin{bmatrix} 0.1 & 0 \\ 0 & 0.2 \end{bmatrix}, \quad \mathbf{A}_{(2,1)(1,1)} = \begin{bmatrix} 0 & 0 \\ 0 & 0.1 \end{bmatrix},$$

$$\mathbf{B}_{(1,1)(2,1)} = \begin{bmatrix} 0 \\ 0.4 \end{bmatrix}, \quad \mathbf{B}_{(2,1)(1,1)} = \begin{bmatrix} 0 \\ 1 \end{bmatrix}.$$

22.3 Description of the Approach

In this section we will describe the control calculation, including the algorithms that gives the nodes to cooperate in order to calculate the control action.

22.3.1 Control Implementation

To implement the control law (22.3) over the network using packets of the form (22.4), we will use communication driven cooperative MPC.

In this work we are assuming that pairs of neighbor nodes in the same column can exchange information. In Algorithm 22.1 we show the cooperation between a pair of nodes. Then, with the CM in (22.2.1) and the Algorithm 22.1, it is possible to calculate a *Feasible Cooperation-Based MPC (FC-MPC)*, as explained in [10].

The calculation of the suboptimal control input, $\mathbf{u}^{*p}_{(i,j)}$, for each iteration p, is performed by solving the FC-MPC problem. So, we will choose the objective function as a linear combination of the individual nodes' objectives, i.e.,

$$J_{(i,j)} = J_{(i^*,j)} = \varpi_{(i,j)} V_{(ij)} + \varpi_{(i^*,j)} V_{(i^*,j)},$$
$$\varpi_{(i,j)}, \varpi_{(i^*,j)} > 0, \varpi_{(i,j)} + \varpi_{(i^*,j)} = 1, i^* \in \{i-1, i+1\}$$

The local objective for each cooperative node depends on the value of

$$\gamma_{(i,j)}^{(i-1,j-1)}(k), \quad \gamma_{(i,j)}^{(i,j-1)}(k) \quad and \quad \gamma_{(i,j)}^{(i+1,j-1)}(k).$$

If at least one of them is equal to one, then we will have the following cost function:

$$J_{(i,j)} = V_{(i,j)}(\mathbf{x}'^{p}_{(i,j)}(k), \mathbf{u}'^{p}_{(i,j)}(k); \mathbf{x}_{(i,j)}(k)) = \sum_{t=k}^{\infty} \left\| \mathbf{x}^{p}_{(i,j)}(t|k) \right\|^2_{\mathbf{Q}} + \left\| \mathbf{u}^{p}_{(i,j)}(t|k) \right\|^2_{\mathbf{R}},$$

where

$$\mathbf{x}'^{p}_{(i,j)}(k) = [\mathbf{x}^{p}_{(i,j)}(k+1|k)^T, \mathbf{x}^{p}_{(i,j)}(k+2|k)^T, ...]^T,$$
$$\mathbf{u}'^{p}_{(i,j)}(k) = [\mathbf{u}^{p}_{(i,j)}(k|k)^T, \mathbf{u}^{p}_{(i,j)}(k+1|k)^T, ...]^T,$$

and $\mathbf{Q} > 0$ and $\mathbf{R} > 0$ are weighting matrices. To calculate these predictions the CM for the node (i, j) has been used, see (22.2).

The notation p indicates the iteration number. During each MPC optimization, the state and input trajectories $(\mathbf{x}'_{(i^*,j)}(k), \mathbf{u}'_{(i^*,j)}(k))$ of the interacting node MPC are not updated, so they remain at $(\mathbf{x}'^{p-1}_{(i^*,j)}(k), \mathbf{u}'^{p-1}_{(i^*,j)}(k))$.

On the other hand, if $\gamma^{(i-1,j-1)}_{(i,j)}(k) = \gamma^{(i,j-1)}_{(i,j)}(k) = \gamma^{(i+1,j-1)}_{(i,j)}(k) = 0$, no information about the state has arrived at node (i, j), so an estimation is used instead. The cost function will be an expected value [14]:

$$J_{(i,j)} = \mathbf{E}\{\sum_{t=k}^{\infty} \mathbf{x}^{p}_{(i,j)}(t|k)^T \mathbf{Q} \mathbf{x}^{p}_{(i,j)}(t|k) + \mathbf{u}^{p}_{(i,j)}(t|k)^T \mathbf{R} \mathbf{u}^{p}_{(i,j)}(t|k)\}$$

$$= \sum_{t=k}^{\infty} \hat{\mathbf{x}}^{p}_{(i,j)}(t|k)^T \mathbf{Q} \hat{\mathbf{x}}^{p}_{(i,j)}(t|k) + \mathbf{tr}\big(\mathbf{Q}\mathbf{P}_{(i,j)}(k)\big) + \mathbf{u}^{p}_{(i,j)}(t|k)^T \mathbf{R} \mathbf{u}^{p}_{(i,j)}(t|k),$$

where $\mathbf{P}_{(i,j)}(k)$ approximates the covariance of $\mathbf{x}_{(i,j)}(k)$ and is calculated as follows:

$$\mathbf{P}_{(i,j)}(k+1) = \mathbf{A}_{(i,j)} \mathbf{P}_{(i,j)}(k) \mathbf{A}^T_{(i,j)} - \Gamma^{(i,j)}(k) \mathbf{P}_{(i,j)}(k) \Gamma^{(i,j)}(k) + \mathbf{D},$$

where

$$\Gamma^{(i,j)}(k) \triangleq \prod_{\substack{\alpha \in \{0,1,\ldots,i-1\} \\ \beta \in \{0,1,\ldots,i-1\}}} \gamma^{(ig^*,jg^*)}_{(\alpha,\beta)}(k)$$

is equal to 1 if and only if $\mathbf{x}_{(i,j)}(k)$ is available at node (i, j) at time $kT + (j-1)\tau$. In the above expression, (ig^*, jg^*) is one of the preceding nodes of (i,j) and is the one that provides the best (the smallest or unique) value of $J_{(i,j)}(k)$.

Remark The objective function $J_{(i,j)}$ is an approximation of:

$$J_{(i,j)} \approx \mathbf{E}\{V_{(i,j)}(\hat{\mathbf{x}}'^{p}_{(i,j)}(k), \mathbf{u}'^{p}_{(i,j)}(k); \hat{\mathbf{x}}_{(i,j)}(k))\},$$

Algorithm 22.1: Cooperative MPC algorithm

Input: $(\bar{\mathbf{u}}^0_{(\iota,j)}, \mathbf{x}_{(\iota,j)}(k))$, $\mathbf{Q} > 0$, $\mathbf{R} > 0$, $\iota \in \{i, i^*\}$, $i^* \in \{i-1, i+1\}$, $p_{\max}(k) \geq 0$, $\epsilon > 0$, $p \leftarrow 1$, $e_{(\iota,j)} \leftarrow \Phi$ and $\Phi >> 1$.

while $e_{(\iota,j)} > \epsilon$ for some $\iota \in \{i, i^*\}$ and $p \leq p_{\max}(k)$ **do**
  $\mathbf{u}^{*p}_{(\iota,j)} \in \arg\min_{(\iota,j)} \; FC-MPC_{(\iota,j)}$
  $\mathbf{u}^{p}_{(\iota,j)} = \varpi_{(\iota,j)}\mathbf{u}^{*p}_{(\iota,j)} + (1-\varpi_{(\iota,j)})\mathbf{u}^{p-1}_{(\iota,j)}$
  $e_{(\iota,j)} = \left\| \mathbf{u}^{p}_{(\iota,j)} - \mathbf{u}^{p-1}_{(\iota,j)} \right\|$
  for each $\iota \in \{i, i^*\}$ **do**
    The node (ι, j) transmits $\mathbf{u}^{p}_{(\iota,j)}$ to its neighbor
  end for
  if $\gamma^{(i-1,j-1)}_{(i,j)}(k) = 1 \vee \gamma^{(i,j-1)}_{(i,j)}(k) = 1 \vee \gamma^{(i+1,j-1)}_{(i,j)}(k) = 1$ **then**
    $\mathbf{x}^{p}_{(\iota,j)} \leftarrow \mathbf{x}^{(\bar{\mathbf{u}}^p_{(i,j)}, \bar{\mathbf{u}}^p_{(i*,j)}; \mathbf{x}_{(\iota,j)})}_{(\iota,j)}$, $\forall \iota \in \{i, i^*\}$
  else
    $\hat{\mathbf{x}}^{p}_{(\iota,j)} \leftarrow \hat{\mathbf{x}}^{(\bar{\mathbf{u}}^p_{(i,j)}, \bar{\mathbf{u}}^p_{(i*,j)}; \hat{\mathbf{x}}_{(\iota,j)})}_{(\iota,j)}$, $\forall \iota \in \{i, i^*\}$
  end if
  $p \leftarrow p + 1$
end while

since $\mathbf{P}_{(i,j)}(k)$ is not the covariance of $\mathbf{x}_{(i,j)}(k)$, but just an approximate value. Therefore the term $\mathbf{tr}\big(\mathbf{QP}_{(i,j)}(k)\big)$ is not exact.

In the Algorithm 22.1, the state sequence generated by the input sequence $\mathbf{u}_{(i,j)}$ and initial state $\mathbf{x}_{(i,j)}$ has been represented by $\mathbf{x}^{(\mathbf{u}_{(i,j)}; \mathbf{x}_{(i,j)})}_{(i,j)}$. Also, the notation $\hat{\mathbf{x}}$ is representing $[\hat{\mathbf{x}}^T_{(i,j)}, \hat{\mathbf{x}}^T_{(i^*,j)}]^T$.

Due to the communication constraints, the maximum number of iterations $p_{\max}$ is limited. It is also possible to loose information during the cooperation. For these reasons only a suboptimal control input $\mathbf{u}^{*p}_{(i,j)}$ will be available.

Notice that $i^* \in \{i-1, i+1\}$, that means that the node (i, j) can communicate with the nodes $(i-1, j)$ and $(i+1, j)$, see Fig. 22.4. Therefore, node (i, j) will solve two cooperative MPC problems and will have two control values. The control value that the node (i, j) transmits will be the one that provides the lowest cost.

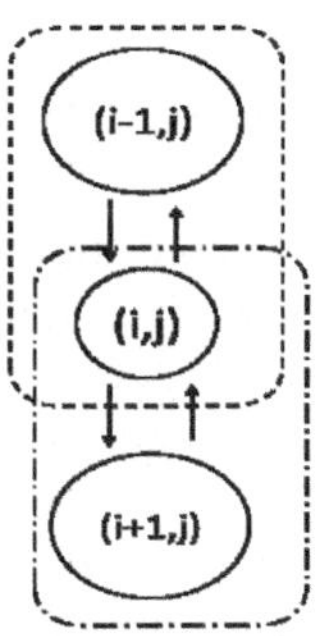

Fig. 22.4 Cooperative nodes for node (i, j)

22.3.2 State Estimation

While only the node $c(k)$ will provide the plant input at instant k, in the present formulation all nodes compute local state estimates, $\hat{\mathbf{x}}_{(i,j)}(k)$, by using the data received from one of the preceding nodes, (ig^*, jg^*). This serves as safeguard for instances when the loop is broken due to dropouts.

Since the nodes don't have full information about the plant, they are only able to calculate a part of the state. That means that $\hat{\mathbf{x}}_{(i,j)}(k)$ is not an estimate of the global state of the plant.

In the sequel, we will focus on situations where acknowledgments of plant inputs are "quite reliable". Thus, the state estimates are simply calculated as

$$\begin{aligned}\hat{\mathbf{x}}_{(i,j)}(k) &= \mathbf{A}_{(i,j)}\hat{\mathbf{x}}_{(i,j)}(k-1)\\ &\quad + \mathbf{B}_{(i,j)}\mathbf{u}_{(i,j)}(k-1) + \mathbf{W}_{(i,j)(i^*,j)}\mathbf{u}_{(i^*,j)}(k-1)\\ &\quad + \mathbf{K}^{(i,j)}(k)\big(\mathbf{x}_{(i,j)}(k) - (\mathbf{A}_{(i,j)}\hat{\mathbf{x}}_{(i,j)}(k-1) + \mathbf{B}_{(i,j)}\mathbf{u}_{(i,j)}(k-1)\\ &\quad + \mathbf{W}_{(i,j)(i^*,j)}\mathbf{u}_{(i^*,j)}(k-1))\big),\end{aligned} \tag{22.5}$$

where $\mathbf{K}^{(i,j)}(k) = \Gamma^{(i,j)}(k)\mathbf{I}$.

In (22.5), $\mathbf{u}_{(i,j)}(k-1)$ and $\mathbf{u}_{(i^*,j)}(k-1)$ are local plant input estimates. In particular, if $\delta^{(i,j)}(k-1) = 1$, then $\mathbf{u}_{(i,j)}(k-1) = \mathbf{u}^A(k-1)$ and $\mathbf{u}_{(i^*,j)}(k-1) = \mathbf{u}^A(k-1)$, where $\mathbf{u}^A(k-1)$ is the applied control signal in time instant $(k-1)$ by the actuator. On the other hand, at instances where $\delta^{(i,j)}(k-1) = 0$, node (i, j) uses the tentative plant input value transmitted in the second field of the previous packet $\mathbf{s}^{(i,j)}(k-1)$ (if non-empty), or otherwise sets $\mathbf{u}_{(i,j)}(k-1)$ and $\mathbf{u}_{(i^*,j)}(k-1)$ as per (22.3), see Algorithm 22.2.

Intuitively, good control performance will be achieved if the state estimation is accurate. Clearly, nodes which are in columns closer to the sensor will have access to more output measurements, see Fig. 22.1. On the other hand, one can expect that nodes which are physically located in columns closer to the actuator node will on average receive more plant input acknowledgments, thus, have better knowledge of plant inputs.

22.3.3 Algorithm for Dynamic Controller Placement

Algorithm 22.2 is run at every node (i, j). Since we assume that acknowledgments from the actuator node are, in general, available, but transmissions of packets $\mathbf{s}^{(i,j)}(k)$ are less reliable, nodes in columns closer to the sensor nodes can be expected to have better state estimates than nodes located in columns further down the network. Therefore, preference is given to forward incoming tentative plant input values.

The sensor node $(i, j) = (0, 0)$ uses as input: $\mathbf{s}^{(0,0)}(k) = (\mathbf{x}(k), \emptyset, \emptyset)$, $\gamma_{(0,0)}(k) = 1$, where $\emptyset$ means that the field is empty.

Algorithm 22.2: Dynamic controller placement

$k \leftarrow 0$, $\hat{\mathbf{x}}_{(i,j)}(0) \leftarrow 0$, $P_0^{(i,j)} \leftarrow P_0$, $m \leftarrow 0$, $i^* = i-1$ or $i^* = i+1$, the cooperative nodes for (i, j).
while $t \geq 0$ **do** $\{t \in \mathbb{R}_{\geq 0}$ is actual time$\}$
  while $t \leq kT + m\tau$ **do** {wait-loop}
    $m \leftarrow m + 1$
  end while
  $\mathbf{P}_{(i,j)}(k+1) \leftarrow \mathbf{A}_{(i,j)}\mathbf{P}_{(i,j)}(k)\mathbf{A}_{(i,j)}^T + \mathbf{D}$
  if $\gamma_{(i,j)}^{(i-1,j-1)}(k) = 0 \wedge \gamma_{(i,j)}^{(i,j-1)}(k) = 0 \wedge \gamma_{(i,j)}^{(i+1,j-1)}(k) = 0$ **then** $\{\mathbf{s}^{(i-1)}(k)$ is dropped$\}$
    if $\delta^{(i,j)}(k-1) = 1$ **then**
      $\mathbf{u}_{(i,j)}(k), \mathbf{u}_{(i^*,j)}(k), J_{(i,j)}(k) \leftarrow$ Algorithm 22.1
      $\mathbf{s}^{(i,j)}(k) \leftarrow \big([\mathbf{x}_{(i,j)}(k)^T, \mathbf{x}_{(i^*,j)}(k)^T]^T, [\mathbf{u}_{(i,j)}(k)^T, \mathbf{u}_{(i^*,j)}(k)^T]^T, J_{(i,j)}(k)\big)$ {a tentative input}
    else
      $\mathbf{s}^{(i,j)}(k) \leftarrow \big(\emptyset, \emptyset, \emptyset\big)$
    end if
  end if
  if $\gamma_{(i,j)}^{(i-1,j-1)}(k) = 1 \vee \gamma_{(i,j)}^{(i,j-1)}(k) = 1 \vee \gamma_{(i,j)}^{(i+1,j-1)}(k) = 1$ **then** $\{\mathbf{s}^{(i-1)}(k)$ is received$\}$
    $\mathbf{S} \leftarrow \mathbf{s}^{(i-1,j-1)}(k)$ and/or $\mathbf{s}^{(i,j-1)}(k)$ and/or $\mathbf{s}^{(i+1,j-1)}(k)$ {**S** is a set containing all the packets received. If all the packets arrive, **S** will contain $\mathbf{s}^{(i-1,j-1)}(k)$, $\mathbf{s}^{(i,j-1)}(k)$ and $\mathbf{s}^{(i+1,j-1)}(k)$}
    $(\mathbf{x}^S, \mathbf{u}^S, J^S) \leftarrow \arg\min_l \; J_l \in \mathbf{S}$
    if $\mathbf{x}^S \neq \emptyset$ **then** $\{\mathbf{x}_{(i,j)}(k)$ is available$\}$
      $\hat{\mathbf{x}}_{(i,j)}(k) \leftarrow \mathbf{x}_{(i,j)}^S$
      $\mathbf{P}_{(i,j)}(k+1) \leftarrow \mathbf{D}$
    end if
    if $\mathbf{u}^S \neq \emptyset$ **then**
      $\mathbf{u}_{(i,j)}(k) = \mathbf{u}_{(i,j)}^S$
      $\mathbf{u}_{(i^*,j)}(k) = \mathbf{u}_{(i^*,j)}^S$
    else
      $\mathbf{u}_{(i,j)}(k), \mathbf{u}_{(i^*,j)}(k), J_{(i,j)}(k) \leftarrow$ Algorithm 22.1
    end if
    if $\mathbf{u}^S = \emptyset \wedge \delta^{(i,j)}(k-1) = 1$ **then**
      $\mathbf{s}^{(i,j)}(k) \leftarrow \big(\mathbf{x}^S, [\mathbf{u}_{(i,j)}(k)^T, \mathbf{u}_{(i^*,j)}(k)^T]^T, J_{(i,j)}(k)\big)$ {a tentative input}
    else
      $\mathbf{s}^{(i,j)}(k) \leftarrow (\mathbf{x}^S, \mathbf{u}^S, J^S)$
    end if
  end if
  while $t < kT + (j+1)\tau$ **do** {wait-loop}
    $m \leftarrow m + 1$
  end while
  transmit $\mathbf{s}^{(i,j)}(k)$
  while $t \leq kT + (M+3)\tau$ **do** {wait-loop}
    $m \leftarrow m + 1$
  end while
  if $\delta^{(i,j)}(k) = 1$ **then**
    $\hat{\mathbf{x}}_{(i,j)}(k+1) \leftarrow \mathbf{A}_{(i,j)}\hat{\mathbf{x}}_{(i,j)}(k) + \mathbf{B}_{(i,j)}\mathbf{u}_{(i,j)}^A(k) + \mathbf{W}_{(i,j)(i^*,j)}\mathbf{u}_{(i^*,j)}^A(k)$
  else
    $\hat{\mathbf{x}}_{(i,j)}(k+1) \leftarrow \mathbf{A}_{(i,j)}\hat{\mathbf{x}}_{(i,j)}(k) + \mathbf{B}_{(i,j)}\mathbf{u}_{(i,j)}(k) + \mathbf{W}_{(i,j)(i^*,j)}\mathbf{u}_{(i^*,j)}(k)$
  end if
  $k \leftarrow k + 1$
end while

This node just passes the information to all the nodes in the first column. The node $(0, 0)$, as the node $(M + 1, M + 1)$, can't calculate control values.

The rest of the nodes in the network can only send information to their three closest neighbors in the following column, except for the lower and uppermost nodes who can only send to two neighbors, see Fig. 22.2. Therefore, the generic node (i, j) can receive zero, one, two or three (if not border node) packets. In the case that it receives more than one packet (as shown in Fig. 22.5), it chooses the one with the minimun value of the cost function J.

The first column of nodes, the nodes (i, j) with $j = 1$, calculate control values cooperating between them per pairs, as explained in Sect. 22.3.1. Each node of that column transmits:

$$\mathbf{s}^{(i,j)}(k) = \left([\mathbf{x}_{(i,j)}(k)^T, \mathbf{x}_{(i*,j)}(k)^T]^T, [\mathbf{u}_{(i,j)}(k)^T, \mathbf{u}_{(i*,j)}(k)^T]^T, J_{(i,j)}(k)\right)$$

to its three closest neighbors in the next column of nodes. Subsequent nodes then relay the arrived packets to the actuator node, choosing the ones with minimum $J_{(i,j)}(k)$.

A new tentative control value has to be calculated only in the following cases:

- No packet has arrived from the previous column,

$$\gamma_{(i,j)}^{(i-1,j-1)}(k) = \gamma_{(i,j)}^{(i,j-1)}(k) = \gamma_{(i,j)}^{(i+1,j-1)}(k) = 0,$$

  but the acknowledgment from the actuator has arrived, $\delta^{(i,j)}(k - 1) = 1$.
- At least one packet has arrived but all of them have the following structure:

$$\mathbf{s} = (\mathbf{x}, \emptyset, \emptyset),$$

  that means that the state is available but there is no information about the control.

Then $\mathbf{u}_{(i,j)}(k)$ is calculated and the following packet is transmitted to the next column: $\mathbf{s}^{(i,j)}(k) = \left([\hat{\mathbf{x}}_{(i,j)}(k)^T, \hat{\mathbf{x}}_{(i*,j)}(k)^T]^T, [\mathbf{u}_{(i,j)}(k)^T, \mathbf{u}_{(i*,j)}(k)^T]^T, J_{(i,j)}(k)\right)$.

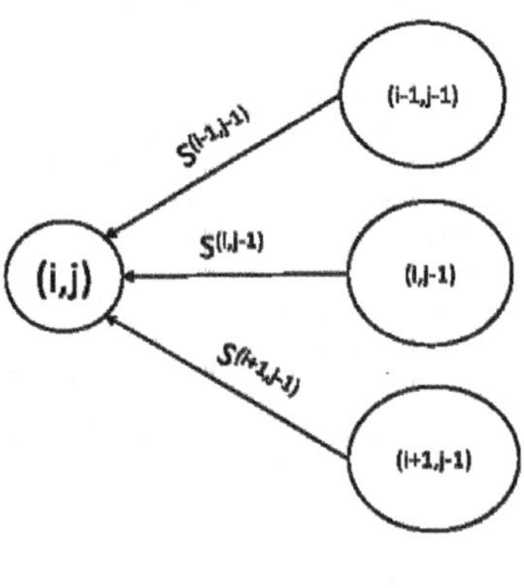

Fig. 22.5 Packets received by node (i, j)

The estimated state $\hat{\mathbf{x}}_{(i,j)}(k)$ is sent to take into account the cases in which the nodes that receive the packet (they don't calculate control value because they have a packet with $\mathbf{u}_{(i,j)}(k) \neq 0$) have some neighbor that is required to calculated a control value, but it can't estimate the whole state by itself.

22.4 Simulation Example

We consider a system with decentralized models

$$\mathbf{A}_{(1,1)(1,1)} = \begin{bmatrix} 1 & 1 \\ 0 & 1 \end{bmatrix}, \quad \mathbf{A}_{(2,1)(2,1)} = \begin{bmatrix} 1 & 1 \\ 0 & 1 \end{bmatrix}, \quad \mathbf{B}_{(1,1)(1,1)} = \begin{bmatrix} 0 \\ 1 \end{bmatrix}, \quad \mathbf{B}_{(2,1)(2,1)} = \begin{bmatrix} 0 \\ 1 \end{bmatrix},$$

and interacting models given by:

$$\mathbf{A}_{(1,1)(2,1)} = \begin{bmatrix} 0 & 0 \\ 0 & 0 \end{bmatrix}, \quad \mathbf{A}_{(2,1)(1,1)} = \begin{bmatrix} 0 & 0 \\ 0 & 0 \end{bmatrix}, \quad \mathbf{B}_{(1,1)(2,1)} = \begin{bmatrix} 0 \\ 1 \end{bmatrix}, \quad \mathbf{B}_{(2,1)(1,1)} = \begin{bmatrix} 0 \\ 1 \end{bmatrix},$$

where we are considering no noise and $\bar{x}_0 = 1$.

The network is as depicted in Fig. 22.3, with i.i.d. transmission processes and success probabilities $\mathbf{Prob}\{\gamma_{(i,j))}^{(i,j-1)}(k) = 1\} = 0.4$ and $\mathbf{Prob}\{\delta^{(i,j)}(1) = 1\} = 1$.

Figure 22.6 shows the empirical distribution of the controller node $c(k)$ obtained by running the algorithm for 100 steps. It is possible to see how 43 % of the times, the controller node is located in the last column of nodes, the one closest to the actuator.

Figures 22.7 and 22.8 compare the plant state trajectory when the algorithm proposed is used with the case in which we locate the controller at the actuator node. The results suggest that the proposed algorithm yields a stable system, while when the controller is at the actuator, the system becomes unstable.

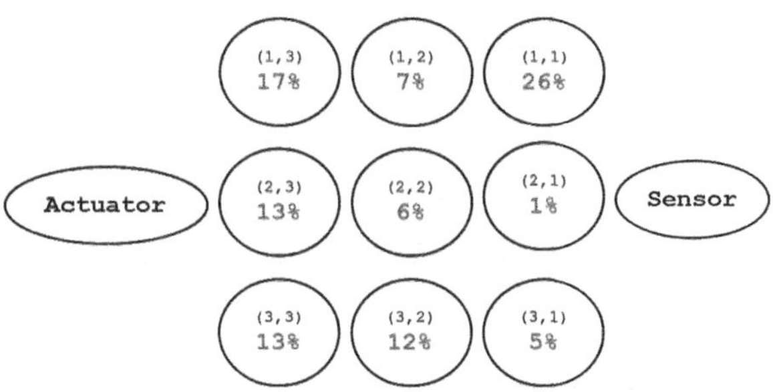

Fig. 22.6 Controller location percentage

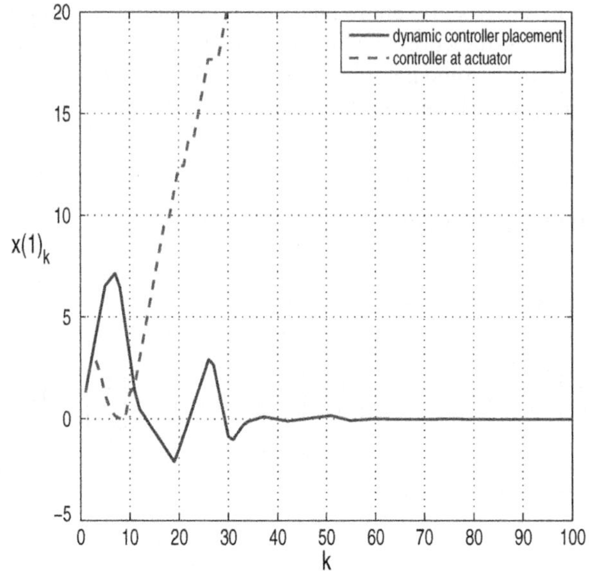

Fig. 22.7 $x(1)$ trajectory

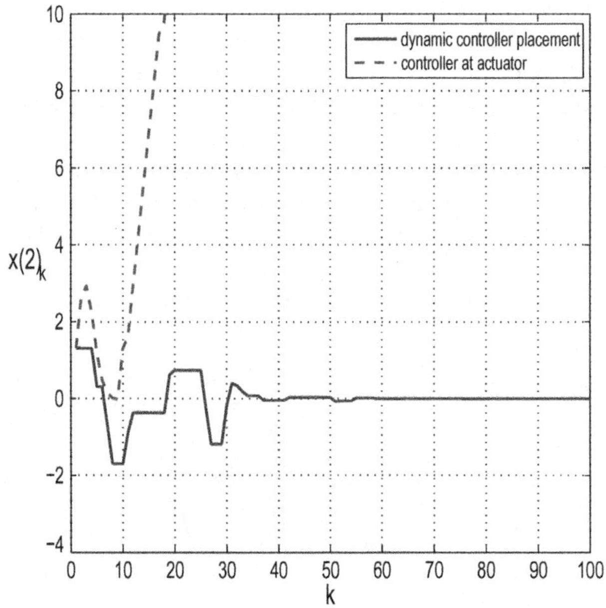

Fig. 22.8 $x(2)$ trajectory

22.5 Conclusions

We have presented a cooperative MPC formulation for NCSs subject to data dropouts. We provide an algorithm that decides which nodes are in charge of the calculation of the the control input, and which ones just relay the received information. This decision depends on transmission outcomes. Once the controller node has been chosen, it interacts with its neighbors over unreliable links solving a cooperative MPC.

Future works may include stability analysis of the proposed architecture. Furthermore, it would be of interest to study a practical application of the proposed method for systems controlled over unreliable networks with time-varying reliability, for example, if there are moving obstacles blocking the nodes [15].

References

1. G.N. Nair, F. Fagnani, S. Zampieri, R.J. Evans, Feedback control under data rate constraints: An overview. Proc. IEEE (Special Issue on Technology of Networked Control Systems) **95**(1), 108–137 (2007)
2. D.E. Quevedo, E.I. Silva, G.C. Goodwin, Subband coding for networked control systems. Int. J. Robust Nonlinear Control **19**(16), 1817–1836 (2009)
3. A.S. Matveev, A.V. Savkin, *Estimation and Control over Communication Networks. Birkäusser*. (Boston, 2009)
4. D. Quevedo, J. Østergaard, D. Nešić, Packetized predictive control of stochastic systems over bit-rate limited channels with packet loss. IEEE Trans. Autom. Control **56**(12), 2854–2868 (2011)
5. E.A. Jorswieck, E.G. Larsson, M. Luise, H.V. Poor, Game theory and the flat-fading Gaussian interference channel. IEEE Signal Process. Mag. (Special Issue on Game Theory in Signal Processing and Communications) **26**(5), 18–27 (2009)
6. E. Altman, T. Boulogne, R. El-Azouzi, T. Jiménez, L. Wynter, A survey on networking games in telecommunications. Comput. Oper. Res. **33**(2), 286–311 (2006)
7. F. Xiao, L. Wang, Consensus protocols for discrete-time multi-agent systems with time-varying delays. Automatica **44**(10), 2577–2582 (2008)
8. J. Chen, K.H. Johansson, S. Olariu, I.C. Paschalidis, I. Stojmenovic, Guest editorial special issue on wireless sensor and actuator networks. IEEE Trans. Autom. Control **56**(10), 2244–2246 (2011)
9. D.E. Quevedo, K.H. Johansson, A. Ahlén, I. Jurado, Dynamic controller allocation for control over erasure channels. in *Proceedings of 3rd IFAC Workshop on Distributed Estimation and Control in Networked Systems*, Santa Barbara, CA, 2012
10. A.N. Venkat, J.B. Rawlings, S.J. Wright. Distributed model predictive control of large-scale systems. in *In Assessment and Future Directions of Nonlinear Model Predictive Control*. (Springer, Berlin, 2007) pp. 591–605
11. C.L. Robinson, P.R. Kumar, Optimizing controller location in networked control systems with packet drops. IEEE J. Sel. Areas Commun. **26**(4), 661–671 (2008)
12. E. Garone, B. Sinopoli, A. Casavola, LQG control over lossy TCP-like networks with probabilistic packet acknowledgements. Int. J. Syst. Control Commun. 2(1,2,3), 55–81 (2010).
13. O.C. Imer, S. Yüksel, T. Basar, Optimal control of LTI systems over unreliable communication links. Automatica **42**(9), 1429–1439 (2006)
14. K.J. Åström, *Introduction to Stochastic Control Theory* (Dover Publications Inc., Mineola, 2006)
15. D.E. Quevedo, A. Ahlén, K.H. Johansson. State estimation over sensor networks with correlated wireless fading channels. IEEE Trans. Automatic Control (Accepted for publication)

Chapter 23
Parallel Implementation of Hybrid MPC

D. Axehill and A. Hansson

Abstract In this chapter parallel implementations of hybrid MPC will be discussed. Different methods for achieving parallelism at different levels of the algorithms will be surveyed. It will be seen that there are many possible ways of obtaining parallelism for hybrid MPC, and it is by no means clear which possibilities that should be utilized to achieve the best possible performance. To answer this question is a challenge for future research.

23.1 Introduction

Speed in numerical computations has increased dramatically for a long period of time. This is partly due to increase of processor performance in computers and partly due to development of more sophisticated algorithms and methods. However, for the last five years single-core processor performance has not significantly increased. To compensate for this multi-core and multi-processor computers have seen an increased use. In addition to this clusters and grids have emerged as another way to speed up computations. Multi-core and multi-processor computers typically have only few cores and processors whereas clusters and grids can be composed of hundreds of processors distributed over a significant number of computers. It is clear that these

---

D. Axehill (✉) · A. Hansson
Division of Automatic Control, Linköping University, Linköping, Sweden
e-mail: axehill@isy.liu.se

A. Hansson
e-mail: hansson@isy.liu.se

J. M. Maestre and R. R. Negenborn (eds.), *Distributed Model Predictive Control Made Easy*, Intelligent Systems, Control and Automation: Science and Engineering 69,
DOI: 10.1007/978-94-007-7006-5_23, © Springer Science+Business Media Dordrecht 2014

new architectures pose new challenges on how algorithms for numerical computations should be designed. In case care is not taken, potential speedup will not happen.

Model Predictive Control (MPC) is a popular control strategy which has been used in many applications for a long time. In later years work has been carried out to generalize MPC to so-called hybrid systems [16]. For these systems the computational demand is even higher. Hybrid systems have applications in, e.g., transportation, logistics, economics, process control, building control, airplane routing and communications.

In recent years there has been a trend in the control community to develop distributed algorithms for control. This type of distributed control has much in common with parallel implementations and a parallel implementation of an optimization algorithm for MPC can be interpreted as a form of a distributed MPC controller. However, it should be stressed that the main objective in this work is computational performance. One of the few references available for the hybrid setup is [11], where a continuous time hybrid optimal control problem is solved using simulations. For other examples it has in [36, 43] been shown that distributed hybrid MPC is suboptimal in general. This chapter will only consider the discrete-time setting of hybrid MPC.

The remaining part of the chapter is organized as follows. First a brief review of hybrid MPC will be given in Sect. 23.2. In Sect. 23.3 different optimization methods for solving hybrid MPC problems will be discussed. Then, in Sect. 23.4, different approaches for parallelization will be reviewed, and their potential applicability to hybrid MPC will be investigated. Finally, in Sect. 23.5, conclusions will be presented together with recommendations for future research.

This chapter is a condensed version of the book chapter in [9], where also explicit MPC is considered.

23.2 Hybrid MPC

In this section, MPC for hybrid systems will be discussed. Furthermore, some commonly used models for hybrid systems that are useful in the MPC context will be reviewed.

23.2.1 Model Predictive Control

The most commonly used variant of MPC is linear MPC, where the dynamics are linear and often a quadratic performance measure similar to the one used in Linear Quadratic (LQ) control is used. A difference compared to LQ control is that it is possible to consider linear inequality constraints on states and control signals. A discrete-time linear time-invariant model on state space form is given by

$$\begin{aligned}\mathbf{x}(k+1) &= \mathbf{A}\mathbf{x}(k) + \mathbf{B}\mathbf{u}(k)\\ \mathbf{y}(k) &= \mathbf{C}\mathbf{x}(k)\end{aligned} \tag{23.1}$$

where $k \in \mathbb{Z}$ is the discrete time, $\mathbf{x}(k) \in \mathbb{R}^{n_x}$ is the state, $\mathbf{u}(k) \in \mathbb{R}^{n_u}$ is the control input and $\mathbf{y}(k) \in \mathbb{R}^{n_y}$ is the controlled output. An example of an objective function, or performance measure, is a quadratic function in the form

$$\begin{aligned}J(k) = &\frac{1}{2}\sum_{l=0}^{N_p-1}\left(\|\mathbf{y}(k+l) - \mathbf{y}_{\text{ref}}(k+l)\|_{\mathbf{Q}}^2 + \|\mathbf{u}(k+l)\|_{\mathbf{R}}^2\right)\\ &+ \frac{1}{2}\left\|\mathbf{y}(k+N_p) - \mathbf{y}_{\text{ref}}(k+N_p)\right\|_{\mathbf{Q}}^2\end{aligned} \tag{23.2}$$

where $\mathbf{Q} \in \mathbb{S}_{+}^{n_x}$ and $\mathbf{R} \in \mathbb{S}_{++}^{n_u}$, and $\mathbf{y}_{\text{ref}}(k) \in \mathbb{R}^{n_y}$ is the reference signal. Other common performance measures for linear MPC are formed by replacing the squared 2-norm in (23.2) with a 1-norm or ∞-norm. Often, the constraints are defined as polyhedral constraints in the form

$$\mathbf{H}_{\mathbf{u}}(k)\mathbf{u}(k) + \mathbf{H}_{\mathbf{x}}(k)\mathbf{x}(k) + \mathbf{h}(k) \leq 0 \tag{23.3}$$

There are several different ways to cast (23.1), (23.2) and (23.3) in the form of a formal optimization problem.

For a more in-depth treatment, see [9, 35]. If the system is linear and the objective function is quadratic, the resulting optimization problem is for a fixed value of the initial state $\mathbf{x}^0$ a Quadratic Programming (QP) problem in the general form

$$\underset{\mathbf{x}}{\text{minisize}} \quad \frac{1}{2}\mathbf{x}^{\mathrm{T}}\mathbf{H}\mathbf{x} + \mathbf{f}^{\mathrm{T}}\mathbf{x} \tag{23.4a}$$

$$\text{subject to} \quad \mathbf{A}_{\mathcal{E}}\mathbf{x} = \mathbf{b}_{\mathcal{E}} \tag{23.4b}$$

$$\mathbf{A}_{\mathcal{I}}\mathbf{x} \leq \mathbf{b}_{\mathcal{I}}, \tag{23.4c}$$

where $\mathbf{x}$ contains the control inputs, states and controlled outputs for the entire prediction horizon stacked. QP problems in general are well-studied problems for which there exist well-developed optimization routines. Similarly, if the system is linear and a 1-norm or an ∞-norm performance measure is used, the resulting optimization problem becomes a Linear Programming (LP) problem. Hence, for linear MPC the optimization problem is considered relatively easy to solve. In order to get closed-loop control, the approach above is used in a receding horizon fashion.

An extension to linear MPC is nonlinear MPC. This extension handles nonlinear systems and a general nonlinear performance measure in the objective function. Unfortunately, the resulting optimization problem is often more difficult to solve compared to the linear case. A special case of nonlinear MPC is control of systems described partly by logic. These are called hybrid systems and provide a framework for describing processes evolving according to continuous dynamics, discrete dynamics and logic rules [16]. This class of systems is especially important when analyzing and controlling systems arising in the growing interaction between physical processes and digital controllers.

23.2.2 Modeling Frameworks for Hybrid Systems

In this subsection, some different modeling frameworks for discrete-time hybrid systems are considered. The most important ones for the purpose of this work are reviewed in more detail and are related to each other.

23.2.2.1 Mixed Logical Dynamical Systems

Mixed Logical Dynamical (MLD) systems is one way of describing an important class of hybrid systems defined by linear dynamic equations subject to linear mixed integer inequalities, i.e., inequalities involving both continuous and binary variables. The MLD description is a very general model class capable of describing a broad range of systems.

In [16], an MPC framework used for systems described by physical laws, logic rules and operating constraints is presented. An important part of this framework consists of the definition of MLD systems. This class of systems includes linear hybrid systems, finite state machines, some classes of discrete event systems, constrained linear systems and nonlinear systems which can be exactly or approximately described by piecewise affine functions. There are many applications for MLD systems reported in the literature. Some illustrative examples can be found in [7, 16, 23]. An MLD system can be described by the following linear relations [16],

$$\begin{aligned}
\mathbf{x}(k+1) &= \mathbf{A}(k)\mathbf{x}(k) + \mathbf{B}_1(k)\mathbf{u}(k) + \mathbf{B}_2(k)(k) + \mathbf{B}_3(k)\mathbf{z}(k) \\
\mathbf{y}(k) &= \mathbf{C}(k)\mathbf{x}(k) + \mathbf{D}_1(k)\mathbf{u}(k) + \mathbf{D}_2(k)(k) + \mathbf{D}_3(k)\mathbf{z}(k) \\
\mathbf{E}_2(k)(k) + \mathbf{E}_3(k)\mathbf{z}(k) &\le \mathbf{E}_1(k)\mathbf{u}(k) + \mathbf{E}_4(k)\mathbf{x}(k) + \mathbf{E}_5(k)
\end{aligned} \tag{23.5}$$

where $k \in \mathbb{Z}$. Furthermore, $\mathbf{y}(k) \in \mathbb{R}^{n_{yc}} \times \{0,1\}^{n_{yl}}$ denotes the controlled outputs, $\mathbf{x}(k) \in \mathbb{R}^{n_{xc}} \times \{0,1\}^{n_{xl}}$ denotes the states of the system, $\mathbf{u}(k) \in \mathbb{R}^{n_{uc}} \times \{0,1\}^{n_{ul}}$ denotes the control inputs, $(k) \in \{0,1\}^{n_\delta}$ denotes the auxiliary binary variables, and $\mathbf{z}(k) \in \{0,1\}^{n_z}$ denotes the auxiliary continuous variables. If the desired finite alphabet is not binary as here, it can be coded using binary variables.

In [16], both optimal control and receding horizon estimation for MLD systems is discussed. The control signal at a state $\mathbf{x}^0$ is found by minimizing either a linear or a quadratic performance measure similar to the one in (23.2) subject to $\mathbf{x}(k) = \mathbf{x}^0$ and the dynamics in (23.5). This MPC problem can be rewritten as an optimization problem in Mixed Integer Quadratic Programming (MIQP) form [16], i.e., in the general form

$$\underset{\mathbf{x}}{\text{minimize}} \quad \frac{1}{2}\mathbf{x}^{\mathrm{T}}\mathbf{H}\mathbf{x} + \mathbf{f}^{\mathrm{T}}\mathbf{x} \tag{23.6a}$$

$$\text{subject to} \quad (22.4\text{b}), (22.4\text{c}) \tag{23.6b}$$

$$x_i \in \{0, 1\}, \forall i \in \mathcal{B}. \tag{23.6c}$$

where $\mathbf{x}$ contains the variables from the system in (23.5) for the entire prediction horizon stacked and $\mathcal{B}$ denotes the set of indices to binary components of $\mathbf{x}$. More details on how to formulate the MPC problem as an optimization problem can be found in [9]. The problem data for the MIQP problem in (23.6) is defined in Sect. 23.3.2.

As for linear MPC, hybrid MPC is implemented in a receding horizon fashion. The difference is that it is much more complicated to find the optimal control signal sequence, since the system is neither linear nor smooth [16]. One way of reducing the computational complexity is to use tailored MIQP solvers. This is further discussed in [2].

23.2.2.2 Piecewise Affine Systems

Piecewise affine (PWA) systems are hybrid systems where the control and state-space is partitioned into different polyhedral regions each implying certain affine dynamics. Mathematically, this can be formulated as [10],

$$\mathbf{x}(k+1) = \mathbf{A}^i\mathbf{x}(k) + \mathbf{B}^i\mathbf{u}(k) + \mathbf{f}^i$$
$$\text{if} \begin{bmatrix} \mathbf{x}(k) \\ \mathbf{u}(k) \end{bmatrix} \in \mathcal{C}^i, i = \{1, \ldots, s\}, \tag{23.7}$$

where $\mathbf{x}(k) \in \mathbb{R}^{n_{xc}} \times \{0, 1\}^{n_{xl}}$ denotes the continuous and binary states, $\mathbf{u}(k) \in \mathbb{R}^{n_{uc}} \times \{0, 1\}^{n_{ul}}$ denotes the continuous and binary control inputs, $\{\mathcal{C}^i\}_{i=1}^{s}$ denotes the polyhedral partition of the set of the state and input space. Often, there are also hard constraints on the inputs and states. These are brought into the framework by adding polyhedral constraints in the form in (23.3). The feasible set for the constrained PWA system is then given by the intersection of the set over which the system is defined in (23.7) and the feasible set of the constraints in (23.3).

The optimal control problem for PWA systems is formulated analogously to the one for MLD problems. An objective function in the form in (23.2) is minimized subject to the constraints in (23.3) and in (23.7). However, in order to bring the optimal control problem into the form of an optimization problem, it is often first reformulated as an MLD system and later solved as an MIQP problem.

23.2.2.3 Discrete Hybrid Automata

Discrete Hybrid Automata (DHA) are composed by the interconnection of a Finite State Machine (FSM), a Switched Affine System (SAS), a Mode Selector (MS), and an Event Generator (EG) [44]. The FSM models the discrete dynamics in the DHA. The state update equation for this subsystem is

$$\mathbf{x_l}(k+1) = \mathbf{f_l}(\mathbf{x_l}(k), \mathbf{u_l}(k), \mathbf{e}(k)) \tag{23.8}$$

where $\mathbf{x}_l(k) \in \{0, 1\}^{n_{xl}}$ is the logic state, $\mathbf{u_l}(k) \in \{0, 1\}^{n_{ul}}$ is the logic input, $\mathbf{e}(k) \in \{0, 1\}^{n_e}$ is a logic event signal from the EG, and $\mathbf{f_l}(\cdot)$ is a Boolean function. The SAS can be represented as

$$\mathbf{x_c}(k+1) = \mathbf{A}_{i(k)}\mathbf{x_c}(k) + \mathbf{B}_{i(k)}\mathbf{u_c}(k) + f_{i(k)} \tag{23.9}$$

where $\mathbf{x_c}(k) \in \mathbb{R}^{n_{xc}}$ is the real-valued state, $\mathbf{u_c}(k) \in \mathbb{R}^{n_{uc}}$ is the real-valued input, and $i(k) \in \{0, 1\}^s$ is an input that is used to select the mode in which the SAS is working. The MS computes the mode of the SAS $i(k)$ based on the state in the finite state machine $\mathbf{x_l}(k)$, the input $\mathbf{u_l}(k)$, and the event signal $\mathbf{e}(k)$ according to

$$i(k) = f_{MS}(\mathbf{x_l}(k), \mathbf{u_l}(k), \mathbf{e}(k)) \tag{23.10}$$

where f_{MS} is a Boolean function. The result from the EG is the binary-valued signal $\mathbf{e}(k)$ which represents if specified linear inequalities are satisfied or not. More specifically,

$$[e_j(k) = 1] \longleftrightarrow [a_j^{\mathrm{T}}\mathbf{x_c}(k) + b_j^{\mathrm{T}}\mathbf{u_c}(k) \leq c_j] \tag{23.11}$$

Optimal control for systems represented as DHA can be performed in at least two ways. First, the DHA model can be transformed into an MLD model and solved as an MPC problem for such systems. Second, the ideas in [14] can be used, where the structure of the DHA is exploited in a more direct way by combining mixed integer optimization with satisfiability solvers and constraint logic programming solvers.

23.2.2.4 Equivalences

As seen in previous sections, there are several different ways of modeling discrete time hybrid systems. Fortunately, it is possible to show equivalences between these under more or less restrictive assumptions. As a result, derived theoretical properties and computational tools can be transferred from one class to another. For optimal control and state estimation, the MLD description is proposed, while most other hybrid techniques are built on a PWA representation [13]. For an in-depth review of equivalences, see e.g. [29].

23.3 Optimization Methods

In this section we will review optimization methods that can be used to solve the optimization problems that were formulated in the previous section. We will focus on on-line methods such as branch and bound (BnB).

23.3.1 Quadratic Programming

In this work, convex QP problems in the form in (23.4a) are considered, where $\mathbf{x} \in \mathbb{R}^n$, $\mathbf{H} \in \mathbb{S}^n$, $\mathbf{f} \in \mathbb{R}^n$ and the rows in $\mathbf{A}_{\mathcal{E}} \in \mathbb{R}^{p\times n}$ are given by the vectors in $\{a_i \in \mathbb{R}^n | i \in \mathcal{E}\}$ and the rows in $\mathbf{A}_{\mathcal{I}} \in \mathbb{R}^{m\times n}$ are given by the vectors in $\{a_i \in \mathbb{R}^n | i \in \mathcal{I}\}$. The column vectors $\mathbf{b}_{\mathcal{E}}$ and $\mathbf{b}_{\mathcal{I}}$ are analogously defined. The sets $\mathcal{I}$ and $\mathcal{E}$ are finite sets of indices where $\mathcal{I} \cap \mathcal{E} = \emptyset$. The problem in (23.4a) can be solved using for example, an active set (AS) method or an interior point (IP) method. If the matrix $\mathbf{H}$ is zero, the problem is an LP problem. These are usually solved either using a simplex method or an IP method. However, the main focus in this work will be on QP problems. More information about QP and how to solve these problems can be found in, e.g. [38].

23.3.2 Mixed Integer Programming

MIQP is a special case of Mixed Integer Non-Linear Programming (MINLP). At a first glance, the MIQP problem looks similar to the ordinary QP problem. There is however one important difference. Some optimization variables are not allowed to be real-valued, but they are constrained to be integer-valued. This seemingly minor modification turns the easily solved QP problem, into an $\mathcal{NP}$-hard problem [49]. A common special case of MIQP is when the integer variables are constrained to be 0 or 1. To use a precise notation, this problem is called a Mixed Binary Quadratic Programming (MBQP) problem. The standard notation for MBQP seems, at least in the control literature, to be MIQP. In what follows, the problem studied will be an MBQP, but to keep the standard notation, it will be denoted MIQP. A survey considering Quadratic Integer Programming (QIP) can be found in [47].

23.3.2.1 Problem Definition

The mathematical formulation of an MIQP problem can be found in (23.6), where $\mathbf{f} \in \mathbb{R}^{n_c+n_b}$ and $\mathbf{H} \in \mathbb{S}_+^{n_c+n_b}$. Furthermore, let $\mathbf{A}_{\mathcal{E}}$, $\mathbf{A}_{\mathcal{I}}$, $\mathbf{b}_{\mathcal{E}}$ and $\mathbf{b}_{\mathcal{I}}$ be defined as in Sect. 23.3.1 with $n = n_c + n_b$. The difference is that n_b optimization variables indexed by the set $\mathcal{B}$ are not real-valued but binary-valued. As a consequence, the problem is no longer convex. Mixed Integer Linear Programming (MILP) can be seen as a special case of MIQP where H is the zero matrix.

Often, this problem is solved using a branch and bound method, where many QP problems in the form in (23.4a) are solved in order to find the optimal solution to the problem in (23.6). The procedure is similar for the MILP case, but the relaxations are of LP type instead of QP type. There also exist other methods for solving these problems. The four most commonly used methods are cutting plane methods, decomposition methods, logic-based methods, and branch and bound methods [16]. Several authors report that branch and bound is the best method for mixed integer programs [16]. In [25], a branch and bound method is compared to Generalized Benders Decomposition (GBD), Outer Approximation (OA) and LP/QP based branch and bound. The conclusion in this reference is that branch and bound is the best method for solving MIQP problems. With a few exceptions, branch and bound is an order of magnitude faster than any of the other methods. An important explanation to this is that the QP subproblems are very cheap to solve. This is not the case for general MINLP, where the subproblems to be solved in the nodes are more complicated. In the MINLP case there exist important problem classes where branch and bound is not the best method. A review of different methods of solving MIQP problems can be found in [47]. There exist several software for solving MIQP problems. For MATLAB, free software like YALMIP[34] or *miqp.m*[15] can be used. A commonly used commercial software is CPLEX.

23.3.2.2 Branch and Bound Methods

If computational burden is not considered, the most straightforward approach to compute the optimal solution to an optimization problem involving binary variables is to enumerate all possible combinations of the binary variables, and for each such combination kept fixed, compute the optimal solution of any real-valued variables also included in the problem. Thereafter, the objective function values are compared and the solution, or solutions, generating the best objective function value is taken as the optimal solution. However, for problems involving many binary variables the computational burden will become overwhelming since the number of combinations of the binary variables is 2^{n_b}. Hence, there is a need for a method that can find the optimal solution without enumerating all possible combinations of the binary variables. One such method is branch and bound, where it for a majority of problems is sufficient to explicitly enumerate only *some* of the possible combinations. Unfortunately, the worst case complexity is still exponential and the number of combinations necessary to enumerate, and solve an optimization problem for, is problem dependent. This classical pessimistic complexity bound is improved in [6].

The basics of a branch and bound method will now be discussed. The main part of the presentation follows the one in, e.g. [49]. The reader is referred to these two references for more details. The general idea of branch and bound is to split the feasible set $\mathcal{S}$ of the optimization problem into K smaller sets such that

$$\mathcal{S} = \bigcup_{i=1}^{K} \mathcal{S}_i \tag{23.12}$$

This partitioning is performed in several steps and it can be represented using a binary tree structure. The topmost node in the tree is called the root node and the nodes at the bottom of the tree are called leaves. The rows of nodes in the tree starting with the root node and ending with the leaves are called levels. An important property of branch and bound is that the entire tree is not known from the beginning and only the parts of the tree explicitly needed in the solution process are further expanded.

An optimal solution over the set $\mathcal{S}$ can be computed by optimizing over the smaller sets separately according to

$$\begin{aligned} z^{i*} &= \underset{\mathbf{x}\in\mathcal{S}_i}{\text{minimize}}\ f_0(\mathbf{x}),\ i \in \{1, \ldots, K\} \\ z^* &= \min_{i\in\{1,\ldots,K\}} \left\{z^{i*}\right\} \end{aligned} \tag{23.13}$$

An optimal solution over $\mathcal{S}$ is found as the optimal solution to a subproblem with the lowest optimal objective function value. Note that the leaves of the tree contain the different combinations of the binary variables that have to be investigated if $\mathcal{S}$ is to be explored by explicit enumeration. Hence, if it is necessary to solve all of the problems represented by the leaves, there is no gain from using the branch and bound method.

The key idea in order to reduce the computational effort is to compute upper and lower bounds on the optimal objective function value for the subproblems in the nodes. Often, these bounds can be used to prune entire subtrees, which means that these subtrees do not have to be explicitly considered, since it can be concluded that the optimal solution cannot be found in any of them. Furthermore, these bounds are supposed to be easily computable. Pruning can be interpreted as an implicit enumeration, and is therefore highly desirable. The tree can be pruned if a relaxation in a node

1. is infeasible. The entire subtree below that node is infeasible.
2. is integer feasible. The optimal value for the entire subtree below that node has been found.
3. has an objective function value that is worse than the best known integer solution so far ("dominance"). The objective function value gets worse as the process proceeds further down in the tree. Hence, there is no use continuing.

To be able to apply the above scheme in practice, it has to be decided how to compute the upper and lower bounds. Usually, upper bounds are found from integer feasible solutions and lower bounds are found from relaxations or duality. In MIQP, often QP relaxations are used, where the integer constraints are relaxed to interval constraints. These relaxations are in the form

$$\begin{aligned}
&\underset{\mathbf{x}}{\text{minimize}} && \frac{1}{2}\mathbf{x}^{\mathrm{T}}\mathbf{H}\mathbf{x} + \mathbf{f}^{\mathrm{T}}\mathbf{x} && \text{(23.14a)}\\
&\text{subject to} && \text{(23.4b), (23.4c)} && \text{(23.14b)}\\
& && 0 \leq x_i \leq 1, \forall i \in \mathcal{B} && \text{(23.14c)}\\
& && x_i = 0, \forall i \in \mathcal{B}_0, \quad x_i = 1, \forall i \in \mathcal{B}_1 && \text{(23.14d)}
\end{aligned}$$

where the original integer constraints have been relaxed to interval constraints. Several relaxations with different choices of the sets $\mathcal{B}_0$ and $\mathcal{B}_1$ (where $\mathcal{B}_0 \cap \mathcal{B}_1 = \emptyset$) are ordered and solved in a structured way in the binary search tree. More about relaxations applicable to branch and bound for MIQP in the hybrid MPC application can be found in [8], where also more advanced relaxations of SDP type are considered. Efficient computational methods for computation of these relaxations in the hybrid MPC application are presented in [4] and in [5] for QP relaxations and SDP relaxations, respectively.

In a branch and bound method, there are several parameters and choices that may affect the performance drastically. One important choice to make is to decide which node to solve next. The three most common choices are depth first, breadth first, and best first. In depth first, the next node to solve is chosen as one of the child nodes of the current node. This process is continued until a node is pruned. In breadth first, all nodes at each level are considered before a node in a new level is considered. In best first, the next problem considered is the one with the lowest lower bound so far.

According to [25], solving the subproblems of QP type using a dual AS method offers the most straightforward way to exploit the structure introduced by the branching procedure. After a branch, the solution to the parent problem is in general infeasible in the child problems. But, a dual feasible starting point for the child problems is directly available from the dual solution of the parent problem. Consequently, it is possible to warm start the AS solver using information from the solution to the parent problem. Also, since a dual AS method is an ascent method generating dual feasible points, it can use an upper bound as a cut-off value for terminating the QP solver prematurely [25]. According to [49], AS methods (the reference considers the LP case) is preferable for solving the relaxed problems in branch and bound. For very large problems, IP methods can be used to solve the first subproblem, but in the subsequent subproblems an AS method should be used.

An important step in a commercial branch and bound code is the preprocessing step. The basic operations in preprocessing is to quickly detect and eliminate redundant constraints and variables, and to tighten bounds if it is possible. A smaller and tighter formulation is preferred, since the number of nodes necessary to consider, and the dimension of the subproblems, might be reduced.

A formal description of a branch and bound algorithm applied to a binary optimization problem P can be found in Algorithm 23.1, where $\bar{z}$ denotes the current upper bound, $\bar{\mathbf{x}}$ denotes the solution associated with the current upper bound, $\underline{z}^i$ denotes the optimal objective function value of the relaxation P_i^R to the problem P_i in node i, and $\underline{\mathbf{x}}^i$ denotes the optimal solution to P_i^R. The feasible set of P_i and P_i^R is denoted $\mathcal{S}_i$ and $\mathcal{S}_i^R$, respectively.

Algorithm 23.1 Sequential branch and bound for binary variables, [49]

$\bar{z} \leftarrow +\infty$
$\bar{\mathbf{x}} \leftarrow$ **void**
Add P to **LIST**.
while length(**LIST**) > 0 **do**
  Pop P_i from **LIST**.
  Solve relaxation $P_i^R \Rightarrow \underline{z}^i$ and $\underline{\mathbf{x}}^i$.
  if $\mathcal{S}_i^R = \emptyset$ **then**
    No feasible solution exists for P_i.
  else if $\underline{z}^i \geq \bar{z}$ **then**
    There exists no feasible solution of P_i which is better than $\bar{\mathbf{x}}$.
  else if $\underline{\mathbf{x}}^i \in \mathcal{S}_i$ **then**
    $\underline{\mathbf{x}}^i$ is integer feasible and is therefore optimal also in P_i.
    $\bar{z} \leftarrow \underline{z}^i$
    $\bar{\mathbf{x}} \leftarrow \underline{\mathbf{x}}^i$
  else
    Split $\mathcal{S}_i$ into $\mathcal{S}_{i0}$ and $\mathcal{S}_{i1}$.
    Push P_{i0} and P_{i1} to **LIST**.
  end if
end while

How subproblems are put on the list and retrieved from the list is decided by the choice of the node selection criterion and the branching priority. If it is possible to easily find an upper bound on the optimal objective function value, this bound can be used to initialize the global upper bound $\bar{z}$.

23.3.2.3 Logic-based Programs

It is possible to modify the branch and bound algorithm for mixed integer programs to logic based programs as described in [14, 28, 39, 45]. The integer variables which are fixed in a node are used to "infer" new knowledge on other integer variables. In this way the best integer solution can potentially be updated faster and hence reduce the computational time. In [14] the reduction is an order of magnitude for a supply chain management example.

23.3.3 Other Methods

Other methods that can be used to solve the optimization problems arising in hybrid MPC are methods such as genetic algorithms, simulated annealing and tabu search. Since these methods in general only provide suboptimal solutions, they will not be discussed in more detail. However, it should be stressed that they can be parallelized fairly easily [40].

23.4 Parallel Implementation

In this section, possibilities for parallelization at different computational levels will be discussed. On the top level, the integer optimization problem is considered directly. The main focus is on branch and bound methods, but also logic-based methods will be considered. In order to solve the MIQP problem of interest using a branch and bound method, QP problems are solved by a convex QP method at an intermediate computational level. At an even lower level, the numerical linear algebra can be found. The reason for the partitioning into levels is that the possibilities and the applicable parallelization methods varies with the level and it is a non-trivial trade-off at what level, or levels, the algorithms should be parallelized. The work should be scheduled such that overhead from, e.g., communication, idle time due to load imbalance, and waiting time for shared resources is minimized [26]. An important aspect that influences parallelism is the way the memory is distributed. Shared address space parallel computers have global memory which all processors can address directly. This is in contrast to message passing computers where each processor has its own local memory. For this latter case the processors can communicate only by sending messages over a network. Another important aspect is that the processors should be "sufficiently loaded" in order to utilize the parallel hardware efficiently according to Amdahl's law. This law states that the speedup from parallelization does not grow linearly with the number of processors for a fixed problem size [26]. This is due to the sequential, non-parallelizable, part of the algorithm which eventually saturates the speedup as the number of processors grows. Finally, it should be stressed that the approaches presented in this work aim at computing a solution with the same properties as a centralized implementation. For example, feasibility properties and the quality of the returned solution is not affect by the fact that the work is distributed on many processors. This in turn implies for the MPC application that, e.g., stability properties are not affected by the parallelization.

23.4.1 Parallel Implementations at High Level

In this section ideas from generic parallel solvers for integer programs and logic-based programs will be reviewed. These implementations have immediate potential for hybrid MPC.

23.4.1.1 Branch and Bound

It is clear from Sect. 23.3.2.2 that the tree used in branch and bound algorithms can easily be portioned into smaller parts that can be searched by different processors. Parallel formulations of branch and bound have shown near linear speedup in many cases [1]. However, because of pruning many of the processors will become idle long

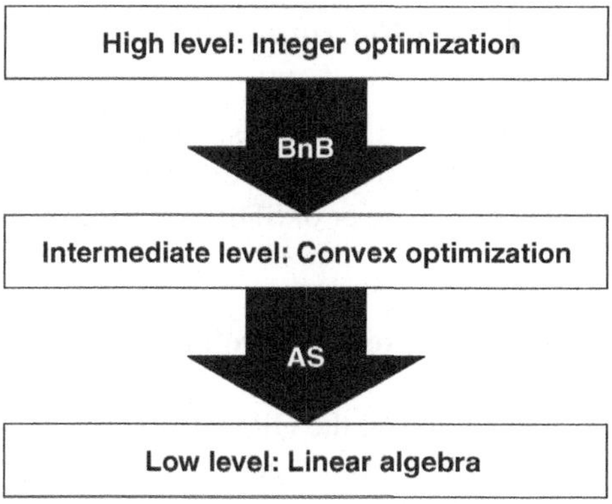

Fig. 23.1 Parallelization for branch and bound can be performed at different computational levels

before the optimal solution is found. This will result in low efficiency, and therefore performing this assignment in a static way is not desirable. Hence there is a need for a dynamic load balancing that minimizes communication and processor idling. In this work, parallelization on different computational levels will be discussed. This is illustrated in Fig. 23.1.

We will only discuss the case of depth-first branch and bound where the list in Algorithm 23.1 implements a stack. This case is commonly used in hybrid MPC and the case that is most easily parallelized. The other cases are discussed in more detail in [26].

For depth-first branch and bound the load balancing is composed of two phases: task portioning and subtask distribution [26]. In the task portioning typically two different techniques can be employed, i.e., so-called stack splitting and node splitting. In stack splitting the stack of a processor is split in two, half of which is kept and half of which is given away to another processor. In node splitting only one node is given away. Node splitting typically results in more work. Subtask distribution involves questions such as when work is split and how it is distributed. This can in both cases be either sender initiated or receiver initiated.

It should be stressed that the best integer solution has to be known to all processors. For shared address space computers this is no problem, but for a message passing computer each processor has to have a local copy of the best integer solution and in case it is updated broadcast it to all other processors.

23.4.1.2 Logic-Based Programming

As explained in the previous section logic-based programming can be used in a branch and bound algorithm to improve performance. Not only the branch and bound algorithm can be parallelized. Also logic-based programming can be parallelized as explained in [37]. For certain problems these algorithms can obtain orders of magnitude in speedup over distributed branch and bound search [37].

23.4.2 Parallel Implementations at Intermediate Level

In this section it is discussed how the subproblems, the relaxations, in branch and bound can be solved more efficiently using parallelization. The relaxed problems are either of QP type or of SDP type. However, the main focus will be on QP problems.

23.4.2.1 The Relaxed Problems in Branch and Bound

As noticed in [2], the QP relaxations used in branch and bound for a hybrid system in MLD form are ordinary linear MPC problems (possibly with an affine system description). This means in principle that parallelization at this level can be performed as for linear MPC. However, this does not mean that this is the most beneficial way, since it might be more beneficial to spend the computational power at another computational level in the algorithm. Since the relaxations are linear MPC problems, also work on distribution of such problems are of interest.

An important key to parallelization of an optimization algorithm is the ability to, in some way, decompose the problem. This topic is by no means new and dates back to the beginning of the 60s. Some fundamental and relevant references on this topic are [17, 18, 20, 22, 32, 33] and the references therein.

23.4.2.2 Tailored Solvers for the QP Subproblems

A tailored solver is a solver that in some way exploits properties of the problem to be solved in order to improve the performance. In [2–4], it was concluded that in order to get good performance for the hybrid MPC problem, tailored, dual active-set-like QP solvers should be used to solve the QP problems in the nodes of branch and bound. AS solvers are preferred over IP solvers, since the former can be efficiently warm-started. Similarly, dual QP solvers can more easily make use of a good initial guess of a (dual) solution to the problem since the geometry of the feasible set in the dual is very simple. The ability to be able to warm-start efficiently is very important in branch and bound since many QP problems are often necessary to solve in order to solve one MIQP problem. In order to overcome the well-known limitation of classical AS methods that one potentially costly iteration is necessary for each addition or removal of a constraint to the working set, a projection method was developed in [4]. In this method, steepest descent directions and Newton directions are projected onto the dual feasible set and the result is a solver which is able to make large changes to the working set, using only a small number of iterations. This is especially important if large disturbances, or large changes in the reference signal, can occur. The tailoring for MPC is in this algorithm performed at the linear algebra level. The steepest descent directions and Newton directions are very efficiently computed with $\mathcal{O}(N)$ computational complexity. Furthermore, the line searches are also exploiting the problem structure. This is further explained in Sect. 23.4.3. Some other variants of

QP solvers tailored for MPC, or at least designed with this application in mind, can be found in [12, 24, 30, 41, 42, 48].

The previously developed algorithms in [2–4] are designed mainly for non-parallel architectures. On parallel architectures there might exist other types of algorithms that potentially can be useful. However, we still believe that some form of dual active-set-like QP solver built on projection of first and second order directions is promising. For these solvers, the number of iterations is kept low but the cost for each iteration is higher. This is a result of that large changes of the working set is common, and hence, updates of factorizations are not that beneficial anymore as in a classical AS solver. Instead, the factorization is in each iteration computed from scratch as in an IP method. A smaller number of computationally heavier iterations seems appealing from a parallel point of view, since it seems more promising to parallelize the computations within one iteration, rather than working on several iterations simultaneously.

23.4.3 Parallel Implementations at Low Level

At the lowest level of the optimization algorithm, parallelization of the linear algebra is the main focus. It is clear that the linear algebra operations in the solver can in principle be performed, more or less, concurrently. See for example [19]. At this level it is interesting to investigate how this can be performed in the best possible way for the hybrid MPC application both in cases where there exists some kind of interconnection structure in the system and when such a structure does not exist.

Today, it is well-known how to tailor most optimization algorithms for the MPC application. Tailoring usually means in this case to exploit the almost block diagonal structure of the KKT system in order to solve this system of equations efficiently. Examples of algorithms for different types of optimization problems related to MPC that exploit this structure can be found in [2–5, 12, 21, 27, 30, 31, 41, 46]. The key to solve these equations efficiently is to factor the coefficient matrix for the KKT system efficiently, which in this case either can be performed using a Riccati recursion or some other generic algorithm that is able to exploit the sparsity in the problem. Using either of these two approaches, the system of equations can be solved with a computational complexity that grows as $\mathcal{O}(N)$ which can be compared to applicable generic dense factorizations which complexity grows as $\mathcal{O}(N^3)$.

A fundamental property of the Riccati recursion is that it factors the coefficient matrix by performing operations on matrices of the size in the order of the state and control signal dimension and it works using recursions backwards and forwards along the prediction horizon. If the number of processors is moderate compared to the state and control signal dimension, it seems reasonable to parallelize the operations that are performed within a single step of the Riccati recursion using standard techniques like those in [19]. However, it is not obvious how to extend the Riccati method to the case where a large number of processors are available, since there is a limit on how much computational power that is necessary for each step in the recursion and the next iteration relies on the result of the previous one. One possibility in order to limit this bottle neck could be to concatenate the dynamics from several time steps and consider a reformulated problem with higher state and control signal dimensions, but with a shorter prediction horizon. For the case with a large number of processors, it might be interesting to search for a completely new way to attack the problem designed to fully exploit parallelism, or to investigate how far it is possible to reach with existing generic parallel methods.

23.5 Conclusions

In this chapter we have discussed parallel implementations of hybrid MPC. Different methods for achieving parallelism at different computational levels have been outlined. At the highest level we have discussed how the branch and bound algorithm for solving hybrid MPC problems can be parallelized. At an intermediate level we have discussed how the relaxed QP problems in branch and bound can be parallelized, and finally at the lowest level we have discussed how the linear algebra can be parallelized. It is clear from what has been said above that there are many possibilities to obtain parallelism in hybrid MPC, and it is not clear which one that is the most efficient one. This is an interesting topic for future research.

References

1. S. Arvindam, V. Kumar, V.N. Rao, Floorplan optimization on multiprocessors, in *Proceedings of the 1989 International Conference on Computer Design*, pp. 109–114, Hyatt Regency Hotel, Cambridge, USA, October 1989
2. D. Axehill, Integer quadratic programming for control and communication, PhD thesis, Linköping University, 2008
3. D. Axehill, A. Hansson, A mixed integer dual quadratic programming algorithm tailored for MPC, in *Proceedings of the 45th IEEE Conference on Decision and Control*, pp. 5693–5698, Manchester Grand Hyatt, San Diego, USA, December 2006
4. D. Axehill, A. Hansson, A dual gradient projection quadratic programming algorithm tailored for model predictive control, in *Proceedings of the 47th IEEE Conference on Decision and Control*, pp. 3057–3064, Fiesta Americana Grand Coral Beach, Cancun, Mexico, December 2008
5. D. Axehill, A. Hansson, L. Vandenberghe, Relaxations applicable to mixed integer predictive control—comparisons and efficient computations, in *Proceedings of the 46th IEEE Conference on Decision and Control*, pp. 4103–4109, Hilton New Orleans Riverside, New Orleans, USA, December 2007
6. D. Axehill, M. Morari, Improved complexity analysis of branch and bound for hybrid MPC, in *Proceedings of the 49th IEEE Conference on Decision and Control*, pp. 4216–4222, Hilton Atlanta, Atlanta, USA, December 2010
7. D. Axehill, J. Sjöberg, Adaptive cruise control for heavy vehicles—hybrid control and MPC, Master's thesis, Linköpings universitet, February 2003
8. D. Axehill, L. Vandenberghe, A. Hansson, Convex relaxations for mixed integer predictive control. Automatica **46**(9), 1540–1545 (2010)

9. D. Axehill, A. Hansson, Towards parallel implementation of hybrid MPC, in *Distributed Decision Making and Control, chapter 14*, ed. by R. Johansson, A. Rantzer (Springer, Berlin, 2011), pp. 313–338
10. M. Baotic, Optimal control of piecewise affine systems—a multi-parametric approach. PhD thesis, ETH, March 2005
11. T. Barth, B. Freisleben, M. Grauer, F. Thilo, Distributed solution of optimal hybrid control problems on networks of workstations, in *Proceedings Second IEEE International Conference on Cluster Computing*, p. 162, Chemnitz, Germany, November 2000
12. R.A. Bartlett, L.T. Biegler, J. Backstrom, V. Gopal, Quadratic programming algorithms for large-scale model predictive control. J. Process Control **12**, 775–795 (2002)
13. A. Bemporad, Efficient conversion of mixed logical dynamical systems into an equivalent piecewise affine form. IEEE Trans. Automat. Control **49**(5), 832–838 (2004)
14. A. Bemporad, N. Giorgetti, Logic-based solution methods for optimal control of hybrid systems. IEEE Trans. Automat. Control **51**(6):963–976 (2006)
15. A. Bemporad, D. Mignone, A Matlab function for solving mixed integer quadratic programs version 1.02 user guide. Technical report, Institut für Automatik, ETH, 2000
16. A. Bemporad, M. Morari, Control of systems integrating logic, dynamics, and constraints. Automatica **35**, 407–427 (1999)
17. J.F. Benders, Partitioning procedures for solving mixed-variables programming problems. Numer. Math. **4**(1), 238–252 (1962)
18. D.P. Bertsekas, J.N. Tsitsiklis, *Parallel and Distributed Computation: Numerical Methods* (Prentice-Hall, Upper Saddle River, 1989)
19. L.S. Blackford, J. Choi, A. Cleary, E. D'Azevedo, J. Demmel, I. Dhillon, J. Dongarra, S. Hammarling, G. Henry, A. Petitet, K. Stanley, D. Walker, R.C. Whaley, *ScaLAPACK Users' Guide* (Society for Industrial and Applied Mathematics, Philadelphia, 1997)
20. G.B. Dantzig, P. Wolfe, Decomposition principle for linear programs. Oper. Res. **8**(1), 101–111 (1960)
21. M. Diehl, H.J. Ferreau, N. Haverbeke, Nonlinear model predictive control, in *Efficient Numerical Methods for Nonlinear MPC and Moving Horizon Estimation* (Springer, Berlin, 2009), pp. 391–417
22. H. Everett, Generalized Lagrange multiplier method for solving problems of optimum allocation of resources. Oper. Res. **11**(3), 399–417 (1963)
23. G. Ferrari-Trecate, D. Mignone, D. Castagnoli, M. Morari, Mixed logical dynamical model of a hydroelectric power plant, in *Proceedings of the 4th International Conference Automation of Mixed Processes: Hybrid Dynamic Systems*, Dortmund, Germany, 2000
24. H.J. Ferreau, H.G. Bock, M. Diehl, An online active set strategy to overcome the limitations of explicit MPC. Int. J. Robust Nonlinear Control **18**(8), 816–830 (2008)
25. R. Fletcher, S. Leyffer, Numerical experience with lower bounds for MIQP branch-and-bound. SIAM J. Optim. **8**(2), 604–616 (May 1998)
26. A.Y. Grama, V. Kumar, A survey of parallel search algorithms for discrete optimization problems. ORSA J. Comput. **7**(4), 365–385 (1995)
27. A. Hansson, A primal-dual interior-point method for robust optimal control of linear discrete-time systems. IEEE Trans. Automat. Control **45**(9):1639–1655 (2000)
28. I. Harjunkoski, V. Jain, I. Grossmann, Hybrid mixedinteger/constraint logic programming strategies for solving scheduling and combinatorical optimization problems. Comput. Chem. Eng. **24**, 337–343 (2000)
29. W.P.M.H. Heemels, B. De Schutter, A. Bemporad, Equivalence of hybrid dynamical models. Automatica **37**, 1085–1091 (2001)
30. H. Jonson, A Newton method for solving non-linear optimal control problems with general constraints, PhD thesis, Linköpings Tekniska Högskola, 1983
31. M. Åkerblad, A. Hansson, Efficient solution of second order cone program for model predictive control. Int. J. Control **77**(1), 55–77 (2004)
32. L.S. Lasdon, *Optimization Theory for Large Systems* (MacMillan, New York, 1970)
33. L.S. Lasdon, *Optimization Theory for Large Systems* (Dover, New York, 2002)

34. J. Löfberg, Yalmip: a toolbox for modeling and optimization in MATLAB, in *Proceedings of the CACSD Conference*, Taipei, Taiwan, 2004
35. B. Lie, M.D. Díez, T.A. Hauge, A comparison of implementation strategies for MPC. Model. Identif. Control **26**(1), 39–50 (2005)
36. E. Mestan, E.M. Turkay, Y. Arkun, Optimization of operations in supply chain systems using hybrid systems approach and model predictive control. Ind. Eng. Chem. Res. **45**, 6493–6503 (2006)
37. P.J. Modi, W.-M. Shen, M. Tambe, Adopt: asynchronous distributed constraint optimization with quality guarantees. Artif. Intell. **161**, 149–180 (2005)
38. J. Nocedal, S.J. Wright, *Numerical Optimization*, 2nd edn (Springer, Berlin, 2006)
39. G. Ottosson, Integration of constraint programming and integer programming for combinatorial optimization, PhD thesis, Computer Science Department, Information Technology, Uppsala, Sweden, 2000
40. P.M. Pardalos, L. Pitsolulis, T. Mavridou, M.G.C. Resende, Parallel search for combinatorial optimization: Genetic algorithms, simulated annealing, tabu search and GRASP, in *Parallel Algorithms for Irregularly Structured Problems*, vol. 980, Lecture Notes in Computer Science, ed. by P. Sanders (Springer, Berlin, 1995), pp. 317–331
41. C.V. Rao, S.J. Wright, J.B. Rawlings, Application of interior-point methods to model predictive control. J. Optim. Theory Appl. **99**(3), 723–757 (1998)
42. S. Richter, C.N. Jones, M. Morari, Real-time input-constrained MPC using fast gradient methods, in *Proceedings of the 48h IEEE Conference on Decision and Control (CDC) held jointly with 2009 28th Chinese Control Conference*, pp. 7387–7393, Shanghai, China, 2009
43. A.N. Tarau, B. de Schutter, J. Hellendoorn, Centralized, decentralized, and distributed model predictive control for route choice in automated baggage handling systems. J. Control Eng. Appl. Inf. **11**(3), 24–31 (2009)
44. F.D. Torrisi, A. Bemporad, HYSDEL—a tool for generating computational hybrid models for analysis and synthesis problems. IEEE Trans. Automat. Control **12**(2):235–249 (2004)
45. E.P.K. Tsang, *Foundations of Constraint Satisfaction* (Academic Press, London, 1993)
46. L. Vandenberghe, S. Boyd, M. Nouralishahi, Robust linear programming and optimal control. Technical report, Department of Electrical Engineering, University of California Los Angeles, 2002, 2002)
47. O.V. Volkovich, V.A. Roshchin, I.V. Sergienko, Models and methods of solution of quadratic integer programming problems. Cybernetics **23**, 289–305 (1987)
48. Y. Wang, S. Boyd, Fast model predictive control using online optimization. IEEE Trans. Control Syst. Technol. **18**(2):267–278 (2010)
49. L.A. Wolsey, *Integer Programming* (John Wiley, New York, 1998)

Chapter 24
A Hierarchical MPC Approach with Guaranteed Feasibility for Dynamically Coupled Linear Systems

M. D. Doan, T. Keviczky and B. De Schutter

Abstract In this chapter we describe an iterative two-layer hierarchical approach to MPC of large-scale linear systems subject to coupled linear constraints. The algorithm uses constraint tightening and applies a primal-dual iterative averaging procedure to provide feasible solutions in every sampling step. This helps overcome typical practical issues related to the asymptotic convergence of dual decomposition based distributed MPC approaches. Bounds on constraint violation and level of suboptimality are provided. The method can be applied to large-scale MPC problems that are feasible in the first sampling step and for which the Slater condition holds (i.e., there exists a solution that strictly satisfies the inequality constraints). Using this method, the controller can generate feasible solutions of the MPC problem even when the dual solution does not reach optimality, and closed-loop stability is also ensured using bounded suboptimality.

24.1 Introduction

When there are couplings among linear subsystems in a distributed MPC problem, dual decomposition is often used in order to divide the computational tasks among the subsystems. A typical requirement of the dual decomposition-based methods is that the dual problem needs to be solved exactly in order to obtain a primal

M. D. Doan · T. Keviczky (✉) · B. Schutter
Delft Center for Systems and Control, Delft University of Technology, Delft, The Netherlands
e-mail: t.keviczky@tudelft.nl

M. D. Doan
e-mail: m.d.doan@tudelft.nl

T. Keviczky
e-mail: t.keviczky@tudelft.nl

B. Schutter
e-mail: b.deschutter@tudelft.nl

J. M. Maestre and R. R. Negenborn (eds.), *Distributed Model Predictive Control Made Easy*, Intelligent Systems, Control and Automation: Science and Engineering 69,
DOI: 10.1007/978-94-007-7006-5_24, © Springer Science+Business Media Dordrecht 2014

feasible solution [1]. However, iterative approaches based on dual decomposition often only converge asymptotically to the optimum, which may not be practical when implementing these approaches in a real-time environment. In this chapter, we apply a dual decomposition technique to the class of MPC problems for linear systems with coupled dynamics and coupled linear constraints, and propose a novel method that is motivated by the use of constraint tightening in robust MPC [5]. This method allows terminating the iterations for the dual problem before reaching convergence while still guaranteeing a feasible primal solution to be found. Moreover, the algorithm also generates a decreasing cost function, leading to closed-loop stability. In summary, the proposed framework guarantees primal feasible solutions and MPC stability using a finite number of iterations with bounded suboptimality.

24.2 Boundary Conditions

Our approach aims at large-scale interconnected systems with constrained discrete-time linear time-invariant dynamics where some of the individual constraints can also be defined over more than one subsystem (coupling constraints). This class includes a wide range of applications, e.g., water and power distribution networks, urban traffic networks, industrial processes, arrays of mechanical actuators, and climate control systems, among others [6–9, 12]. Mathematically, let M be the number of subsystems ($\mathcal{N} = \{1, \ldots, M\}$), and let the dynamical model of each subsystem be represented in the following form:

$$x^i_{k+1} = \sum_{j=1}^{M} \mathbf{A}^{ij} x^j_k + \mathbf{B}^{ij} u^j_k, \quad i \in \mathcal{N} \tag{24.1}$$

The corresponding centralized state-space model is written in a compact form:

$$x_{k+1} = \mathbf{A} x_k + \mathbf{B} u_k \tag{24.2}$$

where $x_k = \left[(x^1_k)^T (x^2_k)^T, \ldots, (x^M_k)^T\right]^T$, $u_k = \left[(u^1_k)^T (u^2_k)^T, \ldots, (u^M_k)^T\right]^T$, $\mathbf{A} = [\mathbf{A}_{ij}]_{i,j \in \mathcal{N}}$ and $\mathbf{B} = [\mathbf{B}_{ij}]_{i,j \in \mathcal{N}}$.

We formulate an MPC problem using a terminal penalty and a terminal constraint set. At a particular time step t the MPC optimization problem is defined as follows:

$$\min_{\mathbf{u},\mathbf{x}} \sum_{k=t}^{t+N_\mathrm{p}-1} \left(x_k^T \mathbf{Q} x_k + u_k^T \mathbf{R} u_k \right) + x_{t+N_\mathrm{p}}^T \mathbf{P} x_{t+N_\mathrm{p}} \tag{24.3}$$

$$\text{s.t.} \quad x^i_{k+1} = \sum_{j \in \mathcal{N}^i} \mathbf{A}^{ij} x^j_k + \mathbf{B}^{ij} u^j_k, \quad i \in \mathcal{N}, \quad k = t, \ldots, t + N_\mathrm{p} - 1 \tag{24.4}$$

$$x_k \in \mathcal{X}, \quad k = t + 1, \ldots, t + N_\mathrm{p} - 1 \tag{24.5}$$

$$x_{t+N_\text{p}} \in \mathcal{X}_\text{f} \subset \mathcal{X} \tag{24.6}$$

$$u_k \in \mathcal{U}, \quad k = t, \ldots, t + N_\text{p} - 1 \tag{24.7}$$

$$u_k^i \in \Omega_i, \quad i \in \mathcal{N}, \quad k = t, \ldots, t + N_\text{p} - 1 \tag{24.8}$$

$$x_t = x(t) \in \mathcal{X} \tag{24.9}$$

where $\mathbf{u} = [u_t^T, \ldots, u_{t+N_\text{p}-1}^T]^T$, $\mathbf{x} = [x_{t+1}^T, \ldots, x_{t+N_\text{p}}^T]^T$, the matrices $\mathbf{Q}$, $\mathbf{P}$, and $\mathbf{R}$ are block-diagonal and positive definite, the constraint sets $\mathcal{U}$, $\mathcal{X}$, and $\mathcal{X}_\text{f}$ are convex polytopes and have nonempty interiors, and each local constraint set Ω_i is a hyperbox. Each subsystem i is assigned a neighborhood, denoted by $\mathcal{N}^i$, containing subsystems that have direct dynamical interactions with subsystem i, including itself. Note that the coupled constraints involving control inputs are represented by the set $\mathcal{U}$ in (24.7), which does not incorporate local constraints that are captured by Ω_i in (24.8). The initial state x_t is the current state at time step t. Notice that the MPC formulation (24.3–24.9) *does not* incorporate a terminal zero-point constraint ($x_{t+N_\text{p}} = 0$). Moreover, the approach proposed in this chapter does not handle any equality constraints on the states except for the dynamical constraints (24.4), since the corresponding constraint sets would have an empty interior, which would prevent constraint tightening, the key idea of this approach.

As $\mathcal{U}$, $\mathcal{X}$, and $\mathcal{X}_\text{f}$ are polytopes, the constraints (24.5) and (24.6) are represented by linear inequalities. Moreover, the state vector $\mathbf{x}$ is affinely dependent on $\mathbf{u}$. Hence, we can eliminate the state variables $x_{t+1}, \ldots, x_{t+N_\text{p}}$ and transform the constraints (24.4)–(24.6) into linear inequalities on the input variable $\mathbf{u}$. Eliminating the state variables in (24.3)–(24.9) leads to an optimization problem in the following form:

$$f_t^* = \min_{\mathbf{u}} \quad f(\mathbf{u}, x_t) \tag{24.10}$$

$$\text{s.t.} \quad g(\mathbf{u}, x_t) \leq 0 \tag{24.11}$$

$$\mathbf{u} \in \tilde{\Omega} \tag{24.12}$$

where f is a convex quadratic function, $g = [g_1, \ldots, g_m]^T$ with $g_i, i = 1, \ldots, m$ affine functions, and $\tilde{\Omega} = \Omega\times, \ldots, \times\Omega$ (N_p times) where $\Omega = \Omega_1\times, \ldots, \times\Omega_M$, is a hyperbox. Note that $f(\mathbf{u}, x_t) > 0, \forall\mathbf{u} \neq 0, x_t \neq 0$, since $\mathbf{Q}$, $\mathbf{P}$, and $\mathbf{R}$ are positive definite.

In addition, our approach also makes use of the following assumptions:

Assumption 24.1 There exists a block-diagonal feedback gain $\mathbf{K}$ such that the matrix $\mathbf{A} + \mathbf{BK}$ is Schur[1] (i.e., $\mathbf{K}$ yields a decentralized stabilizing control law for the unconstrained aggregate system).

Assumption 24.2 The terminal constraint set $\mathcal{X}_\text{f}$ is strictly positively invariant for the closed-loop system with $x_{k+1} = (\mathbf{A} + \mathbf{BK})x_k$, i.e., if $x \in \text{int}(\mathcal{X}_\text{f})$ then $(\mathbf{A} +$

[1] A matrix is Schur if all of its eigenvalues are inside the unit circle.

$\mathbf{BK})x \in \text{int}(\mathcal{X}_\text{f})$. In addition, for any state in $\mathcal{X}_\text{f}$, the control input generated by the terminal controller should satisfy the input constraints, i.e., $-\mathbf{K}x \in \mathcal{U} \cap \Omega,\ \forall x \in \mathcal{X}_\text{f}$.

Assumption 24.3 The Slater condition holds for the constraint (24.11), i.e., there exists a feasible vector that satisfies (24.11) with strict inequality constraints [1]. It is also assumed that prior to computing the control input to be implemented in each time step t, a Slater vector $\bar{\mathbf{u}}_t$ is available, such that

$$\begin{cases} g_j(\bar{\mathbf{u}}_t, x_t) < 0,\ j = 1, \ldots, m \\ \bar{\mathbf{u}}_t \in \tilde{\Omega} \end{cases} \tag{24.13}$$

Assumption 24.4 At each time step t, the following inequality holds:

$$f(\mathbf{u}_{t-1}, x_{t-1}) > f(\bar{\mathbf{u}}_t, x_t) \tag{24.14}$$

For later reference, we define $\Delta_t > 0$ which can be computed before time step t as follows:

$$\Delta_t = f(\mathbf{u}_{t-1}, x_{t-1}) - f(\bar{\mathbf{u}}_t, x_t) \tag{24.15}$$

Assumption 24.5 For each $x_t \in \mathcal{X}$, the Euclidean norm of $g(\mathbf{u}, x_t)$ is bounded:

$$\exists L_t : L_t \geq \|g(\mathbf{u}, x_t)\|_2,\ \forall \mathbf{u} \in \tilde{\Omega} \tag{24.16}$$

Assumptions 24.1 and 24.2 are almost standard in MPC, except that in Assumption 24.2 we require *strict positive invariance* instead of the milder condition of *positive invariance* that is often employed in MPC literature. Assumptions 24.3–24.5 will help us to provide bounds on the constraint violation and the suboptimality of the solution. These assumptions can be satisfied if the search domain is bounded, i.e., there is a lower bound and an upper bound for every control input, which is usually the case in MPC. Assumption 24.4 implies that the decreasing property of the cost function can be obtained directly by using $\bar{\mathbf{u}}_t$ as the solution at time step t, i.e., with the choice $\mathbf{u}_t = \bar{\mathbf{u}}_t$. However, this choice will likely deteriorate the performance of the MPC controller due to the increasing conservativeness of $\bar{\mathbf{u}}_t$ as t increases. Therefore, our approach will make use of $\bar{\mathbf{u}}_t$ only as a starting point, based on which a new solution is computed at every time step t. Moreover, we have developed a method in [2, 4] to find $\bar{\mathbf{u}}_t$ in every time step $t \geq 1$, which makes Assumption 24.4 easy to fulfill. A summary of this method is given later in Remark 24.1.

The communication architecture is assumed to be hierarchical, i.e., there is a coordinator for computing common parameters and performing limited communication with all agents. Each agent only communicates with the coordinator and a small number of other agents (belonging to set $\mathcal{N}^i$) that are considered as its *neighbors*.

24.3 Description of the Approach

Our approach aims at solving a tightened version of problem (24.10)–(24.12) in a hierarchical manner by using a dual decomposition technique. We first construct the tightened problem and its dual problem, and then describe the algorithm.

24.3.1 Initialization

Prior to applying our algorithm to solve the MPC optimization problem at each time step t, there is an initialization which includes two main tasks: formulating the tightened problem to be solved, and determining a sufficient number of iterations to be performed in that time step.

24.3.1.1 Formulation of the Tightened Problem

At the beginning of each MPC step t, a new Slater vector $\bar{\mathbf{u}}_t$ (cf. Assumption 23.3) should be determined. Based on this $\bar{\mathbf{u}}_t$, a value c_t is computed such that:

$$0 < c_t < \min_{j=1,\ldots,m} \{-g_j(\bar{\mathbf{u}}_t, x_t)\} \tag{24.17}$$

Then we construct the tightened problem:

$$f_t'^* = \min_{\mathbf{u}} \quad f(\mathbf{u}, x_t) \tag{24.18}$$

$$\text{s.t.} \quad g'(\mathbf{u}, x_t) \leq 0 \tag{24.19}$$

$$\mathbf{u} \in \tilde{\Omega} \tag{24.20}$$

with the tightened constraint:

$$g'(\mathbf{u}, x_t) \triangleq g(\mathbf{u}, x_t) + \mathbf{1}_m c_t \tag{24.21}$$

where $g'(\mathbf{u}, x_t) = [g_1', \ldots, g_m']^T$, and $\mathbf{1}_m$ is a vector of ones with m elements.

We denote the dual variable as $\mu \in \mathbb{R}_+^m$ and define the Lagrangian function:

$$\tilde{\mathcal{L}}(\mathbf{u}, \mu, x_t) \triangleq f(\mathbf{u}, x_t) + \mu^T g'(\mathbf{u}, x_t) \tag{24.22}$$

A bound L_t on the 2-norm of the constraints g should be given (cf. Assumption 24.5). Using (24.16) and the triangle inequality of the 2-norm, we will get $L_t' = L_t + c_t$ as a norm bound for g', i.e., $L_t' \geq \|g'(\mathbf{u}, x_t)\|_2, \forall \mathbf{u} \in \tilde{\Omega}$.

Using the Slater vector $\bar{\mathbf{u}}_t$, we then compute Δ_t using (24.15). After that we choose two parameters $\alpha_t > 0$ and $\varepsilon_t > 0$ such that $\alpha_t L_t'^2/2 + \varepsilon_t \leq \Delta_t$. Later α_t

will be used as the step size of the algorithm for maximizing the dual function, and ε_t will be the desired suboptimality of the algorithm for minimizing the Lagrangian function.

Remark 24.1 In [2, 4], we describe a method to update L_t and to construct a new Slater vector $\bar{\mathbf{u}}_t$ for each time step $t \geq 1$. In summary, L_t is updated by taking into account the change of the initial state from x_{t-1} to x_t; and the new $\bar{\mathbf{u}}_t$ is constructed by shifting the previous solution $\mathbf{u}_{t-1}$ one-step ahead and adding the terminal state and terminal input, which are feasible due to Assumption 24.2. For the first MPC step, i.e., $t = 0$, these values can be computed offline as follows.

In order to determine L_0, we formulate a maximization problem to find the maximum 2-norm of $g(\mathbf{u}, x_0)$, in which $\mathbf{u} \in \tilde{\Omega}$ and $x_0 \in \mathcal{X}$. This optimization problem has a convex cost function and a convex constraint set, and hence the maximum will occur at one of the vertices of the constraint set. As a consequence, finding the solution of this maximization problem is straightforward by evaluating the 2-norm of g at all the vertices. The maximal function value is then assigned to L_0.

For finding $\bar{\mathbf{u}}_0$, we can start with a guess for c_t, then tighten the constraints using (24.21), and use a centralized solver to find a feasible point of the constraints $g'(\mathbf{u}, x_t) \leq 0$. If this problem is feasible, then the feasible input sequence obtained can be set as $\bar{\mathbf{u}}_0$, and we may increase c_t to find a better Slater vector. If this problem is infeasible, then we reduce c_t until a feasible solution is obtained, which can be used as $\bar{\mathbf{u}}_0$.

24.3.1.2 Determination of the Number of Iterations

The problem (24.18)–(24.20) will be solved by a nested iteration. In the outer loop, the dual function is maximized using a projected gradient method in combination with an averaging scheme that provides bounds for the constraint violation and the cost function. In the inner loop, a hierarchical optimization algorithm is used to provide—with a desired precision—an approximate solution to the minimization of the Lagrangian function. The outer loop is executed for $\tilde{k}_t$ iterations, each of which includes $\tilde{p}_t$ iterations of the inner loop.

The algorithm is simple and is presented in Sect. 24.3.2. Before starting the iterative algorithm, we let a coordinator determine $\tilde{k}_t$ and $\tilde{p}_t$.

As described in [2, 4], we can determine a sufficient minimum number of outer iterations $\tilde{k}_t$ as follows:

$$\tilde{k}_t \geq \frac{1}{\alpha_t c_t}\left(\frac{3}{\gamma_t} f(\bar{\mathbf{u}}_t, x_t) + \frac{\alpha_t L_t'^2}{2\gamma_t} + \alpha_t L_t'\right) \tag{24.23}$$

where $\gamma_t = \min_{j=1,\ldots,m}\{-g_j'(\bar{\mathbf{u}}_t, x_t)\} = \min_{j=1,\ldots,m}\{-g_j(\bar{\mathbf{u}}_t, x_t)\} - c_t$. The number of inner iterations $\tilde{p}_t$ can be computed by:

$$\tilde{p}_t \geq \log_\phi \frac{\varepsilon_t}{\Lambda M \max_i D_i} \tag{24.24}$$

where Λ is the Lipschitz constant of the Lagrangian function $\tilde{\mathcal{L}}$ over the set $\tilde{\Omega}$, D_i is the diameter of the set $\tilde{\Omega}_i$ with respect to the Euclidean norm, i.e., $\|\mathbf{u}_1^i - \mathbf{u}_2^i\|_2 \leq D_i, \forall \mathbf{u}_1^i, \mathbf{u}_2^i \in \tilde{\Omega}_i$, with $\tilde{\Omega}_i \triangleq \Omega_i \times, \ldots, \times \Omega_i$ (N_p times). The value $\phi \in [0, 1)$ is the contraction modulus of the Jacobi algorithm, i.e., the decay rate of the distance from a current iterate to the optimizer, and can be computed by [4]:

$$\phi = \max_i \left\{ \max \left\{ 2\gamma\big(\lambda_{\max}(\mathbf{H}_{ii}) + \sum_{j \neq i} \bar{\sigma}(\mathbf{H}_{ij})\big) - 1, \right.\right.$$
$$\left.\left. 1 - 2\gamma\big(\lambda_{\min}(\mathbf{H}_{ii}) - \sum_{j \neq i} \bar{\sigma}(\mathbf{H}_{ij})\big) \right\}\right\} \tag{24.25}$$

where $\mathbf{H}_{ij}$ with $i, j \in \{1, \ldots, M\}$ denotes the submatrix of $\mathbf{H}$ (the Hessian of the Lagrangian function), containing entries at rows belonging to subsystem i and columns belonging to subsystem j. The notation $\lambda_{\max}(\mathbf{H}_{ii})$ and $\lambda_{\min}(\mathbf{H}_{ii})$ stands for the maximum and minimum eigenvalue of $\mathbf{H}_{ii}$, respectively, while $\bar{\sigma}(\mathbf{H}_{ij})$ denotes the maximum singular value of $\mathbf{H}_{ij}$. Note that depending on the matrices $\mathbf{H}_{ij}$, $i, j \in \{1, \ldots, M\}$, the Jacobi algorithm in the inner loop may or may not converge. When ϕ computed by (24.25) falls into $[0, 1)$, then it means the Jacobi algorithm converges and can be used. Otherwise, another algorithm should be used in the inner loop instead of the Jacobi algorithm; in this case a hierarchical conjugate gradient method can be employed, see [3] for more details.

24.3.2 The Proposed Algorithm

Algorithm (24.1) presents our proposed algorithm for solving the MPC problem at each time step t. Note that prior to applying the algorithm, we need to determine the Lagrangian and the parameters ε_t, α_t as described in the previous section.

Note that the computations (24.27) and (24.28) are performed by the subsystems, with synchronization provided by the coordinator. While solving the local problem (24.28), each subsystem needs to communicate with its neighbors to get the latest update for constructing the proper cost function in the current inner loop iteration.

24.4 Availability of Theoretical Results

In this section, we recall the theoretical guarantees of our method in the form of two propositions.

Algorithm 24.1 Hierarchical Primal Feasible method with Dual Approximate Gradient (HPF-DAG)

1: **Outer loop:** Set $\mu^{(0)} = \mathbf{0}_m$. For $k = 0, \ldots, \tilde{k}_t$, find $\mathbf{u}^{(k)}, \mu^{(k+1)}$ such that:

$$\tilde{\mathcal{L}}(\mathbf{u}^{(k)}, \mu^{(k)}, x_t) \leq \min_{\mathbf{u} \in \tilde{\Omega}} \tilde{\mathcal{L}}(\mathbf{u}, \mu^{(k)}, x_t) + \varepsilon_t \tag{24.26}$$

$$\mu^{(k+1)} = \mathcal{P}_{\mathbb{R}^m_+} \left\{ \mu^{(k)} + \alpha_t d^{(k)} \right\} \tag{24.27}$$

where $\mathcal{P}_{\mathbb{R}^m_+}$ denotes the projection onto the nonnegative orthant, $d^{(k)} = g'(\mathbf{u}^{(k)}, x_t)$ is an ε_t-subgradient of the dual function at $\mu^{(k)}$.
Inner loop:

- Solve problem (24.26) in a distributed way with a Jacobi algorithm. For $p = 0, \ldots, \tilde{p}_t$, every subsystem i computes:

$$\begin{aligned} \mathbf{u}^i(p+1) = \arg \min_{\mathbf{u}^i \in \tilde{\Omega}_i} \tilde{\mathcal{L}}(&\mathbf{u}^1(p), \ldots, \mathbf{u}^{i-1}(p), \mathbf{u}^i, \\ &\mathbf{u}^{i+1}(p), \ldots, \mathbf{u}^M(p), \mu^{(k)}) \end{aligned} \tag{24.28}$$

  where $\tilde{\Omega}_i$ is the local constraint set for control variables of subsystem i.
- Define $\mathbf{u}^{(k)} \triangleq [\mathbf{u}^1(\tilde{p}_t)^T, \ldots, \mathbf{u}^M(\tilde{p}_t)^T]^T$, which is guaranteed to satisfy (24.26).

2: Compute $\hat{\mathbf{u}}^{(\tilde{k}_t)} = \frac{1}{\tilde{k}_t} \sum_{l=0}^{\tilde{k}_t} \mathbf{u}^{(l)}$, and take $\mathbf{u}_t = \hat{\mathbf{u}}^{(\tilde{k}_t)}$ as the solution of (24.10)–(24.12).

Proposition 24.1 *Suppose Assumptions 24.1–24.5 hold. Construct g' as in (24.1). Let the outer loop (24.26)–(24.27) with $\mu^{(0)} = \mathbf{0}_m$ be iterated for $k = 0, \ldots, \tilde{k}_t$. Then $\hat{\mathbf{u}}^{(\tilde{k}_t)}$ computed in Algorithm 24.1 is a feasible solution of (24.10)–(24.12).*

This result is based on the bound of the constraint violation. Let g'^{+} denote the constraint violation, i.e., $g'^{+} = \max\{g', \mathbf{0}_m\}$, then the primal average sequence $\hat{\mathbf{u}}^{(k)} = \frac{1}{k} \sum_{l=0}^{k} \mathbf{u}^{(l)}$ generated at iteration $k \geq 1$ satisfies:

$$\left\| \left[g'\left(\hat{\mathbf{u}}^{(k)}, x_t\right) \right]^+ \right\|_2 \leq \frac{1}{k\tilde{\alpha}_t} \left(\frac{3}{\gamma_t} [f(\bar{\mathbf{u}}_t, x_t) - \tilde{q}'^*_t] + \frac{\tilde{\alpha}_t L'^2_t}{2\gamma_t} + \tilde{\alpha} L'_t \right) \tag{24.29}$$

The proof for (24.29) and Proposition 24.1 can be found in [4]. Proposition 24.1 guarantees that the solution of the algorithm is feasible for implementation.

Let us recall that the optimization problem formulation is motivated by an MPC problem for which Assumption 24.4 holds. The following proposition shows that the cost function of the MPC problem is a decreasing function.

Proposition 24.2 *Suppose Assumptions 24.1–2.5 hold. Then the solution $\hat{\mathbf{u}}^{(\tilde{k}_t)}$ generated by Algorithm 24.1 satisfies the following inequality:*

$$f(\mathbf{u}_t, x_t) < f(\mathbf{u}_{t-1}, x_{t-1}), \quad \forall t \in \mathbb{N}\backslash\{0\} \tag{24.30}$$

This result is made possible using the upper bound on the cost function that is associated to the primal average generated by Algorithm 24.1:

$$f\left(\hat{\mathbf{u}}^{(k)}, x_t\right) \leq f_t'^{*} + \frac{\|\mu^{(0)}\|_2^2}{2k\tilde{\alpha}_t} + \frac{\tilde{\alpha}_t L_t'^2}{2} + \varepsilon_t \tag{24.31}$$

The proof for (24.31) and Proposition 24.2 can be found in [4]. This result leads to closed-loop stability, by using the cost function $f(\mathbf{u}_t, x_t)$ as a Lyapunov function.

The idea of constraint tightening is used often to ensure robustness of MPC, such as when designing robust distributed MPC for the case of decoupled systems [5, 10]. However, to the best of our knowledge it has not been applied for the case of coupled systems. In our proposed framework, we use constraint tightening to provide a feasible solution to the coupled constraints, rather than aiming to achieve an explicit robustness property. We believe that this framework can also be extended for robust hierarchical MPC where coupled constraints are present.

24.5 Availability of Application Results

In this section, we demonstrate an application of the proposed algorithm for a system of irrigation canals, where the objective is to regulate the water flows. Irrigation canals are large-scale systems, consisting of many interacting components, and spanning vast geographical areas. For the most efficient and safe operation of these canals, maintaining the water levels close to pre-specified reference values is crucial, both under normal operating conditions as well as in extreme situations. Manipulation of the water levels and flows in irrigation canals is typically done using devices such as pumps and gates.

The example irrigation canal to be considered is a 4-reach canal system as illustrated in Figure 24.1. In this system, water flows from an upstream reservoir through the reaches, under the control of 4 gates and a pump at the end of the canal system that discharges water. More details about modeling of this system can be found in [2].

The control design is based on the master-slave control paradigm, in which the master controllers compute the flows through the gates, while each slave controller uses the local control actuators to guarantee the flow set by the master controller [11]. We now apply the proposed hierarchical MPC method to design the master controllers.

The Algorithm HPF-DAG is used in this example, and it generates at every time step t a feasible solution with respect to the physical constraints:

$$x_{\min} \leq x_{t+k} \leq x_{\max}, \; k = 1, \ldots, N_{\mathrm{p}} \tag{24.32}$$

$$u_{\min} \leq u_{t+k} \leq u_{\max}, \; k = 0, \ldots, N_{\mathrm{p}} - 1 \tag{24.33}$$

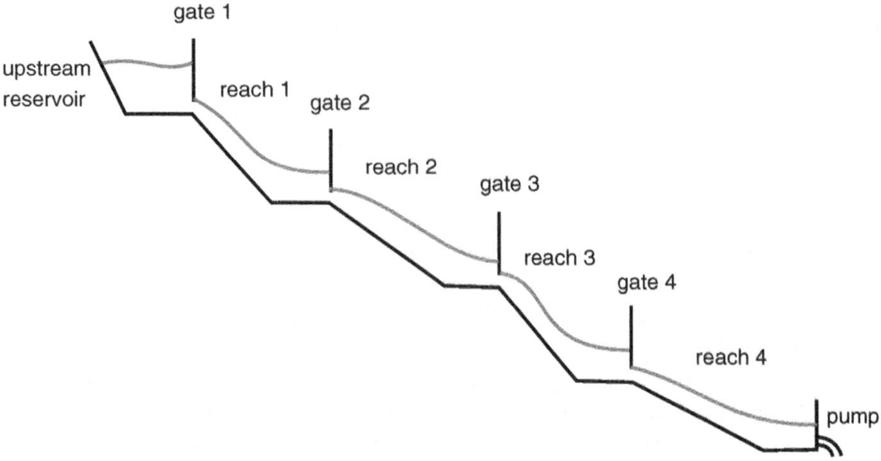

Fig. 24.1 Example of connected canals for water transportation

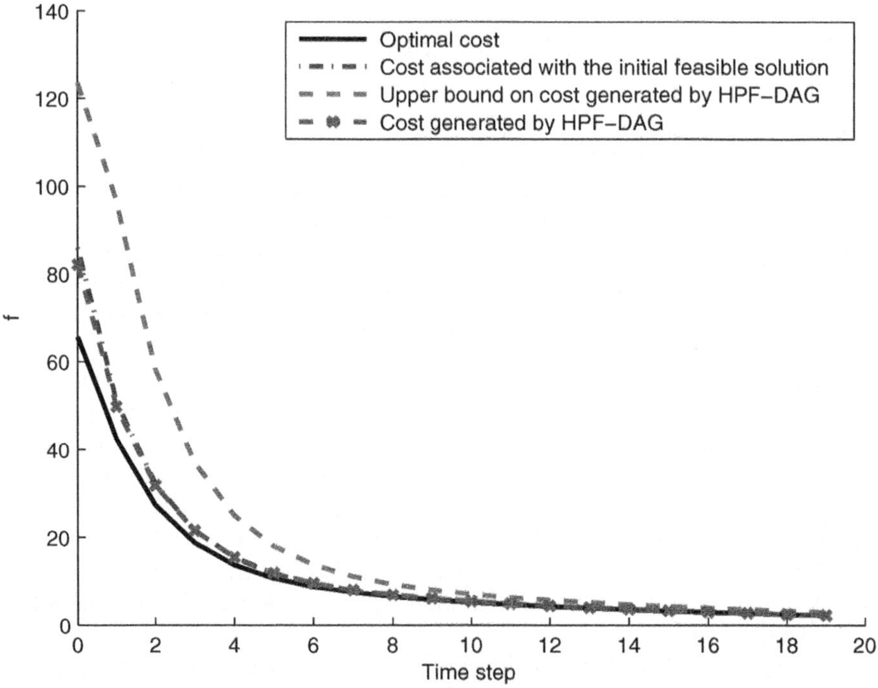

Fig. 24.2 Comparison of cost function evolutions in the simulation example

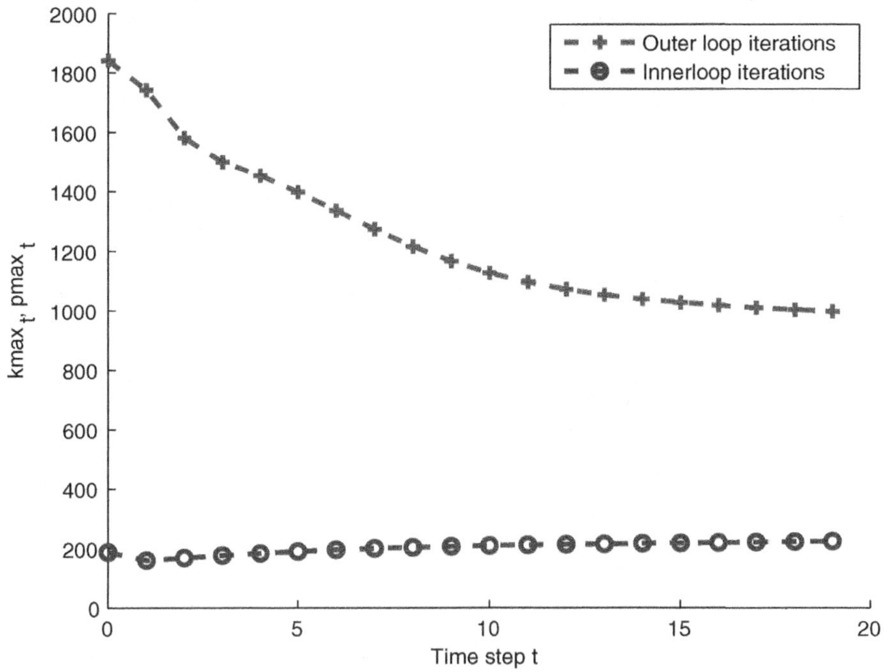

Fig. 24.3 Number of iterations in Algorithm HPF-DAG at each time step

The constraints described in (24.32) are then transformed into the inequality constraint $g(\mathbf{u}, x_t) \geq 0$ after the state variables are eliminated, and the constraints (24.33) are cast into the form $\mathbf{u} \in \tilde{\Omega}$ introduced earlier.

The closed-loop simulation is performed for 20 time steps. In each time step t, Algorithm 24.1 is used to generate a solution to the MPC optimization problem (24.3)–(24.9), with the number of outer loop iterations $\tilde{k}_t$ determined by (24.23), and the number of inner loop iterations $\tilde{p}_t$ determined by (24.24). Then the first control input in the solution generated by Algorithm 24.1 is used to simulate the system, and the routine restarts with the next time step. In every time step, we also solve the same optimization problem by a centralized solver to get the optimal solution and the optimal cost, for comparison purposes.

In Fig. 24.2, we plot the evolutions of cost functions associated with different MPC solutions. At every time step, we compare the optimal cost, the cost associated with the initial feasible solution $\bar{\mathbf{u}}_t$, the cost generated by Algorithm 24.1 (i.e., associated with $\mathbf{u}_t$), and the upper bound of the resulting cost (see 24.31). Although it is easy to obtain closed-loop stability with this example, the comparison of the cost nevertheless confirms that the upper bound of the cost is always respected by the result of the Algorithm HPF-DAG. Moreover, as t increases, the upper bound is closer to the optimal cost, thus the performance of the Algorithm HPF-DAG is also closer to the optimal centralized MPC performance.

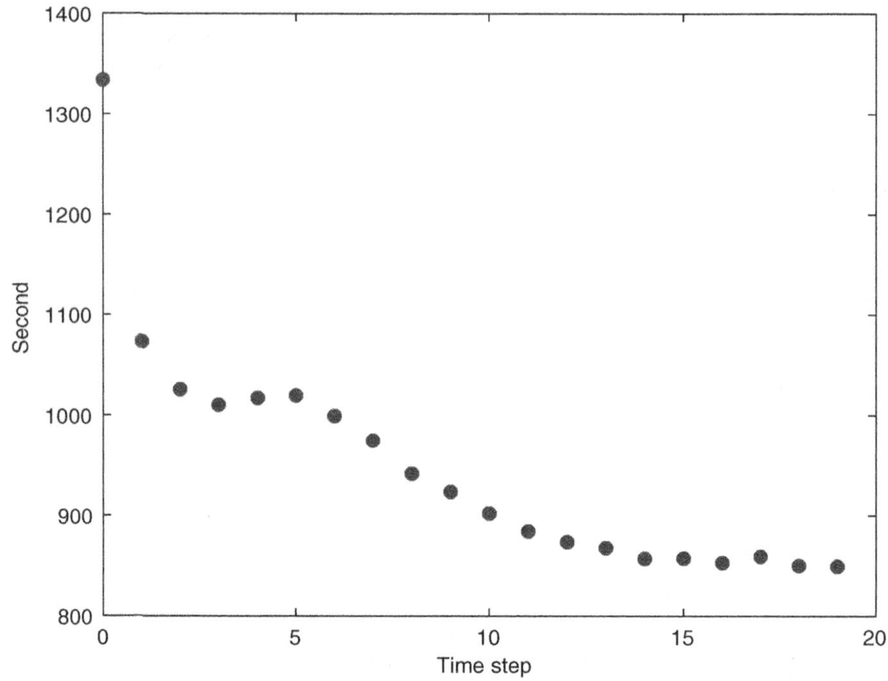

Fig. 24.4 Aggregate computation time over all subsystems for the simulation of Algorithm 24.1 at each time step

In Fig. 24.3, the number of outer loop iterations $\tilde{k}_t$ and the number of inner loop iterations $\tilde{p}_t$, are plotted for each sampling step. The corresponding CPU time required to run Algorithm 24.1 at each sampling step is shown in Fig. 24.4. The CPU time is measured by implementing all the computations in one PC, running MATLAB on Windows with an Intel(R) Core(TM) i7 CPU at 2.30 GHz and with 8 GB RAM. We can see that the overall computation time for the simulation is long, since in every sampling period, there must be $\tilde{k}_t \times \tilde{p}_t \times M$ small optimization problems solved, where M is the number of subsystems (in this example $M = 4$). If the algorithm is simulated in a distributed setting, the computations would be divided between the M subsystems, thus the computation time would be much less than what is indicated by this simulation. More details of the simulation example can be found in [2].

24.6 Conclusions and Future Research

In this chapter, we have presented a constraint tightening approach for solving MPC optimization problems involving large-scale linear systems with coupling in dynamics and constraints. This new approach provides guaranteed feasibility and stability after a finite number of iterations. The method is realized as a hierarchical algorithm, called HPF-DAG, in which the coordinator is in charge of determining the common parameters and the number of iterations, while the main computational tasks are divided up between the subsystems. We have provided the steps for implementation of this algorithm for the case of using a positive definite cost function and linear constraints. Future extensions of this approach can include the consideration of positive semidefinite cost (i.e., some states and inputs are allowed to have zero penalty weights), and more general (nonlinear) convex constraints. These extensions would affect the convergence rate of the Jacobi algorithm used in the inner loop, hence the number of inner iterations would change, or in some cases we would need to replace the Jacobi algorithm with another, more suitable one. Other topics for extensions include the *a posteriori* choice of the solution between either the Slater vector $\bar{\mathbf{u}}_t$ or the primal average $\hat{\mathbf{u}}^{(\bar{k}_t)}$ generated by the algorithm HPF-DAG; or determining optimal α_t, c_t, and ε_t parameters for fastest execution, while taking into account the communication effort.

Acknowledgments The work presented in this chapter has been supported by the European Union Seventh Framework STREP project *Hierarchical and Distributed Model Predictive Control (HD-MPC)* with contract number INFSO-ICT-223854.

References

1. D.P. Bertsekas, *Nonlinear programming* (Athena Scientific, Belmont, 1999)
2. M.D. Doan. *Distributed Model Predictive Controller Design Based on Distributed Optimization*. Ph.D. thesis (Delft University of Technology, Delft, 2012)
3. M.D. Doan, T. Keviczky, B. De Schutter. A dual decomposition-based optimization method with guaranteed primal feasibility for hierarchical MPC problems. in *Proceedings of the 18th IFAC World Congress* (Milan, 2011), pp. 392–397 Aug–Sept 2011
4. M.D. Doan, T. Keviczky, B. De Schutter. A distributed optimization-based approach for hierarchical MPC of large-scale systems with coupled dynamics and constraints. in *Proceedings of the 2011 50th IEEE Conference on Decision and Control and European Control Conference (CDC-ECC)* (Orlando, 2011), pp. 5236–5241 Dec 2011
5. Y. Kuwata, A. Richards, T. Schouwenaars, J.P. How, Distributed robust receding horizon control for multivehicle guidance. IEEE Trans. Control Syst. Technol. **15**(4), 627–641 (2007)
6. M. Mercangoz, F.J. Doyle III. Distributed model predictive control of an experimental four-tank system. J. Process Control **17**(3), 297–308 (2007)
7. P.-D. Morosan, R. Bourdais, D. Dumur, J. Buisson. A distributed MPC strategy based on Benders' decomposition applied to multi-source multi-zone temperature regulation. J. Process Control 21(5), 729–737, (2011). (Special Issue on Hierarchical and Distributed Model Predictive Control)
8. R.R. Negenborn, P.J. van Overloop, T. Keviczky, B. De Schutter, Distributed model predictive control for irrigation canals. Netw. Heterogen. Media **4**(2), 359–380 (2009)
9. V. B. Peccin and E. Camponogara. Distributed model predictive control applied to urban traffic networks: Implementation, experimentation, and analysis. in *2010 IEEE Conference on Automation Science and, Engineering* (2010), pp. 399–405, Aug 2010
10. A.G. Richards, J.P. How, Robust distributed model predictive control. Int. J. Control **80**(9), 1517–1531 (Sept 2007)

11. J. Schuurmans, A. Hof, S. Dijkstra, O.H. Bosgra, R. Brouwer, Simple water level controller for irrigation and drainage canals. J. Irrig. Drainage Eng. **125**(4), 189–195 (1999)
12. A. Venkat, I. Hiskens, J. Rawlings, S. Wright. Distributed MPC strategies with application to power system automatic generation control. IEEE Trans. Control Syst. Technol. (2007)

Chapter 25
Distributed MPC Based on a Team Game

J. M. Maestre, F. J. Muros, F. Fele, D. Muñoz de la Peña and E. F. Camacho

Abstract In this chapter we present a distributed scheme based on a team game for the particular case in which the system is controlled by two agents. The main features of the proposed scheme are the limited amount of global information that the agents share and the low communication burden that it requires. For this reason, this scheme is a good candidate to be implemented in systems with reduced capabilities, for example in wireless sensor and actuator networks.

25.1 Introduction

In this chapter, a distributed model predictive control (DMPC) scheme based on a team game, originally proposed in [8] and formalized in [7], is presented. In this scheme two different agents communicate in order to find a solution to the problem of controlling two constrained linear systems coupled through the inputs; see Fig. 25.1. The assumptions about the amount of global information that the agents have are very restrictive in this scheme. In particular, we assume that each agent only has local model and state information. The only global information that an agent has is how the neighbor's input affects him. Notice that although this assumption can be

J. M. Maestre (✉) · F. J. Muros · F. Fele · D. Muñoz de la Peña · E. F. Camacho
Departamento de Ingeniería de Sistemas y Automática, Universidad de Sevilla, Seville, Spain
e-mail: pepemaestre@cartuja.us.es

F. J. Muros
e-mail: franmuros@gmail.com

F. Fele
e-mail: filiberto@cartuja.us.es

D. Muñoz de la Peña
e-mail: davidmps@cartuja.us.es

E. F. Camacho
e-mail: eduardo@cartuja.us.es

J. M. Maestre and R. R. Negenborn (eds.), *Distributed Model Predictive Control Made Easy*, Intelligent Systems, Control and Automation: Science and Engineering 69,
DOI: 10.1007/978-94-007-7006-5_25, © Springer Science+Business Media Dordrecht 2014

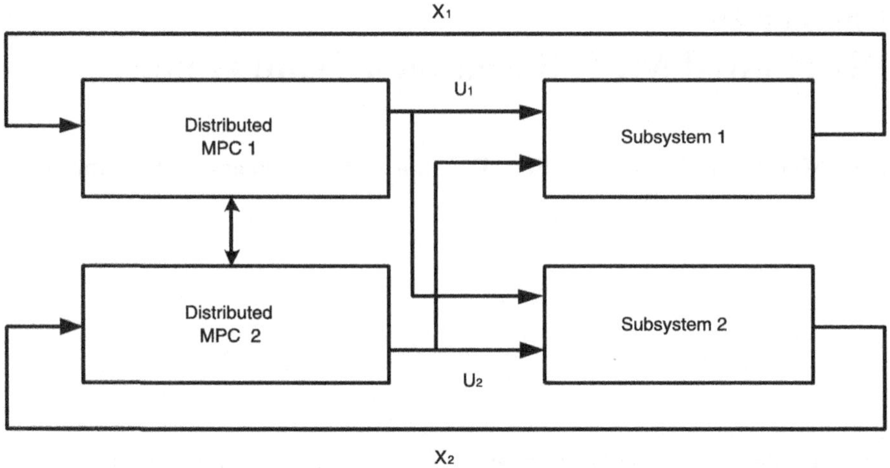

Fig. 25.1 Proposed distributed MPC scheme for two agents

restrictive, it is realistic in the sense that global information about the models and the objectives is not always available. We also assume that the available communication capabilities only allow two communications at each time interval. For this reason, the coordination between the agents is based on a team game that is built in two communication cycles.

25.2 Statement of the Problem and Main Assumptions

Consider the class of distributed linear systems composed by a set of $\mathcal{N} = \{1, 2\}$ interconnected subsystems coupled by the inputs whose dynamics can be described mathematically as:

$$\begin{aligned} \mathbf{x}_1(k+1) &= \mathbf{A}_1\mathbf{x}_1(k) + \mathbf{B}_{12}\mathbf{u}_1(k) + \mathbf{B}_{12}\mathbf{u}_2(k), \\ \mathbf{x}_2(k+1) &= \mathbf{A}_2\mathbf{x}_2(k) + \mathbf{B}_{21}\mathbf{u}_1(k) + \mathbf{B}_{22}\mathbf{u}_2(k), \end{aligned} \tag{25.1}$$

where $\mathbf{x}_i(k) \in \mathbb{R}^{n_{x_i}}$, $i \in \mathcal{N}$ are the states of subsystem i, and $\mathbf{u}_i(k) \in \mathbb{R}^{n_{u_i}}$, $i \in \mathcal{N}$ are their respective inputs. For simplicity we will show the definitions and equations for agent i. In the remaining of the chapter, the neighboring agent will be denoted as ni; for example: $\mathbf{u}_{ni} = \mathbf{u}_2$ if $i = 1$ and $\mathbf{u}_{ni} = \mathbf{u}_1$ if $i = 2$.

States and inputs are constrained into two independent sets defined by a set of linear inequalities

$$\mathbf{x}_i(k) \in \mathcal{X}_i, \quad \mathbf{u}_i(k) \in \mathcal{U}_i, \quad i = 1, 2. \tag{25.2}$$

The agent responsible of subsystem i has complete knowledge of its local model and state $\mathbf{x}_i(k)$, and is able to manipulate the control action $\mathbf{u}_i(k)$. Hence, no agent has full model or state information and communication is required in order to obtain a cooperative solution.

Without loss of generality, we assume that the control objective is to regulate the system to the origin while satisfying the constraints. To this end, we define the following performance index for each agent, which depends on the future evolution of its state and input along a prediction horizon of length N_p:

$$J_i\left(\mathbf{x}_i(k), \mathbf{u}_i(k:k+N_\mathrm{p}-1), \mathbf{u}_{ni}(k:k+N_\mathrm{p}-1)\right) = \sum_{k=0}^{N_\mathrm{p}-1} L_i\left(\mathbf{x}_i(k), \mathbf{u}_i(k)\right) + F_i\left(\mathbf{x}_i(N_\mathrm{p})\right), \quad (25.3)$$

where $L_i(\cdot)$ and $F_i(\cdot)$ with $i \in \mathcal{N}$ are the stage cost and the terminal cost functions respectively, defined as:

$$L_i(\mathbf{x}, \mathbf{u}) = \mathbf{x}_i^\mathrm{T}\mathbf{Q}_i\mathbf{x}_i + \mathbf{u}_i^\mathrm{T}\mathbf{R}_i\mathbf{u}_i, \quad (25.4)$$

$$F_i(\mathbf{x}) = \mathbf{x}_i^\mathrm{T}\mathbf{P}_i\mathbf{x}_i, \quad (25.5)$$

with $\mathbf{Q}_i, \mathbf{P}_i > 0$, $\mathbf{R}_i \geq 0$, and the vector $\mathbf{u}_i(k:k+N_\mathrm{p}-1)$ stands for the future input sequence of agent i:

$$\mathbf{u}_i(k:k+N_\mathrm{p}-1) = \left[\mathbf{u}_i(k)^\mathrm{T}, \mathbf{u}_i(k+1)^\mathrm{T}, \ldots, \mathbf{u}_i(k+N_\mathrm{p}-1)^\mathrm{T}\right]^\mathrm{T}, \quad i \in \mathcal{N}. \quad (25.6)$$

The agents solve a sequence of optimization problems during each sampling interval. These problems are built using local information and assuming a fixed input trajectory for its neighbor. In particular, an agent assumes that its neighbor will behave as in the previous agreed input trajectory, which we will denote with the superscript d, i.e., $\mathbf{u}_i^\mathrm{d}(k-1:k+N_\mathrm{p}-2)$ is the input trajectory that agent i agreed to implement at time $k-1$ for the time interval $[k-1, k+N_\mathrm{p}-2]$.

At this point it is important to remark that the trajectory that agent i assumes for its neighbor must cover the time interval $[k, k+N_\mathrm{p}-1]$. As the last agreed trajectory only provides a trajectory between $[k-1, k+N_\mathrm{p}-2]$ and the first component of that trajectory is already implemented in the system, it is necessary to calculate a new component for the time $k+N_\mathrm{p}-1$. If stability is not an issue in the system, it is possible to complete the sequence simply by adding a 0 vector of the proper size or copying the value of the last known element of the trajectory. In this chapter we assume that each agent updates his decided input trajectory using a feedback gain K_i and the N_p-1 steps ahead prediction of the state assuming. These feedbacks allow us to define the shifted input sequence of agent i at time k as:

$$\mathbf{u}_i^\mathrm{s}(k:k+N_\mathrm{p}-1) = \left[\mathbf{u}_i^\mathrm{d}(k)^\mathrm{T}, \mathbf{u}_i^\mathrm{d}(k+1)^\mathrm{T}, \ldots, \mathbf{u}_i^\mathrm{d}(k+N_\mathrm{p}-2), (\mathbf{K}_i\mathbf{x}_i(k+N_\mathrm{p}-1))^\mathrm{T}\right]^\mathrm{T}, \quad (25.7)$$

where $\mathbf{x}_i(k + N_p - 1)$ is the state of the subsystem $i(N_p - 1)$-steps ahead in the future, which is obtained applying the agreed input trajectories.

25.3 Description of the Approach

Algorithm 25.1 describes the DMPC scheme proposed in this chapter.

From a game theory point of view, at each time step both agents are playing a team game. This game can be synthesized in strategic form by the three by three matrix of Table 25.1. At each time step, the controllers decide among three different options. The shifted optimal input trajectory $\mathbf{u}_i^s(\cdot)$ keeps applying the latest optimal trajectory, so it can be seen like a stable decision. The selfish option $\mathbf{u}_i^*(\cdot)$ provides the best improvement in J_i if the rest of the systems manipulated variables stay unchanged. The altruist option $\mathbf{u}_i^w(\cdot)$ provides the best improvement for the neighbor agent cost function J_{ni} in case it applies its selfish option; i.e., the agent i sacrifices its own welfare in order to improve the overall performance. The cells of the matrix contain the sum of the cost functions of both agents for a particular choice of future inputs, resulting nine possibilities. At each time step, the first option that provides the lowest global cost is chosen. Note that both agents share this information, so they both choose the same option. Some important facts must be remarked about the algorithm:

- The computational burden of the proposed algorithm is in general lower than the one corresponding to the centralized scheme. In particular, the quadratic programming problems solved have less optimization variables.
- Agents operate in parallel in steps 2 and 3, which speeds up the calculation of the input trajectories.
- The minimun number of communication steps that are necessary to obtain a cooperative control scheme is two: in the first one each agent broadcasts its proposals to its neighbors and during the second one feedback about them is received.
- The proposed scheme is cooperative from a game theory point of view because each agent chooses the solution that minimizes a value that depends on the cost of both subsystems. If there were no cooperation between the agents, the solution

Table 25.1 Cost function table used for the decision making

	$\mathbf{u}_2^s(\kappa)$	$\mathbf{u}_2^*(\kappa)$	$\mathbf{u}_2^w(\kappa)$
$\mathbf{u}_1^s(\kappa)$	$J_1(\mathbf{x}_1(k), \mathbf{u}_1^s(\kappa), \mathbf{u}_2^s(\kappa))$ $+ J_2(\mathbf{x}_2(k), \mathbf{u}_2^s(\kappa), \mathbf{u}_1^s(\kappa))$	$J_1(\mathbf{x}_1(k), \mathbf{u}_1^s(\kappa), \mathbf{u}_2^*(\kappa))$ $+ J_2(\mathbf{x}_2(k), \mathbf{u}_2^*(\kappa), \mathbf{u}_1^s(\kappa))$	$J_1(\mathbf{x}_1(k), \mathbf{u}_1^s(\kappa), \mathbf{u}_2^w(\kappa))$ $+ J_2(\mathbf{x}_2(k), \mathbf{u}_2^w(\kappa), \mathbf{u}_1^s(\kappa))$
$\mathbf{u}_1^*(\kappa)$	$J_1(\mathbf{x}_1(k), \mathbf{u}_1^*(\kappa), \mathbf{u}_2^s(\kappa))$ $+ J_2(\mathbf{x}_2(k), \mathbf{u}_2^s(\kappa), \mathbf{u}_1^*(\kappa))$	$J_1(\mathbf{x}_1(k), \mathbf{u}_1^*(\kappa), \mathbf{u}_2^*(\kappa))$ $+ J_2(\mathbf{x}_2(k), \mathbf{u}_2^*(\kappa), \mathbf{u}_1^*(\kappa))$	$J_1(\mathbf{x}_1(k), \mathbf{u}_1^*(\kappa), \mathbf{u}_2^w(\kappa))$ $+ J_2(\mathbf{x}_2(k), \mathbf{u}_2^w(\kappa), \mathbf{u}_1^*(\kappa))$
$\mathbf{u}_1^w(\kappa)$	$J_1(\mathbf{x}_1(k), \mathbf{u}_1^w(\kappa), \mathbf{u}_2^s(\kappa))$ $+ J_2(\mathbf{x}_2(k), \mathbf{u}_2^s(\kappa), \mathbf{u}_1^w(\kappa))$	$J_1(\mathbf{x}_1(k), \mathbf{u}_1^w(\kappa), \mathbf{u}_2^*(\kappa))$ $+ J_2(\mathbf{x}_2(k), \mathbf{u}_2^*(\kappa), \mathbf{u}_1^w(\kappa))$	$J_1(\mathbf{x}_1(k), \mathbf{u}_1^w(\kappa), \mathbf{u}_2^w(\kappa))$ $+ J_2(\mathbf{x}_2(k), \mathbf{u}_2^w(\kappa), \mathbf{u}_1^w(\kappa))$

κ stands for the time interval $[k : k + N_p - 1]$

Algorithm 25.1 DMPC based on a team game

1: At time step k, each agent i measures its state $\mathbf{x}_i(k)$.

2: Each agent i minimizes J_i assuming that the neighbor applies the shifted control sequence calculated during the previous time step and solves the following optimization problem, $i \in \mathcal{N}$

$$\begin{aligned}
&\mathbf{u}_i^*(k:k+N_\mathrm{p}-1) = \\
&\arg \min_{\mathbf{u}_i(k:k+N_\mathrm{p}-1)} J_i\left(\mathbf{x}_i(k), \mathbf{u}_i(k:k+N_\mathrm{p}-1), \mathbf{u}_{ni}^\mathrm{s}(k:k+N_\mathrm{p}-1)\right) \\
&\text{s.t. } \mathbf{x}_i(k+1) = \mathbf{A}_i\mathbf{x}_i(k) + \mathbf{B}_{ii}\mathbf{u}_i(k) + \mathbf{B}_{i,ni}\mathbf{u}_{ni}(k), \\
&\mathbf{x}_i(l) \in \mathcal{X}_i, \quad l = k+1, \ldots, k+N_\mathrm{p}, \\
&\mathbf{u}_i(l) \in \mathcal{U}_i, \quad l = k, \ldots, k+N_\mathrm{p}-1, \\
&\mathbf{x}_i(k+N_\mathrm{p}) \in \Omega_i.
\end{aligned} \tag{25.8}$$

The sets Ω_i, $i \in \mathcal{N}$ define the terminal region constraints necessary to prove closed-loop practical stability following a terminal region/terminal cost approach. If stability is not an issue these constraints can be removed.

3: Each agent i minimizes J_i optimizing the neighbor input sequence $\mathbf{u}_{ni}(\cdot)$ while applying to its own subsystem the input trajectory $\mathbf{u}_i^*(\cdot)$ computed in step 2. This means for agent i solving the following optimization problem, $i \in \mathcal{N}$

$$\begin{aligned}
&\mathbf{u}_{ni}^\mathrm{w}(k:k+N_\mathrm{p}-1) = \\
&\arg \min_{\mathbf{u}_{ni}(k:k+N_\mathrm{p}-1)} J_i\left(\mathbf{x}_i(k), \mathbf{u}_i^*(k:k+N_\mathrm{p}-1), \mathbf{u}_{ni}(k:k+N_\mathrm{p}-1)\right) \\
&\text{s.t. } \mathbf{x}_i(k+1) = \mathbf{A}_i\mathbf{x}_i(k) + \mathbf{B}_{ii}\mathbf{u}_1(k) + \mathbf{B}_{i,ni}\mathbf{u}_{ni}(k), \\
&\mathbf{x}_i(l) \in \mathcal{X}_i, \quad l = k+1, \ldots, N_\mathrm{p}, \\
&\mathbf{u}_{ni}(l) \in \mathcal{U}_{ni}, \quad l = k, \ldots, k+N_\mathrm{p}-1, \\
&\mathbf{x}_i(k+N_\mathrm{p}) \in \Omega_i.
\end{aligned} \tag{25.9}$$

The solution of this optimization problem, $\mathbf{u}_{ni}^\mathrm{w}(k:k+N_\mathrm{p}-1)$, is a proposal for the neighbor (a wished behaviour), i.e., an input trajectory that the neighbor ni can implement to minimize the cost of agent i. Again, the constraint involving Ω_i can be removed if stability is not an issue.

4: Both agents communicate, sending both $\mathbf{u}_i^*(k:k+N_\mathrm{p}-1)$ and $\mathbf{u}_{ni}^\mathrm{w}(k:k+N_\mathrm{p}-1)$ to the other agent, and receiving $\mathbf{u}_{ni}^*(k:k+N_\mathrm{p}-1)$ and $\mathbf{u}_i^\mathrm{w}(k:k+N_\mathrm{p}-1)$.

5: Each agent evaluates its local cost function J_i for each of the nine possible combinations of the input trajectories obtained: $\mathbf{u}_i \in \left\{\mathbf{u}_i^\mathrm{s}(\cdot), \mathbf{u}_i^\mathrm{w}(\cdot), \mathbf{u}_i^*(\cdot)\right\}$, $i \in \mathcal{N}$ (see Table 25.1).

6: Both agents communicate and share the information of the value of local cost function for each possible combination of input trajectories. In this step, both agents receive enough information to take a cooperative decision. In addition, this communication cycle can be used as well to transmit the information regarding the extra component that is necessary to complete the shifted input trajectory vector.

7: Each agent applies the input trajectory that minimizes $J = J_1 + J_2$. Because both agents have access to the same information after the second communication cycle, both agents choose the same optimal input trajectories, i.e., $\mathbf{u}_i^\mathrm{d}(\cdot)$, $i \in \mathcal{N}$.

8: The first input of each optimal sequence is applied and the procedure is repeated at the next time step.

---

attained would converge after several iterations towards the Nash equilibrium of the multi-objective optimization problem defined by the cost functions of the agents.

- Although the outcome chosen by the algorithm is a Pareto optimal of the game that both agents are playing, in general it is not a Pareto optimal of the multi-objective optimization problem defined by the cost functions J_1 and J_2.

- The proposed scheme can be extended to deal with a greater number of agents, i.e., $|\mathcal{N}| = M$. However, the complexity of building the corresponding team game matrix grows exponentially with the number of agents. In order to reduce the complexity, the structure of the system may be exploited taking into account that an input may no affect all the outputs. Also, in general not all the possible cooperation options are employed with the same frequency, so it is possible to reduce further the complexity by not taking into account the less frequent options. In [7], the reader can find a modified distributed scheme which deal with this topic in depth.

25.4 Theoretical Results Availability

In this section we introduce the main theoretical propierties of the proposed DMPC scheme. Please, notice that many theoretical details have been omitted for the sake of clarity. The interested reader is recommended to see [5] for a more rigorous treatment of the topics discussed in this section.

25.4.1 Stability Properties

Controlling a system between two independient agents may lead to an unstable closed-loop system. The resulting closed-loop system is a multiprocess system and studying the stability of this class of systems is in general a difficult task. Following a terminal region/terminal constraint approach, in [5] we provided sufficient conditions that guarantee practical stability of the closed-loop system as well as an optimization based procedure to design the controller so that these conditions are satisfied.

In [5] it is proved that if there exist linear feedbacks $\mathbf{u}_i = \mathbf{K}_i\mathbf{x}_i$, terminal cost functions defined by matrices $\mathbf{P}_i$, and regions Ω_i that satisfy the following conditions for all $i \in \mathcal{N}$, then the system in closed-loop with the proposed controller is ultimately bounded in a region that contains the origin:

$$F_i\left((\mathbf{A}_i + \mathbf{B}_{ii}\mathbf{K}_i)\,\mathbf{x}_i + \mathbf{B}_{i,ni}\mathbf{K}_{ni}\mathbf{x}_{ni}\right) - F_i(\mathbf{x}_i) + L_i\,(\mathbf{x}_i, \mathbf{K}_i\mathbf{x}_i) - d_i \leq 0, \forall \mathbf{x}_{ni} \in \Omega_{ni}, \tag{25.10}$$

$$\mathbf{x}_i \in \Omega_i \rightarrow (\mathbf{A}_i + \mathbf{B}_{ii}\mathbf{K}_i)\,\mathbf{x}_i + \mathbf{B}_{i,ni}\mathbf{K}_{ni}\mathbf{x}_{ni} \in \Omega_i,\ \forall \mathbf{x}_{ni} \in \Omega_{ni}, \tag{25.11}$$

$$\mathbf{K}_i\mathbf{x}_i \in \mathcal{U}_i,\ \forall \mathbf{x}_i \in \Omega_i, \tag{25.12}$$

$$\Omega_i \in \mathcal{X}_i. \tag{25.13}$$

If the aforementioned conditions are satisfied, then it is possible to prove that if $\mathbf{x}_i(0)$ and $\mathbf{u}_i^s(0)$ for all $i \in \mathcal{N}$ are given such that the optimization problem (25.8) is feasible, then the proposed algorithm is feasible for all time steps $k \geq 0$ and system (25.1) in closed-loop with the proposed DMPC controller is ultimately bounded in a region that contains the origin in its interior. Likewise, it is possible to guarantee that the closed-loop system is ultimately bounded in a closed region that contains the origin. Moreover, it is possible to prove that the proposed controller provides asympthotic stability if (25.10) is modified. Specific details about this topic can be found in [5, 7].

25.4.2 Design Procedure

In [5, 7] an optimization based procedure is given to find, for $i \in \mathcal{N}$, local controllers $\mathbf{K}_i$, matrices $\mathbf{P}_i$ and regions Ω_i such that (25.10)–(25.13) holds for a given system. The procedure determines first matrices $\mathbf{K}_i$ and $\mathbf{P}_i$ such that (25.10)–(25.13) hold for any given sets Ω_i solving a linear matrix inequality (LMI) optimization problem. Once the local feedbacks $\mathbf{K}_i$ are fixed, the invariant sets Ω_i are obtained.

The linear feedbacks must guarantee that each system is stable and must provide a certain degree of robustness with respect to the neighbor control input. To certain degree, each local controller assumes that the neighbor input is a bounded disturbance that has to be rejected. This allows us to use well known tools from control of linear uncertain systems in order to determine a local controller such that a given degree of robustness is guaranteed. Constraint (25.10) can be transformed into an LMI and solved using standard techniques. In particular, the following LMI provides a mean for the calculation of $\mathbf{K}_i$ and $\mathbf{P}_i$:

$$\begin{bmatrix} \gamma_i I & 0 & B_{i,ni}^{\mathrm{T}} & 0 & 0 \\ * & \mathbf{W}_i & \mathbf{W}_i\mathbf{A}_{ii}^{\mathrm{T}} + \mathbf{Y}_i^{\mathrm{T}}\mathbf{B}_{ii}^{\mathrm{T}} & \mathbf{W}_i\mathbf{Q}_i^{\frac{1}{2}} & \mathbf{Y}_i^{\mathrm{T}}\mathbf{R}_i^{\frac{1}{2}} \\ * & * & \mathbf{W}_i & 0 & 0 \\ * & * & * & I & 0 \\ * & * & * & * & I \end{bmatrix} > 0. \tag{25.14}$$

where $\mathbf{P}_i = \mathbf{W}_i^{-1}, \mathbf{K}_i = \mathbf{Y}_i\mathbf{W}_i^{-1}$, and γ_i is a constant related with the size of the maximum admissible disturbance for system i: $d_i = \gamma_i \max_{\mathbf{x}\in\Omega_{ni}} (\mathbf{K}_{ni}\mathbf{x})^{\mathrm{T}}\mathbf{K}_{ni}\mathbf{x}$. Notice that the size of the disturbance, given by d_i, depends at the same time on the size of the sets Ω_i. Likewise, note that the conditions presented assume that each input $\mathbf{u}_i$ only depends on the state $\mathbf{x}_i$. This structure can be generalized for local controllers that take into account the full state space.

Once the local controllers and the terminal cost functions are fixed, it is necessary to find sets Ω_i such that (25.10)–(25.13) hold. In general this is a difficult problem because each of the sets depends on the other. The size of the terminal region for agent i is determined by the magnitude of the disturbances induced by its neighbor agent ni and viceversa. The main idea behind the calculation of the invariant sets of both agents is a scalation of their respective input constraints sets by a factor $\lambda_i \leq 1$ with the goal of finding a fair trade off between the disturbances induced to the neighbor and the manipulating capabilities of each agent. There exist several methods to find a set Ω_i that satisfies all the local constraints $\mathcal{X}_i$ and $\lambda_i \mathcal{U}_i$ while coping with the disturbances induced by the neighbor, which depend on $\lambda_{ni} \mathcal{U}_{ni}$; e.g.: in [12] a procedure to find an approximation of the minimal robust positive invariant is given, and in [11] a similar class of invariant systems was studied within the polytopic games framework.

At this point, it is important to stress that the set $\Omega_1 \times \Omega_2$ is an invariant set for the system in closed loop with the linear feedbacks $\mathbf{K}_1, \mathbf{K}_2$; however, the opposite does not hold necessarily, i.e., not all invariant sets satisfy the stability conditions. Note that each set is defined in their corresponding subspace $\mathbf{x}_i$. We denote these sets jointly invariant sets. Note as well that there may exist an infinite number of possible values of λ_{ji} such that these sets exist. In order to choose one, we propose to solve an optimization problem to maximizes the feasibility region of the distributed MPC controller. In [11] it was proved that the feasibility region of this problem is convex. In [9] we prove that the jointly invariant sets Ω_i are polyhedra defined by a set of inequalities whose right hand side can be expressed as an affine combination of the constants γ_{ij}. Using this result, the optimization problem can be cast into a convex optimization problem if the objective function is chosen appropriately, for instance, if the criterium to compare the invariant sets is the radium of a Chebyshev ball inside the invariant region.

Once matrices $\mathbf{K}_i$, $\mathbf{P}_i$, and the sets Ω_i, $i \in \mathcal{N}$ are determined, constants d_i, $i \in \mathcal{N}$ can be calculated in order to obtain an estimation of the set in which the closed-loop system is ultimately bounded.

25.5 Applications of the DMPC Scheme

The algorithm that we propose in this chapter has been tested with simulated and real systems. The applications and the results that are described in this section are explained in detail in [2, 5–8, 10].

25.5.1 Supply Chain Problem

In [5, 6], the proposed controller is applied to a reduced version of the MIT beer game and compared with other distributed control schemes. The MIT beer game is based on the concept of a supply chain, i.e., the set of structures and processes used by an organization to provide a service or a good to a costumer. The original MIT beer game is composed of four agents: retailer, wholesaler, distributor and factory. In our case, a reduced version of the problem with only two agents is considered: the retailer and the supplier, see Fig. 25.2. Notice that there is no loss of generality since the structure of the game is regular: there is a cascade of firms, each maintaining and controlling its stock.

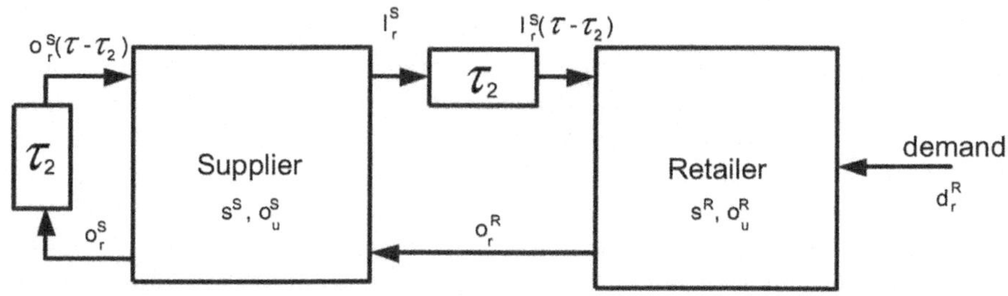

Fig. 25.2 Reduced beer game

The control objective is to regulate the number of stocked beers to a given reference and the manipulated variable is the number of beers that are ordered at each node. The retailer must satisfy the demand of beers and orders new beers to the supplier, which at the same time asks the factory for beers.

The performance comparison between the different schemes considered is based on simulations that are carried out in four different scenarios. Each scenario is defined by a different initial state, a different retailer demand, and a different demand forecast. In general, the proposed algorithm provides a performance in the same order of magnitude than the centralized MPC which, as expected, obtains the best results. The simulations show that the proposed distributed scheme outperforms a noncooperative iterative DMPC scheme with a greater number of communication cycles.

25.5.2 Continuosly Stirred Tank Reactor

In [8, 10] the proposed scheme is applied to a system described by a transfer function: a continuosly stirred tank reactor (CSTR) (taken from [3]). The manipulated variables are respectively the flow rate and the flow of coolant in the jacket. The controlled variables are respectively the effluent concentration and the reactor temperature,

Fig. 25.3 Continuously stirred tank reactor (CSTR)

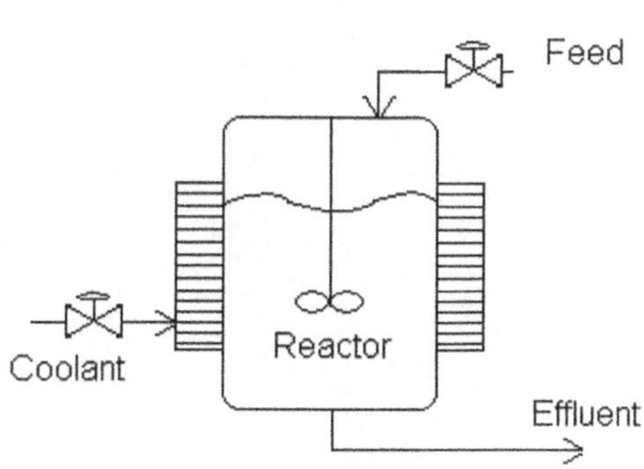

see Fig. 25.3. The control objective is to track a given constant reference from a random initial state. Our results show that the proposed scheme is able to stabilize the closed-loop system while other decentralized schemes fail. On top of that, the rise time and the convergence rate of the overall cost function are determined as a function of the average number of data losses in the communication channel. Hence, this system is used to test the robustness of the algorithm against data losses. Notice that as the reliability of the communication channel decreases, so does the amount of information shared by the agents, and the controller tends to operate in a decentralized manner. In particular, when there are data losses, the agents do not receive $\mathbf{u}_i^{\mathrm{w}}$, which is needed to build the global cost table (Table 25.1). In this case, each agent must decide whether to keep applying the last optimal input trajectory $\mathbf{u}_i^{\mathrm{s}}$, or behave selfishly and try to minimize its local cost function choosing $\mathbf{u}_i^*$. In

Fig. 25.4 The four tank process: the real plant diagram

order to test the robustness of the proposed approach on the worst possible case, we assume that when communication errors occur each controller applies $\mathbf{u}_i^*$. A set of simulations with different average values of data losses can be found in [10]. The simulation results show that the algorithm is able to stabilize the closed-loop system whenever the number of data losses is lower than the 50 %.

25.5.3 The Four Tank Process

In [2, 5] the proposed scheme and other DMPC policies are compared in an real benchmark: a four tank process (see Fig. 25.4), which is one of the benchmarks of the european project HD-MPC. The physical plant is located in facilities of the University of Seville and is described in [1]. It is an educational plant designed to test control techniques with real industrial instrumentation and control devices. The plant is a hydraulic process of four inteconnected tanks inspired by the educational quadruple tank process proposed by Johansson [4]. The main characteristic of this process is that is a simple multivariable system with highly coupled nonlinear dynamics that can exhibit transmission zeros dynamics. The four tank process has proven to be a very interesting system for control education and research.

The objective of the benchmark is to test and compare centralized, decentralized and distributed predictive controllers under similar operation conditions. To this end an experiment is defined in [2, 5], in which the controllers must regulate the levels of tanks 1 and 2 to follow a set of reference changes by manipulating the inlet flows based of the measured levels of the four tanks.

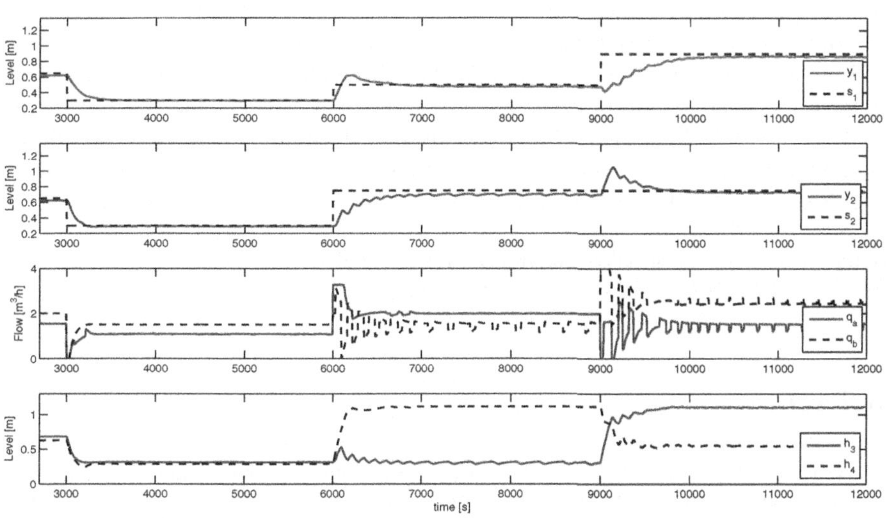

Fig. 25.5 Experimental results on the four tank process

Our results with this plant show a performance index relatively close to the centralized MPC and, moreover, our DMPC scheme outperforms other more complex schemes that have higher communication burdens. Nevertheless our results also show a possible implementation issue for other applications: the resulting input trajectories are not smooth (see Fig. 25.5), which is natural since the controller chooses among nine different modes of operation. Depending on the application, this switching may not be acceptable. Figure 25.5 also shows the cooperative nature of the proposed scheme in the sense that the inputs switch constantly between the two possible trajectories that suit best with the local objectives of each agent.

25.6 Conclusions

We have presented a DMPC scheme based on game theory for a class of systems controlled by two agents. The proposed controller only needs two communication steps in order to provide a cooperative solution for the centralized optimization problem. Each agent solves an optimization problem that depends only on its local model and state information. After sharing information about the cost of the control actions considered, the agents choose the solution that provides the best global performance among a set of possible suboptimal choices (the options are suboptimal because each agent has an incomplete view of the system and they propose the best solutions from a local perspective).

The proposed algorithm guarantees feasibility and stability of the closed-loop system if the feedback laws are designed according to the proposed procedure. On top of that, our results show a good behavior of the control scheme, which is specially remarkable when its low communication and informational requirements are taken into account. Likewise, the robustness of the proposed scheme against failures in the communication channel has been tested in simulation as well.

Finally, it is worthwhile to mention that although it is possible to extend the proposed scheme to control problems with more than 2 agents, the size of the team game that must be built grows exponentially with the number of agents. Therefore it is necessary to reduce the number of options that are proposed and evaluated by the agents such as it is done in [9], where an evolution of this scheme is presented

Acknowledgments Financial support from the HYCON2 EU-project from the ICT-FP7 and MEC-Spain, DPI2008-05818, and F.P.I. grants is gratefully acknowledged.

References

1. I. Alvarado, D. Limón, W.M. Garca, T. Álamo, E.F. Camacho, An educational plant based on the quadruple-tank process, in *Preprints of the 7th IFAC Symposium on Advances in Control Education*, Madrid, Spain, June 2006
2. I. Alvarado, D. Limón, D. Muñoz de la Peña, J.M. Maestre, F. Valencia, H. Scheu, R.R. Negenborn, M.A. Ridao, B. De Schutter, J. Espinosa, W. Marquardt, A comparative analysis

of distributed MPC techniques applied to the HD-MPC four tank benchmark. J. Process Control **21**(5), 800–815 (June 2011)
3. E.F. Camacho, C. Bordóns, *Model Predictive Control in the Process Industry*, 2nd edn. (Springer, London, England, 2004)
4. K.H. Johansson, The quadruple-tank process: a multivariable laboratory process with an adjustable zero. IEEE Trans. Control Syst. Technol. **8**(3), 456–465 (May 2000)
5. J.M. Maestre, *Distributed Model Predictive Control Based on Game Theory*, PhD thesis, Departamento de Ingeniería de Sistemas y Automática, Universidad de Sevilla, Seville, Spain, Nov 2010
6. J.M. Maestre, D. Muñoz de la Peña, E.F. Camacho, Distributed MPC: a supply chain case study, in *Proceedings of the joint 48th IEEE Conference on Decision and Control and 28th Chinese Control Conference*, Dec 2009
7. J.M. Maestre, D. Muñoz de la Peña, E.F. Camacho, Distributed model predictive control based on a cooperative game. Optimal Control Appl. Methods **32**(2), 153–176, Mar/Apr 2011
8. J.M. Maestre, D. Muñoz de la Peña, E.F. Camacho, A distributed MPC scheme with low communication requirements, in *Proceedings of the American Control Conference*, pp. 2797–2802, June 2009
9. J.M. Maestre, D. Muñoz de la Peña, E.F. Camacho, T. Álamo, Distributed model predictive control based on agent negotiation. J. Process Control **21**(5), 685–697 (June 2011)
10. F.J. Muros, J.M. Maestre, E.F. Camacho, *Estudio de robustez frente a retardos y pérdida de datos de una estrategia dmpc basada en pocos ciclos de comunicación* (In Actas de las XXIX Jornadas de Automática, Tarragona, Spain, Sept, 2008)
11. S. Rakovic, E. De Santis, P. Caravani, Invariant equilibria of polytopic games via optimized robust control invariance, in *Proceedings of the 44th IEEE Conference on Decision and Control and the European Control Conference*, pp. 7686–7691, Seville, Spain, Dec 2005
12. S. Rakovic, E. Kerrigan, K. Kouramas, D. Mayne, Robust MPC for nonlinear discrete-time systems. IEEE Trans. Automat. Control **50**, 406–410 (2005)

Chapter 26
Distributed MPC: A Noncooperative Approach Based on Robustness Concepts

G. Betti, M. Farina and R. Scattolini

Abstract The Distributed Predictive Control (DPC) algorithm presented in this chapter has been designed for control of an overall system made by linear discrete-time dynamically interconnected subsystems. It consists of a non-cooperative, non-iterative algorithm where a neighbor-to-neighbor transmission protocol is needed. The DPC algorithm enjoys the following properties: (i) state and input constraints can be considered; (ii) convergence is guaranteed; (iii) it is not necessary for each subsystem to know the dynamical models of the other subsystems; (iv) the transmission of information is limited.

26.1 Introduction

In this chapter the state feedback Distributed Predictive Control (DPC) algorithm originally proposed in [7] is sketched and discussed. The overall system to be controlled is assumed to be composed by a number of interacting subsystems S_i with nonoverlapping states, linear dynamics and possible state and control constraints. The dynamics of each subsystem can depend on the state and input variables of the other subsystems, and joint state and control constraints can be considered. A subsystem S_i is a neighbor of subsystem S_j if the state and/or control variables of S_i influence the dynamics of S_j or if a joint constraint on the states and/or on the inputs of S_i and S_j must be fulfilled.

G. Betti (✉) · M. Farina · R. Scattolini
Dipartimento di Elettronica e Informazione, Politecnico di Milano, Milan, Italy
e-mail: betti@elet.polimi.it

M. Farina
e-mail: farina@elet.polimi.it

R. Scattolini
e-mail: riccardo.scattolini@elet.polimi.it

J. M. Maestre and R. R. Negenborn (eds.), *Distributed Model Predictive Control Made Easy*, Intelligent Systems, Control and Automation: Science and Engineering 69,
DOI: 10.1007/978-94-007-7006-5_26, © Springer Science+Business Media Dordrecht 2014

DPC has been developed with the following rationale: at each sampling time, the subsystem S_i sends to its neighbors information about its future state $\tilde{\mathbf{x}}_i$ and input $\tilde{\mathbf{u}}_i$ reference trajectories, and guarantees that its actual trajectories $\mathbf{x}_i$ and $\mathbf{u}_i$ lie within certain bounds in the neighborhood of the reference ones. Therefore, these reference trajectories are known exogenous variables for the neighboring subsystems to be suitably compensated, while the differences $\mathbf{x}_i - \tilde{\mathbf{x}}_i$, $\mathbf{u}_i - \tilde{\mathbf{u}}_i$ are regarded as unknown bounded disturbances to be rejected. In this way, the control problem is set in the framework of robust MPC, and the "tube-based" approach inspired by [10] is used to formally state and solve a robust MPC problem for each subsystem.

The highlights of DPC are the following.

- It is not necessary for each subsystem to know the dynamical models governing the trajectories of the other subsystems (not even the ones of its neighbors), leading to a non-cooperative approach.
- The transmission of information is limited (i.e., DPC is non-iterative [14] and requires a neighbor-to-neighbor communication network), in that each subsystem needs to know the reference trajectories only of its neighbors.
- Its rationale is similar to the MPC algorithms often employed in industry: reference trajectories tailored on the dynamics of the system under control are used.
- Convergence and stability properties are guaranteed under mild assumptions.
- The algorithm can be extended to cope with the output feedback case and with the tracking problem.

26.2 Statement of the Problem and Main Assumptions

Consider a system composed of $|\mathcal{N}|$ dynamically coupled non-overlapping subsystems, each one described by the following state-space model.

$$\mathbf{x}_i(k+1) = \mathbf{A}_{ii}\,\mathbf{x}_i(k) + \mathbf{B}_{ii}\mathbf{u}_i(k) + \sum_{j=1, j\neq i}^{|\mathcal{N}|} \{\mathbf{A}_{ij}\mathbf{x}_j(k) + \mathbf{B}_{ij}\mathbf{u}_j(k)\} \tag{26.1}$$

$\mathbf{x}_i(k) \in \mathcal{X}_i \subseteq \mathcal{R}^{n_{x_i}}$ and $\mathbf{u}_i(k) \in \mathcal{U}_i \subseteq \mathcal{R}^{n_{u_i}}$ are the state and input vectors of the i-th subsystem S_i ($i = 1, \ldots, |\mathcal{N}|$), and the sets $\mathcal{X}_i$ and $\mathcal{U}_i$ are convex neighborhoods of the origin.

The subsystem S_j is called a *dynamic neighbor* of the subsystem S_i if and only if the state or the input of S_j affects the dynamics of S_i i.e., iff $A_{ij} \neq 0$ or $B_{ij} \neq 0$. The symbol $\mathcal{D}_{Ni}$ denotes the set of dynamic neighbors of S_i (which excludes i).

Letting $\mathbf{x}(k) = [\mathbf{x}_1^T(k), \ldots, \mathbf{x}_{|\mathcal{N}|}^T(k)]^T$, $\mathbf{u}(k) = [\mathbf{u}_1^T(k), \ldots, \mathbf{u}_{|\mathcal{N}|}^T(k)]^T$, the overall collective system can be written as

$$\mathbf{x}(k) = \mathbf{A}\,\mathbf{x}(k) + \mathbf{B}\mathbf{u}(k), \tag{26.2}$$

where the matrices $\mathbf{A}$ and $\mathbf{B}$ have block entries $\mathbf{A}_{ij}$ and $\mathbf{B}_{ij}$ respectively, $\mathbf{x} \in \mathcal{X} = \prod_{i=1}^{|\mathcal{N}|} \mathcal{X}_i \subseteq \mathcal{R}^{n_x}$, $n_x = \sum_{i=1}^{|\mathcal{N}|} n_{x_i}$, $\mathbf{u} \in \mathcal{U} = \prod_{i=1}^{|\mathcal{N}|} \mathcal{U}_i \subseteq \mathcal{R}^{n_u}$, $n_u = \sum_{i=1}^{|\mathcal{N}|} n_{u_i}$, and $\mathcal{X}, \mathcal{U}$ are convex by convexity of $\mathcal{X}_i$ and $\mathcal{U}_i$, respectively.
The states and inputs of the subsystems can be subject to coupling static constraints described in collective form by

$$H_s(\mathbf{x}(k), \mathbf{u}(k)) \leq 0$$

where $s = 1, \ldots, n_c$. H_s is a constraint on S_i if $\mathbf{x}_i$ and/or $\mathbf{u}_i$ are arguments of H_s, while $\mathcal{C}_i = \{s \in \{1, \ldots, n_c\}$: H_s is a constraint on $i\}$ denotes the set of constraints on S_i. Subsystem S_j is a *constraint neighbor* of subsystem S_i if there exists $\bar{s} \in \mathcal{C}_i$ such that $\mathbf{x}_j$ and/or $\mathbf{u}_j$ are arguments of $H_{\bar{s}}$, while $\mathcal{H}_i$ is the set of the constraint neighbors of S_i. Finally, for all $s \in \mathcal{C}_i$, a function is defined $h_{s,i}(\mathbf{x}_i, \mathbf{u}_i, \mathbf{x}, \mathbf{u}) = H_s(\mathbf{x}, \mathbf{u})$, where $\mathbf{x}_i$ and $\mathbf{u}_i$ are not arguments of $h_{s,i}(\mathbf{a}, \mathbf{b}, \cdot, \cdot)$. When $\mathcal{X} = \mathcal{R}^{n_x}, \mathcal{U} = \mathcal{R}^{n_u}$ and $n_c = 0$ the system is unconstrained. In general S_j is called a neighbor of S_i if $j \in \mathcal{D}_{Ni} \cup \mathcal{H}_i$. In line with these definitions, the communication topology which will be assumed from now on is a neighbor-to-neighbor one. Indeed, we require that information is transmitted from subsystem S_j to subsystem S_i if S_j is a neighbor of S_i.

The algorithm proposed in this chapter is based on MPC concepts and aims to solve, in a distributed fashion, the regulation problem for the described network of subsystems, while guaranteeing constraint satisfaction. Towards this aim, the following main assumption on decentralized stabilizability is introduced.

Assumption 26.1 *There exists a block diagonal matrix* $\mathbf{K} = diag(\mathbf{K}_1, \ldots, \mathbf{K}_{|\mathcal{N}|})$, *with* $\mathbf{K}_i \in R^{n_{u_i} \times n_{x_i}}$, $i = 1, \ldots, |\mathcal{N}|$ *such that: (i)* $\mathbf{A} + \mathbf{B}\mathbf{K}$ *is Schur, (ii)* $\mathbf{F}_{ii} = (\mathbf{A}_{ii} + \mathbf{B}_{ii}\mathbf{K}_i)$ *is Schur,* $i = 1, \ldots, |\mathcal{N}|$.

26.3 Description of the Approach

In DPC, at any time instant t, each subsystem S_i transmits to its neighbors its future state and input reference trajectories (to be later specified) $\tilde{\mathbf{x}}_i(k+\nu)$ and $\tilde{\mathbf{u}}_i(k+\nu)$, $\nu = 0, \ldots, N_p - 1$, respectively. Moreover, by adding suitable constraints to its MPC formulation, S_i is able to guarantee that, for all $t \geq 0$, its real trajectories lie in specified time invariant neighborhoods of the reference trajectories, i.e., $\mathbf{x}_i(k) \in \tilde{\mathbf{x}}_i(k) \oplus \mathcal{E}_i$ and $\mathbf{u}_i(k) \in \tilde{\mathbf{u}}_i(k) \oplus \mathcal{E}_i^{\mathcal{U}}$, where $0 \in \mathcal{E}_i$ and $0 \in \mathcal{E}_i^{\mathcal{U}}$. In this way, the dynamics (26.1) of S_i can be written as

$$\mathbf{x}_i(k+1) = \mathbf{A}_{ii}\,\mathbf{x}_i(k) + \mathbf{B}_{ii}\mathbf{u}_i(k) + \sum_{j \in \mathcal{D}_{Ni}} (\mathbf{A}_{ij}\tilde{\mathbf{x}}_j(k) + \mathbf{B}_{ij}\tilde{\mathbf{u}}_j(k)) + \mathbf{w}_i(k) \quad (26.3)$$

where

$$\mathbf{w}_i(k) = \sum_{j \in \mathcal{D}_{Ni}} \left(\mathbf{A}_{ij}(\mathbf{x}_j(k) - \tilde{\mathbf{x}}_j(k)) + \mathbf{B}_{ij}(\mathbf{u}_j(k) - \tilde{\mathbf{u}}_j(k))\right) \in \mathcal{W}_i$$

and $\mathcal{W}_i = \bigoplus_{j \in \mathcal{D}_{Ni}} \{\mathbf{A}_{ij}\mathcal{E}_j \oplus \mathbf{B}_{ij}\mathcal{E}_j^{\mathcal{U}}\}$.
As also discussed in the Introduction, the main idea behind DPC is that each subsystem solves a robust MPC optimization problem considering that its dynamics is given by (26.3), where the term $\sum_{j \in \mathcal{D}_{Ni}} (\mathbf{A}_{ij}\tilde{\mathbf{x}}_j(k+\nu) + \mathbf{B}_{ij}\tilde{\mathbf{u}}_j(k+\nu))$ can be interpreted as an input known in advance over the prediction horizon $\nu = 0, \ldots, N_p - 1$ to be suitably compensated and $\mathbf{w}_i(k)$ is a bounded disturbance to be rejected.
By definition, $\mathbf{w}_i(k)$ represents the uncertainty of the future actions that will be carried out by the dynamic neighbors of subsystem S_i. Therefore the local MPC optimization problem to be solved at each time instant by the controller embedded in subsystem S_i must minimize the cost associated to S_i for any possible uncertainty values, i.e., without having to make any assumption on strategies adopted by the other subsystems, provided that their future trajectories lie in the specified neighborhood of the reference ones. Such *conservative* but *robust* local strategies adopted by each subsystem can be interpreted, from a dynamic non-cooperative game theoretic perspective, as maxmin strategies, i.e., the strategies that maximize "worst case utility" of S_i (for more details see, e.g., [15]).
To solve local robust MPC problems (denoted i-DPC problems), the algorithm proposed in [10] has been selected in view of the facts that no burdensome minmax optimization problem is required to be solved on-line, and that it naturally provides the future reference trajectories $\tilde{\mathbf{x}}_i(k)$ and $\tilde{\mathbf{u}}_i(k)$, as it will be clarified later in this chapter. According to [10], a nominal model of subsystem S_i associated to Eq. (26.3) must be defined to produce predictions

$$\hat{\mathbf{x}}_i(k+1) = \mathbf{A}_{ii}\,\hat{\mathbf{x}}_i(k) + \mathbf{B}_{ii}\hat{\mathbf{u}}_i(k) + \sum_{j \in \mathcal{D}_{Ni}} (\mathbf{A}_{ij}\tilde{\mathbf{x}}_j(k) + \mathbf{B}_{ij}\tilde{\mathbf{u}}_j(k)) \qquad (26.4)$$

while the control law to be used for S_i is

$$\mathbf{u}_i(k) = \hat{\mathbf{u}}_i(k) + \mathbf{K}_i(\mathbf{x}_i(k) - \hat{\mathbf{x}}_i(k)) \qquad (26.5)$$

where $\mathbf{K}_i$ must be chosen to satisfy Assumption 26.1.
Letting $\mathbf{z}_i(k) = \mathbf{x}_i(k) - \hat{\mathbf{x}}_i(k)$, in view of (26.3), (26.4), and (26.5) one has

$$\mathbf{z}_i(k+1) = \mathbf{F}_{ii}\mathbf{z}_i(k) + \mathbf{w}_i(k) \qquad (26.6)$$

where $\mathbf{w}_i(k) \in \mathcal{W}_i$. Since $\mathcal{W}_i$ is bounded and $\mathbf{F}_{ii}$ is Schur, there exists a robust positively invariant (RPI) set $\mathcal{Z}_i$ for (26.6) such that, for all $\mathbf{z}_i(k) \in \mathcal{Z}_i$, then $\mathbf{z}_i(k+1) \in \mathcal{Z}_i$. Given $\mathcal{Z}_i$, define two sets, neighborhoods of the origin, $\Delta\mathcal{E}_i$ and $\Delta\mathcal{U}_i$, $i = 1, \ldots, |\mathcal{N}|$ such that $\Delta\mathcal{E}_i \oplus \mathcal{Z}_i \subseteq \mathcal{E}_i$ and $\Delta\mathcal{U}_i \oplus K_i\mathcal{Z}_i \subseteq \mathcal{E}_i^{\mathcal{U}}$, respectively.

Finally, define the function $\hat{h}_{s,i}$ such that the constraint $\hat{h}_{s,i}(\hat{\mathbf{x}}_i(k), \hat{\mathbf{u}}_i(k), \tilde{\mathbf{x}}(k), \tilde{\mathbf{u}}(k)) \leq 0$ guarantees that $h_{s,i}(\mathbf{x}_i(k), \mathbf{u}_i(k), \mathbf{x}^*(k), \mathbf{u}^*(k)) \leq 0$ for all $\mathbf{x}_i(k) \in \hat{\mathbf{x}}_i(k) \oplus \mathcal{Z}_i$, $\mathbf{u}_i(k) \in \hat{\mathbf{u}}_i(k) \oplus \mathbf{K}_i \mathcal{Z}_i$, $\mathbf{x}^*(k) \in \tilde{\mathbf{x}}(k) \oplus \prod_{i=1}^{|\mathcal{N}|} \mathcal{E}_i$, and $\mathbf{u}^*(k) \in \tilde{\mathbf{u}}(k) \oplus \prod_{i=1}^{|\mathcal{N}|} \mathcal{E}_i^{\mathcal{U}}$.

26.3.1 The Online Phase: The i-DPC Optimization Problems

At any time instant k, we assume that each subsystem S_i knows the future reference trajectories of its neighbors $\tilde{\mathbf{x}}_j(k+\nu)$, $\tilde{\mathbf{u}}_j(k+\nu)$, $\nu = 0, \ldots, N_p - 1$, $j \in \mathcal{D}_{Ni} \cup \mathcal{H}_i \cup \{i\}$ and, with reference to its nominal system (26.4) only, solves the following i-DPC problem.

$$\min_{\hat{\mathbf{x}}_i(k), \hat{\mathbf{u}}_i(k:k+N_p-1)} V_i^N(\hat{\mathbf{x}}_i(k), \hat{\mathbf{u}}_i(k:k+N_p-1)) = \sum_{\nu=0}^{N_p-1} (\|\hat{\mathbf{x}}_i(k+\nu)\|^2_{\mathbf{Q}_i^o} + \|\hat{\mathbf{u}}_i(k+\nu)\|^2_{\mathbf{R}_i^o}) + \|\hat{\mathbf{x}}_i(k+N_p)\|^2_{\mathbf{P}_i^o} \tag{26.7}$$

subject to (26.4),

$$\mathbf{x}_i(k) - \hat{\mathbf{x}}_i(k) \in \mathcal{Z}_i, \tag{26.8}$$

for $\nu = 0, \ldots, N_p - 1$

$$\hat{\mathbf{x}}_i(k+\nu) - \tilde{\mathbf{x}}_i(k+\nu) \in \Delta\mathcal{E}_i \tag{26.9}$$

$$\hat{\mathbf{u}}_i(k+\nu) - \tilde{\mathbf{u}}_i(k+\nu) \in \Delta\mathcal{U}_i \tag{26.10}$$

$$\hat{\mathbf{x}}_i(k+\nu) \in \hat{\mathcal{X}}_i \subseteq \mathcal{X}_i \ominus \mathcal{Z}_i \tag{26.11}$$

$$\hat{\mathbf{u}}_i(k+\nu) \in \hat{\mathcal{U}}_i \subseteq \mathcal{U}_i \ominus \mathbf{K}_i \mathcal{Z}_i \tag{26.12}$$

to the coupling state constraints, for all $s \in \mathcal{C}_i$,

$$\hat{h}_{s,i}(\hat{\mathbf{x}}_i(k+\nu), \hat{\mathbf{u}}_i(k+\nu), \tilde{\mathbf{x}}(k+\nu), \tilde{\mathbf{u}}(k+\nu)) \leq 0 \tag{26.13}$$

and to the terminal constraint

$$\hat{\mathbf{x}}_i(k+N_p) \in \hat{\mathcal{X}}_i^F \tag{26.14}$$

Note that constraints (26.8), (26.9), and (26.10) are used to guarantee the boundedness of the equivalent disturbance $\mathbf{w}_i$. In fact in the i-DPC problem, for $\nu = 0$, constraints (26.8), (26.9), and (26.10) imply that $\mathbf{x}_i(k) - \tilde{\mathbf{x}}_i(k) \in \Delta\mathcal{E}_i \oplus \mathcal{Z}_i \subseteq \mathcal{E}_i$ and $\mathbf{u}_i(k) - \tilde{\mathbf{u}}_i(k) \in \Delta\mathcal{U}_i \oplus \mathbf{K}_i \mathcal{Z}_i \subseteq \mathcal{E}_i^{\mathcal{U}}$, which in turns guarantees that $\mathbf{w}_i(k) \in \mathcal{W}_i$. This, in view of the invariance property of (26.6), implies that $\mathbf{x}_i(k+1) - \hat{\mathbf{x}}_i(k+1) \in \mathcal{Z}_i$ and, since (26.9) and (26.10) are imposed over the whole prediction horizon, it follows by

induction that $\mathbf{w}_i(k+\nu) \in \mathcal{W}_i$ for all $\nu = 0, \ldots, N_p-1$ and $\mathbf{x}_i(k+\nu)-\hat{\mathbf{x}}_i(k+\nu) \in \mathcal{Z}_i$ for all $\nu = 1, \ldots, N_p$.
In (26.7), $\mathbf{Q}_i^o$, $\mathbf{R}_i^o$, and $\mathbf{P}_i^o$ are positive definite matrices and represent design parameters, whose choice is discussed in Sect. 26.3.2 to guarantee stability and convergence properties, while $\hat{\mathcal{X}}_i^F$ in (26.14) is a nominal terminal set which must be chosen to satisfy the following.

Assumption 26.2 *Letting* $\hat{\mathcal{X}} = \prod_{i=1}^{|\mathcal{N}|} \hat{\mathcal{X}}_i$, $\hat{\mathcal{U}} = \prod_{i=1}^{|\mathcal{N}|} \hat{\mathcal{U}}_i$ *and* $\hat{\mathcal{X}}^F = \prod_{i=1}^{|\mathcal{N}|} \hat{\mathcal{X}}_i^F$, *it holds that:*

1. $\hat{H}_{s,i}(\hat{\mathbf{x}}) \leq 0$ *for all* $\hat{\mathbf{x}} \in \hat{\mathcal{X}}^F$, *for all* $s \in \mathcal{C}_i$, *for all* $i = 1, \ldots, |\mathcal{N}|$, *where* $\hat{H}_{s,i}$ *is defined in such a way that* $\hat{H}_{s,i}(\hat{\mathbf{x}}) = \hat{h}_{s,i}(\hat{\mathbf{x}}_i, \mathbf{K}_i\hat{\mathbf{x}}_i, \hat{\mathbf{x}}, \mathbf{K}\hat{\mathbf{x}})$ *for all* $s \in \mathcal{C}_i$, *for all* $i = 1, \ldots, |\mathcal{N}|$.
2. $\hat{\mathcal{X}}^F \subseteq \hat{\mathcal{X}}$ *is an invariant set for* $\hat{\mathbf{x}}(k+1) = (\mathbf{A}+\mathbf{BK})\hat{\mathbf{x}}(k)$;
3. $\hat{\mathbf{u}} = \mathbf{K}\hat{\mathbf{x}} \in \hat{\mathcal{U}}$ *for any* $\hat{\mathbf{x}} \in \hat{\mathcal{X}}^F$;
4. *for all* $\hat{\mathbf{x}} \in \hat{\mathcal{X}}^F$ *and, for a given constant* $\kappa > 0$,

$$\mathbf{V}^F\left(\hat{\mathbf{x}}^+\right) - \mathbf{V}^F\left(\hat{\mathbf{x}}\right) \leq -(1+\kappa)\ell(\hat{\mathbf{x}}, \hat{\mathbf{u}}) \tag{26.15}$$

where $\mathbf{V}^F(\hat{\mathbf{x}}) = \sum_{i=1}^{|\mathcal{N}|} V_i^F(\hat{\mathbf{x}}_i) = \sum_{i=1}^{|\mathcal{N}|} \|\hat{\mathbf{x}}_i\|_{\mathbf{P}_i^o}^2$ *and* $\ell(\hat{\mathbf{x}}, \hat{\mathbf{u}}) = \sum_{i=1}^{|\mathcal{N}|} \ell_i(\hat{\mathbf{x}}_i, \hat{\mathbf{u}}_i)$ $= \sum_{i=1}^{|\mathcal{N}|} (\|\hat{\mathbf{x}}_i\|_{\mathbf{Q}_i^o}^2 + \|\hat{\mathbf{u}}_i\|_{\mathbf{R}_i^o}^2)$.

At time k, let the pair $\hat{\mathbf{x}}_i(k|k)$, $\hat{\mathbf{u}}_i(k : k+N_p-1|k)$ be the solution to the i-DPC problem and define by $\hat{\mathbf{u}}_i(k|k)$ the input to the nominal system (26.4). Then, according to (26.5), the input to the system (26.1) is

$$\mathbf{u}_i(k) = \hat{\mathbf{u}}_i(k|k) + \mathbf{K}_i(\mathbf{x}_i(k) - \hat{\mathbf{x}}_i(k|k)) \tag{26.16}$$

Denoting by $\hat{\mathbf{x}}_i(k+\nu|k)$ the state trajectory of system (26.4) stemming from $\hat{\mathbf{x}}_i(k|k)$ and $\hat{\mathbf{u}}_i(k : k+N_p-1|k)$, at time k it is also possible to compute $\hat{\mathbf{x}}_i(k+N_p|k)$ and $\mathbf{K}_i\hat{\mathbf{x}}_i(k+N_p|k)$. In DPC, these values incrementally define the trajectories of the reference state and input variables to be used at the next time instant $k+1$, that is

$$\tilde{\mathbf{x}}_i(k+N_p) = \hat{\mathbf{x}}_i(k+N_p|k)\,, \quad \tilde{\mathbf{u}}_i(k+N_p) = \mathbf{K}_i\hat{\mathbf{x}}_i(k+N_p|k) \tag{26.17}$$

Note that the only information to be transmitted consists in the reference trajectories updated as in (26.17). More specifically, at time step k, subsystem S_i computes $\tilde{\mathbf{x}}_i(k+N_p)$ and $\tilde{\mathbf{u}}_i(k+N_p)$ according to (26.17) and transmits their values to all the subsystems having S_i as neighbor, before proceeding to the next time step.

26.3.2 Computation of the Design Parameters

The design of the DPC algorithm requires that a number of tuning parameters are properly selected off-line, i.e., the gain matrices $\mathbf{K}_i$ satisfying Assumption 26.1, the sets $\mathcal{E}_i$, $\mathcal{E}_i^{\mathcal{U}}$, $\Delta\mathcal{E}_i$, $\Delta\mathcal{U}_i$, and $\mathcal{Z}_i$, and the weighting matrices $\mathbf{Q}_i^o$, $\mathbf{R}_i^o$, and $\mathbf{P}_i^o$, satisfying Assumption 26.2.
These points are discussed in this section and specific choices of the tuning parameters are presented. A comment is due: in this section some algorithms for the definition of the mentioned tuning parameters are presented. However, they provide sufficient conditions, meaning that other criteria and algorithms can be devised and adopted, which can be even more efficient, especially when applied to specific case studies.

First an exact algorithm for the computation of the sets is presented as Algorithm 26.1, including a sufficient condition that guarantees the applicability of the algorithm (see step 3).

If $\mathcal{M}^P$ defined in (26.19a, 26.19b) is not Hurwitz, Algorithm 26.1 cannot be applied. An "empirical distributed reachability analysis" procedure, which is not guaranteed to provide a solution, but which has proved its effectiveness in many applications, is given as Algorithm 26.2.

Note that the effectiveness of the presented algorithms strongly depends on some arbitrary initial choices, i.e., in Algorithm 26.1 matrices Ξ_i (or equivalently vectors $\mathbf{f}_i$), $i = 1, \ldots, |\mathcal{N}|$, defining the shape of sets $\mathcal{E}_i$ and in Algorithm 26.2 the shapes and dimension of sets $\Delta\mathcal{E}_i$ and $\Delta\mathcal{U}_i$, $i = 1, \ldots, |\mathcal{N}|$. If the selected algorithm results to be inapplicable for a given choice, a trial-and-error procedure is suggested, in order to find a suitable initial choice (which is nevertheless not guaranteed to exist) guaranteeing the applicability of the selected algorithm.
Both with Algorithms 26.1 and 26.2, once $\mathcal{E}_i$, $\mathcal{E}_i^{\mathcal{U}}$, $\mathcal{Z}_i$, $\Delta\mathcal{E}_i$ and $\Delta\mathcal{U}_i$ are derived, the constraint sets $\hat{\mathcal{X}}_i$, $\hat{\mathcal{U}}_i$, $\hat{\mathcal{X}}_i^F$ and $\hat{h}_{s,i}$, $s \in \mathcal{C}_i$ directly follow (recall (26.11)–(26.14)). Eventually the sets $\hat{\mathcal{X}}_i^F$ can be obtained by scaling the sets $\mathcal{Z}_i$, $i = 1, \ldots, |\mathcal{N}|$ by a common factor, which should be sufficiently small to guarantee that (i), (ii), and (iii) of Assumption 26.2 are verified.

The design of the block diagonal matrix $\mathbf{K}$ satisfying Assumption 26.1, together with the computation of a block diagonal matrix $\mathbf{P} = \mathrm{diag}(\mathbf{P}_1^o, \ldots, \mathbf{P}_{|\mathcal{N}|}^o)$, $\mathbf{P}_i^o \in R^{n_{x_i} \times n_{x_i}}$ satisfying the inequality $(\mathbf{A}^T + \mathbf{K}^T\mathbf{B}^T)\mathbf{P}(\mathbf{A} + \mathbf{BK}) - \mathbf{P} \prec 0$, can be performed using Algorithm 26.3.

Algorithm 26.4 is discussed in [7], and requires that a set of control gains $\mathbf{K}_i$, $i = 1, \ldots, |\mathcal{N}|$, verifying Assumption 26.1 are given.

26.3.3 Initialization

The initial reference trajectories are critical tuning parameters, since they strongly affect the initial feasibility. Generally speaking, feasibility can be guaranteed by setting sufficiently high values of N_p and large sets $\mathcal{E}_i$. Algorithm 26.5 is used for

Algorithm 26.1 Computation of the sets

1: Assume that $\mathcal{E}_i$ can be equivalently represented in one of the following two ways:

$$\begin{aligned}\mathcal{E}_i &= \{\varepsilon_i \in \mathcal{R}^{n_{x_i}} \,|\, \varepsilon_i = \Xi_i \mathbf{d}_i \text{ where } \|\mathbf{d}_i\|_\infty \leq l_i\} \\ &= \{\varepsilon_i \in \mathcal{R}^{n_{x_i}} \,|\, \mathbf{f}_{i,r}^T \varepsilon_i \leq l_i \text{ for all } r\}\end{aligned} \tag{26.18}$$

where $\mathbf{d}_i \in \mathcal{R}^{n_{d_i}}$, $\Xi_i \in \mathcal{R}^{n_{x_i} \times n_{d_i}}$, $\mathbf{f}_{i,r} \in \mathcal{R}^{n_{x_i}}$, and $r = 1, \ldots, \bar{r}_i$ for all $i = 1, \ldots, |\mathcal{N}|$. The constants $l_i \in \mathbb{R}_+$, appearing in both equivalent definitions, can be regarded as scaling factors. Define the shape of the polyhedra with a proper setting of matrices Ξ_i and vectors $\mathbf{f}_{i,r}$, $i = 1, \ldots, |\mathcal{N}|$.

2: Assuming that $\mathbf{F}_{ii}$ is diagonalizable for all $i = 1, \ldots, |\mathcal{N}|$ (which is always possible since $\mathbf{K}_i$s are design parameters) define $\mathbf{N}_i$, $i = 1, \ldots, |\mathcal{N}|$, such that $\mathbf{F}_{ii} = \mathbf{N}_i^{-1} \Lambda_i \mathbf{N}_i$, where $\Lambda_i = \text{diag}(\lambda_{i,1}, \ldots, \lambda_{i,n_{x_i}})$, where $\lambda_{i,j}$ is the j-th eigenvalue of $\mathbf{F}_{ii}$. Define also

$$\mathbf{f}_i = \begin{bmatrix} \mathbf{f}_{i,1}^T \\ \vdots \\ \mathbf{f}_{i,\bar{r}_i}^T \end{bmatrix}$$

$i = 1, \ldots, |\mathcal{N}|$. Then, compute the matrix $\mathcal{M}^P \in \mathcal{R}^{|\mathcal{N}| \times |\mathcal{N}|}$ whose entries μ_{ij}^P are

$$\mu_{ii}^P = -1, i = 1, \ldots, |\mathcal{N}| \tag{26.19a}$$

$$\begin{aligned}\mu_{ij}^P &= \|\mathbf{f}_i \mathbf{N}_i^{-1}\|_\infty (\|\mathbf{N}_i \mathbf{A}_{ij} \Xi_j\|_\infty + \|\mathbf{N}_i \mathbf{B}_{ij} \mathbf{K}_j \Xi_j\|_\infty) \frac{1}{1 - \max_{j=1,\ldots,n_{x_i}} |\lambda_{i,j}|}, \\ & i, j = 1, \ldots, |\mathcal{N}| \text{ with } i \neq j\end{aligned} \tag{26.19b}$$

3: If $\mathcal{M}^P$ is Hurwitz, define the values of l_i, $i = 1, \ldots, |\mathcal{N}|$, as the entries of the strictly positive vector $\mathbf{l}$ satisfying $\mathcal{M}^P \mathbf{l} < 0$. For its computation use the following procedure:

- If the system is irreducible [4], $\mathbf{l}$ is the Frobenius eigenvector of matrix $\mathcal{M}^P$.
- If the system is reducible:

 1. Since the system is reducible there exists a permutation matrix $\mathbf{H}$ (where $\mathbf{H}^T = \mathbf{H}^{-1}$) such that $\tilde{\mathcal{M}}^P = \mathbf{H}\mathcal{M}^P\mathbf{H}^T$ is lower block triangular, with block elements $\mathbf{M}_{ij}$, whose diagonal blocks $\mathbf{M}_{ii}$ are irreducible. Let ω_i (strictly positive element-wise) be the Frobenius eigenvector of $\mathbf{M}_{ii}$, associated to the eigenvalue $\lambda_i < 0$.
 2. Set $\alpha_1 = 1$ and define recursively α_i, for $i > 1$, such that $\alpha_i |\lambda_i v_i| > |\sum_{j=1}^{i-1} \alpha_j \mathbf{M}_{ij} v_j|$ element-wise (v_i is the number of dynamic neighbors of subsystem i).
 3. Define $\omega = [\alpha_1 \omega_1^T, \ldots, \alpha_{|\mathcal{N}|} \omega_{|\mathcal{N}|}^T]^T$, and $\mathbf{v}_\mathcal{M} = \mathbf{H}^T \omega$. From the definition of α_is, it follows that $\tilde{\mathcal{M}}^P \omega < 0$, and so $\mathcal{M}^P \mathbf{v}_\mathcal{M} = \mathbf{H}^T \mathbf{H} \mathcal{M}^P \mathbf{H}^T \mathbf{H} \mathbf{v}_\mathcal{M} = \mathbf{H}^T \tilde{\mathcal{M}}^P \omega < 0$.

4: Compute $\mathcal{E}_i^\mathcal{U} = \mathbf{K}_i \mathcal{E}_i$, for all $i = 1, \ldots, |\mathcal{N}|$.

5: Compute $\mathcal{W}_i = \bigoplus_{j \in \mathcal{D}_{Ni}} \{\mathbf{A}_{ij} \mathcal{E}_j \oplus \mathbf{B}_{ij} \mathcal{E}_j^\mathcal{U}\}$, for all $i = 1, \ldots, |\mathcal{N}|$.

6: For all $i = 1, \ldots, |\mathcal{N}|$, compute $\mathcal{Z}_i$ as the polytopic RPI outer δ-approximation of the minimal RPI (mRPI) set for (26.6), as shown in [12].

7: For all $i = 1, \ldots, |\mathcal{N}|$, the sets $\Delta\mathcal{E}_i$, can be taken as any polytope satisfying $\Delta\mathcal{E}_i \oplus \mathcal{Z}_i \subseteq \mathcal{E}_i$, and finally $\Delta\mathcal{U}_i = \mathbf{K}_i \Delta\mathcal{E}_i$.

Algorithm 26.2 Computation of the sets

1: For all $i = 1, \dots, |\mathcal{N}|$, arbitrarily choose sets $\Delta\mathcal{E}_i$ and $\Delta\mathcal{U}_i$.
2: Initialize $\mathcal{Z}_i = \bigoplus_{j \in \mathcal{D}_{Ni}} \{\mathbf{A}_{ij}\Delta\mathcal{E}_j \oplus \mathbf{B}_{ij}\Delta\mathcal{U}_j\}$ for all $i = 1, \dots, |\mathcal{N}|$.
3: For all $i = 1, \dots, |\mathcal{N}|$, compute $\mathcal{Z}_i^+ = \mathbf{F}_{ii}\mathcal{Z}_i \oplus \{\bigoplus_{j \in \mathcal{D}_{Ni}} \{\mathbf{A}_{ij}\mathcal{Z}_j \oplus \mathbf{B}_{ij}\mathbf{K}_j\mathcal{Z}_j\}\} \oplus \{\bigoplus_{j \in \mathcal{D}_{Ni}} \{\mathbf{A}_{ij}\Delta\mathcal{E}_j \oplus \mathbf{B}_{ij}\Delta\mathcal{U}_j\}\}$.
4: If $\mathcal{Z}_i^+ \subseteq \mathcal{Z}_i$ for all $i = 1, \dots, |N|$ then stop: the sets $\mathcal{Z}_i$ actually correspond to the required RPIs. Otherwise set $\mathcal{Z}_i = \mathcal{Z}_i^+$ and repeat step 3.
5: Compute $\mathcal{E}_i = \Delta\mathcal{E}_i \oplus \mathcal{Z}_i$ and $\mathcal{E}_i^{\mathcal{U}} = \Delta\mathcal{U}_i \oplus \mathbf{K}_i\mathcal{Z}_i$.
6: If $\mathcal{Z}_i \subset \mathcal{X}_i$ and $\mathbf{K}_i\mathcal{Z}_i \subset \mathcal{U}_i$ then stop. Otherwise set $\Delta\mathcal{E}_i = \gamma\Delta\mathcal{E}_i$, $\Delta\mathcal{U}_i = \gamma\Delta\mathcal{U}_i$, and $\mathcal{Z}_i = \gamma\mathcal{Z}_i$ for all $i = 1, \dots, |\mathcal{N}|$, where $\gamma \in (0, 1)$ is sufficiently small to guarantee that $\mathcal{Z}_i \subset \mathcal{X}_i$ and $\mathbf{K}_i\mathcal{Z}_i \subset \mathcal{U}_i$ for all $i = 1, \dots, |\mathcal{N}|$.

Algorithm 26.3 Computation of block diagonal matrices $\mathbf{K}$, $\mathbf{P}$, $\mathbf{Q}_i^o$ and $\mathbf{R}_i^o$

1: Define $\mathbf{K} = \mathbf{Y}\mathbf{S}^{-1}$ and $\mathbf{P} = \mathbf{S}^{-1}$, and solve for $\mathbf{Y}$ and $\mathbf{S}$ the following LMI, as discussed in [2]

$$\begin{bmatrix} \mathbf{S} & \mathbf{S}\mathbf{A}^T + \mathbf{Y}^T\mathbf{B}^T \\ \mathbf{A}\mathbf{S} + \mathbf{B}\mathbf{Y} & \mathbf{S} \end{bmatrix} \succ 0 \tag{26.20}$$

with the additional constraints

$$\mathbf{S}_{ij} = \mathbf{0} \ \forall i, j = 1, ..., |\mathcal{N}| \ (i \neq j) \tag{26.21}$$

$$\mathbf{Y}_{ij} = \mathbf{0} \ \forall i, j = 1, ..., |\mathcal{N}| \ (i \neq j) \tag{26.22}$$

where $\mathbf{S}_{ij} \in \mathcal{R}^{(n_{x_i}+n_{u_i})\times(n_{x_j}+n_{u_j})}$, $\mathbf{Y}_{ij} \in \mathcal{R}^{n_{u_i}\times(n_{x_j}+n_{u_j})}$ are the blocks outside the diagonal of $\mathbf{S}$ and $\mathbf{Y}$, respectively. Finally, denoting by $\mathbf{S}_{ii}$ and $\mathbf{Y}_{ii}$ the block diagonal elements of $\mathbf{S}$ and $\mathbf{Y}$, respectively, the requirement that each block $\mathbf{K}_i$ must be stabilizing for its i-th subsystem (recall again Assumption 26.1), translates in the following set of conditions

$$\begin{bmatrix} \mathbf{S}_{ii} & \mathbf{S}_{ii}\mathbf{A}_{ii}^T + \mathbf{Y}_{ii}^T\mathbf{B}_{ii}^T \\ \mathbf{A}_{ii}\mathbf{S}_{ii} + \mathbf{B}_{ii}\mathbf{Y}_{ii} & \mathbf{S}_{ii} \succ 0 \end{bmatrix} \tag{26.23}$$

In conclusion, the computation of $\mathbf{K}$ and $\mathbf{P}$ calls for the solution of the set of LMI's (26.20), (26.21), (26.22) and (26.23), which can be easily found with suitable available software (e.g., YALMIP [9]).
2: Once $\mathbf{K}$ and $\mathbf{P}$ are available the parameters $\mathbf{Q}_i^o$ and $\mathbf{R}_i^o$ must be chosen to satisfy (26.15). To this end, define $\bar{\mathbf{Q}} = \mathbf{P} - (\mathbf{A} + \mathbf{B}\mathbf{K})^T\mathbf{P}(\mathbf{A} + \mathbf{B}\mathbf{K})$, choose an arbitrarily small positive constant κ and two block diagonal matrices $\mathbf{Q} = \text{diag}(\mathbf{Q}_1^o, \dots, \mathbf{Q}_{|\mathcal{N}|}^o)$, $\mathbf{Q}_i^o \succ 0 \in R^{n_{x_i}\times n_{x_i}}$, and $\mathbf{R} = \text{diag}(\mathbf{R}_1^o, \dots, \mathbf{R}_{|\mathcal{N}|}^o)$, $\mathbf{R}_i^o \succ 0 \in R^{n_{u_i}\times n_{u_i}}$. Then proceed as follows:

- If

$$\bar{\mathbf{Q}} - (\mathbf{Q} + \mathbf{K}^T\mathbf{R}\mathbf{K})(1 + \kappa) \succ 0 \tag{26.24}$$

  set $\mathbf{Q}^o = \mathbf{Q}$ and $\mathbf{R}^o = \mathbf{R}$.
- Otherwise set $\mathbf{Q} = \eta\mathbf{Q}$ and $\mathbf{R} = \eta\mathbf{R}$, with $0 < \eta < 1$, and repeat the procedure until (26.24) is fulfilled. Once $\mathbf{Q}$ and $\mathbf{R}$ satisfying (26.24) have been found, set $\mathbf{Q}^o = \mathbf{Q}$ and $\mathbf{R}^o = \mathbf{R}$.

3: Finally, extract from $\mathbf{Q}^o$ and $\mathbf{R}^o$ the submatrices $\mathbf{Q}_i^o$ and $\mathbf{R}_i^o$ of appropriate dimensions.

Algorithm 26.4 Computation of block diagonal matrices $\mathbf{P}$, $\mathbf{Q}_i^o$ and $\mathbf{R}_i^o$

1: Define $\mathbf{F}_{ij} = \mathbf{F}_{ij} + \mathbf{B}_{ij}\mathbf{K}_j, i, j = 1, \ldots, |\mathcal{N}|$, and let ν_i denote the number of dynamic neighbors of subsystem i plus 1. It is well known that, if $\sqrt{\nu_i}\mathbf{F}_{ii}$ is Schur, then for any $\mathbf{Q}_i = \mathbf{Q}_i^T > 0$ there exists a matrix $\mathbf{Q}_i = \mathbf{Q}_i^T > 0$ satisfying

$$\nu_i \mathbf{F}_{ii}^T \mathbf{P}_i \mathbf{F}_{ii} - \mathbf{P}_i = -\mathbf{Q}_i \tag{26.25}$$

Define the matrix $\mathcal{M}^Q \in \mathbb{R}^{|\mathcal{N}|\times|\mathcal{N}|}$ with entries μ_{ij}^Q

$$\mu_{ii}^Q = -\lambda_m(\mathbf{Q}_i), i = 1, \ldots, |\mathcal{N}| \tag{26.26a}$$

$$\mu_{ij}^Q = \nu_j \|\mathbf{F}_{ji}^T \mathbf{P}_j \mathbf{F}_{ji}\|_2, i, j = 1, \ldots, |\mathcal{N}| \text{ with } i \neq j \tag{26.26b}$$

2: If $\mathcal{M}^Q$ is Hurwitz, define the values of $p_i, i = 1, \ldots, |\mathcal{N}|$ as the entries of the strictly positive vector $\mathbf{p}$ satisfying $\mathcal{M}^Q\mathbf{p} < 0$. For its computation see step 3 of Algorithm 26.1.

3: Set $\mathbf{P}_i^o = p_i \mathbf{P}_i$.

4: $\mathbf{Q}_i^o > 0$, $\mathbf{R}_i^o > 0$, and κ_i are chosen in such a way that

$$(1+\kappa_i)(\mathbf{Q}_i^o + (\mathbf{K}_i)^T \mathbf{R}_i^o \mathbf{R}_i) \leq \tilde{\mathbf{Q}}_i \tag{26.27}$$

where

$$\tilde{\mathbf{Q}}_i = p_i \mathbf{Q}_i - \sum_{j=1}^{|\mathcal{N}|} p_j \nu_j \mathbf{F}_{ji}^T \mathbf{P}_j \mathbf{F}_{ji} \tag{26.28}$$

Set $\kappa = \min(\kappa_1, \ldots, \kappa_{|\mathcal{N}|})$.

the definition of the reference trajectories and, at the same time, for the choice of a suitable value for N_p, which proved to be practically effective in many test cases, especially when coupling constraints are not present.

Algorithm 26.5 Initialization

1: For each subsystem i, at time instant $k = 0$, set $N_p = 1$, $\tilde{\mathbf{x}}_i(0) = \mathbf{x}_i(0)$ and a feasible value of $\tilde{\mathbf{u}}_i(0)$.

2: Repeat the following procedure until $\tilde{\mathbf{x}}_i(N_p - 1) \in \hat{\mathcal{X}}_i^F$.

– Solve the i-DPC problem (26.7), subject to constraints (26.8)–(26.13), and additional constraints (26.11), (26.12) for $k = N_p$.

– Set the reference state and input, at time N_p, as $\tilde{\mathbf{x}}_i(N_p) = \hat{\mathbf{x}}_i(N_p|0)$ and $\tilde{\mathbf{u}}_i(N_p) = \mathbf{K}_i\hat{\mathbf{x}}_i(N_p|0)$, respectively.

– Transmit $\tilde{\mathbf{x}}_i(N_p)$ and $\tilde{\mathbf{u}}_j(N_p)$ and receive $\tilde{\mathbf{x}}_j(N_p)$ and $\tilde{\mathbf{u}}_j(N_p)$ from the neighbors $j \in \mathcal{D}_{Ni} \cup \mathcal{H}_i$.

– Set $N_p = N_p + 1$.

26.4 Properties of DPC

In this section the main theoretical properties of DPC will be discussed, together with some promising extensions to the presented approach. In order to enhance readability and clarity, many theoretical details will be omitted. The interested reader can rely on [2, 6, 7] for a more rigorous treatment of the discussed topics.

Optimality issues and convergence. Global optimality of the interconnected closed loop system cannot be guaranteed using DPC. This, on the one hand, is due to the inherent conservativeness of robust algorithms and, on the other hand, is due to the game theoretic characterization of DPC. Namely, as discussed, the provided solution to the control problem can be cast as a maxmin solution of a dynamic non-cooperative game (see, e.g., [15]) where all the involved subsystems aim to optimize local cost functions which are different from each other: therefore, different and possibly conflicting goals inevitably imply suboptimality.
Differently from suboptimal distributed MPC algorithms discussed in [13], whose solutions can be regarded as Nash solutions of non-cooperative games and which possibly lead to instability of the closed-loop system, the convergence of the DPC algorithm can be guaranteed. Specifically, in [7] a convergence analysis of the DPC algorithm is carried out, showing that such fundamental property is guaranteed provided that the tuning parameters and the required sets can be chosen as specified and that the feasibility of the i-DPC problems holds at time step $k = 0$.

Robustness. As discussed in Sect. 26.3, the algorithm presented in this chapter basically relies on the tube-based robust MPC algorithm proposed in [10]. Namely, robustness is here used to cope with uncertainties on the input and state trajectories of the neighboring subsystems. More specifically, the difference between the reference trajectories of the neighboring subsystems and the real ones is regarded as a disturbance, which is known to be bounded in view of suitable constraints imposed in the optimization problem.
It is easy to see that the described approach can be naturally extended for coping also with standard additive disturbances in the interconnected models (26.1), i.e., in case the interconnected perturbed systems are described by the equations

$$\mathbf{x}_i(k+1) = \mathbf{A}_{ii}\,\mathbf{x}_i(k) + \mathbf{B}_{ii}\mathbf{u}_i(k) + \sum_{j=1, j\neq i}^{|\mathcal{N}|} \{\mathbf{A}_{ij}\mathbf{x}_j(k) + \mathbf{B}_{ij}\mathbf{u}_j(k)\} + \mathbf{v}_i(k) \quad (26.29)$$

where $\mathbf{v}_i(k)$ is a bounded unknown disturbance, i.e., $\mathbf{v}_i(k) \in \mathcal{V}_i$ for all $i = 1, \ldots, |\mathcal{N}|$. Now, similarly to the main lines sketched in Sect. 26.3, each subsystem guarantees that, for all $t \geq 0$, its real trajectories lie in specified time-invariant neighborhoods of the reference trajectories, i.e., $\mathbf{x}_i(k) \in \tilde{\mathbf{x}}_i(k) \oplus \mathcal{E}_i$ and $\mathbf{u}_i(k) \in \tilde{\mathbf{u}}_i(k) \oplus \mathcal{E}_i^{\mathcal{U}}$, and (26.29) can be rewritten as (26.3), where now the term $\mathbf{w}_i$ encompasses also the unknown disturbance $\mathbf{v}_i$, i.e.,

$$\mathbf{w}_i(k) = \sum_{j \in \mathcal{D}_{Ni}} \left(\mathbf{A}_{ij}(\mathbf{x}_j(k) - \tilde{\mathbf{x}}_j(k)) + \mathbf{B}_{ij}(\mathbf{u}_j(k) - \tilde{\mathbf{u}}_j(k)) \right) + \mathbf{v}_i(k) \in \mathcal{W}_i$$

where $\mathcal{W}_i = \bigoplus_{j \in \mathcal{D}_{Ni}} \{\mathbf{A}_{ij}\mathcal{E}_j \oplus \mathbf{B}_{ij}\mathcal{E}_j^{\mathcal{U}}\} + \mathcal{V}_i$. From this point on, all the results that follow are basically unchanged with respect to the nominal case discussed in this chapter. Without giving full details of the differences between this case and the nominal one, it is worth remarking that some small differences arise in the definition of the sets (i.e., Algorithms 26.1 and 26.2), where the presence of $\mathcal{V}_i$ must be accounted for.

Output feedback. The approach that has been previously described for copying with unknown exogenous additive disturbances has been employed in [6] for designing a DPC algorithm for output feedback control. Specifically, assume that the input and output equations of the system are the following

$$\begin{aligned} \mathbf{x}_i^o(k+1) &= \mathbf{A}_{ii}\, \mathbf{x}_i^o(k) + \mathbf{B}_{ii}\mathbf{u}_i(k) + \sum_{j=1, j \neq i}^{|\mathcal{N}|} \{\mathbf{A}_{ij}\mathbf{x}_j^o(k) + \mathbf{B}_{ij}\mathbf{u}_j(k)\} \\ \mathbf{y}_i(k) &= \mathbf{C}_i\mathbf{x}_i^o(k) \end{aligned} \tag{26.30}$$

where the state which is not directly available is here denoted as $\mathbf{x}_i^o$ for reasons that will become clearer later on. Denote with $\mathbf{x}_i$ the estimate of $\mathbf{x}_i^o$, for all $i = 1, \ldots, |\mathcal{N}|$. To estimate the state of (26.30) we employ a decentralized Luenberger-like observer of the type

$$\mathbf{x}_i(k+1) = \mathbf{A}_{ii}\, \mathbf{x}_i(k) + \mathbf{B}_{ii}\mathbf{u}_i(k) + \sum_{j=1, j \neq i}^{|\mathcal{N}|} \{\mathbf{A}_{ij}\mathbf{x}_j(k) + \mathbf{B}_{ij}\mathbf{u}_j(k)\} - \mathbf{L}_i(\mathbf{y}_i(k) - \mathbf{C}_i\mathbf{x}_i(k)) \tag{26.31}$$

Assume that the decentralized observer is convergent i.e., $\mathbf{A} + \mathbf{LC}$ is Schur, where $\mathbf{C} = \text{diag}(\mathbf{C}_1, \ldots, \mathbf{C}_{|\mathcal{N}|})$ and $\mathbf{L} = \text{diag}(\mathbf{L}_1, \ldots, \mathbf{L}_{|\mathcal{N}|})$. Under this assumption it is possible to guarantee that the estimation error for each subsystem is bounded, i.e., $\mathbf{x}_i^o(k) - \mathbf{x}_i(k) \in \Sigma_i$ for all $i = 1, \ldots, |\mathcal{N}|$. In this way (26.31) exactly corresponds with the perturbed system (26.29), where $\mathbf{v}_i(k) = -\mathbf{L}_i\mathbf{C}_i(\mathbf{x}_i^o(k) - \mathbf{x}_i(k))$ is regarded as a bounded disturbance, i.e., $\mathbf{v}_i(k) \in \mathcal{V}_i = -\mathbf{L}_i\mathbf{C}_i\Sigma_i$. From this point on, the output feedback control problem is solved as a robust state feedback problem applied to the system (26.29). Details on this approach can be found in [6], where a condition and a constructive algorithm are derived to compute sets Σ_i in such a way that $\Sigma = \prod_{i=1}^{|\mathcal{N}|} \Sigma_i$ is an invariant set for the interconnected observer error.

Tracking. To address the issue of extending the DPC algorithm for tracking desired output signals, the main problem is to characterize the state and input trajectories, for all subsystems, which correspond to the desired output trajectories. To clarify this, consider the case the desired output trajectories are constant and equal to $\bar{\mathbf{y}}_i$, for all $i = 1, \ldots, |\mathcal{N}|$. Under standard assumptions on the system matrices $(\mathbf{A}, \mathbf{B}, \mathbf{C})$ (i.e., that the input/output collective system obtained from (26.30) has no invariant zeros in 1, see [2] for details), the desired setpoint values for the collective state $\mathbf{x}^o$ and input $\mathbf{u}$ can be computed as follows

$$\begin{bmatrix} \bar{\mathbf{x}}^o \\ \bar{\mathbf{u}} \end{bmatrix} = \begin{bmatrix} \mathbf{A} - \mathbf{I}_{n_x} & \mathbf{B} \\ \mathbf{C} & \mathbf{0} \end{bmatrix}^{-1} \begin{bmatrix} 0 \\ \bar{\mathbf{y}} \end{bmatrix} \tag{26.32}$$

where $\bar{\mathbf{y}} = [\bar{\mathbf{y}}_1^T \dots \bar{\mathbf{y}}_{|\mathcal{N}|}^T]^T$. In this way, the setpoint values $\bar{\mathbf{x}}_i^o$ and $\bar{\mathbf{u}}_i$ for $\mathbf{x}_i^o$ and $\mathbf{u}_i$, respectively, are obtained as the vector components of $\bar{\mathbf{x}}^o$ and $\bar{\mathbf{u}}$ of suitable dimensions.

However, the solution to (26.32) requires either a centralized computation or an iterative procedure (to be carried out within a sampling interval). These solutions are not compatible with the proposed approach. To circumvent the issue of computing a solution to (26.32) one can resort to the so-called *velocity form*, see [2] and references therein, which has also the advantage of guaranteeing offset-free tracking in presence of constant perturbations. This implies a transformation of (26.30) into an equivalent one, whose state variable is the pair $[\delta\mathbf{x}_i^T, \varepsilon_i^T]^T$ and whose input variable is $\delta\mathbf{u}_i$, where $\delta\mathbf{x}_i(k) = \mathbf{x}_i^o(k) - \mathbf{x}_i^o(k-1)$, $\varepsilon_i = \mathbf{y}_i(k) - \bar{\mathbf{y}}_i$, and $\delta\mathbf{u}_i(k) = \mathbf{u}_i(k) - \mathbf{u}_i(k-1)$. In this way, the tracking problem for (26.30) is cast as a more standard regulation problem for the velocity form without having to explicitly compute $\bar{\mathbf{x}}_i^o$ and $\bar{\mathbf{u}}_i$. Therefore, the DPC algorithm presented in this chapter can be applied without significant restrictions.

The main issue in the application of DPC for the solution of this problem is on how constraints on $\mathbf{x}_i$ and on $\mathbf{u}_i$ translate into constraints on $\delta\mathbf{x}_i$, ε_i, and $\delta\mathbf{u}_i$, especially as far as the terminal constraints are concerned. A first solution to this is provided in [2]. More sophisticated ones are still under study.

26.5 Applications of DPC

The DPC algorithm has been tested in a number of different simulation test cases. In [7] DPC has been used for control of the reactor/separator process previously considered, e.g., in [8] and depicted in Fig. 26.1. This plant is a consolidated benchmark problem and has been shown to be of particular interest in view of the fact that product recycling induces a feedback interconnection network between the involved subsystems, and that the interconnection strength between them is significant.

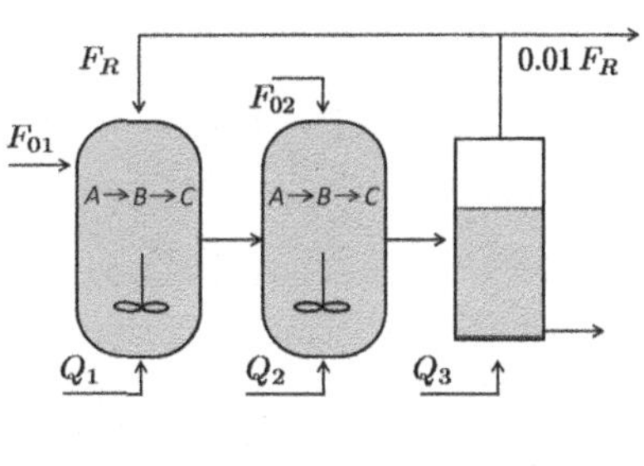

Fig. 26.1 Illustration of the reactor/separator process

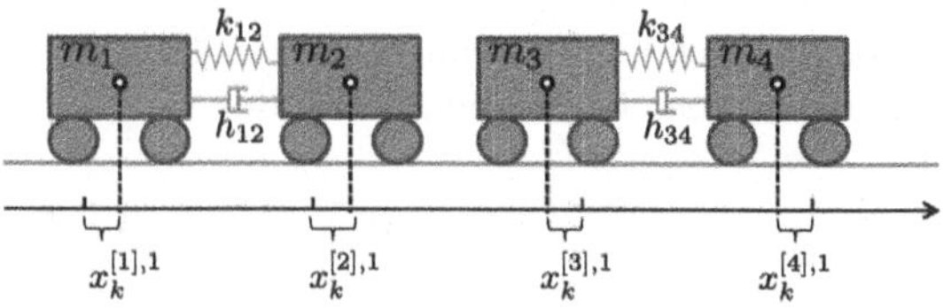

Fig. 26.2 Illustration of the four-truck system

In [6] a different test case is considered. Specifically the system, depicted in Fig. 26.2, consists of four trucks with different masses, each endowed with an engine (exerting bounded input forces). Trucks 1 and 2 (resp. 3 and 4) are dynamically coupled through a spring and a damper. The control problem consists of driving the trucks towards their respective rest positions. Despite its simplicity, this example is of significant interest since it contains all the ingredients of a distributed control problem: dynamic coupling between subsystems, modularity, and presence of both local constraints (i.e., the input forces to each engine are bounded) and collective ones (the distance between two adjacent must not exceed a given value—to prevent breaking of the springs—and must be kept greater than a given threshold—to prevent collisions between trucks). In the paper [2], the considered case study consists of the four-tanks system of Fig. 26.3, used in [1] to test various control algorithms, including distributed approaches.

Finally, the problem of control of an hydro-power valley with DPC has been considered in [11]. The problem and the plant are thoroughly illustrated in [3], where the plant scheme is given. For a description of the system model, as far as the river reaches are concerned, see also [5]. This plant is of particular interest, since it represents a realistic case study, where the order of the system is significant, and where disturbances and linearization are critical issues. Interestingly, the plant is a

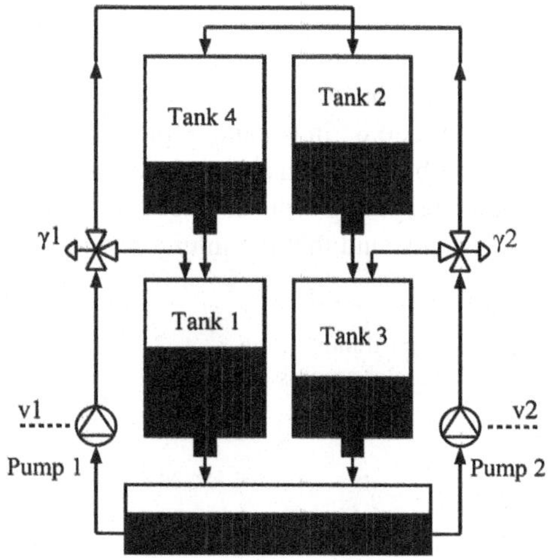

Fig. 26.3 Illustration of the four-tanks system

plain cascaded system and this simplifies the problem of designing a decentralized stabilizing state feedback control law in the DPC framework: in this case, it is also possible to show that the sufficient conditions for application of Algorithms 26.2 and 26.4 for the definition of the sets and of the weighting matrices can always be verified.

Acknowledgments The authors would like to thank Giancarlo Ferrari Trecate, Stefano Riverso and Davide Melzi for fruitful discussions.

References

1. I. Alvarado, D. Limon, D. Muñoz de la Peña, J.M. Maestre, M.A. Ridao, H. Scheu, W. Marquardt, R.R. Negenborn, B. De Schutter, F. Valencia, J. Espinosa, A comparative analysis of distributed MPC techniques applied to the HD-MPC four-tank benchmark. J. Process Control **21**, 800–815 (2011)
2. G. Betti, M. Farina, R. Scattolini, Distributed predictive control for tracking constant references, in *American Control Conference (ACC)*, pp. 6364–6369, (2012)
3. D. Faille, Control specification for hydro power valleys. *Deliverable D.7.2.1 of the European Project HD-MPC*, (2009)
4. L. Farina, S. Rinaldi, *Positive Linear Systems: Theory and Applications* (Wiley-Interscience, New York, 2000)
5. M. Farina, G. Ferrari-Trecate, C. Romani, R. Scattolini, Moving horizon estimation for distributed nonlinear systems with application to cascade river reaches. J. Process Control **21**(5), 767–774 (2011)
6. M. Farina, R. Scattolini, An output feedback distributed predictive control algorithm, in *Decision and Control and European Control Conference (CDC-ECC), 2011 50th IEEE Conference on*, pp. 8139–8144, (2011)
7. M. Farina, R. Scattolini, Distributed predictive control: a non-cooperative algorithm with neighbor-to-neighbor communication for linear systems. Automatica **48**(6), 1088–1096 (June 2012)
8. J. Liu, D. Muñoz de la Peña, P.D. Christofides, Distributed model predictive control of nonlinear process systems. AIChE J. **55**(9), 1171–1184 (2009)
9. J. Löfberg, YALMIP : A toolbox for modeling and optimization in MATLAB, in *Proceedings of the CACSD Conference*, Taipei, Taiwan, (2004)
10. D.Q. Mayne, M.M. Seron, S.V. Rakovic, Robust model predictive control of constrained linear systems with bounded disturbances. Automatica **41**, 219–224 (2005)
11. D. Melzi, Controllo predittivo distribuito di una valle idroelettrica. Master's thesis, Politecnico di Milano, Milan, Italy, (2011)
12. S.V. Rakovic, E.C. Kerrigan, K.I. Kouramas, D.Q. Mayne, Invariant approximations of the minimal robust positively invariant set. IEEE Trans. Autom. Control **50**(3), 406–410 (2005)
13. J.B. Rawlings, D.Q. Mayne, *Model Predictive Control: Theory and Design* (Nob Hill Publishing, Madison, 2009)
14. R. Scattolini, Architectures for distributed and hierarchical model predictive control. J. Process Control **19**, 723–731 (2009)
15. Y. Shoham, K. Leyton-Brown, *Multiagent Systems: Algorithmic, Game-Theoretic, and Logical Foundations* (Cambridge University Press, Cambridge, 2009)

Chapter 27
Decompositions of Augmented Lagrange Formulations for Serial and Parallel Distributed MPC

R. R. Negenborn

Abstract In this chapter we described two distributed MPC schemes for control of interconnected time-invariant discrete-time linear systems: a scheme with serial iterations, and a scheme with parallel iterations. Under the given assumptions, the schemes converge to a solution that a centralized controller would obtain. The schemes have originally been derived from an overall augmented Lagrange formulation in combination with either a block coordinate descent or the auxiliary problem principle. The chapter describes the characteristics of the type of system and control architecture for which the distributed MPC schemes can be used, as well as the actual steps of the schemes, availability of more theoretically oriented extensions, application oriented results, and emerging potential new applications.

27.1 Introduction

We consider a distributed MPC setting in which multiple controllers each control their own subsystem. Due to potential interactions between the subsystems, the controllers need to coordinate their actions. From an MPC point of view: before making decisions, the controllers need to agree with each other in which way they will influence one another's subsystem dynamics over the prediction horizon. Agreement on these influences has to be obtained. In situations in which agreement has to be obtained at each decision step, controllers have to engage in a series of iterations. At each iteration local computations are carried out, taking into account the information available to a controller, followed by communication among the controllers in order to update one another's knowledge about what the controllers want. The iterations can take place in a *serial* or in a *parallel* way. In the serial way, only one controller performs

R. R. Negenborn (✉)
Department of Marine and Transport Technology, Delft University of Technology, Delft, The Netherlands
e-mail: r.r.negenborn@tudelft.nl

J. M. Maestre and R. R. Negenborn (eds.), *Distributed Model Predictive Control Made Easy*, Intelligent Systems, Control and Automation: Science and Engineering 69, DOI: 10.1007/978-94-007-7006-5_27,

its local computations at a time. Typically directly after the local computations the outcomes of these local computations are communicated to other controllers. The other controllers can then take into account the most updated information when it is their turn to do local computations. In the parallel way, all controllers perform local computations simultaneously. Communication then typically takes place after all controllers have completed their local computations. Hence, new information is only taken into account at the next iteration.

In this chapter, we describe the serial scheme and the parallel scheme, first discussed in [12]. The approaches are described from a bottom up perspective, in the sense that we start with the description of the dynamics of an individual subsystem and its controller, and subsequently introduce the presence of other subsystems and controllers. In order to set up the distributed control architecture, no centralized, overall system knowledge is required. It is noted though, that although the description is the schemes is tailored for a bottom up perspective, the original derivation of the schemes followed a bottom up-top down approach. The schemes are derived from the decomposition of an augmented Lagrange formulation of the combined control problem. Due to a non-separable component of the combined control problem obtained when formulating the augmented Lagrange formulation, decomposition of this combined control problem is not straightforward. Different approaches for handling this non-separable term can be considered. The distributed MPC schemes described in this chapter result from using two different approaches: the block coordinate descent approach, or the auxiliary problem principle. The former decomposition approach yields the serial scheme; the latter yields the parallel scheme. Both schemes converge to the overall optimal solution, if the assumptions made are satisfied. In this chapter the ingredients of these schemes are described. For further details (e.g., on the derivation), the interested reader is referred to [14].

This chapter is structured as follows. In Sect. 27.2 the dynamics of the subsystems and the control structure considered are introduced. Section 27.3 introduces the MPC strategy that the controller of a single subsystem considers, without consideration for other controllers. Section 27.4 subsequently describes the serial and parallel MPC strategy when multiple controller are present, influencing the dynamics of one another's subsystems. Section 27.5 points out the availability of more theoretically oriented results. Section 27.6 points out the availability of more application oriented results and an emerging new application domain.

27.2 Boundary Conditions

Consider an overall system composed of n different subsystems. Each subsystem is controlled by a controller. Assumptions on the dynamics of each subsystem and skills and knowledge available to each controller are made as follows.

27.2.1 Dynamics from One Subsystem's Perspective

Let the dynamics of subsystem $i \in \{1, \ldots, n\}$ be given by a deterministic linear discrete-time time-invariant model (possibly obtained after symbolic or numerical linearization of a nonlinear model in combination with discretization):

$$\mathbf{x}_i(k+1) = \mathbf{A}_i\mathbf{x}_i(k) + \mathbf{B}_{1,i}\mathbf{u}_i(k) + \mathbf{B}_{2,i}\mathbf{d}_i(k) + \mathbf{B}_{3,i}\mathbf{v}_i(k) \tag{27.1}$$

$$\mathbf{y}_i(k) = \mathbf{C}_i\mathbf{x}_i(k) + \mathbf{D}_{1,i}\mathbf{u}_i(k) + \mathbf{D}_{2,i}\mathbf{d}_i(k) + \mathbf{D}_{3,i}\mathbf{v}_i(k), \tag{27.2}$$

where at time step k, for subsystem i, $\mathbf{x}_i(k) \in \mathbb{R}^{n_{\mathbf{x}_i}}$ are the local states, $\mathbf{u}_i(k) \in \mathbb{R}^{n_{\mathbf{u}_i}}$ are the local inputs, $\mathbf{d}_i(k) \in \mathbb{R}^{n_{\mathbf{d}_i}}$ are the local known exogenous inputs, $\mathbf{y}_i(k) \in \mathbb{R}^{n_{\mathbf{y}_i}}$ are the local outputs, $\mathbf{v}_i(k) \in \mathbb{R}^{n_{\mathbf{v}}}$ are the remaining variables influencing the local dynamical states and outputs, and $\mathbf{A}_i \in \mathbb{R}^{n_{\mathbf{x}_i} \times n_{\mathbf{x}_i}}$, $\mathbf{B}_{1,i} \in \mathbb{R}^{n_{\mathbf{x}_i} \times n_{\mathbf{u}_i}}$, $\mathbf{B}_{2,i} \in \mathbb{R}^{n_{\mathbf{x}_i} \times n_{\mathbf{d}_i}}$, $\mathbf{B}_{3,i} \in \mathbb{R}^{n_{\mathbf{x}_i} \times n_{\mathbf{v}_i}}$, $\mathbf{C}_i \in \mathbb{R}^{n_{\mathbf{y}_i} \times n_{\mathbf{x}_i}}$, $\mathbf{D}_{1,i} \in \mathbb{R}^{n_{\mathbf{y}_i} \times n_{\mathbf{u}_i}}$, $\mathbf{D}_{2,i} \in \mathbb{R}^{n_{\mathbf{y}_i} \times n_{\mathbf{d}_i}}$, $\mathbf{D}_{3,i} \in \mathbb{R}^{n_{\mathbf{y}_i} \times n_{\mathbf{v}_i}}$ determine how the different variables influence the local states and outputs of subsystem i. The $\mathbf{v}_i(k)$ variables appear due to the fact that a subsystem can be connected to other subsystems. Hence, the $\mathbf{v}_i(k)$ variables represent the influence of other subsystems on subsystem i. If the values of $\mathbf{v}_i(k)$ are fixed, then the dynamics of subsystem i are decoupled from the other subsystems.

27.2.2 Controller of One Subsystem

Subsystem $i \in \{1, \ldots, n\}$ is controlled by a controller i that:

- has a prediction model M_i of the dynamics of subsystem i that matches the subsystem dynamics given by (27.1), (27.2);
- can measure the state $\mathbf{x}_i(k)$ of its subsystem;
- can determine settings $\mathbf{u}_i(k)$ for the actuators of its subsystem;
- can obtain exogenous inputs $\mathbf{d}_i(k+l)$ of its subsystem over a certain horizon of length N_p, for $l = \{0, \ldots, N_\mathrm{p} - 1\}$;
- can communicate with neighboring controllers, i.e., the controllers controlling the subsystems $j \in \mathcal{N}_i$, where $\mathcal{N}_i = \{j_{i,1}, \ldots, j_{i,m_i}\}$ is the set of indices of the m_i subsystems connected to subsystem i, also referred to as the *neighbors* of subsystem or controller i.

27.3 MPC Strategy for One Controller

The controller of a subsystem employs MPC to determine which actions to take. It determines its actions by computing optimal actions over a prediction horizon of N_p time steps according to an objective function, subject to a model of the subsystem, the behavior of the surrounding subsystems, and additional constraints. Assume for

Algorithm 27.1 MPC strategy for one controller

The MPC strategy of controller i at time step k consists of:

1: Measure the initial local state $\bar{\mathbf{x}}_i(k)$,
2: Determine local exogenous inputs over the horizon, $\bar{\mathbf{d}}_i(k+l)$, for $l = \{0,\ldots,N_\mathrm{p}-1\}$,
3: Predict influences of the surrounding subsystems over the prediction horizon, $\bar{\mathbf{v}}_i(k+l)$, for $l = \{0,\ldots,N_\mathrm{p}-1\}$.
4: Solve the following optimization problem:

$$\min_{\tilde{\mathbf{x}}_i(k+1),\tilde{\mathbf{u}}_i(k),\tilde{\mathbf{y}}_i(k)} J_{\mathrm{local,i}}\left(\tilde{\mathbf{x}}_i(k+1),\tilde{\mathbf{u}}_i(k),\tilde{\mathbf{y}}_i(k)\right) = \sum_{l=0}^{N_\mathrm{p}-1} J_{\mathrm{stage,i}}(\mathbf{x}_i(k+1+l),\mathbf{u}_i(k+l),\mathbf{y}_i(k+l)) \tag{27.3}$$

subject to

$$\mathbf{x}_i(k+1+l) = \mathbf{A}_i\mathbf{x}_i(k+l) + \mathbf{B}_{1,i}\mathbf{u}_i(k+l) + \mathbf{B}_{2,i}\bar{\mathbf{d}}_i(k+l) + \mathbf{B}_{3,i}\mathbf{v}_i(k+l) \tag{27.4}$$

$$\mathbf{y}_i(k+l) = \mathbf{C}_i\mathbf{x}_i(k+l) + \mathbf{D}_{1,i}\mathbf{u}_i(k+l) + \mathbf{D}_{2,i}\bar{\mathbf{d}}_i(k+l) + \mathbf{D}_{3,i}\mathbf{v}_i(k+l) \tag{27.5}$$

$$\mathbf{v}_i(k+l) = \bar{\mathbf{v}}_i(k+l) \tag{27.6}$$

$$\text{for } l = 0,\ldots,N_\mathrm{p}-1$$

$$\mathbf{x}_i(k) = \bar{\mathbf{x}}_i(k), \tag{27.7}$$

where $J_{\mathrm{stage,i}}$ is assumed to be twice differentiable function that gives the cost per prediction step given a certain local state, local input, and local output.
5: Implement the actions $\mathbf{u}_i(k)$ until the next time step.
6: Move the time step from k to $k+1$.

now that the controller does not use communication and that it can by itself locally predict the influence of the surrounding subsystems over the prediction horizon, i.e., it knows $\mathbf{v}_i(k+l)$, for $l = 0,\ldots,N_\mathrm{p}-1$. In the following, for notational convenience, we will use the bar over variables to indicate that the values of these variables are known. Moreover, the tilde over variables will be used to denote variables over the prediction horizon, e.g., $\tilde{\mathrm{a}}_i(k) = [\mathbf{a}_i(k)^\mathrm{T},\ldots,\mathbf{a}_i(k+N_\mathrm{p}-1)^\mathrm{T}]^\mathrm{T}$.

The MPC strategy implemented by an individual controller i at each time step k is shown in Algorithm 27.1. A typical choice for the stage cost as used in Algorithm 27.1 is:

$$J_{\mathrm{stage},i}(\mathbf{x}_i(k+1),\mathbf{u}_i(k),\mathbf{y}_i(k)) = \begin{bmatrix}\mathbf{x}_i(k+1)\\ \mathbf{u}_i(k)\\ \mathbf{y}_i(k)\end{bmatrix}^\mathrm{T} \mathbf{Q}_i \begin{bmatrix}\mathbf{x}_i(k+1)\\ \mathbf{u}_i(k)\\ \mathbf{y}_i(k)\end{bmatrix} + \mathbf{f}_i^\mathrm{T}\begin{bmatrix}\mathbf{x}_i(k+1)\\ \mathbf{u}_i(k)\\ \mathbf{y}_i(k)\end{bmatrix}, \tag{27.8}$$

where $\mathbf{Q}_i$ and $\mathbf{f}_i$ are a positive definite weighting matrix and a vector, respectively.

Via (27.6) the assumption of controller i knowing the influence of the surrounding subsystems is made explicit. In most situations, however, controller i cannot know the influence *a priori*, since actions taken by controller i influence the dynamics of its own subsystem and therefore also the dynamics of a neighboring subsystem

$j \in \mathcal{N}_i$, which therefore changes the decision making of neighboring controller j and, hence, changes the actions that controller j chooses, which change the dynamics of subsystem j, and thus changes $\mathbf{v}_i(k+l)$. Therefore, (27.6) can in many cases not be added explicitly. To relax the assumption that this is possible, constraints between control problems have to be introduced. A distributed MPC scheme taking into account these interconnecting constraints should then be employed. Communication between controllers, and iterations of computation and communication are required.

27.4 Distributed MPC for Interconnected MPC Problems

We next discuss formalizing the interconnections among MPC problems, after which we describe a generic distributed MPC algorithm for taking into account the interconnections, and two particular implementations.

27.4.1 Interconnected MPC Problems

Suppose that we have multiple controllers. Each controller controls its subsystem using MPC. The decisions that one controller makes may depend on the decisions that another controller makes, e.g., when the dynamics of the subsystems being controlled are coupled in some way. Hence, the solutions of the MPC problem of one controller will then depend on the solution of the MPC problem of another controller.

We model the interconnections between MPC control problems using *interconnecting variables*. A particular variable of the control problem of controller i is an interconnecting variable with respect to the control problem of controller j if the variable of controller i corresponds to the same physical quantity as a variable in the control problem of controller j. Given the interconnecting variables of two controllers corresponding to the same quantity, it is convenient to define one of these variables as an interconnecting *input* variable and the other as an interconnecting *output* variable. On the one hand, interconnecting input variables $\mathbf{w}_{\text{in},ji}(k)$ of the control problem of controller i with respect to controller j at time step k can be seen as inputs caused by j on the control problem of controller i. On the other hand, interconnecting output variables $\mathbf{w}_{\text{out},ij}(k)$ of the control problem of controller j with respect to the control problem of controller i can be seen as the influence that controller j has on the control problem of controller i. We consider interconnecting variables $\mathbf{w}_{\text{in},ji}(k) \in \mathbb{R}^{n_{\mathbf{w}_{\text{in},ij}}}$ and $\mathbf{w}_{\text{out},ij}(k) \in \mathbb{R}^{n_{\mathbf{w}_{\text{out},ij}}}$.

Define the interconnecting inputs and outputs for the control problem of controller i over a prediction horizon at time step k as:

$$\tilde{\mathbf{w}}_{\text{in},i}(k) = \tilde{\mathbf{v}}_i(k) \tag{27.9}$$

$$\tilde{\mathbf{w}}_{\text{out},i}(k) = \mathbf{K}_i \left[\tilde{\mathbf{x}}_i(k)^{\text{T}} \; \tilde{\mathbf{u}}_i(k)^{\text{T}} \; \tilde{\mathbf{y}}_i(k)^{\text{T}} \right]^{\text{T}}, \tag{27.10}$$

where $\mathbf{K}_i$ is an interconnecting output selection matrix that contains zeros everywhere, except for a single 1 per row corresponding to a local variable that relates to an interconnecting output variable.

The variables $\tilde{\mathbf{w}}_{\text{in},i}(k)$, $\tilde{\mathbf{w}}_{\text{out},i}(k)$ are partitioned such that:

$$\tilde{\mathbf{w}}_{\text{in},i}(k) = \left[\tilde{\mathbf{w}}_{\text{in},j_{i,1}i}(k)^T, \ldots, \tilde{\mathbf{w}}_{\text{in},j_{i,m_i}i}(k)^T \right]^{\text{T}} \tag{27.11}$$

$$\tilde{\mathbf{w}}_{\text{out},i}(k) = \left[\tilde{\mathbf{w}}_{\text{out},j_{i,1}i}(k)^T, \ldots, \tilde{\mathbf{w}}_{\text{out},j_{i,m_i}i}(k)^T \right]^{\text{T}}. \tag{27.12}$$

The interconnecting inputs to the control problem of controller i with respect to controller j must be equal to the interconnecting outputs from the control problem of controller j with respect to controller i, since the variables of both control problems model the same quantity. For controller i this thus gives rise to the following *interconnecting constraints*, for all $j \in \mathcal{N}_i$:

$$\tilde{\mathbf{w}}_{\text{in},ji}(k) = \tilde{\mathbf{w}}_{\text{out},ij}(k) \tag{27.13}$$

$$\tilde{\mathbf{w}}_{\text{out},ji}(k) = \tilde{\mathbf{w}}_{\text{in},ij}(k). \tag{27.14}$$

An interconnecting constraint depends on variables of two different controllers. Therefore, a particular controller will always miss information that it requires to include the interconnecting constraint explicitly in its MPC control problem formulation. Hence, the controller has to use communication with another controller to exchange information that it uses to determine which values it should give to the interconnecting inputs and outputs.

27.4.2 Parallel and Serial Scheme for Distributed MPC

Algorithm 27.2 shows the general distributed MPC scheme that the controllers can employ to obtain agreement on the values of the interconnecting variables. The algorithm involves a number of iterations of communication and computation. These iterations stop when a stopping condition is satisfied. The stopping condition indicates that the controllers have obtained agreement.

Equation (27.16) in Step 5 of the algorithm hints at where the origin of algorithm lies, i.e., dual decomposition using Lagrange multipliers. In the preceding step, Step 4 of the algorithm, in a distributed way a solution to an augmented Lagrange formulation for the overall MPC problem is obtained. As mentioned before, the augmented Lagrange formulation involves a non-separable term. This term can be separated to obtain a solution in a distributed way in two ways:

- **Auxiliary problem principle for parallel implementation** The parallel implementation is the result of using the *auxiliary problem principle* [2, 4, 16] for approximating the non-separable quadratic term in the augmented Lagrange formulation of the combined overall control problem using a linearization with values received from all neighboring controllers at the previous iteration. With

Algorithm 27.2 General distributed MPC scheme

At time step k:

1: For $i = 1, \dots, n$, controller i makes a measurement of the current state of the subsystem $\bar{\mathbf{x}}_i(k) = \mathbf{x}(k)$ and estimates the expected exogenous inputs $\bar{\mathbf{d}}_i(k+l)$, for $l = 0, \dots, N_\mathrm{p} - 1$.

2: Set the iteration counter s to 1 and initialize the Lagrange multipliers $\tilde{\boldsymbol{\lambda}}_{\mathrm{in},ji}(k)^{(s)}$, $\tilde{\boldsymbol{\lambda}}_{\mathrm{out},ij}(k)^{(s)}$ arbitrarily.

3: **repeat**

4: Either serially or in parallel, for $i = 1, \dots, n$, controller i determines $\tilde{\mathbf{x}}_i(k+1)^{(s)}$, $\tilde{\mathbf{u}}_i(k)^{(s)}$, $\tilde{\mathbf{w}}_{\mathrm{in},ji}(k)^{(s)}$, $\tilde{\mathbf{w}}_{\mathrm{out},ij}(k)^{(s)}$, for $j \in \mathcal{N}_i$, by solving:

$$\min_{\substack{\tilde{\mathbf{x}}_i(k+1), \tilde{\mathbf{u}}_i(k), \tilde{\mathbf{y}}_i(k), \\ \tilde{\mathbf{w}}_{\mathrm{in},j_{i,m_i}i}(k), \dots, \tilde{\mathbf{w}}_{\mathrm{in},j_{i,m_i}i}(k), \\ \tilde{\mathbf{w}}_{\mathrm{out},j_{i,1}i}(k), \dots, \tilde{\mathbf{w}}_{\mathrm{out},j_{i,m_i}i}(k)}} \quad J_{\mathrm{local},i}\left(\tilde{\mathbf{x}}_i(k+1), \tilde{\mathbf{u}}_i(k), \tilde{\mathbf{y}}_i(k)\right) + \sum_{j \in \mathcal{N}_i} J_{\mathrm{inter},i}\left(\tilde{\mathbf{w}}_{\mathrm{in},ji}(k), \tilde{\mathbf{w}}_{\mathrm{out},ji}(k), \tilde{\boldsymbol{\lambda}}_{\mathrm{in},ji}(k)^{(s)}, \tilde{\boldsymbol{\lambda}}_{\mathrm{out},ij}(k)^{(s)}\right), \tag{27.15}$$

subject to the local dynamics (27.4)–(27.5) of subsystem i over the horizon and the initial constraint (27.7) of subsystem i.

5: Update the Lagrange multipliers,

$$\tilde{\boldsymbol{\lambda}}_{\mathrm{in},ji}(k)^{(s+1)} = \tilde{\boldsymbol{\lambda}}_{\mathrm{out},ij}(k)^{(s)} + \gamma_\mathrm{c}\left(\tilde{\mathbf{w}}_{\mathrm{in},ji}(k)^{(s)} - \tilde{\mathbf{w}}_{\mathrm{out},ij}(k)^{(s)}\right). \tag{27.16}$$

6: Move on to the next iteration $s+1$.

7: **until** the following stopping condition is satisfied:

$$\left\| \begin{bmatrix} \tilde{\boldsymbol{\lambda}}_{\mathrm{in,j}_{1,1}1}(k)^{(s+1)} - \tilde{\boldsymbol{\lambda}}_{\mathrm{in,j}_{1,1}1}(k)^{(s)} \\ \vdots \\ \tilde{\boldsymbol{\lambda}}_{\mathrm{in,j}n,m_n n}(k)^{(s+1)} - \tilde{\boldsymbol{\lambda}}_{\mathrm{in,j}n,m_n n}(k)^{(s)} \end{bmatrix} \right\|_\infty \leq \gamma_\epsilon, \tag{27.17}$$

where γ_ϵ is a small positive scalar and $\|\cdot\|_\infty$ denotes the infinity norm.

8: The controllers implement the actions until the beginning of the next time step.

9: The next time step is started.

---

this principle, Step 4 of the Algorithm 27.2 will involve a number of parallel iterations in which all controllers perform their local computing step at the same time. In particular, given for $j \in \mathcal{N}_i$, the previous information $\tilde{\mathbf{w}}_{\mathrm{in,prev},ij}(k) = \tilde{\mathbf{w}}_{\mathrm{in},ij}(k)^{(s-1)}$ and $\tilde{\mathbf{w}}_{\mathrm{out,prev},ij}(k) = \tilde{\mathbf{w}}_{\mathrm{out},ji}(k)^{(s-1)}$ *of the last iteration* $s-1$, controller i solves problem (27.15) using the following additional objective function term for the interconnecting constraints:

$$\begin{aligned} &J_{\mathrm{inter},i}\left(\tilde{\mathbf{w}}_{\mathrm{in},ji}(k), \tilde{\mathbf{w}}_{\mathrm{out},ji}(k), \bar{\boldsymbol{\lambda}}_{\mathrm{in},ji}(k)^{(s)}, \bar{\boldsymbol{\lambda}}_{\mathrm{out},ij}(k)^{(s)}\right) \\ &= \begin{bmatrix} \tilde{\boldsymbol{\lambda}}_{\mathrm{in},ji}(k)^{(s)} \\ -\tilde{\boldsymbol{\lambda}}_{\mathrm{out},ij}(k)^{(s)} \end{bmatrix}^\mathrm{T} \begin{bmatrix} \tilde{\mathbf{w}}_{\mathrm{in},ji}(k) \\ \tilde{\mathbf{w}}_{\mathrm{out},ji}(k) \end{bmatrix} + \frac{\gamma_\mathrm{c}}{2} \left\| \begin{bmatrix} \tilde{\mathbf{w}}_{\mathrm{in,prev},ij}(k) - \tilde{\mathbf{w}}_{\mathrm{out},ji}(k) \\ \tilde{\mathbf{w}}_{\mathrm{out,prev},ij}(k) - \tilde{\mathbf{w}}_{\mathrm{in},ji}(k) \end{bmatrix} \right\|_2^2 \\ &\quad + \frac{\gamma_\mathrm{b} - \gamma_\mathrm{c}}{2} \left\| \begin{bmatrix} \tilde{\mathbf{w}}_{\mathrm{in},ji}(k) - \tilde{\mathbf{w}}_{\mathrm{in,prev},ji}(k) \\ \tilde{\mathbf{w}}_{\mathrm{out},ji}(k) - \tilde{\mathbf{w}}_{\mathrm{out,prev},ij}(k) \end{bmatrix} \right\|_2^2. \end{aligned}$$

This scheme uses only information computed during the last iteration $s-1$. The parallel implementation of step 4 of the general distributed MPC scheme therefore consists of the following steps at time step k, iteration s:

4: For all controllers $i \in \{1, \dots, n\}$, *at the same time*, controller i solves the problem (27.15) to determine $\tilde{\mathbf{x}}_i(k+1)^{(s)}$, $\tilde{\mathbf{u}}_i(k)^{(s)}$, $\tilde{\mathbf{w}}_{\mathrm{in},ji}(k)^{(s)}$, $\tilde{\mathbf{w}}_{\mathrm{out},ji}(k)^{(s)}$, and sends to controller $j \in \mathcal{N}_i$ the computed values $\tilde{\mathbf{w}}_{\mathrm{in},ji}(k)^{(s)}$ and $\tilde{\mathbf{w}}_{\mathrm{out},ji}(k)^{(s)}$.

The positive scalar γ_{c} penalizes the deviation from the interconnecting variable iterates that were computed during the last iteration. As additional parameter this scheme uses a positive scalar γ_{b}.

- **Block coordinate descent for serial implementation** The serial implementation that we describe results from using *block coordinate descent* [3, 16] for dealing with the non-separable quadratic term in the augmented Lagrange formulation of the combined overall control problem. This approach minimizes the quadratic term directly, in a serial way, taking for the non-local variables as values those values that have been computed most recently and received from neighboring controllers. Contrarily to the parallel implementation, in the serial implementation one controller after another minimizes its local and interconnecting variables while the other variables stay fixed.
In particular, given the information $\tilde{\mathbf{w}}_{\mathrm{in,prev},ij}(k) = \tilde{\mathbf{w}}_{\mathrm{in},ij}(k)^{(s)}$, $\tilde{\mathbf{w}}_{\mathrm{out,prev},ij}(k) = \tilde{\mathbf{w}}_{\mathrm{out},ij}(k)^{(s)}$ computed at the current iteration s for each controller $j \in \mathcal{N}_i$ that has solved its problem *before* controller i in the *current* iteration s, and given the previous information $\tilde{\mathbf{w}}_{\mathrm{in,prev},ij}(k) = \tilde{\mathbf{w}}_{\mathrm{in},ij}(k)^{(s-1)}$, $\tilde{\mathbf{w}}_{\mathrm{out,prev},ij}(k) = \tilde{\mathbf{w}}_{\mathrm{out},ij}(k)^{(s-1)}$ of the *last* iteration $s-1$ for the other controllers, controller i solves problem (27.15) using the following additional objective function:

$$J_{\mathrm{inter},i}\left(\tilde{\mathbf{w}}_{\mathrm{in},ji}(k), \tilde{\mathbf{w}}_{\mathrm{out},ji}(k), \tilde{\boldsymbol{\lambda}}_{\mathrm{in},ji}(k)^{(s)}, \tilde{\boldsymbol{\lambda}}_{\mathrm{out},ij}(k)^{(s)}\right)$$
$$= \begin{bmatrix} \bar{\boldsymbol{\lambda}}_{\mathrm{in},ji}^{(s)}(k) \\ -\bar{\boldsymbol{\lambda}}_{\mathrm{out},ij}^{(s)}(k) \end{bmatrix}^{\mathrm{T}} \begin{bmatrix} \tilde{\mathbf{w}}_{\mathrm{in},ji}(k) \\ \tilde{\mathbf{w}}_{\mathrm{out},ji}(k) \end{bmatrix} + \frac{\gamma_{\mathrm{c}}}{2} \left\| \begin{bmatrix} \tilde{\mathbf{w}}_{\mathrm{in,prev},ij}(k) - \tilde{\mathbf{w}}_{\mathrm{out},ji}(k) \\ \tilde{\mathbf{w}}_{\mathrm{out,prev},ij}(k) - \tilde{\mathbf{w}}_{\mathrm{in},ji}(k) \end{bmatrix} \right\|_2^2 .$$

Thus, contrarily to the parallel implementation, the serial implementation uses both information from the current iteration and from the last iteration. The serial implementation implements step 4 of the general scheme as follows at time step k, iteration s:

4: For $i = 1, \dots, n$, *one controller after another*, controller i determines $\tilde{\mathbf{x}}_i(k+1)^{(s)}$, $\tilde{\mathbf{u}}_i(k)^{(s)}$, $\tilde{\mathbf{w}}_{\mathrm{in},ji}(k)^{(s)}$, $\tilde{\mathbf{w}}_{\mathrm{out},ji}(k)^{(s)}$ by solving (27.15), and sends to each controller $j \in \mathcal{N}_i$ the computed values $\tilde{\mathbf{w}}_{\mathrm{in},ji}(k)^{(s)}$ and $\tilde{\mathbf{w}}_{\mathrm{out},ji}(k)^{(s)}$.

The role of the scalar γ_{c} is similar as for the parallel implementation, except that for the serial implementation γ_{c} penalizes the deviation from the interconnecting variable iterates that were computed by the controllers before controller i in the current iteration and by the other controllers during the last iteration.

27.5 Theoretical Results Availability

The schemes presented above converge to the optimal solution from an overall point of view, giving a sufficiently tight stopping condition and sufficiently many iterations. Several works focussing on more theoretical sides of the algorithms described above, including the influence of the stopping condition parameter and allowed number of iterations, have been considered:

- Extensive discussion on the influence of values of the parameters required to be set is given in [9, 14].
- Extensions that could be considered in the case that the subsystem dynamics involve both continuous and discrete dynamics (i.e., represent a so-called hybrid system) are proposed in [11].
- Ideas on how to speed up the computation time required for the serial scheme by integrating parallelization are discussed in [13].

27.6 Application Results Availability

27.6.1 Existing Applications

The schemes presented have been used in case studies from application domains such as water systems and power networks:

- **Irrigation canals** are large-scale systems, covering vast geographical areas, and consisting of many interconnected canal reaches that interact with control structures such as pumps and gates. The control of such irrigation canals is currently usually done in a manual way, in which a human operator travels along the irrigation canal to adjust the settings of the gates and pumps in order to obtain a desired water level. In [15] we discuss how the serial distributed MPC scheme can be applied to determine autonomously what the settings of these control structures should be, with as particular application the West-M irrigation canal in Arizona in mind.
- **Urban water supply networks** are large-scale systems that transport potable water over vast geographical areas to millions of consumers. A safe and efficient operation of these networks is crucial, as without it living in today's cities would be impossible. To achieve an adequate operation, these networks are equipped with actuators like pumps and valves, which are used to maintain water pressures and flows within safe margins. Currently, these actuators are controlled in a decentralized way using local controllers that only use local information and that do not take into account the presence of other controllers. As a result, water supply networks regularly experience pressure drops and interruptions of water supply when there is an unexpected increase in water demand. To improve performance the actions of the local controllers should be coordinated. We propose in [5] the application of the parallel distributed MPC schemes for control of urban water supply networks.

Simulations are carried on on model of a part of the urban water supply network of Bogota, the capital of Colombia.

- In [1] physical experiments are performed on a **four-tank water process**. The serial scheme described in this chapter as well as several centralized, decentralized, and distributed MPC schemes for tracking based on cooperative game theory, sensivity-based coordination mechanisms, and bargaining game theory were experimentally assessed, including the serial scheme described in this chapter.
- As the complexity of **power networks** increases, the installation of devices such as High Voltage Direct Current links and Flexible AC Transmission Systems and the use of advanced control techniques can be used to improve network stability. MPC is an example of such an advanced control technique. However, it is often impractical to implement this technique in a centralized manner, as often the problem can be too computationally complex or several independent controllers may be responsible for different subsystems. In [10] the serial distributed MPC scheme is proposed for controlling a multiple link HVDC system using local communications only.

27.6.2 Emerging Applications

Recently, we have started the investigation of the extension of the application of the presented, and related, schemes to the domain of **intermodal and synchromodal transport networks**. Such large-scale networks can often not be controlled from a single point by a single intelligent control agent; instead control has to be performed using multiple intelligent agents. We are focusing on developing a framework for real-time coordinated control of such large-scale transport infrastructures based on the concepts discussed in this and related chapters. Recent developments aimed at application of these concepts are structured along the following lines:

- **Local, microscopic level:**
  Coordinating the actions of equipment inside container terminals
  Container terminals are transport hubs in which containers are moved from one type of transport mode or means to another. An individual container terminal is characterized by a large number of pieces of equipment surrounded by a dynamically operating environment. The transport of a container depends on the actions of multiple pieces of equipment from areas all over the container terminal (Fig. 27.1 illustrates some of the waterside equipment involved). Continuously evolving dynamics of equipment are combined with discrete events, such as the arriving or departing of vessels, trains, and trucks, and the handing over of containers from one piece of equipment to another. We are investigating how to effectively manage the expected commodity volume growth to be handled by container terminals by considering a container terminal as a large-scale system of interacting subsystems. In [17] a first proposal for hybrid MPC of three pieces of equipment is made, as a basis for further study of distributed MPC.

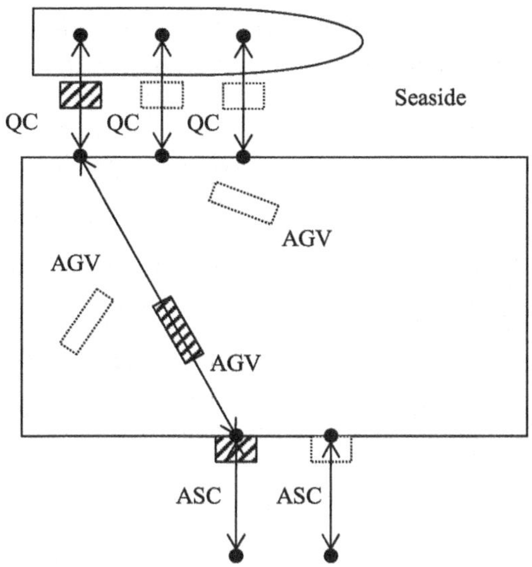

Fig. 27.1 The layout of equipment in the waterside area of an automated container terminal, with quay cranes (QCs), automated guided vehicles (AGVs), and automated stacking cranes (ASCs) [17]

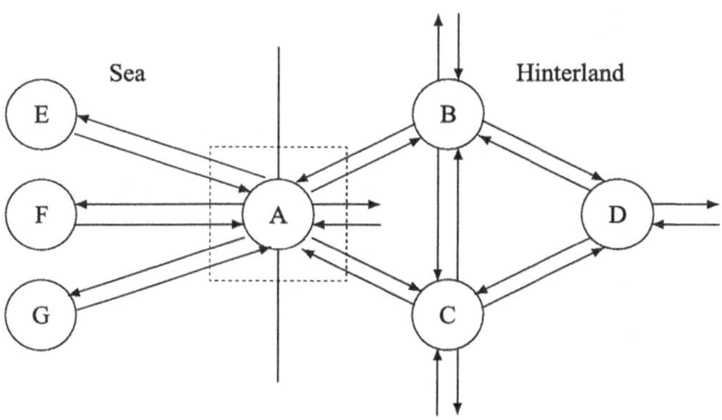

Fig. 27.2 Container terminals and direct surrounding [7]. *Circles* represent terminals and transport connections are indicated by *arrows*

- **Intermediate, mesoscopic, level:**
  Operational control of container terminals and their direct surroundings
  The operations of container terminals are not independent of their surroundings. Container terminals are part of a much larger transport network (Fig. 27.2). At such a network level we consider flows of containers. The increase of international commerce and the expected container vessels capacity with **18,000** TEUs (20-foot equivalent units) will put a considerable pressure on such container hubs. High flow peaks will appear at gateway hubs in the transport network compromis-

ing the cargo transpor towards the hinterland and decreasing the network transport capacity. Moreover, authorities are forcing transport operators to operate in more sustainable ways. For container hubs this is translated into making a preferable choice for barge and train modalities before opting for truck modality. For integrating individual container terminals in a network perspective, we are investigating how to model an individual container terminal in terms of flows. In particular, we propose in [6] a model based on queue theory and container categorization in combination with an MPC controller to solve resource handling allocation problem in a terminal in an optimal way, while respecting maximum resource availability. In [7] we present a framework based on MPC to address the so-called transport modal split problem for the outgoing cargo at container hubs. Giving decision freedom to container hubs using the proposed scheme is a step towards a synchromodal and more flexible transport network. The approach proposed will be extended in

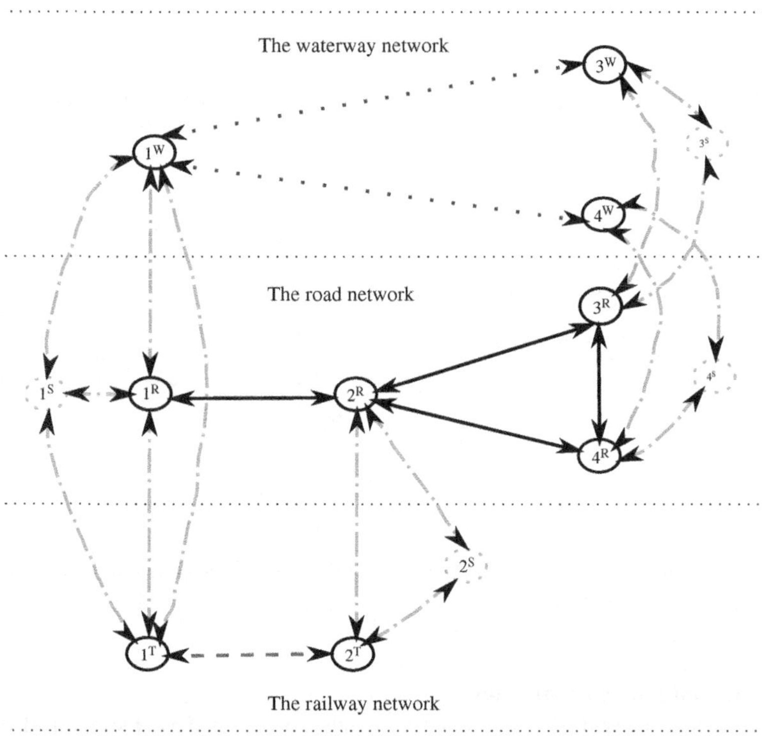

Fig. 27.3 An intermodal transport network model, taken from [8]. The *dotted blue arcs*, the *solid black arcs*, the *dashed red arcs*, and the *dash-dotted green arcs* indicate 4 transport links of the waterway network, 8 transport links of the road network, 2 transport links of the railway network, and 30 transfer links among three different types of transport modes (barges, trucks and trains) in nodes of the intermodal transport network, respectively. The *dashed green nodes* indicate the storage nodes. Each *doubled-headed arc* in the figure represents two directed links with opposite directions.

future work to situations involving multiple terminals.

- **Higher, macroscopic level:**
  Modeling and control of intermodal transport networks
  Intermodal transport networks represent the integration of different types of transport networks, e.g., the road network, the railway network, the waterway network, etc, see Fig. 27.3. The interfaces between these networks are formed by transport hubs (such as container terminals). By integrating and coordinating the use of different transport modes, intermodal transport provides the opportunity to obtain an optimal use of the physical infrastructure so as to guarantee the operating performance of the intermodal transport network as a whole and to provide cost and energy efficient transport services. It is necessary increase the amount of coordination in order to handle the expected increasing amount of commodity to be transported. In [8] a first proposal for modeling such networks with MPC in mind is made. Future research will focus on developing distributed MPC approaches for such large-scale networks.

Acknowledgments This research is supported by the VENI project "Intelligent multi-agent control for flexible coordination of transport hubs" (project 11210) of the Dutch Technology Foundation STW, a subdivision of The Netherlands Organisation for Scientific Research (NWO).

References

1. I. Alvarado, D. Limon, D. Munoz de la Pena, J.M. Maestre, H. Scheu, M.A. Ridao, W. Marquardt, R.R. Negenborn, B. De Schutter, F. Valenciaand, J. Espinosa, A comparative analysis of distributed MPC techniques applied to the HD-MPC four-tank benchmark. J. Process Control **21**(5), 800–815 (2011)
2. J. Batut, A. Renaud, Daily generation scheduling optimization with transmission constraints: a new class of algorithms. IEEE Trans. Power Syst. **7**(3), 982–989 (1992)
3. D.P. Bertsekas, J.N. Tsitsiklis, *Parallel and Distributed Computation: Numerical Methods* (Athena Scientific, New Hampshire, 1997)
4. B.H. Kim, R. Baldick, Coarse-grained distributed optimal power flow. IEEE Trans. Power Syst. **12**(2), 932–939 (1997)
5. S. Leirens, C. Zamora, R.R. Negenborn, B. De Schutter, Coordination in urban water supply networks using distributed model predictive control, in *Proceedings of the American Control Conference 2010*, Baltimore, Maryland, June 2010, pp. 3957–3962
6. J. Lemos Nabais, R.R. Negenborn, M. Ayala Botto, A novel predictive control based framework for optimizing intermodal container terminal operations, in *Proceedings of the 3rd International Conference on Computational Logistics*, Shanghai, China, September 2012, pp. 53–71
7. J. Lemos Nabais, R.R. Negenborn, M. Ayala Botto, Model predictive control for a sustainable transport modal split at intermodal container hubs. in *Proceedings of the 2013 IEEE International Conference on Networking Sensing and Control*, Paris, France, April 2013
8. L. Li, R.R. Negenborn, B. De Schutter, A general framework for modeling intermodal transport networks. in *Proceedings of the 2013 IEEE International Conference on Networking Sensing and Control*, Paris, France, April 2013
9. P. Mc Namara, R.R. Negenborn, B. De Schutter, G. Lightbody, Weight optimisation for iterative distributed model predictive control applied to power networks. Eng. Appl. Artif. Intell. **26**(1), 532–543 (2013)

10. P. Mc Namara, R. R. Negenborn, B. De Schutter, G. Lightbody, Optimal coordination of a multiple HVDC link system using centralised and distributed control. IEEE Trans. Control Syst. Technol. (2013)
11. R.R. Negenborn, B. De Schutter, H. Hellendoorn, Multi-agent model predictive control for transportation networks with continuous and discrete elements, in *Proceedings of the 11th IFAC Symposium on Control in Transportation Systems*, Delft, The Netherlands, August 2006, pp. 609–614
12. R.R. Negenborn, B. De Schutter, J. Hellendoorn, Multi-agent model predictive control for transportation networks: serial versus parallel schemes, in *Proceedings of the 12th IFAC Symposium on Information Control Problems in Manufacturing*, Saint-Etienne, France, May 2006, pp. 339–344
13. R. R. Negenborn, B. De Schutter, J. Hellendoorn, Efficient implementation of serial multi-agent model predictive control by parallelization, in *Proceedings of the 2007 IEEE International Conference on Networking, Sensing, and Control*, London, UK, July 2007, pp. 175–180
14. R.R. Negenborn, B. De Schutter, J. Hellendoorn, Multi-agent model predictive control for transportation networks: serial versus parallel schemes. Eng. Appl. Artif. Intell. **21**(3), 353–366 (2008)
15. R.R. Negenborn, P.J. van Overloop, T. Keviczky, B. De Schutter, Distributed model predictive control for irrigation canals. Netw. Hetrog. Media. **4**(2), 359–380 (2009)
16. C. B. Royo, Generalized unit commitment by the radar multiplier method, PhD thesis, Technical University of Catalonia, Barcelona, Spain, May 2001
17. J. Xin, R.R. Negenborn, G. Lodewijks, Hybrid model predictive control for equipment in an automated container terminal, in *Proceedings of the 2013 IEEE International Conference on Networking Sensing and Control*, Paris, France, April 2013

Chapter 28
A Hierarchical Distributed MPC Approach: A Practical Implementation

A. Zafra-Cabeza and J. M. Maestre

Abstract This chapter presents a hierarchical distributed model predictive control algorithm. Two levels in the problem optimization are presented. At the lower level, a distributed model predictive controller optimizes the operation of the plant manipulating the control variables in order to follow the set-points. The higher level implements a risk management strategy based on the execution of mitigation actions if risk occurrences are expected. In this way it is possible to take into account additional relevant information so that better results are achieved in the optimization of the system.

28.1 Introduction

Optimization plays a very important role in many control systems. However, the materialization of risks that may affect the results of the optimization is rarely considered in the literature. The work presented in this chapter shows a hierarchical control scheme that copes with this issue (see Fig. 28.1). The higher level considers external and internal risk information to improve the overall performance, and the lower level calculates the control actions that are necessary to follow the set-points provided by the higher level. In particular, the higher level uses model predictive control (MPC) to perform optimization and planning of the set-points that are provided to the lower level. The lower level controller is implemented by a distributed MPC (DMPC) controller that ensures the cooperation between the local controllers or agents that are assigned to the different subsystems.

A. Zafra-Cabeza (✉) · J. M. Maestre
System Engineering and Automatic Department, University of Seville, Seville, Spain
e-mail: asunzafra@us.es

J. M. Maestre
e-mail: pepemaestre@cartuja.us.es

J. M. Maestre and R. R. Negenborn (eds.), *Distributed Model Predictive Control Made Easy*, Intelligent Systems, Control and Automation: Science and Engineering 69,
DOI: 10.1007/978-94-007-7006-5_28,

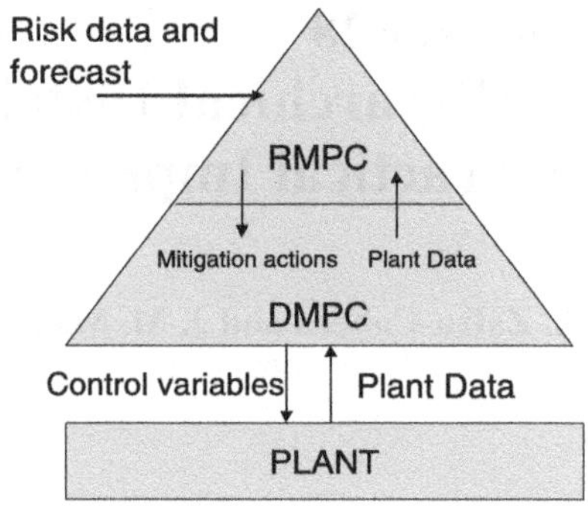

Fig. 28.1 Hierarchical scheme of controllers

The proposed control architecture is called Risk-based DMPC (RDMPC) and is based on the execution of mitigation actions that reduce the impacts of risks. Hence, changes on the set-points of the plant can be seen as mitigation actions that are carried out to reduce the exposure of the aforementioned risks. The introduction of a risk mitigation attitude in control policies improves the reliability of the system and also its robustness with respect to failures or changes in the expected values of the system parameters.

28.2 Problem Statement

In order to clarify the risk mitigation attitude proposed in this chapter, we show in Fig. 28.2 an example of the relationships between risks and actions in a possible strategic plan.

In the first place, we consider a set of parameters $\mathcal{Z} = \{Z_1, Z_2, \ldots\}$ that are monitored in the upper control layer, e.g., economic costs, time delays, demand satisfaction, or delivery times. The system may be associated to some risks, which are identified and grouped into a set $\mathcal{R} = \{R_1, R_2, \ldots\}$. In particular, we define a risk as an event that can happen with a certain probability and that has an impact on one or more of the aforementioned parameters. For example, in Fig. 28.2 it can be seen that the system may be affected by risks R_1 and R_2. On the other hand, risks can be mitigated by actions that can be carried out and that are grouped into a set $\mathcal{A} = \{A_1, A_2, \ldots\}$. Risk R_m is mitigated by the actions A_{p-1} and A_p in Fig. 28.2. Notice that one action can mitigate different risks; for example, in Fig. 28.2 the action A_{p-1} mitigates R_2 and R_m.

In order to illustrate all these relationships, let us consider a tank with a reaction in a solution that needs precise temperature and pressure conditions. If the pressure switches fail (R_1) or the tank is depressurized (R_2) the consequences could be highly dangerous. A maintenance contract of the tank (A_1) and the continuos monitoring of the process (A_2) could prevent this fault, but implementing these actions would increase the system costs as well. Deciding which actions to perform and when to

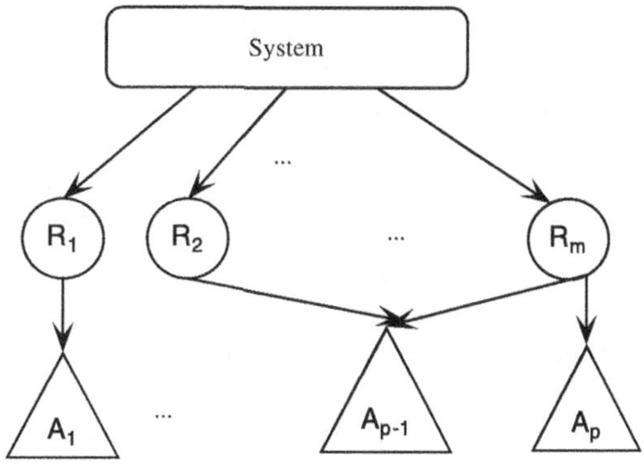

Fig. 28.2 Risk based scheme

run them is a complex optimization problem in which many factors are involved, e.g.: risk probability, impact assessment or costs optimization.

The lower level of the proposed architecture is composed of a DMPC controller that regulates the plant toward the setting points provided by the upper level. We must remark that the control architecture does not depend on any particular DMPC scheme for the lower layer. Hence, any DMPC scheme can be used to this end.

28.3 Risk Based Distributed Model Predictive Control

In this section we describe both the upper and the lower control layers and provide the corresponding algorithms.

28.3.1 Higher Level: Risk-based MPC

At the higher control level a Risk-Based MPC (RMPC) is implemented in order to provide a realistic planning. To this end, a multicriteria objective function that takes into account data from the system and information about external and internal risks is defined as follows:

$$J = \sum_{z \in \mathcal{Z}} \beta_z J_z(\mathbf{u_M}, k) \tag{28.1}$$

where there are $|\mathcal{Z}|$ different parameters to optimize, being J_z the cost associated to the z-th parameter and β_z its corresponding weight. Each of the cost functions J_z is expressed as:

$$J_z(\mathbf{u_M}, k) = \sum_{l=1}^{N_p} (\hat{Z}_z(k+l|k) - Z_{\text{ref},z}(k+l))^2, \tag{28.2}$$

where $\hat{Z}_z$ and $Z_{\text{ref},z}$ are respectively the predicted value of the z-th parameter and its corresponding reference value at time $(k+l)$. N_p is the prediction horizon. The predicted value of Z_z, $\hat{Z}_z$, is calculated as:

$$\hat{Z}_z(k+l|k) = \sum_{r \in \mathcal{R}} RE_{rz}(\mathbf{u_M}, k+l), \tag{28.3}$$

where $RE_{rz}(\mathbf{u_M}, k+l)$ stands for the risk exposure of the risk R_r on parameter Z_z at time step $k+l$. In particular, the risk exposure depends on the probability of the risk, its potential impact, and the implementation of mitigation actions ($\mathbf{u_M}$). More specifically, the risk exposure takes the form:

$$RE_{rz}(\mathbf{u_M}, k) = P_r(k)\big(I_{rz} - \sum_{a \in \mathcal{A}} f_{za}(u_{M_a})\big) + \sum_{a \in \mathcal{A}} g_{za}(u_{M_a}), \tag{28.4}$$

where $P_r(k)$ is the probability assigned to risk R_r at time step k and I_{rz} indicates how the risk R_r affects the parameter Z_z at time step k. The impact I_{rz} of this risk can be reduced by means of mitigation actions. Specifically, an action $a \in \mathcal{A}$ reduces the impact I_{rz} in f_{za} units. Notice that $f_{za} = 0$ in case that the mitigation action a does not reduce the impact I_{rz}. Nevertheless, the execution of the mitigation action a to relieve the risk r over the parameter Z_z comes at a cost, which is represented by the term $g_{za}(u_{M_a})$. Therefore, each mitigation action is described by the following tuple:

$$A_a = \{u_{M_a}, F_a, G_a\} \quad a \in \mathcal{A} \tag{28.5}$$

where u_{M_a} is the decision variable that stands for the mitigation action a; $F_a = \{f_{za} : \mathcal{Z} \to \mathbb{R}\}$ is a function that represents an impact reduction of f_{za} units over the element Z_z when the action A_a is applied; and $G_a = \{g_{za} : \mathcal{Z} \to \mathbb{R}\}$ is the cost of execution of the action. Notice that the intensity of each action has often to be taken into account when deciding how to execute it; this decision will depend on the nature of the mitigation action control variable u_{M_a}, which can either be continuous ($u_{M_a} \in \mathbb{R}$) or discrete ($u_{M_a} \in \mathbb{Z}$).

The outcome of the upper control layer, which is the result of the MPC optimization, is a plan that consider the implementation of control actions to mitigate the risks, as it can be seen in Eqs. (28.1–28.4). Specifically, the planning proposed by the RMPC at each time step is:

$$\mathbf{u_M}^*(k : k+N_\text{p}-1) = \arg\min_{\mathbf{u_M}} \sum_{z \in \mathcal{Z}} \beta_z J_z(\mathbf{u_M}, k), \tag{28.6}$$

that is, the optimal set of mitigation actions (intensity and time instant to be launched) that minimizes the value of the objective function taking into account risk probabilities and the MPC formulation. If some of the involved variables were random, the

Algorithm 28.1 Risk-based MPC (RMPC)

1: Define the set $\mathcal{Z} = \{Z_1, Z_2, \ldots\}$ of the parameters to optimize.
2: Define the set $\mathcal{R} = \{R_1, R_2, \ldots\}$ of the risks that have been identified for the system.
3: For each risk $R_r \in \mathcal{R}$, assign a probability of occurrence $P_r(k)$ and the corresponding impact function $I_{rz}(k)$ at each time instant k $\forall z \in \mathcal{Z}$.
4: Define the set $\mathcal{A} = \{A_1, A_2, \ldots\}$ of the mitigation actions that can be executed in order to reduce impacts.
5: For each mitigation action A_a, define a decision variable, u_{M_a}, and the corresponding functions that determine the risk impact reduction, $F_a(u_{M_a})$, and the additional costs, $G_a(u_{M_a})$.
6: For each risk, define the risk exposure term denoted by RE_{rz} in (28.4) taking into account all the previous information.
7: Define the global objective function (28.1) assigning a value to each weight β_z $\forall z \in \mathcal{Z}$. Notice that the terms J_z should be defined with the information of previous steps.
8: At each time step of the upper control layer, optimize the total objective function J with respect to the mitigation actions decision variables in order to obtain $\mathbf{u}_\mathrm{M}{}^*(k : k + N_\mathrm{p} - 1)$.

optimization would be based on the expected value [2, 5] and properties about the convexity of the objective function could be applied.

The outcome of the optimization problem depends strongly on the weights β_z of the different terms. Likewise, notice that the objective function can be easily enhanced with additional terms in order to consider other goals in the optimization.

At this point it is convenient to remark that the optimization problem solved in (28.6) may be a mixed integer quadratic problem (MIQP) because it is typical that some of the decision variables associated to the implementation of mitigation actions are boolean. This type of problem belongs to the class of NP-complete problems [1] and has a complexity that depends on the number of decision variables and constraints. The computation time required to solve the problem is worst-case exponentially with the problem size. If the problem has n_b binary inputs, the complexity is 2^{n_b} (2^{n_b} QP problems). The number of QP problems to be solved is finite and, therefore, the algorithm finds a feasible solution (if there is one) at a finite time. As a consequence, the time step length of the upper control layer must be large enough to guarantee that the optimization problem can be solved.

The algorithm that details how to implement this control strategy is given as Algorithm 28.1.

Note that depending on the accuracy of available data about risks, the results of the algorithm may change considerably.

28.3.2 Lower Level: Distributed Model Predictive Controller

In this chapter we will use a very simplified version of a DMPC scheme based on agent negotiation, which is proposed in [3]. It assumes that there is a set $\mathcal{N} = \{1, 2, \ldots\}$ of constrained linear subsystems coupled by the inputs, each one controlled by a local controller or agent that has access only to the local model and state information. The goal of this scheme is to minimize a global performance index that is defined as the

Algorithm 28.2 DMPC Based on Agent Negotiation (DMPC-BAN)

1: Initialize the joint decision vector at time k, $\mathbf{u}^{\mathrm{d}}_{\mathcal{N}}(k : k + N_{\mathrm{p}} - 1) = \mathbf{0}$.

2: Each agent i measures its state $\mathbf{x}_i(k)$, solves its optimization problem, and make proposals to its neighbors. A proposal is simply a possible future trajectory for all the system inputs and is built by updating in $\mathbf{u}^{\mathrm{d}}_{\mathcal{N}}(k : k + N_{\mathrm{p}} - 1)$ the trajectories of the inputs that affect the agent, i.e., those in $\mathcal{N}_i$. Hence, the proposal of agent i is:

$$\begin{aligned}
&\mathbf{u}^{i}_{\mathcal{N}}(k : k + N_{\mathrm{p}} - 1) = \arg \min_{\mathbf{u}_{\mathcal{N}}(k:k+N_{\mathrm{p}}-1)} J_i(\mathbf{x}_i(k), \mathbf{u}_{\mathcal{N}_i}(k : k + N_{\mathrm{p}} - 1))\\
&\text{s.t.}\\
&\mathrm{x}_i(l + 1) = \mathbf{A}_i \mathrm{x}_i(l) + \sum_{j \in \mathcal{N}_i} \mathbf{B}_{ij} \mathbf{u}_j(l) \quad \forall i \in \mathcal{N}\\
&\mathbf{u}_j(l) = \mathbf{u}^{\mathrm{d}}_j(l)\ \forall j \notin \mathcal{N}_i, l \in \{k, k + 1, \ldots, k + N_{\mathrm{p}} - 1\}\\
&\mathrm{x}_i(l) \in \mathcal{X}_i,\ l \in \{k + 1, \ldots, k + N_{\mathrm{p}}\} \quad \forall i \in \mathcal{N}\\
&\mathbf{u}_i(l) \in \mathcal{U}_i,\ l \in \{k, k + 1, \ldots, k + N_{\mathrm{p}} - 1\} \quad \forall i \in \mathcal{N}
\end{aligned} \tag{28.8}$$

3: Agent i asks all the agents affected by its proposal if they are free to evaluate it (each agent can only evaluate one proposal at the same time). If all the neighbors acknowledge the petition, the algorithm continues. If not, the agent waits a random time before trying again.

4: Each agent j affected by the proposal evaluates the difference between the cost of the new proposal $\mathbf{u}^{i}_{\mathcal{N}}(k : k + N_{\mathrm{p}} - 1)$ and the cost of the current accepted proposal $\mathbf{u}^{\mathrm{d}}_{\mathcal{N}}(k : k + N_{\mathrm{p}} - 1)$ as

$$\Delta J^{i}_{j} = J_j(x_j(k), \mathbf{u}^{i}_{\mathcal{N}_j}(k : k + N_{\mathrm{p}} - 1)) - J_j(x_j(l), \mathbf{u}^{\mathrm{d}}_{\mathcal{N}_j}(k : k + N_{\mathrm{p}} - 1)) \tag{28.9}$$

This difference ΔJ^{i}_{j} is sent back to the proposer agent i. If the proposal does not satisfy the constraints of the corresponding local optimization problem, an infinite cost increment is assigned. This implies that unfeasible proposals will never be chosen.

5: Agent i receives the local cost increments from all the agents affected by its proposal, so that it can evaluate its impact $\Delta J^{i}(k)$ as

$$\Delta J^{i} = \textstyle\sum_j \Delta J^{i}_{j} \tag{28.10}$$

If ΔJ^{i} is negative, $\mathbf{u}^{\mathrm{d}}_{\mathcal{N}}(k : k + N_{\mathrm{p}} - 1) = \mathbf{u}^{i}_{\mathcal{N}}(k : k + N_{\mathrm{p}} - 1)$ and agent i broadcasts the update in the joint decision vector. Otherwise, the proposal is discarded.

6: Go to Step 2 until the maximum number of proposals have been made or the time available for lacing proposals is over.

7: Apply the first component of $\mathbf{u}^{\mathrm{d}}_{\mathcal{N}}(k : k + N_{\mathrm{p}} - 1)$ and repeat the procedure at the next time step.

---

sum of the local cost functions. Communication between agents is allowed so that they can negotiate within the sample period to take a cooperative decision.

Let $\mathcal{N}_i$ be the set of inputs that affect the dynamics of agent i. The local cost function of agent i at time step k is defined as

$$J_i\left(\mathbf{x}_i(k), \mathbf{u}_{\mathcal{N}_i}(k : k + N_{\mathrm{p}} - 1)\right) = \sum_{l=k+1}^{k+N_{\mathrm{p}}} \|\mathbf{x}_i(l) - \bar{\mathbf{x}}_i\|^2_{\mathbf{Q}_i} + \sum_{l=k}^{k+N_{\mathrm{p}}-1} \sum_{j \in \mathcal{N}_i} \|\mathbf{u}_j(l)\|^2_{\mathbf{R}_{ij}} \tag{28.7}$$

where $\mathbf{u}_{\mathcal{N}_i}(k : k + N_{\mathrm{p}} - 1)$ is the future trajectory of the inputs that affect subsystem i, and $\mathbf{Q}_i > 0$, $\mathbf{R}_{ij} > 0$ are weighting matrices. Next, Algorithm 28.2 details how to implement this control strategy.

28.4 Case Study

The system considered for the case study is an open-canal used for water distribution (for irrigation and drinking water supply), made up of several reaches connected by gates for regulation purposes proposed in [4]. The selected section is Y-shaped (see Fig. 28.3) from the *postrasvase Tajo-Segura* in the south-east of Spain. A main canal split into two with a gate placed at the entry of each of them. The total length of the canals is approximately 24 km. The most important elements in the canals are the main gates that regulate the level of water along the canals and the off-take gates, where farmers take water from the canals for irrigation. Each canal reach has an inflow from an upstream canal reach and an outflow to a downstream canal reach.

Furthermore, other flows are considered as perturbation variables: flows due to rainfall, failure in upstream gate, known off-take outflows by farmers, considered as measurable perturbations,... The controlled variables are the downstream water levels at gates and the manipulated variables are the flow at the head of the canal and the position of the gates.

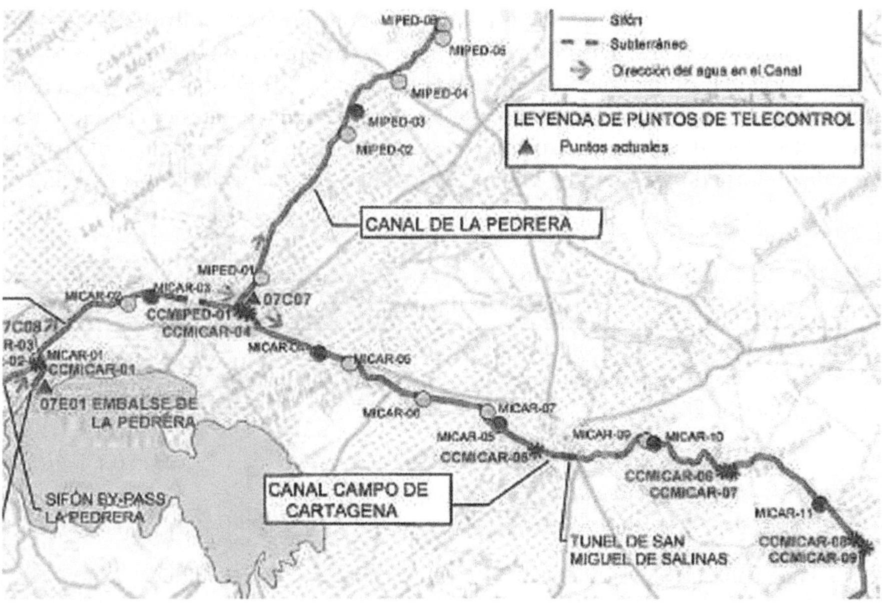

Fig. 28.3 Scheme of the canal

28.4.1 Higher Level

The higher level implements the risk management strategy based on the execution of mitigation actions, which may change the water level set-points if risk occurrences are expected. The main goals of this controller are (a) to minimize cost due to risks and (b) to send the modifications of the water set-points to the DMPC at the lower level if necessary. The list of parameters that we want to optimize is given by the set $\mathcal{Z} = \{Z_1, Z_2\}$, with Z_1 being the cost (euros/day) and Z_2 the variation of the reference level in the reaches (meters). To calculate the predicted output, Eqs. (28.2 and 28.3) are used. The study period is 1 year, sampling time 1 day, prediction horizon, $N_\mathrm{p} = 5$ days and $\beta = [1\ 1]$.

28.4.1.1 Risk Identification

Table 28.1 shows the set of risks, $\mathcal{R}$, that have been considered in this example, their impacts on $\mathcal{Z}$ and probabilities.

Risk R_1 is stated as "Inadequate fresh water quality". If this risk occurs, an impact of 2,000 euros/day is incurred; the probability is constant over time: 0.1. R_2 is associated to failure at gates. The probability of this risk depends on the canal operation: the wear and tear of the gate increases the likelihood of deterioration. Thus, a function θ_1 that depends on the control variables that operate the gates ($\mathbf{u}$) and the time, has been included. As the use of the gates and the time of operation is increasing, the value returned by θ_1 will be greater. R_3 depends on the water levels (h_i) and on

Table 28.1 Risk description (case study)

R_r	Description	Impacts	$P_r(k)$
Internal risks: operation and maintenance			
R_1	Inadequate fresh water quality	$I_{11} = 2{,}000/\ I_{12} = 0$	0.1
R_2	Failure in gates due to wear and tear	$I_{21} = 400/\ I_{22} = 0$	$0.1 + \theta_1(\mathbf{u}, k)$
R_3	Seepage losses	$I_{31} = 10/\ I_{32} = 0$	$0.1 + \theta_2(\mathbf{h}, k)$
External risks: politics and weather			
R_4	Farmers, water demand varies from forecast	$I_{41} = 0/\ I_{42} = +0.15\mathbf{h}(k)$	$P_4(k)$
R_5	Rainfall changes water level of canal, producing water logging of adjacent lands	$I_{51} = 0/\ I_{52} = -RF(k)$	$P_5(k)$
R_6	Changes in politics modify the strategy	$I_{61} = 250/\ I_{62} = 0$	$P_6(k)$
R_7	State policies provide incentives for IC systems	$I_{71} = -2{,}000/\ I_{72} = 0$	0.01
R_8	Uninsured events of force majeure	$I_{81} = 6{,}000/\ I_{82} = 0$	0.01

the time (k). In the same way, a function θ_2 depending on these variables has been included. If farmers take un-authorized, R_4 may appear. This risk is well-known and usually appears in drought season. P_4 is modelled as a normal distribution with mean 0.5 and deviation 0.3 in months June and July. During the rest, this probability is 0. R_4 has impacts on Z_2, and an increment of 15 % on the initial level reference. R_5 states the possibility of water logging. The probability depends on the season. The impact on water level reference consist of an increasing on the initial water level reference. This amount is based on the rainfall forecast of Murcia $RF(k)(l/m^2)$. R_6 has been established as the possibility of the water strategy for the plant being modified by changes in the government, being only 0.1 for the last quarter. R_7 and R_8 are external risks related to incentives for IC plants from government and events of force majeure, respectively. Both of them have an impact on cost and a constant probability.

28.4.1.2 Mitigation Plan

The set $\mathcal{A}$ is composed by six mitigation actions and therefore, $\mathbf{u_M} = \{u_{M_1}^T, \ldots, u_{M_6}^T\}$, where u_{M_5} and u_{M_6} are real and the rest are boolean. A description of the actions used to mitigate risks is shown in Table 28.2. The third column represents functions f_{za} and g_{za} that model the reduction of impacts and cost of execution, respectively. The fourth column is the period of validity of the action (D = Daily, W = Weekly, Y = Yearly, B = Biyearly). That means if an action is executed, it not will be reassessed until past the time of validity. For example, an insurance contract may be executed every 180 days (if estimated) and a water analysis may be undertaken every week.

Some functions f_{za} and g_{za} depend proportionally on the impacts. For example, $f_{11} = 0.7II_1u_{M_1}$ means that the execution of this action reduces the impact by 70 %. Mitigation actions are carried out depending on the setting of the execution frequency. The risk-based structure with links between risks and actions is shown in Fig. 28.4. For example, it can be seen that risk R_1 is mitigated by actions A_1 and A_2 and R_2 is mitigated by A_3.

For risks R_4 and R_5, the discharge from farmers and the rainfall forecast for the city of Murcia have been considered during 2009. Figure 28.5 shows the rainfall forecast in the top panel; the initial level is represented by a dotted green line. This

Table 28.2 Mitigation actions description (case study)

A_a	Description	f_{1a}, g_{1a} on Z_1(cost)	PV
A_1	Periodic water analysis	$f_{11} = 0.7II_1u_{M_1}, g_{11} = 250u_{M_1}$	W
A_2	Control weed growth	$f_{12} = 0.3II_1u_{M_2}, g_{12} = 1{,}500u_{M_2}$	B
A_3	Appropriate monitoring or control over devices	$f_{13} = II_1u_{M_3}, g_{13} = 250u_{M_3}$	W
A_4	Lining irrigation canal	$f_{14} = 0.95II_1u_{M_4}, g_{14} = 2{,}700u_{M_4}$	Y
A_5	Modify set-points of water levels ($u_{M_5} \in \mathbb{R}$)	$f_{15} = 0, g_{15} = 0$	D
A_6	Insurance policy ($u_{M_6} \in \mathbb{R}$)	$f_{16} = 225u_{M_6}, g_{16} = u_{M_6}$	B

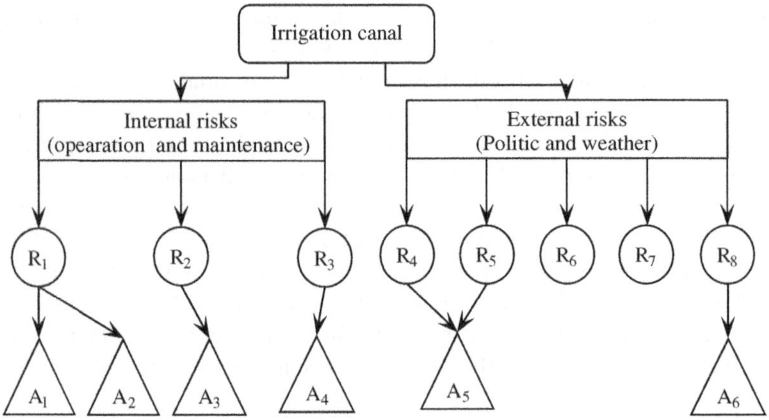

Fig. 28.4 Risk-based structure for the case study

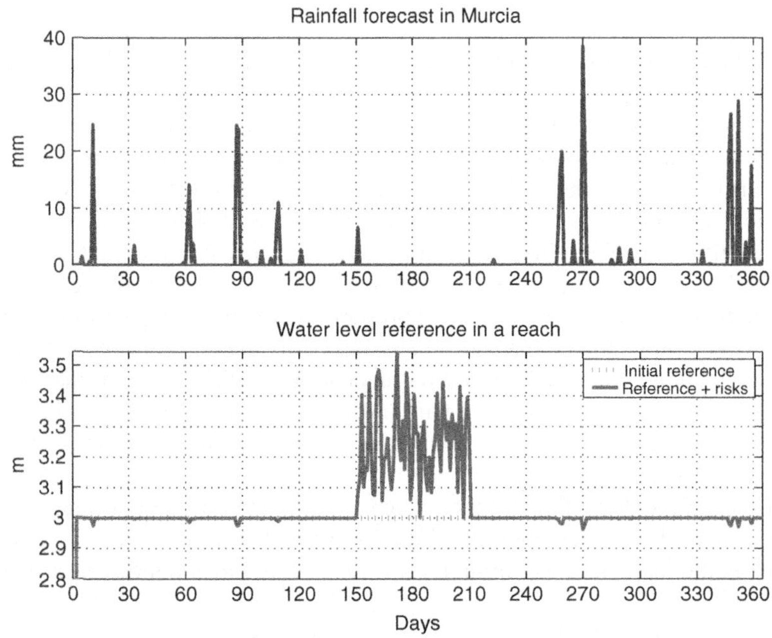

Fig. 28.5 *Top panel* rainfall forecast in Murcia. *Lower panel* level reference in one reach by considering risks R_4, R_5 and action A_5

level is modified by R_4 and R_5, giving rise to an actual level reference shown by the solid red line. A_5 states the modification of water level reference as a consequence of R_4 or R_5. Note that in the summer season the level is increased as farmers may demand more water as a result of drought. A_6 models an insurance contract. The value of u_{M_6} means the cost of the contract and f_{16} the insurance coverage.

28.4.2 Results of Higher Controller

Figure 28.6 shows the following costs obtained with the mitigation:

1. Economic costs with risks but no mitigation (dashed blue line) where impacts are considered but no actions are executed to reduce them.
2. Economic cost with mitigation (solid red line) where mitigation actions are executed to reduce impacts.

The no mitigation case (dashed blue line) is computed considering the accumulative impacts on cost day to day. The mitigation line takes into account the reduction of the impact and the cost of the actions when they are carried out. Note how the no mitigation option reflects the highest cost. As expected, the proposed cost is lower than the no mitigation line. The obtained savings is about 38,000 euros in a mitigation scenario.

The mitigation actions to be executed to reach the optimum costs, are shown in Fig. 28.7; notice that they are undertaken depending on the period of validity shown in Table 28.2 and the risk probabilities. Actions A_1 and A_3 are weekly executed. Action A_2 is carried out every 180 days. The frequency of A_4 is one year; it is only

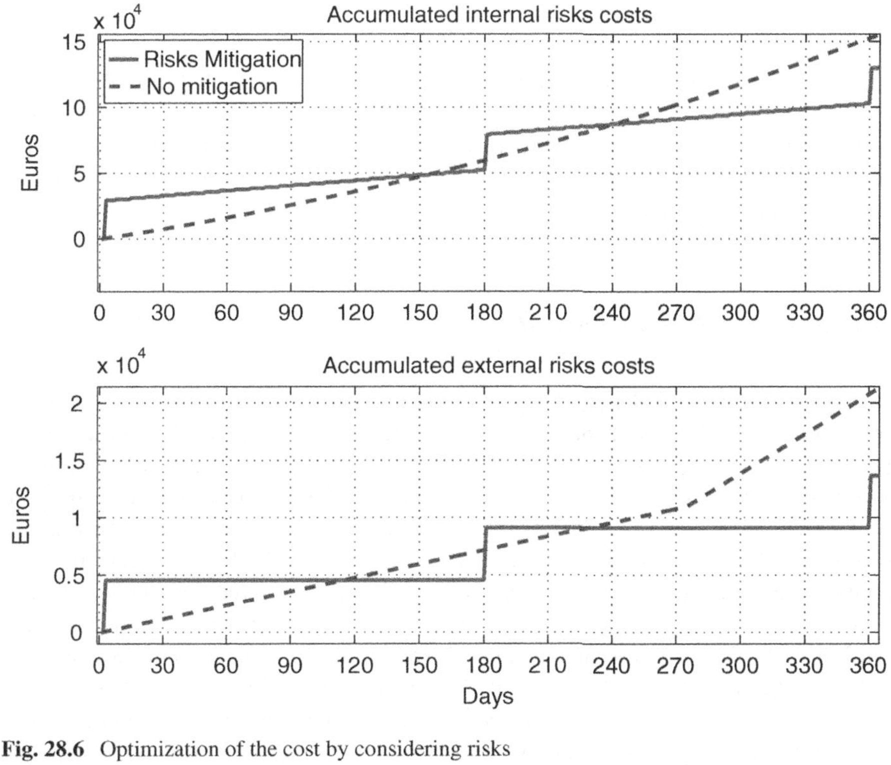

Fig. 28.6 Optimization of the cost by considering risks

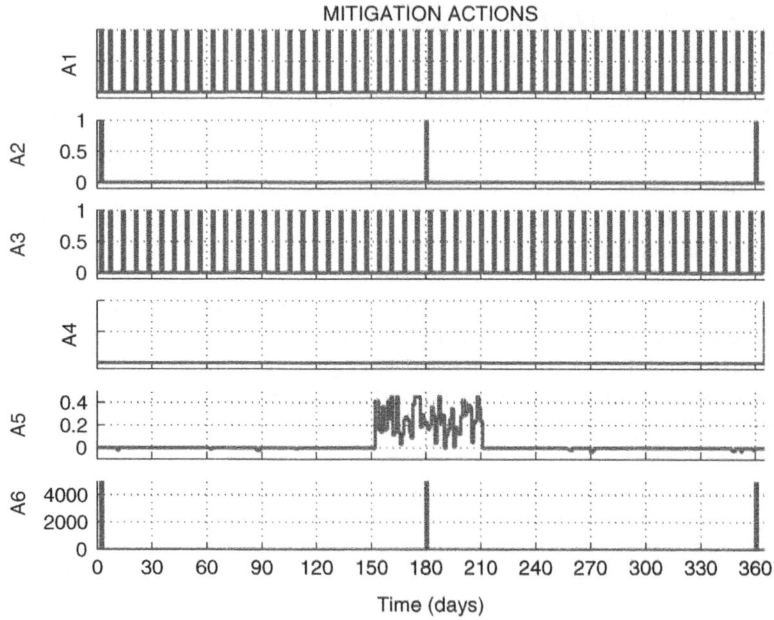

Fig. 28.7 Mitigation actions to be undertaken to reduce risks impacts

carried out at the end of the period due to the increasing of probability P_3. A_5 sets the increasing/decreasing on the water level reference due to risks R_4 and R_5. u_{M_6} is the cost of the insurance contract ($g_{16} = u_{M_6}$). This action takes the value 5,000 because a constraint has been added imposing that the cost of insurance cannot be higher than 5,000 euros.

28.4.3 Lower Level Controller

At the lower level, a distributed model predictive controller optimizes the operation by manipulating flows and gate openings in order to follow the water level set-points provided by the upper control level. There are 7 main gates and 17 off-take gates in the section selected. Hence, we have considered that the system is composed of a set $\mathcal{N} = \{1, 2, \ldots, 7\}$ of subsystems. Each subsystem $i \in \mathcal{N}$ begins at one of the main gates and ends at the next one and is controlled by an agent that can manipulate the upstream gate flow $Q_{\text{in},i}(k) \in \mathbb{R}_+(m^3/s)$ to regulate the water level $h_i(k) \in \mathbb{R}_+(m)$, i.e., $\mathbf{u}_i(k) = Q_{\text{in},i}(k)$ and $h_i(k) \in \mathbf{x}_i(k)\ \forall i \in \mathcal{N}$. Notice that the outflow to the downstream canal reach is the manipulated variable of the downstream neighbor, i.e., $\mathbf{u}_{i+1}(k) = Q_{\text{o},i}(k) \in \mathbb{R}_+(m^3/s)$. Likewise, other flows are considered as perturbation variables: $q_{\text{in},i} \in \mathbb{R}_+(m^3/s)$ models the flows due to rainfall, failures in upstream gate and other unknown disturbances and $q_{\text{o},i} \in \mathbb{R}_+(m^3/s)$ models the

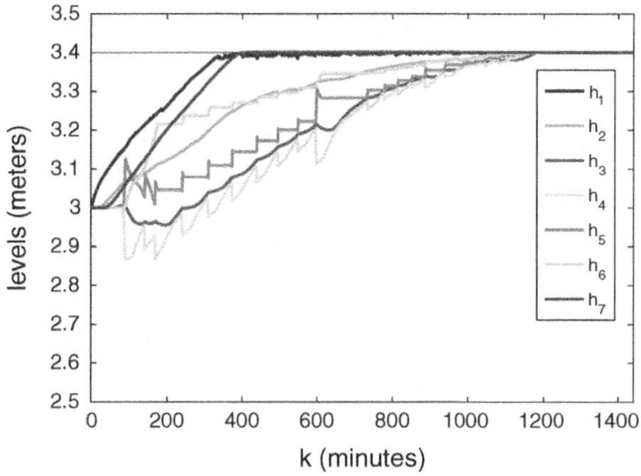

Fig. 28.8 Levels in reaches for day 150 for the nominal case

known off-take outflows from farmers, considered as measurable perturbations. The simplified discrete model that has been considered using the previous variables is:

$$A_i(h_i(k+1) - h_i(k)) = T_d(Q_{\text{in},i}(k - t_{\text{d},i}) + q_{\text{in},i}(k) - Q_{\text{o},i}(k) - q_{\text{o},i}(k))$$

where $T_d = 1$ minute is the time step length, and A_i and $t_{\text{d},i}$ are respectively the surface of the corresponding reach and the water transport delay (the level is measured downstream). The following constraints must be taken into account:

1. Maximum and minimum levels to guarantee that off-take points are submerged.
2. The total amount of water at the head over a determined time period is limited.
3. Maximum and minimum gates opening.

Regarding the controller, the control horizon is set to $N_{\text{c}} = 5$ for all the agents. As the prediction horizon has to be big enough to compensate the transport delay in the calculations, we make it equal to the control horizon plus the delay κ_i of the corresponding reach, that is, $N_{\text{p}}(i) = N_{\text{c}} + \kappa_i$. See [4] for more details about the implementation of the DMPC controller.

Several simulations have been performed for the DMPC controller for a one day period. In these simulations, all the reaches begin with a water level of 3.0 m and there is a change of reference for all the reaches to 3.40 m at time $k = 0$. This change is originated at the higher control level as a function of the risk mitigation policy. In particular, the change of reference corresponds to the day 150 in Fig. 28.5, where the evolution of the references during a 1 year period is depicted. The simulation shown in Fig. 28.8 shows how the reference is followed for all the reaches.

28.5 Conclusions

In this chapter we have presented a RDMPC scheme designed to include in the optimization problem decision variables that are important for the management of the plant. In this way, it is possible to handle explicitly the different risks that may threaten the plant during its life. Therefore, an extra degree of robustness and security can be attained thanks to the implementation of mitigation actions.

Acknowledgments Financial support from the HYCON2 EU-project from the ICT-FP7 and MEC-Spain, DPI2008-05818, and F.P.I. grants is gratefully acknowledged.

References

1. R. Fletcher, S. Leyffer, Numerical experiences with lower bounds for MIQP branch-and-bound. SIAM J. Optim. **8**(2):604–616 (1998)
2. P. Kall, S.W. Wallace, *Stochastic Programming*. (Willey, New York, 1994)
3. J.M. Maestre, D. Muñoz de la Peña, E.F. Camacho, T. Álamo, Distributed model predictive control based on agent negotiation. J. Process Control **21**(5), 685–697 (June 2011)
4. A. Zafra-Cabeza, J.M. Maestre, M.A. Ridao, E.F. Camacho, L. Sánchez, A hierarchical distributed model predictive control approach in irrigation canals: A risk mitigation perspective. J. Process Control **21**(5), 787–799 (2011)
5. A. Zafra-Cabeza, M.A. Ridao, E.F. Camacho, K.G. Kempf, D.E. Rivera, Managing risk in semiconductor manufacturing: a stochastic predictive control approach. Control Eng. Pract. **15**(8), 969–984 (2007)

Chapter 29
Distributed MPC Based on Agent Negotiation

J. M. Maestre, D. Muñoz de la Peña and E. F. Camacho

Abstract In this chapter we propose a distributed model predictive control scheme based on agent negotiation. In particular, we consider the control of several subsystems coupled through the inputs by a set of independent agents that are able to communicate and we assume that each agent has access only to the model and the state of one of the subsystems. This implies that in order to take a cooperative decision, i.e. for the whole system, the agents must negotiate. At each sampling time, following a given protocol, agents make proposals to improve an initial feasible solution on behalf of their local cost function, state and model. These proposals are accepted if the global cost improves the cost corresponding to the current solution. In addition, we study the stability properties of the proposed distributed controller and provide precise conditions that guarantee that the closed-loop system is practically stable along with an optimization based controller and invariant design procedure.

29.1 Introduction

In this chapter we present a distributed model predictive control scheme based on agent negotiation that was presented in [5]. This scheme can be considered an evolution of the DMPC scheme based on a team game proposed in [4], which was developed for the particular case of a system composed by two agents. The bad scal-

J. M. Maestre (✉) · D. Muñoz de la Pēna · E. F. Camacho
Departamento de Ingeniería de Sistemas y Automática,
Universidad de Sevilla, Seville, Spain
e-mail: pepemaestre@cartuja.us.es

D. Muñoz de la Peña
e-mail: davidmps@cartuja.us.es

E. F. Camacho
e-mail: eduardo@cartuja.us.es

J. M. Maestre and R. R. Negenborn (eds.), *Distributed Model Predictive Control Made Easy*, Intelligent Systems, Control and Automation: Science and Engineering 69,
DOI: 10.1007/978-94-007-7006-5_29,

ability properties of the aforementioned scheme required a different approach for a system controlled by multiple agents.

In this scheme we consider the control of several subsystems coupled through the inputs by a set of agents that only have local state and model information. This implies that in order to take a cooperative decision, i.e., from an overall system perspective, the agents must communicate and negotiate. At each time step, agents make proposals to improve an initial feasible solution on behalf of their local cost functions, states and models. These proposals are only accepted if the global cost improves the cost corresponding to the current solution.

One of the main differences between the proposed approach and other cooperative MPC schemes is that the agents do not have a global model of the system. This may be important in some applications in which the centralized model is not available or the agents do not want to share this information with the rest of the subsystems. In addition, there is a potential benefit from this assumption because if a distributed system adds a new subsystem, in the proposed scheme, only those agents affected by this new element would have to be updated, while in other schemes based on global information, the information would have to be broadcasted. One class of systems in which these issues are relevant are transport networks and supply chains, where new consumers/suppliers can appear dynamically. On top of that, it is important to remark that the type of negotiation in which the proposed algorithm is based is independent of the type of system dynamics, i.e., the algorithm presented in this chapter works with distributed systems with more complex dynamics.

29.2 Statement of the Problem and Main Assumptions

Consider a distributed linear system composed of a set of $\mathcal{N} = \{1, 2, \ldots, N\}$ interconnected subsystems coupled by the inputs whose dynamics can be described mathematically as:

$$\mathbf{x}_i(k+1) = \mathbf{A}_i\mathbf{x}_i(k) + \sum_{j\in\mathcal{N}_i} \mathbf{B}_{ij}\mathbf{u}_j(k), \tag{29.1}$$

where $\mathbf{x}_i(k) \in \mathbb{R}^{n_{\mathrm{x},i}}$ is the state of subsystem i and $\mathbf{u}_j(k) \in \mathbb{R}^{n_{\mathrm{u},j}}$ is the input of the agent j with $\mathcal{N}_i$ being the set of agents whose inputs affect the dynamics of agent i, i.e., $\mathcal{N}_i = \{j \in \mathcal{N}|\mathbf{B}_{ij} \neq \mathbf{0}\}$. Likewise, let $\mathcal{M}_j$ be the set of states affected by input j, i.e., $\mathcal{M}_j = \{i \in \mathcal{N}|\mathbf{B}_{ij} \neq \mathbf{0}\}$. At this point it is convenient to remark that in [5] the problem is formulated in a more general framework in which the inputs and the states are in different sets, i.e, the inputs are not associated to any particular subsystem. We decided to adopt the formulation presented in this chapter to simplify the introduction of the scheme and its implementation. Nevertheless, similar results can be obtained with this notation if we assume that there can be agents without state or inputs, i.e., with $n_{\mathrm{x},i} = 0$ or $n_{\mathrm{u},i} = 0$.

States and inputs are constrained into independent sets defined by a set of linear inequalities

$$\mathbf{x}_i(k) \in \mathcal{X}_i, \quad \mathbf{u}_i(k) \in \mathcal{U}_i, \quad i \in \mathcal{N} \tag{29.2}$$

The control objective of the proposed scheme is to minimize a global performance index defined as the sum of each of the local cost functions while satisfying these constraints. The local cost function of agent i based on the predicted trajectories of its state and inputs defined as

$$J_i(\mathbf{x}_i(k), \mathbf{u}_{\mathcal{N}_i}(k : k + N_\mathrm{p} - 1)) = \sum_{l=0}^{N_\mathrm{p}-1} L_i(\mathbf{x}_i(l), \mathbf{u}_{\mathcal{N}_i}(l)) + F_i(\mathbf{x}_i(k + N_\mathrm{p})) \tag{29.3}$$

where $\mathbf{u}_{\mathcal{N}_i}(k : k + N_\mathrm{p} - 1)$ is the future trajectory of the inputs that affect agent i, N_p is the prediction horizon, $L_i(\cdot)$ with $i \in \mathcal{N}$ is the stage cost function defined as

$$L_i(\mathbf{x}_i(k), \mathbf{u}_{\mathcal{N}_i}(k)) = \mathbf{x}_i(k)^\mathrm{T}\mathbf{Q}_i\mathbf{x}_i(k) + \sum_{j\in\mathcal{N}_i} \mathbf{u}_j(k)^\mathrm{T}\mathbf{R}_{ij}\mathbf{u}_j(k) \tag{29.4}$$

with $\mathbf{Q}_i > 0$, $\mathbf{R}_{ij} > 0$ and $F_i(\cdot)$ is the terminal cost defined as

$$F_i(\mathbf{x}_i(k)) = \mathbf{x}_i(k)^\mathrm{T}\mathbf{P}_i\mathbf{x}_i(k) \tag{29.5}$$

with $\mathbf{P}_i > 0$. Note that each of the local cost functions only depends on the trajectories of its state and the inputs that affect it.

At the end of the negotiation rounds, the agents decide a set of input trajectories denoted as $\mathbf{u}^\mathrm{d}_{\mathcal{N}}(k : k+N_\mathrm{p}-1)$. The first input of these trajectories is applied; however, the rest of the trajectories is not discarded but used to generate the initial proposal for the next sampling round, which is given by the shifted future input trajectories of all the inputs $\mathbf{u}^\mathrm{s}_{\mathcal{N}}(k + 1 : k + N_\mathrm{p})$. The last input of each of these trajectories is given by

$$\mathbf{u}^s_j(k + N_\mathrm{p}) = \sum_{p\in\mathcal{M}_j} \mathbf{K}_{jp}\mathbf{x}_p(k + N_\mathrm{p}) \tag{29.6}$$

where $\mathbf{x}_p(k + N_\mathrm{p})$ is the predicted value of the state $\mathbf{x}_p$ after N_p time steps obtained applying $\mathbf{u}^\mathrm{d}_{\mathcal{N}}(k : k+N_\mathrm{p}-1)$ from the initial state $\mathbf{x}_p(k)$, and $\mathbf{K}_{jp}$ are linear feedbacks desined according to a procedure that will be explained in Sect. 29.4. The set of shifted input trajectories will be applied in case the agents do not reach an agreement. This proposal is necessary in order to guarantee closed-loop stability.

29.3 Description of the Approach

Next we present the algorithm that details the implementation of the DMPC scheme proposed in this chapter.

Algorithm 29.1 DMPC Based on Agent Negotiation (DMPC-BAN)

1. Each agent i measures and communicates its state $\mathbf{x}_i(k)$ in order to obtain an initial joint decision vector at time k, $\mathbf{u}^{\mathrm{d}}_{\mathcal{N}}(k : k + N_{\mathrm{p}} - 1)$. To this end, the final joint decision vector at time $k-1$ is shifted so that the components corresponding to its first time step are discarded and new components are aggregated at its end according to Eq. (29.6).
2. Randomly, agents make proposals to their neighbors. A proposal is a possible future trajectory for all the system inputs and is built by updating in $\mathbf{u}^{\mathrm{d}}_{\mathcal{N}}(k : k + N_{\mathrm{p}} - 1)$ the trajectories of some or all the inputs that affect the agent, i.e., those in a set $\mathcal{P} \subseteq \mathcal{N}_i$. Hence, the proposal of agent i is:

$$\begin{aligned}
&\mathbf{u}^{i}_{\mathcal{N}}(k : k + N_{\mathrm{p}} - 1) = \arg \min_{\mathbf{u}_{\mathcal{N}}(k:k+N_{\mathrm{p}}-1)} J_i(\mathbf{x}_i(k), \mathbf{u}_{\mathcal{N}_i}(k : k + N_{\mathrm{p}} - 1))\\
&\text{s.t.}\\
&\mathbf{x}_i(l+1) = \mathbf{A}_i \mathbf{x}_i(l) + \sum_{j \in \mathcal{N}_i} \mathbf{B}_{ij} \mathbf{u}_j(l) \quad \forall i \in \mathcal{N}\\
&\mathbf{u}_j(l) = \mathbf{u}^{\mathrm{d}}_j(l) \quad \forall j \notin \mathcal{P}, l \in \{k, k+1, \ldots, k + N_{\mathrm{p}} - 1\}\\
&\mathbf{x}_i(l) \in \mathcal{X}_i, \ l \in \{k+1, \ldots, k + N_{\mathrm{p}}\} \quad \forall i \in \mathcal{N}\\
&\mathbf{u}_i(l) \in \mathcal{U}_i, \ l \in \{k, k+1, \ldots, k + N_{\mathrm{p}} - 1\} \quad \forall i \in \mathcal{N}
\end{aligned} \tag{29.7}$$

3. Agent i asks all the agents affected by its proposal (i.e., those in the set $\bigcup_{i \in \mathcal{P}} \mathcal{M}_i$) if they are free to evaluate it (each agent can only evaluate one proposal at the time). If all the neighbors acknowledge the petition, the algorithm continues. If not, the agent waits a random time before trying again.
4. Each agent j affected by the proposal (i.e., $j \in \bigcup_{i \in \mathcal{P}} \mathcal{M}_i$) of agent i evaluates the difference between the cost of the new proposal $\mathbf{u}^{i}_{\mathcal{N}}(k : k + N_{\mathrm{p}} - 1)$ and the cost of the current accepted proposal $\mathbf{u}^{\mathrm{d}}_{\mathcal{N}}(k : k + N_{\mathrm{p}} - 1)$ as

$$\Delta J^i_j = J_j(\mathbf{x}_j(k), \mathbf{u}^{i}_{\mathcal{N}_j}(k : k + N_{\mathrm{p}} - 1)) - J_j(\mathbf{x}_j(l), \mathbf{u}^{\mathrm{d}}_{\mathcal{N}_j}(k : k + N_{\mathrm{p}} - 1)) \tag{29.8}$$

   If the proposal does not satisfy the constraints of the corresponding local optimization problem, an infinite cost increment is assigned. This implies that unfeasible proposals will never be chosen. This difference, ΔJ^i_j, is sent back to the proposer agent i.
5. Agent i receives the local cost increments from all the agents affected by its proposal, so that it can evaluate its impact $\Delta J^i(k)$ as

$$\Delta J^i = \textstyle\sum_j \Delta J^i_j \tag{29.9}$$

   If ΔJ^i is negative, $\mathbf{u}^{\mathrm{d}}_{\mathcal{N}}(k : k + N_{\mathrm{p}} - 1) = \mathbf{u}^{i}_{\mathcal{N}}(k : k + N_{\mathrm{p}} - 1)$ and agent i broadcasts the update in the joint decision vector. Otherwise, the proposal is discarded.
6. Go to Step 1 until the maximum number of proposals has been exceeded or the time available for placing proposals is over.

7. Apply the first component of $\mathbf{u}^{\mathrm{d}}_{\mathcal{N}}(k : k + N_{\mathrm{p}} - 1)$ and repeat the procedure in the next sample period.

29.4 Theoretical Results Availability

In this section we introduce the main theoretical propierties of the proposed DMPC scheme. Please, notice that many theoretical details have been omitted for the sake of clarity. The interested reader is recommended to see [1, 5] for a more rigorous treatment of the topics discussed in this section.

29.4.1 Stability

Stability is a major issue in distributed systems. In general, it is a difficult problem because it is not enough to guarantee the stability of each of the subsystems. Actually, stable subsystems may lead to an unstable global system. Hence, although in order to implement the proposed controller, the agents don't need information about the state or the dynamics of the rest of the subsystems, a centralized model of the full system is needed to design the controller so that closed-loop stability is guaranteed.

In order to provide sufficient conditions that guarantee asymptotic stability of the closed-loop system following a standard region/terminal cost approach [6], we make the following assumption: there exist linear feedbacks $\mathbf{u}^{s}_{j}(k) = \sum_{p \in \mathcal{M}_j} \mathbf{K}_{jp}\mathbf{x}_p(k)$ and sets $\Omega_i \subseteq \mathbb{R}^{n_{x,i}}$ such that if $x_i \in \Omega_i$ for all $i \in \mathcal{N}$ then the following conditions hold for all $i \in \mathcal{N}$:

$$\sum_{i \in \mathcal{N}} F_i \left(\mathbf{A}_i \mathbf{x}_i(k) + \sum_{j \in \mathcal{N}_i} \mathbf{B}_{ij} \left(\sum_{p \in \mathcal{M}_j} \mathbf{K}_{jp}\mathbf{x}_p(k) \right) \right) - F_i(\mathbf{x}_i(k))$$

$$+ L_i\left(\mathbf{x}_i(k), \left\{ \sum_{p \in \mathcal{M}_j} \mathbf{K}_{jp}\mathbf{x}_p(k) \right\}_{j \in \mathcal{N}_i}\right) \leq 0 \quad (29.10a)$$

$$\mathbf{A}_i \mathbf{x}_i(k) + \sum_{j \in \mathcal{N}_i} \mathbf{B}_{ij}\mathbf{u}_j(k) \in \Omega_i \quad (29.10b)$$

$$\sum_{p \in \mathcal{M}_j} \mathbf{K}_{jp}\mathbf{x}_p(k) \in \mathcal{U}_j \quad (29.10c)$$

$$\Omega_i \in \mathcal{X}_i \quad (29.10d)$$

The rationale behind our assumption is twofold: in the first place, the local feedbacks must satisfy constraint (29.10a), which implies that the system in closed-loop with these set of local controllers is stable. Second, sets Ω_i such that (29.10b–29.10d) are satisfied must exist. We denote these sets as jointly positive invariants for the system in closed-loop with the controllers defined by matrices $\mathbf{K}_{ij}$. It is important to note that although the cartesian product of these sets is a positive invariant of system, in general it is not possible to obtain the jointly positive invariant sets from an invariant set obtained following standard procedures because each Ω_i must be defined only in the subspace corresponding to state $\mathbf{x}_i$ to ease the separation of the overall control problem. This property is necessary in order to define for each agent a set of constraints that depend only on its state, and hence, only on its model.

If the presented assumption holds and at time step $k = 0$, $\mathbf{u}^{\rm d}_{\mathcal{N}}(0 : N_{\rm p} - 1)$ is given such that each of the $|\mathcal{N}|$ optimization problems (29.7) is feasible, then the proposed algorithm is feasible for all time steps $k \geq 0$ and the system in closed-loop with the proposed distributed MPC controller is asymptotically stable. This statement is demonstrated in [1, 5] following the standard terminal region/terminal constraint approach [6]. Stability is inherited from the set of local controllers defined by matrices $\mathbf{K}_{ij}$ which by (29.10a) are known to stabilize the system. In fact this result is based on the well known idea "feasibility implies stability", see [8].

Finally, we must stress that the stability properties of the proposed scheme rely heavily on the fact that $\mathbf{u}^{\rm d}_{\mathcal{N}}(0 : N_{\rm p} - 1)$ satisfies all the constraints of the optimization problem. This implies, that in the start-up $\mathbf{u}^{\rm d}_{\mathcal{N}}(0 : N_{\rm p} - 1)$ has to be calculated either by a centralized supervisor or in a distributed manner by the agents. Likewise, when applied to a real system in the presence of disturbances and/or possible model errors, if the controller operates close to the state constraints in practice the shifted input trajectory may become unfeasible and it would have to be evaluated again (in a centralized manner or using an appropriate distributed approach). This issue must be taken into account in the implementation procedure of this control strategy.

29.4.2 Controller Design Procedure

The local controllers $\mathbf{K}_{ij}$ must satisfy two necessary conditions. First, the centralized system composed by the $|\mathcal{N}|$ subsystems in closed-loop with the local controllers must be stable. Second, the jointly invariant sets must exist.

The local controllers that depend on each agent; that is, matrices $\mathbf{K}_{ji}$ such that $i \in \mathcal{M}_j$, must be designed in a way such that (29.10a) holds. To take this condition into account, we will use the following centralized model of the system

$$\mathbf{x}(k+1) = \mathbf{A}\mathbf{x}(k) + \mathbf{B}\mathbf{u}(k) \tag{29.11}$$

where

$$\mathbf{x} = [\mathbf{x}_1^{\rm T}, \ldots, \mathbf{x}_N]^{\rm T}, \quad \mathbf{u} = [\mathbf{u}_1^{\rm T}, \ldots, \mathbf{u}_N]^{\rm T} \tag{29.12}$$

and matrices $\mathbf{A}$ and $\mathbf{B}$ are appropriate matrices that depend of the model (29.1) of each subsystem.

In addition, stability of each subsystem in closed-loop with its corresponding local feedback must be guaranteed. A sufficient condition to guarantee stability of each of the subsystems is to require that the cost function defined by the matrices $\mathbf{P}_i$ is a Lyapunov function for the subsystem in closed-loop with its corresponding local feedback. To take into account this condition, we consider the following uncertain model of each of the $|\mathcal{N}|$ subsystems

$$\mathbf{x}_i(k+1) = \mathbf{A}_i\mathbf{x}_i(k) + \mathbf{B}_i\mathbf{v}_i(k) + \mathbf{E}_i\mathbf{w}_i(k) \tag{29.13}$$

where $\mathbf{v}_i$ aggregates the input vectors that depend on $\mathbf{x}_i$ and $\mathbf{w}_i$ aggregates the inputs that depend on the rest of the states when the local controllers are applied; that is,

$$\begin{aligned} \mathbf{B}_i\mathbf{v}_i(k) &= \sum_{j\in\mathcal{N}_i} \mathbf{B}_{ij}\mathbf{K}_{ji}\mathbf{x}_i(k) \\ \mathbf{E}_i\mathbf{w}_i(k) &= \sum_{j\in\mathcal{N}_i} \mathbf{B}_{ij}\left(\sum_{p\in\mathcal{M}_j-\{i\}} \mathbf{K}_{jp}\mathbf{x}_p(k)\right) \end{aligned} \tag{29.14}$$

In this case, the objective is to design a controller $\mathbf{K}_i = \{\mathbf{K}_{ji}\}_{j\in\mathcal{N}_i}$ that stabilizes the subsystem considering $\mathbf{w}_i$ an unknown disturbance. Matrices $\mathbf{B}_i$ and $\mathbf{D}_i$ are appropriate matrices that depend of the model (29.1) of each subsystem.

We provide next a set of linear matrix inequalities (LMI) obtained following standard procedures (see [5] and the references therein) that guarantees that (29.10a) holds and that $\mathbf{K}_i$ stabilizes the subsystem i. In particular, if there exist matrices $\mathbf{W}_i, \mathbf{Y}_i$ with $i = 1, \ldots, N$ such that the following inequalities hold[1]

$$\begin{bmatrix} \Upsilon & \Phi & \Psi & \Xi \\ * & \Upsilon & 0 & 0 \\ * & * & I & 0 \\ * & * & * & I \end{bmatrix} \geq 0 \tag{29.15}$$

with $\mathbf{R}_i = \sum_{j\in n_i} \mathbf{R}_{ij}$, $\mathbf{R} = \text{diag}(\mathbf{R}_1, \ldots, \mathbf{R}_N)$, $\mathbf{K} = [\mathbf{K}_1, \ldots, \mathbf{K}_N]$, $\mathbf{K}_i^{\mathrm{T}} = [\mathbf{K}_{1i}, \ldots, \mathbf{K}_{Ni}]$ and

$$\Phi = \begin{bmatrix} \mathbf{W}_1\mathbf{A}_1^{\mathrm{T}} + \mathbf{Y}_1^{\mathrm{T}}\mathbf{B}_1^{\mathrm{T}} & \mathbf{Y}_1^{\mathrm{T}}\mathbf{B}_2^{\mathrm{T}} & \cdots & \mathbf{Y}_1^{\mathrm{T}}\mathbf{B}_N^{\mathrm{T}} \\ \mathbf{Y}_2^{\mathrm{T}}\mathbf{B}_1^{\mathrm{T}} & \mathbf{W}_2\mathbf{A}_2^{\mathrm{T}} + \mathbf{Y}_2^{\mathrm{T}}\mathbf{B}_2^{\mathrm{T}} & \cdots & \mathbf{Y}_2^{\mathrm{T}}\mathbf{B}_N^{\mathrm{T}} \\ \vdots & \vdots & \ddots & \vdots \\ \mathbf{Y}_N^{\mathrm{T}}\mathbf{B}_1^{\mathrm{T}} & \mathbf{Y}_N^{\mathrm{T}}\mathbf{B}_2^{\mathrm{T}} & \cdots & \mathbf{W}_N\mathbf{A}_N^{\mathrm{T}} + \mathbf{Y}_N^{\mathrm{T}}\mathbf{B}_N^{\mathrm{T}} \end{bmatrix} \tag{29.16}$$

[1] The symbol "$*$" stands for the symmetric part of a matrix.

$$\Upsilon = \begin{bmatrix} \mathbf{W}_1 & \mathbf{0} & \cdots & \mathbf{0} \\ * & \mathbf{W}_2 & \cdots & \mathbf{0} \\ * & * & \ddots & \vdots \\ * & * & * & \mathbf{W}_N \end{bmatrix}, \Xi = \begin{bmatrix} \mathbf{Y}_1^{\mathrm{T}}\mathbf{R}^{\frac{1}{2}} \\ \mathbf{Y}_2^{\mathrm{T}}\mathbf{R}^{\frac{1}{2}} \\ \vdots \\ \mathbf{Y}_N^{\mathrm{T}}\mathbf{R}^{\frac{1}{2}} \end{bmatrix}, \Psi = \begin{bmatrix} \mathbf{W}_1\mathbf{Q}_1^{\frac{1}{2}} & \mathbf{0} & \cdots & \mathbf{0} \\ * & \mathbf{W}_2\mathbf{Q}_2^{\frac{1}{2}} & \cdots & \mathbf{0} \\ * & * & \ddots & \vdots \\ * & * & * & \mathbf{W}_N\mathbf{Q}_N^{\frac{1}{2}} \end{bmatrix} \tag{29.17}$$

and

$$\begin{bmatrix} \mathbf{W}_i & \mathbf{W}_i\mathbf{A}_i^{\mathrm{T}} - \mathbf{Y}_i^{\mathrm{T}}\mathbf{B}_i^{\mathrm{T}} & \mathbf{W}_i\mathbf{Q}_i^{\frac{1}{2}} & \mathbf{Y}_i^{\mathrm{T}}\mathbf{R}_i^{\frac{1}{2}} \\ * & \mathbf{W}_i & \mathbf{0} & \mathbf{0} \\ * & * & I & \mathbf{0} \\ * & * & * & I \end{bmatrix} \geq \mathbf{0} \tag{29.18}$$

for $i \in \mathcal{N}$ then (29.10a) is satisfied for the matrices $\mathbf{P}_i = \mathbf{W}_i^{-1}$, $\mathbf{K}_i = \mathbf{K}_{ji\, j\in\mathcal{N}_i} = \mathbf{Y}_i\mathbf{W}_i^{-1}$ and systems (29.13) are stable in closed-loop with $\mathbf{v}_i = \mathbf{K}_i\mathbf{x}_i$.

Notice that additional constraints can be added to the design procedure so that there is no need to know the state $\mathbf{x}_i$ in order to calculate the input $\mathbf{u}_j$. This is relevant because in order to evaluate the shifted input trajectory, all the subsystems whose state affects a given input must communicate, so in certain cases, it may be desirable to limit these communications.

Once the local controllers and the terminal cost functions are fixed, one needs to find sets Ω_i such that (29.10b–29.10d) hold. In general, this is a difficult problem because each of the sets depends on the others. The size of the terminal region for agent i is determined by the magnitude of the disturbances induced by its neighbor agents and viceversa. A similar class of invariant systems was studied in [7] within the polytopic games framework. In order to obtain sets Ω_i, we will use the uncertain model (29.13) of each agent; that is, each agent assumes that the contribution of its neighbors to the inputs that affect its dynamics are an unknown bounded disturbance. The size of the set in which these disturbances are bounded depend on the size of the sets Ω_i. This implies that finding these sets is in general a complex problem. In order to decouple the design of each set, each agent i limits its contribution to each input j by a factor $\lambda_{ji} \in (0, 1]$ with $\sum_{i\in\mathcal{M}_j} \lambda_{ji} \leq 1$; that is,

$$\mathbf{K}_{ji}\mathbf{x}_i \in \lambda_{ji}\mathcal{U}_j, \ \forall i, j \in \mathcal{N} \tag{29.19}$$

Using the same notation introduced in (29.13), this implies that

$$\mathbf{v}_i \in \mathcal{V}_i(\Lambda), \ \mathbf{w}_i \in \mathcal{W}_i(\Lambda) \tag{29.20}$$

with

$$\begin{aligned} \mathcal{V}_i(\Lambda) &= \lambda_{1i}\mathcal{U}_1 \times \lambda_{2i}\mathcal{U}_2 \times \cdots \times \lambda_{Ni}\mathcal{U}_N \\ \mathcal{W}_i(\Lambda) &= \left(\textstyle\sum_{p\in\mathcal{M}_1-\{i\}} \lambda_{1p}\right)\mathcal{U}_1 \times \left(\textstyle\sum_{p\in\mathcal{M}_2-\{i\}} \lambda_{2p}\right)\mathcal{U}_2 \times \cdots \times \left(\textstyle\sum_{p\in\mathcal{M}_N-\{i\}} \lambda_{Np}\mathcal{U}_N\right) \end{aligned} \tag{29.21}$$

where $\Lambda = \{\lambda_{ij}\}_{\forall i,j}$ is a vector made of all the parameters λ_{ij}. Note that the maximum contribution of a given agent inside Ω_i, is the maximum contribution to the disturbance for the rest of the agents. There exists an infinite number of possible values of λ_{ji} such that the sets Ω_i exist. In order to chose one, we solve the following optimization problem which maximizes the feasibility region of the distributed MPC controller:

$$\begin{aligned} \max_{\lambda_{ji}} \; & f(\Omega_1 \times \Omega_2 ... \times \Omega_N) \\ & \Omega_i = \Omega(\mathbf{A}_i, \mathbf{B}_i, \mathbf{E}_i, \mathcal{X}_i, \mathbf{K}_i, \mathcal{V}_i(\Lambda), \mathcal{W}_i(\Lambda)) \\ & \lambda_{ji} \in (0,1), \;\; \forall j, i \\ & \textstyle\sum_{i \in \mathcal{M}_j} \lambda_{ji} \leq 1, \;\; \forall i \end{aligned} \tag{29.22}$$

where function $f(\cdot)$ is a measure of the size of a polyhedron (for example, its Chebyshev radius) and $\Omega(\cdot)$ stands for the calculation of the invariant set given the corresponding matrices.

Solving problem (29.22) may be difficult in general, however, under certain assumptions it can be posed as a convex problem. In [7] it was proved that the feasibility region of this problem is convex. In [5] we prove that the jointly invariant sets Ω_i are polyhedra defined by a set of inequalities whose right hand side can be expressed as an affine combination of the constants λ_{ij}. This implies, that if an appropriate function $f(\cdot)$ is chosen, problem (29.22) can be cast into a convex optimization problem. For instance, let us suppose that our criterium to compare the invariant sets is the radium of a Chebyshev ball inside the invariant region. In this case we are interested in obtaining the maximum $x^{\mathrm{T}}x$ as function of Λ that verifies all the constraints, which is a convex problem.

29.5 Applications of the DMPC Scheme

The algorithm that we propose in this chapter has been tested with the following simulated systems.

29.5.1 Supply Chain Problem

In [3] an application to the proposed approch to a supply chain system is presented. It is clear that the nodes of a supply chain may not have incentives to share any information beyond the number of units that are ordered upstream or sent downstream. On top of that, supply chain flows usually present three interesting phenomena from the control point of view: oscillation, amplification and phase lag. Due to the material and informational delays, production and inventories overshoot and undershoot the optimal levels. The magnitude of the fluctuations increase as they propagate from the customer to the factory, in what is commonly known as the bullwhip effect. For

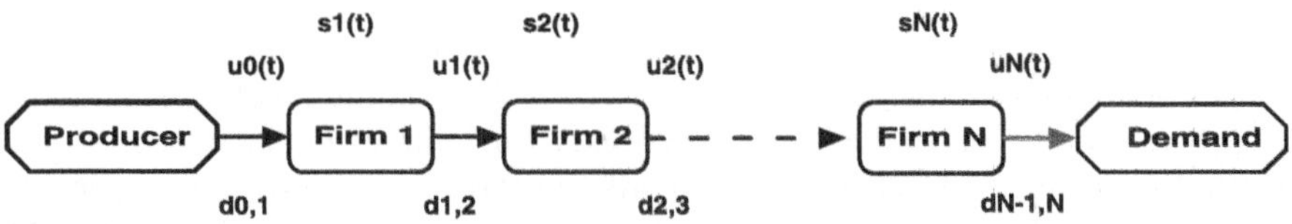

Fig. 29.1 Supply chain

these reasons supply chains dynamics are a very interesting benchark for distributed control schemes.

In this example, we consider a cascade of N firms (see Fig. 29.1). In particular, the discrete time equations that define the dynamics of firm i are given by:

$$s^i(k+1) = s^i(k) + u^{i-1}(k - d_{i-1,i}) - u^i(k) \tag{29.23}$$

The super-scripts $i-1$ and $i+1$ represent, respectively, the dynamics of the upstream and downstream nodes. Variable $s^i(t)$ is the stock level; that is, the number of items available for shipment downstream. The manipulated variable at each stage is $u^i(t)$ which stands for the number of items sent to the downstream node. This is a difference with respect to models in which there is one variable that stands for the order rate and another, which is usually modeled as a disturbance, that stands for the shipment itself. The information flows are assumed to have no time delays and the material flows have a delay modeled by $d_{i,j}$ which corresponds to the time taken by the shipments from node i to node j.

In general, the simulations show that increasing the number of proposals N_{prop} improves the performance of the proposed DMPC scheme. However, communications can be a scarce resource for some systems and it is important to find a trade-off between the number of communications and the performance. In our example it can be seen that a good trade-off happens when N_{prop} is around $5N$ communications, where N is the number of agents. This implies that each agent makes an average of 5 proposals to its neighbors.

29.5.2 Hydro-Power Valley

In [2], a comparison between a decentralized controller and the proposed control architecture on a hydro-power valley (HPV) benchmark is presented. A hydro power network consists of various sub-systems that can transform water flow into energy with the main objective of generating as much energy or financial gain as possible. At the same time, these flows influence water levels in the environmental water systems they belong to. These levels need to be controlled in order to satisfy other objectives such as flood protection or water availability for irrigation and navigation, all with a dynamic, yet predictable behavior. The control infrastructure consists of

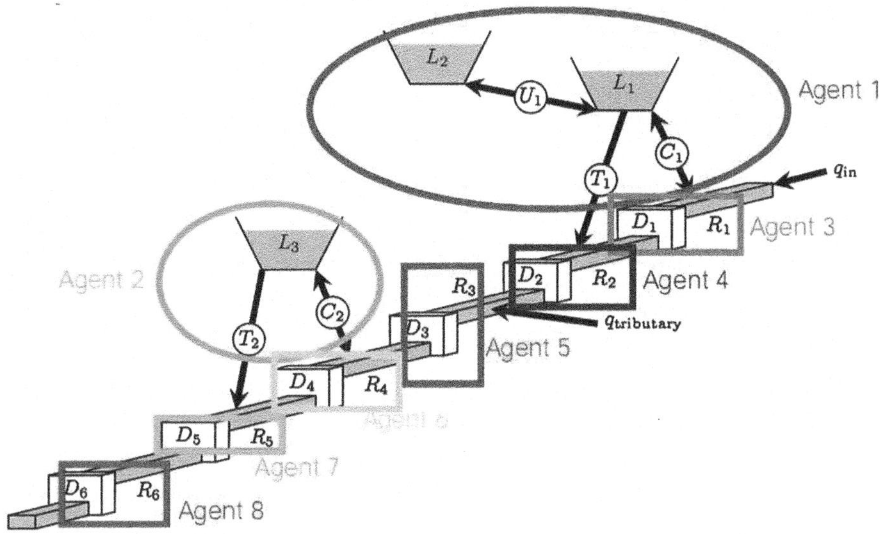

Fig. 29.2 Hydro-power valley (HPV)

pumps, turbines, and gates that are all characterized by limited capacities. As such, the control of a hydro power network can be considered a constrained optimization problem that needs to be solved in real-time

The HPV is composed by three lakes and a river that is divided in six reaches that terminate with dams equipped with turbines for power production; see Fig. 29.2. The lakes and the river reaches are connected in three different ways: by a duct, ducts equipped with a turbine, and ducts equipped with a pump and a turbine. The river is fed by upstream inflows and tributary flows. The test of the controllers is based on the following scenario: the power output of the system must follow a given reference over 24 h while keeping the water levels in the lakes and at the dams as constant as possible. In order to measure the performance of the controllers different indices are considered, e.g.: the mean absolute tracking error, which is the absolute value of the tracking error integrated during the whole simulation; the mean quadratic tracking error, which corresponds to the square error of the tracking error integrated during the whole simulation; and different power reference tracking indices, which provide an economical measurement of the controller performance.

Our results show clearly the advantages derived from the cooperation among the different controllers. In Fig. 29.3 the power produced by the agents hardly follows the reference. On the other hand, it can be seen in the same figure how a joint actuation of the agents allows them to follow the reference. Even when the proposed distributed MPC scheme is tailored for agents that have access only to their local model, state and objective, the negotiation procedure among the agents allows them to calculate an actuation that improves the overall performance of the system as a whole.

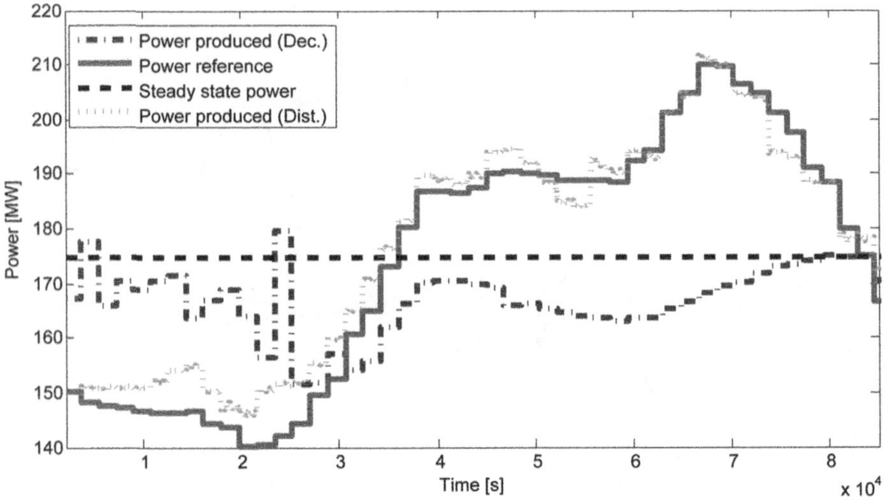

Fig. 29.3 Simulation of a decentralized MPC and the proposed DMPC scheme in the hydro-power valley (HPV)

29.5.3 Control of Irrigation Canals

In [9] we applied the proposed controller to a model of a section of the *postrasvase Tajo-Segura* in the south-east of Spain. The *postrasvase Tajo-Segura* is a set of canals which distribute water coming from the Tajo river in the basin of the Segura river. This water is used for irrigation (78 %) and human consumption (22 %). The selected section is a Y-shape canal, that is, the main canal splits into two canals with a gate placed at the input of each one of them: Canal de la Pedrera, with a total length of 6,680 km, and Canal de Cartagena, with a lenght of 17,444 km. Specifically, the DMPC controller regulates the water level at each of the reaches by manipulating the upstream gate. The time step length was 1 min and the control horizon, N_c, was set to 5. The cost function is designed to include the water traveling from the upstream gates, that is, the states that correspond to the traveling water are also weighted. The weights of the local costs in the canals grow with 2^i, that is, the farther a node is from the beginning, the more important it is. This way of weighting the error facilitates a faster flow of water towards the last canals. Finally, the matrix that weights the control effort R_i was set to zero for simplicity. In [9] it can be seen that with just an average number of 5 communications per agent and time step good regulation results were obtained.

29.6 Conclusions

In this chapter we have presented a distributed MPC algorithm based on negotiation for a class distributed linear systems coupled through the inputs. We assume that each agent has access only to the model and the state of one of the subsystems and that the agents must negotiate in order to reach a cooperative solution. The proposed algorithm has low communication and computational burdens and provides a feasible solution to the centralized problem. In addition, we provide sufficient conditions that guarantee practical stability of the closed-loop system as well as an optimization based procedure to design the controller so that these conditions are satisfied.

Acknowledgments Financial support from the HYCON2 EU-project from the ICT-FP7 and MEC-Spain, DPI2008-05818, and F.P.I. grants is gratefully acknowledged.

References

1. J. M. Maestre, Distributed Model Predictive Control Based on Game Theory. PhD thesis, Departamento de Ingeniería de Sistemas y Automática, Universidad de Sevilla, Nov 2010
2. J.M. Maestre, M.D. Doan, D. Muñoz de la Peña, P.J. van Overloop, T. Keviczky, M. Ridao, B. De Schutter, Benchmarking the operation of a hydro power network through the application of agent-based model predictive controllers, in *Proceedings of the 10th International Conference on Hydroinformatics*, July 2012
3. J.M. Maestre, D. Muñoz de la Peña, E.F. Camacho, Distributed MPC: a supply chain case study, in *Proceedings of the joint 48th IEEE Conference on Decision and Control and 28th Chinese Control Conference*, 2009
4. J.M. Maestre, D. Muñoz de la Peña, E.F. Camacho, Distributed model predictive control based on a cooperative game. Optimal Control Appl. Methods **32**(2), 153–176 (2011)
5. J.M. Maestre, D. Muñoz de la Peña, E.F. Camacho, T. Álamo, Distributed model predictive control based on agent negotiation. J Process Control **21**(5), 685–697 (June 2011)
6. D.Q. Mayne, J.B. Rawlings, C.V. Rao, P.O.M. Scokaert, Constrained model predictive control: Stability and optimality. Automatica **36**, 789–814 (2000)
7. S.V. Rakovic, E. De Santis, P. Caravani. Invariant equilibria of polytopic games via optimized robust control invariance. in *Proceedings of the 44th IEEE Conference on Decision and Control and the European Control Conference*, pp. 7686–7691, Seville, Spain, Dec 2005
8. P.O.M. Scokaert, D.Q. Mayne, J.B. Rawlings, Suboptimal model predictive control (feasibility implies stability). IEEE Trans. Autom. Control **44**, 648–654 (1999)
9. A. Zafra-Cabeza, J.M. Maestre, M.A. Ridao, E.F. Camacho, L. Sánchez, A hierarchical distributed model predictive control approach in irrigation canals: A risk mitigation perspective. J Process Control **21**(5), 787–799 (2011)

Chapter 30
Lyapunov-Based Distributed MPC Schemes: Sequential and Iterative Approaches

J. Liu, D. Muñoz de la Peña and P. D. Christofides

Abstract In this chapter, we focus on two distributed MPC (DMPC) schemes for the control of large-scale nonlinear systems in which several distinct sets of manipulated inputs are used to regulate the system. In the first scheme, the distributed controllers use a one-directional communication strategy, are evaluated in sequence and each controller is evaluated once at each sampling time; in the second scheme, the distributed controllers utilize a bi-directional communication strategy, are evaluated in parallel and iterate to improve closed-loop performance. In the design of the distributed controllers, Lyapunov-based model predictive control techniques are used. To ensure the stability of the closed-loop system, each model predictive controller in both schemes incorporates a stability constraint which is based on a suitable Lyapunov-based controller. We review the properties of the two DMPC schemes from stability, performance, computational complexity points of view. Subsequently, we briefly discuss the applications of the DMPC schemes to chemical processes and renewable energy generation systems.

30.1 Introduction

In this chapter, we focus on distributed MPC (DMPC) of large-scale nonlinear systems in which several distinct sets of manipulated inputs are used to regulate the system. For each set of manipulated inputs, a different model predictive controller,

J. Liu (✉)
Department of Chemical and Materials Engineering, University of Alberta, Edmonton, Canada
e-mail: jinfeng@ualberta.ca

D. Muñoz de la Peña
Departamento de Ingeniería de Sistemas y Automática, Universidad de Sevilla, Sevilla, Spain
e-mail: dmunoz@us.es

P. D. Christofides
Department of Chemical and Biomolecular Engineering and Department of Electrical Engineering, University of California, Los Angeles, CA, USA
e-mail: pdc@seas.ucla.edu

J. M. Maestre and R. R. Negenborn (eds.), *Distributed Model Predictive Control Made Easy*, Intelligent Systems, Control and Automation: Science and Engineering 69,
DOI: 10.1007/978-94-007-7006-5_30,

which is able to communicate with the rest of the controllers in making its decisions, is used to compute the control actions.

It is well known that MPC is not necessarily stabilizing. To achieve closed-loop stability, different approaches have been proposed in the literature. One class of approaches is to use infinite prediction horizons or well-designed terminal penalty terms or to impose stability constraints in the MPC optimization problem; please see [1, 10] for surveys of these approaches. However, the implicit nature of MPC control law makes it very difficult to explicitly characterize, a priori, the admissible initial conditions starting from where the MPC is guaranteed to be feasible and stabilizing. In practice, the initial conditions are usually chosen in an ad hoc fashion and tested through extensive closed-loop simulations. This issue motivated research on Lyapunov-based MPC (LMPC) designs [11, 12] (see also [5, 14]) that allow for an explicit characterization of the stability region of the closed-loop system and lead to a reduced computational complexity of the controller optimization problem. The design of LMPC is based on uniting receding horizon control with explicit Lyapunov-based nonlinear controller design techniques. In order to guarantee the closed-loop stability, contractive constraints based on Lyapunov functions are incorporated in the controller formulations.

Motivated by the above considerations, we adopt the LMPC framework [11, 12] and introduce modifications on the LMPC design both in the optimization problem formulation and in the controller implementation to account for distributed schemes. Specifically, under the assumption that feedback of the state of the process is available to all the distributed controllers at each sampling time and that a model of the plant is available, we present two different DMPC architectures designed via LMPC techniques. In the first architecture, the distributed controllers use a one-directional communication strategy, are evaluated in sequence and each controller is evaluated only once at each sampling time; in the second architecture, the distributed controllers utilize a bi-directional communication strategy, are evaluated in parallel and iterate to improve closed-loop performance. In order to ensure the stability of the closed-loop system, each model predictive controller in both architectures incorporates a stability constraint which is based on a suitable nonlinear control law that can stabilize the closed-loop system.

30.1.1 Notation

Throughout this chapter, the symbol Ω_r is used to denote the set $\Omega_r := \{\mathbf{x} \in \mathbb{R}^{n_x} : V(\mathbf{x}) \le r\}$ where V is a scalar positive definite, continuous differentiable function and $V(0) = 0$. The operator '/' denotes set subtraction, that is, $A/B := \{\mathbf{x} \in \mathbb{R}^{n_x} : \mathbf{x} \in A, \mathbf{x} \notin B\}$. The notation t_0 indicates the initial time instant. The set $\{t_{k \ge 0}\}$ denotes a sequence of synchronous time instants such that $t_k = t_0 + k\Delta$ and $t_{k+i} = t_k + i\Delta$ where Δ is a fixed time interval and i is an integer.

30.2 Problem Formulation

30.2.1 System Description

In this chapter, we consider nonlinear systems described by the following state-space model:

$$\dot{\mathbf{x}}(t) = \mathbf{f}(\mathbf{x}(t)) + \sum_{i \in \mathcal{N}} \mathbf{g}_i(\mathbf{x}(t))\mathbf{u}_i(t) + \mathbf{k}(\mathbf{x}(t))\mathbf{d}(t) \tag{30.1}$$

where $\mathbf{x}(t) \in \mathbb{R}^{n_\mathbf{x}}$ denotes the vector of process state variables, $\mathbf{u}_i(t) \in \mathbb{R}^{n_{\mathbf{u}_i}}$, $i = 1, 2, \dots |\mathcal{N}|$, are the set of control (manipulated) inputs $\mathcal{N}$ and $\mathbf{d}(t) \in \mathbb{R}^{n_\mathbf{w}}$ denotes the vector of disturbance variables. The sets of inputs are restricted to be in nonempty convex sets $U_i \subseteq \mathbb{R}^{n_{\mathbf{u}_i}}$, which are defined as follows:

$$\mathcal{U}_i := \{\mathbf{u}_i \in \mathbb{R}^{n_{\mathbf{u}_i}} : \|\mathbf{u}_i\| \le \mathbf{u}_i^{\max}\}, i \in \mathcal{N} \tag{30.2}$$

where $\mathbf{u}_i^{\max}$, $i \in \mathcal{N}$, are the magnitudes of the input constraints. The disturbance vector is bounded, i.e., $\mathbf{d}(t) \in \mathcal{D}$ where:

$$\mathcal{D} := \{\mathbf{d} \in \mathbb{R}^{n_\mathbf{w}} : \|\mathbf{d}\| \le \theta, \theta > 0\} \tag{30.3}$$

with θ being a known positive real number.

We assume that $\mathbf{f}$, $\mathbf{g}_i$, $i \in \mathcal{N}$, and $\mathbf{k}$ are locally Lipschitz vector, matrix and matrix functions, respectively, and that the origin is an equilibrium of the unforced nominal system (i.e., system (30.1) with $\mathbf{u}_i(t) = 0$, $i \in \mathcal{N}$, $\mathbf{d}(t) = 0$ for all t) which implies that $\mathbf{f}(0) = 0$.

Remark 30.1 In this chapter, in order to account for DMPC designs in which the distributed controllers are evaluated in parallel, we consider nonlinear systems with control inputs entering the system dynamics in an affine fashion. We note that the results presented in Sect. 30.3.1 can be extended to more general nonlinear systems, for example, systems described by the following state-space model:

$$\dot{\mathbf{x}}(t) = \mathbf{f}(\mathbf{x}(t), \mathbf{u}_1(t), \dots, \mathbf{u}_{|\mathcal{N}|}(t), \mathbf{d}(t)). \tag{30.4}$$

30.2.2 Lyapunov-based Control

We assume that there exists a nonlinear control law $\mathbf{h}(\mathbf{x}) = \left[\mathbf{h}_1(\mathbf{x})^T \ \cdots \ \mathbf{h}_{|\mathcal{N}|}(\mathbf{x})^T\right]^T$ with $\mathbf{u}_i = \mathbf{h}_i(\mathbf{x})$, $i \in \mathcal{N}$, which renders (under continuous state feedback) the origin of the nominal closed-loop system asymptotically stable while satisfying the input constraints for all the states $\mathbf{x}$ inside a given stability region. Using converse Lyapunov theorems, this assumption implies that there exist functions $\alpha_i(\cdot)$, $i = 1, 2, 3, 4$ of class $\mathcal{K}$[1] and a continuously differentiable Lyapunov function $V(\mathbf{x})$ for the nominal closed-loop system that satisfy the following inequalities:

$$\alpha_1(\|\mathbf{x}\|) \leq V(\mathbf{x}) \leq \alpha_2(\|\mathbf{x}\|) \tag{30.5}$$

$$\frac{\partial V(\mathbf{x})}{\partial \mathbf{x}}\left(\mathbf{f}(\mathbf{x}) + \sum_{i\in\mathcal{N}} \mathbf{g}_i(\mathbf{x})\mathbf{h}_i(\mathbf{x})\right) \leq -\alpha_3(\|\mathbf{x}\|) \tag{30.6}$$

$$\left\|\frac{\partial V(\mathbf{x})}{\partial \mathbf{x}}\right\| \leq \alpha_4(\|\mathbf{x}\|) \tag{30.7}$$

$$\mathbf{h}_i(\mathbf{x}) \in \mathcal{U}_i, \forall i \in \mathcal{N} \tag{30.8}$$

for all $\mathbf{x} \in O \subseteq \mathbb{R}^{n_x}$ where O is an open neighborhood of the origin. We denote the region $\Omega_\rho \subseteq O$ as the stability region of the closed-loop system under the nonlinear control law $\mathbf{h}(\mathbf{x})$.

30.2.3 Lyapunov-Based Model Predictive Control

In this section, we introduce the LMPC design proposed in [11, 12] that allows for an explicit characterization of the stability region and guarantees controller feasibility and closed-loop stability.

For the predictive control of the system of (30.1), the LMPC is designed based on an existing explicit control law $\mathbf{h}(\mathbf{x})$ that is able to stabilize the closed-loop system and satisfies the conditions of (30.5)–(30.8). The formulation of the LMPC is as follows:

$$\min_{\mathbf{u}_i\in\mathcal{S}(\Delta), i\in\mathcal{N}} \int_{t_k}^{t_{k+N_p}} \left[\|\tilde{\mathbf{x}}(\tau)\|_{Q_c}^2 + \sum_{i\in\mathcal{N}} \|\mathbf{u}_i(\tau)\|_{R_{ci}}^2\right] d\tau \tag{30.9}$$

$$\text{s.t. } \dot{\tilde{\mathbf{x}}}(t) = \mathbf{f}(\tilde{\mathbf{x}}(t)) + \sum_{i\in\mathcal{N}} \mathbf{g}_i(\tilde{\mathbf{x}}(t))\mathbf{u}_i \tag{30.10}$$

$$\mathbf{u}_i(t) \in \mathcal{U}_i, i \in \mathcal{N} \tag{30.11}$$

$$\tilde{\mathbf{x}}(t_k) = \mathbf{x}(t_k) \tag{30.12}$$

$$\sum_{i\in\mathcal{N}} \frac{\partial V(\mathbf{x}(t_k))}{\partial \mathbf{x}} \mathbf{g}_i(\mathbf{x}(t_k))\mathbf{u}_i(t_k) \leq \sum_{i\in\mathcal{N}} \frac{\partial V(\mathbf{x}(t_k))}{\partial \mathbf{x}} \mathbf{g}_i(\mathbf{x}(t_k))\mathbf{h}_i(\mathbf{x}(t_k)) \tag{30.13}$$

where $\mathcal{S}(\Delta)$ is the set of piecewise constant signals with sampling time Δ, $\tilde{\mathbf{x}}$ is the predicted trajectory of the nominal system with $\mathbf{u}_i$, $i \in \mathcal{N}$, the input trajectory computed by this centralized LMPC. The optimal solution to this optimization problem

[1] A continuous function $\alpha : [0, a) \to [0, \infty)$ is said to belong to class $\mathcal{K}$ if it is strictly increasing and satisfies $\alpha(0) = 0$.

is denoted by $\mathbf{u}^*_{ci}(t|t_k)$, $i \in \mathcal{N}$, which is defined for $t \in [t_k, t_{k+N_p})$. The manipulated inputs of the closed-loop system of (30.1) under this centralized LMPC are defined as follows:

$$\mathbf{u}_i(t) = \mathbf{u}^*_{ci}(t|t_k), i \in \mathcal{N}, \forall t \in [t_k, t_{k+1}). \quad (30.14)$$

In what follows, we refer to this controller as the centralized LMPC.

In the LMPC defined by (30.9)–(30.13), the constraint of (30.13) guarantees that the value of the time derivative of the Lyapunov function, $V(\mathbf{x})$, at time t_k is smaller than or equal to the value obtained if the nonlinear control law $\mathbf{u} = \mathbf{h}(\mathbf{x})$ is implemented in the closed-loop system in a sample-and-hold fashion. This is a constraint that allows one to prove (when state measurements are available every synchronous sampling time) that the LMPC inherits the stability and robustness properties of the nonlinear control law $\mathbf{h}(\mathbf{x})$ when it is applied in a sample-and-hold fashion.

One of the main properties of the LMPC of (30.9)–(30.13) is that it possesses the same stability region Ω_ρ as the nonlinear control law $\mathbf{h}(\mathbf{x})$, which implies that the origin of the closed-loop system is guaranteed to be stable and the LMPC is guaranteed to be feasible for any initial state inside Ω_ρ when the sampling time Δ and the disturbance upper bound θ are sufficiently small. Note that the region Ω_ρ can be explicitly characterized. The stability property of the LMPC is inherited from the nonlinear control law $\mathbf{h}(\mathbf{x})$ when it is applied in a sample-and-hold fashion; please see [2, 13] for results on sampled-data systems. The feasibility property of the LMPC is also guaranteed by the nonlinear control law $\mathbf{h}(\mathbf{x})$ since $\mathbf{u} = \mathbf{h}(\mathbf{x})$ is a feasible solution to the optimization problem of (30.9)–(30.13). The main advantage of the LMPC approach with respect to the nonlinear control law $\mathbf{h}(\mathbf{x})$ is that optimality considerations can be taken explicitly into account (as well as constraints on the inputs and the states [12]) in the computation of the control actions within an online optimization framework while improving the closed-loop performance of the system.

30.3 Sequential and Iterative DMPC Designs

The objective of this section is to design DMPC architectures including multiple MPCs for large-scale nonlinear process systems with continuous, synchronous state feedback. Specifically, we will discuss two different DMPC architectures. In the first DMPC architecture, different MPC controllers are evaluated in sequence, only once at each sampling time and require only one-directional communication between consecutive distributed controllers (i.e., the distributed controllers are connected by pairs). In the second architecture, different MPCs are evaluated in parallel, once or more than once at each sampling time depending on the number of iterations, and bi-directional communication among all the distributed controllers (i.e., the distributed controllers are all interconnected) is used.

In each DMPC architecture, we will design $|\mathcal{N}|$ LMPCs to compute $\mathbf{u}_i$, $i \in \mathcal{N}$, and refer to the LMPC computing the input trajectories of $\mathbf{u}_i$ as LMPC i.

Remark 30.2 Note also that the assumption that the full state $\mathbf{x}$ of the system is sampled synchronously is a widely used assumption in the control system design. The control system designs presented in this section can be extended to the case where only part of the state $\mathbf{x}$ is measurable by designing an observer to estimate the whole state vector from output measurements and by designing the control system based on the measured and estimated states. In this case, the stability properties of the resulting output feedback control systems are affected by the convergence of the observer and need to be carefully studied.

30.3.1 Sequential DMPC

A schematic of the architecture considered in this subsection is shown in Fig. 30.1. We first present the implementation strategy of this DMPC architecture and then design the corresponding LMPCs. The implementation strategy of this DMPC architecture is given as Algorithm 30.1.

In this architecture, each LMPC only sends its future input trajectory and the future input trajectories it received to the next LMPC (i.e., LMPC j sends input

Algorithm 30.1: Sequential DMPC

1: At t_k, all the LMPCs receive the state measurement $\mathbf{x}(t_k)$ from the sensors.
2: For $j = |\mathcal{N}|$ to 1

2.1. LMPC j receives the entire future input trajectories of $\mathbf{u}_i$, $i = |\mathcal{N}|, |\mathcal{N}| - 1, \dots, j+1$, from LMPC $j+1$ and evaluates the future input trajectory of $\mathbf{u}_j$ based on $\mathbf{x}(t_k)$ and the received future input trajectories.
2.2. LMPC j sends the first step input value of $\mathbf{u}_j$ to its actuators and the entire future input trajectories of $\mathbf{u}_i$, $i = |\mathcal{N}|, |\mathcal{N}| - 1, \dots, j$, to LMPC $j-1$.

3: When a new measurement is received ($k \leftarrow k+1$), go to Step 1.

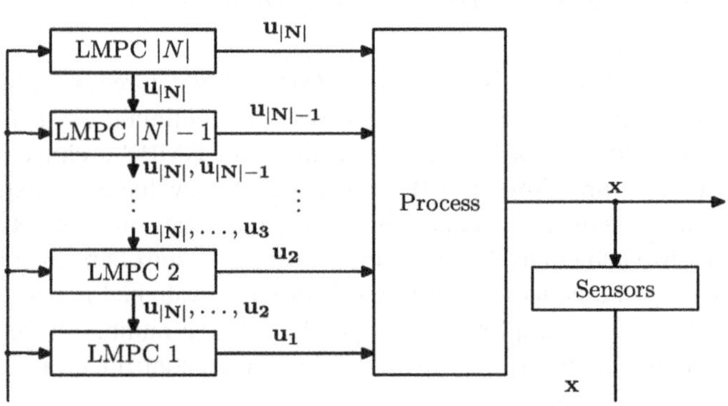

Fig. 30.1 Sequential DMPC architecture

trajectories to LMPC $j-1$). This implies that LMPC j, $j=|\mathcal{N}|, |\mathcal{N}|-1, \dots, 2$, does not have any information about the values that $\mathbf{u}_i$, $i = j-1, j-2, \dots, 1$ will take when the optimization problems of the LMPCs are designed. In order to make a decision, LMPC j, $j=|\mathcal{N}|,|\mathcal{N}|-1,\dots,2$ must assume trajectories for $\mathbf{u}_i$, $i=j-1, j-2, \dots, 1$, along the prediction horizon. To this end, the nonlinear control law $\mathbf{h}(\mathbf{x})$ is used. In order to inherit the stability properties of the controller $\mathbf{h}(\mathbf{x})$, each control input $\mathbf{u}_i$, $i \in \mathcal{N}$ must satisfy a constraint that guarantees a given minimum contribution to the decrease rate of the Lyapunov function $V(\mathbf{x})$. Specifically, the design of LMPC j, $j \in \mathcal{N}$, is based on the following optimization problem:

$$\min_{\mathbf{u}_j \in S(\Delta)} \int_{t_k}^{t_{k+N_p}} \left[\|\tilde{\mathbf{x}}(\tau)\|_{Q_c}^2 + \sum_{i \in \mathcal{N}} \|\mathbf{u}_i(\tau)\|_{R_{ci}}^2 \right] d\tau \tag{30.15}$$

$$\text{s.t. } \dot{\tilde{\mathbf{x}}}(t) = \mathbf{f}(\tilde{\mathbf{x}}(t)) + \sum_{i \in \mathcal{N}} \mathbf{g}_i(\tilde{\mathbf{x}}(t))\mathbf{u}_i \tag{30.16}$$

$$\mathbf{u}_i(t) = \mathbf{h}_i(\tilde{\mathbf{x}}(t_{k+l})), i = 1, \dots, j-1,$$
$$\forall t \in [t_{k+l}, t_{k+l+1}), l = 0, \dots, N_p - 1 \tag{30.17}$$

$$\mathbf{u}_i(t) = \mathbf{u}_{s,i}^*(t|t_k), i = j+1, \dots, |\mathcal{N}| \tag{30.18}$$

$$\mathbf{u}_j(t) \in \mathcal{U}_j \tag{30.19}$$

$$\tilde{\mathbf{x}}(t_k) = \mathbf{x}(t_k) \tag{30.20}$$

$$\frac{\partial V(\mathbf{x}(t_k))}{\partial \mathbf{x}} \mathbf{g}_j(\mathbf{x}(t_k))\mathbf{u}_j(t_k) \le \frac{\partial V(\mathbf{x}(t_k))}{\partial \mathbf{x}} \mathbf{g}_j(\mathbf{x}(t_k))\mathbf{h}_j(\mathbf{x}(t_k)) \tag{30.21}$$

In the optimization problem of (30.15)–(30.21), $\mathbf{u}_{s,i}^*(t|t_k)$ denotes the optimal future input trajectory of $\mathbf{u}_i$ obtained by LMPC i of the form of (30.15)–(30.21) evaluated before LMPC j, $\tilde{\mathbf{x}}$ is the predicted trajectory of the nominal system with $\mathbf{u}_i = \mathbf{u}_{s,i}$, $i = j+1, j+2, \dots, |\mathcal{N}|$, $\mathbf{u}_i$, $i = 1, 2, \dots, j-1$, the corresponding elements of $\mathbf{h}(\mathbf{x})$ applied in a sample-and-hold fashion and $\mathbf{u}_j$ predicted by LMPC j of (30.15)–(30.21). The optimal solution to the optimization problem of (30.15)–(30.21) is denoted as $\mathbf{u}_{s,j}^*(t|t_k)$ which is defined for $t \in [t_k, t_{k+N_p})$.

The constraint of (30.16) is the nominal model of the system of (30.1), which is used to predict the future evolution of the system; the constraint of (30.17) defines the value of the inputs evaluated after $\mathbf{u}_j$ (i.e., $\mathbf{u}_i$ with $i = 1, 2, \dots, j-1$); the constraint of (30.18) defines the value of the inputs evaluated before $\mathbf{u}_j$ (i.e., $\mathbf{u}_i$ with $i = j+1, j+2, \dots, |\mathcal{N}|$); the constraint of (30.19) is the constraint on the manipulated input $\mathbf{u}_j$; the constraint of (30.20) sets the initial state for the optimization problem; the constraint of (30.21) guarantees that the contribution of input $\mathbf{u}_j$ to the decrease rate of the time derivative of the Lyapunov function $V(\mathbf{x})$ at the initial evaluation time (i.e., at t_k), if $\mathbf{u}_j = \mathbf{u}_{s,j}^*(t_k|t_k)$ is applied, is bigger than or equal to the value obtained when $\mathbf{u}_j = \mathbf{h}_j(\mathbf{x}(t_k))$ is applied. This constraint allows proving the closed-loop stability properties of this DMPC.

The manipulated inputs of the system of (30.1) under the DMPC are defined as follows:

$$\mathbf{u}_i(t) = \mathbf{u}^*_{s,i}(t|t_k), i \in \mathcal{N}, \forall t \in [t_k, t_{k+1}). \tag{30.22}$$

In what follows, we refer to this DMPC architecture as the sequential DMPC.

Remark 30.3 Note that, in order to simplify the description of the implementation strategy presented above in this subsection, we do not distinguish LMPC $|\mathcal{N}|$ and LMPC 1 from the others. We note that LMPC $|\mathcal{N}|$ does not receive any information from the other controllers and LMPC 1 does not have to send information to any other controller.

30.3.2 Iterative DMPC

An alternative architecture to the sequential DMPC architecture presented in the previous subsection is to evaluate all the distributed LMPCs in parallel and iterate to improve closed-loop performance. A schematic of this control architecture is shown in Fig. 30.2.

In this architecture, each distributed LMPC must be able to communicate with all the other controllers (i.e., the distributed controllers are all interconnected). More specifically, when a new state measurement is available at a sampling time, each distributed LMPC controller evaluates and obtains its future input trajectory; and then each LMPC controller broadcasts its latest future input trajectory to all the other controllers. Based on the newly received input trajectories, each LMPC controller evaluates its future input trajectory again and this process is repeated until a certain termination condition is satisfied. Specifically, the implementation strategy is given as Algorithm 30.2.

Note that at the initial iteration, all the LMPCs use $\mathbf{h}(\mathbf{x})$ to estimate the input trajectories of all the other controllers. Note also that the number of iterations p can

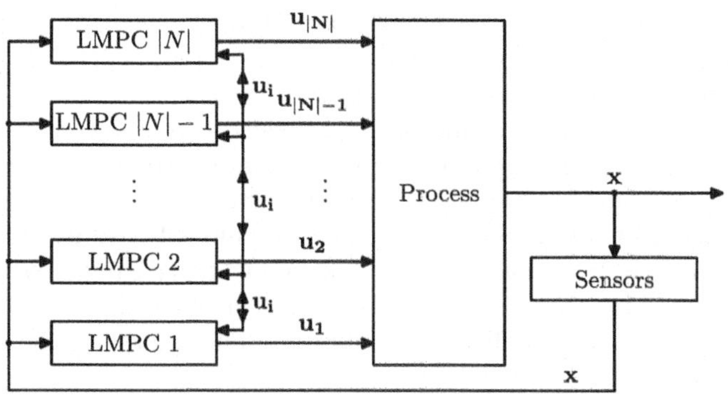

Fig. 30.2 Iterative DMPC architecture

Algorithm 30.2: Iterative DMPC

1: At t_k, all the LMPCs receive the state measurement $\mathbf{x}(t_k)$ from the sensors and then evaluate their future input trajectories in an iterative fashion with initial input guesses generated by $\mathbf{h}(\cdot)$.

2: At iteration p ($p \geq 1$):

2.1. Each LMPC evaluates its own future input trajectory based on $\mathbf{x}(t_k)$ and the latest received input trajectories of all the other LMPCs (when $p = 1$, initial input guesses generated by $\mathbf{h}(\cdot)$ are used).

2.2. The controllers exchange their future input trajectories. Based on all the input trajectories, each controller calculates and stores the value of the cost function.

3: If a termination condition is satisfied, each controller sends its entire future input trajectory corresponding to the smallest value of the cost function to its actuators; if the termination condition is not satisfied, go to Step 2 ($p \leftarrow p + 1$).

4: When a new measurement is received, go to Step 1 ($k \leftarrow k + 1$).

be variable and it does not affect the closed-loop stability of the DMPC architecture presented in this subsection; a point that will be made clear below. For the iterations in this DMPC architecture, there are different choices of the termination condition. For example, the number of iterations p may be restricted to be smaller than a maximum iteration number $p_{\max}$ (i.e., $p \leq p_{\max}$) and/or the iterations may be terminated when the difference of the performance or the solution between two consecutive iterations is smaller than a threshold value and/or the iterations maybe terminated when a maximum computational time is reached.

In order to proceed, we define $\hat{\mathbf{x}}(t|t_k)$ for $t \in [t_k, t_{k+N_p})$ as the nominal sampled trajectory of the system of (30.1) associated with the feedback control law $\mathbf{h}(\mathbf{x})$ and sampling time Δ starting from $\mathbf{x}(t_k)$. This nominal sampled trajectory is obtained by integrating recursively the following differential equation:

$$\begin{aligned} \dot{\hat{\mathbf{x}}}(t|t_k) &= \mathbf{f}(\hat{\mathbf{x}}(t|t_k)) + \sum_{i\in\mathcal{N}} \mathbf{g}_i(\hat{\mathbf{x}}(t|t_k))\mathbf{h}_i(\hat{\mathbf{x}}(t_{k+l}|t_k)), \\ &\qquad \forall \tau \in [t_{k+l}, t_{k+l+1}), l = 0, 1, \ldots, N_p - 1. \end{aligned} \tag{30.23}$$

Based on $\hat{\mathbf{x}}(t|t_k)$, we can define the following variable:

$$\begin{aligned} \mathbf{u}_{n,j}(t|t_k) &= \mathbf{h}_j(\hat{\mathbf{x}}(t_{k+l}|t_k)), \; j \in \mathcal{N}, \\ &\qquad \forall \tau \in [t_{k+l}, t_{k+l+1}), l = 0, 1, \ldots, N_p - 1, \end{aligned} \tag{30.24}$$

which will be used as the initial guess of the trajectory of $\mathbf{u}_j$.

The design of the LMPC j, $j \in \mathcal{N}$, at iteration p is based on the following optimization problem:

$$\min_{\mathbf{u}_j \in S(\Delta)} \int_{t_k}^{t_{k+N_p}} \left[\|\tilde{\mathbf{x}}(\tau)\|_{Q_c}^2 + \sum_{i \in \mathcal{N}} \|\mathbf{u}_i(\tau)\|_{R_{ci}}^2 \right] d\tau \tag{30.25}$$

$$\text{s.t. } \dot{\tilde{\mathbf{x}}}(t) = \mathbf{f}(\tilde{\mathbf{x}}(t)) + \sum_{i \in \mathcal{N}} \mathbf{g}_i(\tilde{\mathbf{x}}(t))\mathbf{u}_i \tag{30.26}$$

$$\mathbf{u}_i(t) = \mathbf{u}_i^{*,p-1}(t|t_k), \forall i \neq j \tag{30.27}$$

$$\mathbf{u}_j(t) \in \mathcal{U}_j \tag{30.28}$$

$$\tilde{\mathbf{x}}(t_k) = \mathbf{x}(t_k) \tag{30.29}$$

$$\frac{\partial V(\mathbf{x}(t_k))}{\partial \mathbf{x}} \mathbf{g}_j(\mathbf{x}(t_k))\mathbf{u}_j(t_k) \le \frac{\partial V(\mathbf{x}(t_k))}{\partial \mathbf{x}} \mathbf{g}_j(\mathbf{x}(t_k))\mathbf{h}_j(\mathbf{x}(t_k)) \tag{30.30}$$

where $\tilde{\mathbf{x}}$ is the predicted trajectory of the nominal system with $\mathbf{u}_k$, the input trajectory, computed by the LMPCs of (30.25)–(30.30) and all the other inputs are the optimal input trajectories at iteration $p-1$ of the rest of distributed controllers (i.e., $\mathbf{u}_i^{*,p-1}(t|t_k)$ for $i \neq j$). The optimal solution to the optimization problem of (30.25)–(30.30) is denoted as $\mathbf{u}_j^{*,p}(t|t_k)$ which is defined for $t \in [t_k, t_{k+N_p})$. Accordingly, we define the final optimal input trajectory of LMPC j (that is, the optimal trajectories computed at the last iteration) as $\mathbf{u}_j^*(t|t_k)$ which is also defined for $t \in [t_k, t_{k+N_p})$.

Note that the first iteration of each distributed LMPC, the input trajectory defined in (30.24) is used as the initial input trajectory guess; that is, $\mathbf{u}_j^{*,0}(t|t_k) = \mathbf{u}_{n,j}(t|t_k)$, $i \in \mathcal{N}$.

The manipulated inputs of the system of (30.1) under this DMPC design with LMPCs of (30.25)–(30.30) are defined as follows:

$$\mathbf{u}_i(t) = \mathbf{u}_i^*(t|t_k), i \in \mathcal{N}, \forall t \in [t_k, t_{k+1}). \tag{30.31}$$

In what follows, we refer to this DMPC architecture as the iterative DMPC.

Remark 30.4 Note that because the manipulated inputs enter the dynamics of the system of (30.1) in an affine manner, the constraints designed in the LMPC optimization problems of (30.15)–(30.21) and (30.25)–(30.30) to guarantee the closed-loop stability can be decoupled for different distributed controllers as in (30.21) and (30.30).

30.4 Theoretical Properties

In [6, 7], it is proved that the two DMPC architectures enforce practical stability in the closed-loop system while improving performance. The sequential DMPC of (30.15)–(30.22) computes the inputs $\mathbf{u}_i$, $i \in \mathcal{N}$, applied to the system of (30.1) in a way such that in the closed-loop system, the value of the Lyapunov function at time instant t_k (i.e., $V(\mathbf{x}(t_k))$) is a decreasing sequence of values with a lower bound. Following Lyapunov arguments, this property guarantees practical stability of the closed-loop system. This is achieved due to the constraint of (30.21). The proof of this property consists of two parts. First it is proved that the optimization problem of (30.15)–(30.21) is feasible for all $j \in \mathcal{N}$, and $\mathbf{x} \in \Omega_\rho$. Then, it is proved that, under the DMPC of (30.15)–(30.22), the state of the system of (30.1) is ultimately bounded in $\Omega_{\rho_{\min}}$. A similar property was derived for iterative controllers.

Note that the DMPC designs have the same stability region Ω_ρ as the one of the nonlinear control law $\mathbf{h}(\mathbf{x})$. When the stability of the nonlinear control law $\mathbf{h}(\mathbf{x})$ is global (i.e., the stability region is the entire state space), then the stability of the DMPC designs is also global. Note also that for any initial condition in Ω_ρ, the DMPC designs are proved to be feasible. The choice of the horizon of the DMPC designs does not affect the stability of the closed-loop system. For any horizon length $N_p \geq 1$, the closed-loop stability is guaranteed by the constraints of (30.21) and (30.30). However, the choice of the horizon does affect the performance of the DMPC designs.

From the performance point of view it is important to remark that for general nonlinear systems, there is no guarantee that the closed-loop performance of one (centralized or distributed) MPC architecture discussed in this section should be superior than the others since the solutions provided by these MPC architectures are proved to be feasible and stabilizing but the superiority of the closed-loop performance of one MPC architecture over another is not established. In applications of these MPC architectures, especially for chemical process control in which non-convex problems is a very common occurrence, simulations should be conducted before making decisions as to which architecture should be used.

With respect to the computational burden, in the sequential DMPC architecture presented in Sect. 30.3.1, the distributed controllers are evaluated in sequence, which implies that the minimal time to obtain a set of solutions to all the LMPCs is the sum of the evaluation times of all the LMPCs; whereas in the iterative DMPC architecture presented in Sect. 30.3.2, the distributed controllers are evaluated in parallel, which implies that the minimal time to obtain a set of solutions to all the LMPCs is the largest evaluation time among all the LMPCs assuming that only one iteration is carried out.

Finally, there is no guaranteed convergence of the optimal cost or solution of a sequential or iterated DMPC to the optimal cost or solution of a centralized MPC for general nonlinear constrained systems because of the non-convexity of the MPC optimization problems and the fact that the DMPC does not solve the centralized LMPC in a distributed fashion due to the way the Lyapunov-based constraint of the centralized LMPC is broken down into constraints imposed on the individual LMPCs. For linear systems, the resulting optimization problems result to be convex and further properties can be derived. The convergence of the solution of a distributed linear or convex MPC design to the solution of a centralized MPC or a Pareto optimal solution is ensured in the context of linear systems (see, for example, [3]).

In addition to the class of systems considered in the previous sections, the DMPC designs have been also extended to include nonlinear systems subject to asynchronous and delayed state feedback [8, 9]. In the case of asynchronous feedback, under the assumption that there is an upper bound on the maximum interval between two consecutive measurements, both DMPC architectures can be extended without further assumptions to take explicitly into account asynchronous feedback. For systems subject to asynchronous measurements that also involve time-delays under the assumption that there exists an upper bound on the maximum feedback delay, however, the schemes have to be modified to include some additional communication between the plant and the controllers. Sufficient conditions under which the proposed distributed control designs guarantee that the states of the closed-loop system are ultimately bounded in regions that contain the origin are provided. The theoretical results were illustrated through a catalytic alkylation of benzene process example.

In [4], we focus on a master-slave type DMPC and discuss how to handle communication disruptions—communication channel noise and data losses—between the distributed controllers. To handle communication disruptions, feasibility problems are incorporated in the DMPC architecture to determine whether the data transmitted through the communication channel is reliable or not. Based on the results of the feasibility problems, the transmitted information is accepted or rejected by the stabilizing MPC. In order to ensure the stability of the closed-loop system under communication disruptions, each model predictive controller utilizes a suitable Lyapunov-based stability constraint.

In [3], we consider the design of a network-based DMPC system using multirate sampling for large-scale nonlinear uncertain systems composed of several coupled subsystems. Specifically, we assume that the states of each local subsystem can be divided into fast sampled states (which are available every sampling time) and slowly sampled states (which are available every several sampling times). The distributed model predictive controllers are connected through a shared communication network and cooperate in an iterative fashion at time instants in which full system state measurements (both fast and slow) are available, to guarantee closed-loop stability. When only fast sampled state feedback is available, the distributed controllers operate in a decentralized fashion to improve closed-loop performance. In this control architecture, the controllers are designed via LMPC techniques taking into account bounded measurement noise, process disturbances and communication noise. Sufficient conditions under which the state of the closed-loop system is ultimately bounded in an invariant region containing the origin are derived.

30.5 Applications

The theoretical results have been illustrated through several high complexity simulated systems; namely a reactor-separator process with recycle, a catalytic alkylation of benzene process and a wind-solar energy generation system.

In [8], the application to a reactor-separator process consisting of two continuously stirred tank reactors (CSTRs) and a flash tank separator was presented. We applied the sequential DMPC to this process and compared its performance with a centralized LMPC. Specifically, we designed two distributed LMPCs. A schematic of the process with the sequential DMPC is shown in Fig. 30.3. From the simulations, we found that both the sequential DMPC and the centralized LMPC designs

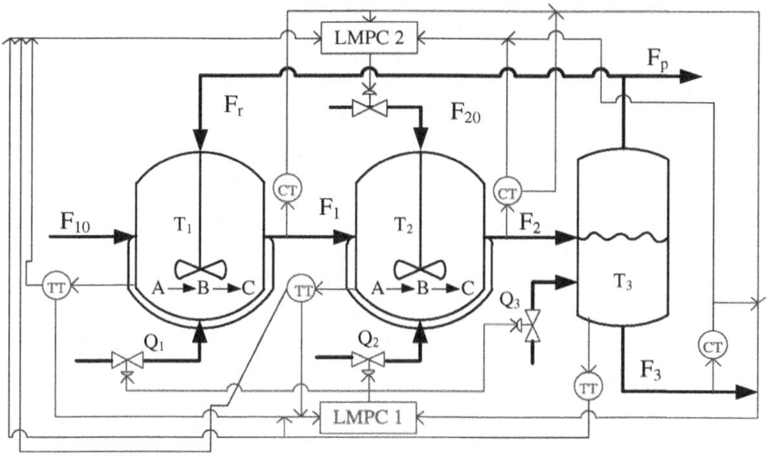

Fig. 30.3 Reactor-separator process with distributed control architecture

stabilize the process and provide a similar closed-loop performance. We also found that the communication between the two distributed LMPCs plays an important role in achieving closed-loop stability and optimal performance.

In [7] the application to a an alkylation of benzene process was presented. The process consists of four CSTRs and a flash tank separator, as shown in Fig. 30.4. In this application, we first demonstrated that the sequential and iterative DMPC schemes can both stabilize the closed-loop system and drive the state of the

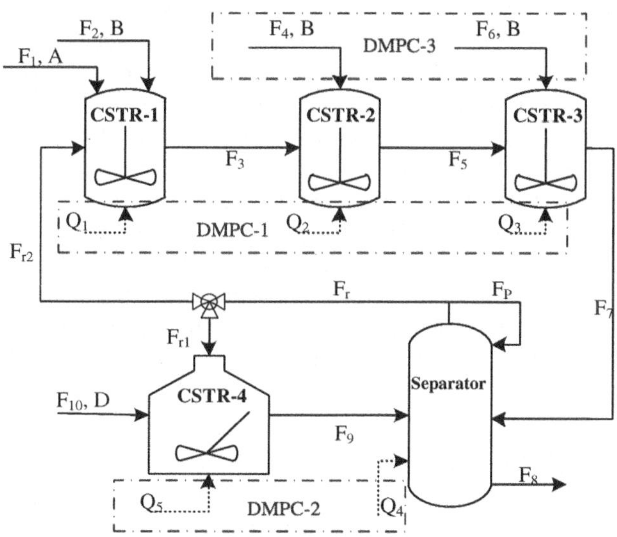

Fig. 30.4 Process flow diagram of alkylation of benzene

closed-loop system to a small region around the operating steady-state. Subsequently, we compared the mean evaluation times of the centralized LMPC optimization problem and the sequential and iterative DMPC optimization problems. From the simulation results, we found that the time needed to solve the centralized LMPC is much larger than the time needed to solve the sequential or iterative DMPCs. This is because the centralized LMPC has to solve a much larger (in terms of decision variables) optimization problem than the DMPCs. We also found that the evaluation time of the centralized LMPC is even larger than the sum of evaluation times of LMPC 1, LMPC 2 and LMPC 3 in the sequential DMPC, and the times needed to solve the DMPCs in both sequential and iterative distributed schemes are of the same order of magnitude. Finally, we compared the centralized LMPC and the two DMPC schemes from a performance index point of view. From the simulation results, we found that the centralized LMPC gives the lowest performance cost, the sequential DMPC gives lower cost than the iterative DMPC when there is no iteration. However, as the iteration number p increases, the performance cost given by the iterative DMPC decreases and converges to the cost of the one corresponding to the centralized LMPC.

In [15], the DMPC schemes were also applied to the design of distributed supervisory MPC systems for the optimal management and operation of distributed wind and solar energy generation systems integrated into the electrical grid . We considered a topology in which a wind subsystem and a solar subsystem are integrated in a direct current (DC) power grid in a distributed fashion, providing electrical power to a local area, and each subsystem is coupled with an energy storage device. Specifically, a sequential and an iterative distributed supervisory MPC architectures which are able to take into account optimality considerations on system operation and battery maintenance were developed to coordinate the actions of the subsystems accordingly. A schematic of the distributed supervisory MPC control system is shown in Fig. 30.5. Simulations of 24-hour system operation were carried out under a centralized and the distributed control architectures to demonstrate the applicability and effectiveness of the distributed supervisory predictive control designs. We first demonstrated that the different control schemes all can coordinate the different subsystems to satisfy the total power demand. Subsequently, we compared the different control schemes from computational time point of view. From the simulations, we found that the distributed supervisory control systems have much smaller computational times compared with the centralized supervisory control system. We also compared the control schemes from a performance point of view. In general, the centralized supervisory control system gives the best performance and the performance provided by the iterative design increases with the number of iterations.

30.6 Conclusions

This chapter presented approaches to distributed predictive control of nonlinear process systems via model predictive control and Lyapunov-based control techniques. These methods provide an explicit characterization for the closed-loop

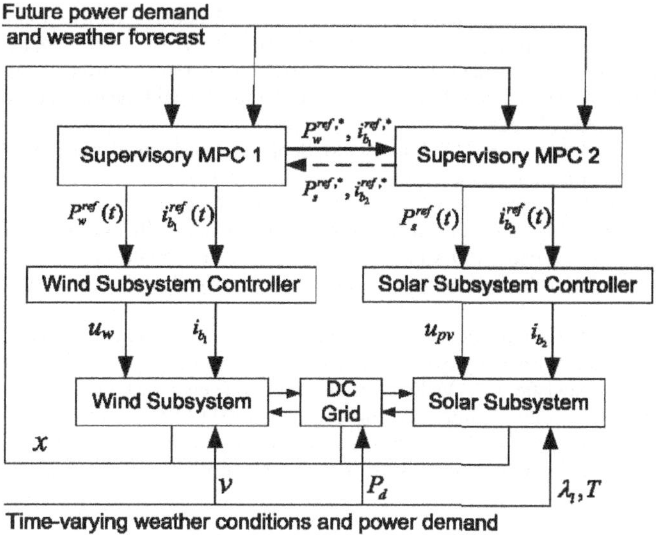

Fig. 30.5 Structure of the distributed supervisory MPC control system. v, P_d, λ_l and T are wind speed, power demand, insolation and cell temperature, respectively

stability region and can be modified to account for the effect of asynchronous feedback and time-varying measurement delays were first developed. The effectiveness and performance of the control approaches were illustrated via applications to nonlinear process networks and wind-solar energy generation systems.

References

1. F. Allgöwer, H. Chen, Nonlinear model predictive control schemes with guaranteed stability, in *NATO ASI on Nonlinear Model Based Process Control*, ed. by R. Berber, C. Kravaris (Kluwer Academic Publishers, Dordrecht, 1998), pp. 465–494
2. F. Clarke, Y. Ledyaev, E. Sontag, Asymtotic controllability implies feedback stabilization. IEEE Trans. Autom. Control **42**(10), 1394–1407 (1997)
3. M. Heidarinejad, J. Liu, D. Muñoz de la Peña, P.D. Christofides, J.F. Davis, Multirate Lyapunov-based distributed model predictive control of nonlinear uncertain systems. J. Process Control **21**(9), 1231–1242 (2011)
4. M. Heidarinejad, J. Liu, D. Muñoz de la Peña, J.F. Davis, P.D. Christofides, Handling communication disruptions in distributed model predictive control of nonlinear systems. J. Process Control **21**, 173–181 (2011)
5. S.L.D. Kothare, M. Morari, Contractive model predictive control for constrained nonlinear systems. IEEE Trans. Autom. Control **45**(6), 1053–1071 (2000)
6. J. Liu, D. Muñoz de la Peña, P.D. Christofides, Distributed model predictive control of nonlinear process systems. AIChE J. **55**(5), 1171–1184 (2009)
7. J. Liu, X. Chen, D. Muñoz de la Peña, P.D. Christofides, Sequential and iterative architectures for distributed model predictive control of nonlinear process systems. AIChE J. **56**(8), 2137–2149 (2010)

8. J. Liu, D. Muñoz de la Peña, P.D. Christofides, Distributed model predictive control of nonlinear systems subject to asynchronous and delayed measurements. Automatica **46**(1), 52–61 (2010)
9. J. Liu, X. Chen, D. Muñoz de la Peña, P.D. Christofides, Iterative distributed model predictive control of nonlinear systems: Handling asynchronous, delayed measurements. IEEE Trans. Autom. Control **57**(2), 528–534 (2012)
10. D.Q. Mayne, J.B. Rawlings, C.V. Rao, P.O.M. Scokaert, Constrained model predictive control: Stability and optimality. Automatica **36**, 789–814 (2000)
11. P. Mhaskar, N.H. El-Farra, P.D. Christofides, Predictive control of switched nonlinear systems with scheduled mode transitions. IEEE Trans. Autom. Control **50**(11), 1670–1680 (2005)
12. P. Mhaskar, N.H. El-Farra, P.D. Christofides, Stabilization of nonlinear systems with state and control constraints using Lyapunov-based predictive control. Syst. Control Lett. **55**(8), 650–659 (2006)
13. D. Nešić, A. Teel, P. Kokotovic, Sufficient conditions for stabilization of sampled-data nonlinear systems via discrete time approximations. Syst. Control Lett. **38**(4–5), 259–270 (1999)
14. J.A. Primbs, V. Nevistic, J.C. Doyle, A receding horizon generalization of pointwise min-norm controllers. IEEE Trans. Autom. Control **45**(5), 898–909 (2000)
15. W. Qi, J. Liu, P.D. Christofides, Distributed supervisory predictive control of distributed wind and solar energy generation systems. IEEE Trans. Control Syst. Technol. **21**, 504–512 (2013) (in press)

Chapter 31
Multi-layer Decentralized MPC of Large-scale Networked Systems

C. Ocampo-Martinez, V. Puig, J. M. Grosso and S. Montes-de-Oca

Abstract In this chapter, a multi-layer decentralized model predictive control (ML-DMPC) approach is proposed and designed for its application to large-scale networked systems (LSNS). This approach is based on the periodic nature of the system disturbance and the availability of both static and dynamic models of the LSNS. Hence, the topology of the controller is structured in two layers. First, an upper layer is in charge of achieving the global objectives from a set $\mathcal{O}$ of control objectives given for the LSNS. This layer works with a sampling time Δt_1, corresponding to the disturbances period. Second, a lower layer, with a sampling time Δt_2, $\Delta t_1 > \Delta t_2$, is in charge of computing the references for the system actuators in order to satisfy the local objectives from the set of control objectives $\mathcal{O}$. A system partitioning allows to establish a hierarchical flow of information between a set $\mathcal{C}$ of controllers designed based on model predictive control (MPC). Therefore, the whole proposed ML-DMPC strategy results in a centralized optimization problem for considering the global control objectives, followed of a decentralized scheme for reaching the local control objectives. The proposed approach is applied to a real case study: the water transport network of Barcelona (Spain). Results obtained with selected simulation scenarios show the effectiveness of the proposed ML-DMPC

C. Ocampo-Martineze(✉) · V. Puig · J. M. Grosso
Institut de Robòtica i Informàtica Industrial (CSIC-UPC), Technical University of Catalonia, Barcelona, Spain
e-mail: cocampo@iri.upc.edu

V. Puig
e-mail: vpuig@iri.upc.edu

J. M. Grosso
e-mail: jgrosso@iri.upc.edu

S. Montes-de-Oca
Automatic Control Department, Technical University of Catalonia, Barcelona, Spain
e-mail: saul.montes.de.oca@upc.edu

J. M. Maestre and R. R. Negenborn (eds.), *Distributed Model Predictive Control Made Easy*, Intelligent Systems, Control and Automation: Science and Engineering 69, DOI: 10.1007/978-94-007-7006-5_31, © Springer Science+Business Media Dordrecht 2014

strategy in terms of system modularity, reduced computational burden and, at the same time, the admissible loss of performance with respect to a centralized MPC (CMPC) strategy.

31.1 Introduction

Large-scale networked systems (LSNS) are very common in the modern societies to transport for example, water, electricity, gas, oil, among others. Thus, their optimal management is a subject of increasing interest that due to its social, economic and environmental impact. The leading control technique for the management of LSNS is model predictive control (MPC) [1, 16]. The main reason for its success is due to after obtaining the network dynamical model, the MPC design just consists in expressing the performance specifications through different control objectives and constraints on system variables (e.g., minima/maxima of selected process variables and/or their rates of change), which are necessary to ensure process safety and asset health. The rest of the MPC design is automatic and follows multiple approaches reported in the literature; see, e.g., [2, 6], among many others.

Traditional MPC procedures assume that all available information is centralized. In fact, a global dynamical model of the system must be available for control design. Moreover, all measurements must be collected in one location to estimate all states and to compute all control actions. However, when considering LSNS, these assumptions usually fail to hold, either because gathering all measurements in one location is not feasible, or because the computational needs of a centralized strategy are too demanding for a real-time implementation. This fact might lead to a lack of scalability. Subsequently, a model change would require the re-tuning of the centralized controller. Thus, the cost of setting up and maintaining the monolithic solution of the control problem is prohibitive. A way of circumventing these issues might be by looking into either *decentralized* or *distributed* MPC techniques, where networked local MPC controllers are in charge of the control of part of the entire system. Those techniques have became one of the hottest topics in control during the early twenty-first century, opening the door to the research towards solving new open issues and related problems of the strategy. Many works have been published in this area; see, e.g., [4, 13, 14, 17], among others.

In order to apply either decentralized or distributed MPC approaches to LSNS, there is a prior problem to be solved: the *system decomposition* into subsystems. The importance of this issue has already been noticed in classic control books addressing the decentralized control of large-scale systems, see, e.g., [15] or [5]. The decomposition of the system in subsystems could be carried out during the modeling of the process by identifying subsystems as parts of the system on the basis of physical insight, intuition or experience. But, when a large-scale complex system with many states, inputs and outputs is considered, it may be difficult, even impossible, to obtain partitions by physical reasoning. A more appealing alternative is to develop systematic methods, which can be used to decompose a given system by extracting information from its structure, which is represented as a graph. Then, this structural information can be analyzed by using methods coming from graph theory. Consequently, the problem of system decomposition into subsystems leads to the problem of graph partitioning, i.e., the decomposition of graph into subgraphs. However, the development of graph partitioning algorithms within the framework of decentralized or distributed MPC is still very incipient and available methods are currently quite limited.

In this chapter, a multi-layer decentralized MPC (ML-DMPC) approach is proposed and designed for its application to LSNS. This approach is based on the assumption that the disturbances affecting the system have a periodic behavior. Moreover, the approach is also based on the availability of both static and dynamic models of the LSNS. Hence, the optimization problem behind the controller is defined to have two layers: the former or *upper* layer, working with a sampling time Δt_1 related to the period of the system disturbances, is in charge of achieving the global objectives from a set $\mathcal{O}$ of control objectives to be fulfilled by the networked system. The latter layer, also named the *lower* layer, with a sampling time Δt_2, $\Delta t_1 > \Delta t_2$, is in charge of computing the references for the system actuators in order to satisfy the local objectives from the set of control objectives $\mathcal{O}$. The system partitioning allows to establish a hierarchical flow of information between the set $\mathcal{C}$ of local MPC controllers. Therefore, the whole proposed ML-DMPC strategy results in a centralized optimization problem for considering the global control objectives, followed by a decentralized scheme for reaching the local control objectives. The advantage of this hierarchical-like DMPC approach is the simplicity of its implementation given the absence of negotiations among controllers. To apply the proposed DMPC approach, the network is decomposed into subsystems by using a novel automatic decomposition algorithm reported in [8], which is based on graph partitioning. The proposed ML-DMPC approach is applied to a real case study: the water transport network of Barcelona (Spain). Results obtained with selected simulation scenarios show the effectiveness of the proposed ML-DMPC strategy in terms of system modularity, reduced computational burden and, at the same time, the admissible loss of performance with respect to a centralized MPC (CMPC) strategy.

This chapter is structured as follows: Sect. 31.2 describes boundary conditions on considered system, control objectives, and constraints. Section 31.3 describes the ML-DMPC approach. Section 31.4 illustrates the proposed approach in the aforementioned case study. Finally, conclusions and some directions for further research are reported in Sect. 31.5.

31.2 Boundary Conditions

31.2.1 Control-Oriented Modeling Framework

Before establishing the fundamentals of the control-oriented modeling framework proposed in this chapter, the statement of the general framework for controlling a LSNS is discussed. The control system architecture of a LSNS may be defined in two

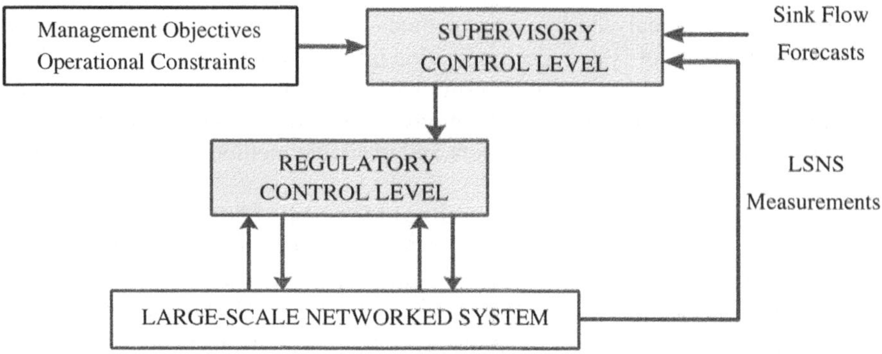

Fig. 31.1 Control architecture for a LSNS

levels as shown in Fig. 31.1. The upper level consists in a supervisory controller that is in charge of the global control of the networked system, establishing references for regulatory controllers (of PID type) at the lower level. Regulatory controllers hide the non-linear behavior of the system to the supervisory controller. This fact may allow the supervisory level to use a control-oriented model.

In a general way, a proper control-oriented model of a given system is defined such that it captures its main behaviors, being as simple as possible in order to save computational burden when such model is used for control design purposes. This chapter considers the use of the control-oriented model with a model-based optimization-based control strategy with constraints. This latter implies not only dynamic and static equations in the mathematical expression of the behavior of the system, but also inequality constraints may be added. In general, these inequalities are associated to bounds in the operational ranges of the physical variables of the system (inputs, states, and outputs). However, some of those inequality constraints may also relate system variables between them together with system disturbances.

The framework of control-oriented modeling of LSNS that is proposed in this chapter relies on the concept of *flow* between or through the constitutive elements of the system. In this framework, the flow is understood in the sense of movement of the raw material related to the use or function of the networked system. In order to have a model structure where the flow concept has sense, it is necessary to define a set of basic elements to be associated with the physical LSNS.

Storage Element: As its name indicates, this element represents the fact of storing the material/data flow, what implies a *volume* given in discrete time by the difference equation

$$x(k+1) = x(k) + \Delta t \, (q_{\text{in}}(k) - q_{\text{out}}(k)), \tag{31.1}$$

where x denotes the stored flow volume, q_{in} and q_{out} denote the net inflow and outflow, respectively; Δt is the considered sampling time and index $k \in \mathbb{Z}_{\geq 0}$ represents the discrete time instant. Notice that (31.1) adds the dynamic nature to the control-oriented model of the whole LSNS. Moreover, this element is not defined to store infinity quantity of flow, what implies a working regime bounded by the storing constraints

$$x_{\min} \leq x(k) \leq x_{\max}, \qquad \forall k, \tag{31.2}$$

where $x_{\min}$ and $x_{\max}$ are the minimum and maximum volume that the element is able to store, respectively.

Node Element: This element, also called *junction*, corresponds to a point where flows are either propagated or merged. Propagation means that the node has one inflow and some outflows. Merging means that two or more inflows are merged into a larger outflow. Thus, two types of nodes may be considered:

- Nodes with one inflow and multiple outputs (splitting nodes), i.e.,

$$q_{\text{in}}(k) = \sum_i q_{\text{out},i}(k). \tag{31.3}$$

- Nodes with multiple inputs and one output (merging nodes), i.e.,

$$\sum_j q_{\text{in},j}(k) = q_{\text{out}}(k). \tag{31.4}$$

Mixed nodes can be described from the basic ones described above, i.e., complex nodes with several inflows and outflows may be defined. Notice that this element would add static relations to the control-oriented model of the whole LSNS. However, some LSNS do not show the behavior modeled by nodes, hence static relations are not always present in the control-oriented model.

Flow source: This element provides the raw material that flows through the network. It may be considered either:

- as an exogenous inflow to the networked system. In that case, constraints such as

$$q_{\min,\Lambda_i} \leq q_{\Lambda_i}(k) \leq q_{\max,\Lambda_i} \tag{31.5}$$

  might be considered, where q_{Λ_i} denotes the inflow from the ith source; $q_{\min,\Lambda_i}$ and $q_{\max,\Lambda_i}$ correspond to minimum and maximum inflow, respectively. For simplicity and compactness of the control-oriented model, constraints in (31.5) are associated to flow handling elements (described below) directly connected to sources;
- or as an external storage element, what implies an expression for its volume $x_\Lambda(k)$ such as in (31.1), with the associated constraint such as in (31.2).

Sink: In this framework, a sink is the element where the flow goes to. From a general point of view, sinks are related to the measured disturbances of the system since they ask for flow according to a given profile. The networked system should be managed in such a way that those elements receive the flow they request.

Link: This element, also called *arc*, represents the general way of connecting two elements which share a flow, e.g., a source with a node, an storage element with a sink, etc. The flow through these elements can be constrained by the range

$$q_{\min} \leq q(k) \leq q_{\max}, \qquad \forall k, \tag{31.6}$$

where $q_{\min}$ and $q_{\max}$ are the minimum and maximum flow through a link, respectively.
Flow Handling Element: In this framework, this element manipulates flow either between storage elements or between a storage element and a node, and viceversa. Hence, flow handling elements are links where the flow is manipulated. Handling elements between storage elements and sinks as well as between nodes and sinks are not considered since the flow handled has to be equal to the flow requested from the sink and, therefore, there is no place for different options. Notice that the flow through these elements is also constrained following (31.6).

Remark 31.1 Regarding storage elements, when their outflow is not manipulated, its expression corresponds with

$$q_{\text{out}}(k) = h(x(k)), \tag{31.7}$$

where h should be determined according to the nature of the particular case study. Notice that this relation can be made more accurate (but also more complex) if h is considered to be nonlinear, thus yielding nonlinear constrained control-oriented model. This latter can be seen considering (31.7) and rewriting the right-hand side of (31.6) as

$$q(k) \leq \min\{q_{\max}, h(x(k))\}, \qquad \forall k. \tag{31.8}$$

Moreover, in the scenario where $x_{\min} \neq 0$ and the outflow of the storage element is manipulated, the left-hand side of (31.6) should be rewritten as

$$\min\{q_{\min}, h(x(k))\} \leq q(k), \qquad \forall k, \tag{31.9}$$

which also implies a non-convex constraint within the control-oriented model of the LSNS. ◇

Consider a given LSNS being represented as the interconnection of n_x storage elements, n_u flow handling elements, n_d sinks and n_q intersection nodes. The n_α sources are considered as inflows. Stating the volume in storage elements as the state variable and the flow through handling elements as the manipulated inputs, an LSNS may be generally described in state-space form by the following linear discrete-time dynamic model:

$$\mathbf{x}(k+1) = \mathbf{A}\mathbf{x}(k) + \boldsymbol{\Gamma}\boldsymbol{v}(k), \tag{31.10a}$$

$$\mathbf{0} = \mathbf{E}_x\mathbf{x}(k) + \mathbf{E}_v\boldsymbol{v}(k), \tag{31.10b}$$

where $\mathbf{x} \in \mathbb{R}^{n_x}$ corresponds to the state vector (stored volumes), $\boldsymbol{\Gamma} \triangleq [\mathbf{B} \;\; \mathbf{B}_p]$, and $\boldsymbol{v}(k) := [\mathbf{u}(k)^{\mathrm{T}} \;\; \mathbf{d}(k)^{\mathrm{T}}]^{\mathrm{T}}$. In turn, $\mathbf{u} \in \mathbb{R}^{n_u}$ is the vector of control inputs (manipulated flows), and $\mathbf{d} \in \mathbb{R}^{n_d}$ corresponds to the vector of measured disturbances (flows to sinks). Moreover, $\mathbf{A} \in \mathbb{R}^{n_x \times n_x}$, $\mathbf{B} \in \mathbb{R}^{n_x \times n_u}$, $\mathbf{B}_p \in \mathbb{R}^{n_x \times n_d}$, are state-space system matrices for mass flow balances in storage elements (31.10a), and $\mathbf{E}_x \in \mathbb{R}^{n_q \times n_x}$ and $\mathbf{E}_{\upsilon} \in \mathbb{R}^{n_q \times (n_u + n_d)}$ are matrices for static flow balances in nodes (31.10b). All vectors and matrices are dictated by the network topology. Notice that $\mathbf{E}_x = 0$ when outflows from storage elements are manipulated. In general, states and control inputs are subject to constraints of the form

$$\mathbf{x}_{\text{min}} \leq \mathbf{x}(k) \leq \mathbf{x}_{\text{max}}, \quad \forall k, \tag{31.11a}$$

$$\mathbf{u}_{\text{min}} \leq \mathbf{u}(k) \leq \mathbf{u}_{\text{max}}, \quad \forall k, \tag{31.11b}$$

where $\mathbf{x}_{\text{min}} \in \mathbb{R}^{n_x}$ and $\mathbf{x}_{\text{max}} \in \mathbb{R}^{n_x}$ denote the vectors of minimum and maximum volumes, respectively, while $\mathbf{u}_{\text{min}} \in \mathbb{R}^{n_u}$ and $\mathbf{u}_{\text{max}} \in \mathbb{R}^{n_u}$ denote the vectors of minimum and maximum flows through flow handling elements, respectively.

Remark 31.2 Notice that manipulated flows may be defined as bidirectional flows. This means that minimum flows of these manipulated links may be negative. In order to cope with this situation, a bidirectional link can be replaced with two separate unidirectional links with null minimum flow, associated with each direction of the original link. Although this approach simplifies the control setup, it might add complexity to the optimization problem related to the optimization-based controller since the number of optimization variables gets higher. ◇

31.2.2 Model Decomposition

Once the control-oriented model is stated, it is important to determine the objective of performing the partition of the networked system no matter what control strategy is followed. In this aspect, the availability of centralized information is fundamental. When all the information about the whole set of system variables is available, the partitioning gains sense from the point of view of modularity of the control architecture and the reduction of computational burden. However, when this global information is not fully available, a control topology based on the partition of the system should be designed given the *physical dispersion* the LSNS might show.

Consider again an LSNS formed by the interconnection of several elements from those proposed beforehand. The way they are interconnected is a key factor for performing the partitioning since it determines the type of variables the resultant subsystems would share. Thus, if for instance the outflow for an storage element from a subsystem S_a is not manipulated and the corresponding flow is just the inflow of a subsystem S_b, then the shared variable corresponds to a system state. Otherwise, it would be a control input.

This chapter considers the partitioning algorithm proposed in [8]. This algorithm, based on graph partitioning, aims at decomposing (31.10) into subsystems. In order to do so, the graph representation of (31.10) is determined by using the system topology, what yields its *incidence matrix* I_M. This matrix describes the connections (edges) between the graph vertices (here represented by storage elements, sources, sinks, and nodes). Once I_M is obtained from the system digraph, the problem of the decomposition into subsystems is formulated in terms of partitioning the corresponding graph into subgraphs such that all subgraphs have nearly the same number of vertices and there exist few connections between subgraphs. These features, motivated by the posterior design of a decentralized/distributed control strategy, guarantee that the obtained subgraphs have a similar size, fact that balances computations between the controllers and allows minimizing communications/interactions between them.

Thus, the overall system (31.10) is assumed to be decomposed in $M \triangleq |\mathcal{N}|$ subsystems collected in the set $\mathcal{N}$, which are not overlapped, output decentralized and input coupled (therefore, $\mathbf{E}_x = 0$). The model of the ith subsystem is stated below for $i \in \{1, \ldots, M\}$ as

$$\mathbf{x}_i(k+1) = \mathbf{A}_i \mathbf{x}_i(k) + \boldsymbol{\Gamma}_i \boldsymbol{v}_i(k) + \mathbf{B}_{\text{sh},i} \boldsymbol{\mu}_i(k), \tag{31.12a}$$

$$\mathbf{0} = \mathbf{E}_{v,i} \boldsymbol{v}_i(k) + \mathbf{E}_{\text{sh},i} \boldsymbol{\mu}_i(k), \tag{31.12b}$$

where $\mathbf{x}_i \in \mathbb{R}^{n_{x_i}}$ and $\boldsymbol{v}_i \in \mathbb{R}^{n_{u_i}+n_{d_i}}$ are the local states and inputs of the subsystem S_i, respectively, and $\boldsymbol{\mu}_i \in \mathbb{R}^{n_{\mu_i}}$ is the vector of shared inputs between S_i and other subsystems. Moreover, $\mathbf{B}_{\text{sh},i}$ and $\mathbf{E}_{\text{sh},i}$ are matrices whose dimensions depend on the number of shared inputs of S_i. The decomposition should assure that $\sum_i n_{x_i} = n_x$, $\sum_i n_{u_i} = n_u$, $\sum_i n_{d_i} = n_d$ and $\sum_i n_{q_i} = n_q$. Matrices $\mathbf{A}_i, \boldsymbol{\Gamma}_i, \mathbf{E}_{v,i}$, are dictated by each subsystem topology. In the same way, the previously defined overall constraints (31.11) are partitioned for each i-th subsystem as

$$\mathbf{x}_{\text{min,i}} \leq \mathbf{x}_i(k) \leq \mathbf{x}_{\text{max,i}}, \quad \forall k, \tag{31.13a}$$

$$\mathbf{u}_{\text{min,i}} \leq \mathbf{u}_i(k) \leq \mathbf{u}_{\text{max,i}}, \quad \forall k. \tag{31.13b}$$

Moreover, it may occur that the n_α flow sources of the LSNS determine the amount of M since the sinks (and therefore storage elements and nodes) related to each subsystem S_j, $j \in \{1, \ldots, n_\alpha\}$ are only supplied by a unique source. Therefore, this topological dependency determines subsystems around a flow source, resulting to be a natural criterion for performing system decomposition. Thus, as seen in Fig. 31.2, the initial LSNS might be decomposed in two stages. In the first stage, subsystems tied with flow sources are determined. From now on, these subsystems are called *anchored subsystems* (AS). It can be seen that there will be as many anchored subsystems as number of sources in the network. Remaining elements are associated in a resultant subsystem namely $\tilde{S}$, where storage elements might be fed from two or more flow sources. In the second stage, subsystems $\tilde{S}$ is now decomposed by following the algorithm proposed in [8]. Notice that, at this point, the shared connections of $\tilde{S}$ that correspond to inflows, my be considered as *pseudo-sources* of $\tilde{S}$. Therefore, depending on the management/control objectives related to the LSNS, it is possible to add some additional criteria to each AS outflow (or $\tilde{S}$ inflow). These criteria can be associated to a weighting factor ω, which is related to

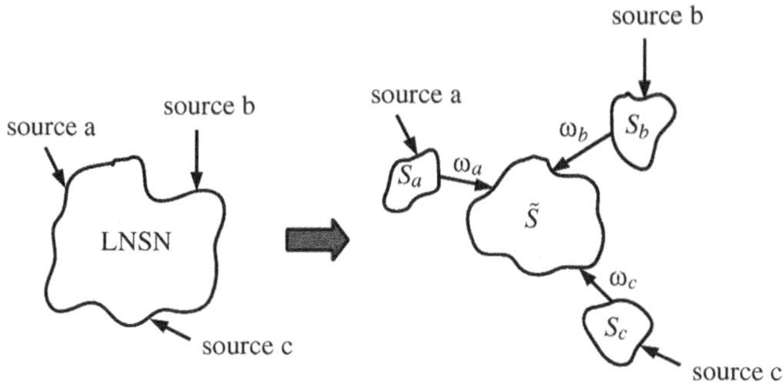

Fig. 31.2 Scheme of LSBS partitioning ($n_\alpha = 3$)

each pseudo-source of $\tilde{S}$ and would be determined within the design of the control strategy for the LSNS (see Sect. 31.3 below). Notice that a second set of pesudo-sources would appear after performing the decomposition of $\tilde{S}$, but their treatment can follow the same procedure considered for the first set of pseudo-sources.

31.2.3 MPC Problem Formulation

From the LSNS model in (31.10), let $\mathbf{u}(k : k+N_p-1)$ be the sequence of control input over a fixed-time prediction horizon N_p. Hence, the following problem is proposed.

Problem 31.1 An MPC controller design is based on the solution of the open-loop multi-objective optimization problem (OOP)

$$\min_{\mathbf{u}^*(k:k+N_p-1)} J(k) \triangleq \sum_{m=1}^{|\mathcal{O}|} \gamma_m J_m(\mathbf{u}(k : k + N_p - 1), \mathbf{x}(k + 1 : k + N_p)), \quad (31.14a)$$

subject to system model (31.10), system constraints (31.11) over N_p, and a set of n_c operative constraints given by management policies of the system and condensed on the form

$$\mathbf{G}_1\mathbf{x}(k + 1 : k + N_p) + \mathbf{G}_2\boldsymbol{v}(k : k + N_p - 1) \leq \mathbf{g}, \quad (31.14b)$$

where $J(\cdot) : \mathbb{R}^{n_u(N_p-1)\times n_x N_p} \mapsto \mathbb{R}$ in (31.14a) is the cost function collecting all control objectives of the set $\mathcal{O}$ and γ_m are positive scalar weights to prioritize the m-th control objective $\mathcal{O}_m \in \mathcal{O}$, particularly represented by J_m within the whole cost function. Moreover, $\mathbf{G}_1 \in \mathbb{R}^{n_c \times n_x N_p}$, $\mathbf{G}_2 \triangleq [\mathbf{G}_{2,u} \quad \mathbf{G}_{2,d}] \in \mathbb{R}^{n_c\times\{n_u(N_p-1)+n_d(N_p-1)\}}$, and $\mathbf{g} \in \mathbb{R}^{n_c N_p}$. □

Assuming that Problem 31.1 is feasible, i.e., there is an optimal solution given by the sequence of control inputs $\mathbf{u}^*(k : k + N_p - 1) \neq \emptyset$, and then the receding horizon philosophy sets

$$\mathbf{u}_{MPC}(\mathbf{x}(k)) \triangleq \mathbf{u}^*(k), \qquad (31.15)$$

and disregards the computed inputs from $k+1$ to $k+N_p-1$, with the whole process repeated at the next time instant $k \in \mathbb{Z}_{\geq 0}$. Expression (31.15) is known in the MPC literature as *the MPC law* [6].

Besides, the decomposition of the original problem leads to design an MPC controller $C_i \in \mathcal{C}$, with $i = \{1, \ldots, M\}$, for each of the M subsystems. This fact also leads to split the cost function (31.14a). Therefore, each subsystem considers the local cost function

$$J_i(k) = \sum_{m=1}^{|\mathcal{O}|} \gamma_{m,i} J_{m,i}(\mathbf{u}_i(k : k + N_p - 1), \mathbf{x}_i(k + 1 : k + N_p)), \qquad (31.16)$$

where $m = \{1, \ldots, |\mathcal{O}|\}$, and $\gamma_{m,i}$ are scalar weights that prioritize local objectives within each subsystem. In the same way, operational constraints may be properly split along the subsystems and expressed as

$$\mathbf{G}_{1,i}\mathbf{x}_i(k + 1 : k + N_p) + \mathbf{G}_{2,i}\upsilon_i(k : k + N_p - 1) \leq \mathbf{g}_i. \qquad (31.17)$$

31.3 Description of the Approach

31.3.1 Preliminary Assumptions

Once the control-oriented model is obtained and decomposed into subsystems, the natural step forward consists in designing the decentralized control strategy considering the given management policies and constraints. Before getting through the proposed methodology for designing such controllers based on predictive control, the following assumptions regarding the LSNS and its management are stated.

Assumption 31.1 All sinks can be supplied by at least one flow source through at least one flow path.[1]

Assumption 31.2 All sinks show a *periodic* flow request, whose period is $T = \Delta t_1$.

Assumption 31.3 The set $\mathcal{O}$ of control objectives is defined as

$$\mathcal{O} = \mathcal{O}_l \cup \mathcal{O}_g, \qquad (31.18)$$

[1] A flow path is formed by a finite set of links, which may connect sources, nodes, sinks, and storage elements.

where $\mathcal{O}_l$ corresponds with the set of *local* control objectives and $\mathcal{O}_g$ with the set of *global* control objectives. Moreover, $m_l \triangleq |\mathcal{O}_l|$, $m_g \triangleq |\mathcal{O}_g|$, and hence $m_l + m_g = |\mathcal{O}|$.

Assumption 31.3 introduces the diversity on the nature of the control objectives of the LSNS. This fact determines the way the decentralized controller is designed since the fulfillment of a global objective from a local point of view should imply information from all the LSNS, fact that is avoided when the system partitioning is performed. Therefore, it is necessary to figure out how to *transform* the formulation of a global objective in a centralized control scheme towards the statement of a set of decentralized controllers $\mathcal{C}$ considering all the control objectives in $\mathcal{O}$ in a suitable way.

First of all, in order to develop this idea, the cost function related to the centralized MPC (CMPC) in Problem 31.1 can be rewritten as

$$J(k) = \sum_{j=1}^{m_g} \gamma_j J_j(\mathbf{u}(k : k + N_p - 1), \mathbf{x}(k + 1 : k + N_p)) + \sum_{m=1}^{m_l} \gamma_m J_m(\mathbf{u}(k : k + N_p - 1), \mathbf{x}(k + 1 : k + N_p)). \tag{31.19}$$

The approach proposed in this chapter consists in designing a decentralized MPC scheme, where each controller $C_i \subset \mathcal{C}$ considers a newer version of (31.16) taking into account the structure of (31.19). Hence, the cost function related to each C_i is written as

$$J_i(k) = \sum_{j=1}^{m_g} \hat{\gamma}_{j,i} \hat{J}_{j,i}(\mathbf{u}_i(k : k + N_p - 1), \mathbf{x}_i(k + 1 : k + N_p)) + \sum_{m=1}^{m_l} \gamma_{m,i} J_{m,i}(\mathbf{u}_i(k : k + N_p - 1), \mathbf{x}_i(k + 1 : k + N_p)), \tag{31.20}$$

where $\hat{J}_{j,i}(\cdot)$ corresponds to the j-th global control objective properly expressed in order to reflect its influence in the local controller. Moreover, $\hat{\gamma}_{j,i}$ is a weight that prioritizes the global objectives that must be filled within the optimization problem.

Thus, the design of the entire control topology gives rise to a twofold optimization problem behind the general MPC topology. This twofold problem consists of two layers operating at different time scales: an *upper layer* works with a sampling time Δt_1, corresponding to the disturbance period. This layer is in charge of achieving the global objectives from a set $\mathcal{O}$ of control objectives given for the networked system. On the other hand, a *lower layer*, with a sampling time Δt_2, $\Delta t_1 > \Delta t_2$, is in charge of computing the references for the system actuators in order to satisfy the local objectives from the set of control objectives $\mathcal{O}$.

31.3.2 Upper Optimization Layer

This layer is designed to take into account the global control objectives in a proper way, i.e., considering information of the entire system in order to fulfill them. This layer is in charge of computing weights ω related to pseudo-sources and discussed in Sect. 31.2.2 (see Fig. 31.2). These weights ω will determine the prioritization weights $\hat{\gamma}_{j,i}$ in (31.20) for the controller design at each subsystem S_i. Therefore, to compute the set of ω, a CMPC problem is stated by considering: (i) a static model of the whole LSNS, and (ii) a cost function that only takes into account the global control objectives associated to the system. Regarding the system static model, the upper optimization layer works with a sampling time Δt_1, corresponding to the periodicity in the flow requested by sinks. Thus, when looking at the volume evolution of storage elements, they show the parallel behavior as the flow to the sinks, i.e., volumes also show a periodic behavior with period Δt_1. For this reason, when modeling the network at sampling time Δt_1, it can be assumed that volumes do not change, i.e., the dynamics of storage elements (31.1) are modified considering $\mathbf{x}(k+1) = \mathbf{x}(k)$. Hence, storage elements behave as nodes and the network dynamic model (31.1) becomes a static model (set of algebraic equations). Having this model and the functional

$$J_{up}(k) = \sum_{j=1}^{m_g} \gamma_j J_j(\mathbf{u}(k : k + N_p - 1), \mathbf{x}(k + 1 : k + N_p)), \tag{31.21}$$

Problem 31.1 is properly formulated in order to obtain the desired weights ω and, indeed, any weight for any arc of any path within the LSNS. To mathematically and systematically find all flow paths in an LSNS, its structure is used by means of node-arc incidence matrices, which represent both the flow balances and the graph structure [3].

31.3.3 Lower Optimization Layer

Having a decentralized predictive controller $C_i \in \mathcal{C}$ for each subsystem $S_i \in \mathcal{N}$ with a cost function as in (31.20), the shared inputs for all subsystems in $\mathcal{N}$ are written as $\boldsymbol{\mu}_{ij}$, whose directionality is defined from S_i to S_j, $i \neq j$. Additionally, $\boldsymbol{\mu}_{ij}$ not only contain values of each component at time step k but also all values over N_p.[2]

The fact of having available this complementary information of the shared variables allows to use predicted values of manipulated flows instead of starting a negotiation procedure between subsystems in order to find their value (following the distributed control philosophy). Besides, the implementation of the hierarchical

[2] This chapter considers $N_u = N_p$, where N_u denotes the control horizon. In the case that $N_u < N_p$, it is still necessary to know the values for shared variables from N_u until N_p, no matter the way they are considered (e.g., keeping constant their value at time instant N_u, make them null, etc.).

DMPC approach requires that subsystem models are modified to coordinate with other subsystems. To introduce such modification, the following concept is introduced.

Definition 31.1 *(Virtual sink).* Consider two subsystems S_1 and S_2, which share a set of manipulated flows $\boldsymbol{\mu}_{12}$. According to the notation employed here, those flows go from S_1 to S_2. If the solution sequence of optimization subproblems—defined by the pre-established hierarchical order—determines that $\boldsymbol{\mu}_{12}$ is computed by C_1, then flows in $\boldsymbol{\mu}_{12}$ are considered as *virtual sinks* in C_2 since their values are now imposed in the same way as the flow to sinks. ■

The *pure* hierarchical control scheme determines a sequence of information propagation among the subsystems, where top-down communication is available from upper to lower level of the hierarchy (see [15]). Note that, despite the subsystem coupling (given by the shared links), the main feature of the pure hierarchical control approach relies on the unidirectionality of the information flow between controllers. However, it may happen that some shared links have defined their flow direction such as bottom-up communications within the hierarchy, which breaks the mentioned unidirectional flow between DMPC controllers. This fact implies that the standard hierarchical control scheme for partitioned LSNS cannot be straight applied. To solve this situation and to design a DMPC strategy, a hierarchical-like DMPC approach, proposed by [10], has been considered and conveniently implemented over the partitioned system. This strategy follows the hierarchical control philosophy and the sequential way of solving the optimization subproblems of the corresponding MPC controllers but also considering the appearance of bidirectional information flows.

The hierarchy defined by the approach of [10] implies that the controller C_i will be allocated in a different level according to the flow request of its corresponding subsystem S_i. Considering the simple topology in Fig. 31.2, this fact means that the controller $C_{\tilde{S}}$ will be at the top of the hierarchy, while controllers C_a, C_b, and C_c will share the bottom level. All controllers work with a sampling time Δt_2 and the computational time spend by the scheme corresponds with the sum of maximum times of each hierarchical level of controllers (e.g., $\tau_{\text{total}} = \tau_{C_{\tilde{S}}} + \max(\tau_{C_a}, \tau_{C_b}, \tau_{C_c})$ for the scheme in Fig. 31.2, where τ denotes the computational time). Special considerations should be done for the treatment of bidirectional shared flows [9, 10].

31.3.4 Interaction of Layers

The sharing of information between layers depends on the nature and features of each application. In general, the computational time that the upper layer spends is quite low with respect to the computational time of the lower layer. This fact is due to the difference in the nature of the models handled by each layer and the interactions given by the DMPC controllers as well as their amount and disposition within the defined hierarchy. Once the optimization problem related to the upper layer is solved, the resultant parameters are properly updated for each optimization problem behind

Algorithm 31.1: ML-DMPC Approach

k=0
loop
  set $x(k)$
  $(\omega, \tilde{\gamma}) \Leftarrow$ **solve** Problem 1 with (31.21)
  while $\frac{k}{\Delta t_1} \notin \mathbb{Z}$
    $u_{\mathrm{MPC},i} \Leftarrow$ **solve** Problem 1 with (31.20) and using $\omega, \tilde{\gamma}$
  end loop
end loop

each $C_i \in C$. This updating is performed with a periodicity Δt_1. Algorithm 31.1 collects the main steps of the proposed ML-DMPC approach.

31.4 Application Results

31.4.1 Case-study Description

The approach presented in this work is assessed with a case study of a real large-scale system, specifically the Barcelona drinking water network (DWN). This network supplies potable water to the Metropolitan Area of Barcelona (Catalunya, Spain). In general the water network operates as a pull interconnected system driven by endogenous and exogenous flow demands; different hydraulic elements are used to collect, store, distribute and serve drinking water to the associated population. For further details about this network, the reader is referred to [7].

31.4.1.1 System Management Criteria

The operational goals in the management of the Barcelona DWN have been provided by AGBAR due their knowledge of the system. These policies are of three kinds: *economic*, *safety*, *smoothness*, and are respectively stated as follows:

1. To provide a reliable water supply in the most economic way, minimizing water production and transport costs,
2. To guarantee the availability of enough water in each reservoir to satisfy its underlying demand, keeping a safety stock to face uncertainties and avoid stock-outs.
3. To operate the transport network under smooth control actions.

These objectives are assessed by minimizing the following performance indices[3]:

[3] The performance indices considered in this work may vary or generalized with the corresponding manipulation to include other control objectives.

$$J_E(k) \triangleq |(\boldsymbol{\alpha}_1 + \boldsymbol{\alpha}_2(k))^{\mathrm{T}} \mathbf{u}(k)|, \tag{31.22a}$$

$$J_S(k) \triangleq \|\boldsymbol{x_i}(k)\|^2, \tag{31.22b}$$

$$J_U(k) \triangleq \|\Delta \mathbf{u}(k)\|^2, \tag{31.22c}$$

where $J_E \in \mathbb{R}_{\geq 0}$ represents the economic cost of network operation taking into account water production cost $\boldsymbol{\alpha}_1 \in \mathbb{R}^{n_u}$ and water pumping cost $\boldsymbol{\alpha}_2 \in \mathbb{R}^{n_u}$ which change every time instant k according to the variable electric tariff; $J_S \in \mathbb{R}_{\geq 0}$ is a performance index which penalizes the amount of volume $\boldsymbol{\xi} \triangleq \min\{\mathbf{0}, \mathbf{x} - \mathbf{x}_s\} \in \mathbb{R}^{n_x}$ that goes down from $\mathbf{x}_s$, a predefined safety volume threshold; $J_U \in \mathbb{R}_{\geq 0}$ represents the penalization of control signal variations $\Delta \mathbf{u}(k) \triangleq \mathbf{u}(k) - \mathbf{u}(k-1)$ to extend actuator life and assure a smooth operation; and $\|\cdot\|$ is the Euclidean norm, i.e., $\|\mathbf{z}\| = \sqrt{\mathbf{z}^{\mathrm{T}}\mathbf{z}}$. More details about the management criteria of this case study can be found in [7].

31.4.1.2 Control-Oriented Modelling

Consider a DWN being represented as the interconnection of n_x tanks, n_u actuators (pumps and valves), n_d sectored demands and n_q intersection nodes; according to Sect. 31.2.1, this system can be generally described in state-space form by (31.10), where $\mathbf{x} \in \mathbb{R}^{n_x}$ is the state vector of water stock volumes in m^3, $\mathbf{u} \in \mathbb{R}^{n_u}$ is the vector of manipulated flows in m^3/s, and $\mathbf{d} \in \mathbb{R}^{n_d}$ corresponds to the vector of disturbances (sectored water demands) in m^3/s. In the particular case of the Barcelona DWN, the outflows from storage elements are manipulated, hence, $\mathbf{E}_x = 0$ in (31.10b).

The states and control inputs are subject to (31.11); this polytopic hard constraints are due to the physical limits in tanks (minimum and maximum volume capacities) and the operational limits in actuators (minimum and maximum flow capacities). For safety and service reliability, in the Barcelona DWN states are also subject to soft constraints

$$\mathbf{x}(k) \geq \mathbf{x}_s(k) - \boldsymbol{x_i}(k) \geq \mathbf{0}, \quad \forall k, \tag{31.23}$$

where $\mathbf{x}_s \in \mathbb{R}^{n_x}$ is a vector of safety volume thresholds in m^3, estimated empirically, above which is desired to keep the reservoirs to avoid stock-outs, and $\boldsymbol{x_i} \in \mathbb{R}^{n_x}$ represents the amount of volume in m^3 that goes down from the desired safety thresholds.

The Barcelona DWN model contains a total amount of 63 tanks and 114 manipulated actuators. Moreover, the network has 88 demand sectors and 17 pipes intersection nodes. Both the demand episodes and the network calibration/simulation set-up are provided by AGBAR. See the aforementioned references for further details of DWN modeling and specific insights related to this case study.

31.4.2 ML-DMPC Setup

This section presents the results of applying the proposed ML-DMPC approach to the partitioned model of the Barcelona DWN developed in [9]. Thus, the overall system is assumed to be composed of six subsystems which are non-overlapped, output-decentralized and input-coupled (see Fig. 31.3). The model of each subsystem is obtained for $i \in \{1, \ldots, 6\}$ following sect. 31.2.2 and expressed by (31.12). In the same way, the hard constraints of the overall DWN are partitioned and expressed by (31.13), while for each i-th subsystem the safety constraints are expressed by

$$\mathbf{x}_i(k) \geq \mathbf{x}_{s,i}(k) - \boldsymbol{x}_{\boldsymbol{i}\,i}(k) \geq \mathbf{0}, \quad \forall k. \tag{31.24}$$

The decomposition of the original problem also leads to split the cost function. Therefore, each subsystem will be solving, at each time step, the following local multi-objective optimization problem:

$$J_i^*(k) \triangleq \min_{\mathbf{u}_i^*(k:k+N_p-1)} \rho_i \left(\gamma_1 J_{E,i} + \gamma_2 J_{S,i} + \gamma_3 J_{U,i}\right), \tag{31.25}$$

where $J_{E,i} \triangleq \sum_{l=0}^{N_c-1} (\boldsymbol{\alpha}_{1,i} + \alpha_{2,i}(k+l))\mathbf{u}_i(k+l)$ is the economic objective, $J_{S,i} \triangleq \sum_{l=1}^{N_p} \|\boldsymbol{x}_{\boldsymbol{i}\,i}(k+l)\|^2$ is the safety objective, $J_{U,i} \triangleq \sum_{l=0}^{N_c-1} \|\Delta\mathbf{u}_i(k+l)\|^2$ is the smoothness objective, N_p, $N_c \in \mathbb{Z}_{\geq 0}$ are the prediction and control horizon respectively, ρ_i is a positive scalar weight to prioritize subsystems, γ_1, γ_2, and γ_3 are positive scalar weights to prioritize each objective in the aggregate local cost function, l is the time step within the receding horizon, and $\mathbf{u}_i$, $x_{i\,i}$ and $\Delta\mathbf{u}_i$ are the ith subsystem local variables previously defined. It can be noticed in Fig. 31.4, in a more compact way, the resulting subsystems and the important couplings between them including their direction. Instead of neglecting the effect of this shared links as classic pure decentralized control schemes do, the multi-layer hierarchical coordination described in Sect. 31.3 is applied here.

The results obtained by applying the ML-DMPC are contrasted with those of applying a CMPC approach and a non multi-layer DMPC strategy proposed in [9]. For this case study, the optimization scheme follows Sect. 31.3, resulting in a *bi-layer* problem which is set up as follows:

- First, the *upper layer* works with a daily time scale and it is in charge of achieving the optimal water source selection. This layer, named *Daily Centralized Control* is a centralized optimization problem with time step $\Delta t = 24$ h, which minimizes the cost function (31.22a) subject to a daily model of the DWN represented by $\mathbf{x}(k+1) = \mathbf{x}(k)$, due to the periodic behavior of states at this time layer, and to constraints (31.11) and (31.23). The objective of this upper layer is to determine and fix in an appropriate way the unitary costs of the critical shared variables that act as sources in the partitioned model, in order to enforce the global economic objective by sequentially coordinating subsystems, allowing them to solve their own problems and achieving the solution of the original system.

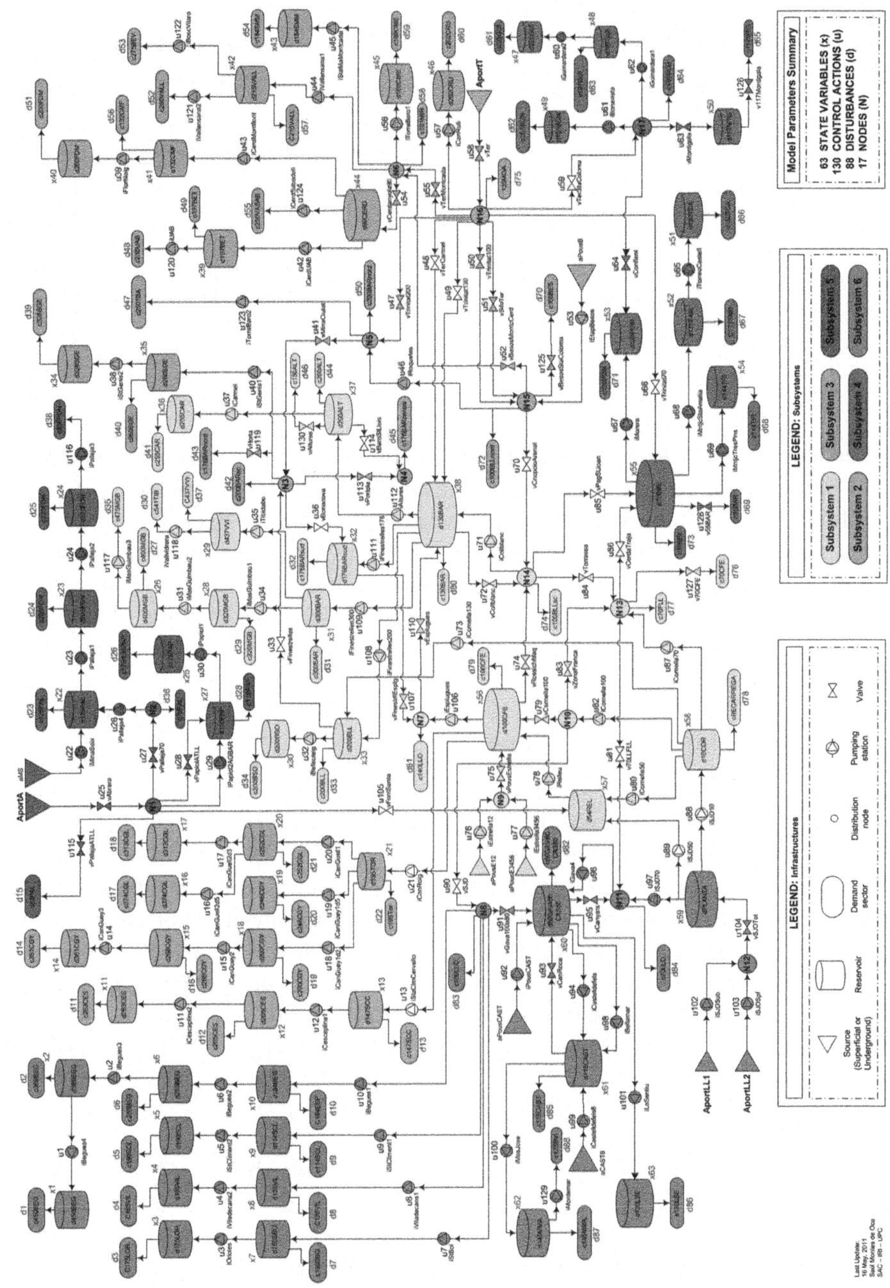

Fig. 31.3 Partition of the Barcelona DWN

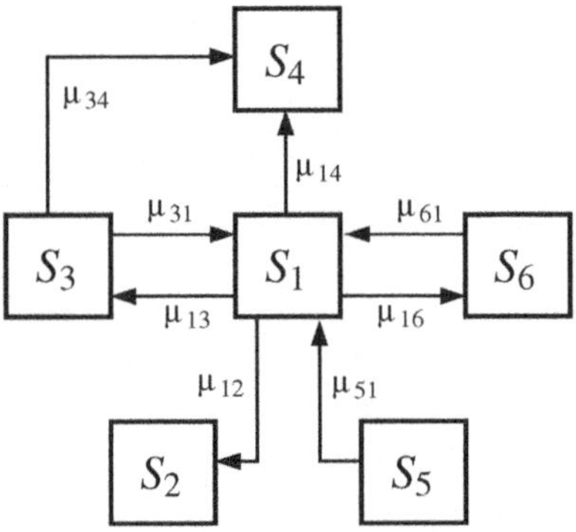

Fig. 31.4 Network subsystems S_i and their sets of shared connections $\boldsymbol{\mu}_{ij}$

- Second, the *lower layer* works with an hourly time scale to cope with the DMPC of the original problem. This layer, named *Hourly Decentralized MPC Control* follows the hierarchical coordination scheme proposed in Sect. 31.3 to perform the minimization of the local cost functions (31.25) subject to (31.12), (31.13), and (31.24), in order to obtain the control policies to operate the DWN and achieve the desired performance. In this hourly layer, following the criterion of the DWN management company, each local MPC controller works with common prediction and control horizons $N_p = N_c = 24$ h. The weights of the cost function (31.25) are $\rho_{1:6} = 1$, $\gamma_1 = 100$, $\gamma_2 = 10$ and $\gamma_3 = 0.005$. See [9] for details on the hierarchical DMPC solution sequence.

The results are obtained for 72 h (July 24-26, 2007). Simulations have been carried out using Matlab® 7.1 (R14SP3). The computer used to run the simulations is a PC Intel® Core™ 2 running at 2.4 GHz with 4 GB of RAM. The tuning of design parameters has been done in a way that the highest priority objective is the economic cost, which should be minimized while maintaining adequate layers of safety volume and control action smoothness. In order to implement the ML-DMPC approach, the demand forecasting algorithms presented in [7, 12] are used to calculate the disturbance vector involved in each control problem. For more details about the twofold-layer optimization problem applied to the Barcelona DWN, the reader is referred to [11].

31.4.3 Main Results

The results of the CMPC, DMPC and ML-DMPC strategies applied to the Barcelona DWN are summarized in Table 31.1 in terms of computational burden and of economic cost as a global management performance indicator. For each MPC approach, the computation time (in seconds) and the water, electric and total cost in economic units (e.u.), is detailed. It can be noticed that an increment of nearly 30 % of the total costs of operation occurs when using the non multi-layer hierarchical DMPC

Table 31.1 Performance comparisons

Index	CMPC	DMPC	ML-DMPC
Water cost	93.01	205.55	97.11
Electric cost	90.31	34.58	87.53
Total cost	183.33	240.13	184.65
CPU time	1143	537	540

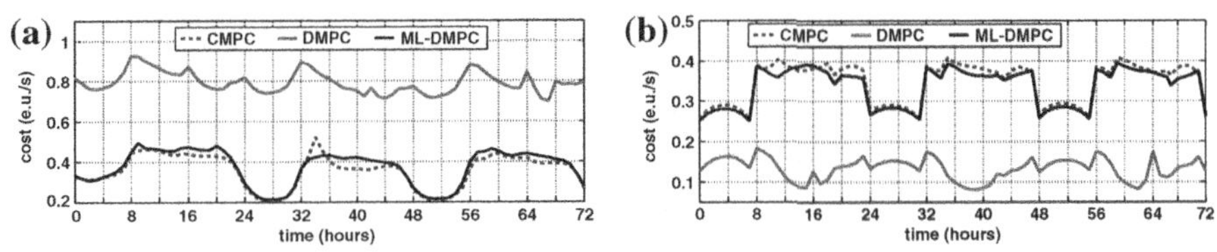

Fig. 31.5 Economic costs of the three MPC strategies. **a** Water cost. **b** Electric cost

strategy with respect to the CMPC baseline. Despite the lower electric costs, the loss of performance in the overall cost is due to the specialized behavior of local MPC controllers to solve their own optimization problems without knowing the real water supply cost of using shared resources with the neighbors. In contrast, the ML-DMPC outperforms the DMPC results by including the bi-layer optimization which allows to propagate the water cost of sources related with neighbor subsystems to the shared links thanks to the daily centralized control layer. With this ML-DMPC approach the level of sub-optimality is very low comparing with the CMPC strategy, i.e., total costs are very similar, but the computational burden is reduced. For this particular application, the computation time of the three approaches is able to satisfy the real-time constraint since the control sampling time is 1 h. Thus, the main motivation for using ML-DMPC is the scalability and easy adaptability of the sub-models if network changes, as well as the modularity of the control policy that leads to face some malfunction/fault without stopping the overall supervisory MPC strategy.

Due the difference of price between water sources and the impact of electric costs on the overall economic performance, the CMPC and ML-DMPC strategies decide to use more water from the Llobregat source despite the consequent pumping of more water through the network (see Fig. 31.6), but achieving a lower total cost, while the hierarchical DMPC decides to exploit in each subsystem their own water source (which could be expensive) and minimize the pumping operation cost. Figure 31.5 shows in detail the evolution of water cost and electric cost, respectively. These results confirm the improvement obtained by including an upper layer optimization to coordinate the local MPCs and face the lack of communication when solving their problems in a tractable way.

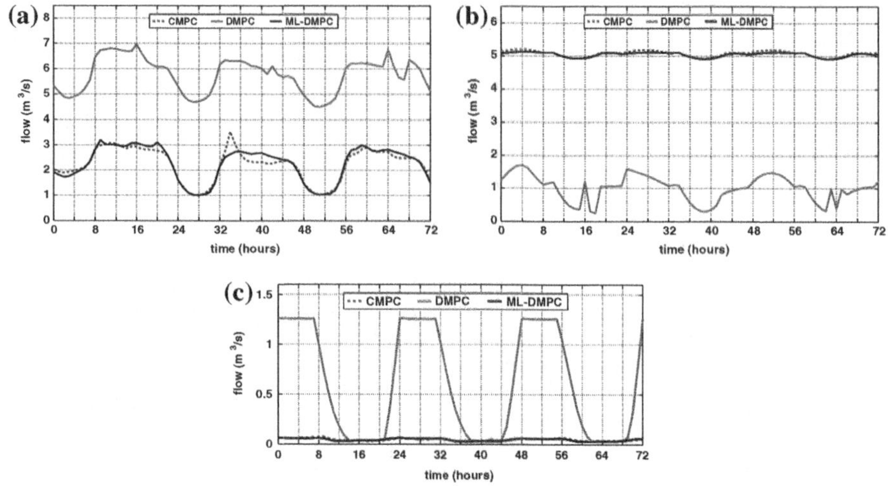

Fig. 31.6 Total flow per water source in the Barcelona DWN. **a** Ter source. **b** Llobregat source. **c** Abrera source

31.5 Conclusions and Future Research

This paper has proposed a multi-layer DMPC approach for large-scale networked systems. The upper layer, working with a larger time scale, is in charge of achieving the global control (in general related to an optimal economic cost). On the other hand, the lower layer is in charge of determining the set-point of the actuators to satisfy the local management/control objectives. The system decomposition is based on graph partitioning theory. Results obtained on selected simulation scenarios has shown the effectiveness of the control strategy in terms of system modularity, reduced computational burden and, at the same time, the reduced loss of performance in contrast to a CMPC strategy and a hierarchical-like DMPC strategy previously presented by the authors. Future work is focused on the formalization of the proposed approach in terms of feasibility, robustness and stability.

Acknowledgments This work has been partially funded by the Spanish Ministry of Science and Technology through the WATMAN project (DPI2009-13744).

References

1. M. Brdys, B. Ulanicki, *Operational Control of Water Systems: Structures, Algorithms and Applications* (Prentice Hall International, UK, 1994)
2. E.F. Camacho, C. Bordons, *Model Predictive Control*, 2nd edn. (Springer, London, 2004)
3. W.C. Cheng, N.S. Hsu, W.M. Cheng, W.W.G. Yeh, A flow path model for regional water distribution optimization. Water Resour. Res. **45**, 1–12 (2009)

4. T. Keviczky, F. Borrelli, G.J. Balas, Decentralized receding horizon control for large scale dynamically decoupled systems. Automatica **42**(12), 2105–2115 (2006)
5. J. Lunze, *Feedback Control of Large-Scale Systems* (Prentice Hall, Great Britain, 1992)
6. J.M. Maciejowski, *Predictive Control with Constraints* (Prentice Hall, Great Britain, 2002)
7. C. Ocampo-Martinez, V. Puig, G. Cembrano, R. Creus, M. Minoves, Improving water management efficiency by using optimization-based control strategies: the Barcelona case study. Water Sci. Technol. Water Supply **9**(5), 565–575 (2009)
8. C. Ocampo-Martinez, S. Bovo, V. Puig, Partitioning approach oriented to the decentralised predictive control of large-scale systems. J. Process Control **21**(5), 775–786 (2011)
9. C. Ocampo-Martinez, V. Puig, S. Bovo, Decentralised MPC based on a graph-partitioning approach applied to Barcelona drinking water network. in *Proceedings of the IFAC World Congress*, Milano (Italy), 2011
10. C. Ocampo-Martinez, D. Barcelli, V. Puig, A. Bemporad, Hierarchical and decentralised model predictive control of drinking water networks: application to the Barcelona case study. IET Control Theory Appl. **6**(1), 62–71 (2012)
11. V. Puig, C. Ocampo-Martinez, S. Montes de Oca, Hierarchical temporal multi-layer decentralised mpc strategy for drinking water networks: application to the Barcelona case study, in *Proceedings of the IEEE Mediterranean Control Conference*, Barcelona (Spain), 2012
12. J. Quevedo, V. Puig, G. Cembrano, J. Blanch, Validation and reconstruction of flow meter data in the Barcelona water distribution network. Control Eng. Pract. **11**(6), 640–651 (2010)
13. J.B. Rawlings, B.T. Stewart, Coordinating multiple optimization-based controllers: new opportunities and challanges. J. Process Control **18**(9), 839–845 (2008)
14. R. Scattolini, Architectures for distributed and hierarchical model predictive control: a review. J. Process Control **19**(5), 723–731 (2009)
15. D.D. Šiljak, *Decentralized Control of Complex Systems* (Academic Press, New York, 1991)
16. P.J. Van Overloop, *Model Predictive Control on Open Water Systems* (Delft University Press, Delft, The Netherlands, 2006)
17. A.N. Venkat, I.A. Hiskens, J.B. Rawlings, S.J. Wright, Distributed MPC strategies with application to power system automatic generation control. IEEE Trans. Control Syst. Technol. **16**(6), 1192–1206 (2008)

Chapter 32
Distributed MPC Using Reinforcement Learning Based Negotiation: Application to Large Scale Systems

B. Morcego, V. Javalera, V. Puig and R. Vito

Abstract This chapter describes a methodology to deal with the interaction (negotiation) between MPC controllers in a distributed MPC architecture. This approach combines ideas from Distributed Artificial Intelligence (DAI) and Reinforcement Learning (RL) in order to provide a controller interaction based on negotiation, cooperation and learning techniques. The aim of this methodology is to provide a general structure to perform optimal control in networked distributed environments, where multiple dependencies between subsystems are found. Those dependencies or connections often correspond to control variables. In that case, the distributed control has to be consistent in each subsystem. One of the main new concepts of this architecture is the negotiator agent. Negotiator agents interact with MPC agents to reach an agreement on the optimal value of the shared control variables. The optimal value of those shared control variables has to accomplish a common goal, probably incompatible with the specific goals of each partition that share the variable. Two cases of study are discussed, a small water distribution network and the Barcelona water network. The results suggest that this approach is a promising strategy when centralized control is not a reasonable choice.

32.1 Introduction

When dealing with LSS and distributed MPC approaches to control them, one of the main concerns is to handle the interactions between subsystems. The goal of the research described in this chapter is to propose and study negotiation methods to solve that problem. Those negotiation methods are based on learning techniques, namely Q-learning, which emerges from the RL theory. That way, it is possible to exploit the attractive features of MPC, such as meaningful objective functions and

B. Morcego (✉) · V. Javalera · V. Puig · R. Vito
Advanced Control Systems Group, Terrassa, Spain
e-mail: bernardo.morcego@upc.edu

J. M. Maestre and R. R. Negenborn (eds.), *Distributed Model Predictive Control Made Easy*, Intelligent Systems, Control and Automation: Science and Engineering 69,
DOI: 10.1007/978-94-007-7006-5_32,

constraints, in a distributed implementation combining learning techniques to perform the negotiation of the variables in a cooperative Multi Agent (MA) environment and over a Multi Agent platform.

Negotiation between distributed controllers in LSS is an open issue. Conventional negotiation techniques are not suitable for many reasons: calculation time, problems handling multiple restrictions and multiple objectives and the impossibility to ensure convergence are the most common reasons. Although there are successful results, there is a need of a methodology that can be used for all kinds of continuous LSS. One of the most accepted techniques is the augmented Lagrangian method. The seminal Tamura coordination method was discussed in the book [2] even before MPC was first introduced. This method is also based on using augmented Lagrangian technique to negotiate values on overlapping sub-networks in distributed LSS. Recent works have applied this method [4, 5, 8]. The main problems with this method are calculation time and the impossibility to ensure convergence. Another negotiation approach is presented in [15], where an iterative, cooperating method for linear discrete-time systems is presented. In particular, the proposed approach guarantees the attainment of the global (Pareto) optimum when the iterative procedure converges, but still ensures closed-loop stability and feasibility if the procedure is stopped at any intermediate iteration. The main problem of this approach is again the calculation time. In [12], an alternative approach to solve the same problem was discussed. The novelty of that approach involves maintaining the distributed structure of all the local controllers, but changing the objective functions so that the local agents cooperate. The main problem with this approach is that there is no systematic method to adjust the objective functions.

All the ideas and the proposed algorithm presented here are integrated in the Multi Agent MPC (MAMPC) Architecture described in [7]. This architecture also provides a methodology that helps in the development process of the distributed MPC controller. The main elements of the MAMPC architecture are MPC Agents and Negotiator Agents.

32.2 Definition of the MAMPC Architecture

32.2.1 System Description and Partitioning

In order to control an LSS in a distributed way, some assumptions have to be made on its dynamics, i.e. on the way the system behaves. Assume first that the system can be divided into n subsystems, where each subsystem consists of a set of nodes and the interconnections between them. The problem of determining the partitions of the network is not addressed here; instead the reader is referred to [9, 10]. The set of partitions should be complete. This means that all system state and control variables should be included at least in one of the partitions.

Definition 32.1 *System partitions.* $\mathcal{N}$ is the set of system partitions and is defined by

$$\mathcal{N} = \{n_1, n_2, \ldots, n_{nn}\} \tag{32.1}$$

where each system partition (subsystem) n_i is described by a deterministic linear time-invariant model that is expressed in discrete-time as follows

$$\begin{aligned} \mathbf{x}_i(k+1) &= \mathbf{A}_i\mathbf{x}_i(k) + \mathbf{B}_{u,i}\mathbf{u}_i(k) + \mathbf{B}_{d,i}\mathbf{d}_i(k) \\ \mathbf{y}_i(k) &= \mathbf{C}_i\mathbf{x}_i(k) + \mathbf{D}_{u,i}\mathbf{u}_i(k) + \mathbf{D}_{d,i}\mathbf{d}_i(k) \end{aligned} \tag{32.2}$$

where variables **x**, **y**, **u** and **d** are the state, output, input and disturbance vectors, respectively; **A**, **B**, **C** and **D** are the state, output, input and direct matrices, respectively. Subindices u and d refer to the type of inputs the matrices model, either control inputs or exogenous inputs (disturbances). Control variables are classified as internal or shared.

Definition 32.2 *Internal Variables.* Internal variables are control variables that appear in the model of only one subsystem in the problem. The set of internal variables of a partition i is defined by

$$\mathcal{U}_i = \{u_1, u_2, \ldots, u_{ni}\} \tag{32.3}$$

Definition 32.3 *Shared Variables.* Shared variables are control variables that appear in the model of at least two subsystems in the problem. Their values should be consistent in the subsystems they appear. They are also called negotiated variables because their values are obtained through a negotiation process. $\mathcal{V}_{ij}$ is the set of negotiated variables between partitions i and j, defined by

$$\mathcal{V}_{ij} = \{v_1, v_2, \ldots, v_{nij}\} \tag{32.4}$$

Each subsystem i is controlled by an MPC controller using:

- the model of the dynamics of subsystem i given by Eq. (32.2);
- the measured state $x_i(k)$ of subsystem i;
- the exogenous inputs $d_i(k+1)$ of subsystem i over a specific horizon of time;

As a result each MPC controller calculates directly the internal control actions, $u_i(k)$, of subsystem i. The shared variables are set by the negotiator agent. There is one agent in charge of each system partition and its duties are to negotiate the shared variables with other agents and to calculate the control actions from the MPC formulation of its partition.

32.2.2 Elements of the MAMPC Architecture

The main elements of the MAMPC architecture are MPC Agents and Negotiator Agents.

Definition 32.4 *MPC Agent.* An MPC Agent is the entity that is in charge of controlling one specific partition of the system. There is one MPC Agent for each system partition. The MPC agent solves an MPC control problem considering the internal variables of the partition and cooperating with one or more negotiator agents to determine the optimal value of the shared variables. $\mathcal{CA}$ is the set of MPC agents and is defined by

$$\mathcal{CA} = \{ca_1, ca_2, \ldots, ca_{nca}\} \tag{32.5}$$

Definition 32.5 *Negotiator Agent.* A Negotiator Agent is the entity that is in charge of determining the value of one or more shared variables between two MPC agents. A negotiator agent exists for every pair of MPC agents that have one or more shared variables in common. Each MPC agent is arranged to cooperate so that the negotiator agent solves the optimization of a common goal by means a reinforcement learning algorithm. This algorithm is based on the Q-learning algorithm, and adapted to be applied in dynamical environments. $\mathcal{NA}$ is the set of negotiator agents and is defined by

$$\mathcal{NA} = \{na_1, na_2, \ldots, na_{nna}\} \tag{32.6}$$

The cooperative behaviour of MPC agents is a basic issue in the proposed approach. Three actions are necessary to behave in such a cooperative way:

- To provide the data required by the negotiator agent.
- To accept the value(s) provided by the negotiator agent of its shared variable(s).
- To solve the MPC control problem of its partition, adjusting the value(s) of its shared control variable(s) in order to coordinate the solution of the negotiation.

Each negotiator agent determines the optimal value of one or more shared variables in the set $\mathcal{V}$. Each shared variable is an optimization problem. This problem is solved as a whole looking for the optimal value of the relation. The method is based on the reinforcements given at each step and on the experience obtained. This experience is stored in a knowledge base (the so called Q-table), one for each negotiation variable. The philosophy of the proposed negotiation algorithm is to consider the shared variables as belonging to a single problem with a single goal, instead of two different problems with conflicting goals. The negotiator agent solves the optimization problem for that variable and communicates the result to the MPC agents at each sampling time. Then, MPC agents set those values as hard constraints in their respective internal control variables and recalculate their multivariable control problem. The optimization algorithm of the negotiator agent is based on its experience and on maximizing the reinforcements received at every action taken in the past on similar situations. The internal architecture of a negotiator agent comprises the following elements:

- A knowledge base (Q-table); a Q-table contains the knowledge acquired by the agent; there is a Q-table for each shared variable because each one can have different behavior and even different goals.
- A communication protocol that allows it to have bi-directional communication with two MPC agents.
- A negotiation algorithm.

Definition 32.6 *Nodes.* A node is the physical device (commonly a computer) in which the agents are located. $\mathcal{W}$ is the set of nodes defined by

$$\mathcal{W} = \{w_1, w_2, \ldots, w_{nw}\} \tag{32.7}$$

There is a node for each MPC Agent. Nodes are communicated via some communication infrastructure (LAN, WAN or Internet).

Definition 32.7 *Agent Platform.* The agent platform works as a virtual machine providing the agents a homogenous medium to communicate and providing the user a way to manage agents. The agent platform is denoted by b. This platform has to be installed and running in all nodes.

Definition 32.8 *MAMPC Architecture.* The MAMPC distributed control architecture is defined as:

$$\mathcal{G} = \{\mathcal{CA}, \mathcal{NA}, \mathcal{N}, \mathcal{W}, \mathcal{V}, \mathcal{U}, b\} \tag{32.8}$$

where: $\mathcal{V}$ is the set formed by all sets of shared variables and $\mathcal{U}$ is the set formed by all sets of internal variables.

32.3 MAMPC Methodology

A methodology has been developed to assign all the elements of the MAMPC architecture, given a system to be controlled. This methodology comprises the analysis and design phases, in which the static elements are defined (agents, nodes, etc.) and the learning and exploitation phases in which the dynamic elements (states, control actions, etc.) are calculated and optimized. A small drinking water network is used to exemplify the proposed MAMPC methodology and its performance. The example was proposed in [1] where a centralized and a decentralized solution was studied and compared. This hypothetical water distribution network has 8 states (tanks) and 11 control variables (valves), see Fig. 32.1.

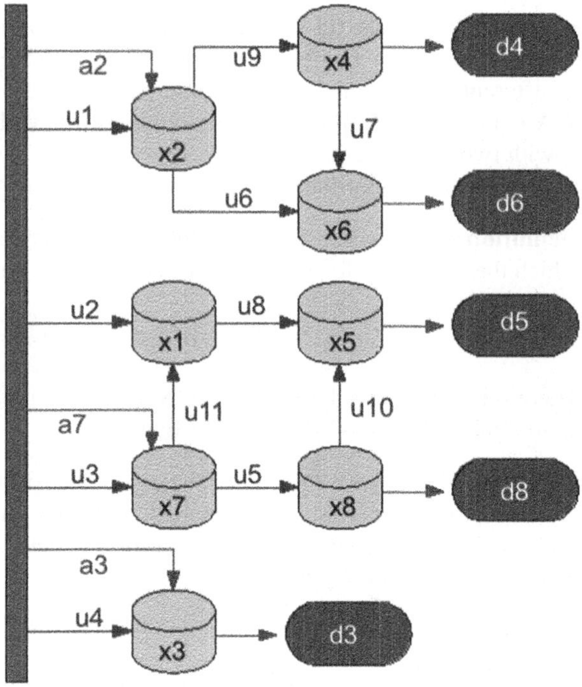

Fig. 32.1 Small water network

32.3.1 Analysis Phase

In the analysis phase, the MAMPC architecture is defined. This phase comprises the following tasks: Partitioning of the network, definition of the architecture and definition of the MPC problem.

32.3.1.1 Partitioning of the Network

The system is decomposed in subsystems. This problem is not addressed here. There are several algorithms to partition a network according to a specific criterion, see [13] for examples. An important step is to check that the partitioning of the plant leads to a complete set of partitions. This is accomplished verifying that all the state variables in the problem are considered in one partition and all the control variables belong to the set of shared or internal variables. The following relation has to be verified:

$$\mathcal{P} = \mathcal{X} \cup \mathcal{V} \cup \mathcal{U} \tag{32.9}$$

In the example, the plant is described by the following set of variables

$$\mathcal{P} = \{x_1, x_2, x_3, x_4, x_5, x_6, x_7, x_8, u_1, u_2, u_3, u_4, u_5, u_6, u_7, u_8, u_9, u_{10}, u_{11}\} \tag{32.10}$$

and the proposed partitioning is

$$\mathcal{X} = \mathcal{X}_1 \cup \mathcal{X}_2, \mathcal{X}_1 = \{x_1, x_2, x_4, x_5, x_6\}, \quad \mathcal{X}_2 = \{x_3, x_7, x_8\} \tag{32.11}$$

$$\mathcal{V}_{12} = \{u_{10}, u_{11}\} \tag{32.12}$$

$$\mathcal{U} = \mathcal{U}_1 \cup \mathcal{U}_2, \ \mathcal{U}_1 = \{u_1, u_2, u_6, u_7, u_8, u_9\}, \ \mathcal{U}_2 = \{u_3, u_4, u_5\} \tag{32.13}$$

which verifies (32.9).

32.3.1.2 Definition of the Architecture

In this step, the MAMPC Architecture is defined. Considering the definition of the architecture in (32.8) and the partition obtained in the previous step it is straightforward.
In the example, two MPC agents and one negotiator agent are needed, the remaining elements are defined as follows:

$$\mathcal{CA} = \{ca_1, ca_2\}, \ \mathcal{NA} = \{na_1\}, \ \mathcal{W} = \{w_1, w_2\} \tag{32.14}$$

32.3.1.3 Definition of the MPC Problem

The optimization goal (or goals) and the problem restrictions are defined on a centralized fashion. The goal and the problem restrictions are global but then they are transferred to the MPC agents domain to set their control goals up. The optimization goal is also the metric used by negotiator agents to guide the learning phase and to determine negotiated values during the exploitation phase.
In the example, the global optimization goal is to keep the water volume in tanks as close as possible to the reference ($x_{ref} = 3\,\text{m}^3$) during the prediction horizon:

$$J = \min \sum_{n=1}^{|\mathcal{X}|} J_i \tag{32.15}$$

The optimization goal of each MPC agent is:

$$J_i = \min \sum_{k=1}^{H_p} \left(\sum_{n \in \mathcal{X}_i} \left(x_n(k) - x_{n,ref}\right)^2 \right) \tag{32.16}$$

The problem restrictions are the following: the maximum volume in tanks is $20\,\mathrm{m}^3$; the control value of the measured variables is from 0.0 to 0.4 except for u_2 that is from 0.0 to 0.1; the sampling time is 1 h and the prediction horizon is $H_p = 24\,\mathrm{h}$; the demands are considered measured perturbations and they typically present a periodic behavior that is repeated daily.

32.3.2 Design Phase

In the design process, the subproblems of every MPC agent and negotiator agent are formulated. The problem formulation is based on the information gathered in the analysis phase which describes the internal structure of MPC Agents. In this step, all the MPC parameters and requirements have to be defined for every agent. They are: the plant; measured, non-measured and manipulated variables; limits and constraints; negotiation variables, set as restrictions; references, optimization goals; prediction horizon; control horizon; initial states and perturbation models. In the example, those parameters corresponding to MPC agent ca_1 are:

- *plant dynamics*:

$$A = C = I_5,\ D = 0,\ B_u = \begin{bmatrix} 0 & 1 & 0 & 0 & -1 & 0 & 0 & 1 \\ 1 & 0 & -1 & 0 & 0 & -1 & 0 & 0 \\ 0 & 0 & 0 & -1 & 0 & 1 & 0 & 0 \\ 0 & 0 & 0 & 0 & 1 & 0 & 1 & 0 \\ 0 & 0 & 1 & 1 & 0 & 0 & 0 & 0 \end{bmatrix},\ B_d = \begin{bmatrix} 0 & 0 & 0 \\ 0 & 0 & 0 \\ 0 & -1 & 0 \\ -1 & 0 & 0 \\ 0 & 0 & -1 \end{bmatrix}$$

- *states*: local states $\{x_1, x_2, x_3, x_4, x_5\}$ correspond to global states $\mathcal{X}_1 = \{x_1, x_2, x_4, x_5, x_6\}$
- *control actions*: local inputs $\{u_1, u_2, u_3, u_4, u_5, u_6\}$ correspond to global inputs $\mathcal{U}_1 = \{u_1, u_2, u_6, u_7, u_8, u_9\}$
- *shared variables*: local inputs $\{u_7, u_8\}$ correspond to global inputs $\mathcal{V}_{12} = \{u_{10}, u_{11}\}$
- *disturbances*: local disturbances $\{d_1, d_2, d_3\}$ correspond to global disturbances $\{d_4, d_6, d_5\}$

The restrictions, references, upper and lower limits, etc. are set according to the considerations mentioned in the analysis phase. The main design element of negotiator agents is their knowledge base, the Q-table. A Q-table is a discrete relation between states and actions. Q-tables maintain the reinforcement gained for each possible state and action. A state represents the global state of each MPC agent subproblem, which is established in terms of the cost function value. The actions are the values assigned to shared variables. In the design phase the size of the Q-table must also be established, i.e. the discretization resolution of states and actions. A simple discretization method is carried out, defining 100 possible values for states and 40 values for actions.

32.3.3 Training Phase

The training phase is an off-line iterative procedure to set up the negotiator agents knowledge bases. The agents' shared variables are assigned according to a training policy in the negotiator in order to gather as much information as possible from the MPC control results, i.e. the objective function values obtained. Q-tables maintain the reinforcement gained for each possible state and action. A state represents the global state of each sub-problem, which is established in terms of the objective function values. The definition of the error of MPC agent i is:

$$\varepsilon_i = J_i \tag{32.17}$$

where J_i is defined in equation (32.16). This is a very general definition of the error. The state value is determined by:

$$s = \frac{1}{2}\left(|\varepsilon_i| + |\varepsilon_j|\right) \tag{32.18}$$

where ε_i is the error of the first agent, and ε_j is the error of the second agent. This state is updated every sampling time. The reward function determines the reward of every action taken by the agent. In this case, the reward function is:

$$r = \rho - s \tag{32.19}$$

where ρ is a positive constant larger than the largest s. Rewards are larger when an action takes the system to a lower state. The negotiation algorithm updates the Q-table according to Algorithm 32.1 where s' is the state attained after applying action a; α is the learning rate, which characterizes the relevance of new rewards over past ones and γ is the discount factor, the relevance of future rewards over previous ones.

The main difference between learning and exploitation phases is the policy used in step 5. The policy in the learning phase is a random or pseudorandom action slection, allowing the negotiator to explore alternative control actions on similar situations.

Algorithm 32.1 Negotiation algorithm

1: **Input:** learning rate α, discount factor γ, $0 \le \alpha, \gamma \le 1$
2: $Q(s,a) \leftarrow 0, \ \forall s \in S, \ a \in A$
3: observe initial state, s
4: **loop**
5: obtain a from policy and Q-table
6: apply a, observe r and s'
7: $Q(s,a) \leftarrow Q(s,a) + \alpha\left[r + \max_{a' \in A} Q(s',a') - Q(s,a)\right]$
8: $s \leftarrow s'$
9: **end loop**

32.3.4 Exploitation Phase

The exploitation phase is the procedure that uses the knowledge obtained in the learning phase to control the LSS in a decentralized way. The shared variables of each MPC agent are assigned according to the Q-learning algorithm given in the previous paragraphs but the policy used by the negotiator agent now tries to minimize the global objective function. Typical policies are greedy or ϵ-greedy policies. The learning rate is also usually reduced to zero or close to zero values.

32.4 Theoretical Results Available

The theoretical foundations of the approach proposed in this chapter are related to Markov Decision Processes (MDP). MDPs are the formal description of discrete time stochastic control problems. Its main components are the state transition probability function, which describes the behavior of the process, the controller policy, which is the sought function that assigns a control action to a state, and the reward function, which evaluates the quality of state transitions. The two areas that have mainly dealt with the algorithms that search solutions of problems formulated as MDPs are Dynamic Programming (DP) and Reinforcement Learning (RL). DP needs an explicit model of the state transition probability function and the reward function while RL does not. RL is a well established theory and there are many contributions that analyze the optimality of its variants. Q-learning, developed by Watkins in 1989 (see [16]) is one of the most celebrated reinforcement learning algorithms. This learning algorithm was proved to be convergent for discounted Markov decision problems [17], and later it was also proved to be convergent in more general cases [6, 14]. Those proofs allow us to consider the negotiation agent as an optimal negotiator given sufficient iterations. The most important step is the appropriate selection of the cost function, which will take the MPC agents to the global optimum solution or elsewhere. Our first attempts of establishing a meaningful cost function were focused on taking the MPC agent error, see Eq. (32.17), as the error associated to the subsystem state directly related to the negotiated variable. That way of managing the state of an agent poses several questions, such as: what to do when a negotiated variable is related to more than one state in the subsystem; what to do when a negotiated variable is weakly but directly related to one particular state and strongly but indirectly related to other state(s); what to do if other optimization objectives (economic costs associated to control variables, for instance) are wanted to be considered. The error as defined in Eq. (32.17) is general enough to give proper answer to all those questions.

32.5 Application Results Available

32.5.1 Small Water Network

First the small drinking water network, already presented in Sect. 32.3, is used to exemplify the proposed algorithm and its integration in the MAMPC architecture.

The demands are measurable and are considered perturbations with a sinusoidal-like behavior throughout the day. Training was carried out off-line using the RL algorithm presented in Sect. 32.3. Control actions calculated from a 48 h scenario of the centralized MPC were used as the starting point to initialize the learning. From that point on, the training continued taking random actions. The reward was calculated for all actions. In the RL exploitation phase, the knowledge acquired in the training phase was used to solve the MPC distributed problem.

The results obtained using the proposed MAMPC Architecture are shown in Fig. 32.2. Each graph presents a 48 h scenario and shows the trajectory of the states (water volumes in tanks) of two representative tanks (x_5 and x_8) in the two agents. The volume of these two tanks is controlled respectively by Agent 1 and Agent 2 through a shared variable u_{10}. The results are contrasted with the centralized MPC solution (dashed line) for each tank. Table 32.1 compares cumulated cost values obtained with the centralized solution and with the proposed distributed MPC solution.

The objective function for the distributed solution is larger than the centralized one. The graphs show that the behaviour of all the signals is reasonable and in some cases the negotiated solution (tanks 5 and 8, Fig. 32.2a, b, respectively) outperforms the centralized optimal solution, at an obvious cost in the other tanks.

32.5.2 Barcelona Drinking Water Network

The second application example is based on the Barcelona drinking water network (DWN). The Barcelona DWN, managed by Aguas de Barcelona, S.A. (AGBAR), supplies drinking water to Barcelona city and also to the metropolitan area. The sources of water are the Ter and Llobregat rivers, which are regulated at their head by dams with an overall capacity of 600 cubic hectometers. There are also underground sources that provide water through pumping stations. Those different water sources currently provide a flow of around 7 m^3/s. The water flow from each source is limited and has different water prices depending on water treatments and legal extraction canons.

Table 32.1 Comparison of cumulated cost values: $J_{centralised}$ versus $J_{distributed}$

J	$J_{centralised}$	$J_{distributed}$
Value	13.3712	14.7201

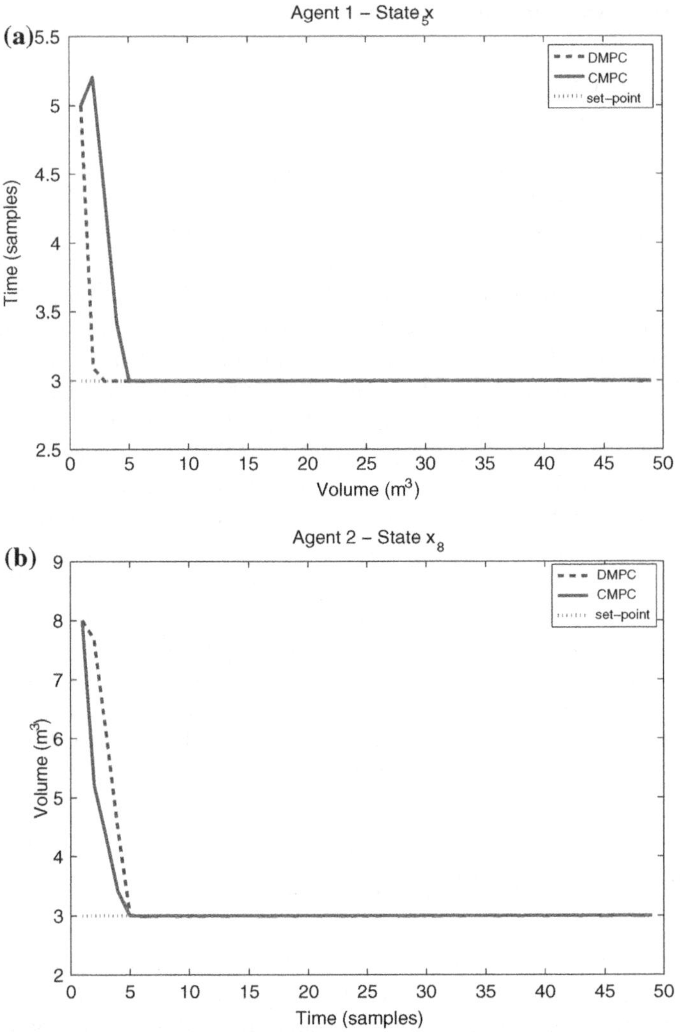

Fig. 32.2 Distributed MAMPC solution (*solid* -) against centralized MPC solution (*dashed* –). Reference (*dot*..): **a** Tank 5 in Agent 1; **b** Tank 8 in Agent 2

The Barcelona DWN is structurally organized in two layers. The upper layer, named *transport network*, which links the water treatment plants with the reservoirs distributed all over the city. The lower layer, named *distribution network* is sectorized in subnetworks. Each subnetwork links a reservoir with each consumer. This application case study is focused on the transport network. Thus, each subnetwork of the distribution network is modeled as a demand sector. The demand of each sector is characterized by a demand pattern, which can be predicted by using a time-series model [11]. The control system of the transport network is also organized in two

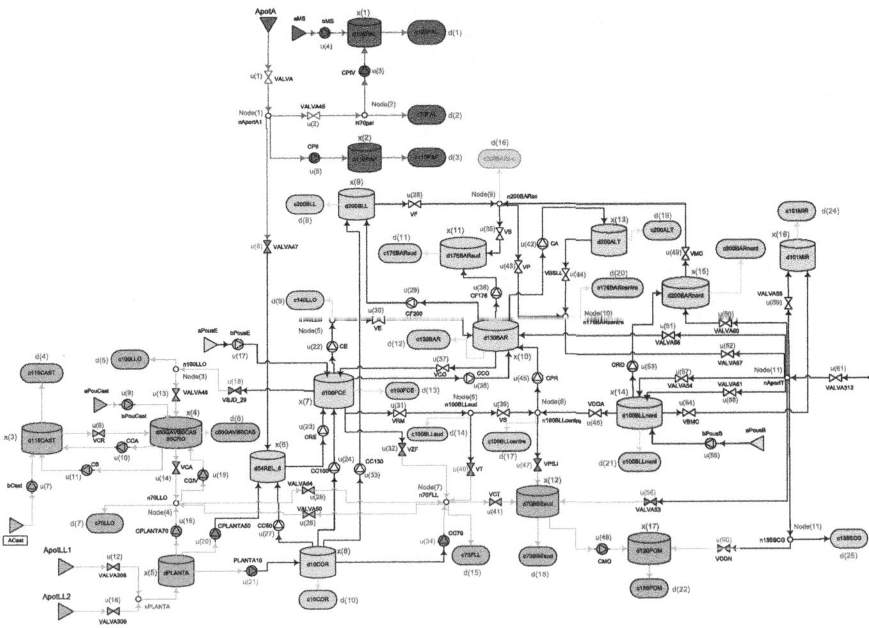

Fig. 32.3 Aggregate case of the Barcelona Drinking Water Network

layers. The upper layer is in charge of the global control of the network, establishing the set-points of the regulatory controllers at the lower layer. Regulatory controllers are of PID type, while the supervisory layer controller is of MPC type. Regulatory controllers hide the network non-linear behaviour to the supervisory controller. This fact allows the MPC supervisory controller to use a control-oriented linear model.

In this chapter, the Barcelona DWN aggregate network in Fig. 32.3 has been used. From this figure, it can be seen that the network comprises 17 tanks (state variables), 61 actuators (26 pumping stations and 35 valves), 11 nodes and 25 main sectors of water demand (model disturbances). The model has been derived using the control oriented modeling methodology proposed in [9]. The obtained model has been compared against real behaviour assessing its validity. The detailed information about physical parameters and other system values are reported in [3].

Using the partitioning obtained in [9], the aggregate model of the Barcelona DWN is decomposed in three subsystems, as depicted in Fig. 32.3 in different colors. Table 32.2 collects the resultant dimensions for each subsystem and the corresponding comparison with the dimensions of the vectors of variables for the entire aggregate network.

The results obtained by means of the proposed DMPC strategy are compared with those obtained when a CMPC strategy is used. The model parameters and measured disturbances (demands) have been supplied by AGBAR. Demands data correspond to the consume of drinking water of the city of Barcelona during 2007.

Table 32.2 Dimension comparison between the subsystems and the whole network

Elements	Subsyst 1	Subsyst 2	Subsyst 3	Whole model
Tanks	2	5	10	17
Actuators	5	22	34	61
Demands	4	9	22	25
Nodes	2	3	6	11

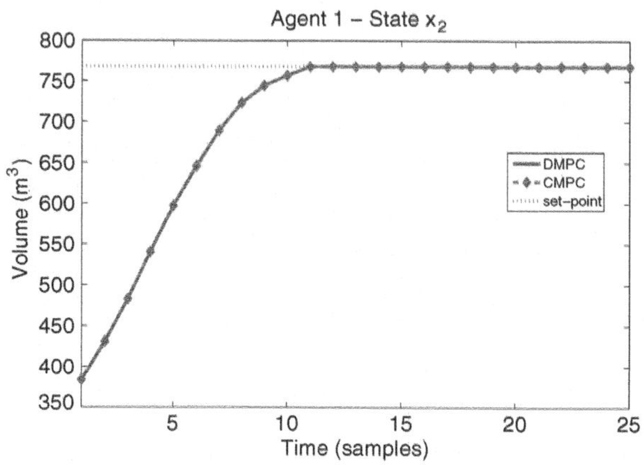

Fig. 32.4 Agent 1—state x_2

Figures 32.4, 32.5 and 32.6 show the control results obtained with the proposed DMPC approach against a CMPC controller. Those figures correspond to the evolution of three state variables (volumes) in the three different agents. In red dotted line results obtained with the CMPC controller are presented, while in blue continuous line the ones obtained with the DMPC approach are presented as well.

It can be seen that the evolution of state x_2 obtained with the CMPC and DMPC approaches are almost not distinguishable since the plots are completely overlapped. In the case of states x_7 and x_{12}, the behaviour is still very similar, with small differences.

Figure 32.7 presents the comparison of the objective function value (32.15) obtained at each iteration using the CMPC and DMPC approaches. It can be noticed that the value obtained with the RL DMPC approach is quite close to the one achieved with the CMPC approach.

32.6 Conclusions and Future Research

The results presented in this chapter suggest that the use of MAMPC architecture based on RL negotiation can converge to the centralized MPC solution with an acceptable degree of approximation but taking advantage from the MAS properties

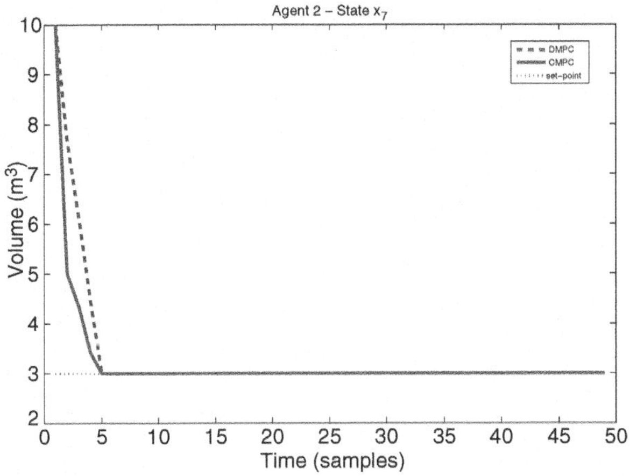

Fig. 32.5 Agent 2—state x_7

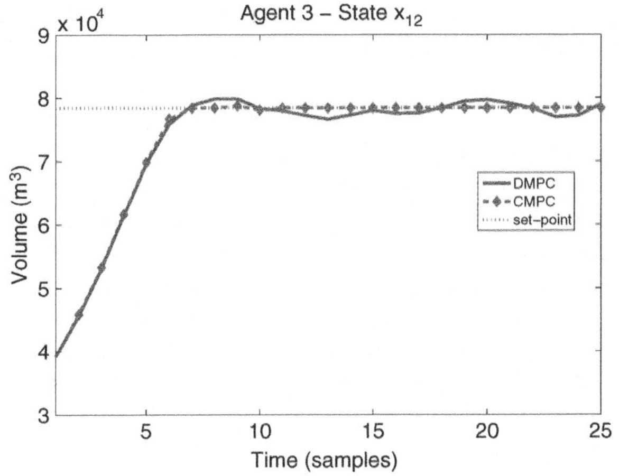

Fig. 32.6 Agent 3—state x_{12}

and the tools that the Agent Oriented Paradigm (AOP) provides for development and implementation. Moreover, the application of learning techniques can provide the Negotiator Agent the ability of prediction. Training of the negotiator can be made directly from a centralized MPC or from human operator driven control. In order to achieve optimization, no model is needed by the negotiator. Data from centralized MPC is recommendable but non essential. The type and quality of the training is a very important issue in order to obtain an efficient optimization. Also the compromise between exploration and exploitation can be implemented on-line to enable the system not just adaptation to the problem but adaptation to changes in time. In this chapter, this capability is not addressed in training but in exploring

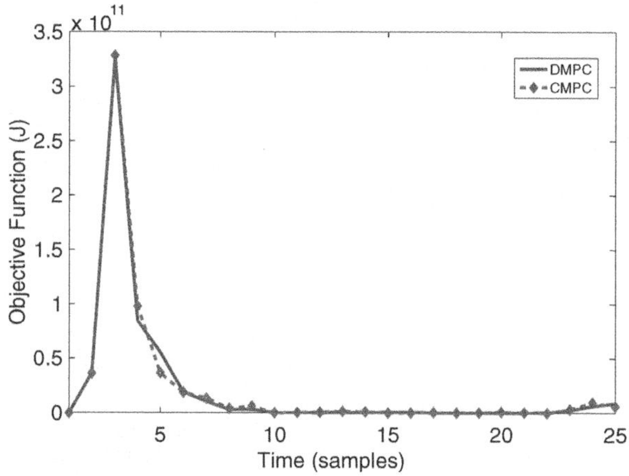

Fig. 32.7 Objective function comparison: DMPC versus CMPC

during the optimization. Communication protocols and coordination methods for MAS have to be studied and tested in a more complex case of study in which many agents interact.

The MAMPC architecture presented in this work is currently being tested on the complete Barcelona water transport network. The Barcelona water network is comprised of 200 sectors with approximately 400 control points. At present, the Barcelona information system receives, in real time, data from 200 control points, mainly through flow meters and a few pressure sensors. This network has been used as a LSS case of study to test several LSS control approaches, see [2]. As a starting point for the application of the MAMPC Architecture, recent work on centralized and decentralized MPC applied to the Barcelona network is being used, as well as the partitioning algorithm developed by [17].

References

1. D. Barcelli, *Decomposizione ottima e controllo predittivo distribuito della rete idrica di Barcellona* (Università di Siena, Facoltà di Ingegneria Informatica, Master's thesis, 2008)
2. M. Brdys, B. Ulanicki, *Operational Control of Water Systems: Structures, Algorithms and Applications* (Prentice Hall International, Hemel Hempstead, Hertfordshire, 1994)
3. V. Fambrini, C. Ocampo-Martinez, *Modelling and decentralized Model Predictive Control of drinking water networks*. Technical Report IRI-TR-04-09, Institut de Robòtica i Informàtica Industrial (CSIC-UPC), April 2009
4. H. El Fawal, D. Georges, and G. Bornard, Optimal control of complex irrigation systems via descomposition-coordination and the use of augmented Lagrangian, in *IEEE International Conference Systems, Man and Cybernetics*, San Diego, 1998, pp. 3874–3879
5. M. Gómez, J. Rodellar, F. Vea, J. Mantecón, and J. Cardona, Decentralized adaptive control for water distribution, in *IEEE International on Systems, Man and Cybernetics*, San Diego, 1998, pp. 1411–1416
6. T. Jaakkola, M.I. Jordan, S.P. Singh, Q-learning. Mach. Learn. **8**, 1185–1201 (1994)

7. V. Javalera, B. Morcego, V. Puig, Distributed MPC for large scale systems using agent-based reinforcement learning, in *IFAC Symposium Large Scale Systems*, Lille 2010
8. R.R. Negenborn, B. De Schutter, J. Hellendoorn, Multi-agent model predictive control for transportation networks: Serial vs. parallel schemes. Eng. Appl. Artif. Intell. **21**(3), 353–366 (April 2008)
9. C. Ocampo-Martinez, D. Barcelli, V. Puig, and A. Bemporad, Hierarchical and decentralised model predictive control of drinking water networks: Application to the barcelona case study. IET Control Theory & Applications, Conditionally accepted, 2011
10. C. Ocampo-Martinez, S. Bovo, V. Puig, Partitioning approach oriented to the decentralised predictive control of large-scale systems. J. Process Control **21**(5), 775–786 (2011)
11. J. Quevedo, V. Puig, G. Cembrano, J. Blanch, Validation and reconstruction of flow meter data in the Barcelona water distribution network. Control Eng. Pract. **11**(6), 640–651 (June 2010)
12. J.B. Rawlings, B. Stewart, Coordinating multiple optimization-based controllers: New opportunities and challenges. J. Process Control **18**(9), 839–845 (2008)
13. D.D. Šiljak, *Decentralized Control of Complex Systems* (Academic Press, New York, 1991)
14. J.N. Tsitsiklis, Asynchronous stochastic approximation and Q-learning. Mach. Learn. **16**, 185–202 (1994)
15. A.N. Venkat, J.B. Rawlings, S.J. Wrigth, Stability and optimality of distributed model predictive control, in *IEEE Conference on Decision and Control and European Control Conference*, Seville, 2005
16. C.I.C.H. Watkins, *Learning from Delayed Rewards*. Doctoral Dissertation (University of Cambridge, Cambridge, 1989)
17. C.I.C.H. Watkins, P. Dayan, Q-learning. Mach. Learn. **8**, 279–292 (1992)

Chapter 33
Hierarchical MPC for Multiple Commodity Transportation Networks

J. L. Nabais, R. R. Negenborn, R. B. Carmona-Benítez, L. F. Mendonça and M. A. Botto

Abstract Transportation networks are large scale complex systems spatially distributed whose objective is to deliver commodities at the agreed time and at the agreed location. These networks appear in different domain fields, such as communication, water distribution, traffic, logistics and transportation. A transportation network has at the macroscopic level storage capability (located in the nodes) and transport delay (along each connection) as main features. Operations management at transportation networks can be seen as a flow assignment problem. The problem dimension to solve grows exponentially with the number of existing commodities, nodes and connections. In this work we present a Hierarchical Model Predictive Control (H-MPC) architecture to determine flow assignments in transportation networks, while minimizing exogenous inputs effects. This approach has the capacity to keep track of commodity types while solving the flow assignment problem. A flow decomposition of the main system into subsystems is proposed to diminish the problem dimension to solve in each time step. Each subsystem is managed by a control

J. L. Nabais(✉)
IDMEC, Department of Informatics and Systems Engineering, Setúbal School of Technology, Polytechnical Institute of Setúbal, Setúbal, Portugal
e-mail: joao.nabais@estsetubal.ips.pt

R. R. Negenborn
Transport Engineering and Logistics, Delft University of Technology, Delft, The Netherlands
e-mail: r.r.negenborn@tudelft.nl

R. B. Carmona-Benítez
School of Business and Economics, Universidad Anáhuac México Norte, Huixquilucan, México
e-mail: rafael.carmona@anahuac.mx

L. F. Mendonça
Department of Marine Engineering, Escola Superior Naútica Infante D. Henrique, Paço d'Arcos, Portugal
e-mail: luismendonca@enautica.pt

M. A. Botto
IDMEC, Instituto Superior Técnico, Technical University of Lisbon, Lisboa, Portugal
e-mail: ayalabotto@ist.utl.pt

J. M. Maestre and R. R. Negenborn (eds.), *Distributed Model Predictive Control Made Easy*, Intelligent Systems, Control and Automation: Science and Engineering 69, DOI: 10.1007/978-94-007-7006-5_33, © Springer Science+Business Media Dordrecht 2014

agent. Control agents solve their problems in a hierarchical way, using a so-called push-pull flow perspective. Further problem dimension reduction is achieved using contracted projection sets. The framework proposed can be easily scaled to network topologies in which hundreds of commodities and connections are present.

33.1 Introduction

In transportation networks (such as cargo transport [2, 4], postal networks, traffic networks [6], water distribution [8, 14], supply chains [10, 17]) the elementary objective is to deliver a certain commodity in the agreed quantity at the agreed time and at the agreed location [13]. The transport need can arise in two different forms: located downstream in the form of customer demand (clients in supply chains and water distribution) or located upstream as clients request to provide a service (deliver letters or containerized cargo). These two types of transport needs are disturbances to the network state and we consider them as exogenous inputs. The main control problem related to transportation networks can be categorized as a tracking control problem and stated as: find the optimal flows inside the network such that the exogenous inputs effects are eliminated and the network follows the desired state over time. In a water distribution application the optimal flows are assured through gate movements in order to keep water depths inside admissible levels in each canal, in cargo transportation the optimal flow is guaranteed by allocating transport capacity such that cargo is delivered at the final destination at the right time and with the exact quantity. The ability to access the stored volume per commodity type at each network node can be used to support operations management towards a more efficient, sustainable, cooperative and reliable transportation. Considering multiple commodities and network nodes a combinatorial issue arises. When realistic applications are considered this becomes a real problem in terms of computation time. Take an example from the freight transport: the *Neuss Trimodal* [5] terminal, recently added as a member of European Gateway Services.[1] This hinterland intermodal container terminal situated at the Rhine river offers connections to the European hinterland through three transport modes: barge, train and truck. With 8 rail tracks it sustains 39 train connections weekly plus 7 inland shipping connections to Rotterdam and Antwerp ports using a quay of 230 m. Adding to these features all kind of container types (hazardous materials, reefer containers and other categories like size, weight and destination) much information has to be captured by the modeling framework. Information can be shared freely over the transportation network, restricted to some subnetworks or

[1] European Gateway Services are a service provided by European Container Terminal (ECT) whose main objective is to create an integrated network of hinterland terminals cooperating to increase the ECT terminals throughput at the Rotterdam port. Neuss Trimodal has been a member of this network since 20 December 2011.

confined to a single component as a consequence of the economical relations between the different parties present at the network. In case of vertical integration, when all parties belong to a same entity, information is usually shared freely.

Using a central model to solve the flow assignment problem in a transportation network with multiple commodities is not an option because the problem dimension to be solved grows exponentially with the number of handled commodities, nodes and connections available. However, some connections can have no transport needs over some time (inactive connections), which means that the optimal solution is partially imposed, and it is also expected that the number of commodity types handled in an active connection (opposed to an inactive connection) is just a subset of all commodity types available at the network. We propose a Hierarchical Model Predictive Control (H-MPC) framework that is able to cope efficiently with this large-scale problem dimension by proposing explicitly measures to face the aspects mentioned above. The framework is based on the following:

- The large-scale system is broken down into smaller subsystems using a decomposition inspired by flows. A subsystem can be related to an arc, path or cycle dependent on the specific network.
- A control agent is assigned to each subsystem and formulates an optimization problem to solve the flow assignment. Control agents will only consider to solve problems related to active subsystems.
- Subproblems will be simplified further by taking into account only the commodity types handled by the subsystem using contracted commodity sets.

One important issue is the order by which the control agents should solve their problems. Following a flow perspective, control agents order is established in a so called push-pull flow perspective [15] based on the exogenous inputs location. If the exogenous input is located downstream, a pull flow perspective is applied and therefore the control agents responsible to move commodities to that downstream node are set to a higher priority. If the exogenous inputs are located upstream, a push flow perspective is applied and control agent responsible to move commodities from the source nodes get a higher priority. A simultaneous push-pull flow perspective is possible. Adding more connections to the network has as a consequence the addition of more control agents. The original problem remains solvable in a reasonable time even for large scale networks with hundreds of commodities, nodes and connections.

In Sect. 33.2 the transportation network model is described to cope with different network topologies. The optimization problem is formulated for the whole network in Sect. 33.3. The H-MPC framework is presented in Sect. 33.3.2 based on a model decomposition inspired by flows and using contraction projected sets. In Sect. 33.4 numerical results are presented, in which the H-MPC architecture is compared with the MPC architecture.

33.2 Multiple Commodity Network Model

33.2.1 Motivation

A transportation network can be represented by a graph $\mathcal{G} = (\mathcal{V}, \mathcal{E})$ where nodes $\mathcal{V}$ represent centers or intersections and arcs $\mathcal{E}$ represent the existing connections between nodes [1]. The assumptions made in this work are intended to produce a general framework able to describe different transportation network topologies that can be found in different research fields such as: traffic networks, water distribution, supply chain, cargo and passenger transportation, postal networks. At a macroscopic level, transportation networks exhibits two major phenomena: storage capability in well-defined areas (where commodities can be produced, manufactured or simply stored) and the transport delay which is the time necessary to transport commodities between different locations. To distinguish these two phenomena inside the transportation network we define two components:

Center Node: is a network node with a significant storage capacity where commodities can be stored temporarily before moving to another network node. The center node degree is always equal of bigger than one. If the center node indegree is zero the node is categorized as a source (upstream) node, if the center node outdegree is zero the node is categorized as a destination (downstream) node. When center nodes have simultaneously the indegree and outdegree bigger than one (see Fig. 33.1a) they are categorized as store nodes;

Connection: is a path between two center nodes and is used to model the transport phenomena. It is composed of a succession of nodes with an indegree and outdegree equal to one, which means that there is only one arc arriving and another arc departing from each node. Connection i is composed of n_{c_i} nodes and $n_{c_i} + 1$ arcs as indicated in Fig. 33.1b. Connections are modeled using a push-pull flow perspective: pushing commodities from the connection upstream node and delivering it to the connection downstream node.

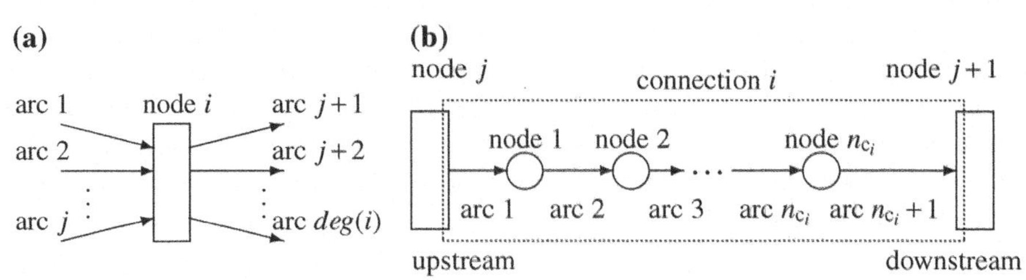

Fig. 33.1 Elementary components in a transportation network ($deg(i)$ stands for node degree). **a** Center node of store type and **b** connection (path) between two center nodes (j and $j + 1$)

All transportation networks are generally composed of center nodes and connections. The complexity of the network model is determined by the following parameters:

- n_t: number of commodity types considered;
- n_c: number of connections existing in the network;
- n_{c_i}: number of nodes belonging exclusively to connection i;
- n_n: number of center nodes in the network that are further divided into source (upstream) nodes n_n^u, destination (downstream) nodes n_n^d and store nodes n_n^s. The store nodes can be distributed through several levels based on similar characteristics, such as distribution centers, consolidation centers;
- n_l: number of levels present in the transportation network, including the source (upstream) and destination (downstream) levels.

For illustration purposes consider the transportation network indicated in Fig. 33.2. The network is composed of 61 nodes ($\sum_{i=1}^{n_c} n_{c_i} = 52$ nodes associated exclusively to connections and $n_n = 9$ center nodes shared by several connections: three source nodes, four store nodes and two destination nodes) and $n_c = 16$ connections with 68 transport flows. This network is divided into four levels ($n_l = 4$), including the source and the destination levels which are level 1 and level 4 respectively. The network topology is generic including: connections between nodes on nonadjacent levels (connections from node 53 to node 58 using the walk 53–1–2–3–4–5–58 and between nodes 55 and node 59 using the walk 55–18–19–20–21–22–59) and cycles between level two and level three (for example cycle formed by nodes 56–26–27–28–59–44–45–46–56). Center nodes can have multiple connections arriving and departing while connections can share limited infrastructure resources to guarantee the desired flows between nodes. Transportation networks are therefore complex systems with coupled dynamics and coupled constraints.

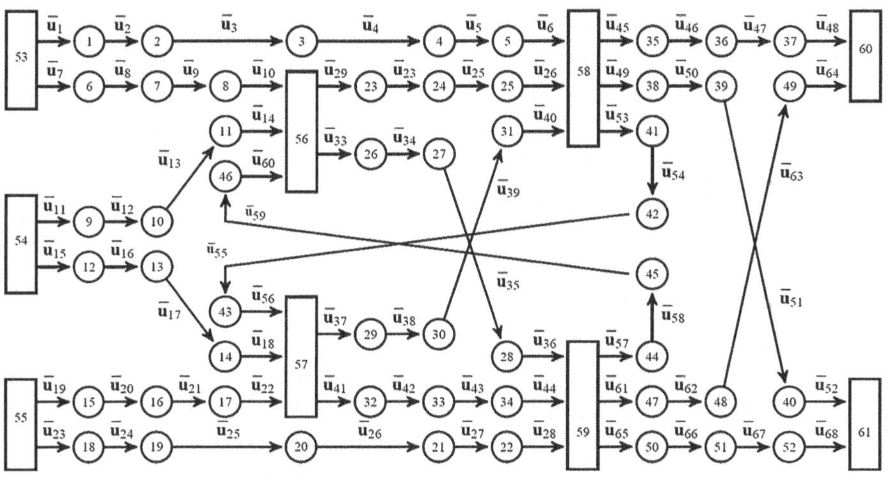

Fig. 33.2 Example of a transportation network

33.2.2 Network Model Framework

A generic framework to model different transportation network is intended but adaptation can be required to accommodate modeling assumptions made for each case scenario. The total number of nodes inside the network is associated with the network topology and is given by,

$$n_{\mathrm{y}} = n_{\mathrm{n}} + \sum_{i=1}^{n_{\mathrm{c}}} n_{\mathrm{c}_i}. \tag{33.1}$$

For each node in the transportation network a state-space vector $\bar{\mathbf{x}}_j(k)$ is defined, and these are merged to form the state-space vector $\mathbf{x}$ of the complete network,

$$\bar{\mathbf{x}}_j(k) = \begin{bmatrix} x_j^1(k) \\ x_j^2(k) \\ \vdots \\ x_j^{n_{\mathrm{t}}}(k) \end{bmatrix}, j = 1, \ldots, n_{\mathrm{y}}, \quad \mathbf{x}(k) = \begin{bmatrix} \bar{\mathbf{x}}_1(k) \\ \bar{\mathbf{x}}_2(k) \\ \vdots \\ \bar{\mathbf{x}}_{n_{\mathrm{y}}}(k) \end{bmatrix}, \tag{33.2}$$

where $x_j^t(k)$ is the volume of commodities of type t at node j at time step k. The state-space dimension $\mathbf{x}$ is given by $n_{\mathrm{t}}n_{\mathrm{y}}$ corresponding to the number of commodity types handled and the number of nodes existing in the network. The state-space vector contains information about the quantity per commodity type not only at the center nodes, with significant storage capacity, but also at connection nodes. The total volume per commodity type inside the network is always accessible through the state-space vector.

The model for the network dynamics can be represented in a state-space form as,

$$\mathbf{x}(k+1) = \mathbf{A}\mathbf{x}(k) + \mathbf{B}_{\mathrm{u}}\mathbf{u}(k) + \mathbf{B}_{\mathrm{d}}\mathbf{d}(k), \tag{33.3}$$

$$\mathbf{y}(k) = \mathbf{C}\mathbf{x}(k), \tag{33.4}$$

$$\mathbf{x}(k) \geq \mathbf{0}, \tag{33.5}$$

$$\mathbf{u}(k) \geq \mathbf{0}, \tag{33.6}$$

$$\mathbf{y}(k) \leq \mathbf{y}_{\max}, \tag{33.7}$$

$$\mathbf{P}_{\mathrm{uu}}\mathbf{u}(k) \leq \mathbf{u}_{\max}, \tag{33.8}$$

$$\mathbf{x}(k) \geq \mathbf{P}_{\mathrm{xu}}\mathbf{u}(k), \tag{33.9}$$

$$\mathbf{x}(k) \in \mathcal{X}, \tag{33.10}$$

$$\mathbf{u}(k) \in \mathcal{U}, \tag{33.11}$$

where $\mathbf{u}$ is the control action vector with length $n_{\mathrm{u}} = n_{\mathrm{t}}(n_{\mathrm{y}} - n_{\mathrm{n}} + n_{\mathrm{c}})$, $\mathbf{d}$ is an exogenous input vector assumed known with dimension $n_{\mathrm{t}}(n_{\mathrm{n}}^{\mathrm{u}} + n_{\mathrm{n}}^{\mathrm{d}})$, $\mathbf{y}$ is the current volume at all nodes with dimension n_{y}, $\mathbf{y}_{\max}$ are the maximum node storage capacities, $\mathbf{u}_{\max}$ the available infrastructure resources according to the network structural layout, $\mathbf{A}$, $\mathbf{B}_{\mathrm{u}}$, $\mathbf{B}_{\mathrm{d}}$ and $\mathbf{C}$ are the state-space matrices, $\mathbf{P}_{\mathrm{xu}}$ is the projection from the control action set $\mathcal{U}$ into the state-space set $\mathcal{X}$ and $\mathbf{P}_{uu}$ is the projection matrix from the control action set $\mathcal{U}$ into the infrastructure resource capacity set $\mathcal{U}_{max}$.

The transportation network state of $\mathbf{x}$ at the next time step, $k+1$, is determined using (33.3) as a function of the current network state $\mathbf{x}$ plus the contribution due to the control action $\mathbf{u}$ and the corresponding exogenous inputs $\mathbf{d}$ capturing the external requests on the transportation network. The control action $\mathbf{u}$ is the flow of commodities between nodes and is imposed through the available infrastructure resources. Inequalities (33.5)–(33.9) are necessary in this framework for imposing the network structural layout and assumptions made:

Nonnegativity of States and Control Actions: negative storage and negative control actions (flows) are not physically possible, imposed by (33.5) and (33.6);

Storage Capacity: each network node has to respect its own storage capacity, this is represented by (33.7);

Maximum Control Actions: the network structural layout in terms of available hardware in quantity and type used to guarantee the desired flows is represented by (33.8);

Feasible Control Actions: not all control actions that satisfy (33.5) and (33.6) are feasible. The control action has to respect the existence of commodity type in the related network node. Inequality (33.9) imposes this relation.

33.2.3 Network Model Decomposition

Taking into account that real transportation networks may serve tens of center nodes and handle hundreds of commodity types it is critical to alleviate the computational burden when considering the sparse central model (33.3)–(33.11) to support operations management. Using a node/arc numbering in a push-flow perspective (from the sources towards the destinations) as indicated in Fig. 33.2, it is possible to obtain a highly structured model without the need for further mathematical manipulations [16]. A connection is by definition the path between two center nodes. The interaction of a single connection into the set of center nodes is done solely at two nodes. To take advantage of the model structure we define a subsystem i as the node collection related to a connection i plus the associated source and destination nodes [11]. The state-space vector $\mathbf{x}_i$ for subsystem i will be composed of the corresponding $\bar{\mathbf{x}}_j$ state-space vectors,

$$\mathbf{x}_i(k) = \begin{bmatrix} \bar{\mathbf{x}}_{nC_i - n_{c_i}+1}(k) \\ \bar{\mathbf{x}}_{nC_i - n_{c_i}+2}(k) \\ \vdots \\ \bar{\mathbf{x}}_{nC_i -1}(k) \\ \bar{\mathbf{x}}_{nC_i}(k) \\ \bar{\mathbf{x}}_i^{in}(k) \\ \bar{\mathbf{x}}_i^{out}(k) \end{bmatrix}, \quad nC_i = \sum_{j=1}^{i} \left(n_{c_j}\right), \quad 1 \le i \le n_c, \tag{33.12}$$

with length $n_t(n_{c_i}+2)$ belonging to state-space set $\mathcal{X}_i$ where $\bar{\mathbf{x}}_i^{\text{in}}$ and $\bar{\mathbf{x}}_i^{\text{out}}$ are the state-space vectors related to the source and destination nodes for connection i respectively. The control action vector $\mathbf{u}_i$ for subsystem i is given by the corresponding $\bar{\mathbf{u}}_j$ state-space vectors,

$$\mathbf{u}_i(k) = \begin{bmatrix} \bar{\mathbf{u}}_{n_{\text{U}_i}-n_{\text{c}_i}}(k) \\ \bar{\mathbf{u}}_{n_{\text{U}_i}-n_{\text{c}_i}+1}(k) \\ \vdots \\ \bar{\mathbf{u}}_{n_{\text{U}_i}-1}(k) \\ \bar{\mathbf{u}}_{n_{\text{U}_i}}(k) \end{bmatrix}, \quad n_{\text{U}_i} = \sum_{j=1}^{i} (n_{\text{c}_j}+1), \quad 1 \le i \le n_{\text{c}}, \tag{33.13}$$

with length $n_t(n_{c_i}+1)$ belonging to control action set $\mathcal{U}_i$. The state-space model for subsystem i is given by,

$$\mathbf{x}_i(k+1) = \mathbf{A}_i\mathbf{x}_i(k) + \mathbf{B}_{\text{u}_i}\mathbf{u}_i(k) + \mathbf{B}_{\text{d}_i}\mathbf{d}_i(k) + \sum_{j=1, j\neq i}^{n_\text{c}} \mathbf{B}_{\text{u}_{i,j}}\mathbf{u}_j(k) \tag{33.14}$$

$$\mathbf{y}_i(k) = \mathbf{C}_i\mathbf{x}_i(k), \tag{33.15}$$

where $\mathbf{y}_i$ is the quantity per commodity type at subsystem i nodes, $\mathbf{d}_i$ is the exogenous input vector associated with subsystem i, $\mathbf{A}_i$, $\mathbf{B}_{\text{u}_i}$, $\mathbf{B}_{\text{u}_{i,j}}$ $\mathbf{B}_{\text{d}_i}$ and $\mathbf{C}_i$ are the state-space matrices for subsystem i. The last term in (33.14) is critical to assure information exchange between control agents, in particular regarding their future behavior, to avoid that two or more control agents respond to the same demand. The complete subsystem i model is obtained including constraints of nonnegativity of states and control actions, storage capacity, maximum control actions, and feasible control actions to the state space model (33.14)–(33.15).

33.3 Hierarchical Network-Flow Control

In this section we present the Hierarchical Model Predictive Control architecture (H-MPC). First, operations management for the transportation network is formulated accordingly to a central MPC approach [3, 9]. After, using a push-pull flow perspective, the large-scale system is decomposed into subsystems to each of which a control agent is assigned. Control agents are associated to a network level if they are delivering commodities to the center nodes located at that level for a pull flow perspective or if they are taking commodities from the center nodes located at that level in a push flow perspective. The problem dimension to be solved at each time step is diminished by considering only control agents related to active connections over the prediction horizon. Further reduction is achieved using the contracted and global commodity sets to formulate the optimization problem considering only the number of commodity types affected over the prediction horizon .

33.3.1 Centralized MPC Formulation

The cost function is defined in accordance to the application domain and is generally a function of the network states, control actions and desired states over the prediction horizon N_p,

$$J(\tilde{\mathbf{x}}_k, \tilde{\mathbf{u}}_k, \tilde{\mathbf{x}}_{\mathrm{ref}}) = \sum_{l=0}^{N_p-1} f(\mathbf{x}(k+1+l), \mathbf{u}(k+l), \mathbf{x}_{\mathrm{ref}}(k+1+l)), \quad (33.16)$$

where $\tilde{\mathbf{x}}_k$ is the vector composed of the state-space vectors for each time step over the prediction horizon $\left[\mathbf{x}^{\mathrm{T}}(k+1), \ldots, \mathbf{x}^{\mathrm{T}}(k+N_p)\right]^{\mathrm{T}}$, $\tilde{\mathbf{u}}_k$ is the vector composed by the control action vectors for each time step over the prediction horizon $\left[\mathbf{u}^{\mathrm{T}}(k), \ldots, \mathbf{u}^{\mathrm{T}}(k+N_p-1)\right]^{\mathrm{T}}$, $\mathbf{x}_{\mathrm{ref}}$ is the state-space reference vector and $\tilde{\mathbf{x}}_{\mathrm{ref}}$ is the vector composed of the state-space reference vectors for each time step over the prediction horizon $\left[\mathbf{x}_{\mathrm{ref}}^{\mathrm{T}}(k+1), \ldots, \mathbf{x}_{\mathrm{ref}}^{\mathrm{T}}(k+N_p)\right]^{\mathrm{T}}$. The weights to be used in the objective function (33.16) are considered time varying to allow changing flow priorities according to the different behaviors desired for the transportation network over time. The MPC problem for the transportation network can be formulated as:

$$\min_{\tilde{\mathbf{u}}_k} \; J(\tilde{\mathbf{x}}_k, \tilde{\mathbf{u}}_k, \tilde{\mathbf{x}}_{\mathrm{ref}}) \quad (33.17)$$

$$\text{subject to} \quad \mathbf{x}(k+1+l) = \mathbf{A}\mathbf{x}(k+l) + \mathbf{B}_{\mathrm{u}}\mathbf{u}(k+l) + \mathbf{B}_{\mathrm{d}}\mathbf{d}(k+l), \quad (33.18)$$

$$\mathbf{y}(k+l) = \mathbf{C}\mathbf{x}(k+l), \quad l = 0, \ldots, N_p - 1, \quad (33.19)$$

$$\mathbf{x}(k+1+l) \geq \mathbf{0}, \quad (33.20)$$

$$\mathbf{u}(k+l) \geq \mathbf{0}, \quad (33.21)$$

$$\mathbf{y}(k+l) \leq \mathbf{y}_{\max}, \quad (33.22)$$

$$\mathbf{P}_{\mathrm{uu}}\mathbf{u}(k+l) \leq \mathbf{u}_{\max}, \quad (33.23)$$

$$\mathbf{x}(k+l) \geq \mathbf{P}_{\mathrm{xu}}\mathbf{u}(k+l), \quad (33.24)$$

$$\mathbf{P}_{\mathrm{dx}}\mathbf{x}(k+1+l) \leq \mathbf{d}_{\mathrm{d}}(k+l), \quad (33.25)$$

where $\mathbf{y}_{\max}$ is the maximum storage capacity by center node, $\mathbf{u}_{\max}$ the maximum available infrastructure resources according to the network structural layout, $\mathbf{d}_{\mathrm{d}}$ is the vector responsible to introduce the exogenous inputs, $\mathbf{P}_{\mathrm{dx}}$ is the projection matrix from the state-space set into the disturbance set and $\mathbf{P}_{\mathrm{uu}}$ is the projection matrix from the control action set into the infrastructure resource set $\mathcal{U}_{\max}$. Constraint (33.25) is included in the MPC problem formulation to introduce the network exogenous inputs prediction.

33.3.2 Hierarchical Formulation

33.3.2.1 MPC Formulation for One Control Agent

The cost function of a control agent is defined in accordance to the application field and is generally a function of the states, control actions and desired states of the subsystem the agent controls over the prediction horizon N_p,

$$J_i\left(\tilde{\mathbf{x}}_{k,i}, \tilde{\mathbf{u}}_{k,i}, \tilde{\mathbf{x}}_{\mathrm{ref},i}\right) = \sum_{l=0}^{N_\mathrm{p}-1} f\left(\mathbf{x}_i(k+1+l), \mathbf{u}_i(k+l), \mathbf{x}_{\mathrm{ref},i}(k+1+l)\right) \tag{33.26}$$

where $\tilde{\mathbf{x}}_{k,i}$ is the vector composed of the state-space vectors for each time step over the prediction horizon $\left[\mathbf{x}_i^\mathrm{T}(k+1)\ ,\ldots,\ \mathbf{x}_i^\mathrm{T}(k+N_\mathrm{p})\right]^\mathrm{T}$ for control agent i, $\tilde{\mathbf{u}}_{k,i}$ is the vector composed by the control action vectors for each time step over the prediction horizon $\left[\mathbf{u}^\mathrm{T}(k)\ ,\ldots,\ \mathbf{u}^\mathrm{T}(k+N_\mathrm{p}-1)\right]^\mathrm{T}$ for control agent i, $\mathbf{x}_{\mathrm{ref},i}$ is the state-space reference vector for control agent i and $\tilde{\mathbf{x}}_{\mathrm{ref},i}$ is the vector composed of the state-space reference vectors for each time step over the prediction horizon $\left[\mathbf{x}_{\mathrm{ref},i}^\mathrm{T}(k+1)\ ,\ldots,\ \mathbf{x}_{\mathrm{ref},i}^\mathrm{T}(k+N_\mathrm{p})\right]^\mathrm{T}$. Transportation networks are large scale systems spatially distributed therefore it is common to have connections with rather different features, in particular the transport delay. For larger transport delays the optimization problem requires a larger prediction horizon in order commodities have enough time to reach the destination node such that this effect is reflected in the cost function. For smaller transport delays smaller prediction horizons can be used at the cost of some performance decrease.

The MPC formulation for control agent i can be stated as:

$$\min_{\tilde{\mathbf{u}}_{k,i}} \quad J_i\left(\tilde{\mathbf{x}}_{k,i}, \tilde{\mathbf{u}}_{k,i}, \tilde{\mathbf{x}}_{\mathrm{ref},i}\right) \tag{33.27}$$

$$\begin{aligned}
\text{subject to}\quad & \mathbf{x}_i(k+1+l) = \mathbf{A}_i\mathbf{x}_i(k+l) + \mathbf{B}_{\mathrm{u}_i}\mathbf{u}_i(k+l) \\
& \qquad + \cdots + \mathbf{B}_{\mathrm{d}_i}\mathbf{d}_i(k+l) + \sum_{j=1, j\neq i}^{n_\mathrm{c}} \mathbf{B}_{\mathrm{u}_{i,j}}\mathbf{u}_j(k+l) && (33.28)\\
& \mathbf{y}_i(k+l) = \mathbf{C}_i\mathbf{x}_i(k+l), \quad l = 0, \ldots, N_\mathrm{p}-1, && (33.29)\\
& \mathbf{x}_i(k+1+l) \geq \mathbf{0}, && (33.30)\\
& \mathbf{u}_i(k+l) \geq \mathbf{0}, && (33.31)\\
& \mathbf{y}_i(k+l) \leq \mathbf{y}_{\mathrm{max},i}, && (33.32)\\
& \mathbf{P}_{\mathrm{uu},i}\mathbf{u}_i(k+l) \leq \mathbf{u}_{\mathrm{max},i}, && (33.33)\\
& \mathbf{x}_i(k+l) \geq \mathbf{P}_{\mathrm{xu},i}\mathbf{u}_i(k+l), && (33.34)\\
& \mathbf{P}_{\mathbf{dx},i}\mathbf{x}_i(k+1+l) \leq \mathbf{d}_{d_i}(k+l), && (33.35)
\end{aligned}$$

where $\mathbf{y}_{\max,i}$ is the maximum capacity for control agent i nodes, $\mathbf{u}_{\max,i}$ the available infrastructure resources according to the network structural layout for agent i, $\mathbf{d}_{d_i}$ is the vector responsible to introduce the exogenous inputs for control agent i, $\mathbf{P}_{\mathrm{uu},i}$ is the projection matrix from the control action set $\mathcal{U}_i$ into the infrastructure resource set for control agent i, $\mathbf{P}_{\mathrm{xu},i}$ is the projection from the control action set $\mathcal{U}_i$ into the state-space set $\mathcal{X}_i$ and $\mathbf{P}_{\mathbf{dx},i}$ is the projection matrix from the state-space set $\mathcal{X}_i$ into the exogenous input set of control agent i.

33.3.2.2 Projections Between Contracted and Global Commodity Sets

Considering the network model as a collection of subsystems reduces the optimization problem dimension to be solved at each time step. It is not expected that each connection in the network is transporting simultaneously all commodity types, therefore further reduction of the problem dimension to be solved in each time step can be made if we consider only the handled commodities over the prediction horizon. Define the following sets:

- $\mathcal{T} := \{1, \ldots, n_{\mathrm{t}}\}$ is the set of all commodity types handled in the transportation network with cardinality $|\mathcal{T}| = n_{\mathrm{t}}$;
- $\mathcal{T}_i(k) = \{1, \ldots, n_i^{\mathrm{ac}}(k)\}$ is the set of the commodity types handled by subsystem i over the prediction horizon at time instant k with cardinality $|\mathcal{T}_i| = n_i^{\mathrm{ac}}$.

It is important to note that the cardinality of $\mathcal{T}_i$ is made time varying to allow different commodity flows over time. The following relation between sets can be derived,

$$\mathcal{T}_i(k) \subset \mathcal{T}. \tag{33.36}$$

The model (33.14)–(33.15) can be written for a new state-space variable $\mathbf{x}_i^{\mathrm{c}}$ and a new control action $\mathbf{u}_i^{\mathrm{c}}$ whose dimensions are a subset of the network commodity set $\mathcal{T}$ by eliminating from the state-space vector $\mathbf{x}_i$ and from the control action vector $\mathbf{u}_i$ all variables related to commodity types that are not included in the contracted commodity set $\mathcal{T}_i$ and for this reason are not expected to change over the prediction horizon. The original state-space representation can be recovered using,

$$\begin{cases} \mathbf{x}_i(k) = \mathbf{P}_{\mathrm{cx},i}(k)\mathbf{x}_i^{\mathrm{c}}(k) \\ \mathbf{u}_i(k) = \mathbf{P}_{\mathrm{cu},i}(k)\mathbf{u}_i^{\mathrm{c}}(k) \end{cases} \tag{33.37}$$

where $\mathbf{P}_{\mathrm{cx},i}$ and $\mathbf{P}_{\mathrm{cu},i}$ are time varying projection matrices between the contracted commodity set $\mathcal{T}_i$ into the global commodity set $\mathcal{T}$ for the state-space and control action vectors respectively. This procedure allows to look for the optimal solution regarding only significant control actions. The control actions associated to the eliminated variables are zero by default.

33.3.2.3 Hierarchical Framework

The order in which the control agents solve their problems at each time step can be fixed over time or depend on the current network state. For setting the control agents order at each time step it is important to recall the transportation network type. Depending on where the network exogenous inputs occur the control agents order is set in a push flow perspective (for upstream exogenous inputs) or in a pull flow perspective (for downstream exogenous inputs). In either case the control agents order is set by groups corresponding each group to a network level and only after the control agents order for a level is determined the order for the control agents associated to the next level is determined.

At the beginning of each time step all control agents update their state using the information about exogenous inputs. After, control agents determine in parallel their expected workload over the prediction horizon to follow the network desired state,

$$c_i(k) = \alpha_i(k) \sum_{l=0}^{N_\mathrm{p}} \left(\mathbf{x}_i^\mathrm{F}(k+1+l) - \mathbf{x}_\mathrm{ref}^\mathrm{F}(k+1+l) \right), \quad i = 1, \dots, n_1, \tag{33.38}$$

where α_i is a time varying penalty term to account commodities transport costs using connection i over time, $\mathbf{x}_i^\mathrm{F}$ and $\mathbf{x}_\mathrm{ref}^\mathrm{F}$ are the state-space and reference vector for the edge node of connection i (upstream center node for a push flow perspective and downstream center node for a pull flow perspective). Each control agent shares its workload information c_i, for the current time step at the network, with the central coordinator that sets the order $\mathbf{o}(k)$ in which the control agents should solve their problems. After analyzing all network levels the complete order $\mathbf{o}(k) = \left[o_1 \; \dots \; o_{n_\mathrm{c}} \right]$ with $1 \leq o_i \leq n_\mathrm{c}$ such that,

$$\underbrace{c_{o_1}(k) > \dots > c_{o_{n_\mathrm{c}}^1}(k)}_{\text{first level}} \quad \cdots \quad \underbrace{c_{o_{n_\mathrm{c}-n_\mathrm{c}^l+1}}(k) > \dots > c_{o_{n_\mathrm{c}}}(k)}_{\text{last level}}, \tag{33.39}$$

where n_c^1 is the number of connections associated to the first network level to be solved and n_c^l is the number of connections associated to the last network level to be solved.

The central coordinator is responsible to set the total amount of infrastructure resources that are available $\theta^0 = \mathbf{u}_\mathrm{max}$ and the current prediction for future decisions set $\mathcal{P}^0 = \{\tilde{\mathbf{u}}_{k-1,o_1}, \dots, \tilde{\mathbf{u}}_{k-1,o_{n_\mathrm{c}}}\}$. The control agent to start (o_1) has all infrastructure resources available. After the initial configuration the iterations are executed in which each control agent o_i ($i = 1, \dots, n_\mathrm{c}$), one after another, performs the following tasks (see Fig. 33.3):

- the maximum admissible resource for control agent o_i is determined as the minimum between the subsystem maximum infrastructure resource $\mathbf{u}_{\mathrm{max},o_i}$ and the infrastructure resources not yet assigned,

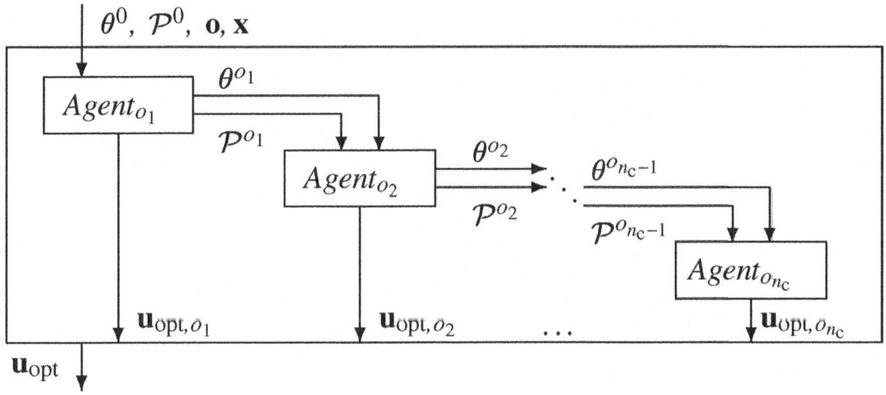

Fig. 33.3 H-MPC schematics at a time step

$$\mathbf{u}_{\max,o_i} = \min\left(\mathbf{P}_{\max,o_i}\theta^{o_{i-1}}; \mathbf{u}_{\max,o_i}\right), \tag{33.40}$$

where $\mathbf{P}_{\max,o_i}$ is the projection matrix from the global infrastructure resource set $\mathcal{U}_{\max}$ to the maximum infrastructure resource set $\mathcal{U}_{\max,o_i}$ for subsystem o_i;

- in case the workload c_{o_i} is zero the optimal control action $\mathbf{u}_{\text{opt},o_i}$ is zero by default. In case the workload c_{o_i} is nonzero the optimal control action $\mathbf{u}_{\text{opt},o_i}$ is found solving the MPC problem (33.27)–(33.35) taking into account the contracted commodity set (33.36). The control agent state-space and control action vectors are recovered using (33.37);
- the available resources to the next control agent o_{i+1} are updated:

$$\theta^{o_{i+1}} = \theta^{o_i} - \mathbf{P}_{\text{mu},o_i}(k)\mathbf{u}_{\text{opt},o_i}(k) \tag{33.41}$$

where $\mathbf{P}_{\text{mu},o_i}(k)$ is the projection matrix from agent o_i infrastructure resources set $\mathcal{U}_{o_i}$ to the control action set $\mathcal{U}_{\max}$;

- the predictions for future decisions are updated and denoted by $\mathcal{P}^{o_{i+1}}$ replacing the control agent initial prediction $\tilde{\mathbf{u}}_{k-1,o_i}$ by the new optimal sequence found $\tilde{\mathbf{u}}_{\text{opt},o_i}$.

The procedure to follow is represented schematically in Fig. 33.3 and is presented in Algorithm 33.1.

33.4 Application Results

The presented H-MPC architecture has been used as a tool to support the resource allocation in a hinterland intermodal container terminal [11] serving three transport modalities: barge, train and truck. In this example the terminal operator perspective is considered and all parties present at the terminal are assumed cooperative in sharing

Algorithm 33.1 Hierarchical Model Predictive Control for Transportation Network

repeat
  control agents determine in parallel the expected workload using (33.39)
  control agents determine their contracted set $\mathcal{T}_{o_i}(k)$ and projections matrices
  central coordinator updates the control agents order as in (33.40)
  central coordinator initialize the infrastructure resource and future decision predictions set
  for $i = 1 \to n_c$ **do**
    update the admissible resources for control agent using (33.41)
    solve optimization problem (33.27)–(33.36) for agent o_i
    recover the global commodity set
    the optimal control action $\mathbf{u}_{\mathrm{opt},o_i}$ is the first component of $\tilde{\mathbf{u}}_{\mathrm{opt},o_i}$
    update the future decision predictions set using $\tilde{\mathbf{u}}_{\mathrm{opt},o_i}$
  end for
  apply the optimal solution $\mathbf{u}_{\mathrm{opt}}$ to the transportation network
  update time step k
until simulation time is reached

information. The container terminal is integrated in a network composed of 4 terminals and faces an average weekly throughput of 16,800 TEU (twenty-foot equivalent units). The terminal allows the berth of two barges simultaneously (Barge A and Barge B), three connections per berth are made on a daily basis. The quay cranes allow a maximum handling capacity of 90 TEU/h, for Barge A this full capacity can be used while for Barge B a maximum rate of 45 TEU/h is possible. As a consequence, Barge A and Barge B will be competing for the same resource at the quay. For the train modality there are two rail tracks (Train A and Train B) that serve four trains each in a single day. The maximum handling capacity for each rail track is 40 TEU/h but the train gate only offers a maximum capacity of 40 TEU/h. In this case both train tracks will be competing for the gate handling capacity.

Although geographically confined to the port area, this network has simultaneously two types of exogenous inputs in the form of request of containers to unload and containers to load to different transports available at the terminal, see Fig. 33.4. The request is presented as the number of containers of each type considered in the network. The containers to unload represent a push of containers towards the *Central Yard* and the containers to load are pulled from the *Central Yard*. Considering that each transport available at the terminal has simultaneously an outflow and inflow we use both flows to define a network path passing through the common node (*Central Yard*). This network path (linking unload and load areas for each available transport at the terminal) will be used to decompose the system into subsystems. In this network, the source and destination nodes are associated exclusively to a single path therefore they were categorized as connection nodes. This is a simple example that shows the need for small adjusts when applying the presented framework.

The considered terminal structural layout is translated into the network graph presented in Fig. 33.4. In this graph there are $n_{c_i} = 5$ exclusive nodes per connection, the containers are categorized into $n_t = 5$ different classes (four destinations A, B, C, D, plus empty containers) and the number of connections that can be served simultaneously at the terminal is $n_c = 5$. The *Central Yard* is a common node to

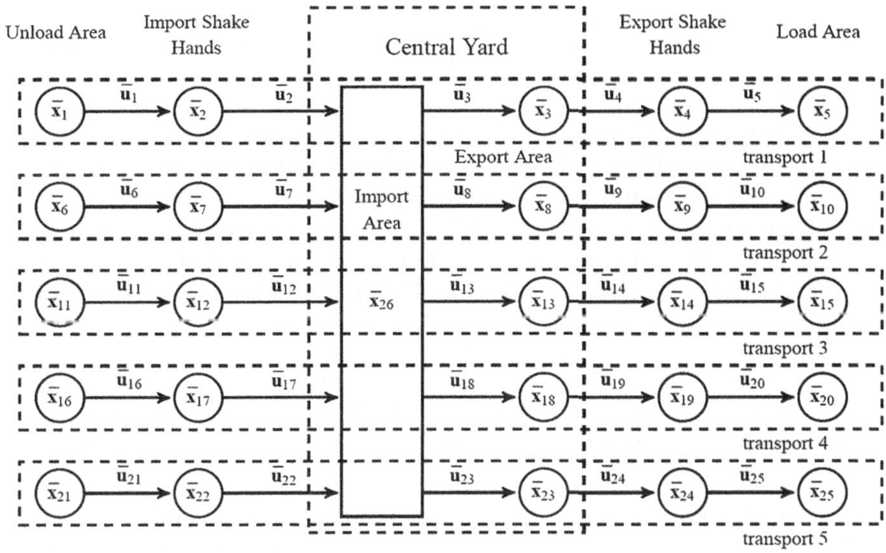

Fig. 33.4 Intermodal container terminal network

all connections and is responsible for the dynamic coupling. A total of 26 nodes are present at the terminal. For this setup the terminal is described by 130 states using the central model (33.3)–(33.11), or by 30 states per subsystem if the decomposed model (33.14)–(33.15) is used. More details about the terminal handling resources and weekly schedules at the terminal are available in [12].

33.4.1 Simulation Setup

The H-MPC architecture and the centralized MPC approach uses the same type of cost function and weights in the optimization problem to allow a fair comparison. It is possible to assign different weights to different terminal nodes, container classes and connections depending on their role in the container terminal dynamics and the desired behavior. The weight assigned to the *Import Area* at the *Central Yard* is zero as it acts as a warehouse for containers between delivery and pick up times. The weights at the *Load Area* are taken negative, such that containers are pulled from the *Import Area*. The main criterion to assign weights is related to the connection priority according to the volume of containers to handle: the higher the volume the higher the priority.

A time step of 1 h is considered. A prediction horizon of $N_\mathrm{p} = 3$ steps is used. To guarantee pulling containers from the *Import Area* to the *Load Area* the weights assigned should respect the relation

$$-(\mathbf{q}_{3+i}+\mathbf{q}_{4+i}) > \sum_{j=1}^{N_\mathrm{p}-2} \mathbf{q}_{5+i} \quad i=0,\ldots,n_\mathrm{c}-1, \tag{33.42}$$

where $\mathbf{q}_3$, $\mathbf{q}_4$ and $\mathbf{q}_5$ are the weights associated to containers located at the *Export Area*, *Export Shake Hands* and *Load Area* for the first connection respectively.

The simulation is performed using MatLab R2009b on a personal computer with a processor Intel(R) Core(TM) i7 at 1.60 GHz with 8 Gb RAM memory in a 64-bit Operating System. The optimization problem is solved at each time step of the simulation using the MPT v2.6.3 toolbox [7] with the CDD Criss–Cross solver for linear programming problems.

33.4.2 Tested Scenario

The scenario presents 1 week. Different criteria to establish the order in which the control agents should be solved in the H-MPC approach were tested; case H-MPC$_1$ the call size $\mathbf{p} = \begin{bmatrix}1 & 1 & 1 & 1 & 1\end{bmatrix}$; case H-MPC$_2$ benefiting from sustainable transport modalities $\mathbf{p} = \begin{bmatrix}2 & 2 & 1 & 1 & 0.5\end{bmatrix}$ and case H-MPC$_3$ inverting the order considered in the MPC strategy $\mathbf{p} = \begin{bmatrix}1 & 1 & 1.5 & 1.5 & 2\end{bmatrix}$. Control strategies are compared using two criteria: 1) the sum of the cost function over the entire simulation and 2) the computation time.

In Fig. 33.5 it is clear that both strategies lead to almost the same terminal behavior over time. This similarity can be confirmed by the cost function performance indicated in Table 33.1. Both strategies achieve a similar performance with a slightly better score for the centralized approach. Interesting to note that all H-MPC strategies tested achieved similar performance. In terms of computation time, the H-MPC approach outperforms the MPC strategy, Table 33.1.

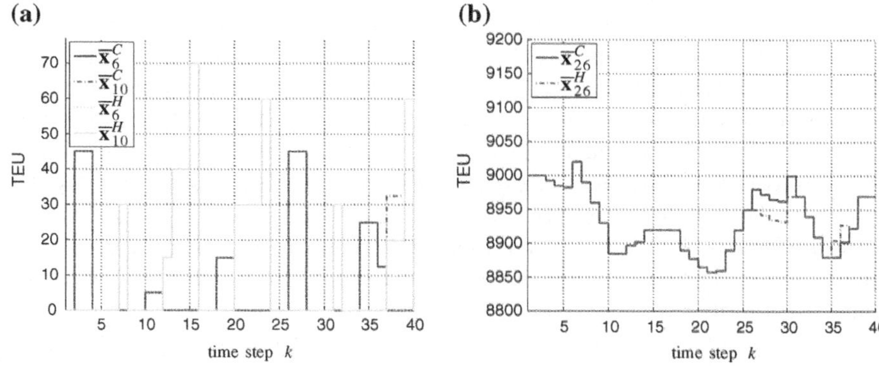

Fig. 33.5 Quantity of containers for the tested scenario (C stands for centralized MPC architecture, H stands for H-MPC architecture). **a** Barge B and **b** *Import Area* at the *Central Yard*

Table 33.1 Control strategies comparison

Strategy	Max (s)	Mean (s)	Stdv (s)	Cost function performance
H-MPC$_1$	4.71	2.66	1.14	-4.660×10^5
H-MPC$_2$	8.28	2.84	1.26	-4.660×10^5
H-MPC$_3$	7.39	2.83	1.21	-4.660×10^5
MPC	367.83	118.16	67.18	-4.766×10^5

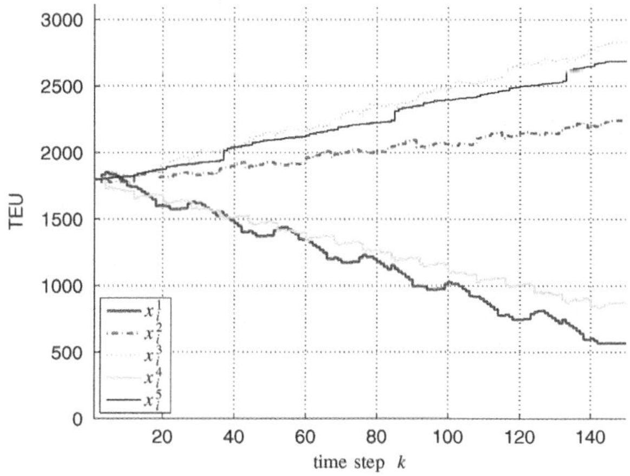

Fig. 33.6 Quantity per container class at the *Central Yard* for the H-MPC framework

Figure 33.6 shows the volume per container class at the *Import Area* in the *Central Yard* over the simulation. This model ability is partially responsible for the large problem dimension. However, when looking to the total volume at the terminal it is almost constant (around 9,000 TEU, Fig. 33.5b). The model complexity is the price to pay to have more information regarding the state of the terminal which is a key element for the transportation network.

33.5 Conclusions and Future Research

This work presents a Hierarchical Model Predictive Control (H-MPC) approach for operations management of large-scale transportation networks with multiple commodity types. The large-scale system is broken down into smaller subsystems, in a flow perspective, to which a control agent is assigned. Based on the network application domain, agents solve their problems in a push-pull flow perspective. The computation burden of considering a sparse central model to support operations management is avoided and a solution is obtained in reasonable time. Further dimension reduction is achieved using contracted and global commodity sets.

Future research will focus on decomposition methods such that the network components or parties are captured into a subsystem. The cooperation mechanism between control agents when multiple parties are present in the transportation network will also be addressed. Extending the approach to more application domains is currently under investigation.

Acknowledgments This work was supported by the Portuguese Government, through Fundação para a Ciência e a Tecnologia, under the project PTDC/EEACRO/102102/2008 - AQUANET, through IDMEC under LAETA and by the VENI project "Intelligent multi-agent control for flexible coordination of transport hubs" (project 11210) of the Dutch Technology Foundation STW, a subdivision of the Netherlands Organisation for Scientific Research (NWO).

References

1. R.K. Ahuja, T.L. Magnanti, J.B. Orlin, *Network Flows* (Prentice Hall, Upper Saddle River, 1993)
2. A. Alessandri, C. Cervellera, M. Cuneo, M. Gaggero, G. Soncin, Modeling and feedback control for resource allocation and performance analysis in container terminals. IEEE Trans. Intell. Transp. Syst. **9**(4), 601–614 (2008)
3. E.F. Camacho, C. Bordons, *Model Predictive Control in the Process Industry* (Springer, Berlin, 1995)
4. T.G. Crainic, K.H. Kim, Intermodal transportation, in *Transportation, Handbooks in Operations Research and Management Science*, ed. by C. Barnhart, G. Laporte (Elsevier, North-Holland, 2007), pp. 467–537
5. ECT Publications. Fast forward 52, Winter 2011
6. A. Hegyi, B. De Schutter, J. Hellendoorn, Optimal coordination of variable speed limits to supress schock waves. IEEE Trans. Intell. Transp. Syst. **11**(1), 102–112 (2005)
7. M. Kvasnica, P. Grieder, M. Baotić. Multi-parametric toolbox (mpt). http://control.ee.ethz.ch/~mpt/
8. S. Leirens, C. Zamora, R.R. Negenborn, B. De Schutter, Coordination in urban water supply networks using distributed model predictive control, in *Proceedings of the 2010 American Control Conference (ACC10)*, pp. 3957–7104, Baltimore, Maryland, June 2010
9. J.M. Maciejowski, *Predictive Control with Constraints* (Prentice-Hall, Harlow, 2002)
10. J.M. Maestre, D. Mu noz de la Pena, E.F. Camacho, Distributed mpc: a supply chain case, in *48th IEEE Conference on Decision and Control and 28th Chinese Control Conference*, pp. 7099–7104, Shanghai, China, Dec 2009
11. J.L. Nabais, R.R. Negenborn, M.A. Botto, Hierarchical model predictive control for optimizing intermodal container terminal operations, Submitted to a conference 2012
12. J.L. Nabais, R.R. Negenborn, M.A. Botto, A novel predictive control based framework for optimizing intermodal container terminal operations, in *Proceedings of the 3rd International Conference on Computational Logistics (ICCL)*, pp. 53–71, Shanghai, China, September 2012
13. R.R. Negenborn, Z. Lukszo, H. Hellendoorn (eds.), *Intelligent Infrastructures* (Springer, Dordrecht, 2010)
14. R.R. Negenborn, P.J. Van Overloop, T. Keviczky, B. De Schutter, Distributed model predictive control of irrigation canals. Netw. Heterogen. Media **4**(1), 359–380 (2009)
15. J. Ottjes, H. Veeke, M. Duinkerken, J. Rijsenbrij, G. Lodewijks, Simulation of a multiterminal system for container handling, in *Container Terminals and Cargo Systems*, edited by K. Hwan Kim, H.-O. Gunther (Springer, Berlin, 2007), pp. 15–36
16. M.E. Sezer, D.D. Šiljak, Decentralized control, in *The Control Handbook*, edited by W.S. Levine (CRC Press, New York, 1996), pp. 779–793
17. C.A. Silva, J.M.C. Sousa, T.A. Runkler, J.M.G. Sá da Costa, Distributed supply chain management using ant colony optimization. Eur. J. Oper. Res. **199**, 349–358 (2009)

Chapter 34
On the Use of Suboptimal Solvers for Efficient Cooperative Distributed Linear MPC

G. Pannocchia, S. J. Wright and J. B. Rawlings

Abstract We address the problem of efficient implementations of distributed Model Predictive Control (MPC) systems for large-scale plants. We explore two possibilities of using suboptimal solvers for the quadratic program associated with the local MPC problems. The first is based on an active set method with early termination. The second is based on Partial Enumeration (PE), an approach that allows one to compute the (sub)optimal solution by using a solution table which stores the information of only a few most recently optimal active sets. The use of quick suboptimal solvers, especially PE, is shown to be beneficial because more cooperative iterations can be performed in the allowed given decision time. By using the available computation time for cooperative iterations rather than local iterations, we can improve the overall optimality of the strategy. We also discuss how input constraints that involve different units (for example, on the summation of common utility consumption) can be handled appropriately. Our main ideas are illustrated with a simulated example comprising three units and a coupled input constraint.

34.1 Introduction

Model predictive control (MPC) has proved to be a highly successful advanced control technique in the process industries [11]. MPC systems are being implemented in more and more plant units. Since these units are often interconnected, interference

G. Pannocchia (✉)
Department of Civil and Industrial Engineering, University of Pisa, Pisa, Italy
e-mail: g.pannocchia@diccism.unipi.it

S. J. Wright
Computer Sciences Department, University of Wisconsin, Madison, WI, USA
e-mail: swright@cs.wisc.edu

J. B. Rawlings
Chemical and Biological Engineering Department, University of Madison, Madison, WI, USA
e-mail: rawlings@engr.wisc.edu

J. M. Maestre and R. R. Negenborn (eds.), *Distributed Model Predictive Control Made Easy*, Intelligent Systems, Control and Automation: Science and Engineering 69,
DOI: 10.1007/978-94-007-7006-5_34,

between MPC controllers is to some extent inevitable, the amount of interference depending on the steady-state and dynamic coupling of the units. These interactions may limit or degrade the overall performance of the control system. From a purely theoretical point of view, the desire for optimality should push practitioners toward control systems with a smaller number of (larger) MPC systems, each encompassing multiple units. In practice, a larger number of (smaller) MPC units may be preferred, for reasons of flexibility in the overall plant-wide control system. Furthermore, for large plants, a global centralized MPC system may be too complex and too demanding in terms of computational resources. For these reasons, researchers are investigating so-called distributed MPC strategies (see [15] for a recent survey), which interconnect different MPC units within a minimal-overhead structure, without increasing the complexity of the online problem solved by each individual MPC unit [2, 7, 14, 16, 17].

Several different types of distributed MPC systems have been considered. The first type is a fully decentralized arrangement, in which each MPC unit optimizes its own objective function, without exchanging information regarding computed input with other MPC units. The second type is a "non-cooperative" distributed MPC arrangement, in which each MPC controller makes its optimal input sequence available to the other controllers. In general, neither of these approaches have proven stability properties in closed loop. In decentralized MPC, the potential for instability comes immediately from the inherent model error induced by neglecting the interactions between different units. In both decentralized and non-cooperative MPC structures, instability may also arise because the different MPC systems optimize over different, possibly conflicting objectives. When the closed-loop system is stable, non-cooperative MPC yields a Nash equilibrium point, which may be arbitrarily far away from the centralized optimum (also known as Pareto equilibrium point).

These issues are extensively discussed by [14, 16, 17], ([13], Chap. 7) who propose the so-called "cooperative" distributed MPC architecture. In this arrangement, each local controller optimizes a common (overall) objective function over its own inputs, and shares its computed optimal input sequence with all other controllers. As discussed by ([13], Chap. 7) this scheme guarantees nominal stability, constraint satisfaction, and convergence towards the optimal centralized MPC solution, provided that no constraint involves coupling of inputs from different units.

In [8], we propose a Partial Enumeration (PE) for centralized MPC systems, that allows fast evaluation of a (possibly suboptimal) solution of the MPC problem. We show in [10] that nominal and inherently robust stability of the origin of the system holds in closed loop, for PE-based MPC. The PE approach has similar features to Explicit MPC [1, 3, 5], but the latter method is best suited to small dimensional systems.

In this chapter, we investigate the use of suboptimal quadratic programming algorithms (PE in particular) for solving local MPC problems, with the goals of (1) increasing the size and complexity of problems that can be addressed efficiently by distributed MPC systems; (2) improving the performance of existing distributed

MPC systems, by allowing faster solution of the local problems, thus allowing time for more cooperative iterations. Furthermore, we consider coupled input constraints, which may limit the achievable performance of distributed MPC systems [14], and outline how they may be handled effectively.

34.2 Boundary Conditions

34.2.1 Overall System, Input Constraints and Local Subsystems

We consider an overall input-constrained discrete time-invariant (plant) system:

$$\mathbf{x}(k+1) = \mathbf{A}\mathbf{x}(k) + \mathbf{B}\mathbf{u}(k), \qquad \mathbf{y}(k) = \mathbf{C}\mathbf{x}(k)\,, \tag{34.1}$$

$$\mathbf{D}\mathbf{u}(k) \leq \mathbf{e}\,, \tag{34.2}$$

in which $\mathbf{x}(k) \in \mathbb{R}^{n_\mathrm{x}}$, $\mathbf{u}(k) \in \mathbb{R}^{n_\mathrm{u}}$ and $\mathbf{y}(k) \in \mathbb{R}^{n_\mathrm{y}}$ are the state, input and output at time k, respectively, and $\mathbf{e} \in \mathbb{R}^{n_\mathrm{e}}$. We assume that the plant is divided into $|\mathcal{N}|$ subsystems (units), and the set of subsystems is denoted by $\mathcal{N}$. Each unit i has n_{y_i} disjoint outputs, $\mathbf{y}_i$, which are affected, possibly, by *all* plant inputs. The objective is to design for each unit a Model Predictive Controller (MPC) that optimizes over a subset of inputs denoted with $\mathbf{u}_i \in \mathbb{R}^{n_{\mathrm{u}i}}$, $i = 1, \ldots, |\mathcal{N}|$. The input vector complementary to $\mathbf{u}_i$ is denoted by $\bar{\mathbf{u}}_i \in \mathbb{R}^{n_\mathrm{u} - n_{\mathrm{u}i}}$. The subvectors $\mathbf{u}_i$ are *not* assumed to be disjoint. We define the selection matrices $\mathbf{T}_i \in \mathbb{R}^{n_{\mathrm{u}i} \times n_\mathrm{u}}$ and $\bar{\mathbf{T}}_i \in \mathbb{R}^{(n_\mathrm{u} - n_{\mathrm{u}i}) \times n_\mathrm{u}}$ to be row submatrices of the identity matrix $\mathbf{I}$ of dimension n_u, such that

$$\mathbf{u}_i = \mathbf{T}_i \mathbf{u}, \qquad \bar{\mathbf{u}}_i = \bar{\mathbf{T}}_i \mathbf{u},$$

and thus (notice that $\mathbf{T}_i^\mathrm{T}\mathbf{T}_i + \bar{\mathbf{T}}_i^\mathrm{T}\bar{\mathbf{T}}_i = \mathbf{I}$)

$$\mathbf{u} = \mathbf{T}_i^\mathrm{T}\mathbf{u}_i + \bar{\mathbf{T}}_i^\mathrm{T}\bar{\mathbf{u}}_i\,.$$

The evolution of each unit $i \in \mathcal{N}$ can thus be described as follows:

$$\mathbf{x}_i(k+1) = \mathbf{A}_i\mathbf{x}_i(k) + \mathbf{B}_i\mathbf{u}_i(k) + \bar{\mathbf{B}}_i\bar{\mathbf{u}}_i(k), \qquad \mathbf{y}_i(k) = \mathbf{C}_i\mathbf{x}_i(k)\,,$$

where we have distinguished the contribution of the inputs that belong to the ith unit from the contribution of the other inputs. The subset of constraints in (34.2) that involve only $\mathbf{u}_i$ can be written as

$$\mathbf{D}_i\mathbf{u}_i(k) \leq \mathbf{e}_i\,, \tag{34.3}$$

with $\mathbf{D}_i$ equal to the non-zero rows of $(\mathbf{D}\mathbf{T}_i^\mathrm{T})$ and with $\mathbf{e}_i$ the corresponding elements of $\mathbf{e}$. Similarly, the subset of constraints in (34.2) that involve only $\bar{\mathbf{u}}_i$ can be written as $\bar{\mathbf{D}}_i \bar{\mathbf{u}}_i(k) \leq \bar{\mathbf{e}}_i$ with $\bar{\mathbf{D}}_i$ equal to the non-zero rows of $(\mathbf{D}\bar{\mathbf{T}}_i^{\mathrm{T}})$ and with $\bar{\mathbf{e}}_i$ the corresponding elements of $\mathbf{e}$.

We consider the following assumptions.

Assumption 34.1 *(Properties of subsystems)* For each subsystem $i \in \mathcal{N}$:

(a) the pair $(\mathbf{A}_i, \mathbf{C}_i)$ is detectable and the pair $(\mathbf{A}_i, \mathbf{B}_i)$ is stabilizable;
(b) the inequality (34.3) represents *all* and *only* the constraints that involve elements of input vector $\mathbf{u}_i$;
(c) the system from the input $\bar{\mathbf{u}}_i$ to $\mathbf{y}_i$ is asymptotically stable.

Remark 34.1 **(Shared inputs)** Notice that Assumption 34.1-b admits the possibility that some inputs belong to more than one subsystem. It does, however, require that all constraints involving any element of $\mathbf{u}_i$ can be written as constraints that do not involve elements of $\bar{\mathbf{u}}_i$. Furthermore Assumption 34.1-c requires that the inputs $\bar{\mathbf{u}}_i$ that are not associated with unit i do not excite any unstable mode of $\mathbf{A}_i$.

An example demonstrates how the input vectors $\mathbf{u}_i$ can be chosen to satisfy our assumptions, even when there is apparent coupling in the input constraints.

Example 34.1 **(Coupled constraints)** Consider an overall system with four inputs and the following input constraint matrix and right-hand-side vector:

$$\mathbf{D} = \left[\begin{array}{cc|cc} 1 & 0 & 0 & 0 \\ 0 & 1 & 0 & 0 \\ 0 & 0 & 1 & 0 \\ 0 & 0 & 0 & 1 \\ -1 & 0 & 0 & 0 \\ 0 & -1 & 0 & 0 \\ 0 & 0 & -1 & 0 \\ 0 & 0 & 0 & -1 \\ \hline 0 & 1 & 1 & 0 \end{array}\right], \qquad \mathbf{e} = \left[\begin{array}{c} 1 \\ 1 \\ 1 \\ 1 \\ 1 \\ 1 \\ 1 \\ 1 \\ \hline 1 \end{array}\right].$$

The first eight rows define upper and lower bounds on the inputs whereas the last row defines an upper bound on the sum of second and third input. Suppose that we want to design two MPCs, one of which optimizes over (u_1, u_2) whereas the other one optimizes over (u_3, u_4). Since the last constraint involves both u_2 and u_3 that, in principle, belong to different units, in order to satisfy Assumption 34.1, we need to include u_3 in the set of inputs for Unit 1 and u_2 in the set of inputs of Unit 2. Thus, for Unit 1 we consider $\mathbf{u}_1 = (u_1, u_2, u_3)$ as inputs, and for Unit 2, we consider $\mathbf{u}_2 = (u_2, u_3, u_4)$ as inputs.

34.2.2 Centralized MPC Problem

Let input, state and output targets be given, and satisfy:

$$\mathbf{x}_{\text{ref}}(k) = \mathbf{A}\mathbf{x}_{\text{ref}}(k) + \mathbf{B}\mathbf{u}_{\text{ref}}(k), \qquad \mathbf{y}_{\text{ref}}(k) = \mathbf{C}\mathbf{x}_{\text{ref}}(k), \qquad \mathbf{D}\mathbf{u}_{\text{ref}}(k) \leq \mathbf{e}.$$

Notice that such targets can be either computed by a plant-wide steady-state optimizer or as the combination of $|\mathcal{N}|$ local steady-state optimizers. We define the predicted *future* deviation of state and input from their *current* target as follows:

$$\mathbf{w}(k+l) = \mathbf{x}(k+l) - \mathbf{x}_{\text{ref}}(k), \qquad \mathbf{v}(k+l) = \mathbf{u}(k+l) - \mathbf{u}_{\text{ref}}(k), \qquad l \geq 0.$$

Given an integer N (the *horizon*), we consider a finite-horizon sequence of deviation inputs $\mathbf{v}_N(k) = \mathbf{v}(k : k+N-1)$, and define the overall cost as:

$$J(\mathbf{w}(k), \mathbf{v}_N(k)) = \frac{1}{2}\sum_{l=0}^{N-1} \mathbf{w}(k+l)\mathbf{Q}\mathbf{w}(k+l) + \mathbf{v}(k+l)^{\mathrm{T}}\mathbf{R}\mathbf{v}(k+l)$$

$$+ \frac{1}{2}\mathbf{w}(k+N)^{\mathrm{T}}\mathbf{P}\mathbf{w}(k+N), \tag{34.4a}$$

$$\text{s. t.} \quad \mathbf{w}(k+l+1) = \mathbf{A}\mathbf{w}(k+l) + \mathbf{B}\mathbf{v}(k+l), \quad l = 0, \ldots, N-1. \tag{34.4b}$$

We make the following assumption.

Assumption 34.2 *(Properties of overall MPC problem)* The matrices $\mathbf{Q}$ and $\mathbf{R}$ are symmetric positive definite. The matrix $\mathbf{P}$ is given by $\mathbf{P} = \mathbf{S}_s^{\mathrm{T}}\boldsymbol{\Pi}\mathbf{S}_s$, where $\boldsymbol{\Pi}$ is the solution to the Lyapunov equation:

$$\boldsymbol{\Pi} = \mathbf{A}_s^{\mathrm{T}}\boldsymbol{\Pi}\mathbf{A}_s + \mathbf{S}_s^{\mathrm{T}}\mathbf{Q}\mathbf{S}_s\,,$$

and $(\mathbf{A}_s, \mathbf{S}_s)$ come from the real Schur decomposition of $\mathbf{A}$:

$$\mathbf{A} = \begin{bmatrix}\mathbf{S}_s & \mathbf{S}_u\end{bmatrix}\begin{bmatrix}\mathbf{A}_s & \mathbf{A}_{su} \\ 0 & \mathbf{A}_u\end{bmatrix}\begin{bmatrix}\mathbf{S}_s^{\mathrm{T}} \\ \mathbf{S}_u^{\mathrm{T}}\end{bmatrix},$$

and $\mathbf{A}_s$ contains all strictly stable eigenvalues of $\mathbf{A}$.

The centralized MPC regulator solves the following problem:

$$\mathbb{P}: \quad \min_{\mathbf{v}_N(k)} J(\mathbf{w}(k), \mathbf{v}_N(k)) \quad \text{s.t.}$$

$$\mathbf{D}\mathbf{v}(k+l) \leq \mathbf{e} - \mathbf{D}\mathbf{u}_{\text{ref}}(k), \quad l = 0, \ldots, N-1, \qquad \mathbf{S}_u^{\mathrm{T}}\mathbf{w}(k+N) = 0. \tag{34.5}$$

Remark 34.2 The constraint $\mathbf{D}\mathbf{v}(k+l) \leq \mathbf{e} - \mathbf{D}\mathbf{u}_{\text{ref}}(k)$ is equivalent to $\mathbf{D}\mathbf{u}(k+l) \leq \mathbf{e}$. The terminal constraint $\mathbf{S}_u^{\mathrm{T}}\mathbf{w}(k+N) = 0$ is present only if the system is open-loop unstable (or integrating) and is needed to zero the unstable modes at the end of the horizon N. Furthermore, the cost function term $\frac{1}{2}\mathbf{w}(k+N)^{\mathrm{T}}\mathbf{P}\mathbf{w}(k+N)$ represents the infinite horizon cost-to-go when $\mathbf{v}(k+l) = 0$ for $l \geq N$.

In the next section, we describe a cooperative Model Predictive Control architecture in which: (1) each local controller optimizes the overall objective function (34.4a) over the sequence of "local" inputs, given the expected sequence of inputs not belonging to its unit; (2) all local controllers communicate the computed solution of local inputs; (3) the previous two steps are repeated (upon convergence or for a finite number iterations).

34.3 Algorithm—Part 1: Cooperative Model Predictive Control

34.3.1 Distributed Cooperative MPC Subproblems

Let $\bar{\mathbf{v}}_{N,i}(k) = \bar{\mathbf{v}}_i(k : k + N - 1)$ be a known sequence vector (in deviation variables) of the inputs that do not belong to Unit i. We define the control problem (conceptually) solved by the ith controller at time k as follows:

$$\min_{\mathbf{v}_N(k)} J(\mathbf{w}(k), \mathbf{v}_N(k)) \quad \text{s. t.} \quad \bar{\mathbf{T}}_{N,i}\mathbf{v}_N(k) = \bar{\mathbf{v}}_{N,i}(k),$$

$$\mathbf{D}\mathbf{v}(k+l) \le \mathbf{e} - \mathbf{D}\mathbf{u}_{\text{ref}}(k), \quad l = 0, \ldots, N-1, \quad \mathbf{S}_u^{\mathrm{T}}\mathbf{w}(k+N) = 0\,, \tag{34.6}$$

in which $\bar{\mathbf{T}}_{N,i} \in \mathbb{R}^{(n_{\mathrm{u}}-n_{\mathrm{u}i})N \times n_{\mathrm{u}}N}$ is the block diagonal matrix containing N blocks, each equal to $\bar{\mathbf{T}}_i$. (Later, we use $\mathbf{T}_{N,i} \in \mathbb{R}^{n_{\mathrm{u}i}N \times n_{\mathrm{u}}N}$ to denote the block diagonal matrix whose N diagonal blocks are equal to $\mathbf{T}_i$.)

Remark 34.3 The first equality constraint enforces the inputs that do not belong to Unit i to be equal to the known value of the sequence vector $\bar{\mathbf{v}}_{N,i}(k)$.

The problem (34.6) contains a large number of decision variables that are fixed, namely, all inputs of the other units. We can eliminate these inputs and reformulate this problem as follows. Let the deviation input sequence $\mathbf{v}_N(k)$ be expressed as

$$\mathbf{v}_N(k) = \mathbf{T}_{N,i}^{\mathrm{T}}\mathbf{v}_{N,i}(k) + \bar{\mathbf{T}}_{N,i}^{\mathrm{T}}\bar{\mathbf{v}}_{N,i}(k)\,, \tag{34.7}$$

in which $\mathbf{v}_{N,i}(k) = \mathbf{v}_i(k : k + N - 1) = \mathbf{T}_{N,i}\mathbf{v}_N(k)$ is the sequence vector of inputs that belong to Unit i, and $\bar{\mathbf{v}}_{N,i}(k) = \bar{\mathbf{T}}_{N,i}\mathbf{v}_N(k)$ is the sequence vector of complementary inputs. We can now write the local control problem as follows.

$$\mathbb{P}_i : \quad \min_{\mathbf{v}_{N,i}(k)} J(\mathbf{w}(k), \mathbf{v}_N(k)) \quad \text{subject to (34.7) and}$$

$$\mathbf{D}_i\mathbf{v}_i(k+l) \le \mathbf{e}_i - \mathbf{D}_i\mathbf{u}_{\text{ref},i}(k), \quad l = 0, \ldots, N-1, \quad \mathbf{S}_u^{\mathrm{T}}\mathbf{w}(k+N) = 0. \tag{34.8}$$

Algorithm 34.1 Cooperative MPC

Input: Targets $(\mathbf{u}_{\text{ref}}(k), \mathbf{x}_{\text{ref}}(k))$, deviation state $\mathbf{w}(k) = \mathbf{x}(k) - \mathbf{x}_{\text{ref}}(k)$, overall initial sequence $\mathbf{v}_N^0(k)$ as in (34.9). Tolerance ρ, maximum number of cooperative iterations $p_{\max}$, convex combination weights $\lambda_i > 0$, $i = 1, \ldots, |\mathcal{N}|$, such that $\sum_{i=1}^{|\mathcal{N}|} \lambda_i = 1$.

Set p $\leftarrow$ 1 and done $\leftarrow$ **false**.
while done=**false do**
    for $i = 1$ **to** $|\mathcal{N}|$ **do**
        Define $\bar{\mathbf{v}}_{N,i}(k) := \bar{\mathbf{T}}_{N,i}\mathbf{v}_N^{p-1}(k)$, solve problem $\mathbb{P}_i$ to obtain $\mathbf{v}_{N,i}(k)$.
        Construct the full solution vector for Unit i from (34.7):
        $\tilde{\mathbf{v}}_{N,i}(k) = \mathbf{T}_{N,i}^{\mathrm{T}}\mathbf{v}_{N,i}(k) + \bar{\mathbf{T}}_{N,i}^{\mathrm{T}}\bar{\mathbf{v}}_{N,i}(k)$.
    end for
    Define the "overall" solution as: $\mathbf{v}_N^p(k) = \sum_{i=1}^{|\mathcal{N}|} \lambda_i \tilde{\mathbf{v}}_{N,i}(k)$.
    if $\left(\frac{\|\mathbf{v}_N^p(k) - \mathbf{v}_N^{p-1}(k)\|}{1 + \|\mathbf{v}_N^{p-1}(k)\|} \le \rho \right.$ **or** $\left. p = p_{\max} \right)$ **then**
        Set $\mathbf{v}_N^*(k) \leftarrow \mathbf{v}_N^p(k)$ and done $\leftarrow$ **true**.
    else
        $p \leftarrow p + 1$.
    end if
end while
return Overall solution $\mathbf{v}_N^*(k)$.

Remark 34.4 In (34.8) we consider only constraints for the inputs of Unit i. Constraints for the other inputs $\bar{\mathbf{D}}_i\bar{\mathbf{v}}_i(k+l) \le \bar{\mathbf{e}}_i - \bar{\mathbf{D}}_i\mathbf{u}_{\text{ref},i}(k)$, $l = 0, \ldots, N-1$, which do not involve $\mathbf{v}_{N,i}(k)$ (by Assumption 34.1-b), are assumed to be satisfied because they are enforced in the other subsystem controllers. The terminal state constraint may also contain equations that are not affected by $\mathbf{v}_{N,i}(k)$, and which can thus be eliminated.

34.3.2 Cooperative MPC Algorithm

In cooperative distributed MPC, each local MPC unit optimizes and communicates its solution with other MPC units, forming a convex combination of the all $|\mathcal{N}|$ unit solutions to obtain an overall solution. If decision time permits, this procedure is repeated iteratively to convergence, or until a specified maximum number of iterations is reached. The distributed MPC algorithm is initiated with an overall input sequence computed from the solution at the previous decision time, as follows:

$$\mathbf{v}_N^0(k) = \left[(\mathbf{u}^*(k) - \mathbf{u}_{\text{ref}}(k))^{\mathrm{T}}, \; \cdots \; , \; (\mathbf{u}^*(k+N-2) - \mathbf{u}_{\text{ref}}(k))^{\mathrm{T}}, \; 0 \right]^{\mathrm{T}}, \quad (34.9)$$

in which we emphasize that the terms $\mathbf{u}^*(k-1+l)$, $l = 1, \ldots, N-1$, are the components of the (sub)optimal vector sequence computed at the *previous* decision time, $k-1$, whereas $\mathbf{u}_{\text{ref}}(k)$ is the input target at the *current* decision time k.

Remark 34.5 The sequence $\mathbf{v}_N^0(k)$ satisfies the input constraint $\mathbf{D}\mathbf{v}(k+l) \leq \mathbf{e} - \mathbf{D}\mathbf{u}_{\text{ref}}(k)$, $l = 0, \ldots, N-1$. This sequence is feasible with respect to the terminal constraint $\mathbf{S}_u^{\mathrm{T}}\mathbf{w}(k+N) = 0$ (when present) provided that the target has not changed, that is, $(\mathbf{x}_{\text{ref}}(k), \mathbf{u}_{\text{ref}}(k)) = (\mathbf{x}_{\text{ref}}(k-1), \mathbf{u}_{\text{ref}}(k-1))$.

The distributed cooperative MPC algorithm is now given in Algorithm 34.1.

Remark 34.6 In Algorithm 34.1, communication takes place in Step 7 when the overall input sequence is computed as a convex combination of the values computed by the local controllers. The control action is sent to actuators after Step 9.

34.4 Algorithm—Part 2: Suboptimal QP Solvers

34.4.1 Introduction

Both the centralized problem $\mathbb{P}$ and each unit's problem $\mathbb{P}_i$ can be written as convex quadratic programs (QP). For problems of small to medium scale systems, they can be solved efficiently using either Active Set Method (ASM) or Interior Point Method (IPM) solvers [4, 6, 12, 18]. As the system dimension increases, however, online solvers may not provide a solution within an acceptable decision time. We consider the use of a suboptimal QP algorithm that produces a suboptimal "solution" satisfying the following conditions:

- The solution is feasible with respect to the constraints;
- The solution has a cost function no-larger than the initial guess.

The above requirements are met by, for example, an active set method with a small number of maximum iterations. In some cases (especially when the initial guess is close to optimal), the QP algorithm may attain optimality in fewer than the specified maximum iterations. In other cases, it may be far from satisfying optimality conditions, especially after a large setpoint change or disturbance.

We also consider a second alternative based on Partial Enumeration (PE) [8]. PE uses a solution table that stores a (small) number of optimal active sets and the associated piecewise linear solution [5]. This approach was applied to large-scale centralized MPC problems in [8] with significant average speed-up factors of 80–200 times over conventional QP solvers, and with small closed-loop suboptimality. Furthermore, it is proved in [10] that nominal and inherent robust stability of the origin of the closed-loop system holds for PE-based MPC. We briefly review PE next, describing the specializations needed to use it for the distributed MPC problem $\mathbb{P}_i$.

34.4.2 Partial Enumeration Algorithm

We start by rewriting the centralized MPC problem $\mathbb{P}$ as a parametric QP:

$$\min_{\mathbf{v}_N} J = \tfrac{1}{2}\mathbf{v}_N^{\mathrm{T}}\mathbf{H}\mathbf{v}_N + \mathbf{v}_N^{\mathrm{T}}\mathbf{G}\boldsymbol{x}_i + \tfrac{1}{2}\boldsymbol{x}_i{}^{\mathrm{T}}\mathbf{Z}\boldsymbol{x}_i \tag{34.10a}$$

$$\text{s.t.} \quad \mathbf{L}\mathbf{v}_N + \mathbf{M}\boldsymbol{x}_i \le \mathbf{e}, \quad \mathbf{E}\mathbf{v}_N + \mathbf{F}\boldsymbol{x}_i = 0, \tag{34.10b}$$

in which $\boldsymbol{x}_i = [\mathbf{w}(k)^{\mathrm{T}},\ \mathbf{u}_{\mathrm{ref}}(k)^{\mathrm{T}}]^{\mathrm{T}}$ (with dimension $n_{x_i} = n_{\mathrm{x}} + n_{\mathrm{u}}$) is the parameter that changes at each decision time point, while all other terms are constant and omitted for the sake of brevity (see, for example, [10]). The equality constraint in (34.10b) is equivalent to the terminal constraint $\mathbf{S}_u^{\mathrm{T}}\mathbf{w}(k+N) = 0$, and thus is present only for open-loop unstable systems.

Explicit control methods [5] perform an offline partitioning of the space of x_i into a number of *regions*, each corresponding to a different active set and each defined by a tuple [10]:

$$(\boldsymbol{\Omega}_P, \boldsymbol{\Omega}_D, \boldsymbol{\Gamma}, \omega_P, \omega_D, \gamma, \mathbf{J}_2, \mathbf{j}_1, J_0). \tag{34.11}$$

The online process is to find the region for which the following conditions on primal and dual feasibility are satisfied by the current parameters:

$$\boldsymbol{\Omega}_P\boldsymbol{x}_i \le \omega_P, \qquad \boldsymbol{\Omega}_D\boldsymbol{x}_i \le \omega_D, \tag{34.12}$$

and then compute the optimal solution $\mathbf{v}_N^*$ from:

$$\mathbf{v}_N^* = \boldsymbol{\Gamma}\boldsymbol{x}_i + \gamma. \tag{34.13}$$

The optimal objective value is then given by:

$$J^* = \tfrac{1}{2}\boldsymbol{x}_i{}^{\mathrm{T}}\mathbf{J}_2\boldsymbol{x}_i + \mathbf{j}_1^{\mathrm{T}}\boldsymbol{x}_i + J_0. \tag{34.14}$$

In PE, we store a table of tuples (34.11) for a fixed number of active sets that were optimal at recent decision time points. Online, we scan the table to see if it contains a tuple for which the optimality conditions (34.12) are satisfied for the current parameter $\boldsymbol{x}_i$. If so, we compute the optimal solution from (34.13). Otherwise, we compute a suboptimal solution (as described below) and inject this solution into the plant. Afterwards, a QP solver is called to compute the optimal solution $\mathbf{v}_N^*$ and derive the tuple (34.11) for the corresponding optimal active set. This entry is inserted into the table as soon as it becomes available. When the table exceeds its maximum size (defined by the user), we delete the entry that was used least recently. By bounding the table size in this way, we ensure a fast table lookup process, while keeping the table entries relevant to the operating conditions for the plant.

Algorithm 34.2 Partial enumeration solver

Input: Table with M entries, each a tuple of the form (34.11); current parameter $\boldsymbol{x}_i$; candidate sequence $\mathbf{v}_N^0$, its cost J^0 if feasible (otherwise $J^0 = \infty$); max. table size $M_{\max}$.

Set: done $\leftarrow$ **false**, $m \leftarrow 0$.
while ($m < M$ **and** done=**false**) **do**
  Set $m \leftarrow m + 1$. Extract the m-th tuple from the table.
  if $\boldsymbol{\Omega}_P \boldsymbol{x}_i \le \boldsymbol{\omega}_P$ **then** {Entry is feasible}
    if $\boldsymbol{\Omega}_D \boldsymbol{x}_i \le \boldsymbol{\omega}_D$ **then** {Entry is optimal}
      Compute optimal solution $\mathbf{v}_N^*$ from (34.13). Move tuple m to first position of the table. Set done $\leftarrow$ **true**.
    else {Entry is suboptimal}
      Compute cost J from (34.14).
      if $J < J^0$ **then**
        Set $\mathbf{v}_N^0 \leftarrow \boldsymbol{\Gamma} \boldsymbol{x}_i + \boldsymbol{\gamma}$, $J^0 \leftarrow J$.
      end if
    end if
  end if
end while
if done=**false then** {No optimal entry found}
  if $J^0 = \infty$ **then** {$\mathbf{v}_N^0$ is infeasible}
    Set $(\hat{\mathbf{L}}, \hat{\mathbf{e}})$ empty matrix/vector.
    while done=**false do**
      Solve the following (equality-constrained) QP:

$$\min_{\mathbf{v}_N} J \quad s.t. \quad \hat{\mathbf{L}} \mathbf{v}_N = \hat{\mathbf{e}}, \; \mathbf{E} \mathbf{v}_N + \mathbf{F} \boldsymbol{x}_i = 0$$

      Let $(\mathbf{v}_N, \boldsymbol{\lambda})$ be minimizer and Lagrange multiplier for $\hat{\mathbf{L}} \mathbf{v}_N = \hat{\mathbf{e}}$, resp.
      if $\mathbf{L} \mathbf{v}_N \le \mathbf{e} - \mathbf{M} \boldsymbol{x}_i$ **then** {Feasible solution found}
        Set done $\leftarrow$ **true**.
      else {Infeasible solution: add/remove constraints}
        Define $(\hat{\mathbf{L}}, \hat{\mathbf{e}})$ as the rows of $(\mathbf{L}, \mathbf{e} - \mathbf{M} \boldsymbol{x}_i)$ for the violated inequalities plus the rows of previous $(\hat{\mathbf{L}}, \hat{\mathbf{e}})$ with nonnegative λ.
      end if
    end while
  else {$\mathbf{v}_N^0$ is feasible}
    Set: $\mathbf{v}_N^* \leftarrow \mathbf{v}_N^0$
  end if
  {Table update, performed after returning $\mathbf{v}_N^*$} Solve QP (34.10) to find the optimal tuple (34.11). If $M = M_{\max}$, delete the final entry in the table. Insert the new entry in first position of the table, set $M \leftarrow \min(M_{\max}, M + 1)$.
end if
return (Sub)optimal sequence $\mathbf{v}_N^*$, updated table.

34.4.3 Application of Partial Enumeration to Cooperative MPC

Each $\mathbb{P}_i$ in (34.8) can be written as the following parametric QP:

$$\min_{\mathbf{v}_{N,i}} \tfrac{1}{2}\mathbf{v}_{N,i}^{\mathrm{T}}\mathbf{H}_i\mathbf{v}_{N,i} + \mathbf{v}_{N,i}^{\mathrm{T}}\mathbf{G}_i\boldsymbol{x}_{ii} + \tfrac{1}{2}\boldsymbol{x}_{ii}\mathbf{Z}_i\boldsymbol{x}_{ii} \tag{34.15a}$$

$$\text{s.t.}\quad \mathbf{L}_i\mathbf{v}_{N,i} + \mathbf{M}_i\boldsymbol{x}_{ii} \leq \mathbf{e}, \quad \mathbf{E}_i\mathbf{v}_{N,i} + \mathbf{F}_i\boldsymbol{x}_{ii} = 0\,, \tag{34.15b}$$

in which $\boldsymbol{x}_{ii} = [\mathbf{w}(k)^{\mathrm{T}},\ \mathbf{u}_{\mathrm{ref}}(k)^{\mathrm{T}},\ \bar{\mathbf{v}}_{N,i}^{\mathrm{T}}]^{\mathrm{T}}$ is the parameter that changes at each decision time. (The difference with centralized MPC problem is that this parameter includes the sequence of inputs that do not belong to Unit i.) For each decision time, the parameter $\boldsymbol{x}_{ii}$ changes at each cooperative iteration because the sequence $\bar{\mathbf{V}}_{N,i}$ is updated. The matrices appearing in (34.15) are:

$$\mathbf{H}_i = \mathbf{T}_{N,i}\mathbf{H}\mathbf{T}_{N,i}^{\mathrm{T}},\ \mathbf{G}_i = \left[\mathbf{T}_{N,i}\mathbf{G}\,\mathbf{T}_{N,i}H\bar{\mathbf{T}}_{N,i}^{\mathrm{T}}\right]\mathbf{Z}_i = \begin{bmatrix}\mathbf{Z} & \mathbf{G}^{\mathrm{T}}\bar{\mathbf{T}}_{N,i}^{\mathrm{T}} \\ \bar{\mathbf{T}}_{N,i}\mathbf{G} & \bar{\mathbf{T}}_{N,i}\mathbf{H}\bar{\mathbf{T}}_{N,i}^{\mathrm{T}}\end{bmatrix},$$

$$\mathbf{L}_i = \mathbf{L}\mathbf{T}_{N,i}^{\mathrm{T}},\ \mathbf{M}_i = \left[\mathbf{M}\ \mathbf{L}\bar{\mathbf{T}}_{N,i}^{\mathrm{T}}\right],\ \mathbf{E}_i = \mathbf{E}\mathbf{T}_i^{\mathrm{T}},\ \mathbf{F}_i = \left[\mathbf{F}\ \mathbf{E}\bar{\mathbf{T}}_{N,i}^{\mathrm{T}}\right].$$

Notice that several rows of $\mathbf{L}_i$ and $\mathbf{E}_i$ are zero—namely, those containing constraints that do not involve the local decision variable $\mathbf{v}_{N,i}$—and can thus be deleted. The QP (34.15) has the same form as (34.10), the main difference being that the parameter $\boldsymbol{x}_i$ is augmented with the known sequence of inputs not belonging to Unit i. Given this significant increase in dimensionality, a fully explicit method is impossible even for small dimensional systems. On the other hand, PE (Algorithm 34.2) can be readily applied to the problem (34.15).

34.5 Theoretical Properties

For simplicity, we assume that the target $(\mathbf{x}_{\mathrm{ref}}(k), \mathbf{u}_{\mathrm{ref}}(k))$ be constant, so that its time argument can be omitted. Further, let $\kappa_N(\mathbf{w}(k)) = \mathbf{v}^*(k)$ be the first component of the (sub)optimal control sequence $\mathbf{v}_N^*(k)$ computed for the initial state $\mathbf{w}(k) = \mathbf{x}(k) - \mathbf{x}_{\mathrm{ref}}$ by any of the methods discussed in this chapter. Notice that $\kappa_N(\cdot)$ is a unique-valued function only for the centralized case based on exact QP solver. In all other cases, $\kappa_N(\cdot)$ is a set-valued map in general. Thus we describe the closed-loop system (in deviation state) as a difference inclusion:

$$\mathbf{w}(k+1) \in \boldsymbol{\Phi}(\mathbf{w}(k)) = \{\mathbf{A}\mathbf{w}(k) + \mathbf{B}\mathbf{v}(k) \mid \mathbf{v}(k) \in \kappa_N(\mathbf{w}(k))\}. \tag{34.16}$$

Let $\mathcal{W}_N \subseteq \mathbb{R}^{n_x}$ be the set for which problem $\mathbb{P}$ in (34.5) admits a solution. Using the theory developed in [9, 10, 16], the following results can be established for all controllers. (We omit the details and proof in the sake of space.)

Theorem 34.1 *The origin of the closed-loop system* $\mathbf{w}(k+1) \in \boldsymbol{\Phi}(\mathbf{w}(k))$ *is exponentially stable on any compact subset of* $\mathcal{W}_N$.

The previous theorem is an extension of the stability results presented in [16] to the case of local suboptimal (rather than optimal) controllers.

Theorem 34.2 *The origin of the closed-loop system* $\mathbf{w}(k+1) \in \boldsymbol{\Phi}(\mathbf{w}(k))$ *is strongly robustly exponentially stable with respect to process and measurement errors on a well-defined compact subset* $\mathcal{C} \subseteq \mathcal{W}_N$.

The previous theorem is an extension of the robust stability results presented in [9, 10] to the case of distributed cooperative (rather than centralized) MPC.

For any initial state deviation $\mathbf{w}(k) \in \mathcal{W}_N$, let $\kappa_N^*(\mathbf{w}(k))$ and $\kappa_N^p(\mathbf{w}(k))$ represent the first component of the solution of the centralized MPC problem (34.5) and of Algorithm 34.1 at the pth iteration, respectively, both computed with an exact QP solver. The following result holds ([16], Lemma 15).

Theorem 34.3 *For any* $\mathbf{w}(k) \in \mathcal{W}_N$, *there holds:* $\lim_{p\to\infty} \kappa_N^p(\mathbf{w}(k)) = \kappa_N^*(\mathbf{w}(k))$.

34.6 Application Example

34.6.1 Overall System and Units Definition

We consider a stable system with 8 inputs, 8 outputs and 48 states. The DC gain of the system is shown below. (Complete details are omitted in the sake of space but are available upon request.)

$$G(0) = \left[\begin{array}{ccc|ccc|cc} -0.259 & -0.783 & 0.347 & 0.221 & 0.029 & -0.309 & -0.18 & 0.700 \\ -0.649 & 0.817 & -0.818 & -0.015 & -0.233 & -0.047 & 0.214 & 0.290 \\ 1.000 & -0.101 & 0.002 & 0.600 & -0.157 & -0.418 & -0.054 & -0.002 \\ \hline 0.217 & -0.162 & -0.036 & -0.218 & 0.260 & 1.000 & 0.085 & 0.444 \\ -0.046 & -0.103 & -0.016 & -0.156 & -0.477 & 0.037 & -0.069 & 0.600 \\ 0.020 & -0.500 & -0.332 & -0.406 & -0.166 & -0.287 & -0.057 & 0.570 \\ \hline 0.063 & -0.078 & 0.199 & 0.369 & 0.700 & -0.170 & -0.177 & -1.000 \\ -0.123 & 0.161 & -0.400 & -0.070 & 0.081 & -0.548 & 0.748 & -0.953 \end{array}\right]. \tag{34.17}$$

Each input of the system is constrained to the interval $[-1, 1]$, and the following coupled input constraint must hold at all times:

$$\begin{bmatrix} 0 & 0 & 0 & 1 & 1 & 0 & 0 & 1 \end{bmatrix} \mathbf{u}(k) \leq 1. \tag{34.18}$$

In the MPC design we use: $\mathbf{Q} = \mathbf{I}$, $\mathbf{R} = \mathbf{I}$, and $N = 30$.

We suppose that this plant is divided into three units. Table 34.1 shows outputs and inputs for each unit, under two possible different distributed MPC designs.

Table 34.1 Outputs and inputs of the three units, according to two design schemes: Design A (existing), Design B (optimal)

	Outputs	Inputs: Design A	Inputs: Design B
MPC Unit 1	(y_1, y_2, y_3)	(u_1, u_2, u_3)	$(u_1, u_2, u_3, u_4, u_8)$
MPC Unit 2	(y_4, y_5, y_6)	(u_4, u_5, u_6)	$(u_3, u_4, u_5, u_6, u_8)$
MPC Unit 3	(y_7, y_8)	(u_7, u_8)	(u_3, u_4, u_7, u_8)

In Design A, which can be regarded as the existing scheme for this plant, no inputs belong to more than one MPC unit at a time. However, because of the coupled constraint (34.18), this input partition scheme fails to satisfy Assumption 34.1-b, so convergence to the optimal centralized solution (Theorem 34.3) cannot be guaranteed. Design B is an alternative input partition scheme in which the inputs (u_3, u_4, u_8), which appear in the coupled constrained, belong to all three MPC units.

34.6.2 Effect of Coupled Constraints

We first investigate the convergence properties for the two distributed MPC architectures. We consider the case that at decision time 10, the input target changes from the origin to $\mathbf{u}_{\text{ref}} = [0, 0, 0, 0.5, 0.2, 0, 0, 0.3]^{\text{T}}$, thus making the coupled constraint active. We report in Fig. 34.1 the closed-loop response of $u_3 + u_4 + u_8$ obtained by three controllers: CMPC is the centralized controller, DMPC-A is the distributed

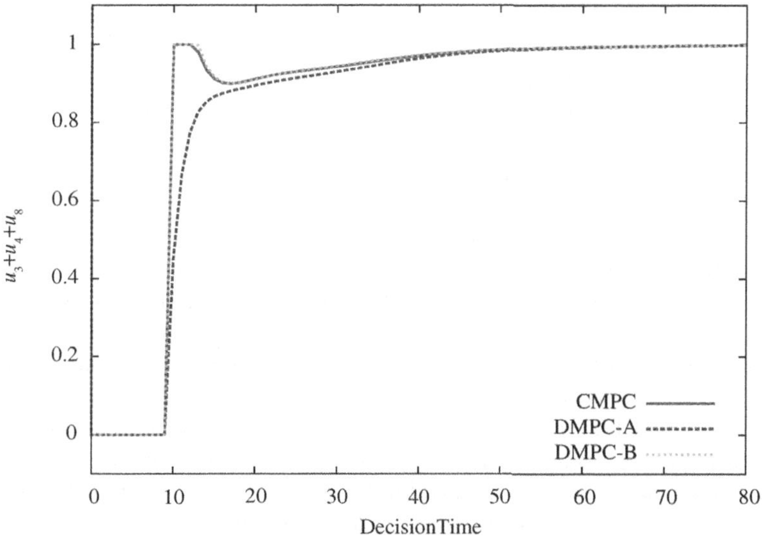

Fig. 34.1 Effect of coupled input constraints: closed-loop response of $u_3 + u_4 + u_8$ for centralized MPC (*CMPC*), distributed cooperative MPC based on Design A (*DMPC-A*), distributed cooperative MPC based on Design B (*DMPC-B*). Both *DMPC-A* and *DMPC-B* make $p = 1$ iteration

Table 34.2 Suboptimality of DMPC-A and DMPC-B for different number of cooperative iterations during a set point change that activates the coupled constraint

DMPC design	S_I			
	$p = 1$	$p = 5$	$p = 10$	$p = 50$
DMPC-A	22.9	0.885	0.682	0.673
DMPC-B	2.29	$5 \cdot 10^{-2}$	$5 \cdot 10^{-4}$	$2 \cdot 10^{-8}$

control structure with $p = 1$ iteration based on Design A, DMPC-B is the distributed control structure with $p = 1$ iteration based on Design B. For this study, we solve the optimal control problems exactly. We report in Table 34.2, the suboptimality of DMPC-A and DMPC-B as the number of iterations p increases, defined by the index:

$$S_I = 100 \cdot \frac{J_{CL} - J_{CL}^*}{J_{CL}},$$

in which J_{CL} is the closed-loop cost for the considered controller and J_{CL}^* is the closed-loop cost for the optimal centralized controller.

As expected, DMPC-B handles the coupled constraint much better than DMPC-A. As the number of iterations increases, DMPC-B converges to the optimal centralized MPC solution, whereas the suboptimality index for DMPC-A does not go to zero.

34.6.3 Comparison of Centralized and Distributed MPC Schemes Based on Different Solvers

We now present the results for cooperative distributed controllers that solve the local MPC problems $\mathbb{P}_i$ in different ways: via an exact (active set) QP solver, via an early terminated (active set) QP solver, and via PE. We are interested in assessing the suboptimality of each approach, as well as its computational efficiency quantified by[1] the average CPU time and the maximum CPU time (in seconds). We consider a closed-loop simulation of 5,000 decision times, in the presence of random output noise, affecting the state estimate and the target at each decision time, and 14 large target changes. When PE is used, each MPC unit deploys an initially empty table of maximum dimension $M_{\max} = 10$. For completeness, we also report the results of the centralized controller based on three solvers: exact QP, early terminated QP, and PE. The results are summarized in Table 34.3.

We notice that for each controller scheme (centralized or decentralized), PE is very effectively, as it achieves nearly the same suboptimality index as exact QP, but with much smaller (average and maximum) CPU times. Moreover, distributed MPC (either based on exact QP or PE) achieves a suboptimality index much less than 1 $p = 5$ iterations. In addition, the early-terminated QP is (obviously) faster than

[1] Computations are performed using GNU Octave on a MacBook Air (1.8 GHz, Intel Core i7).

Table 34.3 Suboptimality and computational efficiency for centralized (CMPC) and distributed (DMPC-B) schemes based on: (1) QP, (2) early terminated QP (max. 10 iterations) and (3) PE

Index	CMPC			DMPC-B ($p = 1$)			DMPC-B ($p = 5$)		
	QP	QP (e.t.)	PE	QP	QP (e.t.)	PE	QP	QP (e.t.)	PE
S_I	0	15.2	0.1	14.7	28.2	14.8	0.60	4.41	0.612
Aver. CPU time	0.126	0.091	0.032	0.036	0.0323	0.0038	0.132	0.127	0.014
Max. CPU time	5.97	0.339	0.513	0.564	0.120	0.0353	1.12	0.462	0.130

the exact QP, at the cost of a much higher suboptimality index. It is also interesting to note that use of an early-terminated QP in a distributed scheme with $p = 5$ cooperative iterations takes similar CPU times to an early-terminated QP in the centralized scheme, but the suboptimality of the distributed scheme is lower. This is due to the fact that each local QP problem is simpler to solve than the centralized one (having fewer decision variables and constraints), so that an early terminated QP is "less suboptimal" for the local problems than for the centralized problem.

34.7 Conclusions

We discussed in this paper efficient implementations for distributed cooperative Model Predictive Control, possibly in the presence of coupled input constraints. The approach is based on suboptimal solutions of the MPC problems solved by each local controller in order to compute each solution more quickly and to (possibly) perform more cooperative iterations within the allowed decision time. We discussed two alternatives: an early terminated active-set QP solver and the Partial Enumeration algorithm. The latter method solves the MPC problem by means of a small solution table, which includes the most recently optimal active sets. If the optimal solution is not found in the table, a quick suboptimal solution is computed for closed-loop control. Meanwhile, the optimal active set is evaluated and inserted into the table, possibly deleting the least recently optimal active set. In this way the size of the table is kept small, thus limiting the required time for scanning it. Finally, we presented a simulation example of an 8 input 8 output plant comprising three units in which we achieved, especially using Partial Enumeration, relevant speedup factors and negligible suboptimality compared to exact QP-based centralized MPC.

References

1. A. Alessio, A. Bemporad, A survey on explicit model predictive control, in *Proceedings of the International Workshop on Assessment and Future Directions of NMPC*, Pavia, Italy, Sept 2008
2. Elvira Marie B. Aske, Stig Strand, Sigurd Skogestad. Coordinator MPC for maximizing plant throughput. Comp. Chem. Eng.**32,** 195–204, (2008)

3. M. Baotic, F. Borrelli, A. Bemporad, M. Morari, Efficient on-line computation of constrained optimal control. SIAM J. Control Optim. **47**, 2470–2489 (2008)
4. R.A. Bartlett, L.T. Biegler, J. Backstrom, V. Gopal, Quadratic programming algorithms for large-scale model predictive control. J. Process Control. **12**(7), 775–795 (2002)
5. A. Bemporad, M. Morari, V. Dua, E.N. Pistikopoulos, The explicit linear quadratic regulator for constrained systems. Automatica **38**, 3–20 (2002)
6. M. Diehl, H.J. Ferreau, N. Haverbeke, Efficient numerical methods for nonlinear MPC and moving horizon estimation, in *Proceedings of International Workshop on Assessment and Future Directions of NMPC*, Pavia, Italy, Sept 2008)
7. W.B. Dunbar, Distributed receding horizon control of dynamically coupled nonlinear systems. IEEE Trans. Auto. Control. **52**, 1249–1263 (2007)
8. G. Pannocchia, J.B. Rawlings, S.J. Wright, Fast, large-scale model predictive control by partial enumeration. Automatica **43**(5), 852–860 (2007)
9. G. Pannocchia, J.B. Rawlings, S.J. Wright, Conditions under which suboptimal nonlinear MPC is inherently robust. Syst. Control Lett. **60**, 747–755 (2011)
10. G. Pannocchia, S.J. Wright, J.B. Rawlings, Partial enumeration MPC: Robust stability results and application to an unstable CSTR. J. Process Control. **21**(10), 1459–1466 (2011)
11. S.J. Qin, T.A. Badgwell, A survey of industrial model predictive control technology. Control Eng. Pract. **11**, 733–764 (2003)
12. C.V. Rao, S.J. Wright, J.B. Rawlings, Application of interior-point methods to model predictive control. J. Optim. Theory Appl. **99**, 723–757 (1998)
13. J.B. Rawlings, D.Q. Mayne, *Model Predictive Control: Theory and Design*, (Nob Hill Publishing, Madison, 2009). pp. 576. ISBN 978-0-9759377-0-9
14. J.B. Rawlings, B.T. Stewart, Coordinating multiple optimization-based controllers: New opportunities and challanges. J. Process Control. **18**, 839–845 (2008)
15. R. Scattolini, A survey on hierarchical and distributed model predictive control. J. Process Control. **19**, 723–731 (2009)
16. B.T. Stewart, A.N. Venkat, J.B. Rawlings, S.J. Wright, G. Pannocchia, Cooperative distributed model predictive control. Syst. Control Lett. **59**, 460–469 (2010)
17. A.N. Venkat, J.B. Rawlings, S.J. Wright, Distributed model predictive control of large-scale systems. in *Assessment and Future Directions of Nonlinear Model Predictive Control*, (Springer, Hidelberg, 2007), pp. 591–605
18. Y. Wang, S. Boyd, Fast model predictive control using online optimization. IEEE Trans. Control Syst. Technol. **18**(2), 267–278 (March 2010)

Chapter 35
Cooperative Distributed MPC Integrating a Steady State Target Optimizer

A. Ferramosca, D. Limon and A. H. González

Abstract In this chapter, a distributed MPC strategy suitable for changing setpoints is described. Based on a cooperative distributed control structure, an extended-cost MPC formulation is proposed, which integrates the problem of computing feasible steady state targets—usually known as Steady State Target Optimizer (SSTO) optimization problem—and the dynamic control problem into a single optimization problem. The proposed controller is able to drive the system to any admissible setpoint in an admissible way, ensuring feasibility under any change of setpoint. It also provides a larger domain of attraction than standard MPC for regulation, due to the particular terminal constraint. Moreover, the controller ensures convergence to the centralized optimum, even in case of coupled constraints. This is possible thanks to the design of the cost function, which integrates the SSTO, and to the *warm start* algorithm used to initialize the optimization algorithm. A numerical simulation illustrates the benefits of the proposal.

35.1 Introduction

This chapter describes a distributed MPC (DMPC) suitable to work in the context of a hierarchical control structure, which is the usual control structure in process industries [3]. At the top of this structure, an economic scheduler and planner decides

A. Ferramosca (✉) · A. H. González
Institute of Technological Development for the Chemical Industry, CONICET-Universidad Nacional del Litoral, Santa Fe, Argentina
e-mail: ferramosca@santafe-conicet.gov.ar

A. H. González
e-mail: alejgon@santafe-conicet.gov.ar

D. Limon
Departamento de Ingeniería de Sistemas y Automática, Universidad de Sevilla, Escuela Superior de Ingenieros, Sevilla, Spain
e-mail: limon@cartuja.us.es

J. M. Maestre and R. R. Negenborn (eds.), *Distributed Model Predictive Control Made Easy*, Intelligent Systems, Control and Automation: Science and Engineering 69,
DOI: 10.1007/978-94-007-7006-5_35, © Springer Science+Business Media Dordrecht 2014

what, when and how much the plant has to produce, taking into account information from the market and from the plant. The output of this layer are production goals, prices, economic cost functions and constraints that are sent to a Real Time Optimizer (RTO). The RTO is a model-based system, operated in closed-loop, whose task is to implement the economic decision in real time. So, it provides the setpoints to the control level. It employs a complex stationary model of the plant and for this reason its sampling time is usually larger than the settling time of the plant. The setpoints calculated by the RTO are sent to the advanced control level, where an advanced control strategy, usually an MPC, calculates the optimal control action to be sent to the plant, in order to regulate the plant as close as possible to the setpoint, taking into account a dynamic model of the plant and constraints and ensuring stability. These control actions are sent to the low-level controllers, which then regulate the plant.

The problem of this hierarchical control structure is that the economic setpoints calculated by the RTO may be inconsistent or unreachable with respect to the dynamic layer, due to the discrepancies between the stationary model of the RTO and the dynamic model used by the MPC, as well as the different time scale of these two layers. Hence, the optimal setpoint calculated by the RTO may not coincide with the dynamic steady state of the system [9].

A way to avoid this problem is the so-called two-layer structure [12]: an upper optimization level is added in between of RTO and MPC. This level, referred as the steady state target optimizer (SSTO), calculates the steady state to which the system has to be stabilized, solving a linear or quadratic programming problem and taking into account information from the RTO. The plant model used in this intermediate level is the same as the MPC one, thus reducing inconsistencies [3]. In [13] the authors analyze the implementation of steady state optimizer with linear MPC.

In [5, 10] the authors propose an MPC for tracking which implements a steady state optimizer into the MPC control problem by means of a so-called *offset cost function* added to the MPC cost function. This controller ensures that under any change of the target provided by the RTO, the closed-loop system maintains the feasibility of the controller and ensures local optimality.

In this chapter, the control formulation proposed in [5, 10] is extended to deal with distributed control schemes. A new formulation of a stabilizing cooperative distributed MPC is presented, which integrates the SSTO problem into the same dynamic control problem and is capable to drive the plant to any changing setpoint provided by the RTO.

35.2 Problem Statement

Consider a system described by a linear invariant discrete time model

$$\begin{aligned}\mathbf{x}(k+1) &= \mathbf{A}\mathbf{x}(k) + \mathbf{B}\mathbf{u}(k) \\ \mathbf{y}(k) &= \mathbf{C}\mathbf{x}(k) + \mathbf{D}\mathbf{u}(k)\end{aligned} \tag{35.1}$$

where $\mathbf{x} \in \mathbb{R}^{n_x}$ is the system state, $\mathbf{u} \in \mathbb{R}^{n_u}$ is the current control vector, $\mathbf{y} \in \mathbb{R}^{n_y}$ is the controlled output. The solution of this system for a given sequence of control inputs $\mathbf{u}(k : k+l)$, $l \in \mathbb{Z}$, and initial state $\mathbf{x}$ is denoted as $\mathbf{x}(j) = p_{h,i}(j; \mathbf{x}, \mathbf{u}(k : k+l))$, $j \in \mathbb{Z}$, where $\mathbf{x} = p_{h,i}(0; \mathbf{x}, \mathbf{u}(k : k+l))$. The state of the system and the control input applied at sampling time k are denoted as $\mathbf{x}(k)$ and $\mathbf{u}(k)$ respectively. The system is subject to hard constraints on state and control:

$$\mathbf{x}(k) \in \mathcal{X}, \quad \mathbf{u}(k) \in \mathcal{U} \tag{35.2}$$

for all $k \geq 0$. $\mathcal{X} \subset \mathbb{R}^{n_x}$ and $\mathcal{U} \subset \mathbb{R}^{n_u}$ are compact convex polyhedra containing the origin in their interior. It is assumed that the following hypothesis hold.

Assumption 35.1 The pair $(\mathbf{A}, \mathbf{B})$ is stabilizable and the state is measured at each sampling time.

The steady state, input and output of the plant $(\mathbf{x}_s, \mathbf{u}_s, \mathbf{y}_s)$ are such that (35.1) is fulfilled, i.e. $\mathbf{x}_s = \mathbf{A}\mathbf{x}_s + \mathbf{B}\mathbf{u}_s$, and $\mathbf{y}_s = \mathbf{C}\mathbf{x}_s + \mathbf{D}\mathbf{u}_s$.

We define the sets of admissible equilibrium states, inputs and outputs as

$$\mathcal{Z}_s = \{(\mathbf{x}, \mathbf{u}) \in \mathcal{X} \times \mathcal{U} \mid \mathbf{x} = \mathbf{A}\mathbf{x} + \mathbf{B}\mathbf{u}\} \tag{35.3}$$

$$\mathcal{X}_s = \{\mathbf{x} \in \mathcal{X} \mid \exists \mathbf{u} \in \mathcal{U} \text{ such that } (\mathbf{x}, \mathbf{u}) \in \mathcal{Z}_s\} \tag{35.4}$$

$$\mathcal{Y}_s = \{\mathbf{y} = \mathbf{C}\mathbf{x} + \mathbf{D}\mathbf{u} \mid (\mathbf{x}, \mathbf{u}) \in \lambda Z_s\} \tag{35.5}$$

where $\lambda \in (0, 1)$ is a parameter chosen arbitrarily close to 1. Notice that $\mathcal{X}_s$ is the projection of $\mathcal{Z}_s$ onto $\mathcal{X}$.

Usually, in practice a steady state target optimizer (SSTO) is added in between the RTO and the MPC, in order to calculate the steady state $(\mathbf{x}_s, \mathbf{u}_s, \mathbf{y}_s)$ to which the system has to be stabilized, solving a linear or quadratic programming problem and taking into account the target operation point $(\mathbf{u}_{\text{ref}}, \mathbf{y}_{\text{ref}})$ provided by the RTO. The steady state optimizer is usually of the form:

$$\begin{aligned} (\mathbf{x}_s, \mathbf{u}_s, \mathbf{y}_s) &= \arg \min_{\mathbf{x},\mathbf{u},\mathbf{y}} \ell_{eco}(\mathbf{u}, \mathbf{y}, \mathbf{u}_{\text{ref}}, \mathbf{y}_{\text{ref}}) \\ \text{s.t.} & \\ & \mathbf{x} \in \mathcal{X}, \ \mathbf{u} \in \mathcal{U} \\ & \mathbf{x} = \mathbf{A}\mathbf{x} + \mathbf{B}\mathbf{u}, \quad \mathbf{y} = \mathbf{C}\mathbf{x} + \mathbf{D}\mathbf{u} \end{aligned} \tag{35.6}$$

where ℓ_{eco} can be a quadratic cost of the form $\|\mathbf{y} - \mathbf{y}_{\text{ref}}\|^2_{\mathbf{Q_t}} + \|\mathbf{u} - \mathbf{u}_{\text{ref}}\|^2_{\mathbf{R_t}}$, for suitable matrices $\mathbf{Q_t}$ and $\mathbf{R_t}$, or even a local convex approximation of the RTO economic function.

In [5, 10] the authors propose a MPC for tracking capable to steer the closed-loop system to any changing setpoint. This MPC for tracking formulation integrates an SSTO in the same MPC problem, thus avoiding the need to solve two optimizations any sample time, and more important, avoiding feasibility problems. Following [8, 15], this formulation will be extended to a distributed control framework.

35.2.1 Characterization of the Steady State of the Plant

The steady conditions of system (35.1) can be determined by a suitable parametrization. Under Assumption 35.1 and *Lemma 1.14* in [14, p. 83], the steady state and input $(\mathbf{x}_s, \mathbf{u}_s)$ of system (35.1) can be parameterized by their associated steady output $\mathbf{y}_s$, that is, every solution of the following equation,

$$\begin{bmatrix} \mathbf{A}-\mathbf{I}_{n_x} & \mathbf{B} & \mathbf{0}_{n_x,n_y} \\ \mathbf{C} & \mathbf{D} & -\mathbf{I}_{n_y} \end{bmatrix} \begin{bmatrix} \mathbf{x}_s \\ \mathbf{u}_s \\ \mathbf{y}_s \end{bmatrix} = \begin{bmatrix} \mathbf{0}_{n_x,1} \\ \mathbf{0}_{n_y,1} \end{bmatrix} \tag{35.7}$$

is given by $(\mathbf{x}_s, \mathbf{u}_s) = \mathbf{M}_y \mathbf{y}_s$, where $\mathbf{M}_y$ is a suitable matrix.
If *Lemma 1.14* in [14, p.83] does not hold, then another parametrization has to be used. In [10], for instance, the authors state that the steady state and input $(\mathbf{x}_s, \mathbf{u}_s)$ are given by a linear combination of a vector $\theta \in \mathbb{R}^{n_u}$, in such a way that $(\mathbf{x}_s, \mathbf{u}_s) = \mathbf{M}_\theta \theta$, where $\mathbf{M}_\theta$ is a suitable matrix.

35.2.2 Distributed Model of the Plant

The plant given by (35.1) can be considered as a collection of coupled subsystems, connected by networks of different nature. These subsystems are considered to be coupled through the inputs. In Fig. 35.1, a schematic plot of the coupled system for the simplest case of two subsystems is shown.

Thus, given model (35.1), without loss of generality, it is considered that $\mathbf{u} = (\mathbf{u}_1, \cdots, \mathbf{u}_{|\mathcal{N}|})$, where $|\mathcal{N}| \leq n_u$. Then, by virtue of [15, Sect. 3.1.1] and [14, Chap. 6, pp. 421–422], model (35.1) is partitioned in $|\mathcal{N}|$ subsystems, coupled by the control inputs and modeled as follows:

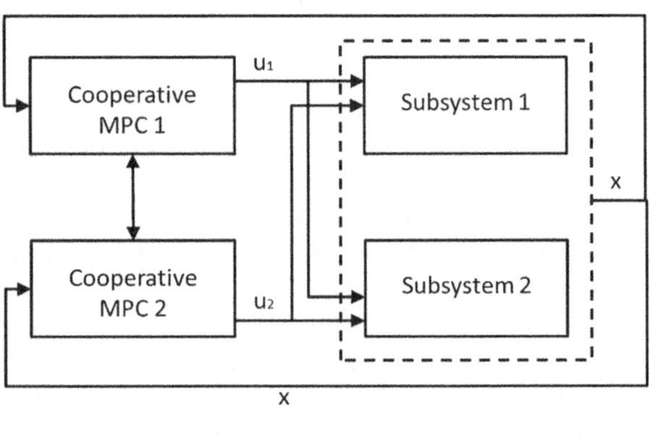

Fig. 35.1 Interaction between subsystems in a cooperative distributed control scheme (two players game)

$$\mathbf{x}_i(k+1) = \mathbf{A}_i\mathbf{x}_i(k) + \sum_{j\in\mathcal{N}} \bar{\mathbf{B}}_{ij}\mathbf{u}_j(k) \qquad (35.8)$$
$$\mathbf{y}_i(k) = \mathbf{C}_i\mathbf{x}_i(k) + \sum_{j\in\mathcal{N}} \bar{\mathbf{D}}_{ij}\mathbf{u}_j(k)$$

where $i \in \mathcal{N}$, $\mathbf{x}_i \in \mathbb{R}^{n_{x_i}}$, $\mathbf{u}_j \in \mathbb{R}^{n_{u_j}}$, $\mathbf{y}_i \in \mathbb{R}^{n_{y_i}}$, $\mathbf{A}_i \in \mathbb{R}^{n_{x_i}\times n_{x_i}}$, $\bar{\mathbf{B}}_{ij} \in \mathbb{R}^{n_{x_i}\times n_{u_j}}$, $\mathbf{C}_i \in \mathbb{R}^{n_{y_i}\times n_{x_i}}$, and $\bar{\mathbf{D}}_i \in \mathbb{R}^{n_{y_i}\times n_{u_j}}$.

The whole system, is hence given by:

$$\mathbf{x}(k+1) = \mathbf{A}\mathbf{x}(k) + \sum_{i\in\mathcal{N}} \mathbf{B}_i\mathbf{u}_i(k) \qquad (35.9)$$
$$\mathbf{y}(k) = \mathbf{C}\mathbf{x}(k) + \sum_{i\in\mathcal{N}} \mathbf{D}_i\mathbf{u}_i(k)$$

where $\mathbf{u}_i \in \mathbb{R}^{n_{u_i}}$, $\mathbf{B}_i \in \mathbb{R}^{n_x\times n_{u_i}}$, and $\mathbf{D}_i \in \mathbb{R}^{n_y\times n_{u_i}}$.

The solution of this system, given the sequences of control inputs $(\mathbf{u}_1(k:k+l),\ldots,\mathbf{u}_{|\mathcal{N}|}(k:k+l))$, $l\in\mathbb{Z}$ and initial state $\mathbf{x}$ is denoted as $\mathbf{x}(j) = p_{h,i}(j;\mathbf{x},\mathbf{u}_1(k:k+l),\ldots,\mathbf{u}_{|\mathcal{N}|}(k:k+l))$, $j\in\mathbb{Z}$, where $\mathbf{x} = p_{h,i}(0;\mathbf{x},\mathbf{u}_1(k:k+l),\ldots,\mathbf{u}_{|\mathcal{N}|}(k:k+l))$.

35.3 Proposed Approach

In this section, the cooperative distributed MPC integrating the SSTO, will be introduce. First of all, notice that in a cooperative game, the players share a common objective, which can be considered to be the overall plant objective. Each i-th agent, minimizes this cost with respect to its own control input $\mathbf{u}_i$, accounting however for "*how its input affects the other players' output as well as its own output*", [14, Chap. 6, p. 433].

The cost function of the controller that we propose here, is given by:

$$\begin{aligned} &J_{N_\mathrm{p}}(\mathbf{x},\mathbf{u}_\mathrm{ref},\mathbf{y}_\mathrm{ref};\mathbf{u}(k:k+N_\mathrm{p}-1),\hat{\mathbf{y}}_s) \\ &= \sum_{l=0}^{N_\mathrm{p}-1} \|\mathbf{x}(k+l)-\hat{\mathbf{x}}_s\|^2_{\mathbf{Q}} + \|\mathbf{u}(k+l)-\hat{\mathbf{u}}_s\|^2_{\mathbf{R}} \\ &+ \|\mathbf{x}(N_\mathrm{p})-\hat{\mathbf{x}}_s\|^2_{\mathbf{P}} + V_O(\hat{\mathbf{y}}_s,\mathbf{u}_\mathrm{ref},\mathbf{y}_\mathrm{ref}) \end{aligned} \qquad (35.10)$$

where $\mathbf{u}(k:k+N_\mathrm{p}-1) = (\mathbf{u}_1(k:k+N_\mathrm{p}-1),\ldots,\mathbf{u}_{|\mathcal{N}|}(k:k+N_\mathrm{p}-1))$ and $(\mathbf{u}_\mathrm{ref},\mathbf{y}_\mathrm{ref})$ defines the operation point provided by an upper level RTO. As in [10], an artificial equilibrium point of the plant $(\hat{\mathbf{x}}_s,\hat{\mathbf{u}}_s,\hat{\mathbf{y}}_s)$, characterized by $\hat{\mathbf{y}}_s$, is added as decision variable. The function $V_O(\hat{\mathbf{y}}_s,\mathbf{u}_\mathrm{ref},\mathbf{y}_\mathrm{ref})$ is the so called *offset cost function* and it is such that

$$V_O(\hat{\mathbf{y}}_s,\mathbf{u}_\mathrm{ref},\mathbf{y}_\mathrm{ref}) = \ell_{eco}(\hat{\mathbf{u}}_s,\hat{\mathbf{y}}_s,\mathbf{u}_\mathrm{ref},\mathbf{y}_\mathrm{ref})$$

Remark 35.1 Notice that the offset cost function can assume different forms depending on the particular control problem: it can involve only output targets, a combination of some inputs and some output targets, a combination of some inputs targets and some output zones (as in [7]), etc. More generally, it can be any convex and positive definite function in $(\hat{\mathbf{y}}_s - \mathbf{y}_{\mathrm{ref}})$, such that it has a unique minimizer given by

$$\mathbf{y}_s = \arg \min_{\hat{\mathbf{y}}_s \in \mathcal{Y}_s} V_O(\hat{\mathbf{y}}_s, \mathbf{u}_{\mathrm{ref}}, \mathbf{y}_{\mathrm{ref}})$$

35.3.1 Main Ingredients

In order to ensure recursive feasibility, convergence and stability, the proposed control algorithm has to fulfill the following Assumptions:

Assumption 35.2

1. Let $\mathbf{R} \in \mathbb{R}^{n_u \times n_u}$ be a positive definite matrix and $\mathbf{Q} \in \mathbb{R}^{n_x \times n_x}$ a positive semi-definite matrix such that the pair $(\mathbf{Q}^{1/2}, \mathbf{A})$ is observable.
2. Let $\mathbf{K} \in \mathbb{R}^{n_u \times n_x}$ be a stabilizing control gain for the centralized system, such that $(\mathbf{A} + \mathbf{BK})$ has all the eigenvalues in the unit cirle.
3. Let $\mathbf{P} \in \mathbb{R}^{n_x \times n_x}$ be a positive definite matrix for the centralized system such that:

$$(A + BK)^T P(A + BK) - P = -(Q + K^T RK) \tag{35.11}$$

4. Let $\Omega_\lambda \subseteq \mathbb{R}^{n_x + n_y}$ be an admissible polyhedral invariant set for tracking for system (35.1) subject to (35.2), for a given gain K. Please, refer to [10] and [4] for a detailed explanation about how to calculate this set.

Remark 35.2 In case of high dimension systems, the computation of Ω_λ can be avoided by using a terminal equality constraint: it is sufficient to impose the last predicted state to be an arbitrary equilibrium point, that is $(\mathbf{x}(N_p), \hat{\mathbf{y}}_s) \in (\mathcal{X}_s \times \mathcal{Y}_s)$, which is invariant (in this case, the term $\|\mathbf{x}(N_p) - \hat{\mathbf{x}}_s\|^2_{\mathbf{P}}$ of the cost function is no longer necessary).

35.3.2 Distributed On-Line Optimization

In a centralized MPC scheme, cost (35.10) is minimized subject to centralized model constraints. In this chapter, however, we adopt a cooperative distributed scheme [15]: at any time k, each i-th agent, in parallel with the other agents, calculates its corresponding couple $(\mathbf{u}_i, \hat{\mathbf{y}}_{s,i})$ by solving an iterative decentralized optimization problem, given an initial guess $(\mathbf{u}^{[0]}(k : k + N_{\mathrm{p}} - 1), \hat{\mathbf{y}}_s^{[0]})$. In this way, two time scales arises: the sampling time, k, which are denoted as the argument of the different variables, and the iteration number p, for a given time k, which is denoted as a superscript.

The optimization problem to be solved at the $(p+1)$-th iteration depends on the complete or centralized state $\mathbf{x}$ (which is precisely the way each agent accounts for the effect of its own input on the other agent's objectives) and on the solution of the p-th iteration $(\mathbf{u}^{[p]}(k:k+N_\mathrm{p}-1), \hat{\mathbf{y}}_s^{[p]})$. This parallel distributed optimization scheme is of the Gauss-Jacobi type [1, pp. 219–223].

This optimization problem, $P_i(\mathbf{x}, \mathbf{u}_\mathrm{ref}, \mathbf{y}_\mathrm{ref}, \mathbf{u}^{[p]}(k:k+N_\mathrm{p}-1), \hat{\mathbf{y}}_s^{[p]})$, is given by:

$$(\mathbf{u}_i^*(k:k+N_\mathrm{p}-1), \hat{\mathbf{y}}_{s,i}^*) = \arg\min_{\mathbf{u}_i(k:k+N_\mathrm{p}-1), \hat{\mathbf{y}}_s} J_{N_p}(\mathbf{x}, \mathbf{u}_\mathrm{ref}, \mathbf{y}_\mathrm{ref}, \mathbf{u}(k.k+N_\mathrm{p}-1)\hat{\mathbf{y}}_s) \tag{35.12a}$$

s.t.

$$\mathbf{x}(k+l+1) = \mathbf{A}\mathbf{x}(k+l) + \sum_{j\in\mathcal{N}} \mathbf{B}_j \mathbf{u}_j(k+l) \tag{35.12b}$$

$$\mathbf{x}(k) = \mathbf{x} \tag{35.12c}$$

$$\mathbf{u}_j(k+l) = \mathbf{u}_j^{[p]}(k+l),\ j \in \mathbb{I}_{1:|\mathcal{N}|},\ j \neq i,\ l \in \mathbb{I}_{0:N_p-1} \tag{35.12d}$$

$$\mathbf{x}(k+l) \in \mathcal{X},\ l \in \mathbb{I}_{0:N_p-1} \tag{35.12e}$$

$$(\mathbf{u}_1(k+l), \ldots, \mathbf{u}_{|\mathcal{N}|}(k+l)) \in \mathcal{U},\ l \in \mathbb{I}_{0:N_p-1} \tag{35.12f}$$

$$(\hat{\mathbf{x}}_s, \hat{\mathbf{u}}_s) = \mathbf{M}_y \hat{\mathbf{y}}_s \tag{35.12g}$$

$$(\mathbf{x}(N_\mathrm{p}), \hat{\mathbf{y}}_s) \in \Omega_\lambda \tag{35.12h}$$

The optimization problem $P_i(\mathbf{x}, \mathbf{u}_\mathrm{ref}, y_\mathrm{ref}, \mathbf{u}^{[p]}(k:k+N_\mathrm{p}-1))$ is a Quadratic Programming (QP) problem. Based on the solution of this optimization problem for each agent, namely $(\mathbf{u}_i^*(k:k+N_\mathrm{p}-1), \hat{y}_{s,i}^*)$, the solution of the $p+1$-iteration is given by the following convex update:

$$\mathbf{u}_i^{[p+1]}(k{:}k+N_\mathrm{p}-1) = w_i \mathbf{u}_i^*(k{:}k+N_\mathrm{p}-1) + (1-w_i)\mathbf{u}_i^{[p]}(k{:}k+N_\mathrm{p}-1),\quad i \in \mathbb{I}_{1:|\mathcal{N}|} \tag{35.13a}$$

$$\hat{y}_s^{[p+1]} = \sum_{i\in\mathcal{N}} w_i \hat{y}_{s,i}^* \quad i \in \mathbb{I}_{1:|\mathcal{N}|} \tag{35.13b}$$

$$\sum_{i\in\mathcal{N}} w_i = 1, \quad w_i > 0,\ i \in \mathbb{I}_{1:|\mathcal{N}|}$$

An agreement between agents is obtained once a given accuracy of the solution is achieved, or the computation time is expired, that is the last iteration $\bar{p}$ is reached. The best available solution is denoted as $(\mathbf{u}^{[\bar{p}]}(k{:}k+N_\mathrm{p}-1), \hat{\mathbf{y}}_s^{[\bar{p}]})$. The first element of the control sequence, that is $\mathbf{u}^{[\bar{p}]}(k)$, is applied to the plant in a receding horizon manner.

35.3.3 Remark on Feasibility

To proceed with the analysis of the proposed controller, we will denote $\mathbf{v} = (\mathbf{u}_1(k{:}k+N_\mathrm{p}-1), \ldots, \mathbf{u}_{|\mathcal{N}|}(k{:}k+N_\mathrm{p}-1), \hat{\mathbf{y}}_s)$. Besides, we will denote $J_{N_\mathrm{p}}(\mathbf{x}, \mathbf{u}_\mathrm{ref}, \mathbf{y}_\mathrm{ref}; \mathbf{v}) = J_{N_p}(\mathbf{x}, \mathbf{u}_\mathrm{ref}, \mathbf{y}_\mathrm{ref}; \mathbf{u}(k : k + N_\mathrm{p} - 1), \hat{\mathbf{y}}_s)$.

A sequence $\mathbf{v}$ is said to be feasible at $\mathbf{x}$ if each optimization problem $P_i(\mathbf{x}; \mathbf{v})$ is feasible for all $i \in \mathbb{I}_{1:|\mathcal{N}|}$. The set of states for which there exists a feasible $\mathbf{v}$, denoted as $\mathcal{X}_{N_\mathrm{p}}$, is given by

$$\begin{aligned}\mathcal{X}_{N_\mathrm{p}} = \{\mathbf{x} \in \mathcal{X} \mid \exists \mathbf{v} \text{ s.t. } (\mathbf{u}_1(k : k + N_\mathrm{p} - 1), \ldots, \mathbf{u}_{|\mathcal{N}|}(k : k + N_\mathrm{p} - 1))\\ \in \mathcal{U}^{N_p}, \hat{\mathbf{y}}_s \in \mathcal{Y}_s, \text{and } (\mathbf{x}, \mathbf{v}) \in \mathcal{Z}_{N_\mathrm{p}}\}\end{aligned}$$

where

$$\begin{aligned}\mathcal{Z}_{N_\mathrm{p}} = \{(\mathbf{x}, \mathbf{v}) \mid (\mathbf{u}_1(k : k + N_\mathrm{p} - 1), \ldots, \mathbf{u}_{|\mathcal{N}|}(k : k + N_\mathrm{p} - 1))\\ \in \mathcal{U}^{N_p}, \hat{\mathbf{y}}_s \in \mathcal{Y}_s, p_{h,i}(l; \mathbf{x}, \mathbf{u}(k : k + N_\mathrm{p} - 1))\\ \in \mathcal{X} \text{ for all } l \in \mathbb{I}_{0:N_\mathrm{p}-1}, p_{h,i}(N_p; \mathbf{x}, \mathbf{u}(k : k + N_\mathrm{p} - 1)) \in \Omega_\lambda\}\end{aligned}$$

Notice that this set is equal to the feasible set of the centralized MPC for tracking [5, 10], i.e. the set of states that can be admissibly steered to $\mathrm{Proj}_x(\Omega_\lambda)$ in N_p steps.

35.3.4 Initialization: The Warm Start Algorithm

In order to precisely define the proposed cooperative control scheme, the initial solution $\mathbf{v}^{[0]} = (\mathbf{u}^{[0]}(k : k + N_\mathrm{p} - 1), \hat{\mathbf{y}}_s^{[0]})$ of the iterative procedure (36.11) must be defined. Since the proposed distributed MPC can be considered as a suboptimal formulation of the centralized MPC, this initialization plays the role of the *warm start* of the suboptimal MPC and determines recursive feasibility and convergence of the control algorithm.

Given the solution $\mathbf{v}(k)$, the objective is to calculate the warm start at sampling time $k + 1$, denoted as

$$\mathbf{v}(k+1)^{[0]} = (\mathbf{u}_1^{[0]}(k+1 : k+N_\mathrm{p}), \ldots, \mathbf{u}_{|\mathcal{N}|}^{[0]}(k+1 : k+N_\mathrm{p}), \hat{\mathbf{y}}_s(k+1)^{[0]})$$

Assuming that the whole system information is available, each i-th agent performs the steps of Algorithm 35.1.

As usual in the suboptimal MPC optimization algorithm, the proposed *warm start* for the first optimization iteration $p = 0$ is given by the previous optimal sequence, shifted one step ahead, with the last control move given by the centralized terminal control law applied to the predicted terminal state of the overall plant and the same

Algorithm 35.1 Warm-start algorithm

1: Each agent defines the first candidate initial solution:

$$\tilde{\mathbf{u}}_i(k+1:k+N_\mathrm{p}) = \{\mathbf{u}_i(1;k), \ldots, \mathbf{u}_i(N_p-1;k), \mathbf{u}_{c,i}(N_p)\}$$

where $\mathbf{u}_c(N_p) = (\mathbf{u}_{c,1}(N_p), \ldots, \mathbf{u}_{c,\mathcal{N}}(N_p)) = \mathbf{K}\mathbf{x}(N_p) + \mathbf{L}\hat{\mathbf{y}}_s(k)$ is the solution given by the centralized terminal control law, and $\mathbf{x}(N_\mathrm{p}) = p_{h,i}(N_\mathrm{p}; \mathbf{x}(k), \mathbf{u}(k:k+N_\mathrm{p}-1))$.

2: Each agent defines the second candidate initial solution:

$$\hat{\mathbf{u}}_i(k+1:k+N_\mathrm{p}) = \{\hat{\mathbf{u}}_{c,i}(0), \ldots, \hat{\mathbf{u}}_{c,i}(N_p-1)\}$$

where $(\hat{\mathbf{u}}_{c,1}(l), \ldots, \hat{\mathbf{u}}_{c,\mathcal{N}}(l)) = \hat{\mathbf{u}}_c(l)$ and

$$\begin{aligned}
\mathbf{x}(0) &= \mathbf{x}(k+1)\\
\mathbf{x}(l+1) &= (\mathbf{A}+\mathbf{B}\mathbf{K})\mathbf{x}(l) + \mathbf{B}\mathbf{L}\hat{\mathbf{y}}_s(k),\ l \in \mathbb{I}_{1:N_\mathrm{p}-2}\\
\hat{\mathbf{u}}_c(l) &= \mathbf{K}\mathbf{x}(l) + \mathbf{L}\hat{\mathbf{y}}_s(k)
\end{aligned}$$

3: **for** Each agent: **do**
4: **if** $(\mathbf{x}(k+1), \hat{\mathbf{y}}_s(k)) \in \Omega_\lambda$ and $J_{N_p}(\mathbf{x}(k+1), \mathbf{u}_{\mathrm{ref}}, \mathbf{y}_{\mathrm{ref}}; \hat{\mathbf{u}}(k+1:k+N_\mathrm{p}), \hat{\mathbf{y}}_s(k)) < J_{N_p}(\mathbf{x}(k+1), \mathbf{u}_{\mathrm{ref}}, \mathbf{y}_{\mathrm{ref}}; \tilde{\mathbf{u}}(k+1:k+N_\mathrm{p}), \hat{\mathbf{y}}_s(k))$ **then**
5: $\mathbf{v}(k+1)^0 = (\hat{\mathbf{u}}_1(k+1:k+N_\mathrm{p}), \ldots, \hat{\mathbf{u}}_{|\mathcal{N}|}(k+1:k+N_\mathrm{p}), \hat{\mathbf{y}}_s(k))$
6: **else**
7: $\mathbf{v}(k+1)^0 = (\tilde{\mathbf{u}}_1(k+1:k+N_\mathrm{p}), \ldots, \tilde{\mathbf{u}}_{|\mathcal{N}|}(k+1:k+N_\mathrm{p}), \hat{\mathbf{y}}_s(k))$
8: **end if**
9: **end for**

artificial steady output, that is $(\tilde{\mathbf{u}}_1(k+1:k+N_\mathrm{p}), \ldots, \tilde{\mathbf{u}}_{|\mathcal{N}|}(k+1:k+N_\mathrm{p}), \hat{\mathbf{y}}_s(k))$. But according to Algorithm 35.1, when the state of the system reaches the invariant set for tracking, that is $(\mathbf{x}(k+1), \hat{\mathbf{y}}_s(k)) \in \Omega_\lambda$, it is desirable that the distributed MPC achieves the same cost than the one provided by using the centralized terminal controller. If this is not possible, i.e., if $J_{N_p}(\mathbf{x}(k+1), \mathbf{y}_{\mathrm{ref}}; \hat{\mathbf{u}}(k+1:k+N_\mathrm{p}), \hat{\mathbf{y}}_s(k)) < J_{N_p}(\mathbf{x}(k+1), \mathbf{y}_{\mathrm{ref}}; \tilde{\mathbf{u}}(k+1:k+N_\mathrm{p}), \hat{\mathbf{y}}_s(k))$, hence the centralized terminal control law is chosen as *warm start*.

Remark 35.3 In this work we are considering suboptimality in the sense that, provided that $\bar{p}$ is finite, the proposed distributed solution is a suboptimal solution of a centralized optimization problem. On the other hand, we assume that the optimal solution of the optimization problem of each agent at each iteration $P_i(\mathbf{x}; \mathbf{v}^{[p]})$ is achieved. In case of suboptimality of this solution, the method proposed in [16] can be used.

Algorithm 35.2 Cooperative distributed control algorithm

1: **if** $k = 0$ **then**
2: Calculate $\mathbf{v}^{[0]}(k)$ solving a Phase I problem [2]
3: **else**
4: Run Algorithm 35.1
5: **end if**
6: Read $\mathbf{x}$
7: **for** $p = 1$ to $p = \bar{p}$ **do**
8: Read $\mathbf{v}^{[p-1]}$
9: Solve $P_i(\mathbf{x}, \mathbf{u}_{\text{ref}}, \mathbf{y}_{\text{ref}}, \mathbf{u}^{[p-1](k:k+N_{\text{p}}-1)}, \hat{\mathbf{y}}_s^{[p-1]})$
10: Calculate the update $(\mathbf{u}_i^{[p]}(k : k + N_{\text{p}} - 1), \hat{\mathbf{y}}_s^{[p]})$ using (36.11)
11: and send it to the other agents
12: **end for**
13: Inject $\mathbf{u}_i(0; k)^{[\bar{p}]}$ into the plant.
14: Go back to 1.

35.3.5 Control Algorithm

At each sample time k, each i-th agent runs control Algorithm 35.2.

35.4 Properties of the Proposed Controller

The controller ensures recursive feasibility, convergence to the centralized optimum and asymptotic stability. Refer to [8] for a detailed proof. Moreover, the controller provides the following properties to the closed-loop system.

- **Enlargement of the domain of attraction.**
  The domain of attraction of the proposed distributed MPC is the set of states that can be admissible steered to Ω_λ in N_{p} steps. Since not a target point, but a set, is used, the domain of attraction of the proposed controller is (potentially) larger than the one of the (distributed) MPC for regulation [10].
- **Changing operating points.**
  Considering that the optimization problem is feasible for any $\mathbf{y}_{\text{ref}}$, the proposed controller is able to track changing operating points maintaining the recursive feasibility and constraint satisfaction.
- **Local optimality.**
  The addition of the artificial reference as decision variable makes that the local optimality property of the controller may be lost even in the case of optimality of the centralized solution. However, if the offset cost function is such that

$$\| V_O(\hat{\mathbf{y}}_s, \mathbf{y}_{\text{ref}}) - V_O(\mathbf{y}_{\text{ref}}, \mathbf{y}_{\text{ref}}) \| \geq \alpha \| \hat{\mathbf{y}}_s - \mathbf{y}_{\text{ref}} \|$$

  then there exists a constant α_{min} such that for all $\alpha > \alpha_{min}$ the local optimality property also holds for this controller [5].

- **Integration of the SSTO.**
  In a distributed control framework, it is natural to think that the SSTO problem has to be solved in a distributed way, that is, each agent has its own SSTO. If the constraints of each subsystem are decoupled, then the distributed SSTO ensures convergence to the centralized optimum. But if the constraints of the problem are coupled, then the optimality of the SSTO problems might be lost, and the controller might fail to steer the plant to the desired setpoint, driving the plant to a suboptimal one. For this reason it is better to consider a centralized SSTO [14, Sect. 6.3.4]. The cooperative MPC for tracking does not have this problem, since each agent consider a centralized offset cost function, which plays the role of the SSTO.
- **Offset cancelation.**
  Model mismatches and unexpected disturbances may produce a steady state offset between the controlled variable and the reference. In order to overcome this problem, a disturbance model can be used to estimate the effect of the disturbance and cancel it [11, 14]. Consider this effect as a constant disturbance on the controlled variable given by the following model:

$$\begin{aligned}\mathbf{x}(k+1) &= \mathbf{A}\mathbf{x}(k) + \mathbf{B}\mathbf{u}(k)\\ \mathbf{y}(k) &= \mathbf{C}\mathbf{x}(k) + \mathbf{D}\mathbf{u}(k) + \mathbf{d}(k)\\ \mathbf{d}(k+1) &= \mathbf{d}(k)\end{aligned}$$

  Then, the disturbance at the current sampling time can be estimated as follows

$$\hat{\mathbf{d}}(k) = \hat{\mathbf{d}}(k-1) + \mathbf{L}_d(\mathbf{y}(k) - (\mathbf{C}\mathbf{x}(k-1) + \mathbf{D}\mathbf{u}(k-1)) - \hat{\mathbf{d}}(k-1))$$

  The gain $\mathbf{L}_d$ must be chosen in such a way that the module of the eigenvalues of $(\mathbf{I}_{n_y} - \mathbf{L}_d)$ is in the interior of the unit circle and sufficiently large to filter the effect of the disturbances on the estimation.
  Consider that the controlled system reaches an equilibrium point $(\mathbf{x}_\infty, \mathbf{u}_\infty)$ and $\mathbf{y}_\infty = \mathbf{C}\mathbf{x}_\infty + \mathbf{D}\mathbf{u}_\infty + \mathbf{d}_\infty$. Assume that the disturbance estimator guarantees that $\hat{\mathbf{d}}_\infty = \mathbf{d}_\infty$. Then the proposed controller ensures that, for a given setpoint $\mathbf{y}_{\text{ref}}$, $\mathbf{C}\mathbf{x}_\infty + \mathbf{D}\mathbf{u}_\infty = \mathbf{y}_{\text{ref}}$. However, the real output is such that $\mathbf{y}_\infty = \mathbf{C}\mathbf{x}_\infty + \mathbf{D}\mathbf{u}_\infty + \mathbf{d}_\infty = \mathbf{y}_{\text{ref}} + \mathbf{d}_\infty$ and hence there exists an offset. This can be easily cancelled if we provided a modified setpoint to the controller in order to compensate the offset, that is, $\hat{\mathbf{y}}_{\text{ref}}(k) = \mathbf{y}_{\text{ref}} - \hat{\mathbf{d}}(k)$.
  The proposed controller ensures feasibility with offset cancellation, for all $\hat{\mathbf{y}}_{\text{ref}}(k)$. Moreover, since $\hat{\mathbf{d}}(k)$ converges to $\mathbf{d}_\infty$, the controlled system converges to an equilibrium point, such that $\mathbf{y}_\infty = \mathbf{C}\mathbf{x}_\infty + \mathbf{D}\mathbf{u}_\infty + \mathbf{d}_\infty = \hat{\mathbf{y}}_{\text{ref},\infty} + \mathbf{d}_\infty = \mathbf{y}_{\text{ref}}$ and hence the controlled output converges to the setpoint.

35.5 Example

In this example, the plant presented in [15] has been considered. This plant consists of two reactors and a separator (see Fig. 36.1). The model for the plant is given by:

$$
\begin{aligned}
\frac{dH_1}{dt} &= \frac{1}{\rho A_1}(F_{f1} + F_R - F_1) \\
\frac{dx_{A1}}{dt} &= \frac{1}{\rho A_1 H_1}(F_{f1}x_{A0} + F_R x_{AR} - F_1 x_{A1}) - k_{A1}x_{A1} \\
\frac{dx_{B1}}{dt} &= \frac{1}{\rho A_1 H_1}(F_R x_{BR} - F_1 x_{B1}) - k_{B1}x_{B1} \\
\frac{dT_1}{dt} &= \frac{1}{\rho A_1 H_1}(F_{f1}T_0 + F_R T_R - F_1 T_1) - \frac{1}{C_p}(k_{A1}x_{A1}\Delta H_A + k_{B1}x_{B1}\Delta H_B) \\
&\quad + \frac{Q_1}{\rho A_1 C_p H_1} \\
\frac{dH_2}{dt} &= \frac{1}{\rho A_2}(F_{f2} + F_1 - F_2) \\
\frac{dx_{A2}}{dt} &= \frac{1}{\rho A_2 H_2}(F_{f2}x_{A0} + F_1 x_{A1} - F_2 x_{A2}) - k_{A2}x_{A2} \\
\frac{dx_{B2}}{dt} &= \frac{1}{\rho A_2 H_2}(F_1 x_{B1} - F_2 x_{B2}) - k_{B2}x_{B2} \\
\frac{dT_2}{dt} &= \frac{1}{\rho A_2 H_2}(F_{f2}T_0 + F_1 T_1 - F_2 T_2) - \frac{1}{C_p}(k_{A2}x_{A2}\Delta H_A + k_{B2}x_{B2}\Delta H_B) \\
&\quad + \frac{Q_2}{\rho A_2 C_p H_2} \\
\frac{dH_3}{dt} &= \frac{1}{\rho A_3}(F_2 - F_D - F_R - F_3) \\
\frac{dx_{A3}}{dt} &= \frac{1}{\rho A_3 H_3}(F_2 x_{A2} - (F_D + F_R)x_{AR} - F_3 x_{A3}) \\
\frac{dx_{B3}}{dt} &= \frac{1}{\rho A_3 H_3}(F_2 x_{B2} - (F_D + F_R)x_{BR} - F_3 x_{B3}) \\
\frac{dT_3}{dt} &= \frac{1}{\rho A_3 H_3}(F_2 T_2 - (F_D + F_R)T_R - F_3 T_3) + \frac{Q_3}{\rho A_3 C_p H_3}
\end{aligned}
$$

in which for all $i \in \mathbb{I}_{1:3}$

$$
F_i = k_{vi} H_i \quad k_{Ai} = k_A \exp\left(-\frac{E_A}{RT_i}\right) \quad k_{Bi} = k_B \exp\left(-\frac{E_B}{RT_i}\right)
$$

The recycle flow and weight percents satisfy

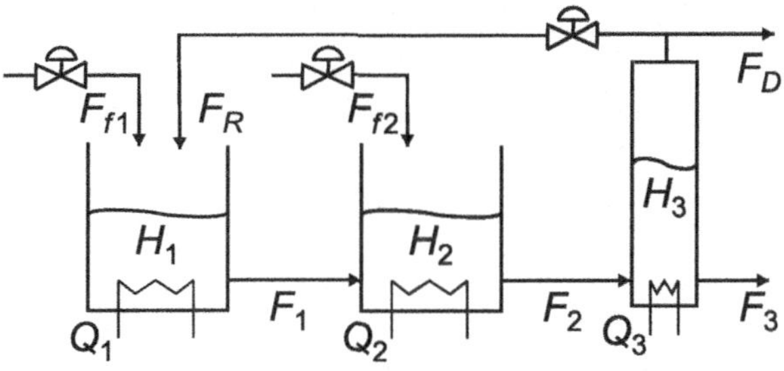

Fig. 35.2 Two reactors in series with separator and recycle [15]

$$F_D = 0.01F_R \quad x_{AR} = \frac{\alpha_A x_{A3}}{\bar{x}_3} \quad x_{BR} = \frac{\alpha_B x_{B3}}{\bar{x}_3}$$
$$\bar{x}_3 = \alpha_A x_{A3} + \alpha_B x_{B3} + \alpha_C x_{C3} \qquad x_{C3} = (1 - x_{A3} - x_{B3})$$

The output and input are denoted, respectively

$$y = \begin{bmatrix} H_1 \; x_{A1} \; x_{B1} \; T_1 \; H_2 \; x_{A2} \; x_{B2} \; T_2 \; H_3 \; x_{A3} \; x_{B3} \; T_3 \end{bmatrix}$$
$$u = \begin{bmatrix} F_{f1} \; Q_1 \; F_{f2} \; Q_2 \; F_R \; Q_3 \end{bmatrix}$$

The plant model has been linearized around the steady state defined by Table 1 of [15] with sampling time $T_s = 1$s. The linearized model obtained has the form of model (35.1).

35.5.1 Distributed Control

The plant has been partitioned in 3 subsystems in order to control each reactor and the separator independently. This partition is given by:

$$y_1 = \begin{bmatrix} H_1 \; x_{A1} \; x_{B1} \; T_1 \end{bmatrix} \qquad y_2 = \begin{bmatrix} H_2 \; x_{A2} \; x_{B2} \; T_2 \end{bmatrix} \qquad y_3 = \begin{bmatrix} H_3 \; x_{A3} \; x_{B3} \; T_3 \end{bmatrix}$$
$$u_1 = \begin{bmatrix} F_{f1} \; Q_1 \end{bmatrix} \qquad u_2 = \begin{bmatrix} F_{f2} \; Q_2 \end{bmatrix} \qquad u_3 = \begin{bmatrix} F_R \; Q_3 \end{bmatrix}$$

The system is constrained in the input. The input constraints are defined in Table 2 of [15]. An MPC with $N_p = 3$ has been considered. The weighting matrices are

$$\mathbf{Q}_{yi} = \text{diag}(1, 0, 0, 0.1) \; \forall i = \mathbb{I}_{1:2} \quad \mathbf{Q}_{y3} = \text{diag}(1, 0, 10^3, 0)$$
$$\mathbf{Q}_i = \mathbf{C}_i^T \mathbf{Q}_{yi} \mathbf{C}_i + 0.001I \quad \mathbf{R}_i = 100I \quad \forall i \in \mathbb{I}_{1:3}$$

Two different formulations of the SSTO cost function have been considered: $\ell_{eco,1}(\hat{u}_s, \hat{\mathbf{y}}_s, u_{\text{ref}}, \mathbf{y}_{\text{ref}}) = \|\hat{\mathbf{y}}_s - \mathbf{y}_{\text{ref}}\|^2_{\mathbf{Q}_t} + \|\hat{u}_s - u_{\text{ref}}\|^2_{\mathbf{R}_t}$, where $\mathbf{T} = 10\mathbf{Q}$ and $\mathbf{R}_t = \mathbf{R}$, and $\ell_{eco,2}(\hat{\mathbf{y}}_s, \mathbf{y}_{\text{ref}}) = \alpha\|\hat{\mathbf{y}}_s - \mathbf{y}_{\text{ref}}\|_1$, where $\alpha = 100$ has been taken following [6]. The initial condition is the steady state given by Table 1 of [15].

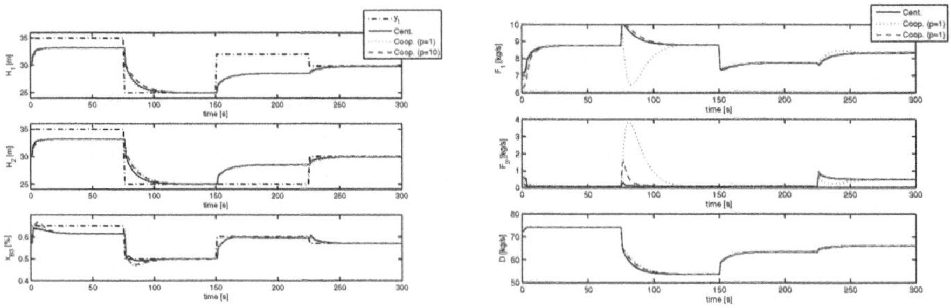

Fig. 35.3 Time evolution of the output and inputs considering $\ell_{eco,1}(.)$

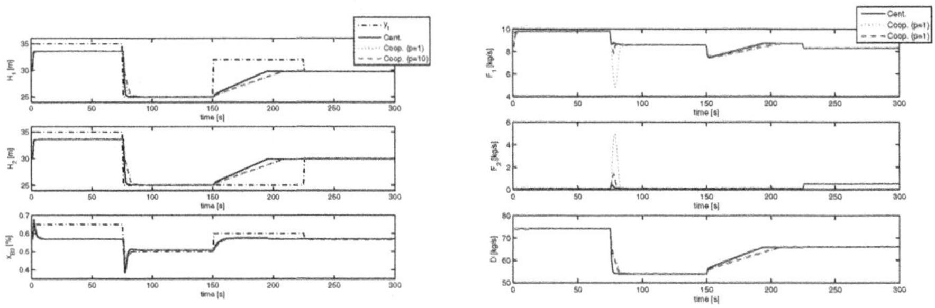

Fig. 35.4 Time evolution of the output and inputs considering $\ell_{eco,2}(.)$

Table 35.1 Performance comparison

	$y_{ref,1}$	$y_{ref,2}$	$y_{ref,3}$	$y_{ref,4}$
Cent. ($\ell_{eco,1}(.)$)	88.54	58.44	260.23	1.94
Coop. ($\bar{p} = 1$, $\ell_{eco,1}(.)$)	377.44	612.99	333.80	3.38
Coop. ($\bar{p} = 10$, $\ell_{eco,1}(.)$)	376.61	606.43	329.93	3.34
Cent. ($\ell_{eco,2}(.)$)	155.03	94.51	485.77	0.0003
Coop. ($\bar{p} = 1$, $\ell_{eco,2}(.)$)	499.37	728.08	583.44	0.0003
Coop. ($\bar{p} = 10$, $\ell_{eco,2}(.)$)	499.34	727.51	582.78	0.0003

Four changes of reference have been considered. In particular three setpoint changes in the level H_1 and H_2 and in the output product weight percent x_{B3} have been simulated. In the last change the system has been brought to the steady state. The setpoints changes are: $\mathbf{y}_{\text{ref},1} \rightarrow (H_1(m) = 35, H_2(m) = 35, x_{B3}(wt.\%) = 0.65)$, $\mathbf{y}_{\text{ref},2} \rightarrow (H_1(m) = 25, H_2(m) = 25, x_{B3}(wt.\%) = 0.50)$, $\mathbf{y}_{\text{ref},3} \rightarrow (H_1(m) = 32, H_2(m) = 22, x_{B3}(wt.\%) = 0.60)$ and $\mathbf{y}_{\text{ref},4} \rightarrow (H_1(m) = 29.8, H_2(m) = 30, x_{B3}(wt.\%) = 0.57)$.

The evolution of outputs and inputs, considering centralized MPC, cooperative MPC with $\bar{p} = 1$, and cooperative MPC with $\bar{p} = 10$ are presented respectively in Figs. 35.3 and 35.4.

See how the distributed controller always drives the system to the centralized optimum. The way the controller approaches the optimum is different in the two cases, due to the two different offset cost functions considered. Notice how the performance of the distributed controller improves with $\bar{p}$ (see Table 35.1). Notice also, that the distributed controller provides a worse performance than the centralized one. This loss of performance appears in the transient, while asymptotically the optimal costs converge. This is the trade-off that one has to face using distributed control: the system is driven to the centralized optimum with a lower computational burden (due to the distributed optimization), even though the performance (in the transient) is worse.

35.6 Conclusions

This chapter presents a general distributed MPC formulation that account for the target calculation problem. By means of an extended cost, a cooperative distributed control scheme was developed, which guarantees stability and recursive feasibility. Under the usual assumption of cooperative control structure, the proposed strategy shows that it works even for a short number of iterations (that corresponds to a short sample time), and although it is in essence a suboptimal MPC controller, it ensures stability and converges to the centralized economic optimum.

Acknowledgments This work has been funded by the National Plan Project (DPI2010-21589-C05-01) of the Spanish Ministry of Science and Innovation, FEDER funds, and ANPCYT, Argentina (PICT 2008, contract number 1833).

References

1. D.P. Bertsekas, J.N. Tsitsiklis, *Parallel and Distributed Computation* (Athena Scientific, Belmont, 1997)
2. S. Boyd, L. Vandenberghe, *Convex Optimization* (Cambridge University Press, Cambridge, 2006)
3. S. Engell, Feedback control for optimal process operation. J. Process Control **17**, 203–219 (2007)
4. A. Ferramosca. Model Predictive Control for Systems with Changing Setpoints. PhD thesis, Université de sevilla de Sevilla, 2011. http://fondosdigitales.us.es/tesis/autores/1537/
5. A. Ferramosca, D. Limon, I. Alvarado, T. Alamo, E.F. Camacho, MPC for tracking with optimal closed-loop performance. Automat. **45**, 1975–1978 (2009)
6. A. Ferramosca, D. Limon, I. Alvarado, T. Alamo, F. Castaño, and E. F. Camacho, Optimal MPC for tracking of constrained linear systems. Int. J. Syst. Sci. **42**(8), Aug (2011). Accepted for publication
7. A. Ferramosca, D. Limon, A.H. González, D. Odloak, E.F. Camacho, MPC for tracking zone regions. J. Process Control **20**, 506–516 (2010)
8. A. Ferramosca, D. Limon, J. B. Rawlings, and E. F. Camacho, Cooperative distributed MPC for tracking, in *Proceedings of the 18th IFAC World Congress* (2011)

9. J. V. Kadam, W. Marquardt, Integration of economical optimization and control for intentionally transient process operation, in *International Workshop on Assessment and Future Direction of Nonlinear Model Predictive Control*, ed. by R. Findeisen, F. Allgöwer, L.T. Biegler (Springer, Berlin, 2007), pp. 419–434
10. D. Limon, I. Alvarado, T. Alamo, E.F. Camacho, MPC for tracking of piece-wise constant references for constrained linear systems. Automat. **44**, 2382–2387 (2008)
11. D. Limon, I. Alvarado, T. Alamo, E.F. Camacho, Robust tube-based MPC for tracking of constrained linear systems with additive disturbances. JPC **20**, 248–260 (2010)
12. K. Muske, Steady-state target optimization in linear model predictive control, in *Proceedings of the ACC* (1997)
13. C.V. Rao, J.B. Rawlings, Steady states and constraints in model predictive control. AIChE J. **45**, 1266–1278 (1999)
14. J.B. Rawlings, D. Q. Mayne, *Model Predictive Control: Theory and Design*, 1st edn. (Nob-Hill Publishing, Madison, 2009)
15. B.T. Stewart, A.N. Venkat, J.B. Rawlings, S.J. Wright, G. Pannocchia, Cooperative distributed model predictive control. Syst. Control Lett. **59**, 460–469 (2010)
16. M.N. Zeilinger, Real-time Model Predictive Control. PhD thesis, ETH Zurich (2011)

Chapter 36
Cooperative MPC with Guaranteed Exponential Stability

A. Ferramosca

Abstract In this chapter, a cooperative distributed MPC is presented. The main features of this control strategy are: constraints satisfaction; cooperation between agents to achieve an agreement; closed-loop stability that is always ensured, even in the case of just one iteration; achieved control actions that are plantwide Pareto optimal and equivalent to the centralized solution; Pareto optimality is achieved also in case of coupled constraints; a coordination layer is not needed. It is proved that cooperative MPC is a particular case of suboptimal MPC; exponential stability is then proved, based on exponential stability of suboptimal centralized MPC.

36.1 Introduction

The aim of this chapter is to present the cooperative Model Predictive Control strategy, developed at Prof. James Rawlings's research group.

The approach accounts for controlling large, networked systems through the integration of subsystem-based MPCs. The distributed scheme is of cooperative type, since the subsystems cooperate in order to achieve an agreement. Such an agreement is the result of a based on iterative optimizations, executed in parallel by each local MPC. At the end of these optimizations, the agents exchange information.

A first comprehensive analysis of cooperative MPC was given by [11]. This thesis provides a great number of results, including exponential closed-loop stability for state and output feedback as well as an extension for asynchronous feedback and optimization. Venkat particularly focuses his attention on applications like power distribution networks [12] and chemical processes [13], which are illustrated in several examples.

A. Ferramosca (✉)
Institute of Technological Development for the Chemical Industry (INTEC), CONICET-Universidad Nacional del Litoral (UNL), Santa Fe, Argentina
e-mail: ferramosca@santafe-conicet.gov.ar

J. M. Maestre and R. R. Negenborn (eds.), *Distributed Model Predictive Control Made Easy*, Intelligent Systems, Control and Automation: Science and Engineering 69, DOI: 10.1007/978-94-007-7006-5_36,

The results proposed by Venkat are later extended and generalized in [8] and [9]. In these works, the stability of cooperative control is extended to a broader class of plants, and optimal convergence of the distributed optimization in presence of coupled constraints, is also shown. Moreover, the asynchronous feedback scheme of Venkat is extended to a general theory of hierarchical cooperative control with asynchronous communication and reduced information exchange. A cooperative MPC scheme for nonlinear systems is also proposed.

In this chapter, we particularly focus on the cooperative MPC for linear systems, presented by Stewart in [8]. The control scheme is analyzed by first describing how to determine the distributed model of a linear plant. Then, the is described by the definition of the cost function to be optimized, the resulting optimization problem, the constraints on it, and the choice of the ingredients necessary to ensure stability. The main properties of the proposed controller are later on summarized, focusing in particular on the fact that the control algorithm can be terminated at any intermediate iterate, without compromising convergence to the Pareto optimum, or closed-loop stability. Finally, the results of some applications of the proposed controller are presented.

In the rest of the chapter, the term iterate is utilized to indicate a set of MPC optimizations executed in parallel (one for each subsystem) followed by an exchange of information among interconnected subsystems. Since going deeper into mathematics is not in the scope of this chapter, the reader is invited to refer to [8] and [5], for detailed proofs of Lemmas and Theorems.

The author would like to remark that the results summarized here, together with interesting discussions about them, with Prof. James B. Rawlings and Dr. Brett T. Stewart, have been the inspiration for the cooperative MPC for tracking algorithm presented in [3].

36.2 Problem Statement

In the discussion of this algorithm, linear invariant discrete time models are considered:

$$\mathbf{x}(k+1) = \mathbf{A}\mathbf{x}(k) + \mathbf{B}\mathbf{u}(k) \tag{36.1}$$

where $\mathbf{x} \in \mathbb{R}^{n_x}$ is the system state, $\mathbf{u} \in \mathbb{R}^{n_u}$ is the current control vector. The solution of this system for a given sequence of control inputs $\mathbf{u}(k : k+l), l \in \mathbb{Z}$, and initial state $\mathbf{x}$ is denoted as $\mathbf{x}(j) = P_{h,i}(j; \mathbf{x}, \mathbf{u}(k : k+l))$, $j \in \mathbb{Z}$, where $\mathbf{x} = P_{h,i}(0; \mathbf{x}, \mathbf{u}(k : k+l))$. The state of the system and the control input applied at sampling time k are denoted as $\mathbf{x}(k)$ and $\mathbf{u}(k)$ respectively.

It is assumed that the following hypothesis hold.

Assumption 36.1 The pair $(\mathbf{A}, \mathbf{B})$ is stabilizable and the state is measured at each time step.

36.2.1 How is the Distributed Model Obtained?

The plant (36.1) is considered as made up by different units, connected in different ways. For each unit, a subsystem i can be defined. These subsystems are considered to be coupled through the inputs. For simplicity, let us consider a 2 player game through the chapter. Following [8, Sect. 3.1.1] and [5, Chapt. 6, pp. 421–422], for each subsystem i there exists a collection of linear models denoting the effects of inputs of subsystem j on the states of subsystem i for all $(i, j) \in \mathbb{I}_{1:2} \times \mathbb{I}_{1:2}$.

$$\mathbf{x}_{ij}(k+1) = \mathbf{A}_{ij}\mathbf{x}_{ij}(k) + \mathbf{B}_{ij}u_j(k)$$

in which $\mathbf{x}_{ij} \in \mathbb{R}^{n_{x_{ij}}}$, $\mathbf{u}_j \in \mathbb{R}^{n_{u_j}}$, $\mathbf{A}_{ij} \in \mathbb{R}^{(n_{x_{ij}} \times n_{x_{ij}})}$, and $\mathbf{B}_{ij} \in \mathbb{R}^{(n_{x_{ij}} \times n_{u_j})}$. For a discussion of the identification of this model choice, see [4]. Moreover, in [8] [Appendix B] it is shown how this models are related to the centralized model. Consider now subsystem 1. Then we can write the model for subsystem 1 as:

$$\begin{bmatrix} \mathbf{x}_{11}(k+1) \\ \mathbf{x}_{12}(k+1) \end{bmatrix} = \begin{bmatrix} \mathbf{A}_{11} & \\ & \mathbf{A}_{12} \end{bmatrix} \begin{bmatrix} \mathbf{x}_{11}(k) \\ \mathbf{x}_{12}(k) \end{bmatrix} + \begin{bmatrix} \mathbf{B}_{11} \\ 0 \end{bmatrix} \mathbf{u}_1(k) + \begin{bmatrix} 0 \\ \mathbf{B}_{12} \end{bmatrix} \mathbf{u}_2(k)$$

which equivalently can be posed as

$$\mathbf{x}_1(k+1) = \mathbf{A}_1\mathbf{x}_1(k) + \bar{\mathbf{B}}_{11}\mathbf{u}_1(k) + \bar{\mathbf{B}}_{12}\mathbf{u}_2(k)$$

for which

$$\mathbf{x}_1 = \begin{bmatrix} \mathbf{x}_{11} \\ \mathbf{x}_{12} \end{bmatrix} \quad \mathbf{A}_1 = \begin{bmatrix} \mathbf{A}_{11} & \\ & \mathbf{A}_{12} \end{bmatrix} \quad \bar{\mathbf{B}}_{11} = \begin{bmatrix} \mathbf{B}_{11} \\ 0 \end{bmatrix} \quad \bar{\mathbf{B}}_{12} = \begin{bmatrix} 0 \\ \mathbf{B}_{12} \end{bmatrix}$$

in which $\mathbf{x}_1 \in \mathbb{R}^{n_{x_1}}$, $\mathbf{A}_1 \in \mathbb{R}^{(n_{x_1} \times n_{x_1})}$, and $\bar{\mathbf{B}}_{1j} \in \mathbb{R}^{(n_{x_1} \times n_{u_j})}$ with $n_{x_1} = n_{x_{11}} + n_{x_{12}}$. Following similar arguments for subsystem 2, the whole plant model can be written as

$$\begin{bmatrix} \mathbf{x}_1(k+1) \\ \mathbf{x}_2(k+1) \end{bmatrix} = \begin{bmatrix} \mathbf{A}_1 & \\ & \mathbf{A}_2 \end{bmatrix} \begin{bmatrix} \mathbf{x}_1(k) \\ \mathbf{x}_2(k) \end{bmatrix} + \begin{bmatrix} \bar{\mathbf{B}}_{11} \\ \bar{\mathbf{B}}_{21} \end{bmatrix} \mathbf{u}_1(k) + \begin{bmatrix} \bar{\mathbf{B}}_{12} \\ \bar{\mathbf{B}}_{22} \end{bmatrix} \mathbf{u}_2(k)$$

The plantwide model notation can be finally simplified to

$$\mathbf{x}(k+1) = \mathbf{A}\mathbf{x}(k) + \mathbf{B}_1\mathbf{u}_1(k) + \mathbf{B}_2\mathbf{u}_2(k) \tag{36.2}$$

for which

$$\mathbf{x} = \begin{bmatrix} \mathbf{x}_1 \\ \mathbf{x}_2 \end{bmatrix} \quad \mathbf{A} = \begin{bmatrix} \mathbf{A}_1 & \\ & \mathbf{A}_2 \end{bmatrix} \quad \mathbf{B}_1 = \begin{bmatrix} \bar{\mathbf{B}}_{11} \\ \bar{\mathbf{B}}_{21} \end{bmatrix} \quad \mathbf{B}_2 = \begin{bmatrix} \bar{\mathbf{B}}_{12} \\ \bar{\mathbf{B}}_{22} \end{bmatrix}$$

Remark 36.1 In case of $|\mathcal{N}|$ subsystems, with $|\mathcal{N}| \leq n_u$, model (36.1) is partitioned as follows:

$$\mathbf{x}_i(k+1) = \mathbf{A}_i\mathbf{x}_i(k) + \sum_{j\in\mathcal{N}} \bar{\mathbf{B}}_{ij}\mathbf{u}_j(k) \tag{36.3}$$

where $i \in \mathcal{N}$, $\mathbf{x}_i \in \mathbb{R}^{n_{x_i}}$, $\mathbf{u}_j \in \mathbb{R}^{n_{u_j}}$, $\mathbf{A}_i \in \mathbb{R}^{n_{x_i}\times n_{x_i}}$ and $\bar{\mathbf{B}}_{ij} \in \mathbb{R}^{n_{x_i}\times n_{u_j}}$. The plantwide model is hence given by:

$$\mathbf{x}(k+1) = \mathbf{A}\mathbf{x}(k) + \sum_{i\in\mathcal{N}} \mathbf{B}_i\mathbf{u}_i(k) \tag{36.4}$$

where $\mathbf{u}_i \in \mathbb{R}^{n_{u_i}}$ and $\mathbf{B}_i \in \mathbb{R}^{n_x\times n_{u_i}}$.

36.3 Design of the Proposed Controller

This section presents the cooperative distributed MPC. The main point of a cooperative game is that players share a common objective, which is the overall plant objective [5, Chap. 6, p. 433].

36.3.1 Objective Functions

Define a quadratic stage cost and terminal penalty functions for subsystem 1:

$$\ell_1(\mathbf{x}_1, \mathbf{u}_1) = \frac{1}{2}(\mathbf{x}_1^T\mathbf{Q}_1\mathbf{x}_1 + \mathbf{u}_1^T\mathbf{R}_1\mathbf{u}_1) \tag{36.5a}$$

$$J_{1f}(\mathbf{x}_1) = \frac{1}{2}\mathbf{x}_1^T\mathbf{P}_{1f}\mathbf{x}_1 \tag{36.5b}$$

where $\mathbf{Q}_1 \in \mathbb{R}^{(n_{x_1}\times n_{x_1})}$, $\mathbf{R}_1 \in \mathbb{R}^{(n_{u_1}\times n_{u_1})}$, and $\mathbf{P}_{1f} \in \mathbb{R}^{(n_{x_1}\times n_{x_1})}$. Then, the objective function for subsystem 1 is given by

$$J_1\big(\mathbf{x}_1(k), \mathbf{u}_1(k:k+N_p-1), \mathbf{u}_2(k:k+N_p-1)\big) = \sum_{l=0}^{N_p-1} \ell_1\big(\mathbf{x}_1(k+l), \mathbf{u}_1(k+l)\big) + J_{1f}\big(\mathbf{x}_1(k+N_p)\big)$$

Notice that J_1 is implicitly a function of both $\mathbf{u}_1(k:k+N_p-1)$ and $\mathbf{u}_2(k:k+N_p-1)$ because $\mathbf{x}_1$ is a function of both $\mathbf{u}_1$ and $\mathbf{u}_2$. Following same arguments for subsystem 2, the plantwide objective function can be expressed as

$$J\big(\mathbf{x}(k),\mathbf{u}(k:k+N_p-1)\big)=\rho_1 J_1\big(\mathbf{x}_1(k),\mathbf{u}_1(k:k+N_p-1),\mathbf{u}_2(k:k+N_p-1)\big) \\ +\rho_2 J_2\big(\mathbf{x}_2(k),\mathbf{u}_1(k:k+N_p-1),\mathbf{u}_2(k:k+N_p-1)\big)$$

where $\mathbf{x}(k) = (\mathbf{x}_1(k), \mathbf{x}_2(k))$, $\mathbf{u}(k:k+N_p-1) = (\mathbf{u}_1(k:k+N_p-1), \mathbf{u}_2(k:k+N_p-1))$, and $\rho_1, \rho_2 > 0$ are relative weights.

36.3.2 Constraints

The inputs have to satisfy

$$\mathbf{u}_1(k+l) \in \mathcal{U}_1 \qquad \mathbf{u}_2(k+l) \in \mathcal{U}_2 \quad l \in \mathbb{I}_{0:N_p-1}$$

with $\mathcal{U}_1$ and $\mathcal{U}_2$ compact and convex sets containing the origin in their interior.

Remark 36.2 In this case, these constraints are considered uncoupled since the feasible region of $\mathbf{u}_1$ is not affected by $\mathbf{u}_2$, and vice-versa.

36.3.2.1 Unstable Modes

In case of unsteable plants, closed-loop stability is ensured by adding a terminal constraint that zeros the unstable modes at the end of the horizon [8]. This terminal constraint can be obtained resorting to the real Schur decomposition of matrices $\mathbf{A}_{ij}$, for each $(i, j) \in \mathbb{I}_{1:2} \times \mathbb{I}_{1:2}$

$$\mathbf{A}_{ij} = \begin{bmatrix} \mathbf{S}^s_{ij} & \mathbf{S}^u_{ij} \end{bmatrix} \begin{bmatrix} \mathbf{A}^s_{ij} & \\ & \mathbf{A}^u_{ij} \end{bmatrix} \begin{bmatrix} \mathbf{S}_{ij}^{sT} \\ \mathbf{S}_{ij}^{uT} \end{bmatrix} \tag{36.6}$$

in which $\mathbf{A}^s_{ij}$ is stable and $\mathbf{A}^u_{ij}$ has all unstable eigenvalues.

36.3.2.2 Terminal Penalty Functions

The terminal penalty functions is another important ingredient for making the closed-loop system controlled by the proposed DMPC, enjoying the property of exponential stability. Following [8] and [5, Chap. 6, pp. 446–448] these functions has to be chosen as the cost to go under zero control:

$$\mathbf{P}_{1f} = \mathbf{S}^s_1 \Sigma_1 {\mathbf{S}^s_1}^T \quad \mathbf{P}_{2f} = \mathbf{S}^s_2 \Sigma_1 {\mathbf{S}^s_2}^T \tag{36.7}$$

where Σ_1 and Σ_2 denote the solution of the Lyapunov equations

$$\mathbf{A}_1^{s\,T}\Sigma_1\mathbf{A}_1^s - \Sigma_1 = -\mathbf{S}_1^{s\,T}\mathbf{Q}_1\mathbf{S}_1^s \qquad \mathbf{A}_2^{s\,T}\Sigma_2\mathbf{A}_2^s - \Sigma_2 = -\mathbf{S}_2^{s\,T}\mathbf{Q}_2 S_2^s \tag{36.8}$$

and

$$\mathbf{S}_i^s = \text{diag}(\mathbf{S}_{i1}^s, \mathbf{S}_{i2}^s) \qquad \mathbf{A}_i^s = \text{diag}(\mathbf{A}_{i1}^s, \mathbf{A}_{i2}^s) \quad \forall i \in \mathbb{I}_{1:2} \tag{36.9a}$$
$$\mathbf{S}_i^u = \text{diag}(\mathbf{S}_{i1}^u, \mathbf{S}_{i2}^u) \qquad \mathbf{A}_i^u = \text{diag}(\mathbf{A}_{i1}^u, \mathbf{A}_{i2}^u) \quad \forall i \in \mathbb{I}_{1:2} \tag{36.9b}$$

came from the Schur decomposition (36.6).

36.3.3 Optimization Scheme

In a distributed MPC scheme, a key point is how the different subsystems execute the optimizations and communicate the results to the other subsystems. In the cooperative MPC scheme proposed in [5, Chap. 6], [8], [11] and [9], the choice is for a parallel optimization scheme. This means that, at any time k, each i-th agent, in parallel with the other agents, solves an iterative optimization problem, and, once the calculation is done, communicates the obtained control sequence $\mathbf{u}_i$ to the other agents. Since the optimization algorithm is iterative, an initial guess $\mathbf{u}_i^{[0]}$ is needed. Notice also that, in the control algorithm there are two time scales to be taken into account: the sampling time, k, and the iteration number p, for a given time k (denoted in what follows as a superscript).

At the $(p+1)$-th iteration, each subsystem solves the following optimization problem:

$$\mathbf{u}_i^*(k{:}k{+}N_\text{p}{-}1) = \arg\min_{\mathbf{u}_i} J(\mathbf{x}_1(k), \mathbf{x}_2(k), \mathbf{u}_1(k{:}k{+}N_\text{p}{-}1), \mathbf{u}_2(k{:}k{+}N_\text{p}{-}1)) \tag{36.10a}$$

subject to

$$\begin{bmatrix} \mathbf{x}_1(k+l+1) \\ \mathbf{x}_2(k+l+1) \end{bmatrix} = \begin{bmatrix} \mathbf{A}_1 & \\ & \mathbf{A}_2 \end{bmatrix} \begin{bmatrix} x_1(k+l) \\ x_2(k+l) \end{bmatrix} + \begin{bmatrix} \bar{\mathbf{B}}_{11} \\ \bar{\mathbf{B}}_{21} \end{bmatrix} \mathbf{u}_1(k+l) + \begin{bmatrix} \bar{\mathbf{B}}_{12} \\ \bar{\mathbf{B}}_{22} \end{bmatrix} \mathbf{u}_2(k+l) \tag{36.10b}$$

$$\mathbf{u}_i(k : k + N_p - 1) \in \mathcal{U}_i^{N_p} \tag{36.10c}$$

$$\|\mathbf{u}_i(k : k + N_p - 1)\| \le d_i \sum_{j \in \mathbb{I}_{1:2}} \|\mathbf{x}_{ji}(k)\| \quad \text{if } x_{ji}(0) \in \mathbb{B}_r \quad \forall j \in \mathbb{I}_{1:2} \tag{36.10d}$$

$$\mathbf{u}_j(k : k + N_p - 1) = \mathbf{u}_j^{[p]}(k : k + N_p - 1) \quad j \in \mathbb{I}_{1:2} \setminus i \tag{36.10e}$$

$$\mathbf{S}_{ji}^{u\,\prime}\mathbf{x}_{ji}(k + N_p) = 0 \quad j \in \mathbb{I}_{1:2} \tag{36.10f}$$

in which the symbol $\mathbb{B}_r$ denotes a closed ball of radius r centered at the origin, defined as $\mathbb{B}_r = \{\mathbf{x} \in \mathbb{R}^n, \|\mathbf{x}\| \leq r\}$, and r, d_1, d_2 are positive scalars.

The solutions to these problems are denoted as

$$\mathbf{u}_1^*(k : k + N_p - 1) \qquad \mathbf{u}_2^*(k : k + N_p - 1)$$

in which the state dependence and the other agent's previous iterate dependence have been omitted to reduce notation.

Given a feasible iterate $(\mathbf{u}_1^{[p]}(k : k + N_p - 1), \mathbf{u}_2^{[p]}(k : k + N_p - 1))$, the next iterate is defined by the following convex update:

$$\mathbf{u}_1^{[p+1]}(k:k+N_\mathrm{p}-1) = w_1\mathbf{u}_1^*(k:k+N_\mathrm{p}-1) + (1 - w_1)\mathbf{u}_1^{[p]}(k:k+N_\mathrm{p}-1), \tag{36.11a}$$

$$\mathbf{u}_2^{[p+1]}(k:k+N_\mathrm{p}-1) = w_2\mathbf{u}_2^*(k:k+N_\mathrm{p}-1) + (1 - w_2)\mathbf{u}_2^{[p]}(k:k+N_\mathrm{p}-1), \tag{36.11b}$$

$$w_1 + w_2 = 1, \quad w_1, w_2 > 0$$

This parallel distributed optimization scheme is of the Gauss-Jacobi type [1, pp. 219–223]. The iterative solution of the optimization problem is stopped once an agreement between agents is achieved. This agreement can be defined as a given accuracy of the solution, or expressed as a maximum number of iterates, $\bar{p}$. At the last iterate $\bar{p}$, the best available solution is defined as $\mathbf{u}^{[\bar{p}]}(k : k + N_\mathrm{p} - 1) = (\mathbf{u}_1^{[\bar{p}]}(k:k+N_\mathrm{p}-1), \mathbf{u}_2^{[\bar{p}]}(k:k+N_\mathrm{p}-1))$. The first element of the control sequence, that is $\mathbf{u}^{[\bar{p}]}(k)$, is then applied to the plant in a receding horizon manner.

36.3.4 Warm-Start: The Initialization of the Iterative Optimizations

As it has been already stated in the previous section, since the optimization problem in solved iteratively, a proper initialization of the optimization algorithm is needed. Recurring to suboptimal MPC notations [7], this initialization is usually referred as warm-start. For each subsystem, the warm-start is given by:

$$\mathbf{u}_1{}^{[0]}(k:k+N_\mathrm{p}-1) = \{\mathbf{u}_1(1; k), \mathbf{u}_1(2; k), \ldots, \mathbf{u}_1(N_p - 1; k), 0\}$$

$$\mathbf{u}_2^{[0]}(k:k+N_\mathrm{p}-1) = \{\mathbf{u}_2(1; k), \mathbf{u}_2(2; k), \ldots, \mathbf{u}_2(N_p - 1; k), 0\}$$

The warm-start $\mathbf{u}_i^{[0]}(k : k + N_\mathrm{p} - 1)$ is used as the initial condition for the cooperative MPC problem in each subsystem i.

Remark 36.3 The term suboptimality in this work refers to the fact that, provided that $\bar{p}$ is finite, the distributed solution obtained with this control algorithm is a

Algorithm 36.1 Cooperative MPC algorithm

1: **if** $k = 0$ **then**
2: Calculate $(\mathbf{u}_1^{[0]}(k), \mathbf{u}_2^{[0]}(k))$ {For instance, by solving a Phase I problem [2]}
3: **else**
4: $\mathbf{u}_1^{[0]}(k\!:\!k+N_\mathrm{p}-1) = \{\mathbf{u}_1(1;k), \mathbf{u}_1(2;k), \ldots, \mathbf{u}_1(N_p-1;k), 0\}$
5: $\mathbf{u}_2^{[0]}(k\!:\!k+N_\mathrm{p}-1) = \{\mathbf{u}_2(1;k), \mathbf{u}_2(2;k), \ldots, \mathbf{u}_2(N_p-1;k), 0\}$
6: **end if**
7: Read $(\mathbf{x}_1, \mathbf{x}_2)$
8: **for** $p = 1$ to $p = \bar{p}$ **do**
9: Read $(\mathbf{u}_1^{[p-1]}(k), \mathbf{u}_2^{[p-1]}(k))$
10: Solve (36.10)
11: Calculate the update $(\mathbf{u}_1^{[p]}(k\!:\!k+N_\mathrm{p}-1), \mathbf{u}_2^{[p]}(k\!:\!k+N_\mathrm{p}-1))$ using (36.11)
12: Broadcast results
13: **end for**
14: Inject $(\mathbf{u}_1(0;k)^{[\bar{p}]}, \mathbf{u}_2(0;k)^{[\bar{p}]})$ into the plant.
15: Go back to 1.

suboptimal solution of a centralized optimization problem. On the other hand, it is assumed that the solution of each agent's optimization problem at each iteration is optimal.

36.3.5 Control Algorithm

The control algorithm executed by the i-th agent at the p-th iterate is depicted as Algorithm 36.1.

36.4 Theoretical Results

In this section, the main theoretical properties of the presented control algorithm will be introduced.

36.4.1 Stability

From the control algorithm described in the previous section, three properties arise.

- **Recursive feasibility** [8, Lemma 5]. Given a feasible initial condition, for all $p \geq 1$,

$$(\mathbf{u}_1^{[p]}(k\!:\!k+N_\mathrm{p}-1), \mathbf{u}_2^{[p]}(k\!:\!k+N_\mathrm{p}-1)) \in \mathcal{U}_1^{N_p} \times \mathcal{U}_2^{N_p}$$

- **Convergence** [8, Lemma 6]. The cost $J(\mathbf{x}_1(k), \mathbf{x}_2(k), \mathbf{u}_1(k : k+N_\mathrm{p}-1), \mathbf{u}_2(k : k+N_\mathrm{p}-1))$ is nonincreasing for each iterate p and converges as $p \to \infty$.
- **Optimality** [8, Lemma 7]. The cost $J(\mathbf{x}_1(k), \mathbf{x}_2(k), \mathbf{u}_1(k : k+N_\mathrm{p}-1), \mathbf{u}_2(k : k+N_\mathrm{p}-1))$ converges to the centralized optimal one, as $p \to \infty$. Moreover, the iterates $(\mathbf{u}_1^{[p]}(k:k+N_\mathrm{p}-1), \mathbf{u}_2^{[p]}(k:k+N_\mathrm{p}-1))$ converge to the Pareto (centralized) solution.

Based on the previous properties, it is possible to state the stability property of the cooperative MPC.

Theorem 36.1 [Exponential stability] [8, Theorem 9] The origin of the closed-loop system (36.2) is exponentially stable on the set $\mathcal{X}_{N_p}$

where $\mathcal{X}_{N_p}$ is the set of all initial x such that there exists a couple $(\mathbf{u}_1(k:k+N_\mathrm{p}-1), \mathbf{u}_2(k:k+N_\mathrm{p}-1)) \in \mathcal{U}_1^{N_p} \times \mathcal{U}_2^{N_p}$ satisfying constraint (36.10f).

A detailed proof of the previous Theorem, as well as of the feasibility, convergence and optimality properties can be found in [8], [9] and [5, Chap. 6].

36.4.1.1 Main Assumptions

In order to enjoy the previous properties the controller has to fulfill the following Assumptions:

Assumption 36.2 1. The couples $(\underline{\mathbf{A}}_i, \underline{\mathbf{B}}_i)$ are stabilizable, where $\underline{\mathbf{A}}_i = \mathrm{diag}(\mathbf{A}_{1i}, \mathbf{A}_{2i})$ and $\underline{\mathbf{B}}_i = \begin{bmatrix} \mathbf{B}_{1i} \\ \mathbf{B}_{2i} \end{bmatrix}$, for every $i \in \mathbb{I}_{1:2}$.

2. Let $\mathbf{R}_i \in \mathbb{R}^{n_{u_i} \times n_{u_i}}$ be a positive definite matrix and $\mathbf{Q}_i \in \mathbb{R}^{n_{x_i} \times n_{x_i}}$ a positive semi-definite matrix such that the pairs $(\mathbf{A}_i, \mathbf{Q}_i)$ are detectable.
3. Let the prediction horizon N_p be such that $N_p \geq \max_{i \in 1:2}(\underline{n}_i^u)$, in which $\underline{n}_i^u$ is the number of unstable modes of $\underline{\mathbf{A}}_i$, that is the number of $\lambda \in \mathrm{eig}(\underline{\mathbf{A}}_i)$ such that $\|\lambda\| \geq 1$.

Remark 36.4 Notice that, according Assumption 36.2.3, the controller has to be designed with an horizon N_p large enough to zero the unstable modes.

36.4.2 Coupled Input Constraints

In the control algorithm presented above, it has been assumed that the constraints are uncoupled, since each input is constrained in a separate feasible region. This assumption however, may not hold in practice, where it is quite common that different subsystems share some resources. In such a case, we talk about coupled input constraints, because each subsystem input depends on how much the other subsystems employ that shared resource, i. e. the other subsystems inputs. Control Algorithm 36.1 may be, then, suboptimal.

In [8] and [9] it is proposed to slightly modify optimization problem (36.10), by considering the control input given by $\mathbf{u} = (\mathbf{u}_1, \mathbf{u}_2, \mathbf{u}_C)$ such that there exists a $\mathcal{U}_1$,$\mathcal{U}_2$, and $\mathcal{U}_C$ that provide

$$\mathcal{U} = \mathcal{U}_1 \times \mathcal{U}_2 \times \mathcal{U}_C$$

and

$$\mathbf{u}_1 \in \mathcal{U}_1, \quad \mathbf{u}_2 \in \mathcal{U}_2, \quad \mathbf{u}_C \in \mathcal{U}_C$$

for which $\mathcal{U}_1$, $\mathcal{U}_2$, and $\mathcal{U}_C$ are compact and convex. Notice that $\mathbf{u}_i$ denotes the uncoupled inputs for subsystems 1 and 2, and $\mathbf{u}_C$ the coupled ones.

Then, by defining the augmented inputs:

$$\hat{\mathbf{u}}_1 = \begin{bmatrix} \mathbf{u}_1 \\ \mathbf{u}_C \end{bmatrix} \quad \hat{\mathbf{u}}_2 = \begin{bmatrix} \mathbf{u}_2 \\ \mathbf{u}_C \end{bmatrix}$$

problem (36.10) can be modified by substituting constraint (36.10c), with

$$\mathbf{u}_i(k : k + N_p - 1) \in \mathcal{U}_i^{N_p} \times \mathcal{U}_C^{N_p} \tag{36.12}$$

It is proved in [8, Lemma 15] that, thanks to modification mentioned above, the cooperative MPC ensures exponential stability and Pareto optimality even in case of coupled constraints.

36.4.3 Hierarchical Cooperative MPC

The control algorithm presented in this chapter assume that all agents in the control network are synchronous and communicate with the other agents at the same time at any optimization iterate. This assumption may be sometimes too strong, since it considers only plants in which information can be easily interchanged, and does not account for subsystems with different time scales.

In [10] and [9] a hierarchical cooperative model predictive control is proposed, in which subsystems are grouped into hierarchies, and optimizations and exchange of information are allowed on any time scale. The main advantage is that, this kind of control enjoys the same properties of hierarchical control schemes without requiring any additional coordinating controller [6].

36.5 Applications

The control algorithm presented here has been tested in several simulation examples, or applications. The goal of this section is to sum up some of these results.

36.5.1 Cooperative MPC Applied to a Chemical Plant

This example appeared in [8]. The considered chemical plant consists of two reactors and a separator. A stream of pure reactant A is added to each reactor and converted to the product B by a first-order reaction. The product is lost by a parallel first-order reaction to side product C. The distillate of the separator is split and partially redirected to the first reactor (see Fig. 36.1).The model for the plant is

$$
\begin{aligned}
\frac{dH_1}{dt} &= \frac{1}{\rho A_1}(F_{f1} + F_R - F_1) \\
\frac{dx_{A1}}{dt} &= \frac{1}{\rho A_1 H_1}(F_{f1}x_{A0} + F_R x_{AR} - F_1 x_{A1}) - k_{A1}x_{A1} \\
\frac{dx_{B1}}{dt} &= \frac{1}{\rho A_1 H_1}(F_R x_{BR} - F_1 x_{B1}) - k_{B1}x_{B1} \\
\frac{dT_1}{dt} &= \frac{1}{\rho A_1 H_1}(F_{f1}T_0 + F_R T_R - F_1 T_1) \\
&\quad - \frac{1}{C_p}(k_{A1}x_{A1}\Delta H_A + k_{B1}x_{B1}\Delta H_B) + \frac{Q_1}{\rho A_1 C_p H_1}
\end{aligned}
$$

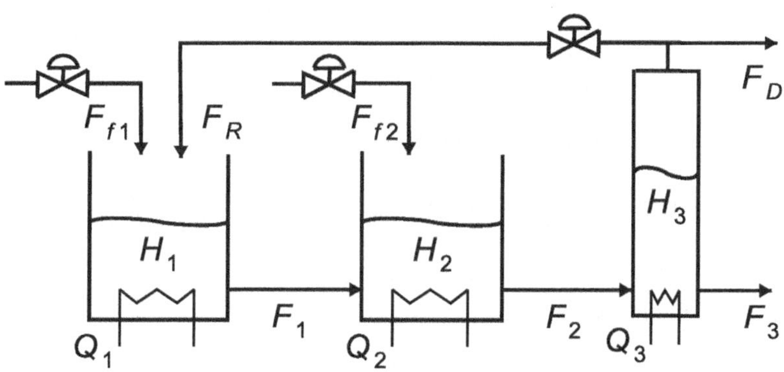

Fig. 36.1 Scheme of the studied plant: two reactors in series with a separator

$$
\begin{aligned}
\frac{dH_2}{dt} &= \frac{1}{\rho A_2}(F_{f2} + F_1 - F_2)\\
\frac{dx_{A2}}{dt} &= \frac{1}{\rho A_2 H_2}(F_{f2} x_{A0} + F_1 x_{A1} - F_2 x_{A2}) - k_{A2} x_{A2}\\
\frac{dx_{B2}}{dt} &= \frac{1}{\rho A_2 H_2}(F_1 x_{B1} - F_2 x_{B2}) - k_{B2} x_{B2}\\
\frac{dT_2}{dt} &= \frac{1}{\rho A_2 H_2}(F_{f2} T_0 + F_1 T_1 - F_2 T_2)\\
&\quad - \frac{1}{C_p}(k_{A2} x_{A2} \Delta H_A + k_{B2} x_{B2} \Delta H_B) + \frac{Q_2}{\rho A_2 C_p H_2}
\end{aligned}
$$

$$
\begin{aligned}
\frac{dH_3}{dt} &= \frac{1}{\rho A_3}(F_2 - F_D - F_R - F_3)\\
\frac{dx_{A3}}{dt} &= \frac{1}{\rho A_3 H_3}(F_2 x_{A2} - (F_D + F_R) x_{AR} - F_3 x_{A3})\\
\frac{dx_{B3}}{dt} &= \frac{1}{\rho A_3 H_3}(F_2 x_{B2} - (F_D + F_R) x_{BR} - F_3 x_{B3})\\
\frac{dT_3}{dt} &= \frac{1}{\rho A_3 H_3}(F_2 T_2 - (F_D + F_R) T_R - F_3 T_3) + \frac{Q_3}{\rho A_3 C_p H_3}
\end{aligned}
$$

in which for all $i \in \mathbb{I}_{1:3}$

$$
F_i = k_{vi} H_i \quad k_{Ai} = k_A \exp\left(-\frac{E_A}{RT_i}\right) \quad k_{Bi} = k_B \exp\left(-\frac{E_B}{RT_i}\right)
$$

The recycle flow and weight percents satisfy

$$
\begin{aligned}
F_D = 0.01 F_R \quad x_{AR} &= \frac{\alpha_A x_{A3}}{\bar{x}_3} \quad x_{BR} = \frac{\alpha_B x_{B3}}{\bar{x}_3}\\
\bar{x}_3 = \alpha_A x_{A3} + \alpha_B x_{B3} + \alpha_C x_{C3} &\quad x_{C3} = (1 - x_{A3} - x_{B3})
\end{aligned}
$$

The output and input are denoted, respectively

$$
\begin{aligned}
y &= \left[\, H_1\ x_{A1}\ x_{B1}\ T_1\ H_2\ x_{A2}\ x_{B2}\ T_2\ H_3\ x_{A3}\ x_{B3}\ T_3 \,\right]\\
u &= \left[\, F_{f1}\ Q_1\ F_{f2}\ Q_2\ F_R\ Q_3 \,\right]
\end{aligned}
$$

The linearized model takes the form:

$$
x^+ = Ax + Bu \quad y = x
$$

Notice that this model is discretized with sampling time $\Delta = 0.1$ s. Please refer to [8] for details on steady state and parameters.

The system is partitioned in three subsystems:

$$y_1 = \left[H_1 \; x_{A1} \; x_{B1} \; T_1 \right]^T \qquad u_1 = \left[F_{f1} \; Q_1 \right]^T$$
$$y_2 = \left[H_2 \; x_{A2} \; x_{B2} \; T_2 \right]^T \qquad u_2 = \left[F_{f2} \; Q_2 \right]^T$$
$$y_3 = \left[H_3 \; x_{A3} \; x_{B3} \; T_3 \right]^T \qquad u_3 = \left[F_R \; Q_3 \right]^T$$

The distributed model is obtained by following [8, Appendix B].

36.5.1.1 Simulation

The tuning parameters for the MPC controllers are

$$Q_{yi} = \mathrm{diag}(1, 0, 0, 0.1) \; \forall i = \mathbb{I}_{1:2} \quad Q_{y3} = \mathrm{diag}(1, 0, 10^3, 0)$$
$$Q_i = C_i^T Q_{yi} C_i + 0.001I \quad R_i = 0.01I \quad \forall i \in \mathbb{I}_{1:3}$$

The inputs are subject to (uncoupled) constraints (see [8]). In the simulation, a set-point change in the output product weight percent x_{B3} at $t = 0.5\,s$ is considered. In Fig. 36.2, the performance of the distributed control strategies are compared to the centralized MPC. Notice that noncooperative control is an improvement over decentralized control. However, cooperative control with just one iteration is significantly better than noncooperative control. Convergence to centralized control speeds up as the number of iteration increases.

36.5.2 Cooperative MPC Applied to Power System Automatic Generation Control

In this section, the application of the cooperative MPC to power systems is taken into account. In particular, the application of the proposed controller to a power system Automatic Generation Control, proposed on [12] will be briefly summarized.

Table 36.1 Performance comparison (from [8])

	Cost	Performance loss (%)
Centralized MPC	1.41	0
Cooperative MPC (10 iterates)	1.64	16.5
Cooperative MPC (1 iterate)	2.59	84.1
Noncooperative MPC	3.88	175
Decentralized MPC	144	1.01e+04

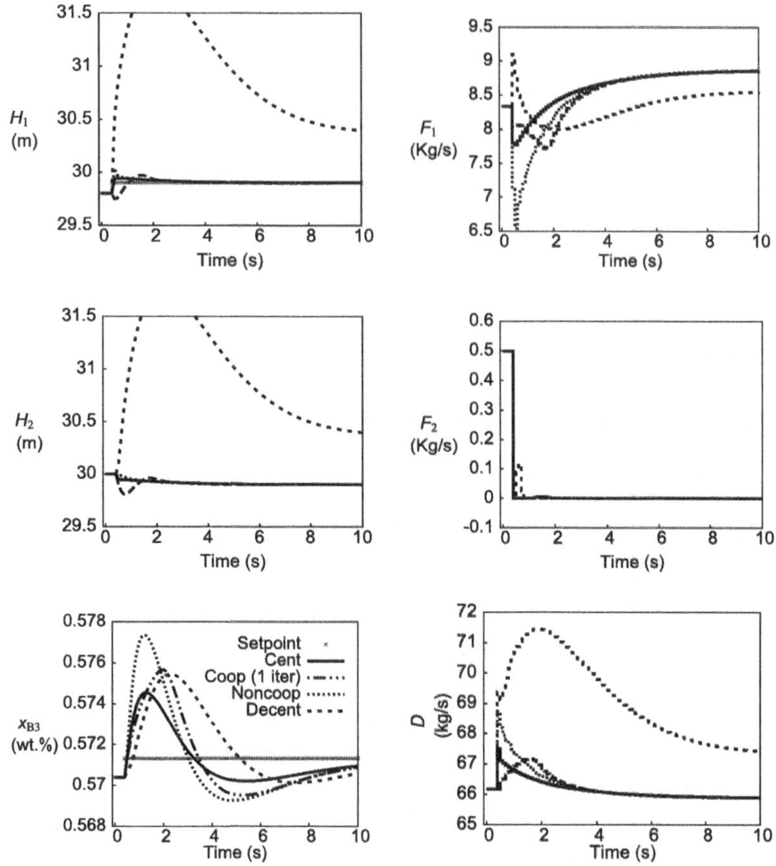

Fig. 36.2 Performance of reactor and separator example (from [8])

Automatic Generation Control (AGC), is a system for adjusting the power output of multiple generators at different power plants, in response to changes in the load. The main idea of AGC is to regulate the real power output of generators throughout the system, by controlling the system frequency and tie-line interchange. Since power systems dimensions tends to expands, and since various limits must be taken into account, including restrictions on the amount and rate of generator power deviation, AGC provides a very relevant example for illustrating the performance of distributed MPC in a power system setting.

In [12] as well as in [11], the cooperative MPC has been tested and compared with centralized MPC and traditional AGC, in three scenarios: a two areas power system network, a four area power system network, and a two area with a FACTS[1] device.

[1] A FACTS is a device employed to manipulate the effective impedance of a tie-line and control power flow between two interconnected control areas.

In all of the three scenarios considered, the cooperative MPC has shown better performance than traditional AGC, as well as performance very close to centralized MPC. Moreover, the cooperative MPC algorithm can be terminated prior to convergence without compromising feasibility or closed-loop stability of the resulting distributed controller. This feature allows one to terminate the algorithm at the end of a sampling interval, even if convergence is not achieved. Since centralized MPC is not a reliable solution for a large scale and geographically extended system like a power system, the results presented in [12] suggest that cooperative distributed MPC, which combines a distributed calculation with centralized performance, can be considered as a suitable control strategy.

36.6 Conclusions

This chapter presented the distributed MPC strategy, proposed by Prof. James Rawlings's Research Group. This strategy has been widely studied and analyzed, in many publications as well as in two doctoral thesis.

The chapter summarized this control strategy underlining its main features and properties. In particular it is important to remark that the control algorithm is of the cooperative distributed MPC type: subsystems optimizes in parallel a certain cost, which can be considered as the overall plant cost. After optimization, an exchange of information between agents takes place, which determines the cooperative nature of the algorithm. This process is iterative and it stops once an agreement - or a certain final iteration - is achieved.

It has been shown that the control algorithm ensures recursive feasibility, convergence, and optimality, in the sense that the centralized optimum is achieved, even in case of coupled constraints. Moreover, the controller enjoys the property of exponential stability.

The proposed controller has been tested in several applications, a couple of which have been briefly described in this chapter.

Acknowledgments The author would like to thank Professor James B. Rawlings and Dr. Brett T. Stewart for helpful discussions and comments, as well as for material provided for the writing of this chapter.

References

1. D.P. Bertsekas, J.N. Tsitsiklis, *Parallel and Distributed Computation* (Athena Scientific, Belmont, 1997)
2. S. Boyd, L. Vandenberghe. *Convex Optimization*. Cambridge University Press, 2006
3. A. Ferramosca, D. Limon, I. Alvarado, E. F. Camacho. Cooperative distributed MPC for tracking. Automatica, 2013 in press

4. R.D. Gudi, J.B. Rawlings, Identification for decentralized model predictive control. AIChE J. **52**(6), 2198–2210 (2006)
5. J. B. Rawlings, D. Q. Mayne. Model Predictive Control: Theory and Design, 1st ed. (Nob-Hill Publishing, 2009)
6. R. Scattolini, Architectures for distributed and hierarchical model predictive control - a review. J. Process Control. **19**(5), 723–731 (2009)
7. P.O.M. Scokaert, D.Q. Mayne, J.B. Rawlings, Suboptimal model predictive control (feasibility implies stability). IEEE Trans. Autom. Control. **44**(3), 648–654 (1999)
8. B.T. Stewart, A.N. Venkat, J.B. Rawlings, S.J. Wright, G. Pannocchia, Cooperative distributed model predictive control. Syst. Control. Lett. **59**, 460–469 (2010)
9. B. T. Stewart. Plantwide Cooperative Distributed Model Predictive Control. Ph.D. thesis, University of Wisconsin, Madison, 2010
10. B. T. Stewart, J. B. Rawlings, S. J. Wright, *Hierarchical cooperative distributed model predictive controlIn Proceedings of the American Control Conference*, Baltimore, June 2010
11. A. N. Venkat. Distributed Model Predictive Control: Theory and Applications. Ph.D. thesis, Univeristy of Wisconsin, Madison, October 2006
12. A.N. Venkat, I.A. Hiskens, J.B. Rawlings, S.J. Wright, Distributed MPC strategies with application to power system automatic generation control. IEEE Ctl. Sys. Tech. **16**(6), 1192–1206 (2008)
13. A. N. Venkat, J. B. Rawlings, S. J. Wright. Distributed model predictive control of large-scale systems. In Assessment and Future Directions of Nonlinear Model Predictive Control, Springer, 2007, pp. 591–605

The manufacturer's authorised representative in the EU is Springer Nature Customer Service Centre GmbH, Europaplatz 3, 69115 Heidelberg, Germany. If you have any concerns regarding our products, please contact ProductSafety@springernature.com

Printed and bound by CPI Group (UK) Ltd, Croydon, CR0 4YY
22/09/2026
02220985-0013